A-Level Year 2

Mathematics

Exam Board: AQA

A-Level Maths is no picnic. In fact, it's the opposite of a picnic. It's like a formal dinner where they serve algebra and calculus instead of dessert.

Fortunately, this CGP Student Book makes it all much easier to digest. It's fully up to date for the latest AQA course, with brilliant study notes, tips, examples and practice questions for every topic.

It even includes a free Online Edition to read on your PC, Mac or tablet!

CGP

How to get your free Online Edition

Go to **cgpbooks.co.uk/extras** and enter this code...

3962 6860 0202 5362

This code will only work once. If someone has used this book before you, they may have already claimed the Online Edition.

Contents

About this Book

In this book you'll find...

Chapter 10 Vectors

1. Vectors in Three Dimensions

In Year 1 you learnt how to represent, calculate with, and model situations using vectors in two dimensions. Now you're going to find out how to do all of that in three dimensions. You might want to look back at your notes on 2D vectors first, to refresh your memory.

Three-dimensional position vectors

- **Three-dimensional vectors** have components in the direction of the x-, y- and z-**axes**. Imagine that the x- and y-axes lie flat on the page. Then imagine a **third axis** sticking straight through the page at right angles to it — this is the z-**axis**.
- The points in three dimensions are given (x, y, z) **coordinates**.
- The **unit vector** in the direction of the z-axis is $\mathbf{k}$, so three-dimensional vectors can be written like this: $x\mathbf{i} + y\mathbf{j} + z\mathbf{k}$ or $\begin{pmatrix} x \\ y \\ z \end{pmatrix}$
- Three-dimensional vectors are used to describe things in **three-dimensional space**, e.g. an aeroplane moving through the sky.
- Calculating with 3D vectors is just the same as with 2D vectors, as the next example shows.

Learning Objectives:
- Understand how to represent vectors in three dimensions, using both unit vectors and column vectors.
- Be able to add and subtract three-dimensional vectors and multiply them by scalars.
- Be able to show that two 3D vectors are parallel, and that three points are collinear.

Example 1

The diagram below shows the position of the points P and Q.

a) Write the position vectors $\overrightarrow{OP}$ and $\overrightarrow{OQ}$ as column vectors.

$\overrightarrow{OP} = 4\mathbf{i} + 3\mathbf{j} + 0\mathbf{k} = \begin{pmatrix} 4 \\ 3 \\ 0 \end{pmatrix}$

$\overrightarrow{OQ} = 2\mathbf{i} + 5\mathbf{j} + 4\mathbf{k} = \begin{pmatrix} 2 \\ 5 \\ 4 \end{pmatrix}$

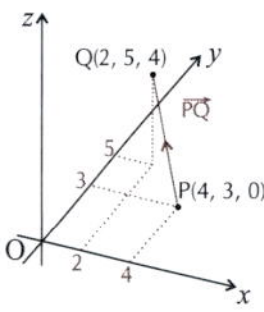

b) Hence find $\overrightarrow{PQ}$ as a column vector.

$\overrightarrow{PQ} = -\overrightarrow{OP} + \overrightarrow{OQ}$

$= -\begin{pmatrix} 4 \\ 3 \\ 0 \end{pmatrix} + \begin{pmatrix} 2 \\ 5 \\ 4 \end{pmatrix} = \begin{pmatrix} 2-4 \\ 5-3 \\ 4-0 \end{pmatrix} = \begin{pmatrix} -2 \\ 2 \\ 4 \end{pmatrix}$

Tip: Don't get confused between a point and its position vector. Points are given by (x, y, z) coordinates and position vectors are the movement needed to get to the point from the origin, given in the form $x\mathbf{i} + y\mathbf{j} + z\mathbf{k}$. You'll need to be able to swap between the forms.

Tip: Just add along each row of the column as you do with 2D column vectors — don't forget that you've multiplied $\overrightarrow{OP}$ by the scalar '–1' so all its entries will change signs.

Chapter 10 Vectors 251

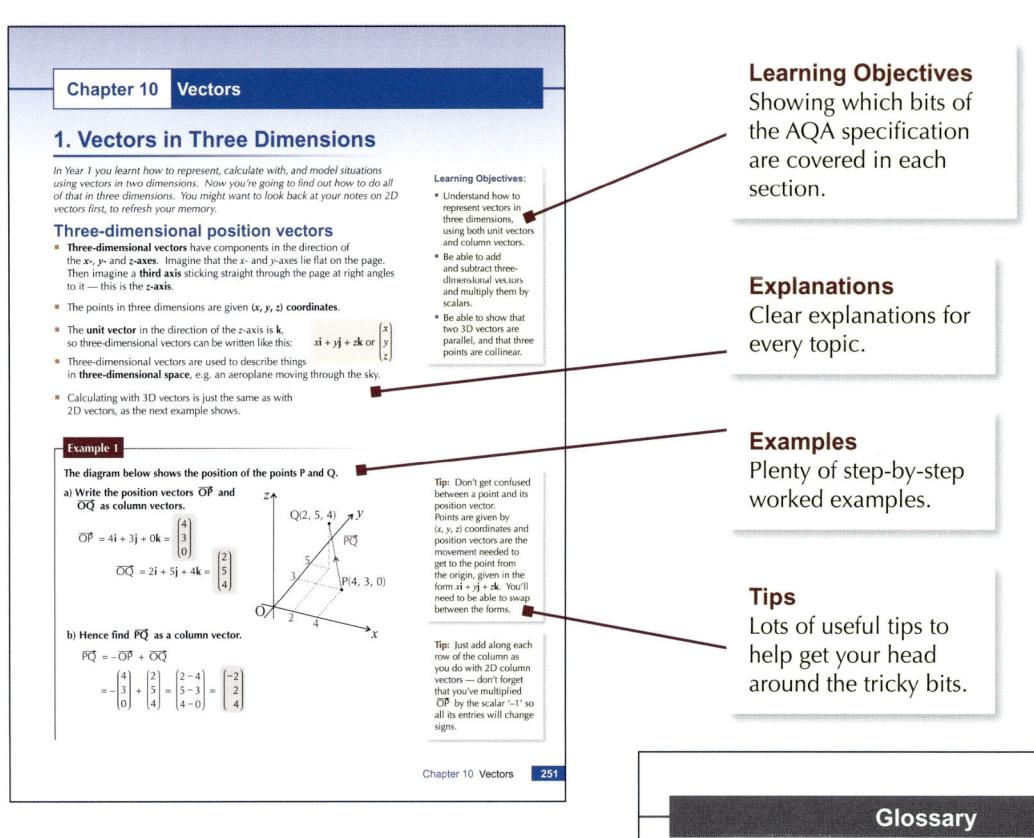

Learning Objectives
Showing which bits of the AQA specification are covered in each section.

Explanations
Clear explanations for every topic.

Examples
Plenty of step-by-step worked examples.

Tips
Lots of useful tips to help get your head around the tricky bits.

Glossary
All the definitions you need to know for the exam, plus other useful words.

Glossary

A

Absolute value
Another name for the **modulus**.

Acceleration
The rate of change of an object's **velocity** with respect to time.

Addition law
A formula linking the probability of the **union** and the probability of the **intersection** of events A and B.

Algebraic division
Dividing one algebraic expression by another.

Algebraic fraction
A fraction made up of algebraic expressions.

Alternative hypothesis
The statement that you will accept instead if you decide to reject the **null hypothesis** in a **hypothesis test.** It gives a range of values for the **parameter** and is usually written H_1.

Arc
The curved edge of a **sector** of a circle.

Arccos
The **inverse** of the cosine function, also written as arccosine or $\cos^{-1}$.

Arcsin
The **inverse** of the sine function, also written as arcsine or $\sin^{-1}$.

Arctan
The **inverse** of the tangent function, also written as arctangent or $\tan^{-1}$.

Arithmetic sequence/series
A **sequence** or **series** where successive terms have a common difference.

Assumption
A simplification of a real-life situation used in a **model**.

B

Beam
A long, **thin**, straight, **rigid** body.

Binomial coefficient
The number of orders in which x objects of one type and $(n - x)$ objects of a different type can be arranged. Equal to $\binom{n}{x} = \frac{n!}{x!(n-x)!}$.

Binomial distribution B(n, p)
A discrete probability distribution which models the number of successes x in n independent trials when the probability of success in each trial is p.

Binomial expansion
A method of expanding functions of the form $(q + px)^n$. Can be used to give a finite expansion or to find an approximation of a value.

C

Cartesian equation
An equation relating the perpendicular axes x and y in 2D (or x, y and z in 3D). Cartesian coordinates are given in the form (x, y) in 2D (or (x, y, z) in 3D).

Chain rule
A method for **differentiating** a function of a function.

Coefficient of friction
A number greater than or equal to zero which measures the effect of **friction** between an object and a surface.

Collinear points
Three or more points are collinear if they all lie on the same straight line.

Component
The effect of a **vector** in a given direction.

Composite function
A combination of two or more functions acting on a value or set of values.

Compression
See **thrust**.

Concave curve
A curve with a negative second derivative — i.e. $f''(x) < 0$ for all x.

Conditional probability
A probability is conditional if it depends on whether or not another event happens.

Constant of integration
A constant term coming from an indefinite **integration** representing any number.

Continuity correction
A correction made in order to approximate a discrete distribution with a continuous distribution.

Convergence
A sequence converges if the terms get closer and closer to a single value.

Convergent sequence/series
A **sequence/series** that tends towards a **limit**.

Convex curve
A curve with a positive second derivative — i.e. $f''(x) > 0$ for all x.

Correlation
A linear relationship between two variables showing that they change together to some extent. (A correlation does not necessarily mean a causal relationship.)

Cosec
The **reciprocal** of the sine function, sometimes written as cosecant.

Cot
The **reciprocal** of the tangent function, sometimes written as cotangent.

Critical region
The set of all values of the **test statistic** that would cause you to reject the **null hypothesis**.

Critical value
The value of the **test statistic** at the edge of the **critical region**.

Glossary 477

Exercise 1.1

Q1 R is the point (4, –5, 1) and S is the point (–3, 0, –1). Write down the position vectors of R and S, giving your answers:

a) as column vectors, and

b) in unit vector form.

Q2 Give $\overrightarrow{GH}$ and $\overrightarrow{HG}$ as column vectors, where

$\overrightarrow{OG} = \begin{pmatrix} 2 \\ -3 \\ 4 \end{pmatrix}$ and $\overrightarrow{OH} = \begin{pmatrix} -1 \\ 4 \\ 9 \end{pmatrix}$.

Q3 Triangle JKL has vertices at the points J (4, 0, –3), K (–1, 3, 0) and L (2, 2, 7). Find the vectors $\overrightarrow{JK}$, $\overrightarrow{KL}$ and $\overrightarrow{LJ}$.

Q3 Hint: The question doesn't mention **i** and **j** components or column vectors so you can answer it using either.

Q4 A 3D printer is being used to make the plastic toy sketched below.

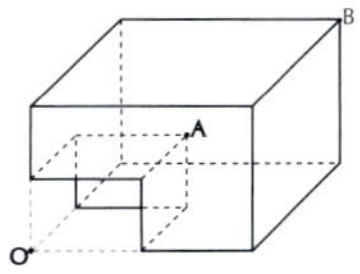

The toy can be modelled as a large cuboid with a smaller cuboid section missing. The large cuboid section is an enlargement of the smaller, 'missing' cuboid, by a scale factor of 2, centred at point O.

a) Prove that $\overrightarrow{AB} = \overrightarrow{OA}$.

b) Find the position vector of point B, if $\overrightarrow{OA} = 3\mathbf{i} + \mathbf{j} + 2\mathbf{k}$.

Q5 M is a point on the line CD, where C has coordinates (–1, 3, –5), M has coordinates (1, 1, –2) and $\overrightarrow{CD} = \begin{pmatrix} 4 \\ -4 \\ 6 \end{pmatrix}$.

Show that M is the midpoint of CD.

Q6 Show that vectors $\mathbf{a} = \frac{3}{4}\mathbf{i} + \frac{1}{3}\mathbf{j} - 2\mathbf{k}$ and $\mathbf{b} = \frac{1}{4}\mathbf{i} + \mathbf{j} - \frac{2}{3}\mathbf{k}$ are not parallel.

Q7 In a 3D board game, players take turns to position counters at points inside a cuboid grid. Players score by forming a straight line with three of their counters, unblocked by their opponent.

Show that counters placed at coordinates (1, 0, 3), (3, 1, 2) and (7, 3, 0) lie on a straight line.

Chapter 10 Vectors 253

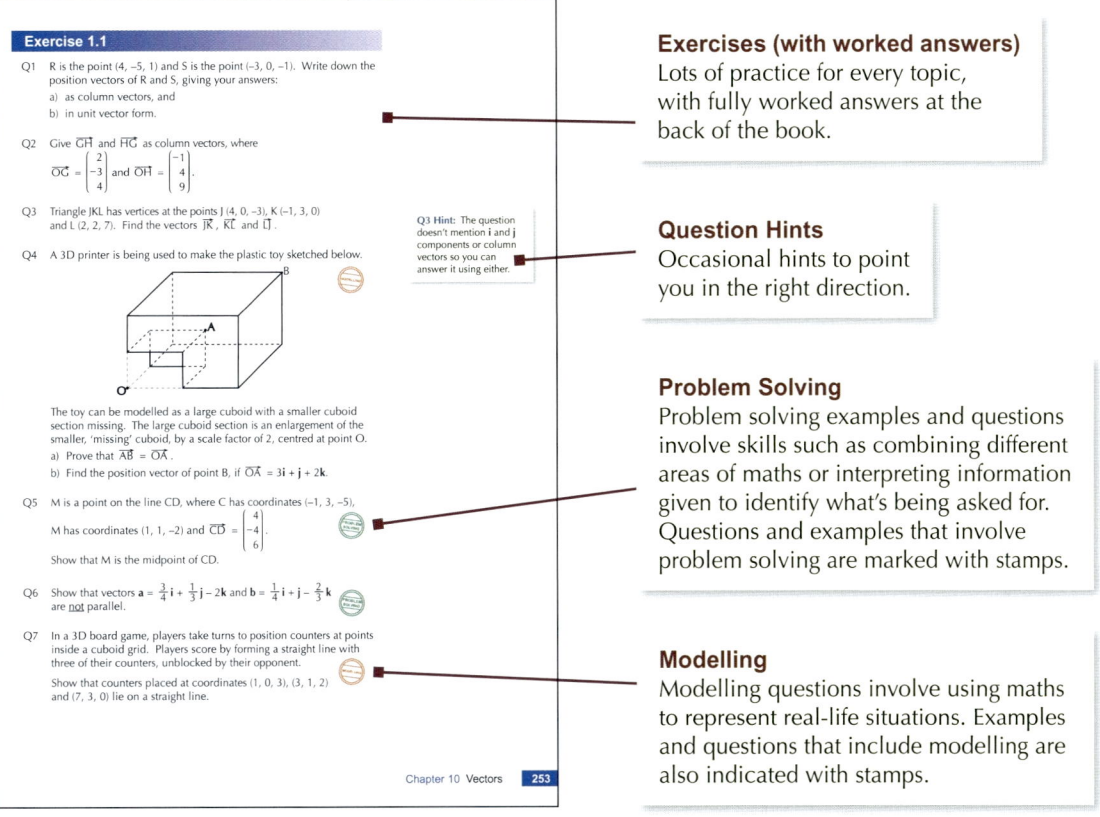

Exercises (with worked answers)

Lots of practice for every topic, with fully worked answers at the back of the book.

Question Hints

Occasional hints to point you in the right direction.

Problem Solving

Problem solving examples and questions involve skills such as combining different areas of maths or interpreting information given to identify what's being asked for. Questions and examples that involve problem solving are marked with stamps.

Modelling

Modelling questions involve using maths to represent real-life situations. Examples and questions that include modelling are also indicated with stamps.

Formula Sheet

Contains all the Year 2 formulas you'll be given in the exams.

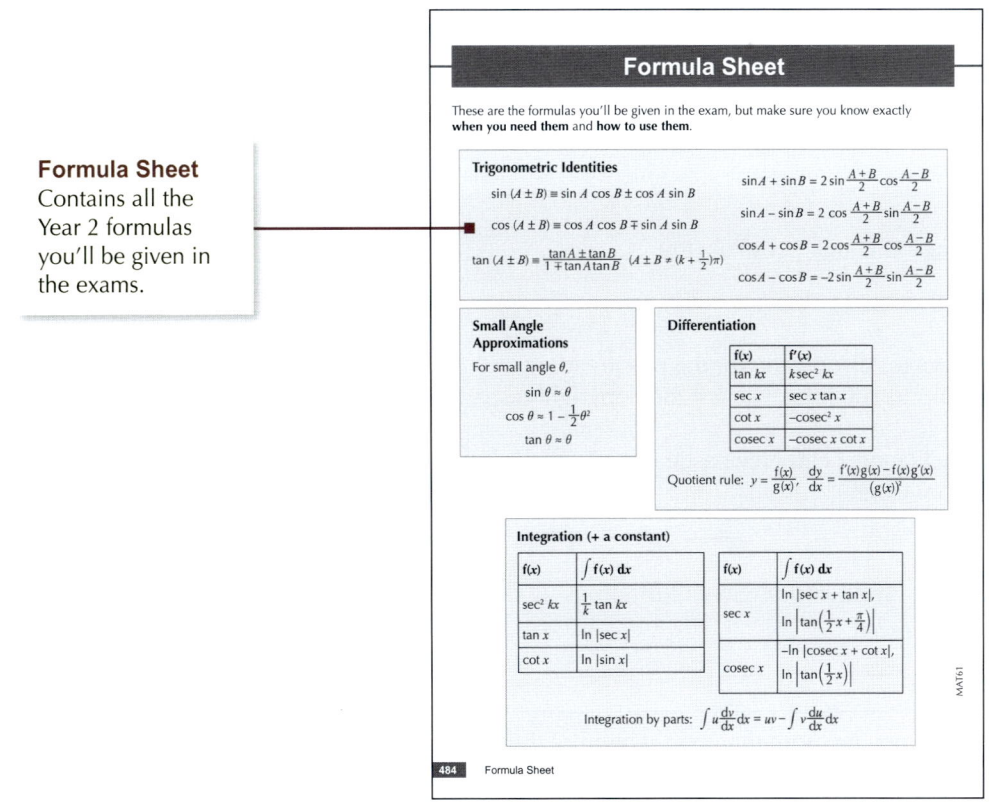

Formula Sheet

These are the formulas you'll be given in the exam, but make sure you know exactly **when you need them** and **how to use them**.

Trigonometric Identities

$\sin(A \pm B) \equiv \sin A \cos B \pm \cos A \sin B$

$\cos(A \pm B) \equiv \cos A \cos B \mp \sin A \sin B$

$\tan(A \pm B) \equiv \frac{\tan A \pm \tan B}{1 \mp \tan A \tan B} \quad (A \pm B \neq (k + \frac{1}{2})\pi)$

$\sin A + \sin B = 2\sin\frac{A+B}{2}\cos\frac{A-B}{2}$

$\sin A - \sin B = 2\cos\frac{A+B}{2}\sin\frac{A-B}{2}$

$\cos A + \cos B = 2\cos\frac{A+B}{2}\cos\frac{A-B}{2}$

$\cos A - \cos B = -2\sin\frac{A+B}{2}\sin\frac{A-B}{2}$

Small Angle Approximations

For small angle θ,

$\sin\theta \approx \theta$

$\cos\theta \approx 1 - \frac{1}{2}\theta^2$

$\tan\theta \approx \theta$

Differentiation

f(x)	f′(x)
$\tan kx$	$k\sec^2 kx$
$\sec x$	$\sec x \tan x$
$\cot x$	$-\text{cosec}^2 x$
$\text{cosec}\, x$	$-\text{cosec}\, x \cot x$

Quotient rule: $y = \frac{f(x)}{g(x)}, \ \frac{dy}{dx} = \frac{f'(x)g(x) - f(x)g'(x)}{(g(x))^2}$

Integration (+ a constant)

f(x)	$\int f(x)\,dx$
$\sec^2 kx$	$\frac{1}{k}\tan kx$
$\tan x$	$\ln\lvert\sec x\rvert$
$\cot x$	$\ln\lvert\sin x\rvert$

f(x)	$\int f(x)\,dx$
$\sec x$	$\ln\lvert\sec x + \tan x\rvert$, $\ln\left\lvert\tan\left(\frac{1}{2}x + \frac{\pi}{4}\right)\right\rvert$
$\text{cosec}\, x$	$-\ln\lvert\text{cosec}\, x + \cot x\rvert$, $\ln\left\lvert\tan\left(\frac{1}{2}x\right)\right\rvert$

Integration by parts: $\int u\frac{dv}{dx}dx = uv - \int v\frac{du}{dx}dx$

MAT61

484 Formula Sheet

Published by CGP

Editors:
Chris Corrall, Sammy El-Bahrawy, Will Garrison, Paul Jordin, Caley Simpson, Ben Train, Dawn Wright.

Contributors:
Andrew Ballard, Jean Blencowe, Katharine Brown, Jane Chow, Michael Coe, Claire Creasor, Margaret Darlington, Anna Gainey, Stephen Green, Dave Harding, Barbara Mascetti, James Nicholson, Charlotte O'Brien, Andy Pierson, Rosemary Rogers, Mike Smith, Janet West.

Cover image © iStock.com/Pobytov

ISBN: 978 1 78294 720 2

With thanks to Alison Palin for the proofreading.
With thanks to Jan Greenway for the copyright research.

Printed by Elanders Ltd, Newcastle upon Tyne.
Clipart from Corel®

Chapter 1 Proof

1. Proof by Contraction

In Year 1, you saw a few different kinds of mathematical proof. There's only one new type to learn here, and that's proof by contradiction.

Learning Objective:

- Be able to use proof by contradiction to show that statements are true.

Types of proof

Proof in mathematics is all about using logical arguments to show that a statement is true or false. So far, you've learned three main types of proof:

- **Proof by deduction**: using known facts and logic to show that the statement must be true.
- **Proof by exhaustion**: breaking a statement down into two or more cases that cover all possible situations, then showing that the statement is true for all of them.
- **Disproof by counter-example**: giving one example that shows that the statement is not true.

The other method that you need to know is **proof by contradiction**.

Tip: Take a look at your Year 1 notes if you can't remember how these proofs work.

Proof by contradiction

To prove a statement by **contradiction**, you start by saying "assume the statement **is not true**...". You then show that this would mean that something **impossible** would have to be true, which means that the initial assumption has to be wrong, so the original statement must be true.

Example 1

Prove the following statement: *"If x^2 is even, then x must be even."*

- You can prove this statement by contradiction. Assume the statement is **not true**. Then there must be an **odd number** x for which x^2 is even.
- If x is odd, then you can write x as $\mathbf{2k + 1}$, where k is an integer.
- Now, $x^2 = (2k + 1)^2 = 4k^2 + 4k + 1$
  $4k^2 + 4k = 2(2k^2 + 2k)$ is **even** because it is 2 × an integer
  $\Rightarrow$ $4k^2 + 4k + 1$ is **odd**.
- But this **isn't possible** if the statement that x^2 is even is true. You've **contradicted** the assumption that there is an odd number x for which x^2 is even.
- So if x^2 is **even**, then x must be **even**, hence the original statement is **true**.

Example 2

Prove that $\sqrt{2}$ is irrational.

- Start by assuming that the statement is **not true**, and that $\sqrt{2}$ can be written as $\frac{a}{b}$ with a and b both **non-zero integers**. You can also assume that a and b do **not** have any **common factors**.

Tip: An irrational number is a real number that can't be written as a fraction $\frac{a}{b}$ (where a and b are both integers and $b \neq 0$). Here, you can assume that a and b have no common factors because if they did, you could simplify the fraction to get a new a and b.

Tip: You can use the same method to prove the irrationality of any surd, although you need to prove the statement *"If x^2 is a multiple of a prime number p, then x must also be a multiple of p"*, which is a bit trickier than the proof in Example 1.

- If $\sqrt{2} = \frac{a}{b}$, then $\sqrt{2}b = a$
- Squaring both sides gives you $2b^2 = a^2$ — so a^2 is an **even** number.
- You saw in the previous example that if a^2 is **even**, then a must be **even** as well. So replace a with $2k$ for some integer k:
  $2b^2 = (2k)^2 = 4k^2 \Rightarrow b^2 = 2k^2$
- Like before, this tells you that b must be **even** (since b^2 is even). However, you assumed at the start that a and b had **no common factors**, so you have **contradicted** your initial assumption.
- Therefore $\sqrt{2}$ **cannot** be written as a fraction $\frac{a}{b}$, so it is **irrational**.

Tip: This example uses proof by contradiction to prove that there is no 'largest number' of a certain type. You assume that there is one, then you can simply add to it to get a bigger number of the same type, which contradicts your initial assumption. You can use this trick in quite a few proofs.

Tip: Some proofs use $P = p_1p_2...p_n + 1$ — but the method is the same.

Example 3

Prove by contradiction that there are infinitely many prime numbers.

- Assume that there are a **finite** number of primes (say n), and list them all:
  $p_1 = 2, p_2 = 3, p_3 = 5, \ldots, p_{n-1}, p_n$.
- Now **multiply** all of these together: $p_1p_2p_3...p_{n-1}p_n$ — call this number P.
- Because of how you defined it, P is a **multiple** of **every** prime number.
- Now think about $P + 1$ — if you **divide** $P + 1$ by p_1, you get:
  $$(P + 1) \div p_1 = (p_1p_2p_3...p_{n-1}p_n + 1) \div p_1$$
  $$= p_2p_3...p_{n-1}p_n \text{ remainder } 1$$
- In fact, dividing $(P + 1)$ by any prime number gives a **remainder** of **1**.
- So $(P + 1)$ **isn't divisible** by **any** of the prime numbers in the list, so either it is **also** a prime number or it is a **product** of some other prime numbers that **aren't** in the list.
- Either way, there is at least one prime number that is **not** on the list, which **contradicts** the assumption that the list contained **all** of them. So there must be **infinitely many** prime numbers.

Exercise 1.1

Q1 Prove, by contradiction, that there is no largest multiple of 3.

Q2 Prove that if x^2 is odd, then x must be odd.

Q3-5 Hint: Remember that every rational number can be written as a fraction $\frac{a}{b}$ where a and b are integers.

Q3 a) Prove that the product of a non-zero rational number and an irrational number is always irrational.

b) Disprove that the product of an irrational number and an irrational number is always irrational.

Q4 Prove that there is no smallest positive rational number.

Q5 Prove that $1 + \sqrt{2}$ is irrational.

Q6 a) Suppose x is an integer. Prove by exhaustion that if x^2 is a multiple of 3, then x must be a multiple of 3. PROBLEM SOLVING

b) Hence prove that $\sqrt{3}$ is irrational.

Chapter 2 Algebra and Functions

1. Simplifying Expressions

Simplifying expressions in this chapter involves a lot of algebraic fractions. You have to factorise, cancel, multiply, divide, add and subtract them. This will come in handy in other parts of maths, so it's a pretty important skill.

Learning Objectives:

- Be able to simplify rational expressions (i.e. algebraic fractions with linear or quadratic denominators) by factorising and cancelling.
- Be able to simplify rational expressions by adding and subtracting algebraic fractions.
- Be able to simplify rational expressions by multiplying and dividing algebraic fractions.
- Be able to simplify algebraic fractions with linear denominators by using algebraic division.

Simplifying algebraic fractions

Algebraic fractions are a lot like normal fractions — and you can treat them in the same way, whether you're adding, subtracting, multiplying or dividing them. All fractions are much easier to deal with when they're in their **simplest form**, so the first thing to do with algebraic fractions is to simplify them as much as possible.

- Look for **common factors** in the numerator and denominator — **factorise** top and bottom and see if there's anything you can **cancel**.
- If there's a **fraction** in the numerator or denominator (e.g. $\frac{1}{x}$), **multiply** the whole algebraic fraction (i.e. top and bottom) by the same factor to get rid of it (for $\frac{1}{x}$, you'd multiply through by x).

Examples

Simplify the following:

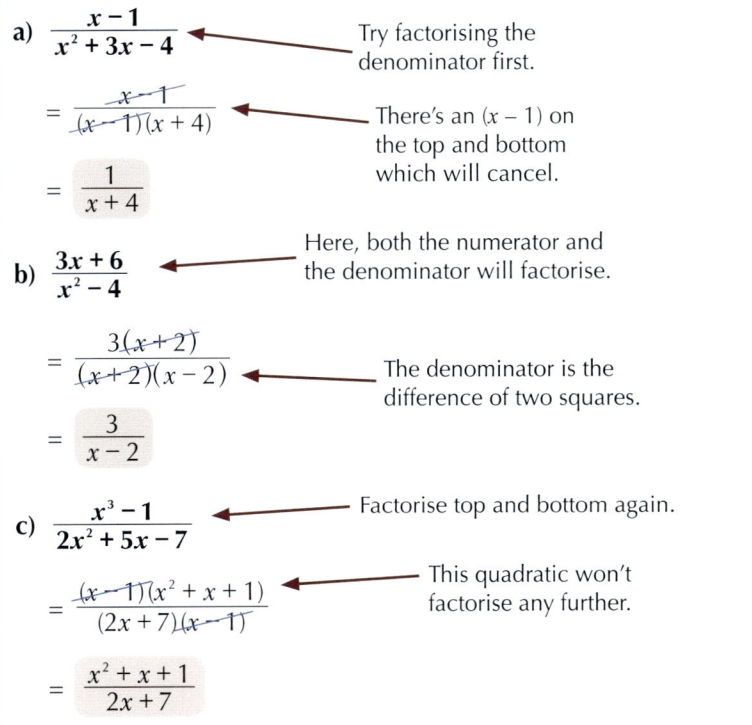

a) $\dfrac{x-1}{x^2+3x-4}$ ← Try factorising the denominator first.

$= \dfrac{\cancel{x-1}}{\cancel{(x-1)}(x+4)}$ ← There's an $(x-1)$ on the top and bottom which will cancel.

$= \dfrac{1}{x+4}$

b) $\dfrac{3x+6}{x^2-4}$ ← Here, both the numerator and the denominator will factorise.

$= \dfrac{3\cancel{(x+2)}}{\cancel{(x+2)}(x-2)}$ ← The denominator is the difference of two squares.

$= \dfrac{3}{x-2}$

c) $\dfrac{x^3-1}{2x^2+5x-7}$ ← Factorise top and bottom again.

$= \dfrac{\cancel{(x-1)}(x^2+x+1)}{(2x+7)\cancel{(x-1)}}$ ← This quadratic won't factorise any further.

$= \dfrac{x^2+x+1}{2x+7}$

Tip: Remember — if the coefficients of a polynomial sum to 0, then $(x-1)$ is a factor.

In part c), $1 + (-1) = 0$, so $(x-1)$ is a factor of the cubic. So write $x^3 - 1 = (x-1)(ax^2+bx+c)$ then expand the brackets and compare the coefficients: $1 = a$

$-1 = (-1) \times c \Rightarrow c = 1$

$0 = b - a \Rightarrow b = 1$

Tip: Take your time with messy expressions and work things out in separate steps.

d) $\dfrac{2+\frac{1}{2x}}{4x^2+x}$ ← Factorise the denominator.

$= \dfrac{2+\frac{1}{2x}}{x(4x+1)}$ ← Get rid of this fraction by multiplying the top and bottom by $2x$.

$= \dfrac{\left(2+\frac{1}{2x}\right)\times 2x}{x(4x+1)\times 2x}$

$= \dfrac{4x+1}{2x^2(4x+1)} = \dfrac{1}{2x^2}$

Exercise 1.1

Simplify the following:

Q1 $\dfrac{4}{2x+10}$

Q2 $\dfrac{5x}{x^2+2x}$

Q3 $\dfrac{6x^2-3x}{3x^2}$

Q4 $\dfrac{4x^3}{x^3+3x^2}$

Q5 $\dfrac{3x+6}{x^2+3x+2}$

Q6 $\dfrac{x^2+3x}{x^2+x-6}$

Q7 $\dfrac{2x-6}{x^2-9}$

Q8 $\dfrac{5x^2-20x}{2x^2-5x-12}$

Q9 $\dfrac{3x^2-7x-6}{2x^2-x-15}$

Q10 $\dfrac{x^3-4x^2-19x-14}{x^2-6x-7}$

Q14 Hint: You need to multiply the top and bottom by the same term to get rid of the fractions. Don't just multiply the top by $2x$ and the bottom by x.

Q11 $\dfrac{x^3-2x^2}{x^3-4x}$

Q12 $\dfrac{1+\frac{1}{x}}{x+1}$

Q13 $\dfrac{3+\frac{1}{x}}{2+\frac{1}{x}}$

Q14 $\dfrac{1+\frac{1}{2x}}{2+\frac{1}{x}}$

Q15 $\dfrac{\frac{1}{3x}-1}{3x^2-x}$

Q16 $\dfrac{2+\frac{1}{x}}{6x^2+3x}$

Q17 $\dfrac{\frac{3x}{x+2}}{\frac{x}{x+2}+\frac{1}{x+2}}$

Q18 $\dfrac{2+\frac{1}{x+1}}{3+\frac{1}{x+1}}$

Q20 Hint: Multiplying each term by x^2 will get rid of all the fractions.

Q19 $\dfrac{1-\frac{2}{x+3}}{x+2}$

Q20 $\dfrac{4-\frac{1}{x^2}}{2-\frac{1}{x}-\frac{1}{x^2}}$

Adding and subtracting algebraic fractions

You'll have come across adding and subtracting fractions before, so here's a little reminder of how to do it:

1. Find the common denominator

- Take all the individual 'bits' from the bottom lines and **multiply** them together.
- Only use each bit **once** unless something on the bottom line is raised to a **power**.

2. Put each fraction over the common denominator

- Multiply both top and bottom of each fraction by the same factor — whichever factor will turn the denominator into the **common denominator**.

3. Combine into one fraction

- Once everything's over the common denominator you can just **add** (or **subtract**) the **numerators**.

Tip: The common denominator should be the lowest common multiple (LCM) of all the denominators.

Examples

a) **Simplify:** $\frac{2}{x-1} - \frac{3}{3x+2}$

- Multiply the denominators to get the **common denominator**:

$$(x-1)(3x+2)$$

- Multiply the top and bottom lines of each fraction by whichever factor changes the denominator into the common denominator:

$$\frac{2\times(3x+2)}{(x-1)\times(3x+2)} - \frac{3\times(x-1)}{(3x+2)\times(x-1)}$$

- All the denominators are the same — so you can just subtract the numerators:

$$\frac{2(3x+2)-3(x-1)}{(3x+2)(x-1)} = \frac{6x+4-3x+3}{(3x+2)(x-1)} = \frac{3x+7}{(3x+2)(x-1)}$$

Tip: Always check if there's any more factorising and cancelling that can be done at the end. Your final answer needs to be fully simplified to get all the marks in an exam question.

b) **Simplify:** $\frac{2y}{x(x+3)} + \frac{1}{y^2(x+3)} - \frac{x}{y}$

The individual 'bits' here are x, $(x + 3)$ and y, but you need to use y^2 because there's a y^2 in the denominator of the second fraction.

- The common denominator is: $xy^2(x+3)$
- Multiply the top and bottom lines of each fraction by whichever factor changes the denominator into the common denominator:

$$\frac{2y\times y^2}{x(x+3)\times y^2} + \frac{1\times x}{y^2(x+3)\times x} - \frac{x\times xy(x+3)}{y\times xy(x+3)}$$

- All the denominators are the same — so you can just add the numerators:

$$= \frac{2y^3+x-x^2y(x+3)}{xy^2(x+3)} = \frac{2y^3+x-x^3y-3x^2y}{xy^2(x+3)}$$

Tip: In theory there's nothing wrong here with having a common denominator of all the denominators multiplied together (i.e. $xy^3(x+3)^2$). You'd still get the same final answer by cancelling down. Being a bit clever about it saves you a lot of effort though, so always try to use the simplest common denominator possible (the LCM).

Exercise 1.2

Simplify the following:

Q1 $\frac{2x}{3}+\frac{x}{5}$

Q2 $\frac{2}{3x}-\frac{1}{5x}$

Q3 $\frac{3}{x^2}+\frac{2}{x}$

Q4 $\frac{x+1}{3}+\frac{x+2}{4}$

Q5 $\frac{2x}{3}+\frac{x-1}{7x}$

Q6 $\frac{3x}{4}-\frac{2x-1}{5x}$

Q7 $\frac{2}{x-1}+\frac{3}{x}$

Q8 $\frac{3}{x+1}+\frac{2}{x+2}$

Q9 $\frac{4}{x-3}-\frac{1}{x+4}$

Q10 $\frac{6}{x+2}+\frac{6}{x-2}$

Q11 $\frac{3}{x-2}-\frac{5}{2x+3}$

Q12 $\frac{3}{x+2}+\frac{x}{x+1}$

Q13 $\frac{5x}{(x+1)^2}-\frac{3}{x+1}$

Q14 $\frac{5}{x(x+3)}+\frac{3}{x+2}$

Q15 $\frac{x}{x^2-4}-\frac{1}{x+2}$

Q16 $\frac{3}{x+1}+\frac{6}{2x^2+x-1}$

Q17 $\frac{2}{x}+\frac{3}{x+1}+\frac{4}{x+2}$

Q18 $\frac{3}{x+4}-\frac{2}{x+1}+\frac{1}{x-2}$

Q19 $2-\frac{3}{x+1}+\frac{4}{(x+1)^2}$

Q20 $\frac{2x^2-x-3}{x^2-1}+\frac{1}{x(x-1)}$

Q2 Hint: Both denominators are a multiple of x. So the common denominator will just be another multiple of x.

Q15 Hint: Factorise the first denominator before you do anything else.

Q19 Hint: To turn the 2 into a fraction just use '1' as the denominator.

Multiplying and dividing algebraic fractions

Multiplying algebraic fractions

You **multiply** algebraic fractions in exactly the same way that you multiply normal fractions — multiply the numerators together, then multiply the denominators. Try to **cancel** any **common factors** before multiplying.

Example 1

Simplify the following:

a) $\frac{x^3}{2y}\times\frac{8y^2}{3}$

$=\frac{x^3}{{}_1\not{2}\not{y}}\times\frac{\not{8}\not{y^2}\,4y}{3}$ ← Cancel all common factors.

$=\frac{x^3\times 4y}{1\times 3}=\frac{4x^3y}{3}$ ← Then multiply top by top and bottom by bottom.

b) $\frac{x^2-2x-15}{2x+8}\times\frac{x^2-16}{x^2+3x}$

$=\frac{\cancel{(x+3)}(x-5)}{2\cancel{(x+4)}}\times\frac{\cancel{(x+4)}(x-4)}{x\cancel{(x+3)}}$ ← Factorise the expressions in both fractions and cancel.

$=\frac{(x-5)(x-4)}{2x}\left(=\frac{x^2-9x+20}{2x}\right)$ ← Then just multiply as before.

Tip: Check whether you can cancel down any further at the end, just in case you missed something before.

Tip: In cases like these, it's a lot easier to do all the factorising and cancelling before you multiply.

Dividing algebraic fractions

To **divide** by an algebraic fraction, you just **multiply** by its reciprocal. The reciprocal is 1 ÷ the original thing — for fractions you just turn the fraction upside down.

Tip: This is the same method that you'd use for normal fractions.

Example 2

Simplify the following:

a) $\frac{8}{5x} \div \frac{12}{x^3}$

Turn the second fraction upside down.

$= \frac{8}{5x} \times \frac{x^3}{12} = \frac{\cancel{8}^{2}}{5\cancel{x}} \times \frac{\cancel{x^3}^{x^2}}{\cancel{12}_{3}}$ ← Cancel all common factors.

$= \frac{2 \times x^2}{5 \times 3} = \frac{2x^2}{15}$ ← Now multiply.

b) $\frac{3x}{5} \div \frac{3x^2 - 9x}{20}$

$= \frac{\cancel{3x}}{\cancel{5}} \times \frac{\cancel{20}^{4}}{\cancel{3x}(x-3)}$ ← Turn the second fraction upside down, and then cancel. Note that the $3x^2 - 9x$ has been factorised first.

$= \frac{4}{x-3}$

Exercise 1.3

Simplify the following:

Q1 a) $\frac{2x}{3} \times \frac{5x}{4}$ b) $\frac{6x^3}{7} \times \frac{2}{x^2}$

c) $\frac{8x^2}{3y^2} \times \frac{x^3}{4y}$ d) $\frac{8x^4}{3y} \times \frac{6y^2}{5x}$

Q1 Hint: Remember — cancelling **before** you multiply will make things a whole lot simpler.

Q2 a) $\frac{x}{3} \div \frac{3}{x}$ b) $\frac{4x^3}{3} \div \frac{x}{2}$

c) $\frac{3}{2x} \div \frac{6}{x^3}$ d) $\frac{2x^3}{3y} \div \frac{4x}{y^2}$

Q3 $\frac{x+2}{4} \times \frac{x}{3x+6}$

Q4 $\frac{4x}{5} \div \frac{4x^2+8x}{15}$

Q5 $\frac{2x^2-2}{x} \times \frac{5x}{3x-3}$

Q6 $\frac{2x^2+8x}{x^2-2x} \times \frac{x-1}{x+4}$

Q5 Hint: Always be on the look out for hidden 'difference of two squares' expressions.

Q7 $\frac{x^2-4}{9} \div \frac{x-2}{3}$

Q8 $\frac{2}{x^2+4x} \div \frac{1}{x+4}$

Q9 $\frac{x^2+4x+3}{x^2+5x+6} \times \frac{x^2+2x}{x+1}$

Q10 $\frac{x^2+5x+6}{x^2-2x-3} \times \frac{3x+3}{x^2+2x}$

Q11 $\frac{x^2-4}{6x-3} \times \frac{2x^2+5x-3}{x^2+2x}$

Q12 $\frac{x^2+7x+6}{4x-4} \div \frac{x^2+8x+12}{x^2-x}$

Q13 $\frac{x^2+4x+4}{x^2-4x+3} \times \frac{x^2-2x-3}{2x^2-2x} \times \frac{4x-4}{x^2+2x}$

Q14 $\frac{x}{6x+12} \div \frac{x^2-x}{x+2} \times \frac{3x-3}{x+1}$

Q15 $\frac{x^2+5x}{2x^2+7x+3} \times \frac{2x+1}{x^3-x^2} \div \frac{x+5}{x^2+x-6}$

Q14 Hint: Turn the fraction you're dividing by upside down and multiply.

Algebraic division

Important terms

There are a few words that come up a lot in algebraic division, so make sure you know what they all mean.

- **Degree** — the highest power of x in the polynomial. For example, the degree of $4x^5 + 6x^2 - 3x - 1$ is 5.
- **Divisor** — this is the thing you're dividing by. For example, if you divide $x^2 + 4x - 3$ by $x + 2$, the divisor is $x + 2$.
- **Quotient** — the bit that you get when you divide by the divisor (not including the **remainder** — see below).

Tip: A **polynomial** is an algebraic expression made up of the sum of constant terms and variables raised to positive integer powers.

For example, $x^3 - 2x + \frac{1}{2}$ is a polynomial, but $x^{-3} - 2x^{\frac{3}{2}}$ is not as it has a negative power and a fractional power of x.

Method 1 — using the formula

There's a handy **formula** you can use to do algebraic division:

A polynomial f(x) can be written in the form:

$$\mathbf{f(x) \equiv q(x)d(x) + r(x)}$$

where: q(x) is the quotient,
d(x) is the divisor,
and r(x) is the remainder.

Tip: The $\equiv$ symbol means it's an identity.

For example:

$$\mathbf{4x^5 - 7x^2 + 3x - 9 \div x^2 - 5x + 8.}$$

This bit is f(x). It has a degree of 5.

This bit is d(x), the divisor. It has a degree of 2.

Tip: The degree of the divisor will always be less than or equal to the degree of the polynomial.

In the exam, you'll only have to divide by a linear factor (i.e. with a degree of 1). Here's a step-by-step guide to using the formula:

- First, you have to work out the **degree** of the **quotient**, which will depend on the degree of the polynomial f(x): **deg q(x) = deg f(x) – 1**. The **remainder** will have degree **0**.
- Write out the division using the formula, but replace q(x) and r(x) with **general polynomials**. For example, a general polynomial of degree 2 is $Ax^2 + Bx + C$, where A, B and C are constants to be found. A general polynomial of degree 0 is just a constant, e.g. D.
- The next step is to work out the values of the **constants** (A, B, etc.). You do this by substituting in values for x to make bits disappear, and by **equating coefficients**.
- It's best to start with the **constant term** and work **backwards** from there.
- Finally, write out the division again, replacing A, B, C, etc. with the values you've found.

Tip: You can use this formula with divisors of higher degrees — the degree of the quotient is deg f(x) – deg d(x), and the degree of the remainder has to be less than the degree of the divisor.

When you're using this method, you might have to use **simultaneous equations** to work out some of the coefficients (have a look back at your Year 1 notes for a reminder of how to do this if you need to). The method looks a bit tricky, but follow through the examples on the next page to see how it works.

Tip: Equating coefficients means comparing the coefficients of each power of x on the left hand side and the right hand side of the identity.

Example 1

Divide $x^4 - 3x^3 - 3x^2 + 10x + 5$ by $x - 2$.

- First, work out the **degrees** of the **quotient** and **remainder**:
  f(x) has degree 4, so the quotient q(x) has degree 4 – 1 = 3.
  The remainder r(x) has degree 0.

- Write out the division in the form **f(x) ≡ q(x)d(x) + r(x)**, replacing q(x) and r(x) with general polynomials of degree 3 and 0:
  $$x^4 - 3x^3 - 3x^2 + 10x + 5 \equiv (Ax^3 + Bx^2 + Cx + D)(x - 2) + E$$

- Substitute $x = 2$ into the identity to make the q(x)d(x) bit disappear:
  $$16 - 3(8) - 3(4) + 10(2) + 5 = 0 + E$$
  $$16 - 24 - 12 + 20 + 5 = E \Rightarrow E = 5$$

  So now the identity looks like this:
  $$x^4 - 3x^3 - 3x^2 + 10x + 5 \equiv (Ax^3 + Bx^2 + Cx + D)(x - 2) + 5$$

- Now substitute $x = 0$ into the identity:
  when $x = 0$, $5 = -2D + 5$

- Solving this gives D = 0. So now you have:
  $$x^4 - 3x^3 - 3x^2 + 10x + 5 \equiv (Ax^3 + Bx^2 + Cx)(x - 2) + 5$$

- For the remaining terms, **equate the coefficients** on both sides. Expanding the brackets on the RHS and collecting terms gives:
  $$x^4 - 3x^3 - 3x^2 + 10x + 5 \equiv Ax^4 + (B - 2A)x^3 + (C - 2B)x^2 - 2Cx + 5$$

- The coefficient of x^4 is A on the RHS and 1 on the LHS, so A = 1. Similarly, comparing the coefficients of x^3 gives B – 2A = –3 (so B = –1) and comparing the coefficients of x gives –2C = 10 (so C = –5). So the identity looks like this:
  $$x^4 - 3x^3 - 3x^2 + 10x + 5 \equiv (x^3 - x^2 - 5x)(x - 2) + 5$$

Tip: After you've done a few of these you'll get used to spotting what the coefficients are going to be in terms of A, B etc. so you won't have to expand the brackets fully each time.

Example 2

Divide $x^3 + 5x^2 - 18x - 18$ by $x - 3$.

- f(x) has degree 3, so q(x) has degree 3 – 1 = 2.

- Write out the division in the form f(x) ≡ q(x)d(x) + r(x):
  $$x^3 + 5x^2 - 18x - 18 \equiv (Ax^2 + Bx + C)(x - 3) + D$$

- Putting $x = 3$ into the identity gives D = 0, so:
  $$x^3 + 5x^2 - 18x - 18 \equiv (Ax^2 + Bx + C)(x - 3)$$

- Now, setting $x = 0$ gives the equation –18 = –3C, so C = 6.
  $$x^3 + 5x^2 - 18x - 18 \equiv (Ax^2 + Bx + 6)(x - 3)$$

- Equating the coefficients of x^3 and x^2 gives A = 1 and B – 3A = 5, so B = 8. So:
  $$x^3 + 5x^2 - 18x - 18 \equiv (x^2 + 8x + 6)(x - 3)$$

Tip: A remainder of zero means it divides exactly (this is from the Factor Theorem).

Tip: If you were asked to solve the equation f(x) = 0, you would set each bracket equal to 0 and find the solutions — $x = 3$ and $x = -4 \pm \sqrt{10}$.

Simply stating the identity at the end doesn't always answer the question. If you're asked to divide one thing by another, then you might need to state the **quotient** and the **remainder** which you've worked out using the formula.

So for Example 1 on the previous page:
$(x^4 - 3x^3 - 3x^2 + 10x + 5) \div (x - 2) = \mathbf{x^3 - x^2 - 5x}$ **remainder 5**.

For Example 2:
$(x^3 + 5x^2 - 18x - 18) \div (x - 3) = \mathbf{x^2 + 8x + 6}$ (i.e. remainder 0).

Tip: You might have come across this method in Year 1.

Method 2 — algebraic long division

You can also use **long division** to divide two algebraic expressions (using the same method you'd use for numbers).

Tip: Note that you only divide each term by the 'x' term, not the '$x - 5$'. The -5 bit is dealt with in the steps in between.

Example 3

Divide $(2x^3 - 7x^2 - 16x + 11)$ by $(x - 5)$.

- Start by dividing the first term in the polynomial by the first term of the divisor: $2x^3 \div x = 2x^2$. Write this answer above the polynomial:

$$\begin{array}{rl} & 2x^2 \\ x-5 & \overline{)\,2x^3 - 7x^2 - 16x + 11} \end{array}$$

- Multiply the divisor $(x - 5)$ by this answer $(2x^2)$ to get $2x^3 - 10x^2$:

$$\begin{array}{rl} & 2x^2 \\ x-5 & \overline{)\,2x^3 - 7x^2 - 16x + 11} \\ & \;\;2x^3 - 10x^2 \end{array}$$

- Subtract this from the main expression to get $3x^2$. Bring down the $-16x$ term just to make things clearer for the next subtraction.

$$\begin{array}{rl} & 2x^2 \\ x-5 & \overline{)\,2x^3 - 7x^2 - 16x + 11} \\ & -\underline{(2x^3 - 10x^2)} \\ & \qquad\qquad 3x^2 - 16x \end{array}$$

- Now divide the first term of the remaining polynomial $(3x^2)$ by the first term of the divisor (x) to get $3x$ (the second term in the answer).

$$\begin{array}{rl} & 2x^2 + 3x \\ x-5 & \overline{)\,2x^3 - 7x^2 - 16x + 11} \\ & -\underline{(2x^3 - 10x^2)} \\ & \qquad\qquad 3x^2 - 16x \end{array}$$

- Multiply $(x - 5)$ by $3x$ to get $3x^2 - 15x$, then subtract again and bring down the $+11$ term.

$$\begin{array}{rl} & 2x^2 + 3x \\ x-5 & \overline{)\,2x^3 - 7x^2 - 16x + 11} \\ & -\underline{(2x^3 - 10x^2)} \\ & \qquad\qquad 3x^2 - 16x \\ & \qquad\quad -\underline{(3x^2 - 15x)} \\ & \qquad\qquad\qquad\quad -x + 11 \end{array}$$

- Divide $-x$ by x to get -1 (the third term in the answer). Then multiply $(x - 5)$ by -1 to get $-x + 5$.

$$
\begin{array}{r}
2x^2 + 3x - 1 \\
x-5 \overline{\big)\, 2x^3 - 7x^2 - 16x + 11} \\
-\ \underline{(2x^3 - 10x^2)} \downarrow \\
3x^2 - 16x \\
-\ \underline{(3x^2 - 15x)} \downarrow \\
-x + 11 \\
-\ \underline{(-x + 5)} \\
6
\end{array}
$$

- After subtracting, this term (6) has a degree that's **less** than the degree of the divisor, $(x - 5)$, so it can't be divided. This is the **remainder**.
- So $(2x^3 - 7x^2 - 16x + 11) \div (x - 5) =$ $2x^2 + 3x - 1$ remainder 6.
- This could also be written as:

$$\frac{2x^3 - 7x^2 - 16x + 11}{x - 5} = 2x^2 + 3x - 1 + \frac{6}{x - 5}$$

Tip: You can multiply the quotient by $(x - 5)$, and then add on the remainder, 6, to check you've got it right.

Exercise 1.4

Q1 Use the formula $f(x) \equiv q(x)d(x) + r(x)$ to divide the following expressions. In each case, state the quotient and remainder.

a) $(x^3 - 14x^2 + 6x + 11) \div (x + 1)$

b) $(2x^3 + 5x^2 - 8x - 17) \div (x - 2)$

c) $(6x^3 + x^2 - 11x - 5) \div (2x + 1)$

Q1 Hint: If you're told which method to use make sure you show all your working clearly to prove that you know how to use the method.

Q2 Write $3x^4 - 8x^3 - 6x - 4$ in the form $(Ax^3 + Bx^2 + Cx + D)(x - 3) + E$, and hence state the result when $3x^4 - 8x^3 - 6x - 4$ is divided by $x - 3$.

Q2 Hint: This is just another way of asking you to use the formula.

Q3 Use long division to divide the following expressions. In each case, state the quotient and remainder.

a) $(x^3 - 14x^2 + 6x + 11) \div (x + 1)$

b) $(x^3 + 10x^2 + 15x - 13) \div (x + 3)$

c) $(2x^3 + 5x^2 - 8x - 17) \div (x - 2)$

d) $(3x^3 - 78x + 9) \div (x + 5)$

e) $(x^4 - 1) \div (x - 1)$

f) $(8x^3 - 6x^2 + x + 10) \div (2x - 3)$

Q3d) Hint: Add in a $0x^2$ term to make sure you don't miss any terms when dividing.

In the following questions you can choose which method to use.

Q4 Divide $10x^3 + 7x^2 - 5x + 21$ by $2x + 1$, stating the quotient and remainder.

Q5 Divide $16x^4$ by $2x - 3$, stating the quotient and remainder.

Q6 Divide $3x^4 + 7x^3 - 22x^2 - 8x$ by $x - 2$, and hence solve the equation $3x^4 + 7x^3 - 22x^2 - 8x = 0$. PROBLEM SOLVING

2. Mappings and Functions

Learning Objectives:

- Understand the definitions of mappings and functions and be able to deduce whether a given mapping is a function.
- Be able to use function notation.
- Be able to identify the domain and range of a given function.
- Be able to restrict the domain of a mapping to make it a function.
- Be able to identify whether a function is one-to-one or many-to-one for a given domain.

A mapping is just a set of instructions that tells you how to get from one value to another, and a function is a special kind of mapping.

Mappings and functions

Mappings

A **mapping** is an operation that takes one number and transforms it into another. For example, 'multiply by 5', 'square root' and 'divide by 7' are all mappings. The set of numbers you start with is called the **domain**, and the set of numbers they become is called the **range**.

Mappings can be drawn as **mapping diagrams**, like the one shown here for 'multiply by 5 and add 1' acting on the domain $\{-1, 0, 1, 2\}$:

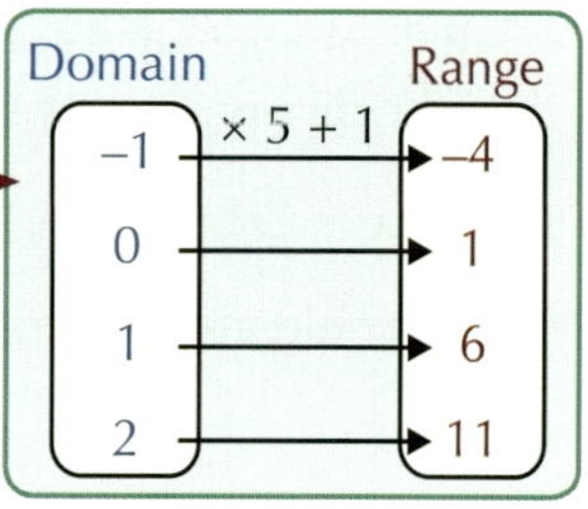

Use the notation $\{1, 2, ...\}$ for the domain and range if they are a **discrete** list of values. If they can take **any value** above or below a limit, use e.g. $x \geq 0$.

The domain and/or range will often be the set of **real numbers**, $\mathbb{R}$. A real number is any positive or negative number (or 0) — including fractions, decimals, integers and surds. If x can take any real value, it's usually written as $x \in \mathbb{R}$. Other sets of numbers include $\mathbb{Z}$, the set of **integers**, and $\mathbb{N}$, the set of **natural numbers** (positive integers, not including 0).

Tip: The symbol $\in$ means "belongs to" or "is a member of the set"— so $x \in \mathbb{R}$ means that x belongs to the set of real numbers. You can also give domains and ranges using set notation, e.g. $\{x : x > 0\}$ or $\{x : x \in \mathbb{Z}\}$.

Tip: Another set of numbers is $\mathbb{C}$, the complex numbers, made up of 'imaginary' numbers — but you don't meet these in A-level Maths.

You might have to work out the range of a mapping from the domain you're given. For example, $y = x^2$, $x \in \mathbb{R}$ has the range $y \geq 0$, as the squares of all real numbers are positive (or zero).

Functions

Some mappings take every number in the domain to exactly **one** number in the range. These mappings are called **functions**. Functions are written using the following notation:

$$f(x) = 5x + 1 \quad \text{or} \quad f : x \to 5x + 1$$

You've probably seen at least the f(x) notation before, but you need to be able to understand and use both.

You can substitute values for x into a function to find the value of the function at that point, as shown in the examples below.

Example 1

a) Give the value of f(–2) for the function $f(x) = x^2 - 1$.

Just replace each x in the function with –2 and calculate the answer:

$f(-2) = (-2)^2 - 1 = 4 - 1 = 3$

b) Find the value of x for which $f(x) = 12$ for the function $f : x \to 2x - 3$.

Solve this like a normal equation:

$2x - 3 = 12 \Rightarrow 2x = 15 \Rightarrow x = 7.5$

Functions can also be given in **several parts** (known as '**piecewise**' functions). Each part of the function will act over a different domain. For example:

$$f(x) = \begin{cases} 2x + 3 & x \leq 0 \\ x^2 & x > 0 \end{cases}$$

So f(2) is $2^2 = 4$ (because $x > 0$), but f(–2) is $2(-2) + 3 = -1$ (because $x \leq 0$).

Tip: Functions don't always use the letter 'f' — you'll see different letters used for functions over the next few pages.

If a mapping takes a number from the domain to **more than one** number in the range (or if it isn't mapped to any number in the range), it's **not** a function.

Example 2

The mapping shown here **is a function**, because any value of x in the domain maps to **only one** value in the range.

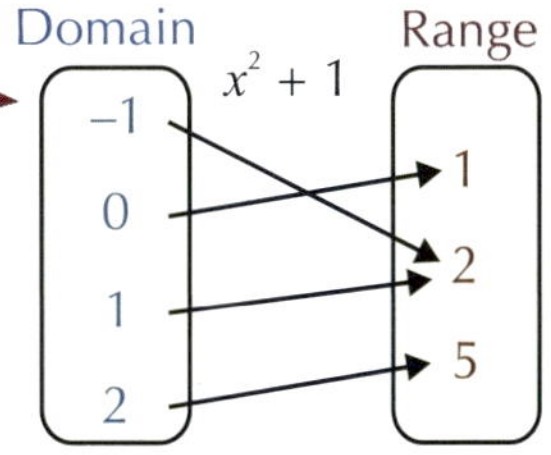

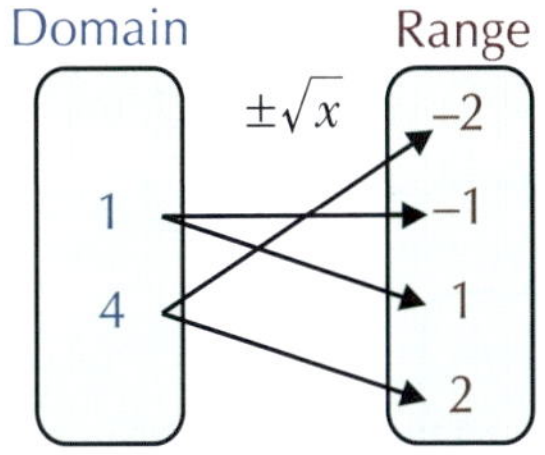

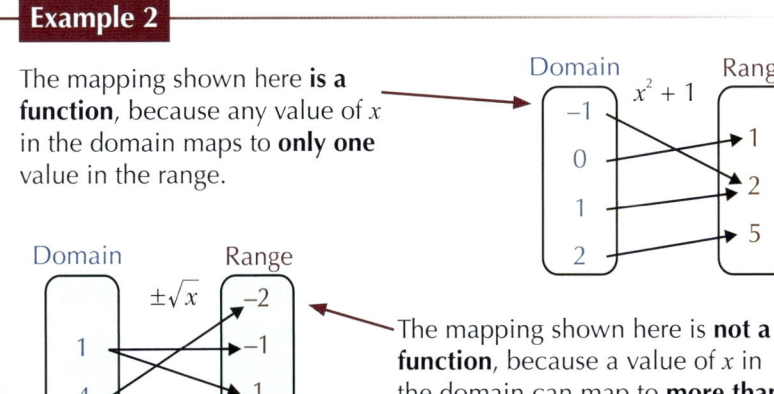

The mapping shown here is **not a function**, because a value of x in the domain can map to **more than one** value in the range.

Tip: Although each value in the domain only maps to one value in the range, the reverse is not true. This means it's a 'many to one' function — there's more about these on page 18.

Exercise 2.1

Q1 Draw a mapping diagram for the map "multiply by 6" acting on the domain {1, 2, 3, 4}.

Q2 $y = x + 4$ is a map with domain $\{x : x \in \mathbb{N}, x \leq 7\}$. Draw the mapping diagram.

Q2 Hint: $x \in \mathbb{N}$ just means that x is in the set of natural numbers (positive integers, not including 0). The $x \leq 7$ means that the domain must be the integers 1-7.

Q3 Complete the mapping diagrams below:

a)

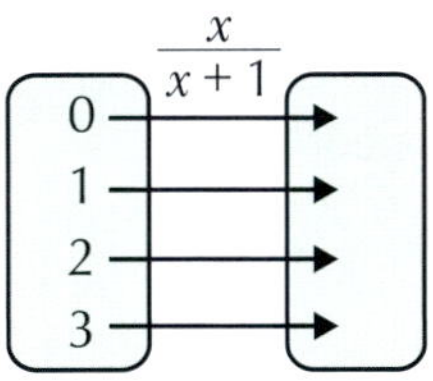

b)

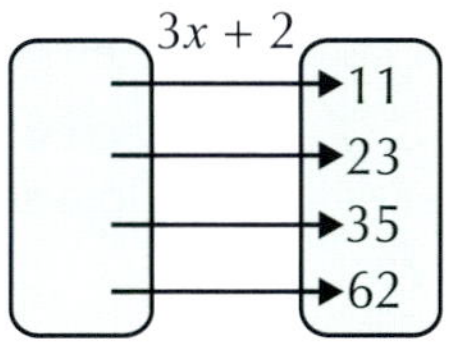

Q4 For the function $g : x \to \frac{1}{2x+1}$, $x > -\frac{1}{2}$, evaluate g(0) and g(2).

Q4 Hint: 'Evaluate' is just another way of asking you to 'find the value of'.

Q5 f defines a function $f : x \to \frac{1}{2 + \log_{10}x}$ for the domain $x > 0.01$. Evaluate f(1) and f(100).

Q6 a) Find the range of the function $h(x) = \sin x$, $0° \leq x \leq 180°$.

b) Find the range of $j(x) = \cos x$ on the same domain.

Q7 State the largest possible domain and range of each function:

a) $f(x) = 3^x - 1$

b) $g(x) = (\ln x)^2$

Q7 Hint: To find the largest possible domain, think about what values of x need to be excluded to make the function valid.

Q8 State whether or not each of the mapping diagrams below shows a function, and if not, explain why.

a) 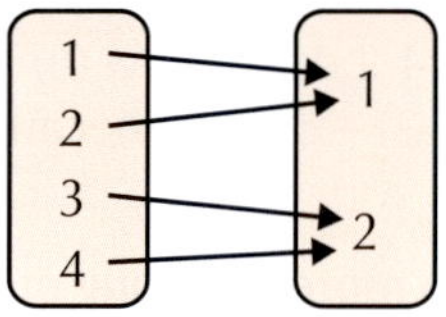

b) 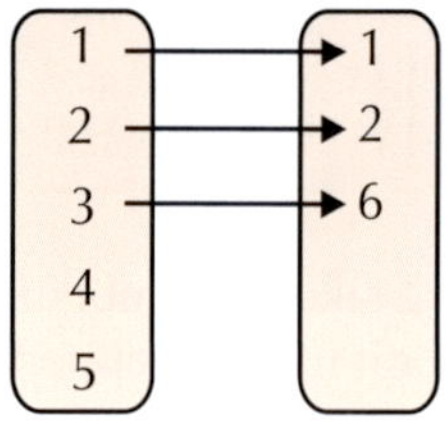

c) 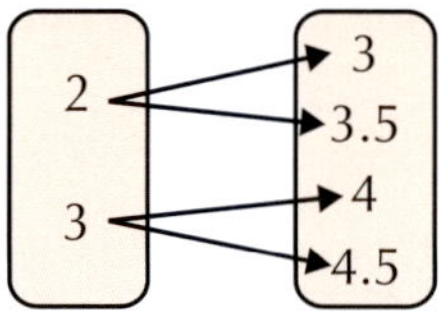

Graphs of functions

Mappings and functions with a **continuous** domain (such as $x \in \mathbb{R}$, i.e. not a discrete set of values) can be drawn as **graphs**. Drawing a graph of, say, $f(x) = x^2$ is exactly the same as drawing a graph of $y = x^2$. For each value of x in the **domain** (which goes along the horizontal x-axis) you can plot the corresponding value of $f(x)$ in the **range** (up the vertical y-axis):

Tip: If you've got a discrete set of values that x can take (e.g. $x \in \{1, 2, 3\}$) then draw a mapping diagram instead of a graph.

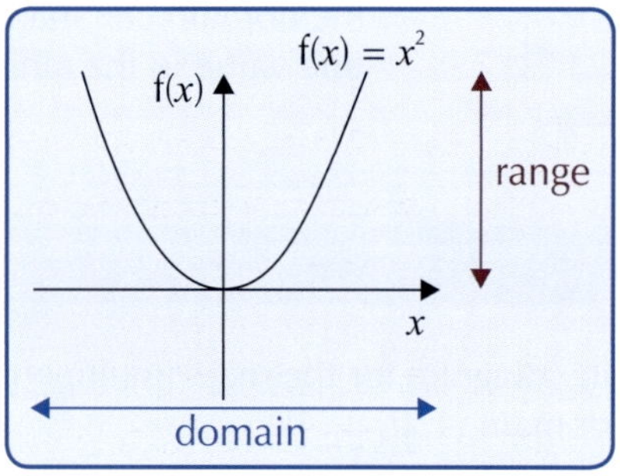

Drawing graphs can make it easier to **identify functions**, as shown below.

Example 1

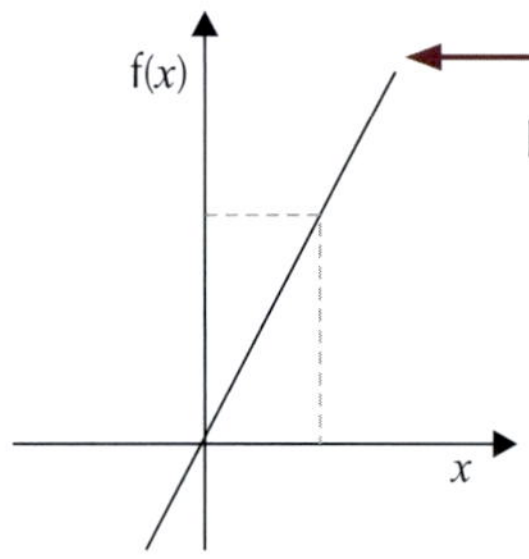

Functions

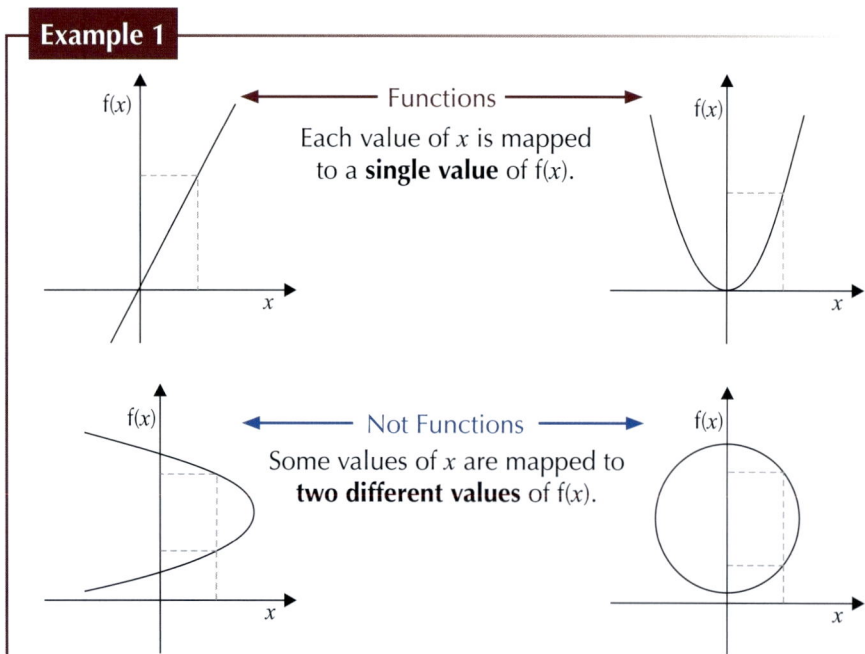

Each value of x is mapped to a **single value** of $f(x)$.

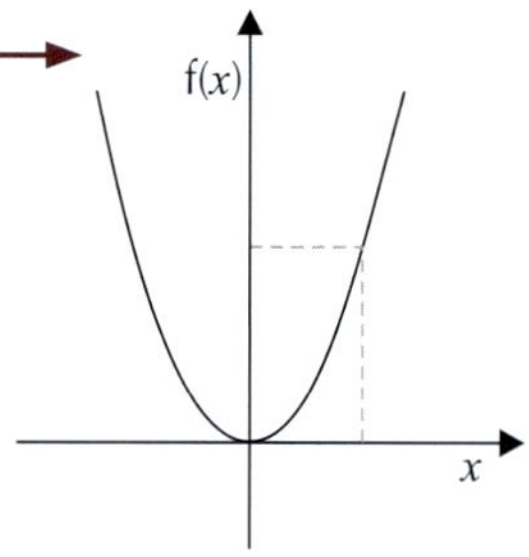

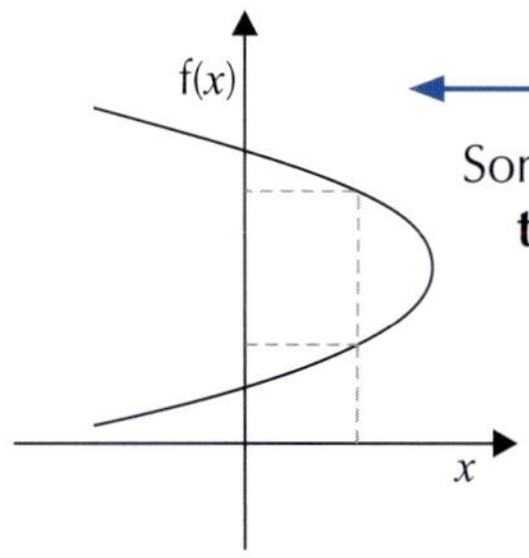

Not Functions

Some values of x are mapped to **two different values** of $f(x)$.

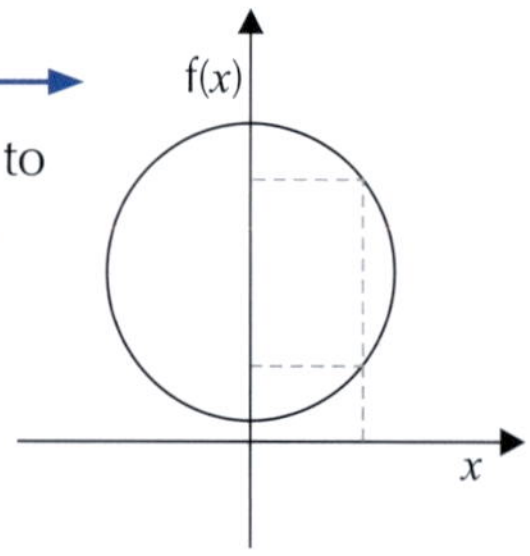

The graph on the right isn't a function for $x \in \mathbb{R}$ because f(x) is **not defined** for $x < 0$.
This just means that when x is negative there is no real value that f(x) can take.

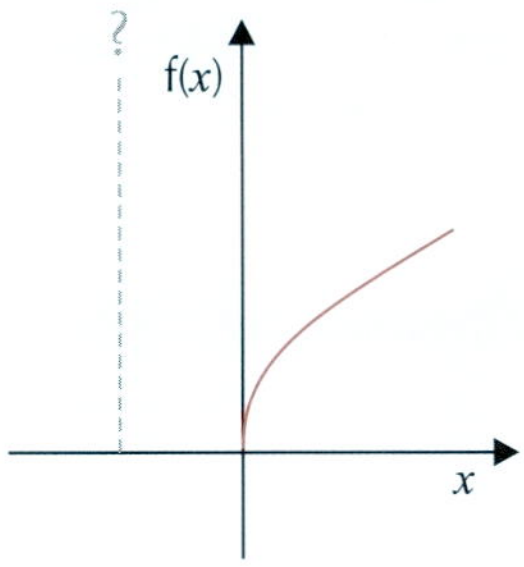

Tip: This could be turned into a function by restricting the domain to $x \geq 0$ — see page 16.

Finding ranges and domains using graphs

Sketching a graph can also be really useful when trying to find limits for the domain and range of a function.

Example 2

a) State the range for the function $f(x) = x^2 - 5$, $x \in \mathbb{R}$.

- The smallest possible value of x^2 is 0.
- So the smallest possible value of $x^2 - 5$ must be –5.
- So the range is $f(x) \geq -5$.
- This can be shown clearly by sketching a graph of $y = f(x)$:

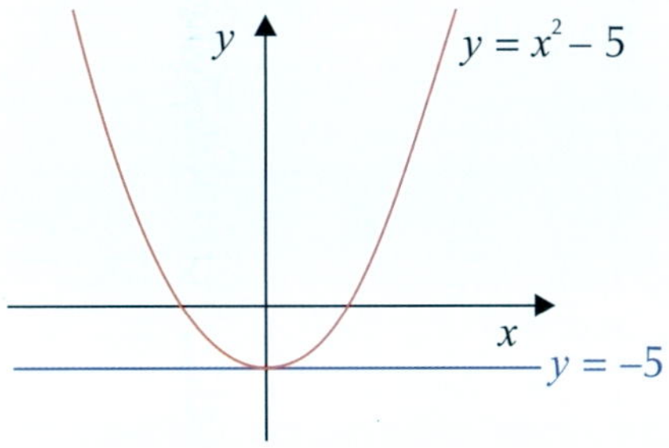

b) State the domain for $f(x) = \sqrt{(x-4)}$, giving your answer in set notation.

- There are no **real** solutions for the square root of a negative number.
- This means there is a limit on the domain so that $x - 4 \geq 0$.
- This gives a domain of $x \geq 4$ — in set notation, this is $\{x : x \geq 4\}$
- Again, this can be demonstrated by sketching a graph of $y = f(x)$:

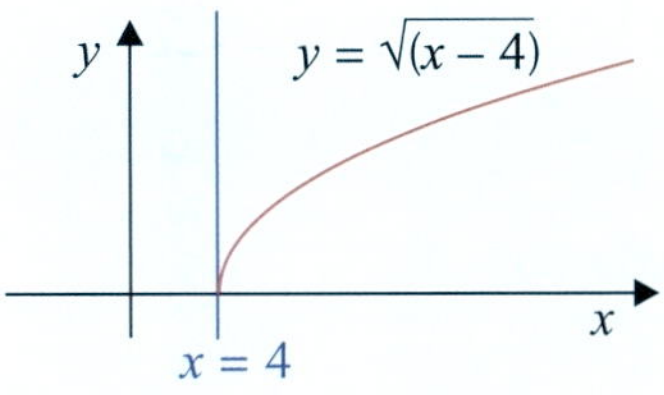

Tip: Remember, the : means "such that", so this is the set of x, such that x is greater than or equal to 4.
The range for this function, in set notation, is $\{f(x) : f(x) \geq 0\}$.

Turning mappings into functions

Some mappings that aren't functions can be turned into functions by **restricting their domain**.

For example, consider the graph of the mapping $y = \frac{1}{x-1}$ for $x \in \mathbb{R}$:

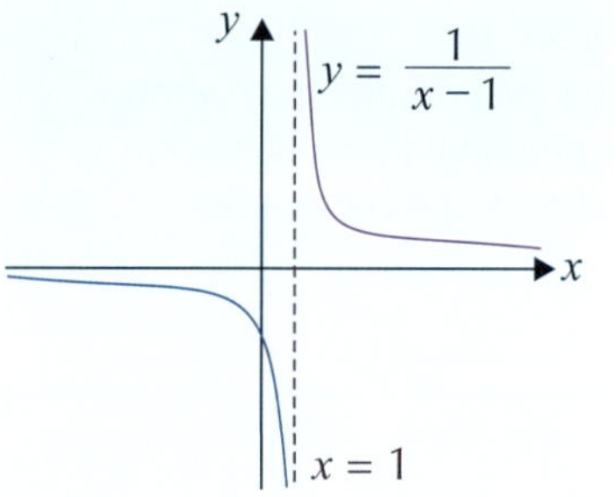

The mapping $y = \frac{1}{x-1}$ for $x \in \mathbb{R}$ is **not** a function, because it's not defined at $x = 1$.

Tip: You will often see asymptotes on a graph when a mapping is undefined at a certain value. Look back at your Year 1 notes if you need a reminder of what asymptotes are.

But if you change the domain to $x > 1$, the mapping is now a function, as shown:

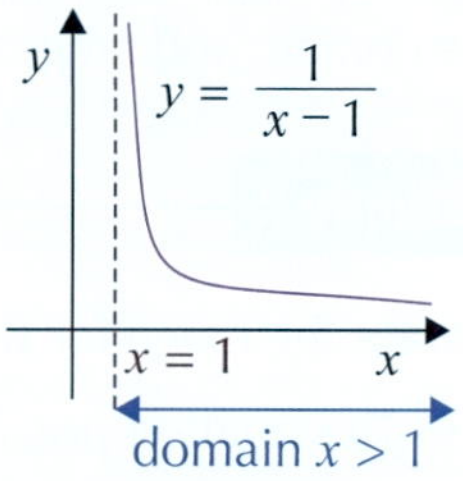

You could also restrict the domain by giving values that x can't be equal to, e.g. $x \in \mathbb{R}, x \neq 1$. In this case the graph would be in two parts like in the first diagram.

Exercise 2.2

Q1 State whether or not each of the graphs below shows a function, and if not, explain why.

a)

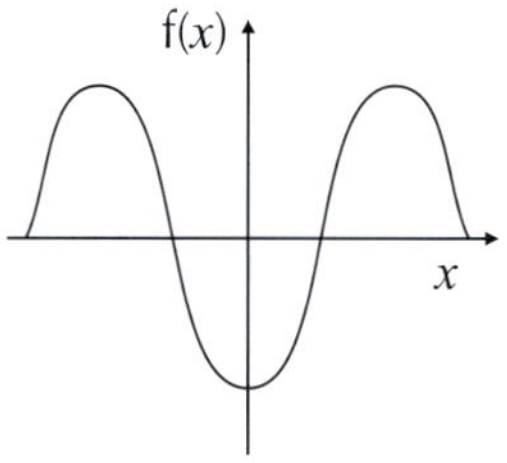

b)

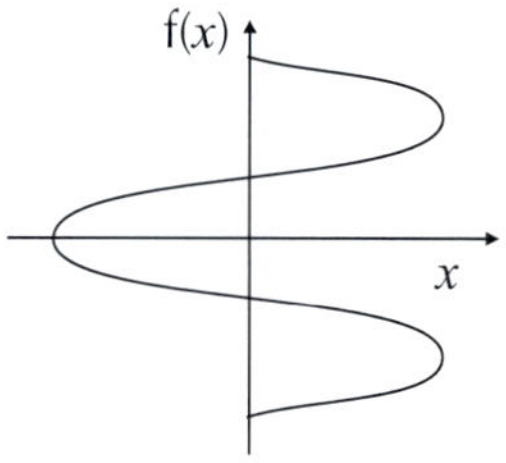

Q2 For each of the following functions, sketch the graph of the function for the given domain, marking relevant points on the axes, and state the range.

a) $f(x) = 3x + 1 \quad x \geq -1$

b) $f(x) = x^2 + 2 \quad -3 \leq x \leq 3$

c) $f(x) = \cos x \quad 0° \leq x \leq 360°$

d) $f(x) = \begin{cases} 5 - x & 0 \leq x < 5 \\ x - 5 & 5 \leq x \leq 10 \end{cases}$

Q3 State the domain and range for the following functions, giving your answers in set notation:

Q3 Hint: Use the given functions to work out the domain from the given range, or the range from the given domain.

a)

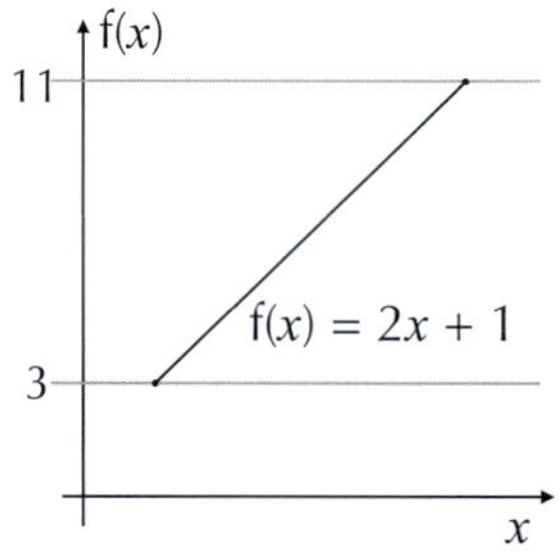

b)

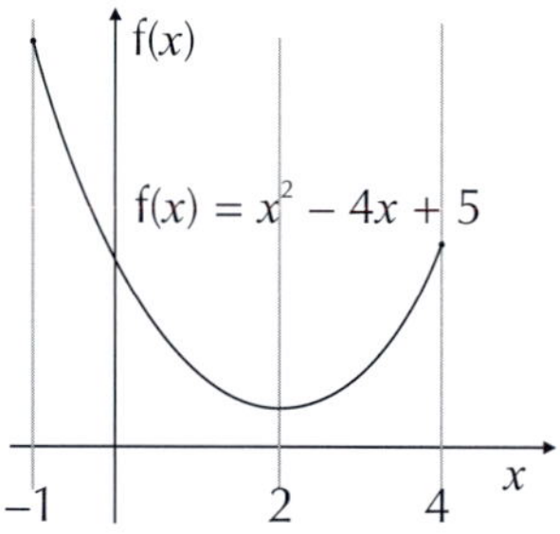

Q4 The graph below shows the function $f(x) = \frac{x+2}{x+1}$, defined for the domain $x \geq 0$. State the range.

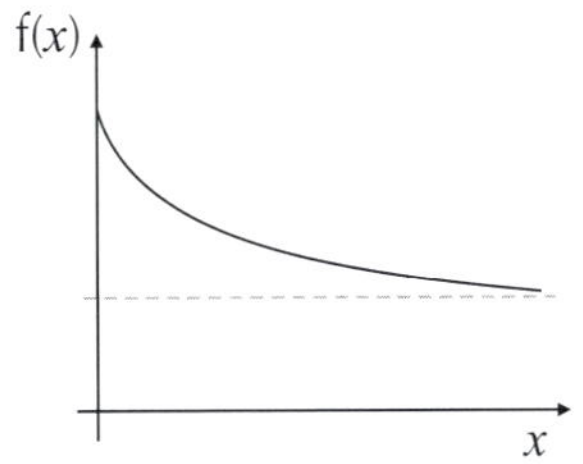

Q5 The diagram shows the function $f(x) = \frac{1}{x-2}$ drawn over the domain $x > a$. State the value of a.

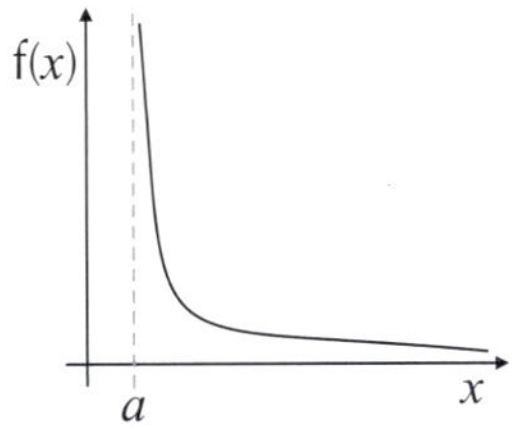

Q6 The diagram shows the function $f(x) = \sqrt{9-x^2}$ for $x \in \mathbb{R}$, $a \leq x \leq b$. State the values of a and b.

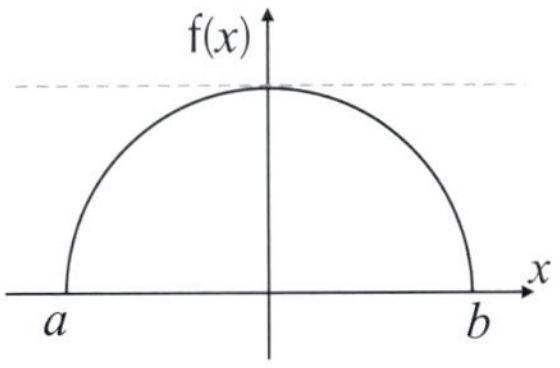

Q7 $h(x) = \sqrt{x+1}$, $x \in \mathbb{R}$. Using set notation, give a restricted domain which would make h a function.

Q8 $k : x \to \tan x$, $x \in \mathbb{R}$.
Give an example of a domain which would make k a function.

Q9 $m(x) = \frac{1}{x^2-4}$.

What is the largest continuous domain which would make $m(x)$ a function?

Q10 The diagram on the right shows the graph of $y = f(x)$.

a) Explain why f is not a function on the domain $x \in \mathbb{R}$.

b) State the largest possible domain that would make f a function.

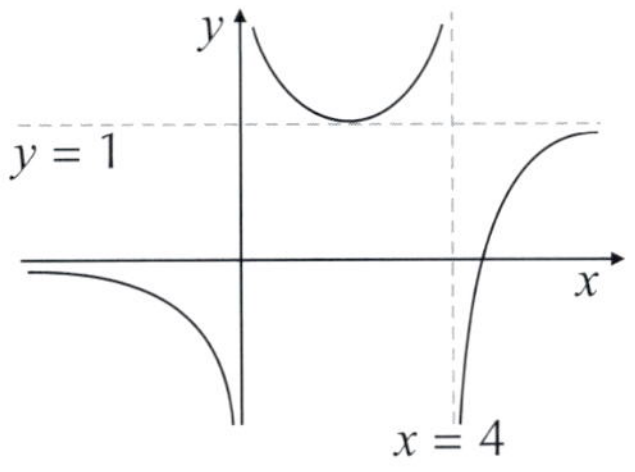

Q4-5 Hint: Use the functions to work out where the asymptotes lie. The range (or domain) will lie on one side of the asymptote.

Q6 Hint: $9 - x^2$ cannot be negative, as you can't take the square root of a negative number, so work out the values of x for which $9 - x^2 \geq 0$ and use these as the domain.

Q8-9 Hint: Sketch a graph of each one first and identify where any asymptotes might be.

Types of function

One-to-one functions

Tip: Sketching a graph is a good way to help you identify the type of function.

A function is **one-to-one** if each value in the **range** corresponds to **exactly one** value in the **domain**.

Example 1

The function $f : x \to 2x,\ x \in \mathbb{R}$ is one-to-one, as only one value of x in the domain is mapped to each value in the range (the range is also $\mathbb{R}$).

You can see this clearly on a sketch of the function:

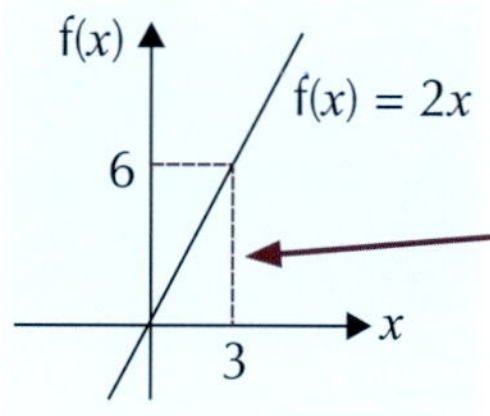

Only 3 in the domain is mapped to 6 in the range.

Many-to-one functions

Tip: There are also mappings known as 'one-to-many' and 'many-to-many', but neither of these types are functions.

A function is **many-to-one** if some values in the **range** correspond to **more than one (many)** values in the **domain**.

Remember that no element in the domain can map to more than one element in the range, otherwise it wouldn't be a function.

Example 2

The function $f(x) = x^2,\ x \in \mathbb{R}$ is a many-to-one function, as two elements in the domain map to the same element in the range, as shown:

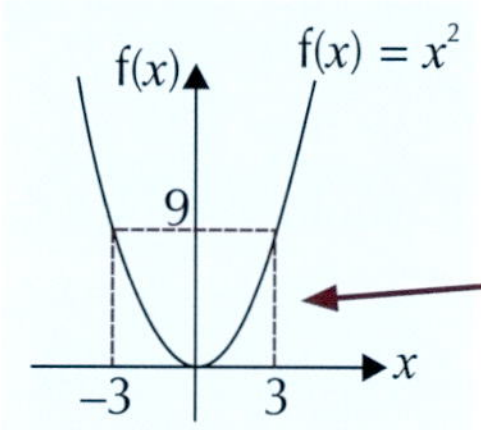

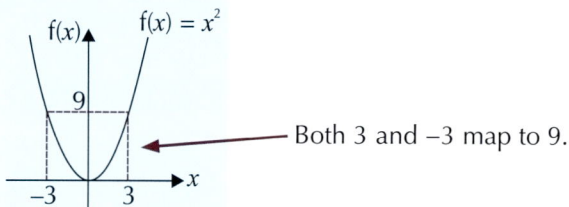

Both 3 and –3 map to 9.

Exercise 2.3

Q1 State whether each function below is one-to-one or many-to-one.

a) $f(x) = x^3 \qquad x \in \mathbb{R}$

b) $f : x \to \sin 2x \qquad -180° < x \leq 180°$

c) $f(x) = \log_{10} x \qquad x > 0$

d) $f(x) = \begin{cases} x + 2 & -2 \leq x < 0 \\ 2 - x & 0 \leq x \leq 2 \end{cases}$

e) $f(x) = \begin{cases} 2^x & x \geq 0 \\ 1 & x < 0 \end{cases}$

3. Composite Functions

When one function is applied to another it makes a different function. This is known as a composite function.

Learning Objectives:

- Be able to combine two or more functions into one composite function.
- Know that fg means 'do g first, then f'.
- Be able to solve equations involving composite functions.

Composite functions

- If you have two functions f and g, you can combine them (do one followed by the other) to make a new function. This is called a **composite function**.
- Composite functions are written **fg(x)**. This means 'do **g first**, then **f**'. If it helps, put brackets in until you get used to it, so fg(x) = f(g(x)).
- The **order** is really important — usually fg(x) ≠ gf(x). If you get a composite function that's written $f^2(x)$, it means ff(x). This just means you have to do f **twice**.

Tip: Composite functions made up of three or more functions work in exactly the same way — just make sure you get the order right.

Example 1

If f(x) = x – 2 and g(x) = 3x, then find:

a) fg(6):

First substitute 6 into g(x).
Then substitute the value that comes out into f(x), as shown below:

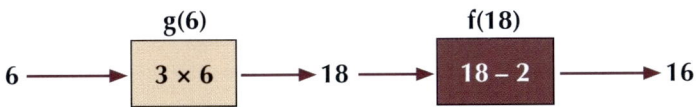

6 → [g(6): 3 × 6] → 18 → [f(18): 18 – 2] → 16

So fg(6) = 16

b) gf(6):

This time substitute 6 into f(x) first.
Then substitute the value that comes out into g(x):

6 → [f(6): 6 – 2] → 4 → [g(4): 3 × 4] → 12

So gf(6) = 12

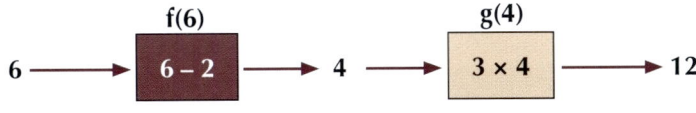

Tip: Comparing the answers to a) and b) you can see that fg(x) ≠ gf(x).

c) fg(x):

This time leave everything in terms of x. Do g first, then f:

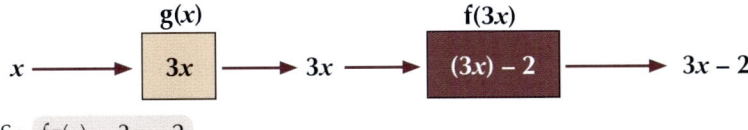

x → [g(x): 3x] → 3x → [f(3x): (3x) – 2] → 3x – 2

So fg(x) = 3x – 2

d) gf(x):

x → [f(x): x – 2] → x – 2 → [g(x – 2): 3(x – 2)] → 3(x – 2) or 3x – 6

So gf(x) = 3x – 6

The key to composite functions is to work things out in steps. Set out your working for composite functions as shown in the examples below.

Example 2

For the functions f : $x \rightarrow 2x^3$, $x \in \mathbb{R}$ and g : $x \rightarrow x - 3$, $x \in \mathbb{R}$, find:

a) fg(4) b) fg(0) c) gf(0) d) fg(x) e) gf(x) f) $f^2(x)$.

a) $fg(4) = f(g(4))$
$= f(4 - 3) = f(1)$
$= 2 \times 1^3 = 2$

b) $fg(0) = f(g(0))$
$= f(0 - 3) = f(-3)$
$= 2 \times (-3)^3 = 2 \times -27$
$= -54$

c) $gf(0) = g(f(0))$
$= g(2 \times 0^3) = g(0)$
$= 0 - 3 = -3$

d) $fg(x) = f(g(x))$
$= f(x - 3)$
$= 2(x - 3)^3$

e) $gf(x) = g(f(x))$
$= g(2x^3)$
$= 2x^3 - 3$

f) $f^2(x) = f(f(x))$
$= f(2x^3)$
$= 2(2x^3)^3 = 16x^9$

Tip: Don't forget the 2^3 when expanding $(2x^3)^3$ in part f).

Domain and range of composite functions

Two functions with given domains and ranges may form a composite function with a **different** domain and range.

Example 3

Give the domain and range of the composite function fg(x), where:
$f(x) = 2x^2 + 1$, domain $x \in \mathbb{R}$, range $f(x) \geq 1$
$g(x) = \frac{1}{x+3}$, domain $x > -3$, range $g(x) > 0$

- First work out the composite function in terms of x:

$$fg(x) = f(g(x))$$
$$= f\left(\frac{1}{x+3}\right) = 2\left(\frac{1}{x+3}\right)^2 + 1$$

- Next, consider the graph of the composite function over the domain and range of the original functions:

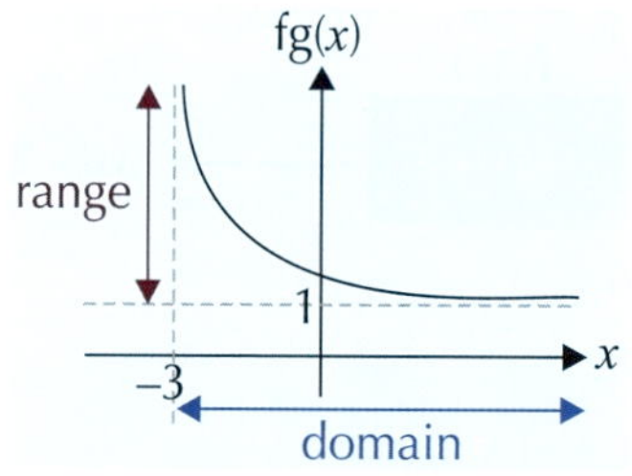

As g(x) is restricted to $x > -3$, so the domain of fg(x) is also restricted to $x > -3$.

Tip: Working out the domains and ranges of composite functions can be tricky — but sketching a graph always helps.

Since $\frac{1}{x+3}$ is always > 0 for the domain $x > -3$,

$2\left(\frac{1}{x+3}\right)^2 + 1$ must be > 1. So the range is $fg(x) > 1$.

For a composite function fg(x), the domain can also be found by putting the **range** of **g(x)** into **f(x)**. If the domain of f(x) does not fully include the range of g(x) then the domain of g(x) will have to be **restricted further.**

Example 4

Find the domain of fg(x), where f(x) = $\sqrt{x}$, $x \geq 0$, and g(x) = $x + 5$, $x \in \mathbb{R}$

The range of g(x) is g(x) $\in \mathbb{R}$. This is bigger than the domain of f(x), so the domain of g(x) will need to be restricted.

The input into f needs to be ≥ 0.
Since for fg(x) the input into f is g(x) (i.e. $x + 5$):

$x + 5 \geq 0 \Rightarrow x \geq -5$.

So the largest possible domain for fg(x) is $x \geq -5$.

If the domain was not restricted, fg(x) would be undefined in places, e.g. fg(–6) = f(–6 + 5) = f(–1) = $\sqrt{-1}$ (which is undefined).

Exercise 3.1

Q1 $f : x \to x^2$, $x \in \mathbb{R}$ and $g : x \to 2x + 1$, $x \in \mathbb{R}$. Find the values of:
a) fg(3) b) gf(3) c) $f^2(5)$ d) $g^2(2)$

Q2 f(x) = sin x for $\{x : x \in \mathbb{R}\}$ and g(x) = $2x$ for $\{x : x \in \mathbb{R}\}$.
Evaluate fg(90°) and gf(90°).

Q3 $f : x \to \frac{3}{x+2}$, $x > -2$ and $g : x \to 2x$, $x \in \mathbb{R}$.
a) Find the values of gf(1), fg(1) and $f^2(4)$.
b) Explain why fg(–1) is undefined.

Q3b) Hint: Try to find the value of fg(–1), or consider the domains and ranges of f(x) and g(x).

Q4 f(x) = cos x, $x \in \mathbb{R}$ and g(x) = $2x$, $x \in \mathbb{R}$. Find the functions:
a) fg(x) b) gf(x)

Q5 f(x) = $2x - 1$, $x \in \mathbb{R}$ and g(x) = 2^x, $x \in \mathbb{R}$. Find the functions:
a) fg(x) b) gf(x) c) $f^2(x)$

Q6 f(x) = $\frac{2}{x-1}$ for $\{x : x > 1\}$ and g(x) = $x + 4$ for $\{x : x \in \mathbb{R}\}$.
Find the functions fg(x) and gf(x), writing them as single fractions in their simplest forms.

Q7 f(x) = $\frac{x}{1-x}$ for $\{x : x \in \mathbb{R}\}$, $x \neq 1$ and g(x) = x^2 for $\{x : x \in \mathbb{R}\}$.
Find $f^2(x)$ and gfg(x).

Q8 f(x) = x^2 with domain $x \in \mathbb{R}$, and g(x) = $2x - 3$ also with domain $x \in \mathbb{R}$.
a) Find fg(x) and write down its range.
b) Find gf(x) and write down its range.

Q9 f(x) = $\frac{1}{x}$ and g(x) = ln (x + 1), both with domain $\{x : x > 0\}$.
a) Find gf(x) and write down its range and largest possible domain.
b) Find fg(x) and write down its range and largest possible domain.

Q8-9 Hint: Sketching the graphs of the composite functions will help you find the ranges and domains.

Q10 Given that f(x) = $3x + 2$, g(x) = $5x - 1$, and h(x) = $x^2 + 1$ (all with domain $\{x : x \in \mathbb{R}\}$), find fgh(x).

Solving composite function equations

If you're asked to **solve** an equation such as $fg(x) = 8$, the best way to do it is to work out what $fg(x)$ is, then **rearrange** $fg(x) = 8$ to make x the subject.

Tip: $\sqrt{x}$ means the positive root of x.

Tip: Be careful with the domains and ranges of composite functions. Have a look back at pages 20-21 for more on how to find them.

Example 1

For the functions $f : x \to \sqrt{x}$ with domain $\{x : x \geq 0\}$ and $g : x \to \frac{1}{x-1}$ with domain $\{x : x > 1\}$, solve the equation $fg(x) = \frac{1}{2}$ and state the range of $fg(x)$.

- First, find $fg(x)$: $fg(x) = f\left(\frac{1}{x-1}\right) = \sqrt{\frac{1}{x-1}} = \frac{1}{\sqrt{x-1}}$
- So $\frac{1}{\sqrt{x-1}} = \frac{1}{2}$
- Rearrange this equation to find x:
  $\frac{1}{\sqrt{x-1}} = \frac{1}{2} \Rightarrow \sqrt{x-1} = 2 \Rightarrow x - 1 = 4 \Rightarrow x = 5$
- To find the range, draw the graph of $fg(x)$.
- From the graph you can see that the domain of $fg(x)$ is $\{x : x > 1\}$ (though the question doesn't ask for this) and the range is $\{fg(x) : fg(x) > 0\}$.

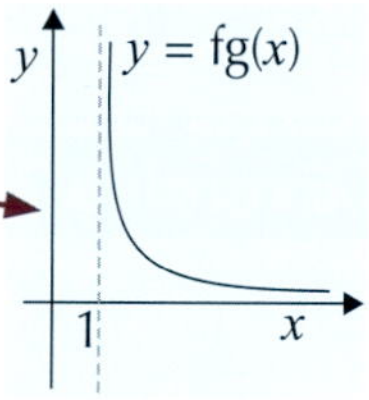

Example 2

For the functions $f : x \to 2x + 1, x \in \mathbb{R}$ and $g : x \to x^2, x \in \mathbb{R}$, solve $gf(x) = 16$.

- Find $gf(x)$: $gf(x) = g(2x + 1) = (2x + 1)^2$
- Now solve $gf(x) = 16$: $(2x + 1)^2 = 16$
  $(2x + 1) = 4$ or -4
  $2x = 3$ or -5
- So $x = \frac{3}{2}$ or $x = -\frac{5}{2}$

Exercise 3.2

Q1 For the following functions, solve the given equations:

a) $f(x) = 2x + 1, x \in \mathbb{R}$ $\quad g(x) = 3x - 4, x \in \mathbb{R}$ $\quad fg(x) = 23$

b) $f(x) = \frac{1}{x}$ for $\{x : x \neq 0\}$ $\quad g(x) = 2x + 5$ for $\{x : x \in \mathbb{R}\}$ $\quad gf(x) = 6$

c) $f(x) = x^2$ for $\{x : x \in \mathbb{R}\}$ $\quad g(x) = \frac{x}{x-3}$ for $\{x : x \neq 3\}$ $\quad gf(x) = 4$

d) $f(x) = x^2 + 1, x \in \mathbb{R}$ $\quad g(x) = 3x - 2, x \in \mathbb{R}$ $\quad fg(x) = 50$

e) $f(x) = 2x + 1, x \in \mathbb{R}$ $\quad g(x) = \sqrt{x}, x \geq 0$ $\quad fg(x) = 17$

f) $f(x) = \log_{10} x, x > 0$ $\quad g(x) = 3 - x, x \in \mathbb{R}$ $\quad fg(x) = 0$

g) $f(x) = 2^x$ for $\{x : x \in \mathbb{R}\}$ $\quad g(x) = x^2 + 2x$ for $\{x : x \in \mathbb{R}\}$ $\quad fg(x) = 8$

h) $f(x) = \frac{x}{x+1}, x \neq -1$ $\quad g(x) = 2x - 1, x \in \mathbb{R}$ $\quad fg(x) = gf(x)$

Q2 $f : x \to x^2 + b, x \in \mathbb{R}$ and $g : x \to b - 3x, x \in \mathbb{R}$ (b is a constant).

a) Find fg and gf and give the range of each in terms of b.

b) Given that $gf(2) = -8$, find the value of $fg(2)$.

4. Inverse Functions

Inverse functions 'undo' functions. So if a function tells you to do a certain thing to x, the inverse of that function tells you how to get back to the start.

Learning Objectives:

- Understand which functions will have inverses.
- Know that $f^{-1}f(x) = ff^{-1}(x) = x$.
- Be able to find the inverse of a function, and find its domain and range.
- Be able to draw and interpret graphs of functions and their inverses.

Inverse functions and their graphs

- An **inverse function** does the **opposite** to the function. So if the function was '+ 1', the inverse would be '– 1', if the function was '× 2', the inverse would be '÷ 2' etc.
- The inverse for a function f(x) is written $\mathbf{f^{-1}(x)}$.
- An inverse function maps an element in the **range** to an element in the **domain** — the opposite of a function. This means that only **one-to-one** functions have inverses, as the inverse of a many-to-one function would be one-to-many, which isn't a function.

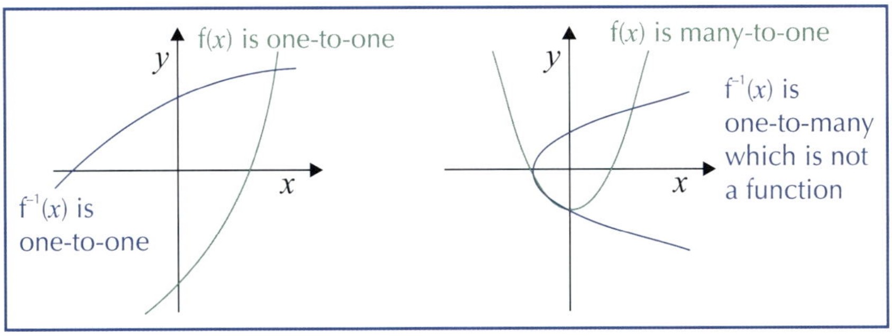

Tip: See page 18 for more on the different types of functions.

- For any inverse $f^{-1}(x)$:

$$f^{-1}f(x) = x = ff^{-1}(x)$$

Doing the function and then the inverse...

...is the same as doing the inverse then doing the function — both just give you x.

Tip: $f^{-1}f(x)$ is a composite function. (see page 19). It just means 'do f then f^{-1}'.

- The **domain** of the inverse is the **range** of the function, and the **range** of the inverse is the **domain** of the function.

Example 1

A function $f(x) = x + 7$ has domain $x \geq 0$ and range $f(x) \geq 7$.
State whether the function has an inverse.
If so, find the inverse, and give its domain and range.

The function $f(x) = x + 7$ is one-to-one, so it does have an inverse.

The inverse of +7 is –7, so $f^{-1}(x) = x - 7$.

$f^{-1}(x)$ has domain $x \geq 7$ (the range of f(x)).

It has range $f^{-1}(x) \geq 0$ (the domain of f(x)).

Tip: Check that this is correct by seeing if $f^{-1}f(x) = x$:
$f^{-1}f(x) = f^{-1}(x + 7)$
$= (x + 7) - 7 = x$
So it's correct.

For simple functions (like the one in the example above), it's easy to work out what the inverse is just by looking at it. But for more complex functions, you need to **rearrange** the original function to **change the subject**.

Tip: It's easier to work with y than $f(x)$.

Finding the inverse of a function

Here's a general method for finding the inverse of a given function:

- Replace $f(x)$ with y to get an equation for **y in terms of x**.
- **Rearrange** the equation to make x the subject.
- Replace x with $f^{-1}(x)$ and y with x — this is the **inverse function**.
- **Swap** round the **domain** and **range** of the function.

Tip: Breaking it into steps like this means you're less likely to go wrong. It's worth doing it this way even for easier functions.

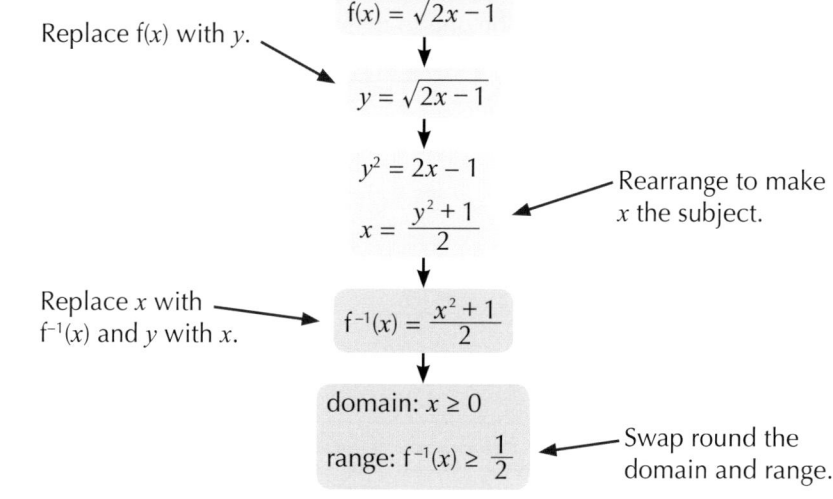

Example 2

Find the inverse of the function $f(x) = \sqrt{2x-1}$, with domain $x \geq \frac{1}{2}$ and range $f(x) \geq 0$. State the domain and the range of the inverse.

$$f(x) = \sqrt{2x-1}$$

Replace $f(x)$ with y.

$$y = \sqrt{2x-1}$$

$$y^2 = 2x - 1$$

$$x = \frac{y^2+1}{2}$$

Rearrange to make x the subject.

Replace x with $f^{-1}(x)$ and y with x.

$$f^{-1}(x) = \frac{x^2+1}{2}$$

domain: $x \geq 0$

range: $f^{-1}(x) \geq \frac{1}{2}$

Swap round the domain and range.

Tip: You have to use the laws of logs in this example — take a look back at your Year 1 notes if you need a reminder.

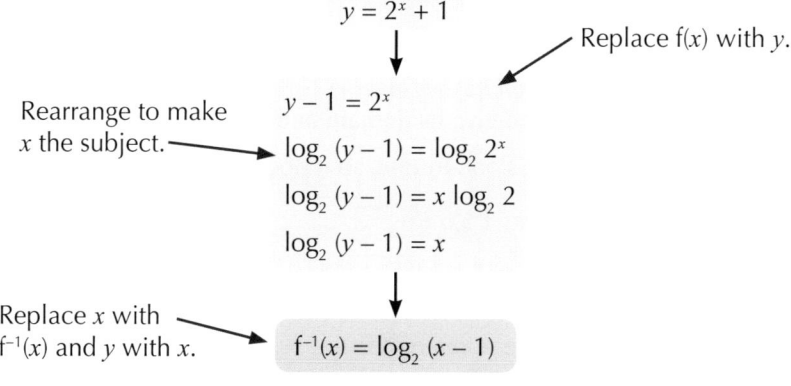

Example 3

Find the inverse of the function $f(x) = 2^x + 1$ with domain $\{x : x \geq 0\}$, and state its domain and range.

$$y = 2^x + 1$$

Replace $f(x)$ with y.

Rearrange to make x the subject.

$$y - 1 = 2^x$$

$$\log_2 (y-1) = \log_2 2^x$$

$$\log_2 (y-1) = x \log_2 2$$

$$\log_2 (y-1) = x$$

Replace x with $f^{-1}(x)$ and y with x.

$$f^{-1}(x) = \log_2 (x-1)$$

Tip: If you're not given the domain and / or the range of the function you'll need to work it out. In this example, x is always at least 0, so $f(x)$ must always be at least $2^0 + 1 = 2$.

The range of $f(x)$ is $f(x) \geq 2$, so $f^{-1}(x)$ has domain $\{x : x \geq 2\}$.

The domain of $f(x)$ is $x \geq 0$ and so the inverse has range $\{f^{-1}(x) : f^{-1}(x) \geq 0\}$.

Graphs of inverse functions

The inverse of a function is its **reflection** in the line $y = x$.

Example 4

Sketch the graph of the inverse of the function $f(x) = x^2 - 8$ with domain $x \geq 0$.

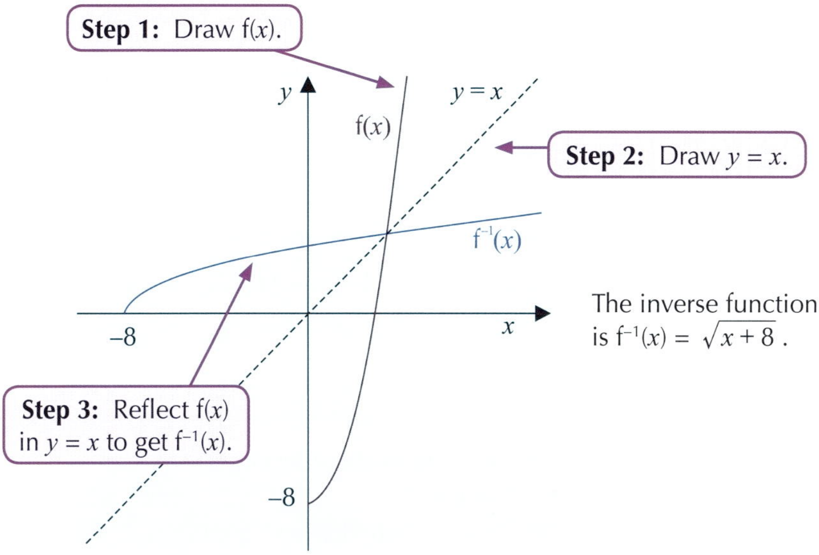

The inverse function is $f^{-1}(x) = \sqrt{x+8}$.

It's easy to see what the domains and ranges are from the graph — $f(x)$ has domain $x \geq 0$ and range $f(x) \geq -8$, and $f^{-1}(x)$ has domain $x \geq -8$ and range $f^{-1}(x) \geq 0$.

Tip: Only sketch the function over the given domain and range. Otherwise you won't be able to see the correct domain and range for the graph of the inverse when you do the reflection.

Exercise 4.1

Q1 Do the functions shown in the diagrams below have inverses? Justify your answers.

a) b)

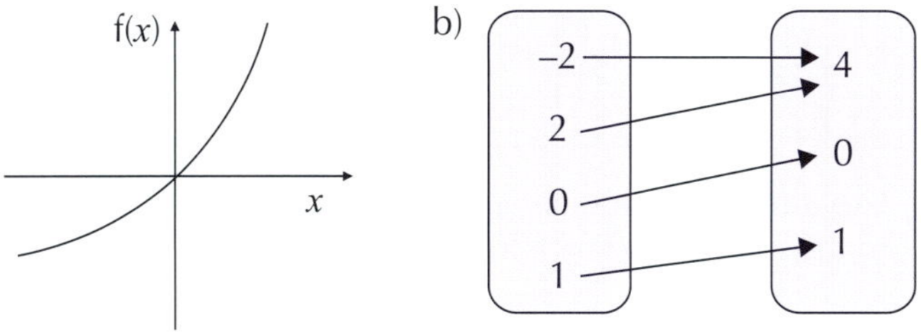

Q1 Hint: Work out which type of function is shown in each of the diagrams.

Q2 For the following functions, explain whether or not an inverse f^{-1} exists:

a) $f(x) = \sin x,\ x \in \mathbb{R}$ b) $f(x) = x^2 + 3,\ x \in \mathbb{R}$

c) $f(x) = (x - 4)^2$ for $\{x : x \geq 4\}$

Q2-6 Hint: If in doubt, sketch a graph of the function to check domains and ranges.

Q3 Find the inverse of each of the following functions, stating the domain and range:

a) $f(x) = 3x + 4,\ x \in \mathbb{R}$ b) $f(x) = 5(x - 2),\ x \in \mathbb{R}$

c) $f(x) = \frac{1}{x+2},\ x > -2$ d) $f(x) = x^2 + 3$ for $\{x : x > 0\}$

Q4-5 Hint: The range of f(x) is quite tricky to find — you might find it helpful to think about what happens to f(x) as $x \to \infty$ and sketch the graph.

Q4 $f(x) = \frac{3x}{x+1}$, $x > -1$

a) Find $f^{-1}(x)$, stating the domain and range.

b) Evaluate $f^{-1}(2)$.

c) Evaluate $f^{-1}\left(\frac{1}{2}\right)$.

Q5 $f(x) = \frac{x-4}{x+3}$ for $\{x : x > -3\}$

a) Find $f^{-1}(x)$, stating the domain and range.

b) Evaluate $f^{-1}(0)$.

c) Evaluate $f^{-1}\left(-\frac{2}{5}\right)$.

Q6 Find the domain and range of $f^{-1}(x)$ for the following functions:

a) $f(x) = \log_{10}(x-3)$, $x > 3$

b) $f(x) = 4x - 2$, $1 \leq x \leq 7$

c) $f(x) = \frac{x}{x-2}$ for $\{x : x < 2\}$

d) $f(x) = 3^{x-1}$ for $\{x : x \geq 2\}$

e) $f(x) = \tan x$, $0° \leq x < 90°$

f) $f(x) = \ln(x^2)$ for $\{x : 3 \leq x \leq 4\}$

Q7 Find the inverse $f^{-1}(x)$ for the following functions, giving the domain and range:

a) $f(x) = e^{x+1}$, $x \in \mathbb{R}$

b) $f(x) = x^3$, $x < 0$

c) $f(x) = 2 - \log_2(x)$, $x \geq 1$

d) $f(x) = \frac{1}{x-2}$ for $\{x : x \neq 2\}$

Q8 $f(x) = 2x + 3$, $x \in \mathbb{R}$. Sketch $y = f(x)$ and $y = f^{-1}(x)$ on the same set of axes, marking the points where the functions cross the axes.

Q9 $f(x) = x^2 + 3$, $x > 0$.

a) Sketch the graphs of $f(x)$ and $f^{-1}(x)$ on the same set of axes.

b) State the domain and range of $f^{-1}(x)$.

Q10 Hint: The functions are equal where the two graphs cross.

Q10 $f(x) = \frac{1}{x+1}$ for $\{x : x > -1\}$.

a) Sketch the graphs of $f(x)$ and $f^{-1}(x)$ on the same set of axes.

b) Explain how your diagram shows that there is just one solution to the equation $f(x) = f^{-1}(x)$.

Q11 $f(x) = \frac{1}{x-3}$, $x > 3$.

a) Find $f^{-1}(x)$ and state its domain and range.

b) Sketch $f(x)$ and $f^{-1}(x)$ on the same set of axes.

c) How many solutions are there to the equation $f(x) = f^{-1}(x)$?

d) Solve $f(x) = f^{-1}(x)$.

5. Modulus

Sometimes in maths you want to work with numbers or functions without having to deal with negative values. The modulus function lets you do this.

The modulus function

Modulus of a number

The **modulus** of a number is its **size** — it doesn't matter if it's positive or negative. So for a positive number, the modulus is just the same as the number itself, but for a negative number, the modulus is its numerical value without the minus sign.

> The modulus of a number, x, is written $|x|$.
>
> In general terms, for $x \geq 0$, $|x| = x$ and for $x < 0$, $|x| = -x$.

For example, the modulus of 8 is 8, and the modulus of –8 is also 8. This is written $|8| = |-8| = 8$.

Modulus of a function

Functions can have a modulus too — the modulus of a function f(x) is just f(x) but with any negative values that it can take turned positive. Suppose $f(x) = -6$, then $|f(x)| = 6$. In general terms:

> $|f(x)| = f(x)$ when $f(x) \geq 0$ and
>
> $|f(x)| = -f(x)$ when $f(x) < 0$.

If the modulus is inside the brackets in the form $f(|x|)$, then you make the x-value positive **before** applying the function. So $f(|-2|) = f(2)$.

Modulus graphs

$y = |f(x)|$

- For the graph of $y = |f(x)|$, any **negative** values of f(x) are made **positive** by **reflecting** them in the ***x*-axis**.
- This **restricts** the **range** of the modulus function to $|f(x)| \geq 0$ (or some subset within $|f(x)| \geq 0$, e.g. $|f(x)| \geq 1$).
- The easiest way to draw a graph of $y = |f(x)|$ is to initially draw $y = f(x)$, then **reflect** the negative part in the ***x*-axis**.

$y = f(|x|)$

- For the graph of $y = f(|x|)$, the **negative** x-values produce the same result as the corresponding **positive** x-values. So the graph of f(x) for $x \geq 0$ is **reflected** in the ***y*-axis** for the negative x-values.
- The range of $f(|x|)$ will be the **same** as the range of f(x) for values of $x \geq 0$.
- To draw a graph of $y = f(|x|)$, first draw the graph of $y = f(x)$ for **positive** values of x, then **reflect** this in the ***y*-axis** to form the rest of the graph.

Learning Objectives:

- Understand the meaning of the modulus, including modulus notation.
- Be able to write the modulus of a number or function.
- Be able to sketch the graph of $y = |ax + b|$.
- Be able to sketch the graph of $y = |f(x)|$ given the graph of $y = f(x)$.
- Be able to solve equations and inequalities involving the modulus.

Tip: The modulus is sometimes called the **absolute value**.

Tip: Don't get the two types of modulus graph mixed up. For some functions $y = |f(x)|$ looks very different to $y = f(|x|)$ — as you'll see in the example on the next page.

Example

Draw the graphs of $y = |f(x)|$ and $y = f(|x|)$ for $f(x) = 5x - 5$. State the range of each.

> **Tip:** Always mark on any key points such as the places that the graph touches or crosses the axes.

$y = |f(x)|$

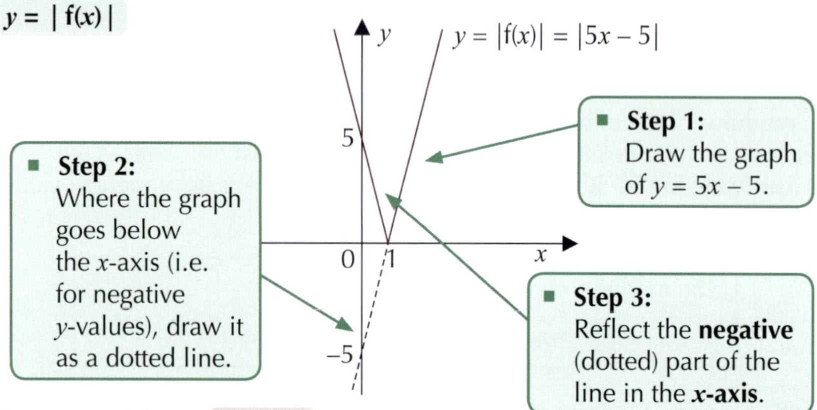

y
y = |f(x)| = |5x – 5|
5
0
1
x
–5
Step 1: Draw the graph of y = 5x – 5.
Step 2: Where the graph goes below the x-axis (i.e. for negative y-values), draw it as a dotted line.
Step 3: Reflect the negative (dotted) part of the line in the x-axis.

The range of this is $|f(x)| \geq 0$

> **Tip:** For $y = |f(x)|$ graphs you reflect the **dotted** line in the ***x*-axis**, but for these $y = f(|x|)$ graphs you reflect the **solid** line in the ***y*-axis**.

$y = f(|x|)$

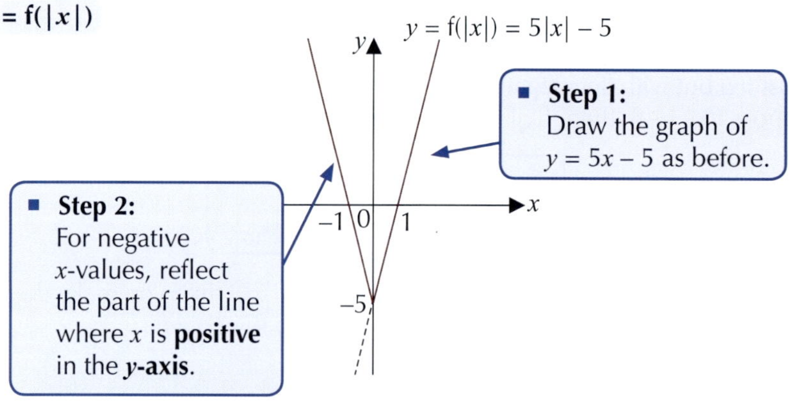

y
y = f(|x|) = 5|x| – 5
x
–1
0
1
–5
Step 1: Draw the graph of y = 5x – 5 as before.
Step 2: For negative x-values, reflect the part of the line where x is positive in the y-axis.

The range of this is $f(|x|) \geq -5$

Exercise 5.1

> **Q1-2 Hint:** Don't forget to label the key points such as where the graphs cross the axes.

Q1 Sketch the following graphs and state the range of each:

a) $y = |x + 3|$ b) $y = |5 - x|$ c) $y = |3x - 1|$

d) $y = |x| - 9$ e) $y = 2|x| + 5$

Q2 For each of the following functions, sketch the graph of $y = |f(x)|$:

a) $f(x) = 2x + 3$ b) $f(x) = 4 - 3x$ c) $f(x) = -4x$

d) $f(x) = 7 - \frac{1}{2}x$ e) $f(x) = -(x + 2)$

> **Q3 Hint:** Try sketching the graphs from the given functions to see what shape they should be, then compare them in terms of where they cross the x and y axes.

Q3 Match up each graph (1-4) with its correct equation (a-d):

a) $y = |x| + 4$

b) $y = |2x - 10|$

c) $y = |x + 1|$

d) $y = |2x| - 2$

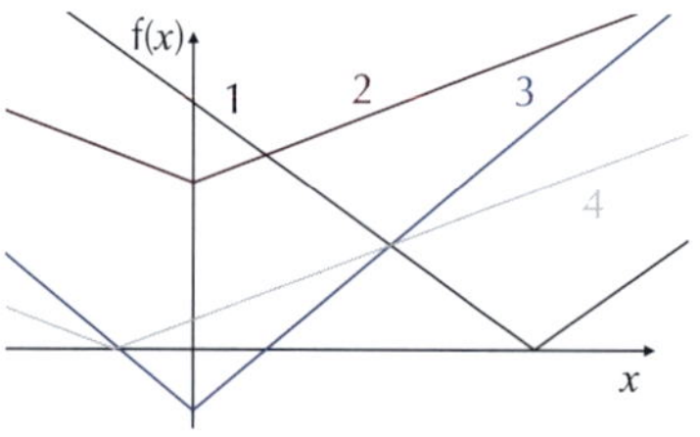

f(x)
1
2
3
4
x

Q4 Draw the graph of the function $f(x) = \begin{cases} |2x+4| & x < 0 \\ |x-4| & x \geq 0 \end{cases}$

Q5 For the function $f(x) = 3x - 5$:

a) Draw, on the same axes, the graphs of $y = f(x)$ and $y = |f(x)|$.

b) How many solutions are there to the equation $|3x - 5| = 2$?

Q6 For the function $f(x) = 4x + 1$:

a) Draw accurately the graph of $y = |f(x)|$.

b) Use your graph to solve the equation $|f(x)| = 3$.

Q5b) Hint: Read across from 2 on the y-axis to find the number of values of x for which $|3x - 5| = 2$.

Solving modulus equations and inequalities

You might be asked to substitute a modulus value into an expression to find the possible values that the expression could take.

Example 1

a) If $|x| = 2$, what are the possible values of $5x - 3$?

If $|x| = 2$, then either $x = 2$ or $x = -2$.
So substitute these two values into the expression.

- If $x = 2$, then $5x - 3 = 5(2) - 3 = 10 - 3 = 7$
- If $x = -2$, then $5x - 3 = 5(-2) - 3 = -10 - 3 = -13$

b) Find all of the possible values of $|3x - 4|$ when $|x| = 6$.

- $|x| = 6$ means $x = 6$ or $x = -6$.
- Substituting $x = 6$ into $|3x - 4|$ gives: $|3(6) - 4| = |18 - 4| = |14| = 14$
- Substituting $x = -6$ gives: $|3(-6) - 4| = |-18 - 4| = |-22| = 22$

|f(x)| = n and |f(x)| = g(x)

The method for solving equations of the form $|f(x)| = n$ is shown below. Solving $|f(x)| = g(x)$ is exactly the same — just replace n with $g(x)$.

- **Step 1:** Sketch the functions $y = |f(x)|$ and $y = n$ on the same axes. The solutions you're trying to find are where they **intersect**.
- **Step 2:** From the graph, work out the ranges of x for which $f(x) \geq 0$ and $f(x) < 0$: e.g. $f(x) \geq 0$ for $x \leq a$ or $x \geq b$ and $f(x) < 0$ for $a < x < b$. These ranges should 'fit together' to cover **all** possible x-values.
- **Step 3:** Use this to write **two new equations**, one true for each range of x:

  ① $f(x) = n$ for $x \leq a$ or $x \geq b$

  ② $-f(x) = n$ for $a < x < b$
- **Step 4:** Solve each equation and check that any solutions are **valid**. Get rid of any solutions outside the range of x you have for that equation.
- **Step 5:** Look at the graph and **check** that your solutions look right.

Tip: Sketching the graphs will show you where the solutions should be — it's easy to accidentally find solutions of $f(x) = g(x)$ that **don't** satisfy $|f(x)| = g(x)$ if you don't draw the graphs first.

Tip: The original equation $|f(x)| = n$ becomes $f(x) = n$ in the range where $f(x) \geq 0$, and it becomes $-f(x) = n$ in the range where $f(x) < 0$.

Example 2

Solve $|2x - 4| = 5 - x$. ← This is an example of $|f(x)| = g(x)$, where $f(x) = 2x - 4$ and $g(x) = 5 - x$.

- Sketch $y = |2x - 4|$ and $y = 5 - x$. The graphs cross twice.

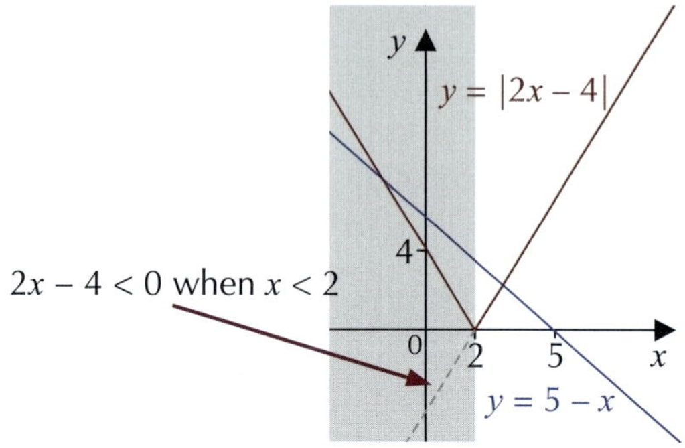

- Looking at where $f(x) \geq 0$ and where $f(x) < 0$ gives:

  ① $2x - 4 = 5 - x$ for $x \geq 2$
  ② $-(2x - 4) = 5 - x$ for $x < 2$.

- Solving these gives:

  ① $3x = 9 \Rightarrow x = 3$ ← valid because $x \geq 2$

  and

  ② $-x = 1 \Rightarrow x = -1$ ← valid because $x < 2$

- Checking against the graph, there are two solutions and they're where we expected.

Tip: You can see from the graph that there will be one solution in the shaded area where $-f(x) = g(x)$, and one solution in the positive region ($x \geq 2$) where $f(x) = g(x)$.

Tip: Make sure that you check whether your solutions are within the valid range for each equation.

$|f(x)| = |g(x)|$

When using **graphs** to solve functions of the form $|f(x)| = |g(x)|$ you have to do a bit more work at the start to identify the different areas of the graph. There could be regions where:

- $f(x)$ and $g(x)$ are **both** positive or **both** negative — for solutions in these regions you need to solve the equation $\mathbf{f(x) = g(x)}$.
- One function is **positive** and the other is **negative** — for solutions in these regions you need to solve the equation $\mathbf{-f(x) = g(x)}$.

There is also an **algebraic** method for solving equations of this type:

$$|a| = |b| \Leftrightarrow a^2 = b^2$$
$$\text{So } |f(x)| = |g(x)| \Leftrightarrow [f(x)]^2 = [g(x)]^2$$

This is true because squaring gives the **same answer** whether the value is **positive** or **negative**. You'll usually be left with a **quadratic** to solve, but in some cases this might be easier than using a graphical method.

The following example shows how you could use either method to solve the same equation.

Tip: Solving $-f(x) = -g(x)$ is the same as solving $f(x) = g(x)$, and solving $f(x) = -g(x)$ is the same as solving $-f(x) = g(x)$.

Tip: The squaring method works for solving $|f(x)| = n$ as well — write $[f(x)]^2 = n^2$, then rearrange and solve.

Example 3

Solve $|x - 2| = |3x + 4|$.

- $|x - 2| = |3x + 4|$
  Square both sides to give:
  $$(x-2)^2 = (3x+4)^2$$
  $$x^2 - 4x + 4 = 9x^2 + 24x + 16$$
  $$8x^2 + 28x + 12 = 0$$
  $$2x^2 + 7x + 3 = 0$$
- Factorise and solve: $(2x + 1)(x + 3) = 0$
  $$x = -\frac{1}{2} \text{ and } x = -3$$
- You can check these solutions by sketching the graphs:
- There are two intersections — one where f(x) is negative but g(x) is positive (shaded grey) and the other where f(x) and g(x) are both negative (i.e. where $x < -1\frac{1}{3}$).

  These correspond to the two solutions $x = -\frac{1}{2}$ and -3.

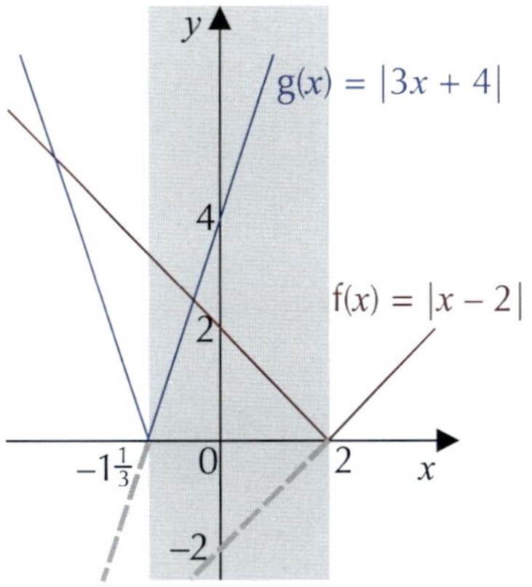

Tip: For this example the algebraic method involves less work than using graphs.

Tip: You can use the quadratic formula to solve it if it won't easily factorise.

Solving inequalities

Inequalities with a modulus can be a bit nasty to solve. Just remember that $|x| < 5$ means that $x < 5$ and $-x < 5$ (which is the same as $x > -5$). So $|x| < 5 \Rightarrow -5 < x < 5$. Similarly, if $|x| > 5$, you'd end up with $x > 5$ and $x < -5$. In general, for $a > 0$:

- $|x| < a \Rightarrow -a < x < a$
- $|x| > a \Rightarrow x > a$ **or** $x < -a$

Using this, you can **rearrange** more complicated inequalities like $|x - a| \leq b$. From the method above, this means that $-b \leq x - a \leq b$, so **adding a** to **each bit** of the inequality gives $a - b \leq x \leq a + b$.

Example 4

Solve $|x - 4| < 7$

- Using the theory above:
  $$|x - 4| < 7 \Rightarrow -7 < x - 4 < 7$$
- Now add 4 to each bit:
  $$-7 + 4 < x < 7 + 4, \text{ so } -3 < x < 11$$

For more complicated modulus inequalities, it's often helpful to draw the graph — see the example on the next page.

Example 5

Solve $|2x - 4| > 5 - x$.

- Sketch the graphs of $y = |2x - 4|$ and $y = 5 - x$ (if you're thinking that this looks familiar, you're dead right — this is the graph you drew on p.30).
- You solved the equation $|2x - 4| = 5 - x$ on p.30, so you know that the graphs cross at $x = 3$ and $x = -1$.
- Now highlight the areas where the graph of $|2x - 4|$ (the red line) is above the graph of $5 - x$ (the blue line):

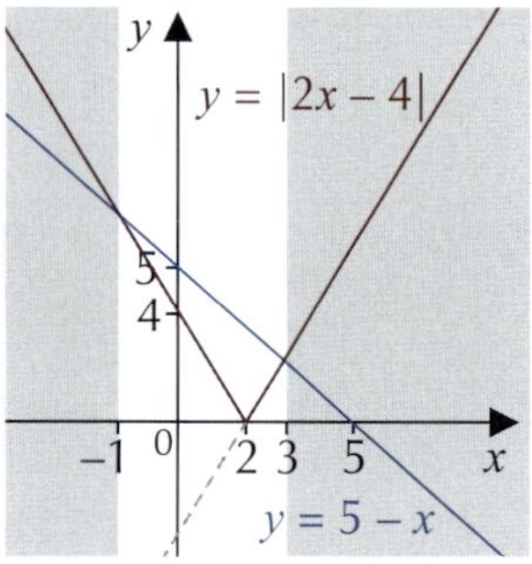

- You can see that there are two regions that satisfy the inequality — the region to the left of $x = -1$ and the region to the right of $x = 3$.
- So the solutions of the inequality are:

  $x < -1$ or $x > 3$

Tip: In set notation, the answer would be: $\{x : x < -1\} \cup \{x : x > 3\}$

Exercise 5.2

Q1 Solve the following equations:

a) $|x - 2| = 6$ b) $|4x + 2| = 10$ c) $2 - |3x - 4| = 1$

Q2 If $|x| = 5$, find the possible values of $|3x + 2|$

Q3 a) On the same axes sketch the graphs of $|f(x)|$ and $g(x)$, where:

$f(x) = 2x + 3$ and $g(x) = x - 1$

b) Hence find any solutions of the equation $|f(x)| = g(x)$.

Q4 a) On the same axes sketch the graphs of $|f(x)|$ and $|g(x)|$, where:

$f(x) = 5x + 10$ and $g(x) = x + 1$

b) Hence find any solutions of the equation $|f(x)| = |g(x)|$.

Q5 Solve, either graphically or otherwise:

a) $|x + 2| = |2x|$ b) $|4x - 1| = |2x + 3|$ c) $|3x - 6| = |10 - 5x|$

Q5 Hint: You can solve these equations algebraically if you prefer, using the squaring method.

Q6 If $|4x + 1| = 3$, find the possible values of $2|x - 1| + 3$ PROBLEM SOLVING

Q7 Solve the following inequalities:

a) $|x| < 8$ b) $|x| \geq 5$ c) $|2x| > 12$

d) $|4x + 2| \leq 6$ e) $3 \geq |3x - 3|$ f) $6 - 2|x + 4| < 0$

g) $3x + 8 < |x|$ h) $2|x - 4| \geq x$ i) $|x - 3| \geq |2x + 3|$

Q8 Give the solutions of $x + 6 \leq |3x + 2|$ in set notation.

Q9 Find the possible values of $|5x + 4|$, given that $|1 + 2x| \leq 3$. PROBLEM SOLVING

6. Transformations of Graphs

You should be familiar with the basic graph transformations. Now you have to know how to put them all together to form combinations of transformations.

Transformations of graphs

The four transformations

The transformations you've met before are translations (a vertical or horizontal shift), stretches (either vertical or horizontal) and reflections in the x- or y- axis. Here's a quick reminder of what each one does:

$y = f(x + a)$

For $a > 0$:

- $f(x + a)$ is $f(x)$ translated a **left**,
- $f(x - a)$ is $f(x)$ translated a **right**.

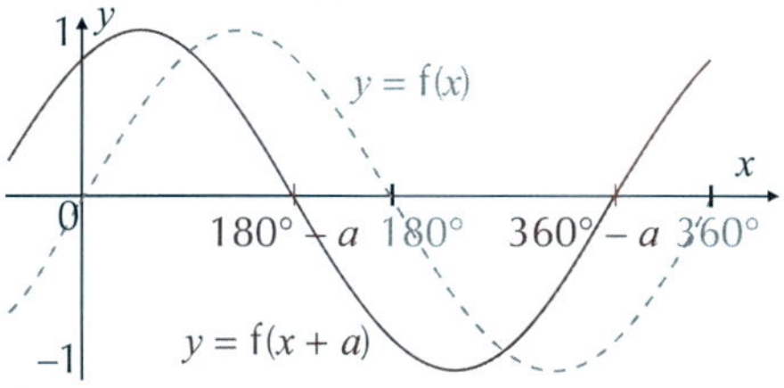

$y = f(x) + a$

For $a > 0$:

- $f(x) + a$ is $f(x)$ translated a **up**,
- $f(x) - a$ is $f(x)$ translated a **down**.

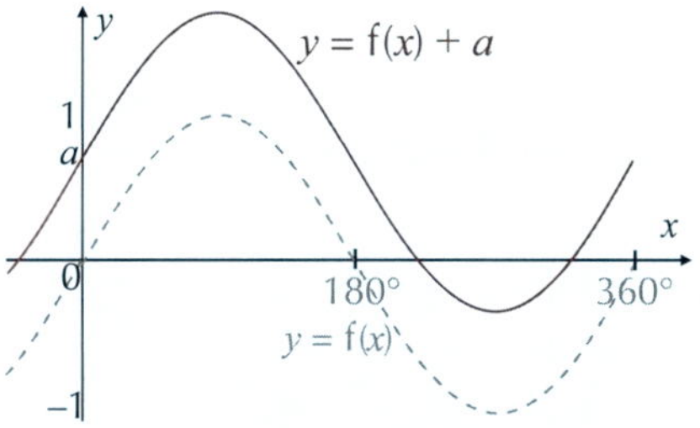

$y = af(x)$

- The graph of $af(x)$ is $f(x)$ **stretched** parallel to the **y-axis** (i.e. vertically) by a factor of a.
- And if $a < 0$, the graph is also **reflected** in the **x-axis**.

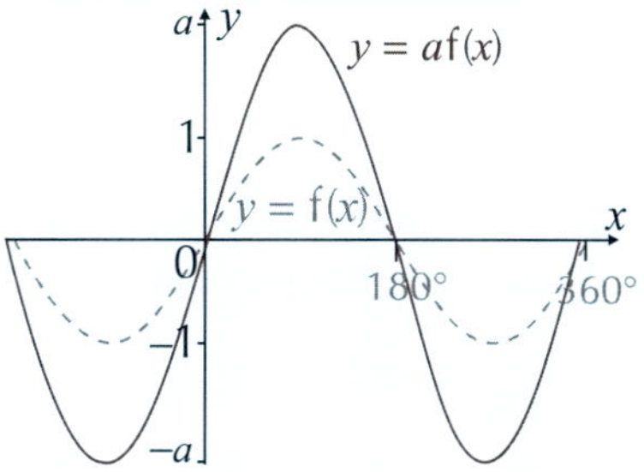

$y = f(ax)$

- The graph of $f(ax)$ is $f(x)$ **stretched** parallel to the **x-axis** (i.e. horizontally) by a factor of $\frac{1}{a}$.
- And if $a < 0$, the graph is also **reflected** in the **y-axis**.

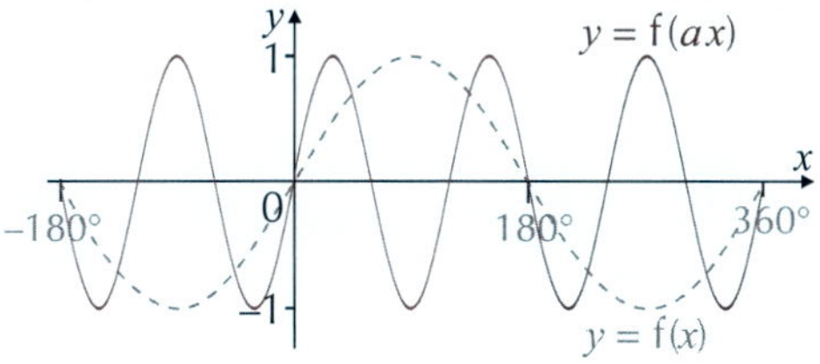

Learning Objectives:

- Be able to sketch graphs when $y = f(x)$ has been affected by a combination of these transformations: $y = f(x + a)$, $y = f(x) + a$, $y = af(x)$, and $y = f(ax)$.
- Be able to interpret transformed graphs, including finding coordinates of points.

Tip: All these graphs use $f(x) = \sin x$.

Tip: Translations can also be given as column vectors. For example, the vector $\begin{pmatrix} 2 \\ -3 \end{pmatrix}$ is a translation of 2 to the right and 3 down.

Tip: It might be easier to think of a stretch by a scale factor **$0 < a < 1$** as a 'squash' — but make sure you use the word stretch in the exam.

Tip: In this case, the transformation is a stretch when **$0 < a < 1$** and a 'squash' when **$a > 1$**.

Combinations of transformations

Combinations of transformations can look a bit tricky, but if you take them one step at a time they're not too bad. Don't try and do all the transformations at once — break it up into the separate bits shown on the previous page and draw a graph for each stage.

> **Tip:** Make sure you do the transformations the right way round — you should do the bit in the brackets first.

Example 1

The graph below shows the function $y = f(x)$. Draw the graph of $y = 3f(x + 2)$, showing the coordinates of the turning points.

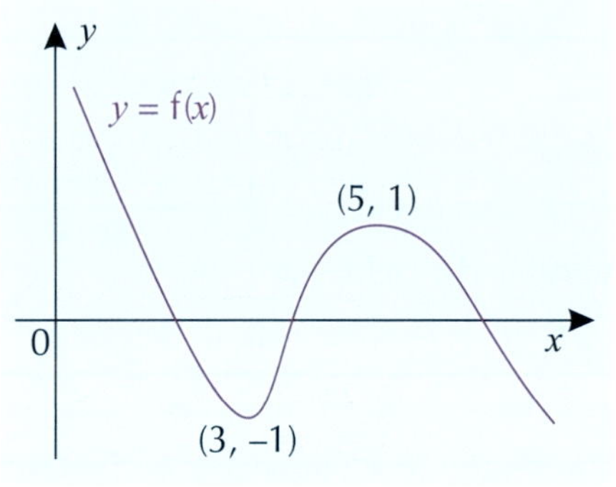

y
y = f(x)
(5, 1)
0
x
(3, –1)

- Don't try to do everything at once. First draw the graph of $y = f(x + 2)$ and work out the coordinates of the turning points:

> **Tip:** Remember — $y = f(x + a)$ is $f(x)$ translated a to the left.

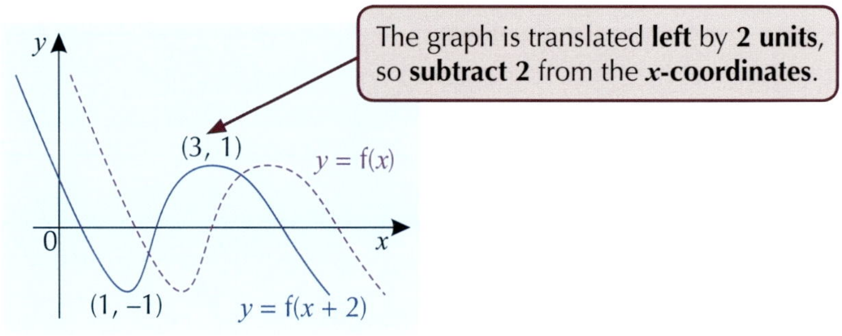

The graph is translated **left** by **2 units**, so **subtract 2** from the ***x*-coordinates**.
y
(3, 1)
y = f(x)
0
x
(1, –1)
y = f(x + 2)

- Now use your graph of $y = f(x + 2)$ to draw the graph of $y = 3f(x + 2)$:

> **Tip:** $y = af(x)$ is $f(x)$ stretched vertically by a factor of a.

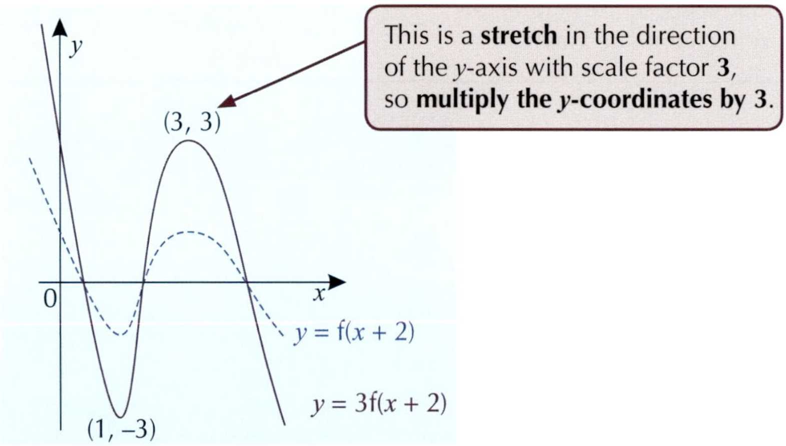

This is a **stretch** in the direction of the y-axis with scale factor **3**, so **multiply the *y*-coordinates by 3**.
y
(3, 3)
0
x
y = f(x + 2)
y = 3f(x + 2)
(1, –3)

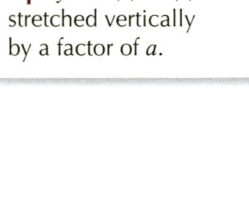

Example 2

The graph shows the function f(x) = $|x|$.

a) Draw the graph of f(x) after a translation by $\begin{pmatrix}-1\\4\end{pmatrix}$ and give its equation.

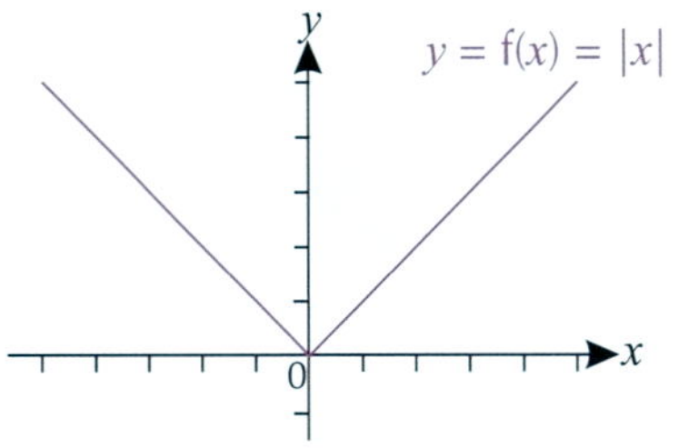

Tip: Have a look back at pages 27-28 to refresh your memory on the modulus function $y = |x|$ and its graph.

- You need to do a **horizontal translation left** by **1**...

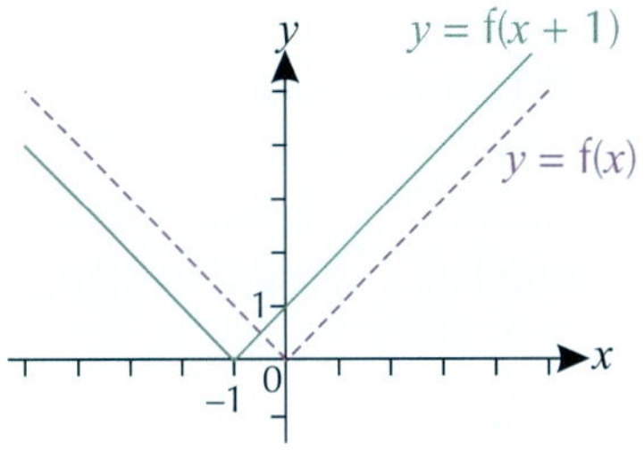

- ...followed by a **vertical translation** of **4 upwards**.
- The equation of the new graph is

  $y = f(x + 1) + 4$
- Replacing f(x) with the original function $|x|$ gives the answer:

  $y = |x + 1| + 4$

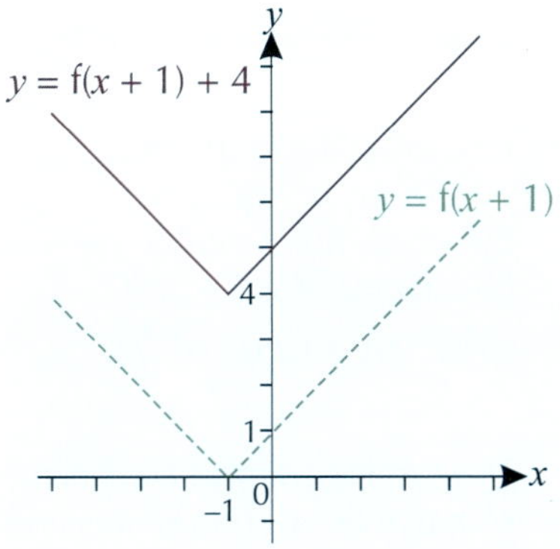

Tip: If you're not sure whether your graph is correct, try putting some numbers into the function and checking them against coordinates on the graph.

b) The function g(x) is a translation of f(x). Give the translation vector for this transformation, and the equation of g(x).

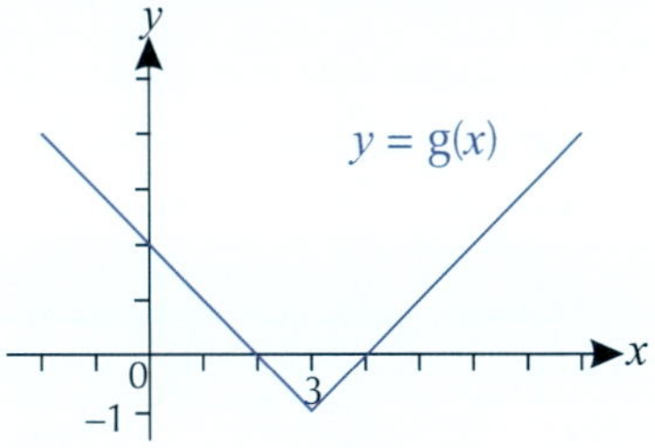

- The transformation is a **horizontal** translation of 3 to the **right**, and a **vertical** translation of 1 **downwards**.
- This gives a **translation vector** of $\begin{pmatrix}3\\-1\end{pmatrix}$
- To find the equation of g(x), use the equations for each translation:
  Horizontal translation of 3 to the right = f(x – 3)
  Vertical translation of 1 downwards = f(x – 3) – 1
- Now replace f with the actual function given above:

  $g(x) = f(x - 3) - 1$

  $g(x) = |x - 3| - 1$

Example 3

The graph below shows the function $y = \sin x$, $0° \le x \le 360°$.
Draw the graph of $y = 2 - \sin 2x$, $0° \le x \le 360°$.

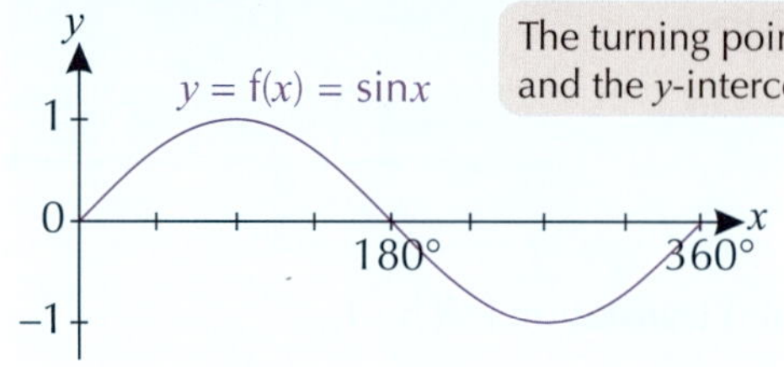

The turning points are at (90°, 1) and (270°, –1), and the y-intercept is (0, 0).

- This is a lot easier to deal with if you rearrange the function from $y = 2 - \sin 2x$ to $\boldsymbol{y = -\sin 2x + 2}$. This gets it in the form $y = -f(2x) + 2$. So we need a **horizontal stretch** by a factor of $\frac{1}{2}$, followed by a **vertical stretch** by a factor of **–1**, followed by a **vertical translation** by **2 up** (in the positive y-direction).

Tip: Always try to break it down like this before you start drawing lots of graphs.

- First draw the graph of $y = \sin 2x$, by squashing the graph horizontally by a factor of 2 (i.e. a stretch by a factor of $\frac{1}{2}$).

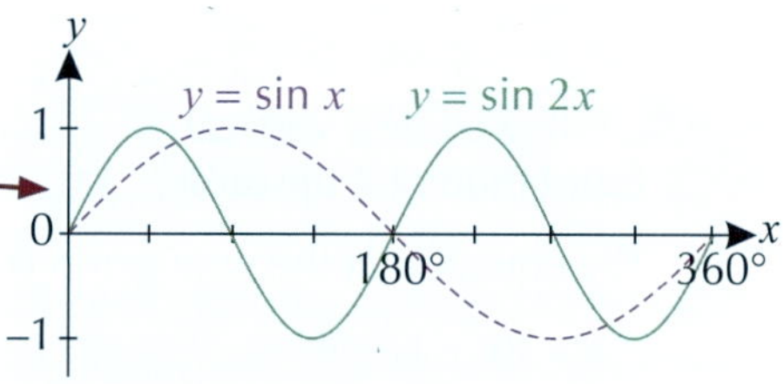

The turning points have been squashed up in the x-direction, so halve the x-coordinates: (45°, 1) and (135°, –1).

There are also now an extra two within the domain, each one occurring a further 90° along the x-axis: (225°, 1) and (315°, –1).

Tip: A stretch with a factor of –1 doesn't change the size of the graph, you just have to reflect in the x-axis.

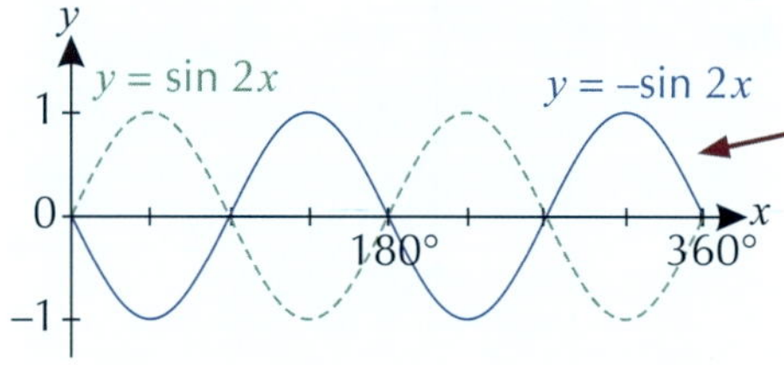

- From there, draw the graph of $y = -\sin 2x$, by reflecting in the x-axis.

This transformation flips the turning points, so multiply the y-coordinates by –1. They're now at (45°, –1), (135°, 1), (225°, –1) and (315°, 1).

- Finally, translate the graph of $y = -\sin 2x$ up by 2 to get the graph of $y = -\sin 2x + 2$ (or $y = 2 - \sin 2x$).

  Add 2 to the y-coordinates of the turning points to give (45°, 1), (135°, 3), (225°, 1) and (315°, 3).

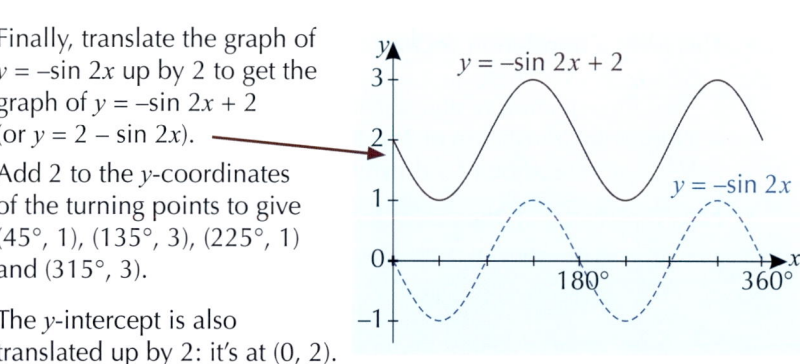

  The y-intercept is also translated up by 2: it's at (0, 2).

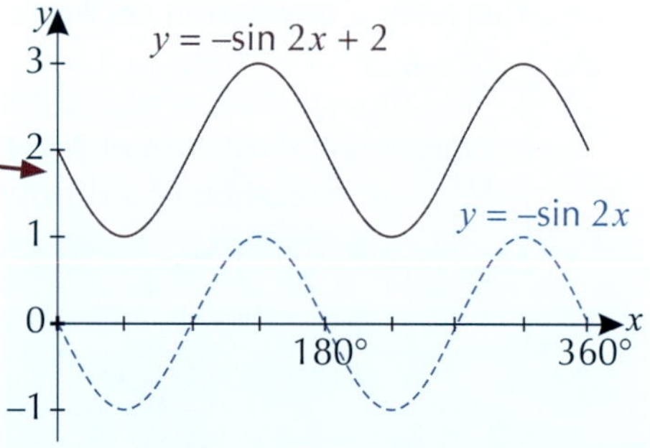

Tip: Having clearly labelled axes makes it easier to read off key points at the end of all the transformations and check your answer.

Exercise 6.1

Q1 Given that $f(x) = x^2$, sketch the following graphs on the same axes:

a) $y = f(x)$ b) $y = f(x) + 3$ c) $y = f(x - 2)$ d) $y = f(x + 4) - 1$

In each case, write down the coordinates of the turning point, and in parts b)-d), give the translation as a column vector.

Q2 The graph of $f(x) = x^3$ is translated to form the graph of $g(x) = f(x - 1) + 4$.

a) Sketch the graphs of $y = f(x)$ and $y = g(x)$.

b) Give this translation as a column vector.

c) What is the equation of $g(x)$?

Q2a) Hint: Do this step by step.

Q3 Given that $f(x) = |x|$, sketch the following graphs on the same axes:

a) $y = f(x)$ b) $y = f(x) + 2$ c) $y = f(x - 4)$ d) $y = 2f(x + 1)$

In parts b)-d) describe the transformation from $y = f(x)$ in words.

Q4 Let $f(x) = |2x - 6|$. On the same axes sketch the graphs of:

a) $y = f(x)$ b) $y = f(-x)$ c) $y = f(-x) + 2$

Q5 Let $f(x) = \frac{1}{x}$. On the same axes sketch the graphs of:

a) $y = f(x)$ b) $y = -f(x)$ c) $y = -f(x) - 3$

Q6 a) Let $f(x) = \cos x$. Sketch the graph $y = f(x)$ for $0° \leq x \leq 360°$.

b) On the same axes sketch the graph of $y = f(2x)$.

c) On the same axes sketch the graph of $y = 1 + f(2x)$.

d) State the coordinates of the minimum point(s) of the graph $y = \cos 2x + 1$, in the interval $0° \leq x \leq 360°$.

Q7 Complete the following table for the function $f(x) = \sin x$ ($0° \leq x \leq 360°$).

Transformed function	New equation	Maximum value of transformed function	Minimum value of transformed function
$f(x) + 2$			
$f(x - 90°)$			
$f(3x)$			
$4f(x)$			

Q7 Hint: Try sketching the graphs of the transformed functions first.

Q8 Complete the following table for the function $f(x) = x^3$:

Transformed function	New equation	Coordinates of point of inflection
$f(x) + 1$		
$f(x - 2)$		
$-f(x) - 3$		
$f(-x) + 4$		

Q8 Hint: The point of inflection is the stationary point. For $y = x^3$ the point of inflection is at (0, 0)

Q9 The graph $y = \cos x$ is translated by the vector $\begin{pmatrix} 90° \\ 0 \end{pmatrix}$ and stretched by scale factor $\frac{1}{2}$ parallel to the y-axis.

a) Sketch the new graph for $0 \leq x \leq 360°$.

b) Write down its equation.

Q10 Hint: You can also write g(x) as $-\frac{1}{x} + 3$.

Q10 a) Sketch the graph of $y = f(x)$ where $f(x) = \frac{1}{x}$.

b) Write down the sequence of transformations needed to map f(x) on to $g(x) = 3 - \frac{1}{x}$.

c) Sketch the graph of $y = g(x)$.

Q11 Complete the following table:

Original graph	New graph	Sequence of transformations
$y = x^3$	$y = (x - 4)^3 + 5$	
$y = 4^x$	$y = 4^{3x} - 1$	
$y = \|x + 1\|$	$y = 1 - \|2x + 1\|$	
$y = \sin x$	$y = -3\sin 2x + 1$	

Q12 Hint: Try taking out a common factor (the value of a) before completing the square.

Q12 a) Write $y = 2x^2 - 4x + 6$ in the form $y = a[(x + b)^2 + c]$.

b) Hence list the sequence of transformations that will map $y = x^2$ on to $y = 2x^2 - 4x + 6$.

c) Sketch the graph of $y = 2x^2 - 4x + 6$.

d) Write down the coordinates of the minimum point of the graph.

Q13 Starting with the curve $y = \cos x$, state the sequence of transformations which could be used to sketch the following curves:

a) $y = 4 \cos 3x$ b) $y = 4 - \cos 2x$ c) $y = 2 \cos(x - 60°)$

Q14 The diagram shows $y = f(x)$ with a minimum point, P, at (2, –3).

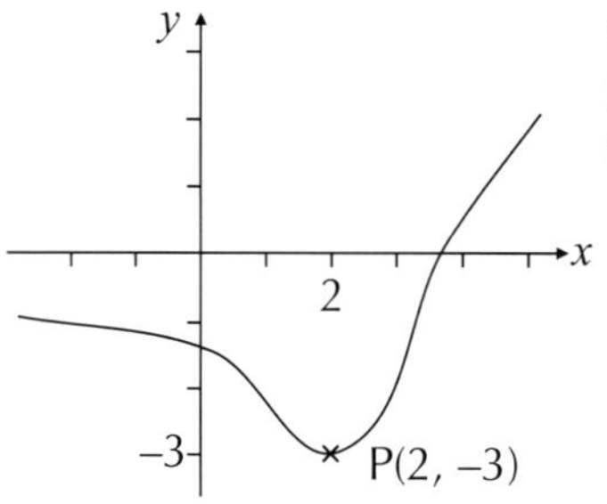

Copy the diagram and sketch each of the following graphs. In each case state the new coordinates of the point P.

a) $y = f(x) + 5$

b) $y = f(x + 4)$

c) $y = -f(x)$

Q15 The diagram shows $y = f(x)$ with a minimum point, Q, at (–1, –3) and a maximum point, P, at (2, 4).

Copy the diagram and sketch each of the following graphs. In each case state the new coordinates of the points P and Q.

a) $y = f(x - 1) + 3$

b) $y = -f(2x)$

c) $y = |f(x + 2)|$

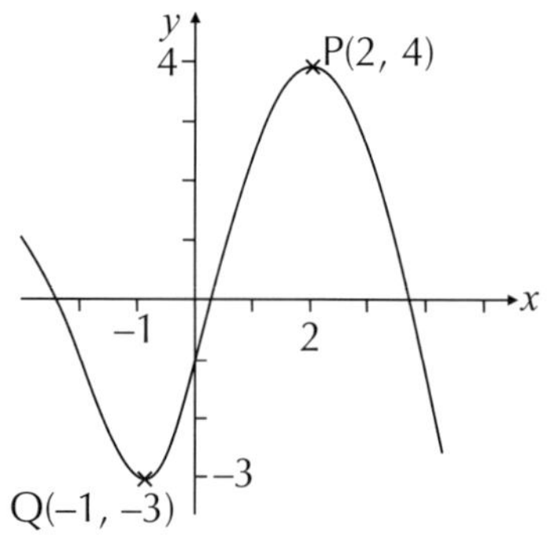

7. Partial Fractions

Sometimes an algebraic fraction with a complicated denominator can be split into a sum of simpler fractions. In this chapter you'll see a couple of methods you can use to do this, depending on what type of denominator you have.

Learning Objectives:

- Be able to write algebraic fractions with linear or constant numerators as partial fractions, including fractions with denominators of the form $(ax + b)(cx + d)(ex + f)$ and $(ax + b)(cx + d)^2$.

Expressing in partial fractions

- You can split a fraction with **more than one linear factor** in the denominator into **partial fractions**.
- This means writing it as a **sum** of two or more **simpler fractions**.
- The **denominators** of these simpler fractions will be **factors** of the denominator of the original fraction.
- This is useful in lots of areas of maths, such as **integration** (see p.219-220) and **binomial expansions** (p.132-133).

Example 1

- $\frac{7x-7}{(2x+1)(x-3)}$ can be written as partial fractions of the form $\frac{A}{(2x+1)}+\frac{B}{(x-3)}$.
- $\frac{7x-1}{(x-3)(x-1)(x+2)}$ can be written as partial fractions of the form $\frac{A}{(x-3)}+\frac{B}{(x-1)}+\frac{C}{(x+2)}$.
- $\frac{21x-2}{9x^2-4}$ can be written as partial fractions of the form $\frac{A}{(3x-2)}+\frac{B}{(3x+2)}$.

Tip: If you're asked to write an algebraic fraction as partial fractions, start by writing the partial fractions out with A, B and C as numerators as shown here. You might have to factorise the denominator first, like in the last example.

The tricky bit is **working out** what ***A*, *B* and *C*** are. Follow this method:

- **Write out** the expression as an identity, e.g.

$$\frac{7x-7}{(2x+1)(x-3)} \equiv \frac{A}{(2x+1)}+\frac{B}{(x-3)}$$

- **Add** the partial fractions together, i.e. write them over a **common denominator**.
- **Cancel** the denominators from both sides (they'll be the same).
- This will give you an **identity** for A and B, for example:

$$7x - 7 \equiv A(x - 3) + B(2x + 1)$$

- Use the **Substitution** method or the **Equating Coefficients** method:

Substitution
Substitute a number for x to leave you with just one constant on the right hand side.

Equating Coefficients
Equate the constant terms, coefficients of x and coefficients of x^2, then solve the equations simultaneously.

Example 2

Express $\frac{7x-1}{(x-3)(x-1)(x+2)}$ **in partial fractions.**

- Write it out as an **identity**:

$$\frac{7x-1}{(x-3)(x-1)(x+2)} \equiv \frac{A}{(x-3)} + \frac{B}{(x-1)} + \frac{C}{(x+2)}$$

- **Add** the partial fractions — this means writing them over a **common denominator**:

$$\frac{A}{(x-3)} + \frac{B}{(x-1)} + \frac{C}{(x+2)} \equiv \frac{A(x-1)(x+2) + B(x-3)(x+2) + C(x-3)(x-1)}{(x-3)(x-1)(x+2)}$$

- **Cancel** the denominators from both sides of the original identity, so the numerators are **equal**:

$$7x - 1 \equiv A(x-1)(x+2) + B(x-3)(x+2) + C(x-3)(x-1)$$

Substitution Method

Substitute values of x which make one of the expressions in brackets equal zero to get rid of all but one of A, B and C.

- Substituting $x = 3$ gets rid of B and C:

$$21 - 1 = A(3-1)(3+2) + 0 + 0$$
$$20 = 10A \quad \Rightarrow \mathbf{A = 2}$$

- Substituting $x = 1$ gets rid of A and C:

$$7 - 1 = 0 + B(1-3)(1+2) + 0$$
$$6 = -6B \quad \Rightarrow \mathbf{B = -1}$$

- Substituting $x = -2$ gets rid of A and B:

$$-14 - 1 = 0 + 0 + C(-2-3)(-2-1)$$
$$-15 = 15C \quad \Rightarrow \mathbf{C = -1}$$

Equating Coefficients Method

- Compare coefficients in the numerators:

$$\begin{aligned} 7x - 1 &\equiv A(x-1)(x+2) + B(x-3)(x+2) + C(x-3)(x-1) \\ &\equiv A(x^2 + x - 2) + B(x^2 - x - 6) + C(x^2 - 4x + 3) \\ &\equiv (A + B + C)x^2 + (A - B - 4C)x + (-2A - 6B + 3C) \end{aligned}$$

Equating x^2 coefficients: $0 = A + B + C$
Equating x coefficients: $7 = A - B - 4C$
Equating constant terms: $-1 = -2A - 6B + 3C$

- Solving these equations simultaneously gives:
$A = 2$, $B = -1$ and **$C = -1$** (the same as the substitution method).

- Finally, **replace** A, B and C in the original identity:

$$\frac{7x-1}{(x-3)(x-1)(x+2)} \equiv \frac{2}{(x-3)} - \frac{1}{(x-1)} - \frac{1}{(x+2)}$$

Tip: The adding step can be a bit fiddly, so you should always check that each term will cancel to produce the original fraction. Another method is to multiply both sides of the original identity by the left hand denominator to get the identity you want.

Tip: For some questions one method will be easier than the other — if one seems too tricky try the other. Sometimes you might want to use a combination of both methods.

Tip: Generally it's easier to use the substitution method first. In this example, it's actually quite tricky to solve the simultaneous equations that you get by equating coefficients, so you're better off using substitution.

Tip: Don't forget to write out your solution like this once you've done all the working.

Example 3

Express $\frac{3-x}{x^2+x}$ in partial fractions.

- First you need to **factorise** the denominator: $\frac{3-x}{x^2+x} \equiv \frac{3-x}{x(x+1)}$.
- Now write as an **identity** with partial fractions: $\frac{3-x}{x(x+1)} \equiv \frac{A}{x} + \frac{B}{x+1}$.
- **Add** the partial fractions and **cancel** the denominators from both sides:

$$\frac{3-x}{x(x+1)} \equiv \frac{A(x+1)+Bx}{x(x+1)} \quad \Rightarrow \quad 3 - x \equiv A(x+1) + Bx$$

- In this example it's easier to **equate coefficients** because A is the only letter that appears in the constant term on the right hand side.
- **Compare coefficients** in $3 - x \equiv A(x+1) + Bx$:

  Equating constant terms: $\mathbf{3 = A}$

  Equating x coefficients: $-1 = A + B \Rightarrow -1 = 3 + B \Rightarrow \mathbf{B = -4}$
- **Replace** A and B in the identity: $\frac{3-x}{x^2+x} \equiv \frac{3}{x} - \frac{4}{(x+1)}$

Tip: In this example equating coefficients is a good bet because you get one of the constants without any work.

Exercise 7.1

Q1 Express $\frac{3x+3}{(x-1)(x-4)}$ in the form $\frac{A}{x-1} + \frac{B}{x-4}$.

Q2 Express $\frac{5x-1}{x(2x+1)}$ in the form $\frac{A}{x} + \frac{B}{2x+1}$.

Q1 & 2 Hint: Use the substitution method for Q1, and equate coefficients for Q2.

Q3 Find the values of the constants A and B in the identity $\frac{3x-2}{x^2+x-12} \equiv \frac{A}{x+4} + \frac{B}{x-3}$.

Q4 Write $\frac{2}{x^2-16}$ in partial fractions.

Q4 Hint: Look out for the difference of two squares.

Q5 Factorise $x^2 - x - 6$ and hence express $\frac{5}{x^2-x-6}$ in partial fractions.

Q6 Write $\frac{11x}{2x^2+5x-12}$ in partial fractions.

Q7 a) Factorise $x^3 - 9x$ fully.

b) Hence write $\frac{12x+18}{x^3-9x}$ in partial fractions.

Q8 Write $\frac{3x+9}{x^3-36x}$ in the form $\frac{A}{x} + \frac{B}{x+6} + \frac{C}{x-6}$.

Q9 a) Use the Factor Theorem to fully factorise $x^3 - 7x - 6$.

b) Hence write $\frac{6x+2}{x^3-7x-6}$ in partial fractions.

Q9a) Hint: Have a look back at your Year 1 notes for a reminder of the Factor Theorem.

Q10 Express $\frac{6x+4}{(x+4)(x-1)(x+1)}$ in partial fractions.

Q11 Express $\frac{15x-27}{x^3-6x^2+3x+10}$ in partial fractions.

Repeated factors

If the denominator of an algebraic fraction has **repeated linear factors** the partial fractions will take a slightly **different form**, as shown in the examples below.

- The **power** of the repeated factor tells you **how many** times that factor should appear in the partial fractions.

$$\frac{7x-3}{(x+1)^2(x-4)} \text{ is written as } \frac{A}{(x+1)}+\frac{B}{(x+1)^2}+\frac{C}{(x-4)}$$

- A factor that's **squared** in the original denominator will appear in the denominator of **two** of your partial fractions — once squared and once just as it is.

$$\frac{32x-14}{x^2(2x+7)} \text{ is written as } \frac{A}{x}+\frac{B}{x^2}+\frac{C}{(2x+7)}$$

Tip: Make sure you don't miss repeated factors like x^2.

Tip: A factor that's **cubed** will appear **three** times — once cubed, once squared and once just as it is.

Example 1

Express $\frac{5x+12}{x^2(x-3)}$ in partial fractions.

- x is a **repeated factor** so the answer will be of the form $\frac{A}{x}+\frac{B}{x^2}+\frac{C}{(x-3)}$.
- Write it out as an **identity**: $\frac{5x+12}{x^2(x-3)} \equiv \frac{A}{x}+\frac{B}{x^2}+\frac{C}{(x-3)}$
- **Add** the partial fractions: $\frac{5x+12}{x^2(x-3)} \equiv \frac{Ax(x-3)+B(x-3)+Cx^2}{x^2(x-3)}$
- **Cancel** the denominators from both sides, so the numerators are **equal**:

$$5x+12 \equiv Ax(x-3)+B(x-3)+Cx^2$$

- **Substituting** $x = 3$ gets rid of A and B:

$$15+12 = 0+0+C(3^2)$$
$$27 = 9C \Rightarrow \mathbf{C = 3}$$

- **Substituting** $x = 0$ gets rid of A and C:

$$0+12 = 0+B(0-3)+0$$
$$12 = -3B \Rightarrow \mathbf{B = -4}$$

- There's no value of x you can substitute to get rid of B and C and just leave A, so **equate coefficients** of x^2:

Coefficients of x^2 are: $0 = A + C$
You know $C = 3$, so: $0 = A + 3 \Rightarrow \mathbf{A = -3}$

- **Replace** A, B and C in the identity:

$$\frac{5x+12}{x^2(x-3)} \equiv -\frac{3}{x}-\frac{4}{x^2}+\frac{3}{(x-3)}$$

Tip: You don't need to multiply through by every denominator to add these fractions together.

Tip: There's no x^2 term on the LHS, so the coefficient of x^2 is just 0.

Tip: You could equate coefficients of x instead to find A, but you can't equate constant terms because A only appears as a coefficient of x or x^2 in the identity.

Example 2

Express $\dfrac{4x+15}{(x+2)^2(3x-1)}$ in partial fractions.

- Write the **identity**: $\dfrac{4x+15}{(x+2)^2(3x-1)} \equiv \dfrac{A}{(x+2)} + \dfrac{B}{(x+2)^2} + \dfrac{C}{(3x-1)}$
- **Add** the partial fractions:
  $\dfrac{A}{(x+2)} + \dfrac{B}{(x+2)^2} + \dfrac{C}{(3x-1)} \equiv \dfrac{A(x+2)(3x-1)+B(3x-1)+C(x+2)^2}{(x+2)^2(3x-1)}$
- **Cancel** the denominators from both sides:
  $4x + 15 \equiv A(x + 2)(3x - 1) + B(3x - 1) + C(x + 2)^2$.
- **Substituting** $x = -2$ gets rid of A and C:
  $-8 + 15 = 0 + B(-6 - 1) + 0$
  $7 = -7B \Rightarrow \mathbf{B = -1}$
- **Substituting** $x = \frac{1}{3}$ gets rid of A and B:
  $\frac{4}{3} + 15 = 0 + 0 + C\left(\frac{1}{3}+2\right)^2$
  $\frac{49}{3} = \frac{49}{9}C \Rightarrow \mathbf{C = 3}$
- There's no value of x you can substitute to get rid of B and C to just leave A, so try **equating coefficients** instead:
  Equate coefficients of x^2: $0 = 3A + C$
  You know $C = 3$, so: $-3 = 3A \Rightarrow \mathbf{A = -1}$
- **Replace** A, B and C in the original identity:

$$\frac{4x+15}{(x+2)^2(3x-1)} \equiv -\frac{1}{(x+2)} - \frac{1}{(x+2)^2} + \frac{3}{(3x-1)}$$

Tip: Remember, you won't need to multiply through by all of the denominators. E.g. for the second fraction you just need to multiply the top and bottom by $(3x - 1)$.

Tip: Another method you can use when there's no value of x which will get rid of B and C is to substitute in any simple value of x, e.g. $x = 1$, and the values that you have calculated for B and C to work out A.

Exercise 7.2

Q1 Express $\dfrac{3x}{(x+5)^2}$ in the form $\dfrac{A}{(x+5)} + \dfrac{B}{(x+5)^2}$.

Q2 Write $\dfrac{5x+2}{x^2(x+1)}$ in the form $\dfrac{A}{x} + \dfrac{B}{x^2} + \dfrac{C}{(x+1)}$.

Q3 Write the following in partial fractions.

a) $\dfrac{2x-7}{(x-3)^2}$ b) $\dfrac{6x+7}{(2x+3)^2}$ c) $\dfrac{7x}{(x+4)^2(x-3)}$ d) $\dfrac{11x-10}{x(x-5)^2}$

Q4 Express $\dfrac{5x+10}{x^3-10x^2+25x}$ in partial fractions.

Q5 Express $\dfrac{3x+2}{(x-2)(x^2-4)}$ in partial fractions.

Q4-5 Hint: Factorise the denominator to get a repeated factor.

Q6 Find the value of c such that: PROBLEM SOLVING

$$\frac{x+17}{(x+1)(x+c)^2} = \frac{1}{x+1} - \frac{1}{x+c} + \frac{5}{(x+c)^2}$$

Chapter 3 Trigonometry

1. Arcs and Sectors

Learning Objectives:

- Be able to use radians as the measure for the size of an angle.
- Be able to use the formula $s = r\theta$ to work out the arc length of a sector of a circle.
- Be able to use the formula $A = \frac{1}{2}r^2\theta$ to work out the area of a sector.

You'll be familiar with angles being measured in degrees. Radians are another unit of measurement for angles. They can be easier to use than degrees when measuring things like the arc length of a sector of a circle or its area, and they come up a lot in trigonometry and throughout the course.

Radians

A **radian** (rad) is just another unit of measurement for an angle.

1 radian is the angle formed in a **sector** that has an **arc length** that is the same as the **radius**.

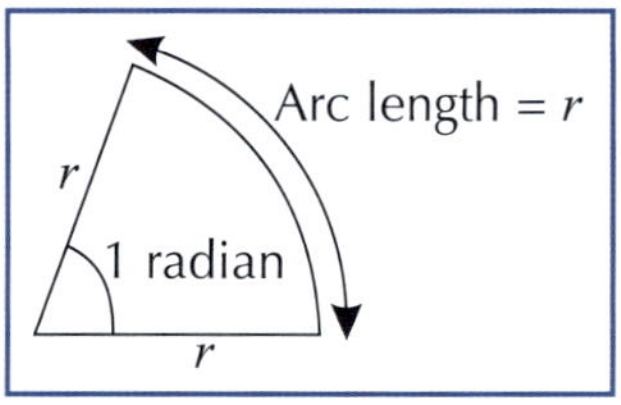

In other words, if you have a **sector** with an angle of **1 radian**, then the **length** of the **arc** will be exactly the **same length** as the **radius** r.

It's important to know how **radians relate to degrees**:

- 360 degrees (a complete circle) = 2π radians
- 180 degrees = π radians
- 1 radian is about 57 degrees

Tip: Radians are sometimes shown by 'rad' after the number, or a 'c' (e.g. $4\pi^c$), but most of the time angles in radians are given without a symbol.

You also need to know how to **convert** between the two units. The table below shows you how:

Converting angles	
Radians to degrees: Divide by π, multiply by 180.	**Degrees to radians:** Divide by 180, multiply by π.

Here's a table of some of the **common angles** you're going to need, in degrees and radians:

Degrees	0	30	45	60	90	120	180	270	360
Radians	0	$\frac{\pi}{6}$	$\frac{\pi}{4}$	$\frac{\pi}{3}$	$\frac{\pi}{2}$	$\frac{2\pi}{3}$	π	$\frac{3\pi}{2}$	2π

Tip: It's a good idea to learn these common angles — they come up a lot.

Examples

a) **Convert $\frac{\pi}{15}$ into degrees.**

- To convert from radians to degrees:
- Divide by π... $\frac{\pi}{15} \div \pi = \frac{1}{15}$
- ... then multiply by 180... $\frac{1}{15} \times 180 = 12°$

Tip: Notice how the angle was given without a symbol — if this happens, you can just assume it's in radians.

b) Convert 120 degrees into radians.

- You could smugly use the table on the last page to find the answer, but to show that the rule for converting from degrees to radians works...
- ... divide by 180 and then multiply by π.

$$\frac{120}{180} \times \pi = \frac{2\pi}{3}$$

c) Convert 297 degrees into radians.

- To convert from degrees to radians, divide by 180 and then multiply by π.

$$\frac{297}{180} \times \pi = 1.65\pi \text{ or } 5.18 \text{ rad (3 s.f.)}$$

Tip: In the exam, you can give your answer to part c) in terms of π or as a rounded number, unless the question states otherwise. Generally though, it's better to just keep it in terms of π, and if the question asks for an exact answer you have to do this.

Exercise 1.1

Q1 Convert the angles below into radians. Give your answers in terms of π.

a) 180° b) 135° c) 270°

d) 70° e) 150° f) 75°

Q2 Convert the angles below into degrees.

a) $\frac{\pi}{4}$ b) $\frac{\pi}{2}$ c) $\frac{\pi}{3}$

d) $\frac{5\pi}{2}$ e) $\frac{3\pi}{4}$ f) $\frac{7\pi}{3}$

Arc length and sector area

A **sector** is part of a circle formed by **two radii** and part of the **circumference**. The **arc** of a sector is the **curved** edge of the sector. You can work out the **length** of the **arc**, or the **area** of the sector — as long as you know the **angle** at the **centre** (**θ**) and the **length** of the **radius** (**r**). When working out arc length and sector area you **always** work in radians.

Arc length

For a circle with **radius r**, a sector with **angle θ** (measured in **radians**) has **arc length s**, given by:

$$s = r\theta$$

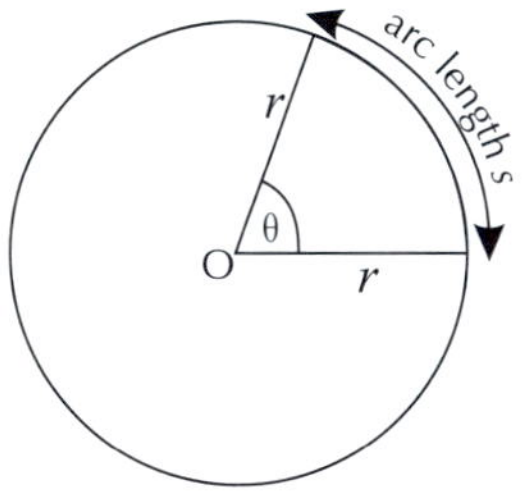

- If you put $\theta = 2\pi$ in this formula (and so make the sector equal to the whole circle), you find that the distance all the way round the outside of the circle is $s = 2\pi r$.
- This is just the normal **circumference** formula.

Sector area

For a circle with **radius r**, a sector with **angle θ** (measured in **radians**) has **area A**, given by:

$$A = \frac{1}{2}r^2\theta$$

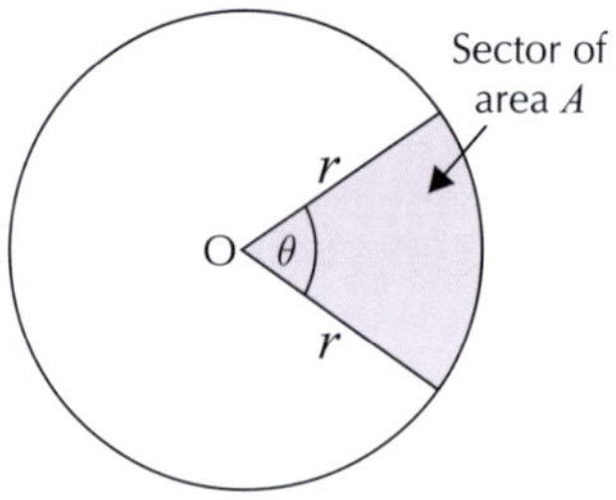

- Again, if you put $\theta = 2\pi$ in the formula, you find that the area of the whole circle is $A = \frac{1}{2}r^2 \times 2\pi = \pi r^2$.
- This is just the normal '**area of a circle**' formula.

Example 1

Find the exact length L and area A in the diagram to the right.

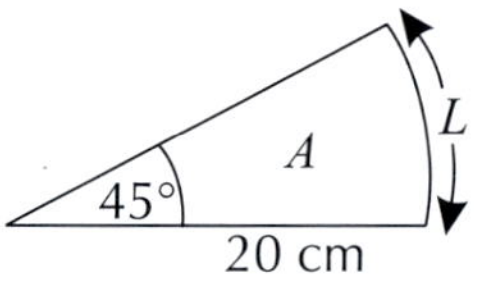

- It's asking for an arc length and sector area, so you need the angle in **radians**.

$$45° = \frac{45 \times \pi}{180} = \frac{\pi}{4} \text{ radians}$$

Tip: Or you could just quote this if you've learnt the stuff on page 44 off by heart.

- Now put everything in your formulas:

$$L = r\theta = 20 \times \frac{\pi}{4} = 5\pi \text{ cm}$$

$$A = \frac{1}{2}r^2\theta = \frac{1}{2} \times 20^2 \times \frac{\pi}{4} = 50\pi \text{ cm}^2$$

Example 2

Find the area of the shaded part of the symbol.

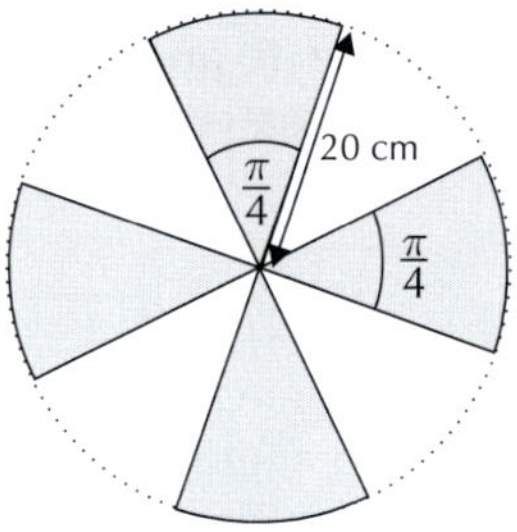

- Each 'leaf' has area:

$$\frac{1}{2} \times 20^2 \times \frac{\pi}{4} = 50\pi \text{ cm}^2$$

- So the area of the whole symbol is:

$$4 \times 50\pi = 200\pi \text{ cm}^2$$

Tip: Instead, you could use the total angle of all the shaded sectors (π).

Example 3

Find the exact value of θ in the diagram to the right.

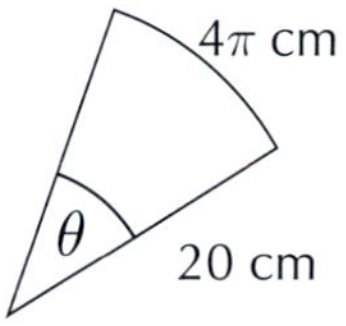

Use the formula for the arc length:

$$s = r\theta \Rightarrow 4\pi = 20\theta$$

$$\theta = \frac{4\pi}{20} = \frac{\pi}{5} \text{ radians}$$

Example 4

The sector shown has an area of 6π cm². Find the arc length, s.

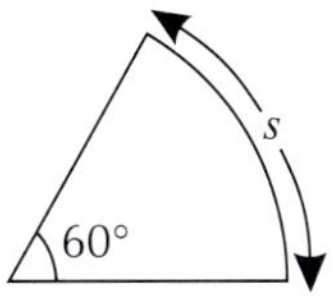

- First, get the angle in **radians**:

$$60° = \frac{60 \times \pi}{180} = \frac{\pi}{3} \text{ radians}$$

- You want to find the arc length, but you don't know the sector's radius.
- But you know its area, so use the area formula to **work out** the **radius**:

$$A = \frac{1}{2}r^2\theta \Rightarrow 6\pi = \frac{1}{2} \times r^2 \times \frac{\pi}{3}$$
$$36 = r^2$$
$$r = 6$$

- Putting this value of r into the equation for arc length gives:

$$s = r\theta = 6 \times \frac{\pi}{3} = 2\pi \text{ cm}$$

Tip: This is one of the common angles from the table on page 44 — it's definitely worth learning them.

Exercise 1.2

Q1 The diagram below shows a sector OAB. The centre is at O and the radius is 6 cm. The angle AOB is 2 radians.

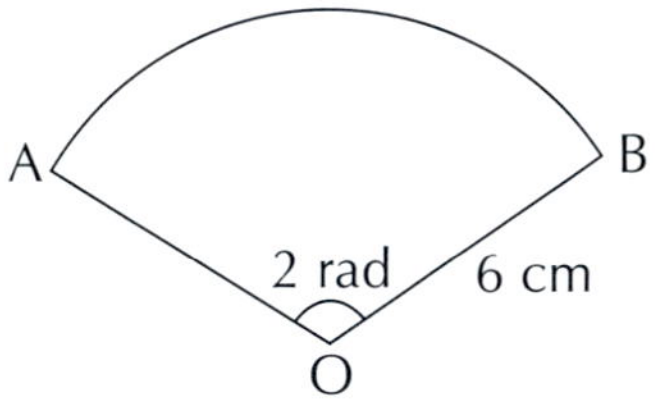

Find the arc length and area of this sector.

Q2 The diagram below shows a sector OAB. The centre is at O and the radius is 8 cm. The angle AOB is 46°.

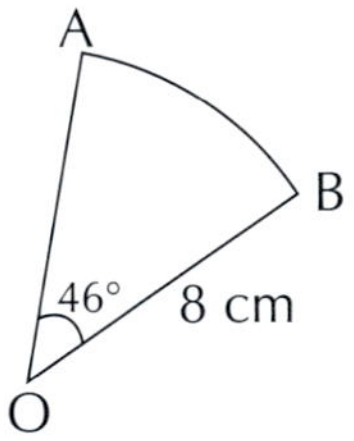

Find the arc length and area of this sector to 1 d.p.

Q2 Hint: Remember to convert 46° to radians first.

Q3 A sector of a circle of radius 4 cm has an area of 6π cm². Find the exact value of the angle θ.

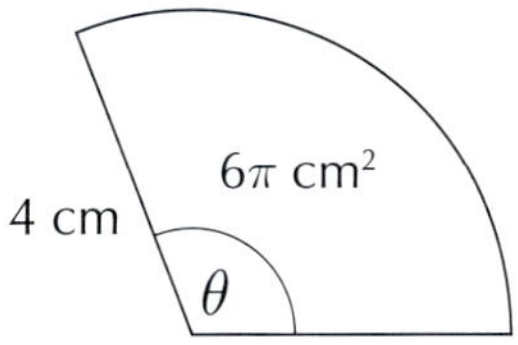

Q4 The diagram below shows a sector of a circle with a centre O and radius r cm. The angle AOB shown is θ.

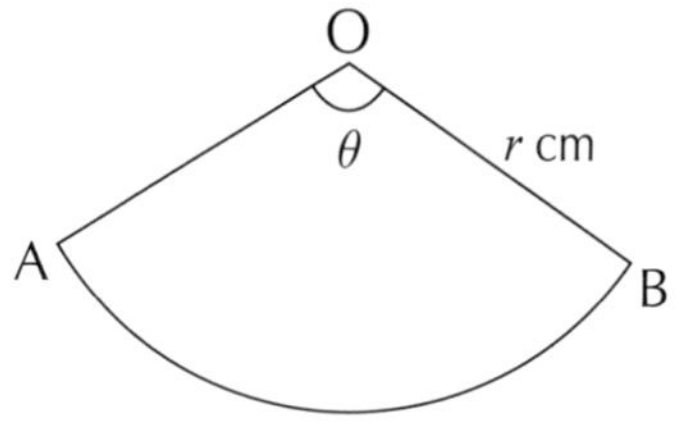

For each of the following values of θ and r, give the arc length and the area of the sector. Where appropriate give your answers to 3 s.f.

a) $\theta = 1.2$ radians, $r = 5$ cm

b) $\theta = 0.6$ radians, $r = 4$ cm

c) $\theta = 80°$, $r = 9$ cm

d) $\theta = \frac{5\pi}{12}$, $r = 4$ cm

Q5 The diagram to the right shows a sector ABC of a circle, where the angle BAC is 0.9 radians.

Given that the area of the sector is 16.2 cm^2, find the arc length s.

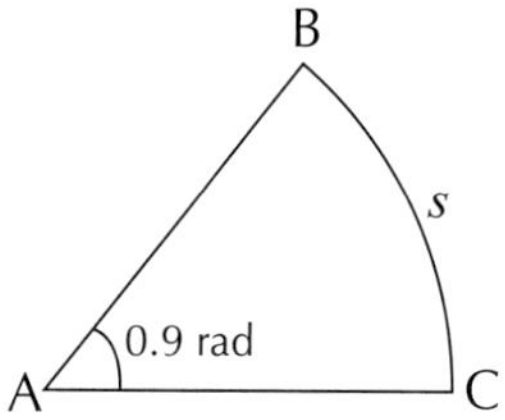

Q6 A circle C has a radius of length 3 cm with centre O. A sector of this circle is given by angle AOB which is 20°.

Find the length of the arc AB and the area of the sector. Give your answer in terms of π.

Q7 The sector shown has an arc length of 7 cm. The angle BAC is 1.4 rad. Find the area of the sector.

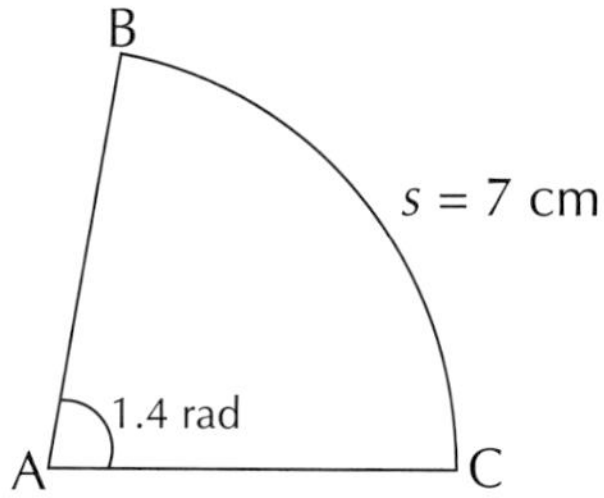

Q8 A circle of radius r contains a sector of area 80π cm^2. Given that the arc length of the sector is 16π cm, find the angle of the sector (θ) and the value of r, giving your answers to 3 s.f.

PROBLEM SOLVING

Q9 The diagram below shows a semicircle of radius 2 cm, with a smaller sector of radius 1 cm removed.

PROBLEM SOLVING

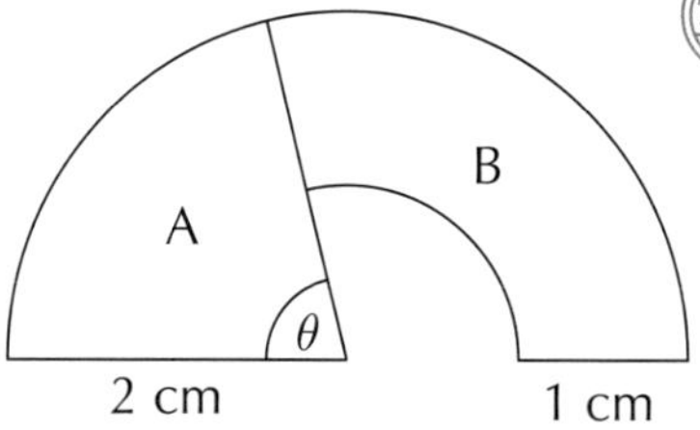

Given that the area of the sector A and the area of B are equal, find the exact value of θ.

2. Small Angle Approximations

When you're working in radians, there are some very handy approximations for the sin, cos and tan values of a really small angle. They're not perfect, and they become less accurate with larger angles, but they're important for finding the derivatives of the trig functions (which you'll see in Chapter 7).

Learning Objectives:
- Be able to use the small angle approximations for sin θ, cos θ and tan θ.
- Be able to use a combination of these to approximate functions for small θ.

The small angle approximations

When θ (measured in radians) is very small, you can approximate the value of sin θ, cos θ and tan θ using the **small angle approximations**:

$$\sin\theta \approx \theta \qquad \cos\theta \approx 1 - \frac{1}{2}\theta^2 \qquad \tan\theta \approx \theta$$

You don't need to know where they come from, but it might help you to understand why they work. Sin θ, cos θ and tan θ can each be written as an **infinite series** (a sum of terms — see Chapter 5 for more on series and summation notation) known as the **Maclaurin series**:

$$\sin\theta = \theta - \frac{1}{6}\theta^3 + \frac{1}{120}\theta^5 - ... = \sum_{n=0}^{\infty}(-1)^n\frac{\theta^{2n+1}}{(2n+1)!}$$

$$\cos\theta = 1 - \frac{1}{2}\theta^2 + \frac{1}{24}\theta^4 - ... = \sum_{n=0}^{\infty}(-1)^n\frac{\theta^{2n}}{(2n)!}$$

$$\tan\theta = \theta + \frac{1}{3}\theta^3 + \frac{2}{15}\theta^5 - ...$$

Tip: Don't worry, you don't need to know anything about the Maclaurin series for now (but you might come across it if you do Further Maths).

Tip: The Maclaurin series for tan θ is a bit messy in sum notation, so it's been left out here.

When θ is small ($\theta < 1$), θ^n gets smaller and smaller as n increases, so most of the terms in each series will be **tiny** (almost 0). Choosing to **ignore** all of the terms with θ^3 or a higher power leaves you with something **close** to the actual answer — so $\sin\theta \approx \theta$, $\cos\theta \approx 1 - \frac{1}{2}\theta^2$ and $\tan\theta \approx \theta$.

Tip: These approximations only work when $\theta < 1$ — above that, the θ^n terms get bigger instead of smaller, so you can't ignore them.

Remember, θ must be in **radians** for these to work, not in degrees — you may have to **convert** before you use the approximations.

Example 1

Give an approximation for cos 0.2.

- The small angle approximation for cos is:
  $$\cos\theta \approx 1 - \frac{1}{2}\theta^2$$
- Set $\theta = 0.2$:
  $$\cos 0.2 \approx 1 - \frac{1}{2}(0.2)^2$$
  $$\approx 1 - 0.02 = 0.98$$

Tip: The actual value of cos 0.2 is 0.980067 (to 6 d.p.), so the approximation is pretty accurate. Try it with some other small values of θ and compare the approximation to the actual answer.

Approximating functions

These approximations are most useful for approximating more **complicated functions**, which could involve sin, cos and tan of **multiples** of θ (when θ is small, you can assume that multiples of θ are also small). Make sure that you apply the approximation to **everything** inside the trig function — for example:

$$\tan 4\theta \approx 4\theta \qquad \sin\frac{1}{2}\theta \approx \frac{1}{2}\theta \qquad \cos 3\theta \approx 1 - \frac{1}{2}(3\theta)^2$$

You might have to use more than one formula in a question.

Example 2

Find an approximation for $f(\theta) = 4\cos\theta\tan 3\theta$ when θ is small.

Replace cos and tan with the small angle approximations:

$$f(\theta) \approx 4 \times (1 - \tfrac{1}{2}\theta^2) \times 3\theta$$

$\approx \cos\theta$ $\approx \tan 3\theta$

$$= (4 - 2\theta^2) \times 3\theta$$

$$= 12\theta - 6\theta^3 \text{ (or } 6\theta(2 - \theta^2))$$

Example 3

Show that $\dfrac{2\theta\sin 2\theta}{1-\cos 5\theta} \approx \dfrac{8}{25}$ when θ is small.

Use the small angle approximations for each trig function:

$$f(\theta) = \frac{2\theta\sin 2\theta}{1-\cos 5\theta}$$

$$\approx \frac{2\theta(2\theta)}{1-\left(1-\frac{1}{2}(5\theta)^2\right)}$$

$$= \frac{4\theta^2}{\frac{25}{2}\theta^2} = \frac{8}{25} \text{ as required}$$

Exercise 2.1

Q1 Use the small angle approximations to estimate the following values, then find the actual values on a calculator:

a) sin 0.23 b) cos 0.01 c) tan 0.18

Q2 For the values of θ below, use the small angle approximations to estimate the value of $f(\theta) = \sin\theta + \cos\theta$, then use a calculator to find the actual answer:

a) $\theta = 0.3$ b) $\theta = 0.5$ c) $\theta = 0.25$ d) $\theta = 0.01$

Q3 Find an approximation for the following expressions when θ is small:

a) $\sin\theta\cos\theta$ b) $\theta\tan 5\theta\sin\theta$ c) $\dfrac{\sin 4\theta\cos 3\theta}{2\theta}$

d) $3\tan\theta + \cos 2\theta$ e) $\sin\frac{1}{2}\theta - \cos\theta$ f) $\dfrac{\cos\theta - \cos 2\theta}{1-(\cos 3\theta + 3\sin\theta\tan\theta)}$

Q4 A pendulum of length 6 cm follows the arc of a circle. Its straight-line displacement as a vector is given by:
$\mathbf{d} = 6\sin\theta\,\mathbf{i} + 6(1 - \cos\theta)\mathbf{j}$.

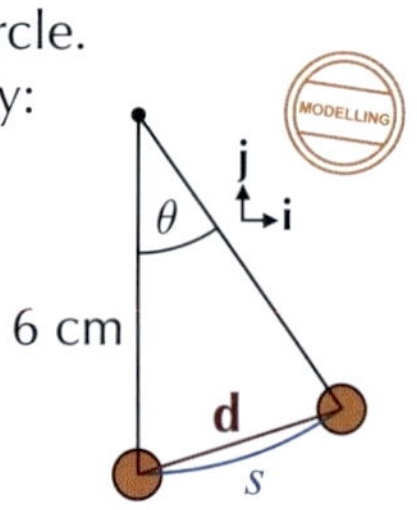

a) Show that the magnitude of the displacement is $6\sqrt{2(1-\cos\theta)}$.

b) Show that, when θ is small, the magnitude of the displacement can be approximated by the arc length s.

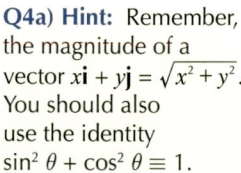

Q4a) Hint: Remember, the magnitude of a vector $x\mathbf{i} + y\mathbf{j} = \sqrt{x^2+y^2}$. You should also use the identity $\sin^2\theta + \cos^2\theta \equiv 1$.

3. Inverse Trig Functions

In Chapter 2 you saw that some functions have inverses, which reverse the effect of the function. The trig functions have inverses too.

Learning Objectives:

- Know that the inverse of the trig functions sin, cos and tan are arcsin, arccos and arctan.
- Recognise and be able to sketch the graphs of arcsin, arccos and arctan, including their restricted domains and ranges.
- Be able to evaluate the inverse trig functions.

Arcsin, arccos and arctan

The inverse trig functions

- **Arcsin** is the inverse of **sin**.
  You might see it written as arcsine or $\sin^{-1}$.
- **Arccos** is the inverse of **cos**.
  You might see it written as arccosine or $\cos^{-1}$.
- **Arctan** is the inverse of **tan**.
  You might see it written as arctangent or $\tan^{-1}$.

The inverse trig functions **reverse** the effect of sin, cos and tan. For example, $\sin 30° = 0.5$, so $\arcsin 0.5 = 30°$. You should have buttons for doing arcsin, arccos and arctan on your calculator — they'll probably be labelled $\sin^{-1}$, $\cos^{-1}$ and $\tan^{-1}$.

Graphs of the inverse trig functions

The functions sine, cosine and tangent **aren't one-to-one** mappings (see p.18). This means that more than one value of x gives the same value for $\sin x$, $\cos x$ or $\tan x$. For example: $\cos 0 = \cos 2\pi = \cos 4\pi = 1$, and $\tan 0 = \tan \pi = \tan 2\pi = 0$.

If you want the inverses to be **functions**, you have to **restrict the domains** of the trigonometric functions to make them **one-to-one**.

The graphs of the inverse functions are the **reflections** of the sin, cos and tan graphs in the line $y = x$.

Tip: Functions are mappings which have just one y-value for every x-value. There's more on functions on pages 12-18, and on inverse functions on pages 23-25.

Arcsin

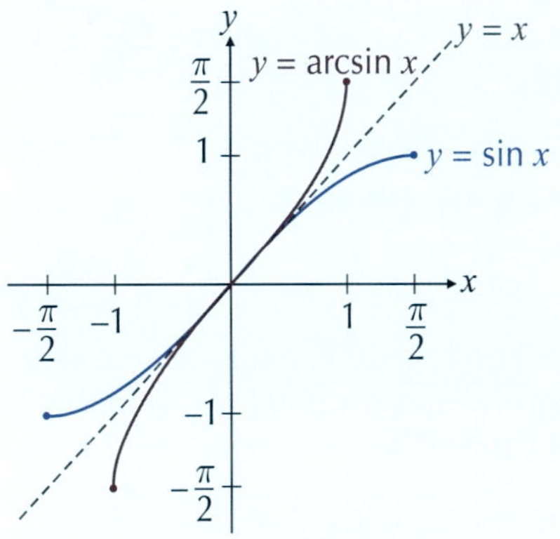

- For arcsin, limit the **domain** of $\sin x$ to $-\frac{\pi}{2} \leq x \leq \frac{\pi}{2}$ (the range of $\sin x$ is still $\mathbf{-1 \leq \sin x \leq 1}$).
- So the **domain** of $\arcsin x$ is $\mathbf{-1 \leq x \leq 1}$.
- The **range** of $\arcsin x$ is $\mathbf{-\frac{\pi}{2} \leq \arcsin x \leq \frac{\pi}{2}}$.
- The graph of $\mathbf{y = \arcsin x}$ goes through the **origin**.
- The coordinates of its **endpoints** are $\left(-1, -\frac{\pi}{2}\right)$ and $\left(1, \frac{\pi}{2}\right)$.

Tip: These graphs show values in radians, but you might have to use angles in degrees too. Remember that π radians is 180°.

Arccos

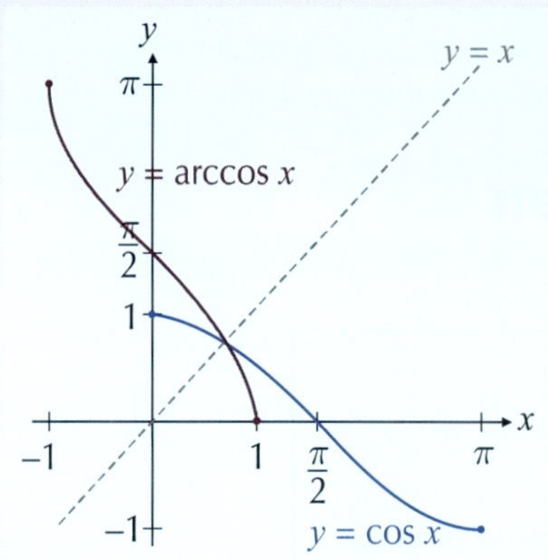

- For arccos, limit the **domain** of $\cos x$ to $\mathbf{0 \le x \le \pi}$ (the range of $\cos x$ is still $\mathbf{-1 \le \cos x \le 1}$.)
- So the **domain** of arccos x is $\mathbf{-1 \le x \le 1}$.
- The **range** of arccos x is $\mathbf{0 \le \arccos x \le \pi}$.
- The graph of $y = \arccos x$ crosses the **y-axis** at $\left(0, \frac{\pi}{2}\right)$.
- The coordinates of its **endpoints** are $(-1, \pi)$ and $(1, 0)$.

Tip: Learn the key features of the inverse trig graphs — you might be asked to transform them in different ways. There's more on transformations of graphs on pages 33-36.

Arctan

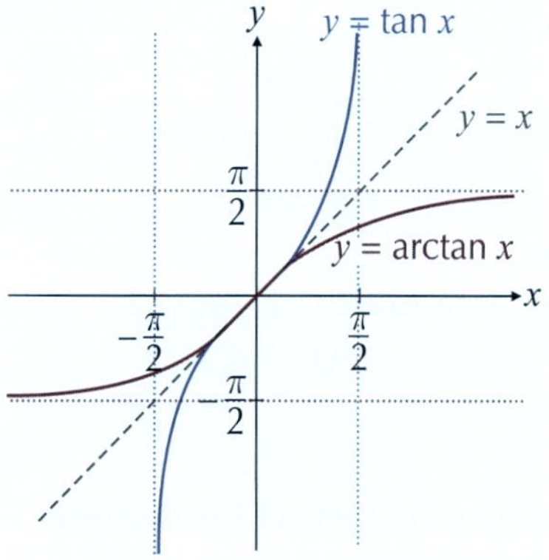

- For arctan, limit the **domain** of $\tan x$ to $\mathbf{-\frac{\pi}{2} < x < \frac{\pi}{2}}$ (this **doesn't limit** the range of $\tan x$.)
- This means that the **domain** of arctan x **isn't** limited (it's $x \in \mathbb{R}$).
- The **range** of arctan x is $\mathbf{-\frac{\pi}{2} < \arctan x < \frac{\pi}{2}}$.
- The graph of $y = \arctan x$ goes through the **origin**.
- It has **asymptotes** at $y = \frac{\pi}{2}$ and $y = -\frac{\pi}{2}$.

Tip: Note that there are no endpoints marked on the graph as the domain is not restricted.

Evaluating arcsin, arccos and arctan

If a is an angle within the interval $-\frac{\pi}{2} \le a \le \frac{\pi}{2}$ (or $-90° \le a \le 90°$) such that $\mathbf{\sin a = x}$, then $\mathbf{\arcsin x = a}$. So to evaluate **arcsin x** you need to find the angle a in this interval such that $\mathbf{\sin a = x}$. Using a calculator, this will be the answer you get when you enter "**$\sin^{-1} x$**" (for a given value of x).

Similarly, to find **arccos x**, you need to find the angle a within the interval $\mathbf{0 \le a \le \pi}$ (or $0° \le a \le 180°$) such that $\mathbf{\cos a = x}$.
And **arctan x** is the angle a in the interval $-\frac{\pi}{2} < a < \frac{\pi}{2}$ (or $-90° < a < 90°$) such that $\mathbf{\tan a = x}$.

When evaluating inverse trig functions, it'll be helpful if you know the **sine**, **cosine** and **tangent** of some **common angles**. Here's a quick recap of the method of drawing triangles — and **SOH CAH TOA**.

Remember: SOH CAH TOA...

$\sin x = \frac{\text{opp}}{\text{hyp}}$ $\quad \cos x = \frac{\text{adj}}{\text{hyp}}$ $\quad \tan x = \frac{\text{opp}}{\text{adj}}$

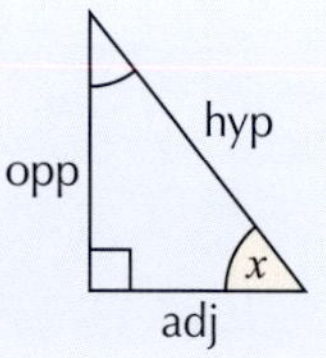

The sin, cos and tan of 30°, 45° and 60° can be found by drawing the following triangles and using **SOH CAH TOA.**

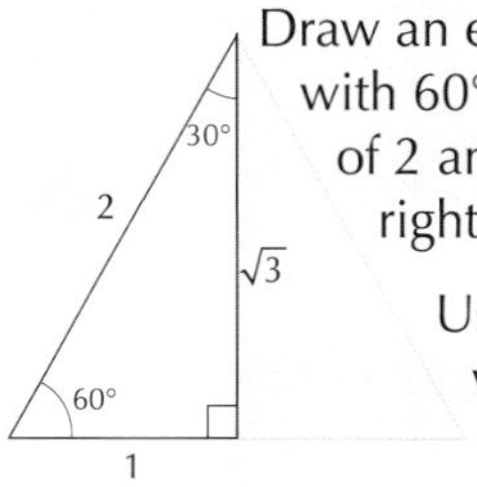

Draw an equilateral triangle with 60° angles and sides of 2 and split it to make a right-angled triangle.

Use Pythagoras to work out the length of the third side.

Draw a right-angled triangle where the edges adjacent to the right angle have length 1. Use Pythagoras to work out the length of the third side.

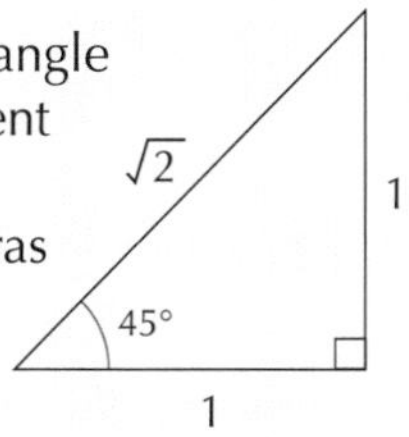

You should already know the sin, cos and tan of 90° and 180°, so you can work out all the values in this table:

$x°$	x (rad)	$\sin x$	$\cos x$	$\tan x$
0	0	0	1	0
30	$\frac{\pi}{6}$	$\frac{1}{2}$	$\frac{\sqrt{3}}{2}$	$\frac{1}{\sqrt{3}}$
45	$\frac{\pi}{4}$	$\frac{1}{\sqrt{2}}$	$\frac{1}{\sqrt{2}}$	1
60	$\frac{\pi}{3}$	$\frac{\sqrt{3}}{2}$	$\frac{1}{2}$	$\sqrt{3}$
90	$\frac{\pi}{2}$	1	0	—
180	π	0	–1	0

Be careful though — the first solution you find might **not** lie within the appropriate domain for the inverse function (see the graphs on pages 51 and 52). To find a solution that **does** lie in the correct domain, you need to use the **graphs** of the functions, or the **CAST diagram** that was introduced in the first year of this course. The following examples show how to use these methods.

Tip: You might find it useful to look back at your Year 1 notes on solving trig equations in a given interval.

Examples

a) Evaluate, without using a calculator, arccos 0.5. Give your answer in degrees.

- First, work out the angle a for which $\cos a = 0.5$.
  Since you're expected to do this without a calculator, it will be a common angle you can find using a right-angled triangle:

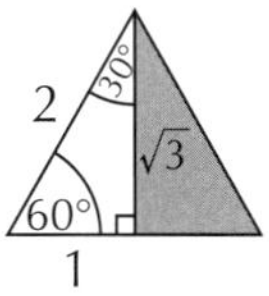

0.5 is $\frac{1}{2}$ which is either sin 30° (if using $\frac{\text{opp}}{\text{hyp}}$ on the triangle) or cos 60° (if using $\frac{\text{adj}}{\text{hyp}}$).

Tip: For all of these examples, the values that you need are in the table — it really helps to learn it.

- We are looking for the inverse of **cos**, so $a = 60°$.
- Next check that this answer lies in the appropriate domain for cos:
  $0 \le a \le \pi$ in radians, which is $0° \le a \le 180°$.
  60° lies within this domain, so arccos 0.5 = 60°.

b) Evaluate, without using a calculator, arctan –1. Give your answer in radians.

Tip: You could also use the CAST diagram (as shown in the example on the next page) to find any negative solutions — for tan they lie in the 'S' and 'C' quadrants.

- Work out the angle a for which $\tan a = -1$, over $-\frac{\pi}{2} < a < \frac{\pi}{2}$.

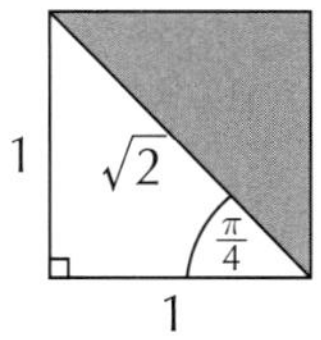

- Using this triangle you can see that $\tan \frac{\pi}{4} = 1$.
  But you need to look at the symmetry of the tan x graph to find the solution for $\tan a = -1$:

- The graph shows that if $\tan\frac{\pi}{4} = 1$ then $\tan -\frac{\pi}{4} = -1$.
- So $\arctan -1 = -\frac{\pi}{4}$.

  This answer lies in the appropriate domain for tan, i.e. $-\frac{\pi}{2} < a < \frac{\pi}{2}$.

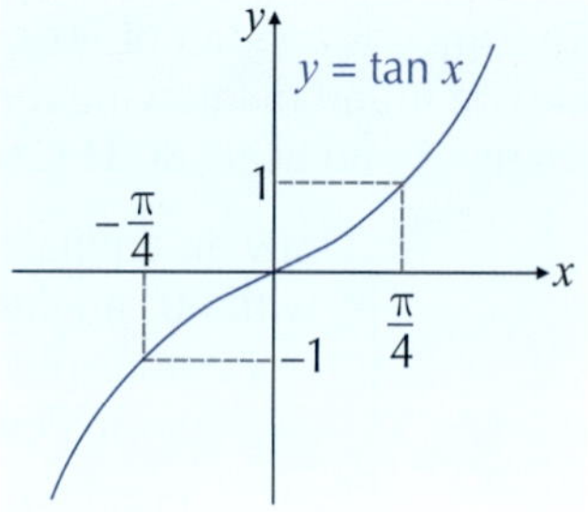

c) Evaluate $\arcsin -\frac{1}{\sqrt{2}}$ without using your calculator. Give your answer in radians.

- Using the triangle from the previous example: $\sin\frac{\pi}{4} = \frac{1}{\sqrt{2}}$.
- To find the angle a such that $\sin a = -\frac{1}{\sqrt{2}}$, look at the **CAST diagram**:

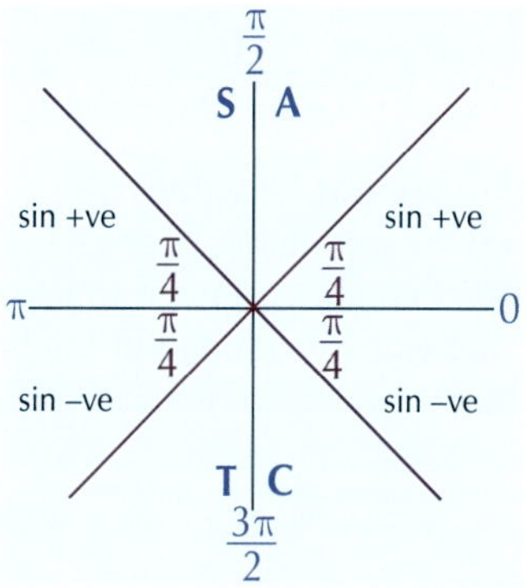

- Positive solutions for $\sin\frac{\pi}{4}$ are found in the quadrants labelled 'S' (for 'sin positive') and 'A' (for 'all positive'). This means that the desired **negative** solutions will be in the 'T' and 'C' quadrants:
  so $a = \pi + \frac{\pi}{4} = \frac{5\pi}{4}$, or $0 - \frac{\pi}{4} = -\frac{\pi}{4}$.
- Only the second of these answers lies within the appropriate domain $\left(-\frac{\pi}{2} \le a \le \frac{\pi}{2}\right)$, so $\arcsin -\frac{1}{\sqrt{2}} = -\frac{\pi}{4}$.

Tip: If you're not confident with using the CAST diagram you could sketch the sin graph instead and look at the symmetry (as in the previous example).

Exercise 3.1

Q1 Evaluate the following, giving your answer in radians.

a) arccos 1 b) $\arcsin\frac{\sqrt{3}}{2}$ c) $\arctan\sqrt{3}$

Q2 a) Sketch the graph of $y = 2\arccos x$ for $-1 \le x \le 1$.

b) Sketch the graph of $y = \frac{1}{2}\arctan x$ and state the range.

Q3 By drawing the graphs of $y = \frac{x}{2}$ and $y = \cos^{-1} x$, determine the number of real roots of the equation $\cos^{-1} x = \frac{x}{2}$.

Q3 Hint: Remember $\cos^{-1} x$ is just another name for arccos x.

Q4 Evaluate the following, giving your answers in radians:

a) $\sin^{-1}(-1)$ b) $\cos^{-1}\left(-\frac{\sqrt{3}}{2}\right)$

Q5 Evaluate the following:

a) $\tan\left(\arcsin\frac{1}{2}\right)$ b) $\cos^{-1}\left(\cos\frac{2\pi}{3}\right)$ c) $\cos\left(\arcsin\frac{1}{2}\right)$

Q6 $f(x) = 1 + \sin 2x$. Find an expression for $f^{-1}(x)$. PROBLEM SOLVING

4. Cosec, Sec and Cot

There are a few more trigonometric functions to learn. This time it's the reciprocals of sin, cos and tan — cosec, sec and cot.

Graphs of cosec, sec and cot

When you take the **reciprocal** of the three main trig functions, sin, cos and tan, you get three new trig functions — **cosecant** (or **cosec**), **secant** (or **sec**) and **cotangent** (or **cot**).

$$\text{cosec}\ \theta \equiv \frac{1}{\sin\theta} \qquad \sec\theta \equiv \frac{1}{\cos\theta} \qquad \cot\theta \equiv \frac{1}{\tan\theta}$$

Since $\tan\theta \equiv \frac{\sin\theta}{\cos\theta}$, you can also think of **cot θ** as being $\frac{\cos\theta}{\sin\theta}$.

Examples

Write the following in terms of sin and cos only:

a) cosec 20° $\qquad \text{cosec}\ 20° = \frac{1}{\sin 20°}$

b) sec π $\qquad \sec\pi = \frac{1}{\cos\pi}$

c) cot $\frac{\pi}{6}$ $\qquad \cot\frac{\pi}{6} = \frac{1}{\tan\frac{\pi}{6}} = \frac{\cos\frac{\pi}{6}}{\sin\frac{\pi}{6}}$

Graph of cosec

This is the graph of $y = \text{cosec}\ x$:

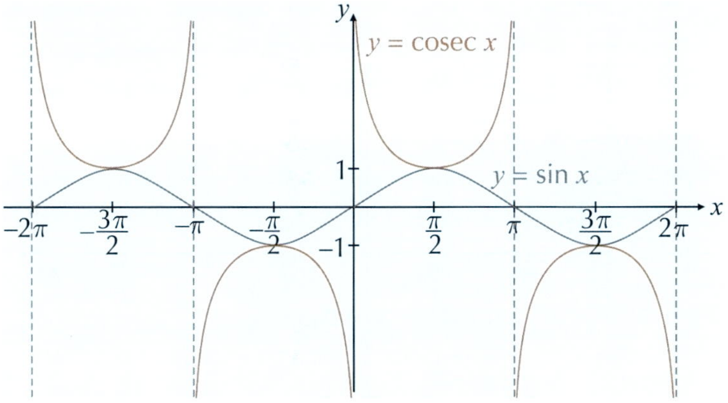

- Since $\text{cosec}\ x = \frac{1}{\sin x}$, $y = \text{cosec}\ x$ is **undefined** at any point where $\sin x = 0$.
- So $y = \text{cosec}\ x$ has **vertical asymptotes** at $x = n\pi$ (where n is any integer).
- The graph $y = \text{cosec}\ x$ has **minimum** points at $x = ..., -\frac{3\pi}{2}, \frac{\pi}{2}, \frac{5\pi}{2}, ...$ (wherever the graph $y = \sin x$ has a **maximum**). At these points, $y = 1$.
- It has **maximum** points at $x = ..., -\frac{\pi}{2}, \frac{3\pi}{2}, \frac{7\pi}{2}, ...$ (wherever the graph $y = \sin x$ has a **minimum**). At these points, $y = -1$.

Learning Objectives:

- Know that the reciprocals of sin, cos and tan are cosec, sec and cot.
- Recognise and be able to sketch the graphs of cosec, sec and cot.
- Be able to evaluate the reciprocal trig functions.
- Be able to simplify expressions involving the reciprocal trig functions.
- Be able to solve equations involving the reciprocal trig functions.

Tip: The trick for remembering which is which is to look at the third letter — cosec (1/sin), sec (1/cos) and cot (1/tan).

Tip: The x-coordinates of the turning points for cosec x are the same as for $\sin x$ — but remember a maximum on $\sin x$ becomes a minimum on cosec x, and vice versa.

Graph of sec

This is the graph of $y = \sec x$:

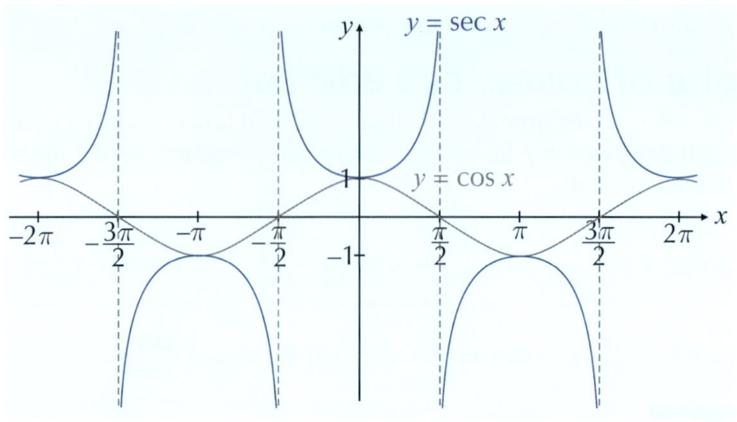

> **Tip:** Just like the graphs of $\sin x$ and $\cos x$, the graphs of $\operatorname{cosec} x$ and $\sec x$ have a **period** of 2π radians — this just means they repeat themselves every 2π (or 360°).

- As $\sec x = \frac{1}{\cos x}$, $y = \sec x$ is **undefined** at any point where $\cos x = 0$.
  So $y = \sec x$ has **vertical asymptotes** at $\boldsymbol{x = \left(n\pi + \frac{\pi}{2}\right)}$ (where n is any integer).
- The graph of $y = \sec x$ has **minimum** points at $x = 0, \pm 2\pi, \pm 4\pi, ...$ (wherever the graph of $y = \cos x$ has a **maximum**). At these points, $\boldsymbol{y = 1}$.
- It has **maximum** points at $x = \pm\pi, \pm 3\pi, ...$ (wherever the graph of $y = \cos x$ has a **minimum**). At these points, $\boldsymbol{y = -1}$.

> **Tip:** The integer n can be negative too, so the asymptotes are at $x = -\frac{\pi}{2}, -\frac{3\pi}{2}, -\frac{5\pi}{2}$... etc, as well as $\frac{\pi}{2}, \frac{3\pi}{2}, \frac{5\pi}{2}$...

Graph of cot

This is the graph of $y = \cot x$:

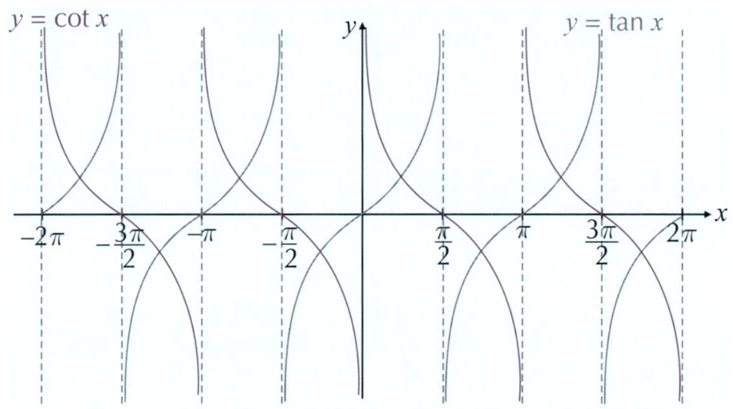

> **Tip:** The graphs of $\tan x$ and $\cot x$ both have a period of π radians.

- Since $\cot x = \frac{1}{\tan x}$, $y = \cot x$ is **undefined** at any point where $\tan x = 0$.
- So $y = \cot x$ has **vertical asymptotes** at $\boldsymbol{x = n\pi}$ (where n is any integer).
- $y = \cot x$ **crosses the x-axis** at every place where the graph of $\tan x$ has an asymptote. This is any point with the coordinates $\left(\left(n\pi + \frac{\pi}{2}\right), 0\right)$.

> **Tip:** The asymptotes for $\cot x$ are the $\tan x$ asymptotes shifted horizontally by $\frac{\pi}{2}$.

Transformations of cosec, sec and cot

The graphs of the cosec, sec and cot functions can be transformed in the same way as other functions.

Examples

a) Sketch the graph of $y = \cot 2x$ over the interval $-\pi \leq x \leq \pi$.

If $f(x) = \cot x$, then $y = f(2x)$. This transformation is a **horizontal stretch** by a factor of $\frac{1}{2}$ (i.e. the graph is squashed up in the x-direction by a factor of 2).

Tip: Look back at pages 33-36 for more on transformations of graphs.

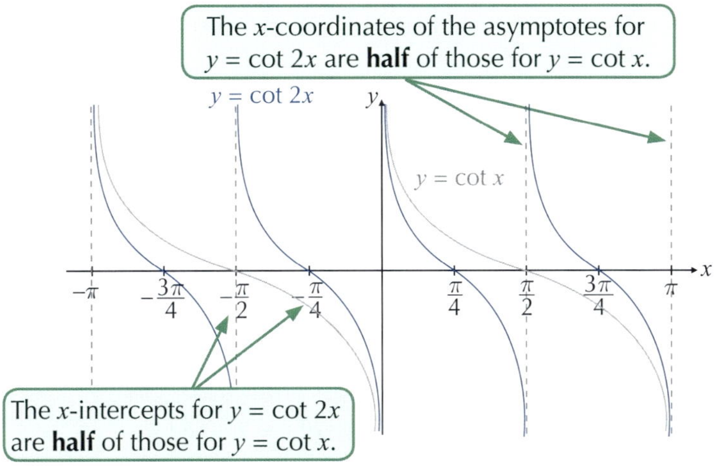

The x-coordinates of the asymptotes for y = cot 2x are half of those for y = cot x.
y = cot 2x
y
y = cot x
x
−π
−3π/4
−π/2
−π/4
π/4
π/2
3π/4
π
The x-intercepts for y = cot 2x are half of those for y = cot x.

The **period** of the graph is also halved:
$y = \cot 2x$ repeats itself every $\frac{\pi}{2}$ radians.

b) Give the coordinates of the maximum point on the graph of $y = \sec(x - 30°) + 1$, between 0° and 360°.

- The graph of $f(x) = \sec x$ has a maximum point at **(180°, –1)**.
- $y = f(x - 30°) + 1$. This transformation is a **horizontal translation right** by **30°**, followed by a **vertical translation up** by **1**.
- The coordinates of the maximum point will be affected by this transformation — the x-coordinate will be **increased by 30°** and the y-coordinate will be **increased by 1**.
- So the coordinates of the maximum point on the transformed graph will be: (210°, 0).

Tip: Make sure you know all the key coordinates on the graphs in both radians and degrees.

c) Describe the position of the asymptotes on the graph of $y = \text{cosec}\left(x + \frac{\pi}{3}\right)$.

- The graph of $f(x) = \text{cosec } x$ has asymptotes at $x = n\pi$ (where n is any integer).
- $y = f\left(x + \frac{\pi}{3}\right)$. This transformation is a **horizontal translation left** by $\frac{\pi}{3}$.
- Each asymptote will be translated left by $\frac{\pi}{3}$.

  So there will be asymptotes at $x = n\pi - \frac{\pi}{3}$ (where n is any integer).

Tip: The position of a vertical asymptote is only affected by transformations in the x-direction.

Exercise 4.1

> **Q1d) Hint:** Think about which values are **not** included in the range.

Q1 a) Sketch the graph of $y = \sec x$ for $-2\pi \le x \le 2\pi$.
b) Give the coordinates of the minimum points within this interval.
c) Give the coordinates of the maximum points within this interval.
d) State the range of $y = \sec x$.

Q2 a) Sketch the graph of $y = \text{cosec } x$ for $0 < x < 2\pi$.
b) Give the coordinates of any maximum and minimum points within this interval.
c) State the domain and range of $y = \text{cosec } x$.

> **Q3 Hint:** Compare the graphs you've drawn for Q1 and Q2.

Q3 Describe the transformation that maps $y = \sec x$ onto $y = \text{cosec } x$.

Q4 a) Describe the transformation that maps $y = \cot x$ onto $y = \cot \frac{x}{4}$.
b) What is the period, in degrees, of the graph $y = \cot \frac{x}{4}$?
c) Sketch the graph of $y = \cot \frac{x}{4}$ for $0° \le x \le 360°$.

Q5 a) Sketch the graph of $y = 2 + \sec x$ for $-2\pi \le x \le 2\pi$.
b) Give the coordinates of any maximum and minimum points within this interval.
c) State the domain and range of $y = 2 + \sec x$.

Q6 a) Sketch the graph of $y = 2 \text{ cosec } 2x$ for $0° \le x \le 360°$.
b) Give the coordinates of the minimum points within this interval.
c) Give the coordinates of the maximum points within this interval.
d) For what values of x in this interval is $y = 2 \text{ cosec } 2x$ undefined?

> **Q7 Hint:** Work out the transformations from the graph of $y = \text{cosec } x$ first and see how they affect the graph's properties.

Q7 a) Describe the position of the asymptotes on the graph of $y = 2 + 3 \text{ cosec } x$.
b) What is the period, in degrees, of the graph $y = 2 + 3 \text{ cosec } x$?
c) Sketch the graph of $y = 2 + 3 \text{ cosec } x$ for $-180° < x < 180°$.
d) State the range of $y = 2 + 3 \text{ cosec } x$.

Evaluating cosec, sec and cot

To **evaluate** cosec, sec or cot of a number, first evaluate sin, cos or tan then work out the **reciprocal** of the answer.

Examples

a) Evaluate 2 sec(–20°) + 5, giving your answer to 3 significant figures.

- First write out the expression in terms of sin, cos or tan.

$$\sec x = \frac{1}{\cos x}, \text{ so } 2\sec(-20°) + 5 = \frac{2}{\cos(-20°)} + 5.$$

- Now use a calculator to find the answer:

$$\frac{2}{\cos(-20°)} + 5 = \frac{2}{0.93969...} + 5 = 7.13 \text{ to 3 s.f.}$$

b) Evaluate cosec $\frac{\pi}{4}$ without a calculator. Give your answer in surd form.

- $\text{cosec } x = \frac{1}{\sin x}$, so $\text{cosec } \frac{\pi}{4} = \frac{1}{\sin\frac{\pi}{4}}$
- Using the triangle on the right, $\sin\frac{\pi}{4} = \frac{1}{\sqrt{2}}$.
- So $\text{cosec } \frac{\pi}{4} = \frac{1}{\left(\frac{1}{\sqrt{2}}\right)} = \sqrt{2}$

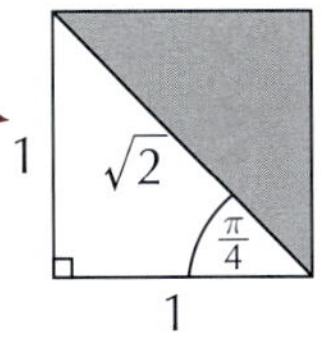

Tip: If you're asked for an answer in surd form, or to give an exact answer, it means you probably have to do it without a calculator. You should be able to solve it by considering angles of 0, $\frac{\pi}{6}$, $\frac{\pi}{4}$, $\frac{\pi}{3}$, $\frac{\pi}{2}$, π or 2π — see p.53.

c) Find the exact value of cot $\left(-\frac{\pi}{6}\right)$.

- $\cot x = \frac{1}{\tan x}$, so $\cot\left(-\frac{\pi}{6}\right) = \frac{1}{\tan\left(-\frac{\pi}{6}\right)}$
- Using the triangle, you can see that $\tan\frac{\pi}{6} = \frac{1}{\sqrt{3}}$.

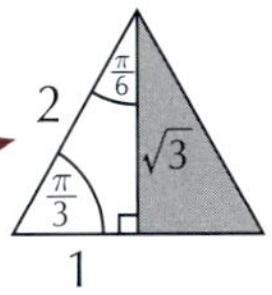

- The graph of $y = \tan x$ below shows that if $\tan\frac{\pi}{6} = \frac{1}{\sqrt{3}}$

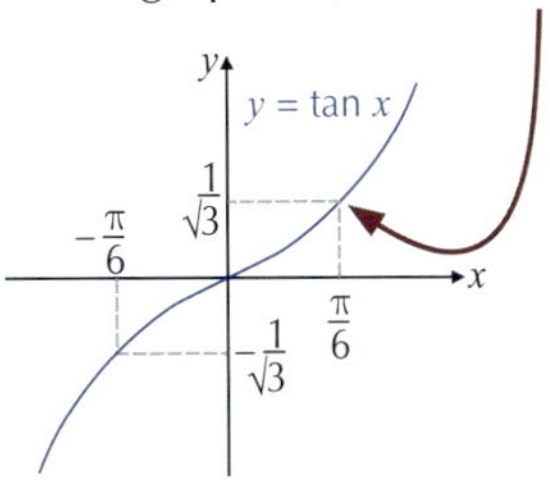

then $\tan\left(-\frac{\pi}{6}\right) = -\frac{1}{\sqrt{3}}$.

- So $\cot\left(-\frac{\pi}{6}\right) = \frac{1}{\left(-\frac{1}{\sqrt{3}}\right)} = -\sqrt{3}$

Tip: You could use a CAST diagram for this too. $-\frac{\pi}{6}$ is in a negative quadrant for tan, so it is the same as $-\tan\frac{\pi}{6}$.

d) Find cosec 300° without using a calculator.

- $\text{cosec } 300° = \frac{1}{\sin 300°} = \frac{1}{\sin(360° - 60°)}$.
- The CAST diagram shows that sin 300° is the same size as sin 60°, but it lies in a quadrant where sin is negative.

  So $\sin 300° = -\sin 60° = -\frac{\sqrt{3}}{2}$.

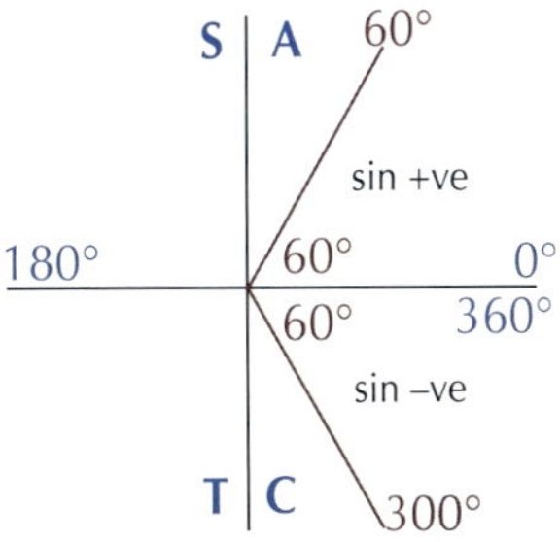

- So $\text{cosec } 300° = \frac{1}{\left(-\frac{\sqrt{3}}{2}\right)} = -\frac{2}{\sqrt{3}}$

Tip: Get used to spotting angles that are 30°, 45° and 60° either side of 180° and 360° so that you can use the CAST diagram to help find them without a calculator.

Exercise 4.2

Q1 Evaluate the following, giving your answers to 2 decimal places:

a) cosec 80° b) sec 75° c) cot 30°

d) sec(–70°) e) 3 – cot 250° f) 2 cosec 25°

Q2 Evaluate the following, giving your answers to 3 significant figures:

a) sec 3 b) cot 0.6 c) cosec 1.8 d) sec(–1)

e) $\text{cosec } \frac{\pi}{8}$ f) $8 + \cot\frac{\pi}{8}$ g) $\frac{1}{1 + \sec\frac{\pi}{10}}$ h) $\frac{1}{6 + \cot\frac{\pi}{5}}$

Q1-2 Hint: If you have to give an answer rounded to a certain accuracy it means you'll have to use a calculator for part of it. Don't forget to switch between radians and degrees where needed.

Q3 Using the table of common angles on p.54, find the exact values of:

a) $\sec 60°$ b) $\text{cosec } 30°$ c) $\cot 45°$ d) $\text{cosec } \frac{\pi}{3}$

e) $\sec (-180°)$ f) $\text{cosec } 135°$ g) $\cot 330°$ h) $\sec \frac{5\pi}{4}$

i) $\text{cosec } \frac{5\pi}{3}$ j) $\text{cosec } \frac{2\pi}{3}$ k) $3 - \cot \frac{3\pi}{4}$ l) $\frac{\sqrt{3}}{\cot \frac{\pi}{6}}$

Q4 Hint: Use the table of common values on page 53 if you need to.

Q4 Find, without a calculator, the exact values of:

a) $\frac{1}{1 + \sec 60°}$ b) $\frac{2}{6 + \cot 315°}$ c) $\frac{1}{\sqrt{3} - \sec 30°}$

d) $1 + \cot 420°$ e) $\frac{2}{7 + \sqrt{3}\cot 150°}$

Simplifying expressions and solving equations

Simplifying expressions

You can use the cosec, sec and cot relationships to **simplify expressions**. This can make it a lot easier to **solve** trig equations.

Examples

a) Simplify $\cot^2 x \tan x$.

$\cot x = \frac{1}{\tan x}$, so:

$$\cot^2 x \tan x = \left(\frac{1}{\tan^2 x}\right) \tan x = \frac{1}{\tan x} = \cot x$$

Tip: To 'show that' one thing is the same as another, you need to rearrange the expression on one side of the identity until it's the same as the other.

b) Show that $\frac{\cot x \sec x}{\text{cosec}^2 x} \equiv \sin x$.

$\cot x = \frac{\cos x}{\sin x}$, so:

$$\frac{\cot x \sec x}{\text{cosec}^2 x} = \frac{\left(\frac{\cos x}{\sin x}\right)\left(\frac{1}{\cos x}\right)}{\left(\frac{1}{\sin^2 x}\right)} = \frac{\left(\frac{1}{\sin x}\right)}{\left(\frac{1}{\sin^2 x}\right)} = \sin x \text{ as required}$$

c) Write the expression $(\text{cosec } x + 1)(\sin x - 1)$ as a single fraction in terms of $\sin x$ only.

- First expand the brackets:

$$(\text{cosec } x + 1)(\sin x - 1) = \text{cosec } x \sin x + \sin x - \text{cosec } x - 1$$

- $\text{cosec } x \sin x = \left(\frac{1}{\sin x}\right)\sin x = 1$, so:

$$1 + \sin x - \text{cosec } x - 1 = \sin x - \text{cosec } x$$

- Using $\text{cosec } x = \frac{1}{\sin x}$ the expression becomes:

$$= \sin x - \frac{1}{\sin x}$$

$$= \frac{\sin^2 x - 1}{\sin x}$$

Tip: You usually need to write an expression in terms of one type of trig function in order to solve an equation.

Solving equations

You can **solve** equations involving cosec, sec and cot by **rewriting** them in terms of sin, cos or tan and solving as usual (often you'll need to get them all in terms of the same trig function, e.g. all sin). You may also need to use the **CAST diagram** (or the graph of the trig function) to find all the solutions in a given interval.

Examples

a) Solve $\sec x = \sqrt{2}$ in the interval $0 \leq x \leq 2\pi$.

- $\sec x = \sqrt{2}$, so $\cos x = \frac{1}{\sqrt{2}}$ ← Write in terms of cos x by giving the reciprocal.
- The triangle on the right shows that one solution to $\cos x = \frac{1}{\sqrt{2}}$ is $x = \frac{\pi}{4}$.

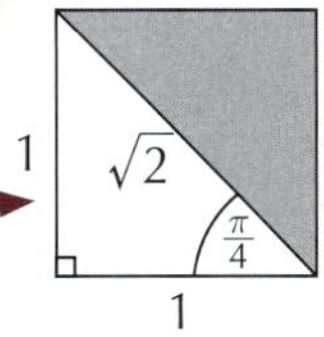

- Use the **CAST diagram** to find the **other** solution in the interval:

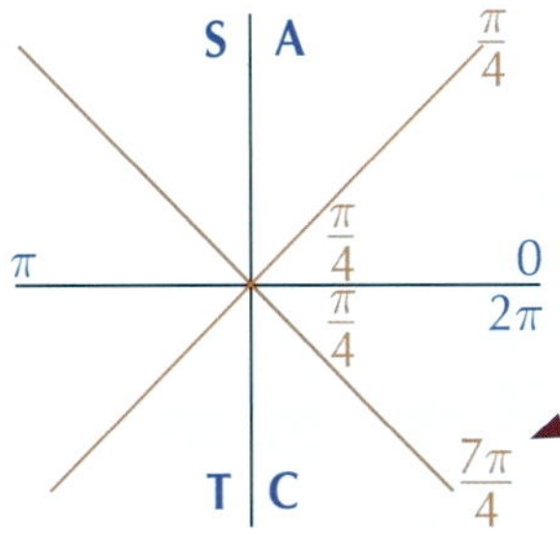

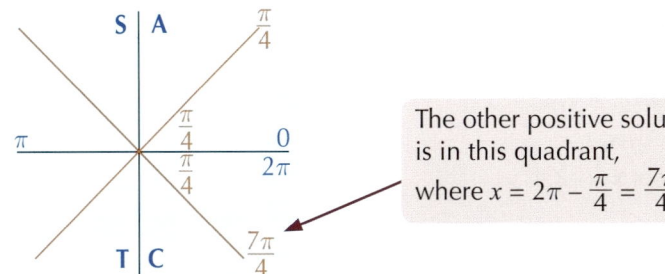

The other positive solution is in this quadrant, where $x = 2\pi - \frac{\pi}{4} = \frac{7\pi}{4}$.

- So the two solutions are $x = \frac{\pi}{4}$ and $x = \frac{7\pi}{4}$.

b) Solve $\text{cosec}^2 x - 3\,\text{cosec}\, x + 2 = 0$ in the interval $-180° \leq x \leq 180°$.

- First you need to spot that this is a **quadratic equation** in cosec x, which can be factorised as follows: PROBLEM SOLVING

$$(\text{cosec}\, x - 1)(\text{cosec}\, x - 2) = 0$$

- This gives **two** equations to solve:

$$\text{cosec}\, x - 1 = 0 \qquad \text{cosec}\, x - 2 = 0$$
$$\text{cosec}\, x = 1 \qquad \text{cosec}\, x = 2$$
$$\sin x = 1 \qquad \sin x = \frac{1}{2}$$

One solution to this is $x = 90°$. (for $\sin x = 1$)

One solution to this is $x = 30°$. (for $\sin x = \frac{1}{2}$)

- Look at the **graph** of $y = \sin x$ over the interval $-180° \leq x \leq 180°$:

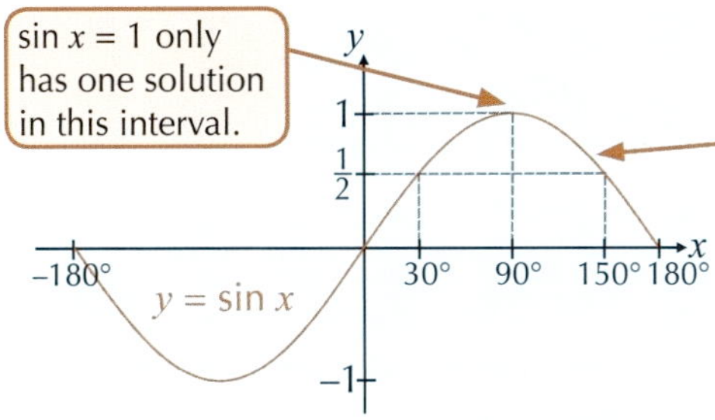

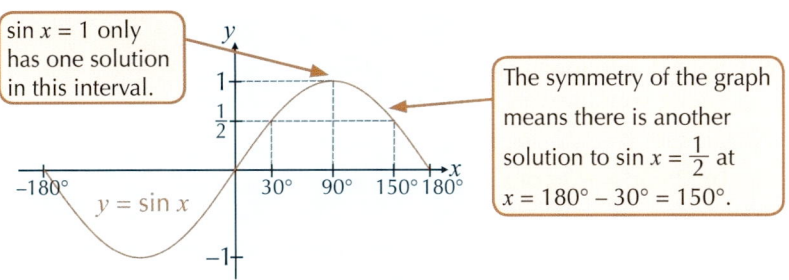

$\sin x = 1$ only has one solution in this interval.

The symmetry of the graph means there is another solution to $\sin x = \frac{1}{2}$ at $x = 180° - 30° = 150°$.

- So the three solutions are $x = 30°$, $x = 90°$ and $x = 150°$.

Tip: If you've learnt the values of the sin, cos and tan of common angles on p.53, you won't need to keep drawing these triangles.

Tip: Make a substitution of $y = \text{cosec}\, x$ and solve as a quadratic in y if you're struggling here.

Exercise 4.3

Q2 Hint: You might find the identity $\sin^2 x + \cos^2 x \equiv 1$ useful.

Q3 Hint: Use a calculator to find one solution, but then look at the graph or the CAST diagram to find any other solutions in the interval.

Q3-4 Hint: For questions where you're looking for solutions for a multiple of x, such as $2x$, the interval that you'll need to look in will be different. E.g. If $a \le x \le b$ then $2a \le 2x \le 2b$.

Q1 Simplify the following expressions:

a) $\sec x + \frac{1}{\cos x}$

b) $(\text{cosec}^2 x)(\sin^2 x)$

c) $2 \cot x + \frac{1}{\tan x}$

d) $\frac{\sec x}{\text{cosec}\, x}$

e) $(\cos x)(\text{cosec}\, x)$

f) $\frac{\text{cosec}^2 x}{\cot x}$

Q2 Show that:

a) $\sin x \cot x \equiv \cos x$

b) $\sec x - \cos x \equiv \tan x \sin x$

c) $\tan x \,\text{cosec}\, x \equiv \sec x$

d) $\frac{(\tan^2 x)(\text{cosec}\, x)}{\sin x} \equiv \sec^2 x$

Q3 Solve these equations for $0° \le x \le 360°$. Give your answers in degrees to one decimal place.

a) $\sec x = 1.9$

b) $\cot x = 2.4$

c) $\text{cosec}\, x = -2$

d) $\sec x = -1.3$

e) $\cot x = -2.4$

f) $4 \sec 2x = -7$

Q4 Solve these equations for $0 \le x \le 2\pi$, giving your answers in radians in terms of π.

a) $\sec x = 2$

b) $\text{cosec}\, x = -2$

c) $\cot 2x = 1$

d) $\sec 5x = -1$

Q5 Solve the equation $\cot 2x - 4 = -5$ in the interval $0 \le x \le 2\pi$. Give your answers in terms of π.

Q6 Solve for $0° \le x \le 360°$: $2 \,\text{cosec}\, 2x = 3$. Give your answers to 1 decimal place.

Q7 Find, for $0 \le x \le 2\pi$, all the solutions of the equation $-2 \sec x = 4$. Give your answers in terms of π.

Q8 Solve $\sqrt{3}\, \text{cosec}\, 3x = 2$ for $0 \le x \le 2\pi$. Give your answers in terms of π.

Q9 Solve the following for $0° \le x \le 180°$, giving your answers to 1 d.p. where appropriate: PROBLEM SOLVING

a) $\sec^2 x - 2\sqrt{2} \sec x + 2 = 0$

b) $2 \cot^2 x + 3 \cot x - 2 = 0$

Q10 Solve the equation $(\text{cosec}\, x - 3)(2 \tan x + 1) = 0$ for $0° \le x \le 360°$. Give your answers to 1 decimal place.

5. Identities Involving Cosec, Sec and Cot

An identity is an equation that's true for all values of a variable. You've met some trig identities — you can build on these to include cosec, sec and cot.

Learning Objective:
- Know and be able to use the following identities:
$\sec^2\theta \equiv 1 + \tan^2\theta$
$\text{cosec}^2\theta \equiv 1 + \cot^2\theta$

Deriving the identities

By now you should be familiar with the following **trig identities**:

$$\cos^2\theta + \sin^2\theta \equiv 1 \qquad \tan\theta \equiv \frac{\sin\theta}{\cos\theta}$$

You can use them to produce a couple of other identities:

$$\sec^2\theta \equiv 1 + \tan^2\theta \qquad \text{cosec}^2\theta \equiv 1 + \cot^2\theta$$

Tip: The $\equiv$ sign tells you that this is true for all values of θ, rather than just certain values.

You need to know how to **derive** these identities from the ones you already know.

Deriving $\sec^2\theta \equiv 1 + \tan^2\theta$

Start with the identity $\cos^2\theta + \sin^2\theta \equiv 1$ and divide through by $\cos^2\theta$.

$$\frac{\cos^2\theta}{\cos^2\theta} + \frac{\sin^2\theta}{\cos^2\theta} \equiv \frac{1}{\cos^2\theta}$$

$$\tan\theta \equiv \frac{\sin\theta}{\cos\theta}, \text{ so } \frac{\sin^2\theta}{\cos^2\theta} = \tan^2\theta$$

$$1 + \tan^2\theta \equiv \frac{1}{\cos^2\theta}$$

The definition of $\sec\theta = \frac{1}{\cos\theta}$, so replace $\frac{1}{\cos^2\theta}$ with $\sec^2\theta$.

$$1 + \tan^2\theta \equiv \sec^2\theta$$

Rearrange slightly... $\sec^2\theta \equiv 1 + \tan^2\theta$

Tip: Remember that $\cos^2\theta = (\cos\theta)^2$.

Deriving $\text{cosec}^2\theta \equiv 1 + \cot^2\theta$

Start again with $\cos^2\theta + \sin^2\theta \equiv 1$ but this time divide through by $\sin^2\theta$.

$$\frac{\cos^2\theta}{\sin^2\theta} + \frac{\sin^2\theta}{\sin^2\theta} \equiv \frac{1}{\sin^2\theta}$$

$$\tan\theta \equiv \frac{\sin\theta}{\cos\theta}, \text{ so } \frac{\cos^2\theta}{\sin^2\theta} = \frac{1}{\tan^2\theta}.$$

$$\frac{1}{\tan^2\theta} + 1 \equiv \frac{1}{\sin^2\theta}$$

The definition of **cot θ** = $\frac{\mathbf{1}}{\mathbf{tan}\,\boldsymbol{\theta}}$, so replace $\frac{1}{\tan^2\theta}$ with $\cot^2\theta$.

$$\cot^2\theta + 1 \equiv \frac{1}{\sin^2\theta}$$

The definition of **cosec θ** = $\frac{\mathbf{1}}{\mathbf{sin}\,\boldsymbol{\theta}}$, so replace $\frac{1}{\sin^2\theta}$ with $\mathrm{cosec}^2\,\theta$.

$$\cot^2\theta + 1 \equiv \mathrm{cosec}^2\,\theta$$

Rearrange slightly... $\boldsymbol{\mathrm{cosec}^2\,\theta \equiv 1 + \cot^2\theta}$

Tip: These derivations are examples of proof by deduction, where known or accepted facts are used to prove that other relationships are true. In this case, the identities $\cos^2\theta + \sin^2\theta \equiv 1$ and $\tan\theta \equiv \frac{\sin\theta}{\cos\theta}$ are the known facts.

Using the identities

You can use identities to get rid of any trig functions that are making an equation difficult to solve.

Example 1

Simplify the expression 3 tan x + sec²x + 1.

- Use $\boldsymbol{\sec^2\theta \equiv 1 + \tan^2\theta}$ to swap $\sec^2 x$ for $1 + \tan^2 x$:

$$3\tan x + 1 + \tan^2 x + 1$$

- Now rearrange:

$$\tan^2 x + 3\tan x + 2$$

- This is a quadratic in $\tan x$ which will **factorise**:

$$(\tan x + 1)(\tan x + 2)$$

Tip: It's usually best to get the expression all in terms of one thing — in this case $\tan x$. You'll often be asked to simplify an expression in order to then solve an equation involving that expression.

Example 2

Solve the equation cot²x + 5 = 4 cosec x in the interval 0° ≤ x ≤ 360°.

PROBLEM SOLVING

- You can't solve this while it has both cot and cosec in it, so use $\boldsymbol{\mathrm{cosec}^2\,\theta \equiv 1 + \cot^2\theta}$ to swap $\cot^2 x$ for $\mathrm{cosec}^2 x - 1$.

$$\mathrm{cosec}^2 x - 1 + 5 = 4\,\mathrm{cosec}\,x$$

$$\mathrm{cosec}^2 x + 4 = 4\,\mathrm{cosec}\,x$$

$$\mathrm{cosec}^2 x - 4\,\mathrm{cosec}\,x + 4 = 0$$

Rearrange so that one side is zero.

- So you've got a quadratic in cosec x which will **factorise**.

$$(\mathrm{cosec}\,x - 2)(\mathrm{cosec}\,x - 2) = 0$$

- One of the brackets must be equal to zero — here they're both the same, so you only get one **equation**:

$$(\mathrm{cosec}\,x - 2) = 0$$

$$\Rightarrow \mathrm{cosec}\,x = 2$$

Tip: If it helps, think of this as $y^2 - 4y + 4 = 0$. Factorise it, and then replace the y with cosec x.

- Now you can convert this into **sin x**, and **solve** it:

$$\sin x = \frac{1}{2}$$
$$\Rightarrow x = 30°$$

- To find the other values of x, draw a quick sketch of the sin curve:

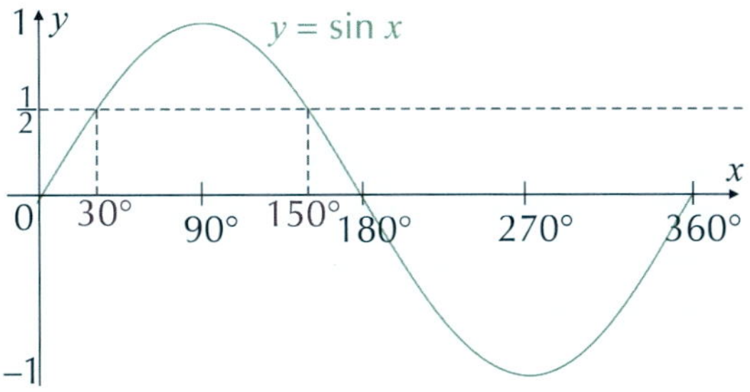

- From the graph, you can see that sin x takes the value of $\frac{1}{2}$ **twice** in the given interval, once at $x = 30°$ and once at $x = 180 - 30 = 150°$.

Example 3

Given that $\cot x = \sqrt{8}$, where $0° \le x \le 180°$, show how you can use the identity $\text{cosec}^2\,\theta \equiv 1 + \cot^2\theta$ to find the exact value of $\sin x$. Use Pythagoras' Theorem to confirm the result.

PROBLEM SOLVING

- If $\cot x = \sqrt{8}$, then $\mathbf{\cot^2 x = 8}$.
- $\text{cosec}^2\,\theta \equiv 1 + \cot^2\theta$ so: $\mathbf{\text{cosec}^2\,x = 8 + 1 = 9}$.

$$\mathbf{\text{cosec}\,x = \pm 3}$$

- Since $\text{cosec}\,x = \frac{1}{\sin x}$, $\mathbf{\sin x = \pm\frac{1}{3}}$.
- You're told that $\mathbf{0 \le x \le 180°}$, and sin x is **positive** over this interval, so: $\sin x = \frac{1}{3}$

Look at the graph of sin x to see why.

- To confirm this using **Pythagoras' Theorem**:
$\cot x = \sqrt{8} \Rightarrow \tan x = \frac{1}{\cot x} = \frac{1}{\sqrt{8}}$.
- Now draw a right-angled triangle for which $\tan x = \frac{1}{\sqrt{8}}$
(i.e. the opposite has a length of 1 and the adjacent has a length of $\sqrt{8}$):

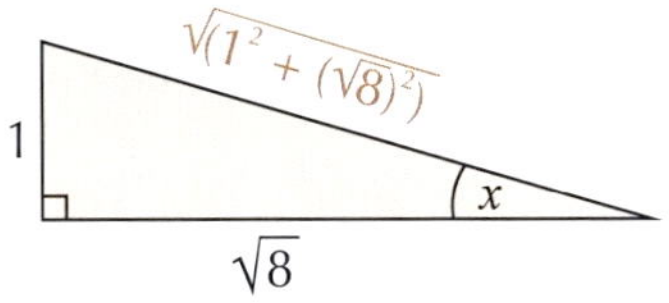

- Using Pythagoras' Theorem, the hypotenuse will have a length of $\sqrt{1^2 + (\sqrt{8})^2} = 3$.
- So $\sin x = \frac{\text{opp}}{\text{hyp}} = \frac{1}{3}$

Tip: In this example, you're told which identity to use, but this won't always be the case. It takes practice to quickly spot which identity will work best.

Exercise 5.1

Q1 Express $\operatorname{cosec}^2 x + 2\cot^2 x$ in terms of $\operatorname{cosec} x$ only.

Q2 Simplify the following expression: $\tan^2 x - \frac{1}{\cos^2 x}$.

Q3 Given that $x = \sec\theta + \tan\theta$, show that $x + \frac{1}{x} = 2\sec\theta$.

Q4 Hint: The 'hence' in part b) means that you should use the result of part a) and solve $\sec^2 x - 2\sec x - 3 = 0$. It's a quadratic in $\sec x$.

Q4 a) Show that the equation $\tan^2 x = 2\sec x + 2$ can be written as $\sec^2 x - 2\sec x - 3 = 0$.

b) Hence solve $\tan^2 x = 2\sec x + 2$ over the interval $0° \leq x \leq 360°$, giving your answers in degrees to 1 decimal place.

Q5 a) Show that the equation $2\operatorname{cosec}^2 x = 5 - 5\cot x$ can be written as $2\cot^2 x + 5\cot x - 3 = 0$.

b) Hence solve $2\operatorname{cosec}^2 x = 5 - 5\cot x$ over the interval $-\pi \leq x \leq \pi$, giving your answers in radians to 2 decimal places.

Q6 a) Show that the equation $2\cot^2 A + 5\operatorname{cosec} A = 10$ can be written as $2\operatorname{cosec}^2 A + 5\operatorname{cosec} A - 12 = 0$.

b) Hence solve $2\cot^2 A + 5\operatorname{cosec} A = 10$ over the interval $0° \leq x \leq 360°$, giving your answers in degrees to 1 decimal place.

Q7 Hint: First write out the equation in terms of $\tan x$ only.

Q7 Solve the equation $\sec^2 x + \tan x = 1$ for $0 \leq x \leq 2\pi$, giving exact answers.

Q8 a) Given that $\operatorname{cosec}^2\theta + 2\cot^2\theta = 2$, find the possible values of $\sin\theta$.

b) Hence solve the equation $\operatorname{cosec}^2\theta + 2\cot^2\theta = 2$ in the interval $0° \leq \theta \leq 180°$.

Q9 Solve the equation $\sec^2 x = 3 + \tan x$ in the interval $0° \leq x \leq 360°$, giving your answers to 1 decimal place.

Q10 Solve the equation $\cot^2 x + \operatorname{cosec}^2 x = 7$, giving all the solutions in the interval $0 \leq x \leq 2\pi$ in terms of π.

Q11 Solve the equation $\tan^2 x + 5\sec x + 7 = 0$, giving all the solutions in the interval $0 \leq x \leq 2\pi$ to 2 decimal places.

Q12 Given that $\tan\theta = \frac{60}{11}$, and $180° \leq \theta \leq 270°$, find the exact value of: a) $\sin\theta$ b) $\sec\theta$ c) $\operatorname{cosec}\theta$

PROBLEM SOLVING

Q14 Hint: You're told to use this identity so you need to show your working out for this — you can't get away with stating the answer after using a different method.

Q13 Given that $\operatorname{cosec}\theta = -\frac{17}{15}$, and $180° \leq \theta \leq 270°$, find the exact value of: a) $\cos\theta$ b) $\sec\theta$ c) $\cot\theta$

PROBLEM SOLVING

Q14 Given that $\cos x = \frac{1}{6}$, use the identity $\sec^2\theta = 1 + \tan^2\theta$ to find the two possible exact values of $\tan x$.

Proving other identities

You can also use identities to prove that two trig expressions are the same, as shown in the examples below. You just need to take one side of the identity and play about with it until you get what's on the other side.

Examples

PROBLEM SOLVING

a) Show that $\frac{\tan^2 x}{\sec x} \equiv \sec x - \cos x$.

- Start by looking at the **left-hand side** of the identity: $\frac{\tan^2 x}{\sec x}$
- Try replacing $\tan^2 x$ with $\sec^2 x - 1$:

$$\frac{\tan^2 x}{\sec x} \equiv \frac{\sec^2 x - 1}{\sec x} \equiv \frac{\sec^2 x}{\sec x} - \frac{1}{\sec x} \equiv \sec x - \cos x$$

Do some **rearranging**...

...until you get the **right-hand side** of the identity.

b) Prove the identity $\frac{\tan^2 x}{\sec x + 1} \equiv \sec x - \cos^2 x - \sin^2 x$.

- As before, replace $\tan^2 x$ with $\sec^2 x - 1$...

$$\frac{\tan^2 x}{\sec x + 1} \equiv \frac{\sec^2 x - 1}{\sec x + 1} \equiv \frac{(\sec x + 1)(\sec x - 1)}{\sec x + 1}$$

...But factorise the $\sec^2 x - 1$ as it's the **difference of two squares**.

$$\equiv \sec x - 1$$

$$\equiv \sec x - (\cos^2 x + \sin^2 x)$$

$$\equiv \sec x - \cos^2 x - \sin^2 x$$

The right-hand side of the identity has a $\cos^2 x$ and $\sin^2 x$. So use $\cos^2 x + \sin^2 x \equiv 1$ to replace the '1' here.

Tip: Keep checking that you're getting closer to the right-hand side of the identity. As well as using the known identities, there are lots of little tricks you can use — such as looking for the 'difference of two squares', and multiplying the top and bottom of a fraction by the same expression.

Exercise 5.2

PROBLEM SOLVING

Q1 a) Show that $\sec^2 \theta - \text{cosec}^2 \theta \equiv \tan^2 \theta - \cot^2 \theta$.

b) Hence prove that $(\sec \theta + \text{cosec } \theta)(\sec \theta - \text{cosec } \theta) \equiv (\tan \theta + \cot \theta)(\tan \theta - \cot \theta)$.

Q2 Prove the identity $(\tan x + \cot x)^2 \equiv \sec^2 x + \text{cosec}^2 x$.

Q3 Prove the identity $\cot^2 x + \sin^2 x \equiv (\text{cosec } x + \cos x)(\text{cosec } x - \cos x)$.

Q4 Prove the identity $\frac{(\sec x - \tan x)(\tan x + \sec x)}{\text{cosec } x - \cot x} \equiv \cot x + \text{cosec } x$.

Q5 Prove that $\frac{\cot x}{1 + \text{cosec } x} + \frac{1 + \text{cosec } x}{\cot x} \equiv 2 \sec x$.

Q6 Prove the identity $\frac{\text{cosec } x + 1}{\text{cosec } x - 1} \equiv 2 \sec^2 x + 2 \tan x \sec x - 1$.

6. The Addition Formulas

Learning Objectives:

- Know and be able to use the formulas for $\sin(A \pm B)$, $\cos(A \pm B)$ and $\tan(A \pm B)$.
- Understand the geometric proof of the addition formulas.

The addition formulas are a special set of trig identities that can be used to simplify trig expressions where there are two different angles, or where there is a sum of angles.

Finding exact values

The identities shown below are known as the **addition formulas**. You can use the addition formulas to find the **sin**, **cos** or **tan** of the **sum** or **difference** of two angles, and to 'expand the brackets' in expressions such as $\sin(x + 60°)$ or $\cos\left(n - \frac{\pi}{2}\right)$.

$$\sin(A \pm B) \equiv \sin A \cos B \pm \cos A \sin B$$

$$\cos(A \pm B) \equiv \cos A \cos B \mp \sin A \sin B$$

$$\tan(A \pm B) \equiv \frac{\tan A \pm \tan B}{1 \mp \tan A \tan B}$$

Watch out for the $\pm$ and $\mp$ signs in the formulas — especially for cos and tan. If you use the sign on the **top** on the **left-hand side** of the identity, you have to use the sign on the **top** on the **right-hand side** too. So $\cos(A + B) = \cos A \cos B - \sin A \sin B$.

Proving the addition formulas

You need to understand the **geometric proof** of these formulas, and although it looks rather complicated, it only uses basic trigonometry.

- Start with a right-angled triangle with a hypotenuse of 1, where one of the angles is $A + B$.
- This is a right-angled triangle, so $\sin(A + B) = \frac{\text{opp}}{\text{hyp}}$ and $\cos(A + B) = \frac{\text{adj}}{\text{hyp}}$, and since the hypotenuse is 1, you can write down the lengths of each side: opp = $\sin(A + B)$ and adj = $\cos(A + B)$.
- Add these labels to the diagram as shown:

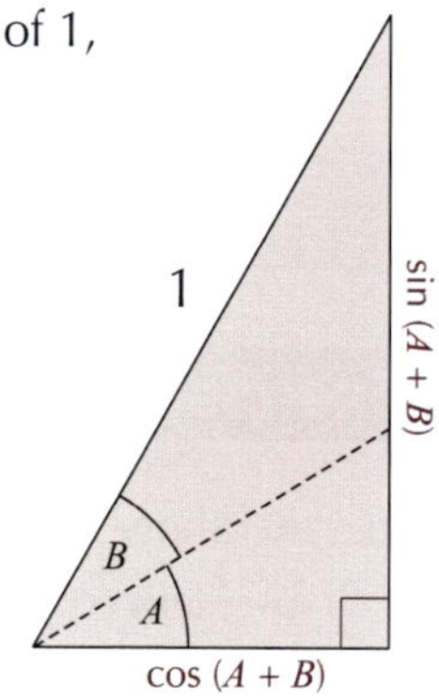

- If you extend the dotted line, you can form another right-angled triangle with a hypotenuse of 1. This time, the angle is B.
- You can find the lengths of the sides of this triangle in the same way: write out $\sin B = \frac{\text{opp}}{\text{hyp}}$ and $\cos B = \frac{\text{adj}}{\text{hyp}}$, then use the fact that the hypotenuse is 1 to get opp = $\sin B$ and adj = $\cos B$.
- Then add these to the diagram as well.

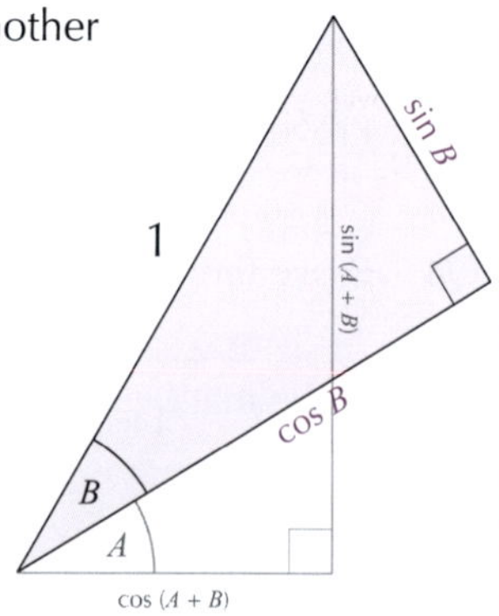

- Now form another right-angled triangle using angle A as shown. This one has a hypotenuse with length $\cos B$.
- So this time, when you rearrange the equations $\sin A = \frac{\text{opp}}{\text{hyp}}$ and $\cos A = \frac{\text{adj}}{\text{hyp}}$, use hyp = $\cos B$, which gives opp = $\sin A \cos B$ and adj = $\cos A \cos B$.
- Label these sides, and make sure it's clear which length is $\cos (A + B)$ and which is $\cos A \cos B$.

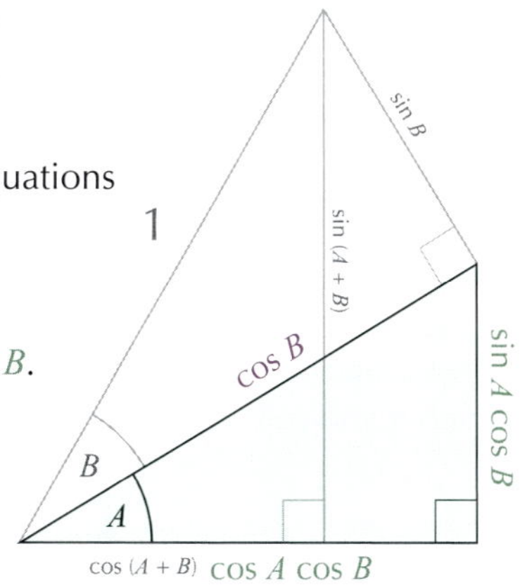

- The last triangle you need to draw has a hypotenuse of $\sin B$ and an angle of A. If you're not sure where this angle comes from, have a look at the diagram.
- So $\sin A = \frac{\text{opp}}{\text{hyp}}$ and $\cos A = \frac{\text{adj}}{\text{hyp}}$, and since hyp = $\sin B$, opp = $\sin A \sin B$ and adj = $\cos A \sin B$.
- Label these sides on the diagram.

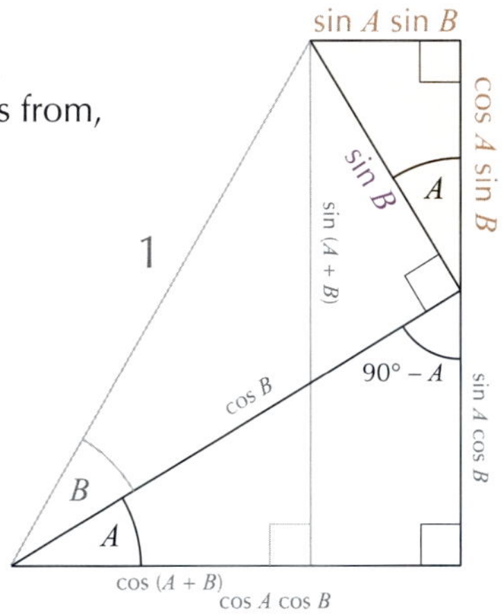

- You've now labelled enough sides to prove the formulas — here's the diagram again, with the labels for $\sin (A + B)$ and $\cos (A + B)$ moved to make things a bit clearer.
- Using the height of the diagram, you can see that

$$\sin (A + B) = \sin A \cos B + \cos A \sin B$$

- The width of the diagram shows you that

$$\cos A \cos B = \cos (A + B) + \sin A \sin B$$

$$\Rightarrow \cos (A + B) = \cos A \cos B - \sin A \sin B$$

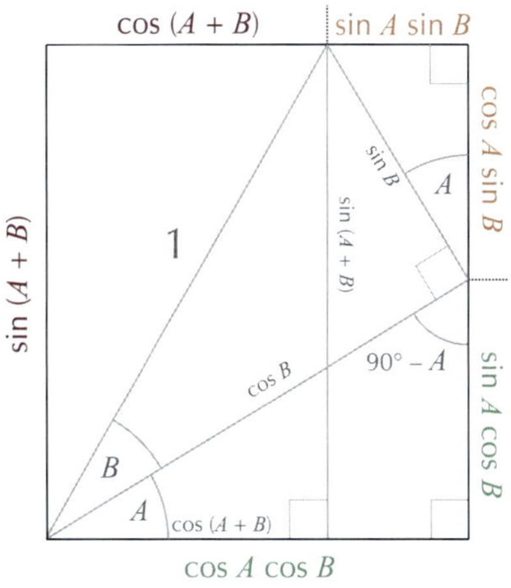

Tip: This is the proof of the sin and cos formulas. See p.73 for a proof of the tan formula.

Tip: Have a look at the graphs of $y = \sin x$ and $y = \cos x$ if you want to check these facts.

You can work out the subtraction formulas from the addition ones, by using the fact that **sin –*B* = –sin *B*** and **cos –*B* = cos *B***:

$$\sin (A - B) = \sin A \cos -B + \cos A \sin -B$$
$$= \sin A \cos B - \cos A \sin B$$

$$\cos (A - B) = \cos A \cos -B - \sin A \sin -B$$
$$= \cos A \cos B + \sin A \sin B$$

You can also prove them geometrically using a similar method as on p.68-69. You'll end up with the diagram below — try and work through it yourself to make sure you know where the labels come from.

Tip: Start with the triangle with angle $A - B$ and hypotenuse 1, and then add on the triangle with angle B below it.

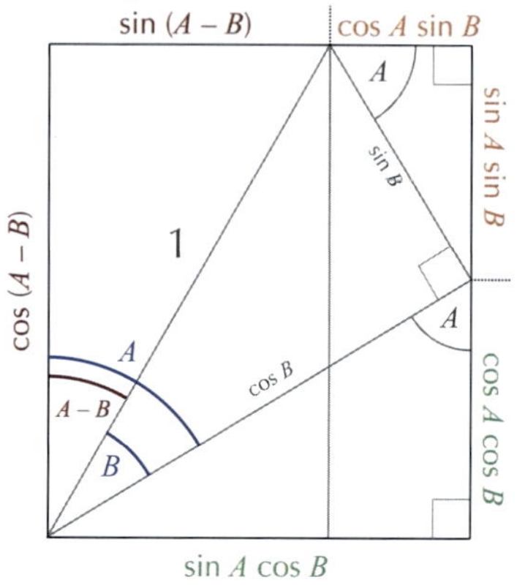

Example 1

a) Find the exact value of sin 18° cos 12° + cos 18° sin 12°.

- Using the **sin** addition formula:

$\sin A \cos B + \cos A \sin B \equiv \sin (A + B)$. ← *A* is 18° and *B* is 12°.

$\sin 18° \cos 12° + \cos 18° \sin 12° = \sin (18° + 12°) = \sin 30° = \frac{1}{2}$

b) Write $\frac{\tan 5x - \tan 2x}{1 + \tan 5x \tan 2x}$ as a single trigonometric ratio.

Tip: Note that it's $\tan A - \tan B$ (rather than +) on the top line, which means it's the $\tan (A - B)$ formula.

- Use the **tan** addition formula, with $A = 5x$ and $B = 2x$:

$$\frac{\tan 5x - \tan 2x}{1 + \tan 5x \tan 2x} = \tan (5x - 2x) = \tan 3x$$

c) Find cos $(x + y)$ if sin $x = \frac{4}{5}$ and sin $y = \frac{15}{17}$. Both x and y are acute. Give an exact answer.

Tip: You could also use the identity $\cos^2 \theta + \sin^2 \theta \equiv 1$ to work out $\cos x$ and $\cos y$ here.

- In order to use the **cos** addition formula, first find **cos *x*** and **cos *y***.
- You know x and y are acute, so draw triangles and use **SOH CAH TOA** and Pythagoras to work out $\cos x$ and $\cos y$...

$\cos x = \frac{\text{adj}}{\text{hyp}} = \frac{3}{5}$ →

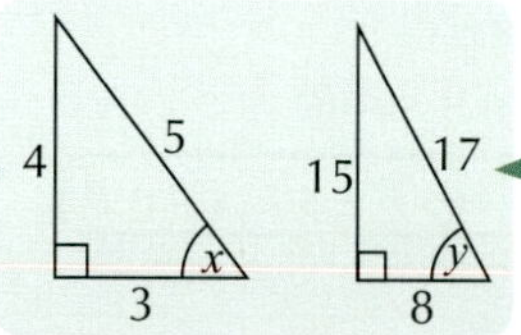

← $\cos y = \frac{\text{adj}}{\text{hyp}} = \frac{8}{17}$

- $\cos (x + y) = \cos x \cos y - \sin x \sin y$

$$= \left(\frac{3}{5} \times \frac{8}{17}\right) - \left(\frac{4}{5} \times \frac{15}{17}\right) = -\frac{36}{85}$$

You should know the value of sin, cos and tan for **common angles**, in degrees and radians. You can use your knowledge of these angles, along with the addition formulas, to find the **exact value** of sin, cos or tan for **other** angles.

Find a **pair** of common angles which **add or subtract** to give the angle you're after. Then plug them into the addition formula, and work it through.

Tip: If you need a reminder of the sin, cos and tan of common angles, see the table on p.53.

Example 2

Using the addition formula for tangent, show that $\tan 15° = 2 - \sqrt{3}$.

- Pick two angles that **add or subtract** to give **15°**, and put them into the **tan** addition formula. It's easiest to use **tan 60°** and **tan 45°** here, since neither of them are fractions:

$$\tan(A - B) = \frac{\tan A - \tan B}{1 + \tan A \tan B}$$

$$\tan 15° = \tan(60° - 45°) = \frac{\tan 60° - \tan 45°}{1 + \tan 60° \tan 45°}$$

- **Substitute** the values for tan 60° (= $\sqrt{3}$) and tan 45° (= 1) into the equation.

$$= \frac{\sqrt{3} - 1}{1 + (\sqrt{3} \times 1)} = \frac{\sqrt{3} - 1}{\sqrt{3} + 1}$$

- Now **rationalise the denominator** of the fraction to get rid of the $\sqrt{3}$.

$$= \frac{\sqrt{3} - 1}{\sqrt{3} + 1} \times \frac{\sqrt{3} - 1}{\sqrt{3} - 1} = \frac{3 - 2\sqrt{3} + 1}{3 - \sqrt{3} + \sqrt{3} - 1}$$

- Then **simplify** the expression...

$$= \frac{4 - 2\sqrt{3}}{2} = 2 - \sqrt{3}$$

...and there's the **right-hand side**.

Tip: If you can't remember how to rationalise the denominator have a look at your Year 1 notes.

Exercise 6.1

Q1 Use the addition formulas to find the exact values of the following:

a) $\cos 72° \cos 12° + \sin 72° \sin 12°$

b) $\cos 13° \cos 17° - \sin 13° \sin 17°$

c) $\dfrac{\tan 12° + \tan 18°}{1 - \tan 12° \tan 18°}$

d) $\dfrac{\tan 500° - \tan 140°}{1 + \tan 500° \tan 140°}$

e) $\sin 35° \cos 10° + \cos 35° \sin 10°$

f) $\sin 69° \cos 9° - \cos 69° \sin 9°$

Q2 Use the addition formulas to find the exact values of the following:

a) $\sin \frac{2\pi}{3} \cos \frac{\pi}{2} - \cos \frac{2\pi}{3} \sin \frac{\pi}{2}$

b) $\cos 4\pi \cos 3\pi + \sin 4\pi \sin 3\pi$

c) $\dfrac{\tan \frac{5\pi}{12} + \tan \frac{5\pi}{4}}{1 - \tan \frac{5\pi}{12} \tan \frac{5\pi}{4}}$

Q1-2 Hint: You've been asked for exact values, which is a big clue that they'll have something to do with the common angles you should know.

Q3 Write the following expressions as a single trigonometric ratio:

a) $\sin 5x \cos 2x - \cos 5x \sin 2x$

b) $\cos 4x \cos 6x - \sin 4x \sin 6x$

c) $\dfrac{\tan 7x + \tan 3x}{1 - \tan 7x \tan 3x}$

d) $5 \sin 2x \cos 3x + 5 \cos 2x \sin 3x$

e) $8 \cos 7x \cos 5x + 8 \sin 7x \sin 5x$

Q4 Hint: You'll need to work out $\cos x$ and $\sin y$ before you can answer parts a)-d). You can use the triangle method or the identity $\cos^2 \theta + \sin^2 \theta \equiv 1$ to work them out.

Q4 $\sin x = \frac{3}{4}$ and $\cos y = \frac{3}{\sqrt{10}}$, where x and y are both acute angles. Calculate the exact value of:

a) $\sin (x + y)$

b) $\cos (x - y)$

c) $\operatorname{cosec} (x + y)$

d) $\sec (x - y)$

Q5 Using the addition formula for cos, show that $\cos \frac{\pi}{12} = \frac{\sqrt{6} + \sqrt{2}}{4}$.

Q6 Using the addition formula for sin, show that $\sin 75° = \frac{\sqrt{6} + \sqrt{2}}{4}$.

Q7 Using the addition formula for tan, show that $\tan 75° = \frac{\sqrt{3} + 1}{\sqrt{3} - 1}$.

Simplifying, solving equations and proving identities

You might be asked to use the addition formulas to **prove an identity**. All you need to do is put the numbers and variables from the left-hand side into the addition formulas and simplify until you get the expression you're after.

Example 1

Prove that $\cos (a + 60°) + \sin (a + 30°) \equiv \cos a$.

- Put the numbers from the question into the addition formulas:

$$\cos(a + 60°) + \sin(a + 30°)$$
$$\equiv (\cos a \cos 60° - \sin a \sin 60°) + (\sin a \cos 30° + \cos a \sin 30°)$$

- Now substitute in any sin and cos values that you know...

Tip: Be careful with the + and – signs here.

$$= \frac{1}{2} \cos a - \frac{\sqrt{3}}{2} \sin a + \frac{\sqrt{3}}{2} \sin a + \frac{1}{2} \cos a$$

cos 60° | sin 60° | cos 30° | sin 30°

- ..and simplify:

$$= \frac{1}{2} \cos a + \frac{1}{2} \cos a = \cos a$$

Example 2

Use the sine and cosine addition formulas to prove that $\tan(A + B) \equiv \frac{\tan A + \tan B}{1 - \tan A \tan B}$.

$$\tan(A+B) \equiv \frac{\sin(A+B)}{\cos(A+B)}$$

Start with the identity $\tan\theta \equiv \frac{\sin\theta}{\cos\theta}$.

$$\equiv \frac{\sin A\cos B + \cos A\sin B}{\cos A\cos B - \sin A\sin B}$$

Replace $\sin(A + B)$ and $\cos(A + B)$ with the addition formulas for each.

$$\equiv \frac{\frac{\sin A\cancel{\cos B}}{\cos A\cancel{\cos B}} + \frac{\cancel{\cos A}\sin B}{\cancel{\cos A}\cos B}}{\frac{\cancel{\cos A\cos B}}{\cancel{\cos A\cos B}} - \frac{\sin A\sin B}{\cos A\cos B}}$$

Divide each part of the fraction by $\cos A\cos B$ and cancel where possible.

$$\equiv \frac{\frac{\sin A}{\cos A} + \frac{\sin B}{\cos B}}{1 - \left(\frac{\sin A}{\cos A}\right)\left(\frac{\sin B}{\cos B}\right)} \equiv \frac{\tan A + \tan B}{1 - \tan A\tan B}$$

Tip: You know that the first term on the denominator needs to be '1' — so divide through by whatever this term is and you'll get the 1 in the right place.

Example 3

PROBLEM SOLVING

Use the addition formulas to show that $\cos A + \cos B \equiv 2\cos\left(\frac{A+B}{2}\right)\cos\left(\frac{A-B}{2}\right)$.

- Start with the **cos addition formulas**:

  $\cos(x + y) \equiv \cos x\cos y - \sin x\sin y$

  $\cos(x - y) \equiv \cos x\cos y + \sin x\sin y$

- **Add them together** to get:

  $\cos(x + y) + \cos(x - y)$

  $\equiv \cos x\cos y - \sin x\sin y + \cos x\cos y + \sin x\sin y \equiv 2\cos x\cos y$

- Now **substitute** in $A = x + y$ and $B = x - y$.

  Subtracting these gives $A - B = x + y - (x - y) = 2y$, so $y = \frac{A-B}{2}$

  Adding gives $A + B = x + y + (x - y) = 2x$, so $x = \frac{A+B}{2}$

- So $\cos A + \cos B \equiv 2\cos\left(\frac{A+B}{2}\right)\cos\left(\frac{A-B}{2}\right)$

You can also use the addition formulas to **solve** complicated trig equations.

Example 4

Solve $\sin\left(x + \frac{\pi}{2}\right) = \sin x$ in the interval $0 \le x \le 2\pi$.

- First replace $\sin\left(x + \frac{\pi}{2}\right)$ using the **sin** addition formula:

  $\sin x\cos\frac{\pi}{2} + \cos x\sin\frac{\pi}{2} = \sin x \Rightarrow 0 + \cos x = \sin x$

- **Divide** through by $\cos x$: $\frac{\cos x}{\cos x} = \frac{\sin x}{\cos x}$

- Replace $\frac{\sin x}{\cos x}$ with $\tan x$: $\frac{\cancel{\cos x}}{\cancel{\cos x}} = \tan x \Rightarrow \tan x = 1$

- Solve for $0 \le x \le 2\pi$: $x = \frac{\pi}{4}$ and $\frac{5\pi}{4}$

Tip: Using the table of common values on page 53, $\cos\frac{\pi}{2} = 0$ and $\sin\frac{\pi}{2} = 1$.

Tip: Remember — to solve a trig equation you need to get it all in terms of sin, cos or tan. You can use any of the identities you've learnt so far to do this.

Exercise 6.2

Q1 Hint: Look at the proof of tan $(A + B)$ on the previous page.

Q1 Use the sine and cosine addition formulas to prove that

$\tan(A - B) \equiv \frac{\tan A - \tan B}{1 + \tan A \tan B}$.

Q2 Prove the following identities:

a) $\frac{\cos(A-B) - \cos(A+B)}{\cos A \sin B} \equiv 2\tan A$

b) $\frac{1}{2}[\cos(A - B) - \cos(A + B)] \equiv \sin A \sin B$

c) $\sin(x + 90°) \equiv \cos x$

Q3 Solve $4\sin\left(x - \frac{\pi}{3}\right) = \cos x$ in the interval $-\pi \le x \le \pi$. Give your answers in radians to 2 decimal places.

Q4 a) Show that $\tan\left(-\frac{\pi}{12}\right) = \sqrt{3} - 2$.

b) Use your answer to a) to solve the equation $\cos x = \cos\left(x + \frac{\pi}{6}\right)$ in the interval $0 \le x \le \pi$. Give your answer in terms of π.

Q5 Show that $2\sin(x + 30°) \equiv \sqrt{3}\sin x + \cos x$.

Q6 Write an expression for $\tan\left(\frac{\pi}{3} - x\right)$ in terms of $\tan x$ only.

Q7 $\tan A = \frac{3}{8}$ and $\tan(A + B) = \frac{1}{4}$. Find the exact value of $\tan B$.

Q8 Show that $\sin A + \sin B \equiv 2\sin\left(\frac{A+B}{2}\right)\cos\left(\frac{A-B}{2}\right)$. PROBLEM SOLVING

Q9 a) Given that $\sin(x + y) = 4\cos(x - y)$, write an expression for $\tan x$ in terms of $\tan y$.

b) Use your answer to a) to solve $\sin\left(x + \frac{\pi}{4}\right) = 4\cos\left(x - \frac{\pi}{4}\right)$ in the interval $0 \le x \le 2\pi$.

Q10 Solve the following equations in the given interval. Give your answers to 2 decimal places.

a) $\sqrt{2}\sin(\theta + 45°) = 3\cos\theta$, $\quad 0° \le \theta \le 360°$

b) $2\cos\left(\theta - \frac{2\pi}{3}\right) - 5\sin\theta = 0$, $\quad 0 \le \theta \le 2\pi$

c) $\sin(\theta - 30°) - \cos(\theta + 60°) = 0$, $\quad 0° \le \theta \le 360°$

Q11 Use the sin addition formula to show that PROBLEM SOLVING

$\sin\left(x + \frac{\pi}{6}\right) \approx \frac{1}{2} + \frac{\sqrt{3}}{2}x - \frac{1}{4}x^2$ when x is small.

7. The Double Angle Formulas

You can use the addition formulas to get the double angle formulas — you'll find them really handy throughout the course.

Learning Objective:
- Know and be able to use the double angle formulas for sin $2A$, cos $2A$ and tan $2A$.

Deriving the double angle formulas

Double angle formulas are a special case of the addition formulas, using $(A + A)$ instead of $(A + B)$. They're called "double angle" formulas because they take an expression with a $2x$ term (a double angle) inside a trig function, and change it into an expression with only single x's inside the trig functions.

You need to know the double angle formulas for sin, cos and tan. Their derivations are given below.

Tip: You can also prove these geometrically in the same way as on pages 68-69 — just replace B with A.

$$\sin 2A \equiv 2\sin A \cos A$$

- Start with the **sin addition formula** (see p.68), but replace 'B' with 'A':

$$\sin (A + A) \equiv \sin A \cos A + \cos A \sin A$$

- Sin $(A + A)$ can be written as sin $2A$, and so:

$$\sin 2A \equiv \sin A \cos A + \cos A \sin A \equiv 2 \sin A \cos A$$

$$\cos 2A \equiv \cos^2 A - \sin^2 A$$

- Start with the **cos addition formula**, but again replace 'B' with 'A':

$$\cos (A + A) \equiv \cos A \cos A - \sin A \sin A$$

$$\Rightarrow \cos 2A \equiv \cos^2 A - \sin^2 A$$

- You can then use $\cos^2 A + \sin^2 A \equiv 1$ to get:

$\cos 2A \equiv \cos^2 A - (1 - \cos^2 A)$ and $\cos 2A \equiv (1 - \sin^2 A) - \sin^2 A$

$$\cos 2A \equiv 2\cos^2 A - 1 \qquad \cos 2A \equiv 1 - 2\sin^2 A$$

Tip: The double angle formula for cos has three different forms which are all very useful — but you can work out the second two from the general one as shown.

$$\tan 2A \equiv \frac{2\tan A}{1 - \tan^2 A}$$

- Start with the **tan addition formula**, and again replace 'B' with 'A':

$$\tan (A + A) \equiv \frac{\tan A + \tan A}{1 - \tan A \tan A}$$

- Simplifying this gives:

$$\tan 2A \equiv \frac{2\tan A}{1 - \tan^2 A}.$$

Using the double angle formulas

Like the other trig identities covered in this chapter, the double angle formulas are useful when you need to find an exact value.

Tip: Watch out for sneaky double angles (e.g. $\sin x \cos x$ is just $\frac{1}{2}\sin 2x$).

Tip: You won't usually be told which identity to use, so work on being able to spot the clues. If you're asked for an 'exact value' you should be thinking of your common angles. 15° is half of a common angle, so this should get you thinking about double angle formulas.

Example 1

a) Use a double angle formula to work out the exact value of $\sin 15° \cos 15°$.

- This looks the most like the **sin** double angle formula, $\sin 2A \equiv 2 \sin A \cos A$, but it needs to be **rearranged** slightly:

$$\sin 2A \equiv 2 \sin A \cos A \Rightarrow \sin A \cos A \equiv \frac{1}{2}\sin 2A$$

- Now put in the **numbers** from the question:

$$\sin 15° \cos 15° = \frac{1}{2}\sin 30° = \frac{1}{2} \times \frac{1}{2} = \frac{1}{4}$$

b) $\sin x = \frac{2}{3}$, where x is acute. Find the exact value of $\cos 2x$ and $\sin 2x$.

- For **$\cos 2x$**, use the **cos** double angle formula in terms of **sin**:

$$\cos 2A \equiv 1 - 2\sin^2 A$$

$$\Rightarrow \cos 2x = 1 - 2\left(\frac{2}{3}\right)^2 = \frac{1}{9}$$

- For **$\sin 2x$**, use the **sin** double angle formula.

$$\sin 2A \equiv 2 \sin A \cos A$$

- To use this, first work out **$\cos x$** from $\sin x$ using the triangle method:

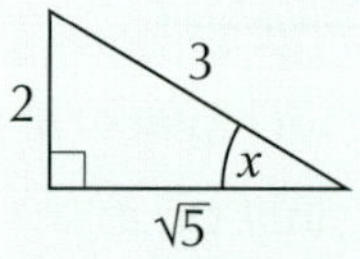

$$\cos x = \frac{\text{adj}}{\text{hyp}} = \frac{\sqrt{5}}{3}$$

- Now put the values into the sin double angle formula as usual:

$$\sin 2x = 2 \sin x \cos x = 2 \times \frac{2}{3} \times \frac{\sqrt{5}}{3} = \frac{4\sqrt{5}}{9}$$

The double angle formulas are also handy for **simplifying expressions** in order to solve equations.

Example 2

Write $1 - 2\sin^2\left(\frac{3x}{2}\right)$ as a single trigonometric ratio.

- Look for an identity that is similar to this expression, containing a '$\sin^2$'. The **cos** double angle formula (in terms of **sin**) looks best:

$$\cos 2A \equiv 1 - 2\sin^2 A$$

- Comparing the expression with the right-hand side of the identity, we need to use $A = \frac{3x}{2}$, and so $2A = 3x$.
- Putting this into the identity gives:

$$1 - 2\sin^2 \frac{3x}{2} \equiv \cos 3x$$

Exercise 7.1

Q1 Use the double angle formulas to write down the exact values of:

a) $4\sin\frac{\pi}{12}\cos\frac{\pi}{12}$ b) $\cos\frac{2\pi}{3}$ c) $\frac{\sin 120°}{2}$

d) $\frac{\tan 15°}{2-2\tan^2 15°}$ e) $2\sin^2 15° - 1$

Q1 Hint: For some of these there are other ways to find the answer, but if you've been asked to use a certain method then show your working using that method.

Q2 An acute angle x has $\sin x = \frac{1}{6}$. Find the exact values of:

a) $\cos 2x$ b) $\sin 2x$ c) $\tan 2x$

Q3 Angle x has $\sin x = -\frac{1}{4}$, and $\pi \le x \le \frac{3\pi}{2}$. Find the exact values of:

a) $\cos 2x$ b) $\sin 2x$ c) $\tan 2x$

Q3 Hint: Angle x lies in the 3rd quadrant of the CAST diagram, so $\sin x$ and $\cos x$ are negative but $\tan x$ is positive.

Q4 Write the following expressions as a single trigonometric ratio:

a) $\frac{\sin 3\theta \cos 3\theta}{3}$ b) $\sin^2\left(\frac{2y}{3}\right) - \cos^2\left(\frac{2y}{3}\right)$

c) $\frac{1-\tan^2\left(\frac{x}{2}\right)}{2\tan\left(\frac{x}{2}\right)}$

Solving equations and proving identities

If an equation has a mixture of $\sin x$ and $\sin 2x$ terms in it, there's not much that you can do with it in that state. But you can use one of the double angle formulas to simplify it, and then solve it.

Example 1

PROBLEM SOLVING

Solve the equation $\cos 2x - 5\cos x = 2$ in the interval $0 \le x \le 2\pi$.

- First use the **cos double angle formula** to get rid of $\cos 2x$:

$$\cos 2A \equiv 2\cos^2 A - 1$$

$$\Rightarrow 2\cos^2 x - 1 - 5\cos x = 2$$

Tip: Use this version of the formula so that you don't end up with a mix of sin and cos terms.

- **Simplify** so you have zero on one side...

$$2\cos^2 x - 5\cos x - 3 = 0$$

- ...then **factorise** and **solve** the **quadratic** that you've made:

$$(2\cos x + 1)(\cos x - 3) = 0$$

$$\Rightarrow (2\cos x + 1) = 0 \text{ or } (\cos x - 3) = 0$$

Tip: Let $y = \cos x$ and write as a quadratic in y if it helps.

- The second bracket gives you...

$$\cos x = 3$$

...which has **no solutions** since $-1 \le \cos x \le 1$.

- So all that's left is to solve the first bracket to find x:

$$2\cos x + 1 = 0$$

$$\cos x = -\frac{1}{2}$$

- You know that $\cos x = \frac{1}{2}$ for $x = \frac{\pi}{3}$ so using the symmetry of the graph below you get:

$$x = \frac{2\pi}{3} \text{ or } x = \frac{4\pi}{3}$$

- Remember — you can sketch the **graph** of $\cos x$ to find all values of x in the given interval:

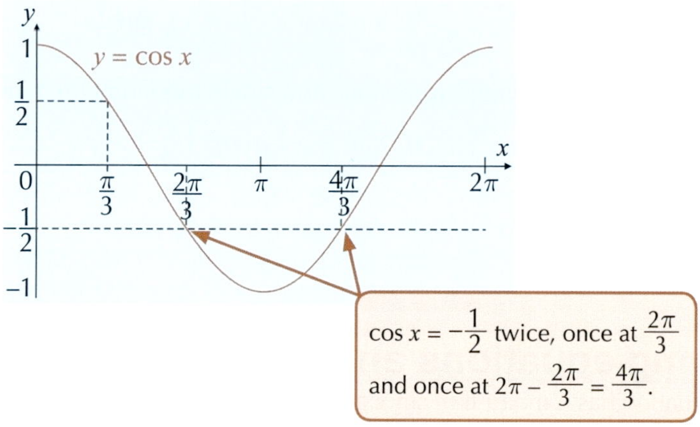

Tip: You can also use the CAST diagram.

The following examples show how the double angle formulas can be used to **prove** other identities.

Example 2

Prove that $2\cot\frac{x}{2}\left(1 - \cos^2\frac{x}{2}\right) \equiv \sin x$. PROBLEM SOLVING

- Use the identity $\sin^2\theta \equiv 1 - \cos^2\theta$ to replace the $1 - \cos^2\frac{x}{2}$ on the left-hand side, so you have:

Left-hand side: $2\cot\frac{x}{2}\sin^2\frac{x}{2}$

- Now write $\cot\theta$ as $\frac{\cos\theta}{\sin\theta}$:

$$\frac{2\cos\frac{x}{2}\sin^2\frac{x}{2}}{\sin\frac{x}{2}} \equiv 2\cos\frac{x}{2}\sin\frac{x}{2}$$

- Now use the **sin double angle formula**, using $A = \frac{x}{2}$, to write...

$$2\cos\frac{x}{2}\sin\frac{x}{2} \equiv \sin x$$

...which gives you the right-hand side.

Tip: cot is just $\frac{1}{\tan}$, and tan is $\frac{\sin}{\cos}$, so cot is $\frac{\cos}{\sin}$. Look back at pages 55-56 if you need to refresh your memory on cot, cosec and sec.

Example 3

Show that $\cos 3\theta \equiv 4\cos^3\theta - 3\cos\theta$. PROBLEM SOLVING

- First, write $\cos 3\theta$ as $\cos(2\theta + \theta)$.
- Now use the **cos addition formula**:

$$\cos 3\theta \equiv \cos(2\theta + \theta) \equiv \cos 2\theta \cos\theta - \sin 2\theta \sin\theta.$$

- Now use the **cos** and **sin double angle formulas** to get rid of the 2θ:

$$\cos 2\theta \cos\theta - \sin 2\theta \sin\theta \equiv (2\cos^2\theta - 1)\cos\theta - (2\sin\theta\cos\theta)\sin\theta$$

cos double angle formula — sin double angle formula

- **Tidy this up** by expanding the brackets and using $\sin^2\theta \equiv 1 - \cos^2\theta$:

$$\equiv 2\cos^3\theta - \cos\theta - 2\sin^2\theta\cos\theta$$

$$\equiv 2\cos^3\theta - \cos\theta - 2(1 - \cos^2\theta)\cos\theta$$

$$\equiv 2\cos^3\theta - \cos\theta - 2\cos\theta + 2\cos^3\theta$$

$$\equiv 4\cos^3\theta - 3\cos\theta \text{ as required}$$

Tip: Clever tricks like splitting up the angle so you can use the addition formulas can really help if you're stuck on a trig identity question.

Tip: You can use a similar method to show that $\sin 3\theta \equiv 3\sin\theta - 4\sin^3\theta$.

The half angle formulas

The double angle formulas for cos can be **rearranged** to give another three useful identities known as the **half angle formulas**.

The following examples show how you can derive them.

Example 4

a) **Show that $\cos^2\left(\frac{\theta}{2}\right) \equiv \frac{1}{2}(1 + \cos\theta)$.**

- Start with the **double angle** formula for **cos**: $\cos 2A \equiv 2\cos^2 A - 1$
- Replace A with $\frac{\theta}{2}$: → $\cos\theta \equiv 2\cos^2\left(\frac{\theta}{2}\right) - 1$
- Now **rearrange** to get the **half angle** formula for **cos**: → $\cos^2\left(\frac{\theta}{2}\right) \equiv \frac{1}{2}(1 + \cos\theta)$

b) **Show that $\sin^2\left(\frac{\theta}{2}\right) \equiv \frac{1}{2}(1 - \cos\theta)$.**

- This time, start with the **double angle** formula for **cos** that contains **sin**: → $\cos 2A \equiv 1 - 2\sin^2 A$
- Again, replace A with $\frac{\theta}{2}$: → $\cos\theta \equiv 1 - 2\sin^2\left(\frac{\theta}{2}\right)$
- Now **rearrange** to get the **half angle** formula for **sin**: → $\sin^2\left(\frac{\theta}{2}\right) \equiv \frac{1}{2}(1 - \cos\theta)$

Tip: You don't need to know these derivations for your exam, but they're good examples of using identities.

Tip: That 'hence' in the question tells you to use your results for sin and cos. Any time you have to use sin and cos to prove something for tan you should be thinking of the identity $\tan\theta \equiv \frac{\sin\theta}{\cos\theta}$.

c) Hence show that $\tan^2\left(\frac{\theta}{2}\right) \equiv \frac{1-\cos\theta}{1+\cos\theta}$.

- Start with the identity $\tan x \equiv \frac{\sin x}{\cos x}$: → $\tan\left(\frac{\theta}{2}\right) \equiv \frac{\sin\left(\frac{\theta}{2}\right)}{\cos\left(\frac{\theta}{2}\right)}$
- **Square** both sides: → $\tan^2\left(\frac{\theta}{2}\right) \equiv \frac{\sin^2\left(\frac{\theta}{2}\right)}{\cos^2\left(\frac{\theta}{2}\right)}$
- Replace $\sin^2\left(\frac{\theta}{2}\right)$ and $\cos^2\left(\frac{\theta}{2}\right)$ with their **half angle formulas** from examples a) and b). → $\tan^2\left(\frac{\theta}{2}\right) \equiv \frac{\frac{1}{2}(1-\cos\theta)}{\frac{1}{2}(1+\cos\theta)}$
- Now **simplify** to get the **half angle** formula for **tan**: → $\tan^2\left(\frac{\theta}{2}\right) \equiv \frac{1-\cos\theta}{1+\cos\theta}$

Exercise 7.2

Q1d) Hint: There should be 7 solutions in the given interval, but two of them are easy to miss...

Q1 Solve the equations below in the interval $0 \le x \le 360°$. Give your answers to 1 decimal place.

a) $4\cos 2x = 14\sin x$ b) $5\cos 2x + 9\cos x = -7$

c) $4\cot 2x + \cot x = 5$ d) $\tan x - 5\sin 2x = 0$

Q2 Solve the equations below in the interval $0 \le x \le 2\pi$. Give your answers to 3 significant figures.

a) $4\cos 2x - 10\cos x + 1 = 0$ b) $\frac{\cos 2x - 3}{2\sin^2 x - 1} = 3$

Q3b) Hint: Try writing $\sin x$ as $\sin 2\left(\frac{x}{2}\right)$ and using the sin double angle formula.

Q3 Solve the equations below in the interval $0 \le x \le 2\pi$. Give your answers in terms of π.

a) $\cos 2x + 7\cos x = -4$ b) $\sin x + \cos\frac{x}{2} = 0$

Q4 Use the double angle formulas to prove each of the identities below.

a) $\sin 2x \sec^2 x \equiv 2\tan x$ b) $\frac{2}{1+\cos 2x} \equiv \sec^2 x$

c) $\cot x - 2\cot 2x \equiv \tan x$ d) $\tan 2x + \cot 2x \equiv 2\operatorname{cosec} 4x$

Q4d) Hint: Write the left-hand side in terms of sin and cos first.

Q5 a) Show that $\frac{1+\cos 2x}{\sin 2x} \equiv \cot x$.

b) Use your answer to a) to solve $\frac{1+\cos 4\theta}{\sin 4\theta} = 7$ in the interval $0 \le \theta \le 360°$. Give your answers to 1 d.p.

Q6a) Hint: Write $\operatorname{cosec} x$ as $\frac{1}{\sin x}$ and then use $\sin x = \sin 2\left(\frac{x}{2}\right)$.

Q6 a) Show that $\operatorname{cosec} x - \cot\frac{x}{2} \equiv -\cot x$.

b) Use your answer to a) to solve $\operatorname{cosec} y = \cot\frac{y}{2} - 2$ in the interval $-\pi \le y \le \pi$. Give your answers to 3 s.f.

Q7 Given that $\sin\theta = \frac{5}{13}$, and that θ is acute, find

a) (i) $\cos\left(\frac{\theta}{2}\right)$ (ii) $\sin\left(\frac{\theta}{2}\right)$

b) Hence find $\tan\left(\frac{\theta}{2}\right)$.

8. The R Addition Formulas

The R addition formulas are used to help solve equations which contain a mix of cos and sin terms.

Learning Objective:

- Know and be able to use expressions for $a \cos \theta + b \sin \theta$ in the equivalent forms of: $R \cos (\theta \pm \alpha)$ or $R \sin (\theta \pm \alpha)$.

Expressions of the form $a \cos \theta + b \sin \theta$

If you're solving an equation that contains both $\sin \theta$ and $\cos \theta$ terms, e.g. $3 \sin \theta + 4 \cos \theta = 1$, you need to rewrite it so that it only contains one trig function. The formulas that you use to do that are known as the ***R* formulas**:

One set for **sine**: $$a \sin \theta \pm b \cos \theta \equiv R \sin (\theta \pm \alpha)$$

And one set for **cosine**: $$a \cos \theta \pm b \sin \theta \equiv R \cos (\theta \mp \alpha)$$

where a, b and R are **positive**, and α is **acute**.

You need to be careful with the + and – signs in the cosine formula. If you have $a \cos \theta$ **+** $b \sin \theta$ then use $R \cos (\theta - \alpha)$.

Using the R formulas

- You'll start with an identity like $2 \sin x + 5 \cos x \equiv R \sin (x + \alpha)$, where R and α need to be found.
- First, **expand** the right hand side using the **addition formulas** (see p.68): $2 \sin x + 5 \cos x \equiv R \sin x \cos \alpha + R \cos x \sin \alpha$.
- **Equate the coefficients** of $\sin x$ and $\cos x$. You'll get two equations: ① $R \cos \alpha = 2$ and ② $R \sin \alpha = 5$
- To find α, **divide** equation ② by equation ①, (because $\frac{R \sin \alpha}{R \cos \alpha} = \tan \alpha$) then take $\mathbf{tan^{-1}}$ of the result.
- To find R, **square** equations ① and ② and **add** them together, then take the **square root** of the answer. This works because: $(R \sin \alpha)^2 + (R \cos \alpha)^2 \equiv R^2 (\sin^2 \alpha + \cos^2 \alpha) \equiv R^2$ (using the identity $\sin^2 \alpha + \cos^2 \alpha \equiv 1$).

Tip: This method looks a bit scary, but follow through the next example and it should make more sense.

Example 1

Express $4 \cos x + 5 \sin x$ in the form $R \cos (x \pm \alpha)$.

- First you need to get the **sign** right in the formula. For this use:

$$4 \cos x + 5 \sin x \equiv R \cos (x - \alpha)$$

- Now **expand** the right hand side using the **cos addition formula**:

$$4 \cos x + 5 \sin x \equiv R \cos x \cos \alpha + R \sin x \sin \alpha.$$

Tip: The addition formula used here is: $\cos (A - B) = \cos A \cos B + \sin A \sin B$.

Tip: Equating coefficients is a useful technique when you are working with identities like this — you know that the two sides are identical for all x, so the coefficients of $\sin x$ and $\cos x$ will be the same on each side.

- **Equating the coefficients** of $\cos x$ gives:

$$① \; R\cos\alpha = 4$$

and equating the coefficients of $\sin x$ gives:

$$② \; R\sin\alpha = 5$$

- Dividing ② by ① gives:

$$\tan\alpha = \frac{5}{4} \Rightarrow \alpha = 51.3° \text{ (1 d.p.)}$$

- Squaring ① and ② gives:

$$①^2: \; R^2\cos^2\alpha = 16$$
$$②^2: \; R^2\sin^2\alpha = 25$$

$$①^2 + ②^2: \; R^2\cos^2\alpha + R^2\sin^2\alpha = 16 + 25$$
$$\Rightarrow R^2(\cos^2\alpha + \sin^2\alpha) = 41$$
$$\Rightarrow R^2 = 41$$
$$\Rightarrow R = \sqrt{41}$$

Tip: Remember that R is always positive (so take the positive root) and α is always acute (so take the angle in the first quadrant of the CAST diagram when solving).

- Finally, put the values for α and R back into the identity to give:

$$4\cos x + 5\sin x \equiv \sqrt{41}\cos(x - 51.3°)$$

Example 2

a) Show that $5\sin x - 5\sqrt{3}\cos x \equiv 10\sin\left(x - \frac{\pi}{3}\right)$.

- As before, pick a formula and find values for R and α — then you can show that they are the **same** as in the right-hand side of the given identity. Using the **sin addition formula**:

$$5\sin x - 5\sqrt{3}\cos x \equiv R\sin(x - \alpha)$$
$$\equiv R\sin x\cos\alpha - R\cos x\sin\alpha$$

Tip: $\sin(A - B) = \sin A\cos B - \cos A\sin B$.

- Equate coefficients of $\sin x$ and $\cos x$ to get:

$$① \; R\cos\alpha = 5 \quad \text{and} \quad ② \; R\sin\alpha = 5\sqrt{3}$$

$$② \div ①: \; \frac{R\sin\alpha}{R\cos\alpha} = \tan\alpha = \frac{5\sqrt{3}}{5} = \sqrt{3}$$
$$\Rightarrow \alpha = \frac{\pi}{3}$$

$$①^2 + ②^2: \; R^2\cos^2\alpha + R^2\sin^2\alpha = 5^2 + (5\sqrt{3})^2$$
$$\Rightarrow R^2(1) = 100$$
$$\Rightarrow R = 10$$

- So, putting the values for α and R back into the identity gives...

$$5\sin x - 5\sqrt{3}\cos x \equiv 10\sin\left(x - \frac{\pi}{3}\right)$$

...which is the **right-hand side** of the identity you're trying to prove.

b) Hence sketch the graph of $y = 5 \sin x - 5\sqrt{3} \cos x$ in the interval $-\pi \le x \le \pi$.

Writing $y = 5 \sin x - 5\sqrt{3} \cos x$ as $y = 10 \sin\left(x - \frac{\pi}{3}\right)$ makes it a lot easier to sketch the graph — just **transform** the graph of $y = \sin x$ as appropriate:

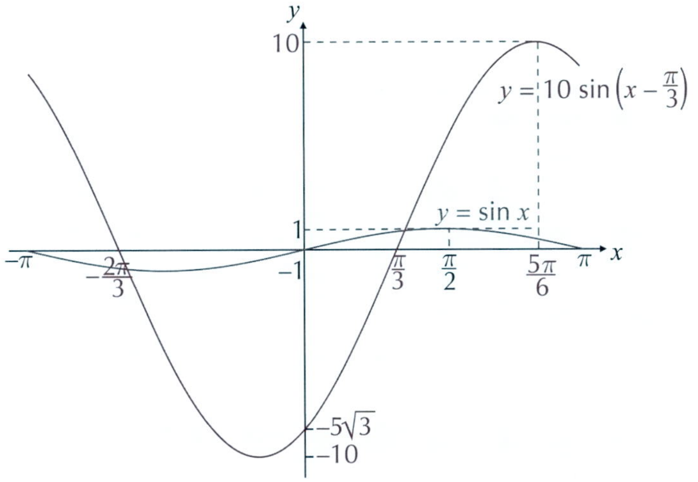

You can transform the **maximum** and **minimum** points of the graph in the same way. E.g. the maximum on $y = \sin x$ at $\left(\frac{\pi}{2}, 1\right)$ gets translated right by $\frac{\pi}{3}$ (so add $\frac{\pi}{3}$ to the x-coordinate) and stretched vertically by 10 (so multiply the y-coordinate by 10), to become $\left(\frac{5\pi}{6}, 10\right)$.

Tip: Look back at pages 33-36 for a reminder about transformations of graphs.

Tip: This transformation is just a translation of the graph of $y = \sin x$ horizontally right by $\frac{\pi}{3}$ followed by a stretch vertically by a scale factor of 10.

Exercise 8.1

Q1 Express $3 \sin x - 2 \cos x$ in the form $R \sin (x - \alpha)$.
Give R in surd form and α in degrees to 1 decimal place.

Q2 Express $6 \cos x - 5 \sin x$ in the form $R \cos (x + \alpha)$.
Give R in surd form and α in degrees to 1 decimal place.

Q3 Express $\sin x + \sqrt{7} \cos x$ in the form $R \sin (x + \alpha)$.
Give R in the form $m\sqrt{2}$ and α in radians to 3 significant figures.

Q4 Show that $\sqrt{2} \sin x - \cos x \equiv \sqrt{3} \sin (x - \alpha)$, where $\tan \alpha = \frac{1}{\sqrt{2}}$.

Q5 Show that $3 \cos 2x + 5 \sin 2x \equiv \sqrt{34} \cos (2x - \alpha)$, where $\tan \alpha = \frac{5}{3}$.

Q5 Hint: Treat the $2x$ just the same as an x.

Q6 a) Express $\sqrt{3} \sin x + \cos x$ in the form $R \sin (x + \alpha)$.
Give R and α as exact answers and α in radians in terms of π.

b) Hence sketch the graph of $y = \sqrt{3} \sin x + \cos x$ in the interval $-\pi \le x \le \pi$.

c) State the coordinates of any maximum and minimum points and intersections with the axes of the graph in b).

Q6c) Hint: Write down their coordinates on the graph of $y = \sin x$ first, then apply the transformations to them.

Applying the R addition formulas

To solve equations of the form $a \sin \theta + b \cos \theta = c$, it's best to work things out in different stages — first writing out the equation in the form of one of the R formulas, then solving it. The example below shows how.

Tip: Questions like these are sometimes already written in stages. You might be asked for something else too, like the maximum and minimum values of the function.

Example

a) Solve $2 \sin x - 3 \cos x = 1$ in the interval $0° \le x \le 360°$.

- Before you can **solve** this equation you need to get $2 \sin x - 3 \cos x$ in the form $\boldsymbol{R} \textbf{ sin } \boldsymbol{(x - \alpha)}$:

$$2 \sin x - 3 \cos x \equiv R \sin (x - \alpha)$$
$$2 \sin x - 3 \cos x \equiv R \sin x \cos \alpha - R \cos x \sin \alpha$$

- Equating coefficients gives the equations:

$$R \cos \alpha = 2 \text{ and } R \sin \alpha = 3$$

- Solving for α:

$$\frac{R \sin \alpha}{R \cos \alpha} = \tan \alpha = \frac{3}{2}$$
$$\Rightarrow \alpha = \tan^{-1} 1.5 = 56.31° \text{ (2 d.p.)}$$

- Solving for R:

$$R^2 \cos^2 \alpha + R^2 \sin^2 \alpha = 2^2 + 3^2$$
$$\Rightarrow R^2 = 13 \Rightarrow R = \sqrt{13}$$

- So $2 \sin x - 3 \cos x = \sqrt{13} \sin (x - 56.31°)$

- Now **solve the equation.** If $2 \sin x - 3 \cos x = 1$, then:

$$\sqrt{13} \sin (x - 56.31°) = 1$$
$$\Rightarrow \sin (x - 56.31°) = \frac{1}{\sqrt{13}}$$

- Since $0° \le x \le 360°$, you should be looking for solutions in the interval $\boldsymbol{-56.31° \le (x - 56.31°) \le 303.69°}$ (just take 56.31 away from the original interval).

- Solve the equation:

$$x - 56.31° = \sin^{-1}\left(\frac{1}{\sqrt{13}}\right) = 16.10°,$$
$$\text{or } 180 - 16.10 = 163.90° \text{ (2 d.p.)}$$

There are no solutions in the range $-56.31° \le (x - 56.31°) \le 0°$ since $\sin x$ is negative in the range $-90° \le x \le 0°$.

Tip: The other positive solution for sin is in the second quadrant of the CAST diagram, at 180° – 16.10° (the angle in the first quadrant).

- So $x = 16.10 + 56.31 = 72.4°$ (1 d.p.)

or $x = 163.90 + 56.31 = 220.2°$ (1 d.p.)

b) What are the maximum and minimum values of $2 \sin x - 3 \cos x$?

The maximum and minimum values of the **sin** (and cos) function are ± 1, so the maximum and minimum values of $\boldsymbol{R} \textbf{ sin } \boldsymbol{(x - \alpha)}$ are $\boldsymbol{\pm R}$.

As $2 \sin x - 3 \cos x = \sqrt{13} \sin (x - 56.31°)$, $R = \sqrt{13}$,

so the maximum and minimum values are $\pm\sqrt{13}$.

Tip: This is similar to being asked for the coordinates of the maximum and minimum points on the graph, so you can use a graph to help if you prefer.

Exercise 8.2

Q1 a) Express $5 \cos \theta - 12 \sin \theta$ in the form $R \cos (\theta + \alpha)$, where $R > 0$ and α is an acute angle (in degrees, to 1 decimal place).

b) Hence solve $5 \cos \theta - 12 \sin \theta = 4$ in the interval $0° \leq \theta \leq 360°$.

c) State the maximum and minimum values of $5 \cos \theta - 12 \sin \theta$.

Q2 a) Express $2 \sin 2\theta + 3 \cos 2\theta$ in the form $R \sin (2\theta + \alpha)$, where $R > 0$ (given in surd form) and $0 < \alpha < \frac{\pi}{2}$ (to 3 significant figures).

b) Hence solve $2 \sin 2\theta + 3 \cos 2\theta = 1$ in the interval $0 \leq \theta \leq 2\pi$.

Q2b) Hint: Take care with the interval here to make sure you get all the correct solutions for θ.

Q3 a) Express $3 \sin \theta - 2\sqrt{5} \cos \theta$ in the form $R \sin (\theta - \alpha)$. Give R in surd form and α in degrees to 1 decimal place.

b) Hence solve $3 \sin \theta - 2\sqrt{5} \cos \theta = 5$ in the interval $0° \leq \theta \leq 360°$.

c) Find the maximum value of $f(x) = 3 \sin x - 2\sqrt{5} \cos x$ and the smallest positive value of x at which it occurs.

Q4 $f(x) = 3 \sin x + \cos x$.

a) Express $f(x)$ in the form $R \sin (x + \alpha)$ where $R > 0$ (given in surd form) and $0° < \alpha < 90°$ (to 1 d.p.).

b) Hence solve the equation $f(x) = 2$ in the interval $0° \leq x \leq 360°$.

c) State the maximum and minimum values of $f(x)$.

Q5 a) Express $4 \sin x + \cos x$ in the form $R \sin (x + \alpha)$, where $R > 0$ (given in surd form) and $0 < \alpha < \frac{\pi}{2}$ (to 3 significant figures).

b) Hence find the greatest value of $(4 \sin x + \cos x)^4$.

c) Solve the equation $4 \sin x + \cos x = 1$ for values of x in the interval $0 \leq x \leq \pi$.

Q6 $f(x) = 8 \cos x + 15 \sin x$.

a) Write $f(x)$ in the form $R \cos(x - \alpha)$, where $R > 0$ and $0 < \alpha < \frac{\pi}{2}$.

b) Solve the equation $f(x) = 5$ in the interval $0 \leq x \leq 2\pi$.

c) Find the minimum value of $g(x) = (8 \cos x + 15 \sin x)^2$ and the smallest positive value of x at which it occurs.

Q6c) Hint: Think about what happens to the negative values when you square a function.

Q7 The function g is given by $g(x) = 2 \cos x + \sin x$, $x \in \mathbb{R}$. $g(x)$ can be written as $R \cos (x - \alpha)$, where $R > 0$ and $0° < \alpha < 90°$.

a) Show that $R = \sqrt{5}$, and find the value of α (to 3 s.f.).

b) Hence state the range of $g(x)$.

Q7b) Hint: This is just another way of asking for the maximum and minimum values of the function.

Q8 Express $3 \sin \theta - \frac{3}{2} \cos \theta$ in the form $R \sin (\theta - \alpha)$, where $R > 0$ and $0 < \alpha < \frac{\pi}{2}$, and hence solve the equation $3 \sin \theta - \frac{3}{2} \cos \theta = 3$ for values of θ in the interval $0 \leq \theta \leq 2\pi$.

Q9 Solve the equation $4 \sin 2\theta + 3 \cos 2\theta = 2$ for values of θ in the interval $0 \leq \theta \leq \pi$.

9. Modelling with Trig Functions

Learning Objective:
- Be able to use trigonometry to model real-life problems.

Physicists and engineers use trig functions a lot — the fact that they repeat every 2π makes them really useful for modelling situations where motion repeats periodically, such as bouncing or swinging. You'll need to use all the trigonometry you've learned so far.

MODELLING

Trigonometry in modelling

Trigonometric functions show up a lot in modelling problems. Things that happen in a cycle, like the motion of a child on a swing, or tidal patterns, could be modelled with sin and cos.

When tackling these questions, remember to check whether you should be working in degrees or radians. Make sure that you give your answers to a suitable degree of accuracy (3 s.f. is usually fine).

Example 1

The height of an object bouncing on a spring is modelled by the equation

$$h = 5 + 2\sin\left(5t + \frac{\pi}{3}\right)$$

where t is the time in seconds and h is the height in cm.
Find the first time at which $h = 4$ cm.

Tip: Remember — if you see π in a question it means it's in radians.

- You need to solve the equation $5 + 2\sin\left(5t + \frac{\pi}{3}\right) = 4$:

$$2\sin\left(5t + \frac{\pi}{3}\right) = -1$$

$$\sin\left(5t + \frac{\pi}{3}\right) = -\frac{1}{2}$$

$$\left(5t + \frac{\pi}{3}\right) = \sin^{-1}\left(-\frac{1}{2}\right) = -\frac{\pi}{6}, \frac{7\pi}{6}, \frac{11\pi}{6}, \text{... etc.}$$

Tip: Be careful when using inverse trig functions to solve equations — you usually get multiple solutions so you have to decide which one(s) to use.

- Time can't be negative, which means that $t \geq 0 \Rightarrow 5t + \frac{\pi}{3} \geq \frac{\pi}{3}$
- The first solution that satisfies this inequality is $\frac{7\pi}{6}$.

$$5t + \frac{\pi}{3} = \frac{7\pi}{6} \Rightarrow 5t = \frac{5\pi}{6} \Rightarrow t = \frac{\pi}{6}$$

- So the first time it has a height of 4 cm is at $t = \frac{\pi}{6}$ = 0.524 s (3 s.f.)

Example 2

Two sound waves are modelled by the functions f(θ) and g(θ), where f(θ) = 12 sin θ and g(θ) = $4\sqrt{3}$ cos θ.
The two waves combine to produce a new wave given by h(θ) = f(θ) + g(θ).

a) Write the equation for this new wave in the form $R\sin(\theta + \alpha)$, where α is measured in radians, giving your answers as exact values.

- $h(\theta) = 12\sin\theta + 4\sqrt{3}\cos\theta = R\sin(\theta + \alpha)$
- Expand $R\sin(\theta + \alpha)$ using the sin addition formula and equate coefficients:

$$R\sin(\theta + \alpha) = R\sin\theta\cos\alpha + R\cos\theta\sin\alpha$$

$$\Rightarrow R\cos\alpha = 12 \text{ and } R\sin\alpha = 4\sqrt{3}$$

Tip: This uses the method from page 81.

- Now solve for R and α:

$$R = \sqrt{12^2 + (4\sqrt{3})^2} = \sqrt{144 + 48} = \sqrt{192} = 8\sqrt{3}$$

$$\tan \alpha = \frac{4\sqrt{3}}{12} = \frac{\sqrt{3}}{3} = \frac{1}{\sqrt{3}} \Rightarrow \alpha = \frac{\pi}{6}$$

- So $h(\theta) = 8\sqrt{3} \sin\left(\theta + \frac{\pi}{6}\right)$

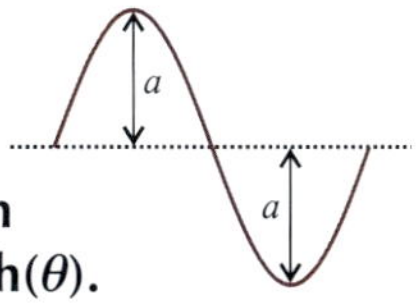

b) The amplitude, *a*, of a sound wave is equivalent to half the distance between the maximum and minimum values of the wave (as shown). Find the amplitude of h(θ).

- $\sin \theta$ has a maximum of 1 and a minimum of –1, so its amplitude is $\frac{1}{2}(1 - (-1)) = 1$.
- To transform $\sin \theta$ into h(θ), you do a horizontal translation by $\frac{\pi}{6}$, and a vertical stretch by a scale factor of $8\sqrt{3}$.
- The maximum and minimum values of h(θ) are $\pm R$ — i.e. $8\sqrt{3}$ and $-8\sqrt{3}$, so its amplitude is $\frac{1}{2}(8\sqrt{3} - (-8\sqrt{3})) = 8\sqrt{3}$

Tip: The horizontal translation won't affect the amplitude, since it doesn't change the height of the maximum or minimum of the function.

Exercise 9.1

MODELLING

Q1 A circular plot of land with a radius of 20 m is separated into three gardens. Each garden is a sector of the circle, with angles of 120°, 144° and 96° respectively. Calculate the area and perimeter of each garden (to 3 s.f.).

Q2 Adam wants to form a function to model the hours of daylight in his town throughout the year. He knows that the function should be of the form $f(t) = A + B \cos t$, where A and B are positive constants and the time, t, is in radians.

a) The daylight hours vary from 7 to 17. Find the values of A and B.

b) He now wants to adjust the model so that t is measured in months. He rewrites his function as $g(t) = A + B \cos (Ct + D)$. Find the values of C and D, such that the longest day of the year occurs at $t = 0$ and the shortest day of the year occurs at $t = 6$.

Q2 Hint: You might find it helpful to think of the graph of $y = \cos x$, and what transformations you could apply to it to make it fit the model.

Q3 The height of a buoy floating in a harbour, measured in metres, is modelled by the function $h(t) = 14 + 5 (\sin t + \cos t)$, where t is time in hours. By writing h(t) in the form $14 + R \cos (t - \alpha)$, find the maximum and minimum height of the buoy (to 1 d.p.).

Q4 Antonia goes on a fairground ride. She is strapped into a small spinning disc, which is attached to a large rotating wheel.

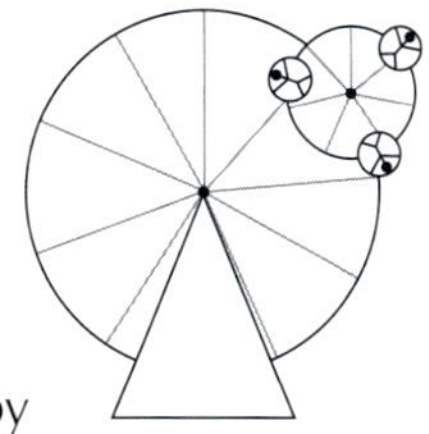

a) The height, in metres, of the centre of the disc above the ground after t seconds is given by $H = 10 + \frac{7}{2}(\sin t - \sqrt{3} \cos t)$. Write this in the form $H = 10 + R \sin (t - \alpha)$, where $R > 0$ and $0 < \alpha \leq \frac{\pi}{2}$.

b) Antonia's height above the ground, h, is given by $h = H - \cos 2(t - \alpha)$. Use the double angle formulas to show that $h = A + R \sin (t - \alpha) + B \sin^2 (t - \alpha)$, and give the values of A and B.

c) Hence find the first time when Antonia is 13 m above the ground.

Q4 Hint: In part b), you might find it easier to replace $(t - \alpha)$ with a single variable (e.g. x), and do the same for $\sin (t - \alpha)$ in part c) — then you have a quadratic that you can solve.

1. Parametric Equations of Curves

Learning Objectives:

- Be able to calculate Cartesian coordinates for a curve given in parametric form.
- Be able to find the coordinates of intersection points between a parametric curve and other lines.

Tip: Parametric equations can be used to model the movement of particles in kinematics — see Chapter 14.

Tip: You won't necessarily be expected to sketch a curve from its parametric equations in the exam — but finding the Cartesian coordinates like this can make it easier to picture the graph.

Parametric equations are ones where you have x and y in separate equations, both defined in terms of another variable. It sounds complicated, but it can often make things a lot easier, as you'll see in this chapter.

Finding coordinates from parametric equations

- Normally, graphs in the (x, y) plane are described using a **Cartesian equation** — a single equation linking x and y. Sometimes, particularly for more complicated graphs, it's easier to have two linked equations, called **parametric equations**.
- In parametric equations, x and y are each defined separately in terms of a **third variable**, called a **parameter**. The parameter is usually either t or θ.
- Parametric equations are often used to model the path of a moving particle, where its **position** (given by x and y) depends on time, t.

Example 1

Sketch the graph given by the parametric equations $y = t^3 - 1$ and $x = t + 1$.

- Start by making a **table of coordinates**.
Choose some values for t and calculate x and y at these values:

t	–2	–1	0	1	2
x	–1	0	1	2	3
y	–9	–2	–1	0	7

$x = -2 + 1 = -1$, and $y = (-2)^3 - 1 = -9$.

$x = 1 + 1 = 2$, and $y = 1^3 - 1 = 0$.

- Now plot the **Cartesian (x, y) coordinates** on a set of axes as usual:

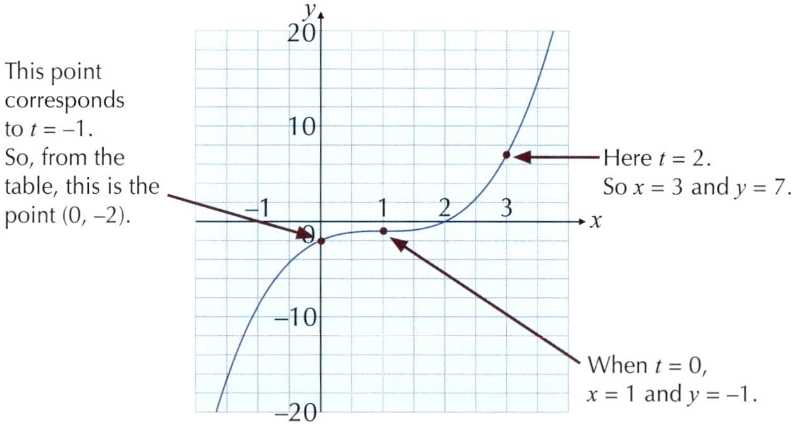
y
20
10
–10
–20
–1
1
2
3
x
This point corresponds to t = –1. So, from the table, this is the point (0, –2).
Here t = 2. So x = 3 and y = 7.
When t = 0, x = 1 and y = –1.

Example 2

Sketch the graph given by $x = \cos\theta$ and $y = \sin\theta + 1$.

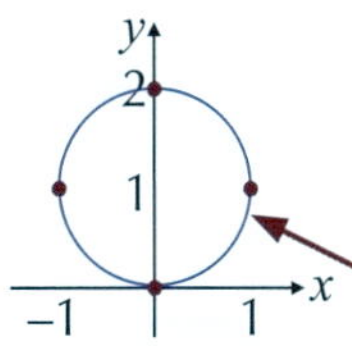

- Make a **table** as before with values for θ over a suitable interval (such as $0 \le \theta \le 2\pi$).
- Plot the points — you should spot that they lie on a **circle** with centre **(0, 1)** and **radius 1**.

θ	x	y
0	1	1
$\frac{\pi}{2}$	0	2
π	–1	1
$\frac{3\pi}{2}$	0	0
2π	1	1

Tip: $x = a + r\cos\theta$, $y = b + r\sin\theta$ are the general parametric equations for a circle with centre (a, b). If r takes different values in the x and y equations, the curve is an ellipse.

You can use the parametric equations to find **coordinates** of points on a graph, and to find the value of the **parameter** for given x- or y-coordinates.

Example 3

A flying disc is thrown from the point (0, 0). After t seconds, it has travelled x m horizontally and y m vertically, modelled by the parametric equations $x = t^2 + 2t$ and $y = 6t - t^2$ $(0 \le t \le 6)$.

MODELLING

a) Find the x- and y- values of the position of the disc after 2.5 seconds.

Just **substitute** $t = 2.5$ into the equations for x and y:

When $t = 2.5$: $x = 2.5^2 + 2 \times 2.5 = 6.25 + 5 = 11.25$

$y = 6 \times 2.5 - 2.5^2 = 15 - 6.25 = 8.75$

b) After how many seconds does the disc reach a height of 5 metres?

Now you want the value of t when $y = 5$:

$5 = 6t - t^2 \Rightarrow t^2 - 6t + 5 = 0 \Rightarrow (t - 1)(t - 5) = 0$

$\Rightarrow t = 1$ s and $t = 5$ s

c) What is the value of y when the disc reaches the point $x = 24$ m?

- Use the equation for x to find t first:

  $24 = t^2 + 2t \Rightarrow t^2 + 2t - 24 = 0 \Rightarrow (t + 6)(t - 4) = 0$

  $\Rightarrow t = -6$ or $t = 4$, but t is restricted to $0 \le t \le 6$, so $t = 4$ s

- Then use it in the other equation to find y:

  $y = 6 \times 4 - 4^2 = 24 - 16 = 8$

Tip: There's often a limit on the **domain** of the parameter — see pages 12-16 for more about domains of functions. For example, the parameter can't take a value that would set the denominator of any fraction to zero.

Exercise 1.1

Q1 A curve is defined by the parametric equations $x = 3t$, $y = t^2$.
- a) Find the coordinates of the point where $t = 5$.
- b) Find the value of t at the point where $x = 18$.
- c) Find the possible values of x at the point where $y = 36$.

Q2 A curve is defined by the parametric equations $x = 2t - 1$, $y = 4 - t^2$.
- a) Find the coordinates of the point where $t = 7$.
- b) Find the value of t at the point where $x = 15$.
- c) Find the possible values of x at the point where $y = -5$.

Q3b) & c) Hint: There is only one acute value of θ possible in b) ($0 < \theta < \frac{\pi}{2}$), and only one obtuse value of θ possible in c) ($\frac{\pi}{2} < \theta < \pi$).

Q3 A curve has parametric equations $x = 2 + \sin\theta$, $y = -3 + \cos\theta$.

a) Find the coordinates of the point where $\theta = \frac{\pi}{4}$.

b) Find the acute value of θ at the point where $x = \frac{4+\sqrt{3}}{2}$.

c) Find the obtuse value of θ at the point where $y = -\frac{7}{2}$.

Q4 Complete the table below, and hence sketch the curve represented by the parametric equations $x = 5t$, $y = \frac{2}{t}$, for $t \neq 0$.

t	–5	–4	–3	–2	–1	1	2	3	4	5
x										
y										

Q5 Hint: Make sure your calculator's set to radians.

Q5 Sketch the curve represented by the parametric equations $x = 1 + \sin\theta$, $y = 2 + \cos\theta$ for the values $0 \leq \theta \leq 2\pi$.
Use the table below to help you, and give your answers to 2 d.p.

θ	0	$\frac{\pi}{4}$	$\frac{\pi}{3}$	$\frac{\pi}{2}$	$\frac{2\pi}{3}$	$\frac{3\pi}{4}$	π	$\frac{4\pi}{3}$	$\frac{3\pi}{2}$	$\frac{5\pi}{3}$	2π
x											
y											

Q6 The orbit of a comet around the Sun is modelled by the parametric equations $x = 3\sin\theta$, $y = 7 + 9\cos\theta$, for $0 \leq \theta \leq 2\pi$, where the point (0, 0) represents the Sun, and where 1 unit on the x- and y-axes represents 1 astronomical unit (AU, 1 AU $\approx$ 150 million km). MODELLING

How far, in AU, is the comet from the Sun when:

a) $\theta = 0$, b) $\theta = \frac{\pi}{2}$?

Q6b) Hint: You'll need to use Pythagoras here.

Q7 The path of a toy plane thrown from a tower is modelled by the parametric equations $x = t^2 + 4t$, $y = 25 - t^2$ for $0 \leq t \leq 5$, where t is the time taken in seconds, and x and y are the horizontal and vertical distances in metres to the plane from the point at ground level at the foot of the tower. MODELLING

a) How far does the plane travel in the horizontal direction in the first 2 seconds?

b) Sonia is standing 21 m from the base of the tower, in line with the path of the plane. At what height above the ground does the toy plane pass over Sonia's head?

Finding intersections

A lot of parametric equations questions involve identifying points on the curve defined by the equations. You'll often be given the parametric equations of a curve, and asked to find the coordinates of the **points of intersection** of this curve with another line (such as the x- or y-axis).

Use the information in the question to **solve for t** at the intersection point(s). Then **substitute the value(s) of t** into the parametric equations to work out the **x and y values** (i.e. the **coordinates**) at the intersection point(s).

The example on the next page shows how to tackle a question like this.

Example

The curve shown has the parametric equations $y = t^3 - t$ and $x = 4t^2 - 1$.

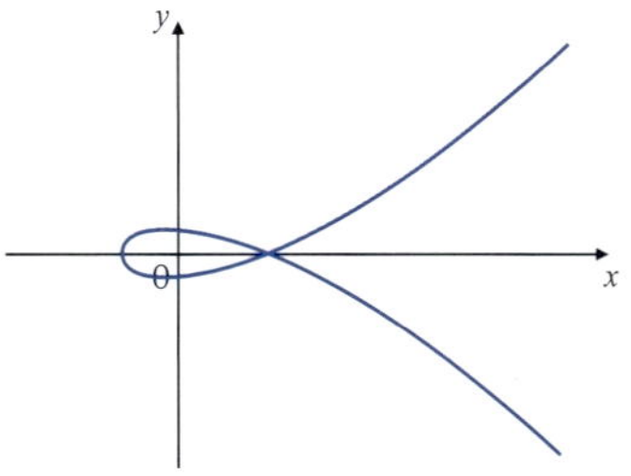

Find the coordinates of the points where the graph crosses:
a) the x-axis, b) the y-axis, and c) the line $8y = 3x + 3$.

a) On the x-axis, $y = 0$.

- Use the parametric equation for y to find the values of t where the graph crosses the x-axis. You need to factorise and solve the cubic:

  $0 = t^3 - t \Rightarrow t(t^2 - 1) = 0 \Rightarrow t(t + 1)(t - 1) = 0$
  $\Rightarrow t = 0, t = -1, t = 1$

- Now use those values of t to find the x-coordinates:

  When $t = 0$: $x = 4(0)^2 - 1 = -1$
  When $t = -1$: $x = 4(-1)^2 - 1 = 3$
  When $t = 1$: $x = 4(1)^2 - 1 = 3$

- So the graph crosses the x-axis at the points $(-1, 0)$ and $(3, 0)$.

Tip: $t = -1$ and $t = 1$ give the same coordinates — that's where the curve crosses over itself.

b) On the y-axis, $x = 0$.

So: $0 = 4t^2 - 1 \Rightarrow t^2 = \frac{1}{4} \Rightarrow t = \pm\frac{1}{2}$

When $t = \frac{1}{2}$: $y = \left(\frac{1}{2}\right)^3 - \frac{1}{2} = -\frac{3}{8}$

When $t = -\frac{1}{2}$ $y = \left(-\frac{1}{2}\right)^3 - \left(-\frac{1}{2}\right) = \frac{3}{8}$

So the graph crosses the y-axis at the points $\left(0, -\frac{3}{8}\right)$ and $\left(0, \frac{3}{8}\right)$.

Tip: The sketch shows there are two points where the graph crosses each axis.

c) Part c) is just a little trickier.

- First, **substitute** the **parametric equations** into $8y = 3x + 3$:

  $8y = 3x + 3 \Rightarrow 8(t^3 - t) = 3(4t^2 - 1) + 3$

- **Rearrange** and **factorise** to find the values of t you need:

  $8t^3 - 8t = 12t^2 \Rightarrow 8t^3 - 12t^2 - 8t = 0 \Rightarrow t(2t + 1)(t - 2) = 0$
  $\Rightarrow t = 0, t = -\frac{1}{2}, t = 2$

- Go back to the parametric equations to find the x- and y-coordinates:

  When $t = 0$: $x = -1, y = 0$

  When $t = -\frac{1}{2}$: $x = 4\left(\frac{1}{4}\right) - 1 = 0, y = \left(-\frac{1}{2}\right)^3 + \frac{1}{2} = \frac{3}{8}$

  When $t = 2$: $x = 4(4) - 1 = 15, y = 2^3 - 2 = 6$

- So the graph crosses the line $8y = 3x + 3$ at the following points:

  $(-1, 0), \left(0, \frac{3}{8}\right), (15, 6)$

Tip: You can check the answers by sticking these values back into $8y = 3x + 3$.

Exercise 1.2

Q1 Hint: A lies on the x-axis and B lies on the y-axis, so in each case you know one of the coordinates.

Q1 The curve with parametric equations $x = 3 + t$, $y = -2 + t$ meets the x-axis at the point A and the y-axis at the point B. Find the coordinates of A and B.

Q2 The curve C has parametric equations $x = 2t^2 - 50$, $y = 3t^3 - 24$.

a) Find the value of t where the curve meets the x-axis.

b) Find the values of t where the curve meets the y-axis.

Q3 The curve with parametric equations $x = 64 - t^3$, $y = \frac{1}{t}$, for $t \neq 0$, meets the y-axis at the point P. Find the coordinates of the point P.

Q4 Hint: Replace the x and y in the Cartesian equation with the parametric equations.

Q4 Find the coordinates of the point of intersection, P, of the line $y = x - 3$ and the curve with parametric equations $x = 2t + 1$, $y = 4t$.

Q5 Find the coordinates of the point(s) of intersection of the curve $y = x^2 + 32$ and the curve with parametric equations $x = 2t$, $y = 6t^2$.

Q6 Find the points of intersection of the circle $x^2 + y^2 = 32$ and the curve with parametric equations $x = t^2$, $y = 2t$.

Q7 The curve with parametric equations $x = a(t - 2)$, $y = 2at^2 + 3$ (where $a \neq 0$), meets the y-axis at the point (0, 4).

a) Find the value of the constant a.

b) Hence determine whether the curve meets the x-axis.

Q8 A curve has parametric equations $x = \frac{2}{t}$, $y = t^2 - 9$, for $t \neq 0$.

a) Find the point(s) at which the curve crosses the x-axis.

b) Does the curve meet the y-axis? Explain your answer.

c) Find the coordinates of the point(s) at which this curve meets the curve $y = \frac{10}{x} - 3$.

Q9 A curve has parametric equations $x = 3 \sin t$, $y = 5 \cos t$ and is defined for the domain $0 \leq t \leq 2\pi$.

a) Determine the coordinates at which this curve meets the x- and y-axes.

b) Find the points where the curve meets the line $y = \left(\frac{5\sqrt{3}}{9}\right)x$.

Q10 A simulation models the paths of two ships using parametric equations. The ships are modelled as points, with no width or length. The path taken by the first ship is given by $x_1 = 24 - t$, $y_1 = 10 + 3t$. The path taken by the second ship is given by $x_2 = t + 10$, $y_2 = 12 + 2t - 0.1t^2$. For both sets of equations, $0 \leq t \leq 30$, where t is the time in hours since the start of the simulation, and x and y are measured in miles East and North respectively. According to the simulation, will the two ships collide?

MODELLING

PROBLEM SOLVING

2. Parametric and Cartesian Equations

As well as finding Cartesian coordinates for a curve given in parametric form, you can also work out the Cartesian equation from the parametric equations.

Learning Objective:
- Be able to convert parametric equations into Cartesian form.

Converting parametric equations to Cartesian equations

Some parametric equations can be **converted** into **Cartesian equations**. There are two main ways to do this:

- **Rearrange** one of the equations to make the **parameter** the subject, then **substitute** the result into the **other** equation.

- If your equations involve **trig functions**, use **trig identities** to **eliminate** the parameter.

Tip: You'll have come across loads of trig identities in A-Level maths already. Look back at your notes for Year 1 and see pages 63, 68 and 75 of this book if you need to refresh your memory.

The examples below show how the first method works.

Example 1

Give the Cartesian equations, in the form $y = f(x)$, of the curves represented by the following pairs of parametric equations:

a) $y = t^3 - 1$ and $x = t + 1$,

b) $y = \frac{1}{3t}$ and $x = 2t - 3$, $t \neq 0$.

Tip: The equations in part a) are the same as in the example on p.88.

a) You want the answer in the form $y = f(x)$, so leave y alone for now, and **rearrange** the equation for x to **make t the subject**:

$$x = t + 1 \Rightarrow t = x - 1$$

Now you can **eliminate t** from the equation for y:

$$y = t^3 - 1$$

$$\Rightarrow y = (x - 1)^3 - 1 = (x - 1)(x^2 - 2x + 1) - 1$$
$$\Rightarrow y = x^3 - 2x^2 + x - x^2 + 2x - 1 - 1$$
$$\Rightarrow y = x^3 - 3x^2 + 3x - 2$$

So the Cartesian equation is $y = x^3 - 3x^2 + 3x - 2$.

Tip: Just replace every 't' in the equation for y with '$x - 1$'.

Tip: You could use the binomial theorem or Pascal's triangle to find the coefficients in the expansion of $(x - 1)^3$.

b) Use the same method as above:

$$x = 2t - 3 \Rightarrow t = \frac{x+3}{2}$$

$$\text{So } y = \frac{1}{3t} \Rightarrow y = \frac{1}{3\left(\frac{x+3}{2}\right)}$$

$$\Rightarrow y = \frac{1}{\left(\frac{3(x+3)}{2}\right)} \Rightarrow y = \frac{2}{3x+9}$$

Trigonometric functions

Things get a little trickier when the likes of sin and cos decide to put in an appearance. For trig functions you need to use **trig identities**.

Example 2

A curve has parametric equations $x = 1 + \sin\theta$, $y = 1 - \cos 2\theta$.

Give the Cartesian equation of the curve in the form $y = f(x)$.

- If you try to make θ the subject of these equations, things will just get messy. The trick is to find a way to get both x and y in terms of the **same trig function**. You can get $\sin\theta$ into the equation for y using the identity $\cos 2\theta \equiv 1 - 2\sin^2\theta$:

$$y = 1 - \cos 2\theta = 1 - (1 - 2\sin^2\theta) = 2\sin^2\theta$$

- **Rearranging** the equation for x gives: $\sin\theta = x - 1$
- **Replace** '$\sin\theta$' in the equation for y with '$x - 1$' to get y in terms of x:

$$y = 2\sin^2\theta \Rightarrow y = 2(x-1)^2 = 2x^2 - 4x + 2$$

- So the **Cartesian equation** is $y = 2x^2 - 4x + 2$.

Tip: If one of the parametric equations includes $\cos 2\theta$ or $\sin 2\theta$, that's probably the one you need to substitute — so make sure you know the **double angle formulas** from page 75.

Example 3

A curve is defined parametrically by $x = 4\sec\theta$, $y = 4\tan\theta$. Give the equation of the curve in the form $y^2 = f(x)$, and hence determine whether the curve intersects the line $y + 3x = 4$.

- Start by **rearranging** each equation to make the **trig function** the subject:

$$x = 4\sec\theta \Rightarrow \sec\theta = \frac{x}{4} \quad \text{and} \quad y = 4\tan\theta \Rightarrow \tan\theta = \frac{y}{4}$$

- Find an **identity** that contains both **sec θ** and **tan θ**:

$$\sec^2\theta \equiv 1 + \tan^2\theta$$

- **Substitute** the trig functions with the x and y terms:

$$\left(\frac{x}{4}\right)^2 = 1 + \left(\frac{y}{4}\right)^2$$

- **Rearrange** to get an equation in the right form:

$$x^2 = 16 + y^2 \Rightarrow y^2 = x^2 - 16$$

- To see whether the curve intersects $y + 3x = 4$, **rearrange** to $y = 4 - 3x$ and **replace** in the equation of the curve:

$$(4 - 3x)^2 = x^2 - 16$$
$$\Rightarrow 16 - 24x + 9x^2 = x^2 - 16$$
$$\Rightarrow 8x^2 - 24x + 32 = 0$$
$$\Rightarrow x^2 - 3x + 4 = 0$$

- There are **no real roots** to this equation ($b^2 - 4ac < 0$), so the line and the curve **do not intersect**.

Tip: Note that you're asked for the equation in the form $y^2 = f(x)$ not $y = f(x)$.

Tip: The last bit comes from the **quadratic formula**. For a quadratic $ax^2 + bx + c = 0$, there are only real solutions when $b^2 - 4ac \geq 0$, because of the $\sqrt{b^2 - 4ac}$ bit in the formula.

Exercise 2.1

Q1 For each of the following parametrically-defined curves, find the Cartesian equation of the curve in an appropriate form.

a) $x = t + 3,\ y = t^2$

b) $x = 3t,\ y = \frac{6}{t},\ t \neq 0$

c) $x = 2t^3,\ y = t^2$

d) $x = t + 7,\ y = 12 - 2t$

e) $x = t + 4,\ y = t^2 - 9$

f) $x = \sin\theta,\ y = \cos\theta$

g) $x = 1 + \sin\theta,\ y = 2 + \cos\theta$

h) $x = \sin\theta,\ y = \cos 2\theta$

i) $x = \cos\theta,\ y = \cos 2\theta$

j) $x = \cos\theta - 5,\ y = \cos 2\theta$

Q2 By eliminating the parameter θ, express the curve defined by the parametric equations $x = \tan\theta$, $y = \sec\theta$ in the form $y^2 = f(x)$.

Q3 Write the curve $x = 2\cot\theta$, $y = 3\operatorname{cosec}\theta$ in the form $y^2 = f(x)$.

Q4 A circle is defined by the parametric equations $x = 5 + \sin\theta$, $y = -3 + \cos\theta$.

a) Find the coordinates of the centre of the circle, and the radius of the circle.

b) Write the equation of the curve in Cartesian form.

Q5 A curve has parametric equations $x = \frac{1 + 2t}{t}$, $y = \frac{3 + t}{t^2}$, $t \neq 0$.

a) Express t in terms of x.

b) Hence show that the Cartesian equation of the curve is: $y = (3x - 5)(x - 2)$.

c) Sketch the curve.

Q6 Express $x = \frac{2 - 3t}{1 + t}$, $y = \frac{5 - t}{4t + 1}$ $(t \neq -1,\ t \neq -0.25)$, in Cartesian form.

Q7 Find the Cartesian equation of the curve defined by the parametric equations $x = 5\sin^2\theta$, $y = \cos\theta$.
Express your answer in the form $y^2 = f(x)$.

Q8 a) Express $x = a\sin\theta$, $y = b\cos\theta$ in Cartesian form.

b) Use your answer to a) to sketch the curve.

c) What type of curve has the form $x = a\sin\theta$, $y = b\cos\theta$?

Q8b) Hint: Find the x- and y-intercepts first — it will give you an idea of the graph's shape.

Q9 A curve has parametric equations $x = 3t^2$, $y = 2t - 1$.

a) Show that the Cartesian equation of the curve is $x = \frac{3}{4}(y + 1)^2$.

b) Hence find the point(s) of intersection of this curve with the line $y = 4x - 3$.

Q10 Find the Cartesian equation of the curve $x = 7t + 2$, $y = \frac{5}{t}$, $t \neq 0$, in the form $y = f(x)$ and hence sketch the curve, labelling any asymptotes and points of intersection with the axes clearly.

Q10 Hint: If you need a recap on transformations of curves, look back at pages 33-36. It's much easier to sketch if you can transform a standard curve shape.

Chapter 5 Sequences and Series

1. Sequences

Learning Objectives:

- Be able to use an nth term formula to generate terms in a sequence, or find the position of a term with a given value.
- Be able to show that a sequence is increasing, decreasing or neither.
- Be able to write recurrence relations and use them to generate sequences.

A sequence is a list of numbers that follow a certain pattern — e.g. 2, 4, 6, 8..., −5, 2, −5, 2, −5, ... or 1, 4, 9, 16... There are two main ways of describing sequences — from an nth term formula and from a recurrence relation.

n^{th} term

Before we get going with this section, there's some **notation** to learn:

> a_n just means the **n^{th} term** of the sequence — e.g. a_4 is the 4th term, and a_{n+1} is the term after a_n.

The idea behind the nth term is that you can use a formula to generate any term in a sequence from its **position**, n, in the sequence.

Example 1

A sequence has n^{th} term $a_n = 4n + 1$.

a) **Find the value of a_{10}.**

Just substitute 10 for n in the nth term expression: $4(10) + 1 = 41$

So a_{10} = 41.

b) **A term in the sequence is 33. Find the position of this term.**

- The position of the term is n where $4n + 1 = 33$.
- So $n = (33 - 1) \div 4 = 8$

  This means $a_8 = 33$

Tip: This sequence is an example of the 'common difference' type — the first few terms are 5, 9, 13, 17, ..., which have a common difference of 4. These are actually known as arithmetic sequences (see pages 101-102).

Example 2

A sequence has the nth term $an^2 + b$, where a and b are constants.

a) **If the 3rd term is 7 and the 5th term is 23, find the nth term formula.**

- Form equations using the information given in the question.

  For the 3rd term $n = 3$: $a(3^2) + b = 9a + b = 7$

  For the 5th term $n = 5$: $a(5^2) + b = 25a + b = 23$

- Solve the equations simultaneously to find the values of a and b.

  Subtract the equation on the left from the equation on the right:

$$\begin{array}{r} 25a + b = 23 \\ -\,(9a + b = \ \ 7) \\ \hline 16a \qquad = 16 \\ a = 1 \end{array}$$

  Then use $a = 1$ to find b using the left (or right) hand side equation.

  $9(1) + b = 7 \Rightarrow b = -2$

  So the nth term formula is $n^2 - 2$.

b) Is 35 a term in the sequence?

For questions like this, you need to form and solve an equation in n and see if you get a positive whole number:

$$n^2 - 2 = 35 \Rightarrow n^2 = 37 \Rightarrow n = \sqrt{37}$$

$\sqrt{37}$ is not an integer, so 35 is **not** in the sequence.

Tip: If your value for n had been a positive integer, then that number would have been a term in the sequence. E.g. 34 is in the sequence, because solving $n^2 - 2 = 34$ gives $n = 6$, so 34 is the 6th term in the sequence.

Increasing and decreasing sequences

There are a few types of sequences that you need to know:

- In an **increasing sequence**, each term is larger than the previous term, so $a_{k+1} > a_k$ for all terms — e.g. the square numbers 1, 4, 9, 16, 25, ...
- In a **decreasing sequence**, each term is smaller than the previous term, so $a_{k+1} < a_k$ for all terms — e.g. the sequence 16, 13, 10, 7, 4, ...

Tip: A **finite** sequence is one that has a 'last term' (see p.103), while an **infinite** sequence just keeps on going forever.

Example 3

A sequence has the nth term $7n - 16$. Show that the sequence is increasing.

- Use the nth term formula to find a_k and a_{k+1}

  $a_k = 7k - 16$

  $a_{k+1} = 7(k + 1) - 16 = 7k - 9$

- Show that $a_{k+1} > a_k$ for all k

  $7k - 9 > 7k - 16 \Rightarrow -9 > -16$

  This is true so it is an increasing sequence.

Tip: If a sequence is decreasing, you can show that $a_{k+1} < a_k$ is true for all values of k.

You may also come across **periodic sequences**, where the terms **repeat** in a cycle. The number of repeated terms is known as the **order**. For example, the sequence 1, 0, 1, 0, 1, 0... is periodic with order 2.

Watch out — some sequences **aren't** increasing, decreasing **or** periodic.

Tip: If you can show $a_{k+1} > a_k$ for some values of k and $a_{k+1} < a_k$ for some other values of k, then the sequence is neither increasing nor decreasing.

Exercise 1.1

Q1 A sequence has n^{th} term $a_n = 3n - 5$. Find the value of a_{20}.

Q2 Find the 4th term of the sequence with n^{th} term $n(n + 2)$.

Q3 Find the first 5 terms of the sequence with n^{th} term $(n - 1)(n + 1)$.

Q4 The kth term of a sequence is 29.
The nth term of this sequence is $4n - 3$. Find the value of k.

Q5 A sequence has the nth term $13 - 6n$.
Show that the sequence is decreasing.

Q6 A sequence has n^{th} term $= an^2 + b$, where a and b are constants.
If the 2nd term is 15, and the 5th term is 99, find a and b.

Q7 A sequence starts 9, 20, 37, ... Its nth term $= en^2 + fn + g$, where e, f and g are constants. Find the values of e, f and g. PROBLEM SOLVING

Q8 The nth term of the sequence is given by $(n - 1)^2$. A term in the sequence is 49. Find the position of this term.

Q9 How many terms of the sequence with n^{th} term $15 - 2n$ are positive?

Recurrence relations

A **recurrence relation** is another way to describe a sequence.

> A recurrence relation tells you how to work out a term in a sequence from the previous term — i.e. $\boldsymbol{a_{k+1} = f(a_k)}$ for some function f.

Tip: The notation using subscripts (see p.96) is used in recurrence relations, so you need to get used to it. Any letter can be used, e.g. $x_n = (n - 1)(n + 1)$. It just avoids having to repeat "with n^{th} term...".

For example, if each term in the sequence is **2 more** than the previous term:

$$a_{k+1} = a_k + 2$$

So, if $k = 5$, this says that $a_6 = a_5 + 2$, that is, the 6th term is equal to the 5th term + 2.

This recurrence relation will be true for loads of sequences, e.g. 1, 3, 5, 7..., and 4, 6, 8, 10... So to describe a **particular sequence** you also have to give one term. E.g. the sequence 1, 3, 5, 7... is described by:

$$a_{k+1} = a_k + 2, \ a_1 = 1$$

a_1 stands for the 1st term.

Example 1

Find the recurrence relation of the sequence 5, 8, 11, 14, 17, ...

- Each term in this sequence equals the one before it, plus 3.
- The recurrence relation is written like this:

$$a_{k+1} = a_k + 3$$

- **BUT**, as you saw above $a_{k+1} = a_k + 3$ on its own isn't enough to describe 5, 8, 11, 14, 17, ... For example, the sequence 87, 90, 93, 96, ... also has each term being 3 more than the one before.
- The description needs to be more specific, so you've got to give **one term** in the sequence, as well as the recurrence relation.
- Putting all of this together gives 5, 8, 11, 14, 17, ... as:

$$a_{k+1} = a_k + 3, \ a_1 = 5$$

Tip: You usually give the first value, a_1 — but using $a_2 = 8$ would also describe the sequence 5, 8, 11, 14, 17.

Example 2

A sequence is given by the recurrence relation $\boldsymbol{a_{k+1} = a_k - 4, \ a_1 = 20}$. Find the first five terms of this sequence.

- You're given the first term (a_1) — it's 20.
- Now you need to find the second term (a_2). If $a_1 = a_k$, then $a_2 = a_{k+1}$.

$a_{k+1} = a_k - 4$

$\Rightarrow a_2 = a_1 - 4 = 20 - 4 = 16$ ← This is 4 less than the first term.

- Repeat this to find the third, fourth and fifth terms.

$a_3 = a_2 - 4 = 16 - 4 = 12$

$a_4 = a_3 - 4 = 12 - 4 = 8$

$a_5 = a_4 - 4 = 8 - 4 = 4$

So the first five terms of the sequence are 20, 16, 12, 8, 4.

Tip: When you've got the hang of recurrence relations, you'll be able to see straight away that $a_{k+1} = a_k - 4$ just means the sequence decreases by 4 each time.

Example 3

A sequence is generated by the recurrence relation:

$$\boldsymbol{u_{n+1} = 3u_n + k, \; u_1 = 2.}$$

a) Find u_3 in terms of k.

You're given the first term, u_1, so use this to generate the second term, u_2, in terms of k, then use the second term to generate the third term, u_3.

$u_{n+1} = 3u_n + k$

$\Rightarrow u_2 = 3u_1 + k = 3(2) + k = u_2 = 6 + k$

Substitute in the expression for u_2 that you found above.

$\Rightarrow u_3 = 3u_2 + k = u_3 = 3(6 + k) + k = 18 + 4k$

Tip: This is the downside of recurrence relations — you have to go through each term to work out the one you're after. It'd take ages to find, say, the 100th term.

b) Given that $u_4 = 28$, find k.

- Form an expression for u_4 using the recurrence relation:

  $u_4 = 3u_3 + k = 3(18 + 4k) + k = 54 + 13k$

  Substitute in the expression for u_3 from part a).

- The question tells you that $u_4 = 28$, so form an equation and solve to find k.

  $28 = 54 + 13k \Rightarrow 13k = -26 \Rightarrow k = -2$

The next example's a bit harder, as there's an n^2 in there. But the method works in exactly the same way — you just end up with a slightly more complicated formula for the recurrence relation.

Example 4

A sequence has the general term $x_n = n^2$.
Write down a recurrence relation which generates the sequence.

- Start by finding the first few terms of the sequence (i.e. $n = 1, 2, 3, 4...$)

  $x_1 = 1^2 = \mathbf{1}$ $\quad x_2 = 2^2 = \mathbf{4}$ $\quad x_3 = 3^2 = \mathbf{9}$ $\quad x_4 = 4^2 = \mathbf{16}$

  So the first four terms are: 1, 4, 9, 16

- You're now at the same point as you were at the start of Example 1. You use the same method to form the recurrence relation, but it's a little trickier as n in the general term is squared.

Tip: It's sometimes helpful to draw some little diagrams where n^2 is involved. E.g. these diagrams let you see that it's $2k + 1$ added each time.

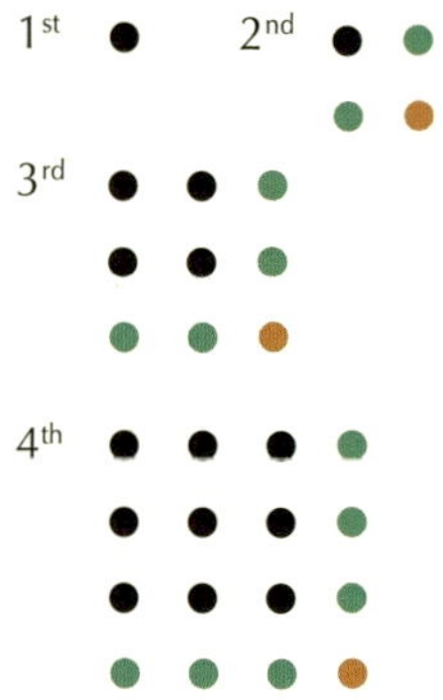

- Look at the difference between each term:

$k =$	1	2	3	4
$x_k =$	1	4	9	16
	+3	+5	+7	
	(2×1) + 1	(2×2) + 1	(2×3) + 1	

- To get from term x_k to the next term, x_{k+1}, you take the value of term x_k, and add $2k + 1$ to it. E.g. 3rd term = 2nd term + 2(2) + 1.
So the recurrence relation is: $\mathbf{x_{k+1} = x_k + 2k + 1}$

- To generate the sequence you need to give **one term** in the sequence, and there's no reason not to stick to convention and use the first term:

$\mathbf{x_{k+1} = x_k + 2k + 1, x_1 = 1}$ ← Try substituting in different k values to check your recurrence relation does generate the sequence.

Exercise 1.2

Q1 A sequence is defined for $n \geq 1$ by $u_{n+1} = 3u_n$ and $u_1 = 10$. Find the first 5 terms of the sequence.

Q2 Find the first 4 terms of the sequence in which $u_1 = 2$ and $u_{n+1} = u_n^2$ for $n \geq 1$.

Q4b) Hint: Most of the sequences you'll see are infinite — they go on forever. But you can have finite sequences which stop after a certain number of terms.

Q3 Write down a recurrence relation which produces the sequence 3, 6, 12, 24, 48, ...

Q4 a) Write down a recurrence relation which produces the sequence 12, 16, 20, 24, 28, ...
b) The sequence is finite and ends at 100. Find the number of terms.

Q5 Hint: You need to find an operation that goes back and forth between 4 and 7. There's more than one possible answer — think about one of these number facts:
7 + 4 = 11 or
7 × 4 = 28.

Q5 Find a recurrence relation which generates the sequence 7, 4, 7, 4, 7, ...

Q6 In a sequence $u_1 = 4$ and $u_{n+1} = 3u_n - 1$ for $n \geq 1$. Find the value of k if $u_k = 95$.

Q7 In a sequence $x_1 = 9$ and $x_{n+1} = (x_n + 1) \div 2$ for $n \geq 1$. Find the value of r if $x_r = \frac{5}{4}$.

Q8 Find the first 5 terms of the sequence in which $u_1 = 7$ and $u_{n+1} = u_n + n$ for $n \geq 1$.

Q9 Hint: Form all the equations you can from the information, and then solve them.

Q9 In a sequence $u_1 = 6$, $u_2 = 7$ and $u_3 = 8.5$. If the recurrence relation is of the form $u_{n+1} = au_n + b$, find the values of the constants a and b.

Q10 A sequence is generated by $u_1 = 8$ and $u_{n+1} = \frac{1}{2}u_n$ for $n \geq 1$. Find the first 5 terms and a formula for u_n in terms of n.

2. Arithmetic Sequences

Now it's time to look at a particular kind of sequence in greater detail. When the terms of a sequence progress by adding a fixed amount each time, this is called an arithmetic sequence.

Learning Objectives:

- Be able to find and use the formula for the nth term of an arithmetic sequence.

Finding the nth term

Here are some examples of arithmetic sequences:
5, 7, 9, 11... (add 2 each time); 20, 17, 14, 11... (add –3 each time).

The formula for the nth term of an arithmetic sequence is:

$$u_n = a + (n - 1)d$$

where:
a is the **first term** of the sequence.
d is the amount you add each time — the **common difference**.
n is the **position** of any term in the sequence.

Tip: Arithmetic sequences are often referred to as arithmetic progressions — it's exactly the same thing.

This box shows you how the formula is derived:

Term	n	
1st	1	$u_1 = a$
2nd	2	$u_2 = u_1 + d = a + d$
3rd	3	$u_3 = u_2 + d = (a + d) + d = a + 2d$
4th	4	$u_4 = u_3 + d = (a + 2d) + d = a + 3d$
⋮	⋮	⋮
n^{th}	n	$u_n = a + (n - 1)d$

This is the formula to find the nth term.

Tip: Each term is made up of the previous one, plus d. It's a recurrence relation.

Example 1

For the arithmetic sequence 2, 5, 8, 11, ... find u_{20} and the formula for u_n.

- The nth term of a sequence is $u_n = a + (n - 1)d$.
- For this sequence, $a = 2$ and $d = 3$.
- Plug the numbers into the nth term formula:

  $u_{20} = 2 + (20 - 1) \times 3 = 2 + 19 \times 3 = 59$

  So $u_{20} = 59$.
- u_n is the general term, i.e. $a + (n - 1)d$.
  Just substitute in the a and d values and simplify:

  $$u_n = 2 + (n - 1)3$$
  $$u_n = 3n - 1$$
- Finally, check the formula works with a couple of values of n:
  $n = 1$ gives $3(1) - 1 = 2$ ✓
  $n = 2$ gives $3(2) - 1 = 5$ ✓

Tip: You should always check your nth term formula is correct by sticking in some values for n and seeing if it produces the terms of the sequence.

You only actually need to know **two terms** of an arithmetic sequence (and their positions) — then you can work out any other term.

Example 2

The 2nd term of an arithmetic sequence is 21, and the 9th term is –7. Find the 23rd term of this sequence.

- Set up an equation for each of the known terms:

  2nd term = 21, so $a + (2 - 1)d = 21$
  $a + d = 21$

  9th term = –7, so $a + (9 - 1)d = -7$
  $a + 8d = -7$

- You've now got two **simultaneous equations** — so solve them to find a and d:

  $a + d = 21$ — ①

  $a + 8d = -7$ — ②

  ①–②: $-7d = 28 \Rightarrow d = -4$

  ①: $a + d = 21 \Rightarrow a - 4 = 21 \Rightarrow a = 25$

- Write the nth term formula... n^{th} term $= a + (n - 1)d$
  $= 25 + (n - 1) \times -4$
  $= 29 - 4n$

- ... and use it to find the 23rd term ($n = 23$): 23rd term $= 29 - 4 \times 23$
  $= -63$

Tip: The common difference and the first term have also been found along the way here (questions sometimes ask you to find these).

Exercise 2.1

Q1 An arithmetic progression has first term 7 and common difference 5. Find its 10th term.

Q2 Find the nth term for each of the following sequences:

a) 6, 9, 12, 15, ...

b) 4, 9, 14, 19, ...

c) 12, 8, 4, 0, ...

d) 1.5, 3.5, 5.5, 7.5 ...

Q3 In an arithmetic sequence, the fourth term is 19 and the tenth term is 43. Find the first term and common difference.

Q4 In an arithmetic progression, $u_7 = 8$ and $u_{11} = 10$. Find u_3.

Q5 In an arithmetic sequence, $u_3 = 15$ and $u_7 = 27$. Find the value of k if $u_k = 66$.

Q6 In an arithmetic sequence the first three terms are $\ln(x)$, $\ln(x + 8)$, $\ln(x + 48)$. Find the value of x and the next term in the sequence. PROBLEM SOLVING

3. Arithmetic Series

Series and sequences are very similar and quite easy to confuse. Remember that a sequence is a just list of terms that follow a pattern. You'll often want to add these terms together — when you do this, it becomes a series.

Learning Objectives:

- Be able to use the nth term formula to solve arithmetic series problems.
- Be able to find the sum of the first n terms of an arithmetic series.
- Use sigma notation (Σ) to refer to the sum of a series.
- Be able to find the sum of the first n natural numbers.

Sequences and series

Here is an arithmetic sequence. It's an infinite sequence — it goes on forever.

5, 8, 11, 14, 17, 20, ...

Now suppose you wanted to find the sum of the first 5 terms of this sequence. You'd write this by replacing the commas with '+' signs like this:

5 + 8 + 11 + 14 + 17

This is now an **arithmetic series**. It's a finite series with 5 terms. And if you actually added up the numbers you'd find that the **sum** for this series is 55.

So sequences become series when you add up their terms to find sums.

Sum of the first n terms

It would very quickly stop being fun if you had to find the sum of a 100-term series manually. Instead, you can use one of these **two formulas**.

S_n represents the **sum of the first n terms**

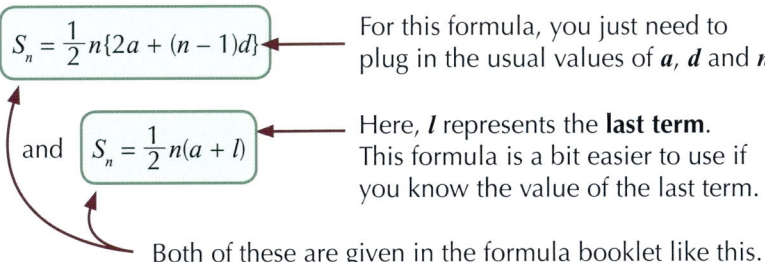

$$S_n = \frac{1}{2}n\{2a + (n-1)d\}$$

For this formula, you just need to plug in the usual values of a, d and n

and

$$S_n = \frac{1}{2}n(a + l)$$

Here, l represents the **last term**. This formula is a bit easier to use if you know the value of the last term.

Both of these are given in the formula booklet like this.

Tip: You can work out a, d and the nth term for a series, just as you would for a sequence. So in the 5-term series above, $a = 5$, $d = 3$ and nth term $= 3n + 2$ (for $1 \leq n \leq 5$). Also, because the series is finite, you can state its last term, which is 17.

There's a nice little proof for these formulas:

For any series, you can express S_n as:

$$S_n = a + (a + d) + (a + 2d) + \ldots + (a + (n-3)d) + (a + (n-2)d) + (a + (n-1)d)$$

Now, if you reverse the order of the terms you can write it as:

$$S_n = (a + (n-1)d) + (a + (n-2)d) + (a + (n-3)d) + \ldots + (a + 2d) + (a + d) + a$$

Adding the two expressions for S_n gives:

$$2S_n = (2a + (n-1)d) + (2a + (n-1)d) + (2a + (n-1)d) + \ldots + (2a + (n-1)d)$$

So we've now got the term "$(2a + (n-1)d)$" repeated n times, which is:

$$2S_n = n \times (2a + (n-1)d) \Rightarrow S_n = \frac{1}{2}n\{2a + (n-1)d\}$$

So we've derived the first formula.
Now to get the second, just replace "$a + (n-1)d$" with l:

$$S_n = \frac{1}{2}n\{a + a + (n-1)d\}, \text{ so } S_n = \frac{1}{2}n(a + l)$$

Now it's time to try out the sum formulas in some worked examples.

Example 1

Find the sum of the series with first term 3, last term 87 and common difference 4.

- You're told the last term, so use the S_n formula with l in: $S_n = \frac{1}{2}n(a + l)$
- You know a (3) and l (87), but you don't know n yet.
- Find n by putting what you do know into the 'n^{th} term' formula:

$$a + (n - 1)d = 87$$

$$3 + (n - 1)4 = 87$$
$$4n - 1 = 87$$
$$n = 22$$

- You're now all set to plug the values for a, l and n into the S_n formula:

$n = 22$ means that there are 22 terms in the series.

$$S_n = \frac{1}{2}n(a + l)$$

$$S_{22} = \frac{1}{2} \times 22 \times (3 + 87) = 11 \times 90$$
$$= 990$$

The sum of the series is **990**.

Example 2

Tip: This question could just have easily been "about the series $-5 + -2 + 1 + 4 + 7 + ...$" — the working would have been exactly the same. Don't worry too much about the distinction between sequences and series in this type of question.

This question is about the sequence –5, –2, 1, 4, 7...

a) Is 67 a term in the sequence? If it is, give its position.

First, find the formula for the nth term of the sequence:

$$n^{th} \text{ term} = a + (n - 1)d$$
$$= -5 + (n - 1)3 \qquad a = -5 \text{ and } d = 3$$
$$= 3n - 8$$

Put 67 into the formula and see if this gives a whole number for n:

$$67 = 3n - 8$$
$$3n = 75 \Rightarrow n = 25$$

67 **is** a term in the sequence. It's the **25th** term.

b) Find the sum of the first 20 terms.

We know $a = -5$, $d = 3$ and $n = 20$, so plug these values into the formula $S_n = \frac{n}{2}[2a + (n - 1)d]$:

$$S_{20} = \frac{20}{2}[2(-5) + (20 - 1)3]$$

$$S_{20} = 10[-10 + 19 \times 3]$$

$$S_{20} = 470$$

The sum of the first 20 terms is **470.**

Example 3

Find the possible numbers of terms in the arithmetic series starting 21 + 18 + 15... if the sum of the series is 75.

- You're told the first term, a, is 21 and the sum of the series, S_n, is 75. You can see that the common difference, d, is –3. What you need to work out is n, the number of terms in the series.
- You can use the S_n formula for this: $S_n = \frac{n}{2}[2a + (n-1)d]$

$$S_n = \frac{n}{2}[2(21) + (n-1) \times -3]$$

$$75 = \frac{n}{2}[42 - 3n + 3]$$

$$75 = \frac{45n}{2} - \frac{3n^2}{2}$$

$$-3n^2 + 45n - 150 = 0$$

Divide through by –3 to simplify the quadratic

$$n^2 - 15n + 50 = 0$$

Solve this quadratic equation

$$(n-5)(n-10) = 0$$

$$n = 5 \text{ or } n = 10$$

There are **5 or 10 terms** in the series.

Tip: There are two answers to this one — the series goes into negative numbers, so the sum of 75 is reached twice. Look at the first 10 terms of the series written out in full and you'll see what I mean: 21 + 18 + 15 +12 + 9 + 6 + 3 + 0 + (–3) + (–6).

Sigma notation

So far, the letter S has been used for the sum. The Greeks did a lot of work on this — their capital letter for S is Σ or **sigma**. This is used today, together with the general term, to mean the sum of the series.

For example, the following means the sum of the series with n^{th} term $2n + 3$.

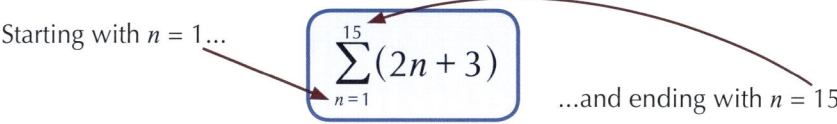

Starting with $n = 1$...

$$\sum_{n=1}^{15}(2n+3)$$

...and ending with $n = 15$

Example 4

Find $\sum_{n=1}^{15}(2n+3)$.

- The first term ($n = 1$) is 5, the second term ($n = 2$) is 7, the third is 9, ... and the last term ($n = 15$) is 33. So in other words, you need to find 5 + 7 + 9 + ... + 33. This gives $a = 5$, $d = 2$, $n = 15$ and $l = 33$.
- You know all of a, d, n and l, so you can use either formula:

$$S_n = \frac{1}{2}n(a+l)$$

$$S_{15} = \frac{1}{2} \times 15(5+33)$$

$$S_{15} = \frac{1}{2} \times 15 \times 38$$

$$S_{15} = 285$$

It makes no difference which formula you use.

$$S_n = \frac{n}{2}[2a + (n-1)d]$$

$$S_{15} = \frac{15}{2}[2 \times 5 + 14 \times 2]$$

$$S_{15} = \frac{15}{2}[10 + 28]$$

$$S_{15} = 285$$

Exercise 3.1

Q1 An arithmetic series has first term 8 and common difference 3. Find the 10th term and the sum of the first 10 terms.

Q2 In an arithmetic series $u_2 = 16$ and $u_5 = 10$. Find a, d and S_8.

Q3 In an arithmetic series $a = 12$ and $d = 6$. Find u_{100} and S_{100}.

Q4 Find $\sum_{n=1}^{12}(5n-2)$.

Q5 Find $\sum_{n=1}^{9}(20-2n)$.

Q6 In an arithmetic series $a = 3$ and $d = 2$. Find n if $S_n = 960$.

Q7 Given that $\sum_{n=1}^{k}(5n+2) = 553$, show that the value of k is 14. PROBLEM SOLVING

Q8 An arithmetic sequence begins $x + 11$, $4x + 4$, $9x + 5$, ... Find the sum of the first 11 terms.

Sum of the first *n* natural numbers

The **natural numbers** are the positive whole numbers, i.e. 1, 2, 3, 4...

They form a simple arithmetic progression with $a = 1$ and $d = 1$.

The sum of the first n natural numbers is: $S_n = \frac{1}{2}n(n+1)$

This formula can be derived from the previous sum formulas by plugging in values:

The sum of the first n natural numbers looks like this:

$$S_n = 1 + 2 + 3 + \ldots + (n-2) + (n-1) + n$$

So $a = 1$, $l = n$ and also $n = n$.

$$S_n = \frac{1}{2}n(a+l) \longrightarrow S_n = \frac{1}{2}n(n+1)$$

You can also derive the formula from first principles — the proof is almost identical to the one for a general arithmetic series on page 103.

- $S_n = 1 + 2 + 3 + \ldots + (n-2) + (n-1) + n$ ①
- Rewrite ① with the terms reversed: $S_n = n + (n-1) + (n-2) + \ldots + 3 + 2 + 1$ ②
- ① + ② gives:

  $2S_n = (n+1) + (n+1) + (n+1) + \ldots + (n+1) + (n+1) + (n+1)$

  $\Rightarrow 2S_n = n(n+1) \Rightarrow \mathbf{S_n = \frac{1}{2}n(n+1)}$

Example 1

Find the sum of the first 100 natural numbers.

$S_n = \frac{1}{2}n(n+1)$

$S_{100} = \frac{1}{2} \times 100 \times 101 = \mathbf{5050}$

Sum of the first 100 natural numbers = 5050

Example 2

The sum of the first k natural numbers is 861. Find the value of k.

- Form an equation in k:

$$\frac{1}{2}k(k+1) = 861$$

- Expand the brackets and rearrange:

$$k^2 + k = 1722$$
$$k^2 + k - 1722 = 0$$

- So we have a quadratic in k to solve.
  We're looking for a whole number for k, so it should factorise.

$$k^2 + k - 1722 = 0$$
$$(k \quad\;)(k \quad\;) = 0$$

- It looks tricky to factorise, but notice that '$b = 1$', so we're looking for two numbers that are 1 apart and multiply to 1722.

$$(k + 42)(k - 41) = 0$$
$$k = -42 \text{ or } k = 41$$

- We can ignore the negative solution here, so the answer is $k = 41$.

Tip: In this question we have k numbers added together, so an answer of $k = -42$ wouldn't make any sense.

Exercise 3.2

Q1 Find the sum of the first:
a) 10, b) 2000 natural numbers.

Q2 Find $\sum_{n=1}^{32} n$.

Q3 Find $\sum_{n=1}^{10} n$ and $\sum_{n=1}^{20} n$. Hence find $\sum_{n=11}^{20} n$.

Q3 Hint: This question uses a handy little trick for finding series sums that don't start from $n = 1$.

Q4 The sum of the first n natural numbers is 66. Find n.

Q5 Find k if $\sum_{n=1}^{k} n = 120$.

Q6 Find the sum of the series 16 + 17 + 18 + ... + 35.

Q7 What is the first natural number k for which $\sum_{n=1}^{k} n$ is greater than 1 000 000?

4. Geometric Sequences and Series

Learning Objectives:

- Be able to recognise geometric sequences and series.
- Know and be able to use the formula for the general term of a geometric sequence or series.
- Know and be able to use the formula for the sum of the first n terms of a geometric sequence or series.
- Be able to recognise convergent geometric series and find their sum to infinity.

On pages 101-107 you saw how to find the general term and how to find the sum of a number of terms for an arithmetic sequence. Now you'll do the same thing for geometric sequences and series.

Geometric sequences

Remember, with **arithmetic** sequences you get from one term to the next by **adding** a fixed amount each time.

They have a **first term** (a), and the amount you add to get from one term to the next is called the **common difference** (d).

With **geometric sequences**, rather than adding, you get from one term to the next by **multiplying** by a **constant** called the **common ratio** (r).

Tip: Geometric sequences are also called geometric progressions.

- This is a **geometric sequence** where you find each term by **multiplying** the previous term by 2:

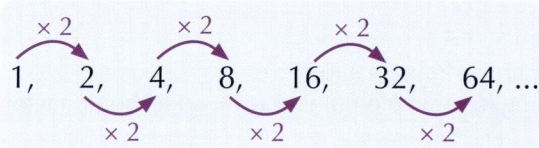

$$1, \quad 2, \quad 4, \quad 8, \quad 16, \quad 32, \quad 64, \ldots$$

(× 2 between each pair of consecutive terms)

- If the common ratio is **negative**, the signs of the sequence will **alternate**. For this geometric sequence the common ratio is –3.

$$2, -6, 18, -54, 162, -486, \ldots$$

- The common ratio might **not** be a **whole number**. Here, it's $\frac{3}{4}$.

$$16, 12, 9, \frac{27}{4}, \frac{81}{16}, \frac{243}{64}, \ldots$$

You get each term by multiplying the first term by the common ratio some number of times. In other words, each term is the **first term** multiplied by **some power** of the **common ratio**.

This is how you describe geometric sequences using **algebra**:

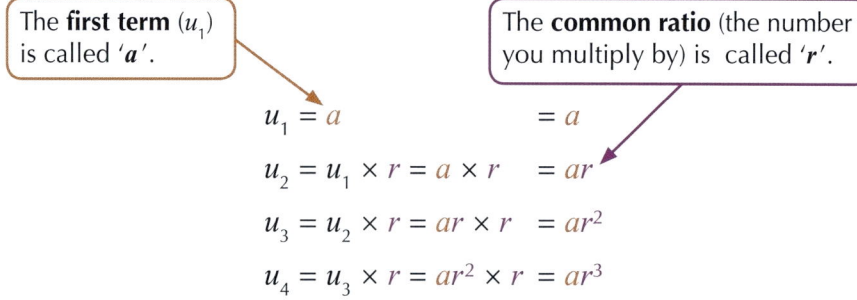

The **first term** (u_1) is called ***'a'***.

The **common ratio** (the number you multiply by) is called ***'r'***.

$$u_1 = a \qquad = a$$

$$u_2 = u_1 \times r = a \times r \qquad = ar$$

$$u_3 = u_2 \times r = ar \times r \qquad = ar^2$$

$$u_4 = u_3 \times r = ar^2 \times r \qquad = ar^3$$

Tip: In the first sequence at the top of the page (1, 2, 4, 8, ...) $a = 1$ and $r = 2$.

So the **formula** that describes **any term** in a geometric sequence is:

$$\boldsymbol{u_n = ar^{n-1}}$$

Tip: The term u_n is called the 'n^{th} term', or 'general term' of the sequence. You need to learn this formula.

If you know the values of a and r, you can substitute them into the **general formula** to find an expression that describes the whole sequence:

Example 1

A chessboard has a 1p piece on the first square, 2p on the second square, 4p on the third, 8p on the fourth and so on until the board is full. Find a formula for the amount of money on each square.

- This is a **geometric sequence**: $u_1 = 1$, $u_2 = 2$, $u_3 = 4$, $u_4 = 8$, ...
- You get each term in the sequence by **multiplying** the previous one by **2**.
- So $a = 1$ (because you start with 1p on the first square) and $r = 2$.
- Then the **formula** for the amount of money (in pence) on each square is:

$$u_n = ar^{n-1} = 1 \times (2^{n-1}) = 2^{n-1}$$

$$u_n = 2^{n-1}$$

Tip: The trick with questions like this is to recognise that you're being asked about a geometric sequence — there's more about this on page 117.

You can also use the formula to find the **first term** a, the **common ratio** r or a **particular term** in the sequence, given other information about the sequence:

Example 2

a) Find the 5th term in the geometric sequence 1, 3, 9, ...

- First find the common ratio r. Each term is the previous term multiplied by r, so you find the common ratio by dividing consecutive terms:

$$\text{second term} = \text{first term} \times r \Rightarrow r = \frac{\text{second term}}{\text{first term}} = \frac{3}{1} = 3$$

- Then find the 5th term. The 3rd term is 9, so:

$$4^{\text{th}} \text{ term} = 3^{\text{rd}} \text{ term} \times r = 9 \times 3 = 27$$

$$5^{\text{th}} \text{ term} = 4^{\text{th}} \text{ term} \times r = 27 \times 3 = 81$$

Tip: You can choose any two consecutive terms — e.g. dividing the third term by the second will also give you r.

Tip: Working up from the last term you know is a good method for finding a couple more terms in the sequence, but when you're asked for a higher term you're better off using the method in example b).

b) A geometric sequence has first term 2 and common ratio 1.2. Find the 15th term in the sequence to 3 significant figures.

$a = 2$ and $r = 1.2$, so the nth term is given by $u_n = ar^{n-1} = 2 \times (1.2)^{n-1}$

Then the 15th term is $u_{15} = 2 \times (1.2)^{14} = 2 \times 12.839... = 25.7$ to 3 s.f.

c) A geometric sequence has first term 25 and 10th term 80. Calculate the common ratio. Give your answer to 3 significant figures.

$a = 25$, so the nth term is given by $u_n = ar^{n-1} = 25r^{n-1}$.

The 10th term is $80 = u_{10} = 25r^9$

$$80 = 25r^9 \Rightarrow r^9 = \frac{80}{25} \Rightarrow r = \sqrt[9]{\frac{80}{25}} = 1.137... = 1.14 \text{ to 3 s.f.}$$

Tip: Make sure you know how to find the nth root on your calculator — and remember, $\sqrt[n]{x}$ is the same as $x^{\frac{1}{n}}$.

d) The 8th term of a geometric sequence is 4374 and the common ratio is 3. What is the first term?

The common ratio $r = 3$, so the nth term is $u_n = ar^{n-1} = a(3)^{n-1}$.

Then the 8th term is $4374 = a(3)^7 = 2187a \Rightarrow a = \frac{4374}{2187} = 2$

Exercise 4.1

Q1 Find the seventh term in the geometric progression 2, 3, 4.5, 6.75 ...

Q2 The sixth and seventh terms of a geometric sequence are 2187 and 6561 respectively.
What is the first term?

Q3 A geometric sequence is 24, 12, 6, ... What is the 9th term?

Q4 The 14th term of a geometric progression is 9216.
The first term is 1.125.
Calculate the common ratio.

Q5 Hint: Write yourself an equation or inequality and then solve it using logs. Look back at your Year 1 notes if you need a reminder of how to use logs.

Q5 The first and second terms of a geometric progression are 1 and 1.1 respectively.
How many terms in this sequence are less than 4?

Q6 A geometric progression has a common ratio of 0.6. If the first term is 5, what is the difference between the 10th term and the 15th term? Give your answer to 5 d.p.

Q7 A geometric sequence has a first term of 25 000 and a common ratio of 0.8.
Which term is the first to be below 1000?

Q8 A geometric sequence is 5, –5, 5, –5, 5, ...
Give the common ratio.

Q9 The first three terms of a geometric progression are $\frac{1}{4}$, $\frac{3}{16}$ and $\frac{9}{64}$.

a) Calculate the common ratio.

b) Find the 8th term. Give your answer as a fraction.

Q10 The 7th term of a geometric sequence is 196.608 and the common ratio is 0.8. What is the first term?

Q11 3, –2.4, 1.92,... is a geometric progression.

a) What is the common ratio?

b) How many terms are there in the sequence before you reach a term with modulus less than 1? PROBLEM SOLVING

Geometric series

A **sequence** becomes a **series** when you **add** the terms to find the **sum**. Geometric series work just like geometric sequences (they have a **first term** and a **common ratio**), but they're written as a **sum of terms** rather than a list:

geometric sequence:	geometric series:
3, 6, 12, 24, 48, ...	3 + 6 + 12 + 24 + 48 + ...

Sometimes you'll need to find the **sum** of the **first few terms** of a geometric series:

- The sum of the **first n terms** is called S_n.
- S_n can be written in terms of the first term a and the common ratio r:
  $$S_n = u_1 + u_2 + u_3 + u_4 + \dots u_n = a + ar + ar^2 + ar^3 + \dots + ar^{n-1}$$
- There's a nice **formula** for finding S_n that doesn't involve loads of adding. Here's the **proof** of the formula:

For any geometric sequence:

$$S_n = a + ar + ar^2 + ar^3 + \dots ar^{n-2} + ar^{n-1} \quad ①$$

Multiplying this by r gives:

$$rS_n = ar + ar^2 + ar^3 + \dots ar^{n-2} + ar^{n-1} + ar^n \quad ②$$

Subtract equation ② from equation ①: $S_n - rS_n = a - ar^n$

Factorise both sides: $(1 - r)S_n = a(1 - r^n)$

Then divide through by $(1 - r)$: $S_n = \dfrac{a(1-r^n)}{1-r}$

Tip: Geometric series can be infinite (i.e. they can go on forever). We're just adding up bits of them for now, but summing an infinite series is covered on page 115.

Tip: You could also subtract equation ① from equation ② to get:

$$S_n = \frac{a(r^n-1)}{r-1}$$

Both versions are correct.

So the sum of the first n terms of a geometric series is:

$$S_n = \frac{a(1-r^n)}{1-r}$$

This formula is given in the formula booklet.

Example 1

a) A geometric series has first term 3.5 and common ratio 5. Find the sum of the first 6 terms.

You're told that $a = 3.5$ and $r = 5$, and you're looking for the sum of the first 6 terms, so just stick these values into the formula for S_6:

$$S_6 = \frac{a(1-r^6)}{1-r} = \frac{3.5(1-5^6)}{1-5} = 13\,671$$

b) The first two terms in a geometric series are 20, 22. To 2 decimal places, the sum of the first k terms of the series is 271.59. Find k.

$a = 20$, $r = \dfrac{\text{second term}}{\text{first term}} = \dfrac{22}{20} = 1.1$, so put these into the sum formula:

$$S_k = \frac{a(1-r^k)}{1-r} = \frac{20(1-(1.1)^k)}{1-1.1} = -200(1-(1.1)^k)$$

Tip: You're looking for a number of terms so the answer must be a positive integer.

So $271.59 = -200(1 - (1.1)^k) \Rightarrow -\frac{271.59}{200} - 1 = -(1.1)^k$

$\Rightarrow -2.35795 = -(1.1)^k$

$\Rightarrow 2.35795 = 1.1^k$

$\Rightarrow \log(2.35795) = k\log(1.1)$

$\Rightarrow k = \frac{\log(2.35795)}{\log(1.1)} = 9$

Sigma notation

You saw on page 105 that the **sum** of the first ***n*** **terms** of a series (S_n) can also be written using **sigma** (Σ) **notation**.
For geometric series, sigma notation looks like this:

$$S_n = u_1 + u_2 + u_3 + \ldots + u_n = a + ar + ar^2 + \ldots + ar^{n-1} = \sum_{k=0}^{n-1} ar^k$$

Tip: Remember, Σ just means sum (it's the Greek letter for S). In this case, it's the sum of ar^k from $k = 0$ to $k = n - 1$. Be careful with the limits — it's $n - 1$ on top of the Σ, but the sum is S_n.

So, using the formula from the previous page, the sum of the first n terms can be written:

$$\sum_{k=0}^{n-1} ar^k = \frac{a(1-r^n)}{1-r}$$

Example 2

$a + ar + ar^2 + \ldots$ is a geometric series, and $\sum_{k=0}^{4} ar^k = 85.2672$.

Given that $r = -1.8$, find the first term a.

You've got the sum of the first 5 terms:

$$85.2672 = \sum_{k=0}^{4} ar^k = S_5 = \frac{a(1-r^5)}{1-r}$$

So plug the value of r into the formula:

$$85.2672 = \frac{a(1-r^5)}{1-r} = \frac{a(1-(-1.8)^5)}{1-(-1.8)} = a\frac{19.89568}{2.8} = 7.1056a$$

$$\Rightarrow a = \frac{85.2672}{7.1056} = 12$$

Exercise 4.2

Q1 The first term of a geometric sequence is 8 and the common ratio is 1.2. Find the sum of the first 15 terms.

Q2 A geometric series has first term $a = 25$ and common ratio $r = 0.7$. Find $\sum_{k=0}^{9} 25(0.7)^k$.

Q3 The sum of the first n terms of a geometric series is 196 605. The common ratio of the series is 2 and the first term is 3. Find n.

Q4 A geometric progression starts with 4, 5, 6.25. The first x terms add up to 103.2 to 4 significant figures. Find x.

Q5 The 3rd term of a geometric series is 6 and the 8th term is 192. Find:

a) the common ratio b) the first term

c) the sum of the first 15 terms

Q6 $m + 10$, m, $2m - 21$, ... is a geometric progression, m is a positive constant.

a) Show that $m^2 - m - 210 = 0$.

b) Hence show that $m = 15$.

c) Find the common ratio of this series.

d) Find the sum of the first 10 terms.

Q6a) Hint: The ratio of the first term to the second is the same as the ratio of the second term to the third.

Q7 The first three terms of a geometric series are $1, x, x^2$. The sum of these terms is 3 and each term has a different value.

a) Find x. b) Calculate the sum of the first 7 terms.

Q8 $a, ar, ar^2, ar^3, \ldots$ is a geometric progression. Given that $a = 7.2$ and $r = 0.38$, find $\sum_{k=0}^{9} ar^k$.

Q9 The sum of the first eight terms of a geometric series is 1.2. Find the first term of the series, given that the common ratio is $-\frac{1}{3}$.

Q10 $a, -2a, 4a, -8a, \ldots$ is a geometric sequence. Given that $\sum_{k=0}^{12} a(-2)^k = -5735.1$, find a.

Convergent geometric series

Convergent sequences

Some geometric sequences **tend towards zero** — in other words, they get closer and closer to zero (but they never actually reach it). For example:

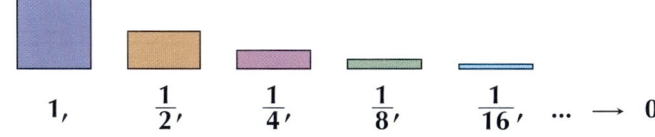

$$1, \frac{1}{2}, \frac{1}{4}, \frac{1}{8}, \frac{1}{16}, \ldots \longrightarrow 0$$

Tip: The arrow here means 'tends to'.

- Sequences like this are called **convergent** — the terms **converge** (get closer and closer) to a **limit** (the number they get close to).
- Geometric sequences either **converge to zero** or **don't converge** at all.
- A sequence that doesn't converge is called **divergent**.
- A geometric sequence $a, ar, ar^2, ar^3, \ldots$ will converge to **zero** if each term is **closer** to zero than the one before.
- This happens when $-1 < r < 1$. You can write this as $|r| < 1$, where $|r|$ is the modulus of r (see p.27), so:

$$a, ar, ar^2, ar^3, \ldots \rightarrow 0 \text{ when } |r| < 1$$

Tip: You ignore the sign of r because you can still have a convergent sequence when r is negative. In that case the terms will alternate between > 0 and < 0, but they'll still be getting closer and closer to zero.

Convergent series

- When you **sum** a sequence that tends to zero you get a **convergent series**.
- Because each term is getting closer and closer to zero, you're **adding smaller and smaller** amounts each time. So the sum gets **closer and closer** to a certain number, but never reaches it — this is the **limit** of the series.

For example, the **sum** of the **convergent sequence** $1, \frac{1}{2}, \frac{1}{4}, \frac{1}{8}, \frac{1}{16}, \ldots$ gets closer and closer to 2:

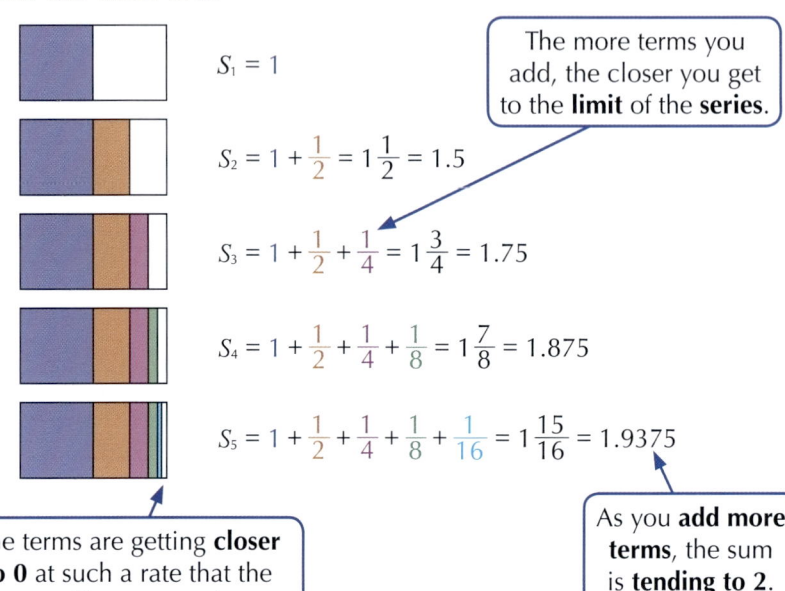

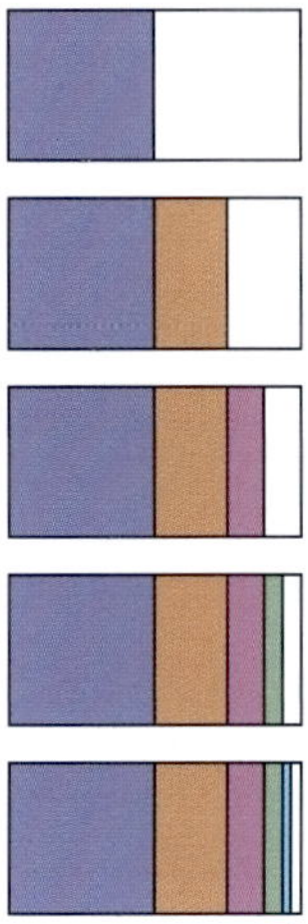

$S_1 = 1$

$S_2 = 1 + \frac{1}{2} = 1\frac{1}{2} = 1.5$

$S_3 = 1 + \frac{1}{2} + \frac{1}{4} = 1\frac{3}{4} = 1.75$

$S_4 = 1 + \frac{1}{2} + \frac{1}{4} + \frac{1}{8} = 1\frac{7}{8} = 1.875$

$S_5 = 1 + \frac{1}{2} + \frac{1}{4} + \frac{1}{8} + \frac{1}{16} = 1\frac{15}{16} = 1.9375$

The more terms you add, the closer you get to the **limit** of the **series**.

The terms are getting **closer to 0** at such a rate that the sum will never reach 2.

As you **add more terms**, the sum is **tending to 2**.

$$1 + \frac{1}{2} + \frac{1}{4} + \frac{1}{8} + \frac{1}{16} + \ldots = 2$$

Tip: This is the sum of the sequence from the previous page.

Tip: Not all sequences that tend to zero produce a convergent series — this rule is only true for geometric progressions.
For example, the series $1 + \frac{1}{2} + \frac{1}{3} + \frac{1}{4} + \frac{1}{5} + \ldots$ diverges.

- So when the **sequence** $a, ar, ar^2, ar^3, \ldots$ **converges** to zero, the **series** $a + ar + ar^2 + ar^3 + \ldots$ **converges** to a **limit**.
- Like sequences, series converge when $|r| < 1$.

Geometric series
$a + ar + ar^2 + ar^3 + \ldots$
with $|r| < 1$ are **convergent**

Geometric series
$a + ar + ar^2 + ar^3 + \ldots$
with $|r| \geq 1$ are **divergent**

Example 1

Determine whether or not the following sequences are convergent.

a) 1, 2, 4, 8, 16, ...

$r = \frac{2^{\text{nd}}\text{ term}}{1^{\text{st}}\text{ term}} = \frac{2}{1} = 2$, $|r| = |2| = 2 > 1$, so the series is not convergent

b) 81, –27, 9, –3, 1, ...

$r = \frac{2^{\text{nd}}\text{ term}}{1^{\text{st}}\text{ term}} = \frac{-27}{81} = -\frac{1}{3}$, $|r| = \left|-\frac{1}{3}\right| = \frac{1}{3} < 1$, the series is convergent

Tip: You can usually spot straight away if a geometric series is convergent — the terms will be getting closer and closer to zero. But you still need to check $|r| < 1$ to prove it.

Summing to infinity

When a series is **convergent** you can find its **sum to infinity**.

- The sum to infinity is called S_∞ — it's the **limit** of S_n as $n \to \infty$.
- This just means that the sum of the first n terms of the series (S_n) gets closer and closer to S_∞ the more terms you add (the bigger n gets).
- In other words, it's the **number** that the **series converges to**.

Example 2

If $a = 2$ and $r = \frac{1}{2}$, find the sum to infinity of the geometric series.

$u_1 = 2 \longrightarrow S_1 = 2$

$u_2 = 2 \times \frac{1}{2} = 1 \longrightarrow S_2 = 2 + 1 = 3$

$u_3 = 1 \times \frac{1}{2} = \frac{1}{2} \longrightarrow S_3 = 2 + 1 + \frac{1}{2} = 3\frac{1}{2}$

$u_4 = \frac{1}{2} \times \frac{1}{2} = \frac{1}{4} \longrightarrow S_4 = 2 + 1 + \frac{1}{2} + \frac{1}{4} = 3\frac{3}{4}$

$u_5 = \frac{1}{4} \times \frac{1}{2} = \frac{1}{8} \longrightarrow S_5 = 2 + 1 + \frac{1}{2} + \frac{1}{4} + \frac{1}{8} = 3\frac{7}{8}$

These values are getting closer (**converging**) to 4. So the sum to infinity is 4.

These values are getting **smaller** each time.

You can show this **graphically**. The line on the graph is getting **closer and closer** to 4, but it'll never actually get there.

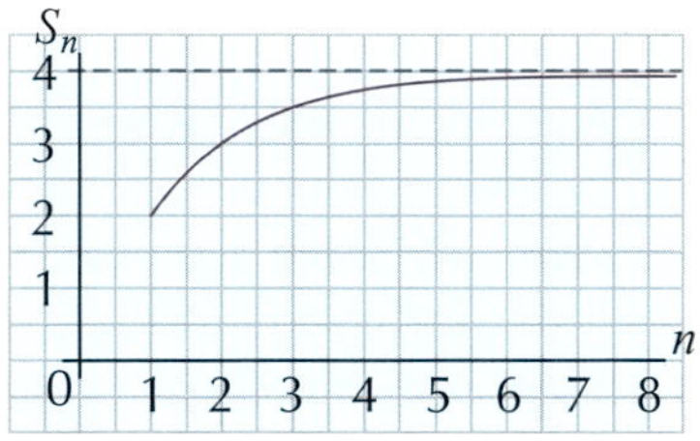

Tip: Luckily you don't have to find a list of sums like this to get the sum to infinity — there's a nifty formula for working it out further down the page. This example is just to show you what's going on when a series converges.

There's a **formula** you use to work out the **sum to infinity** of a geometric series.

- The sum of the **first n terms** of a geometric series is $S_n = \frac{a(1-r^n)}{1-r}$
- If $|r| < 1$ and n is very, very big, then r^n will be very, very **small**, i.e. $r^n \to 0$ as $n \to \infty$.
- This means $(1 - r^n)$ will get really **close to 1**, so $(1 - r^n) \to 1$ as $n \to \infty$.
- Putting this back into the sum formula gives $S_n \to \frac{a \times 1}{1-r} = \frac{a}{1-r}$ as $n \to \infty$.
- So: $S_\infty = \frac{a}{1-r}$

Tip: Remember, $|r| < 1$ just means $-1 < r < 1$.

Tip: The sum of the first n terms formula is on p.111. Both of these formulas are given in the formula booklet.

Example 3

a) If $a = 2$ and $r = \frac{1}{2}$, find the sum to infinity of the geometric series.

$|r| = \left|\frac{1}{2}\right| = \frac{1}{2} < 1$, so the series converges and you can find its sum to infinity using the formula for S_∞:

$$S_\infty = \frac{a}{1-r} = \frac{2}{1-\frac{1}{2}} = \frac{2}{\frac{1}{2}} = 4$$

Tip: This is the same as the example above, but this time the sum to infinity is worked out using the formula.

b) Find the sum to infinity of the geometric series 8 + 2 + 0.5 + 0.125 + ...

First find a and r as before: $a = 8$, $r = \frac{2^{nd}\text{ term}}{1^{st}\text{ term}} = \frac{2}{8} = 0.25$

Again, $|r| < 1$, so the series converges.

Now find the sum to infinity: $S_\infty = \frac{a}{1-r} = \frac{8}{1-0.25} = \frac{32}{3} = 10\frac{2}{3}$

Divergent series **don't** have a **sum to infinity**.

Because the terms aren't tending to zero, the size of the sum will just keep increasing as you add more terms, so there is **no limit** to the sum.

Exercise 4.3

Q1 Hint: If $r = 1$, the sequence is just the same term repeated, so it diverges. If $r = -1$, the sequence alternates between two terms forever.

Q1 State which of these sequences will converge and which will not.

a) 1, 1.1, 1.21, 1.331, ... b) $0.8, 0.8^2, 0.8^3, \ldots$

c) $1, \frac{1}{4}, \frac{1}{16}, \frac{1}{64}, \ldots$ d) $3, \frac{9}{2}, \frac{27}{4}, \ldots$

e) $1, -\frac{1}{2}, \frac{1}{4}, -\frac{1}{8}, \frac{1}{16}, \ldots$ f) 5, 5, 5, 5, 5, ...

Q2 A geometric series is 9 + 8.1 + 7.29 +... Calculate the sum to infinity.

Q3 $a, ar, ar^2, \ldots$ is a geometric sequence. Given that $S_\infty = 2a$, find r.

Q4 The sum to infinity of a geometric progression is 13.5 and the first 3 terms add up to 13.

a) Find the common ratio r

b) Find the first term a.

Q5 $a + ar + ar^2 + \ldots$ is a geometric series. $ra = 3$ and $S_\infty = 12$. Find r and a.

Q6 The sum to infinity of a geometric series is 10 and the first term is 6.

a) Find the common ratio.

b) What is the 5th term?

Q7 The 2nd term of a geometric progression is –48 and the 5th term is 0.75. Find:

a) the common ratio

b) the first term

c) the sum to infinity

Q8 The sum of the terms after the 10th term of a convergent geometric series is less than 1% of the sum to infinity. The first term is positive. Show that the common ratio $|r| < 0.631$. PROBLEM SOLVING

Q9 In a convergent geometric series $S_\infty = \frac{9}{8} \times S_4$. PROBLEM SOLVING
Find the value of r, given that r is positive and real.

5. Modelling Problems

Sometimes a sequence or series is disguised by a 'real life' situation. You'll have to spot whether the sequence or series is arithmetic or geometric and turn the wordy question into the right maths.

MODELLING

Learning Objectives:
- Be able to answer modelling problems involving arithmetic and geometric sequences and series.

Real life problems

Modelling problems might not say the words sequence, series, arithmetic or geometric in the question — you have to decide what you're dealing with.

Look for questions that have a time period (e.g. each year) and describe how values increase or decrease over that time period.

Example 1

Mo is training for a 10 km running race. On the first day he runs 2 km. He schedules his training to increase by 0.5 km each day, so that he runs 2.5 km on the second day and 3 km on the third day and so on. This continues until he reaches the maximum distance of 10 km on the seventeenth day. The distance he runs each day then remains at 10 km until the race. There are 20 days before the race.

What is the total distance Mo will run in training?

- The question describes the sequence 2, 2.5, 3, ... , 10 which is arithmetic because each term increases by 0.5.
- You are asked for the total distance, so use the series formula. $S_n = \frac{n}{2}(2a + (d - 1)d)$
- The series contains 17 terms and you know that $a = 2$ and $d = 0.5$ so $S_{17} = \frac{17}{2}(2 \times 2 + (17 - 1) \times 0.5) = \frac{17}{2}(4 + 8) = 102$ km
- There are also 3 extra days where he will run 10 km: $3 \times 10 = 30$ km. So the total distance Mo will run is $102 + 30 =$ 132 km

Tip: You also know $l = 10$ for the series so you could use the formula $S_n = \frac{1}{2}n(a + l)$ to get the same answer.

Example 2

When a baby is born, £3000 is invested in an account with a fixed interest rate of 4% per year.

a) What will the account be worth at the start of the seventh year?

- Start by working out the first few terms.

  The first term is: $u_1 = a = 3000$

  The second term is: $u_2 = 3000 + (4\% \text{ of } 3000)$ ← This is the interest.
  $= 3000 + (0.04 \times 3000)$
  $= 3000(1 + 0.04)$
  $= 3000 \times 1.04$

  The third term is: $u_3 = u_2 \times 1.04 = (3000 \times 1.04) \times 1.04$
  $= 3000 \times (1.04)^2$

  And so on — this is a geometric sequence with $r = 1.04$
- So the nth term of the sequence is: $u_n = ar^{n-1} = 3000 \times (1.04)^{n-1}$

Tip: You might recognise that the sequence is geometric with $r = 1.04$ straight away, if you know that a 4% interest rate means you multiply by 1.04.

- The value of the account at the start of the first year is the 1st term, so the value of the account at the start of the seventh year is the 7th term:

  $u_7 = ar^6 = 3000 \times (1.04)^6 = 3795.957...$

  So it's £3795.96 (to the nearest penny)

b) After how many full years will the account have doubled in value?

- You need to know when $u_n > 3000 \times 2 = 6000$
- From part **a)** you know $u_n = 3000 \times (1.04)^{n-1}$

  So $3000 \times (1.04)^{n-1} > 6000 \Rightarrow (1.04)^{n-1} > 2$
- To complete this you need to use logs:

$$\Rightarrow \log(1.04)^{n-1} > \log 2$$
$$\Rightarrow (n-1)\log(1.04) > \log 2$$
$$\Rightarrow n - 1 > \frac{\log 2}{\log 1.04}$$
$$\Rightarrow n - 1 > 17.67$$
$$\Rightarrow n > 18.67 \quad \text{(to 2 d.p.)}$$

This is true because 1.04 > 1 and the log of numbers greater than 1 is positive.

- So u_n is more than 6000 when n is more than 18.67. Then u_{19} (the amount at the start of the 19th year) will be more than double the original amount. After 18 years, the account will have doubled in value.

Tip: It's OK to take logs of both sides of an inequality because logs are increasing functions, so if $x > y$, $\log x > \log y$.

Tip: If you have to give an answer in years, make sure you think through which term belongs to which year carefully — it's easy to give the wrong answer even if you've done all the maths right.

Exercise 5.1

MODELLING

Q1 Morag starts a new job. In the first week she is paid £60, but this rises by £3 per week, so she earns £63 in the second week and £66 in the third week. How much does she earn in her 12th week?

Q2 A collector has 8 china dolls that fit inside each other. The smallest doll is 3 cm high and each doll is 25% taller than the previous one. If he lines them up in height order (shortest to tallest), how tall is the 8th doll?

Q3 Mario opens a sandwich shop. On the first day he sells 40 sandwiches. As people hear about the shop, sales increase and on the second day he sells 45 sandwiches. Daily sales rise in an arithmetic sequence. On which day will he sell 80 sandwiches?

Q4 A retro cassette player is launched into the market. In the first month after launch, the product takes £300 000 of revenue. It takes £270 000 in the second month and £240 000 in the third. If this pattern continues, when would you expect monthly sales to fall below £50 000?

Q5 Hint: 'Depreciates' just means 'decreases in value'.

Q5 A car depreciates by 15% each year. The value of the car after each year forms a geometric sequence. After 10 years from new the car is valued at £2362. How much was the car when new?

Q6 A fishing licence cost £120 in 2011.
The cost rose 3% each year for the next 5 years.

a) How much was a fishing licence in 2012?

b) Nigel bought a fishing licence every year between 2011 and 2016 (including 2011 and 2016). How much in total did he spend?

Q7 "Cornflake Collector" magazine sells 6000 copies in its first month of publication, 8000 in its second month and 10 000 in its third month. If this pattern continues, how many copies will it sell in the first year of publication?

Q8 Ron is growing his prize leeks for the Village Show. The bag of compost he's using on his leeks says it will increase their height by 15% every 2 days. After 4 weeks the leeks' height has increased from 5 cm to 25 cm. Has the compost done what it claimed?

Q9 It's predicted that garden gnome value will go up by 2% each year, forming a geometric progression.
Jean-Claude has a garden gnome currently valued at £80 000.
If the predicted rate of inflation is correct:

a) What will Jean-Claude's gnome be worth after 1 year?

b) What is the common ratio of the geometric progression?

c) What will Jean-Claude's gnome be worth after 10 years?

d) It will take k years for the value of Jean-Claude's gnome to exceed £120 000. Find k.

Q10 Frazer draws one dot in the first square of his calendar for July, two dots in the second square, and so on up to 31 dots in the last day of the month. How many dots does he draw in total?

Q11 The thickness of a piece of paper is 0.01 cm. The Moon is 384 000 km from the Earth. The piece of paper is on the Earth. Assuming you can fold the piece of paper as many times as you like, how many times would you have to fold it for it to reach the Moon?

Q11 Hint: Have a go — it really works!*

*Disclaimer: it gets a bit tricky after 7 folds...

Q12 An athlete is preparing for an important event and sets herself a target of running 3% more each day. On day 1 she runs 12 miles.

a) How far does she run on day 10?

b) Her training schedule lasts for 20 days.
To the nearest mile, how far does she run altogether?

Q13 Laura puts 1p in her jar on the first day, 2p in on the second day, 3p in on the third day, etc. How many days will it take her to collect over £10?

Q14 Chardonnay wants to invest her savings for the next 10 years.
She wants her investment to double during this time.
If interest is added annually, what interest rate does she need?

Q14 Hint: Call the initial value of her savings a.

Chapter 6 The Binomial Expansion

1. The Binomial Expansion

Learning Objectives:

- Be able to expand $(p + qx)^n$ for any rational n using the general formula for binomial expansions.
- Be able to state the validity of the expansion — i.e. the values of x for which the expansion is valid.

You might well recognise binomial expansions from Year 1. This chapter should refresh your memory on the formula and take it a step further — including how to raise an expression to a fractional or negative power.

Expansions where n is a positive integer

The **binomial expansion** is a way to raise a given expression to any power.

For simpler cases it's basically a fancy way of multiplying out brackets. You can also use it to approximate more complicated expressions.

In Year 1, you met the formula for the binomial expansion of $(1 + x)^n$:

$$(1+x)^n = 1 + nx + \frac{n(n-1)}{1\times 2}x^2 + \ldots + \frac{n(n-1)\ldots(n-r+1)}{1\times 2\times\ldots\times r}x^r + \ldots$$

From the formula, it looks like the expansion always goes on forever. But if ***n* is a positive integer**, the binomial expansion is **finite**.

Example 1

Give the binomial expansion of $(1 + x)^5$.

Use the formula and plug in $n = 5$:

($n = 5$, $n(n-1)$)

$$(1+x)^5 = 1 + 5x + \frac{5(5-1)}{1\times 2}x^2 + \frac{5(5-1)(5-2)}{1\times 2\times 3}x^3 + \frac{5(5-1)(5-2)(5-3)}{1\times 2\times 3\times 4}x^4 + \frac{5(5-1)(5-2)(5-3)(5-4)}{1\times 2\times 3\times 4\times 5}x^5 + \frac{5(5-1)(5-2)(5-3)(5-4)(5-5)}{1\times 2\times 3\times 4\times 5\times 6}x^6 + \ldots$$

$$= 1 + 5x + \frac{5\times 4}{1\times 2}x^2 + \frac{5\times 4\times 3}{1\times 2\times 3}x^3 + \frac{5\times 4\times 3\times 2}{1\times 2\times 3\times 4}x^4 + \frac{5\times 4\times 3\times 2\times 1}{1\times 2\times 3\times 4\times 5}x^5 + \frac{5\times 4\times 3\times 2\times 1\times 0}{1\times 2\times 3\times 4\times 5\times 6}x^6 + \ldots$$

You can stop here — all the terms after this one are **zero**.

$$= 1 + 5x + \frac{20}{2}x^2 + \frac{60}{6}x^3 + \frac{120}{24}x^4 + \frac{120}{120}x^5 + \frac{0}{720}x^6 + \ldots$$

$$= 1 + 5x + 10x^2 + 10x^3 + 5x^4 + x^5$$

Tip: It may seem a bit dull to write all the steps out in this way for simple expansions, but it's important for more complex ones to make sure you don't miss anything out.

Tip: The expansion is finite because at some point you introduce an $(n - n)$ (i.e. zero) term in the numerator which then appears in every coefficient from that point on, making them all zero.

The formula still works if the coefficient of x **isn't 1**, i.e. $(1 + ax)^n$ — just **replace** each 'x' in the formula with **(ax)**. The **'a'** should be raised to the **same power** as the 'x' in each term, and included in the coefficient when you simplify at the end. This next example shows you how it's done.

Example 2

Give the binomial expansion of $(1 - 3x)^4$.

Use the **formula** with $n = 4$, but replace every x with $-3x$. Think of this as $(1 + (-3x))^4$ — put the **minus** into the formula as well as the $3x$.

$n = 4$ → $n(n-1)$

$$(1-3x)^4 = 1 + 4(-3x) + \frac{4 \times 3}{1 \times 2}(-3x)^2 + \frac{4 \times \cancel{3} \times \cancel{2}}{1 \times \cancel{2} \times \cancel{3}}(-3x)^3 + \frac{\cancel{4} \times \cancel{3} \times \cancel{2} \times \cancel{1}}{\cancel{1} \times \cancel{2} \times \cancel{3} \times \cancel{4}}(-3x)^4 + \frac{\cancel{4} \times \cancel{3} \times \cancel{2} \times \cancel{1} \times 0}{\cancel{1} \times \cancel{2} \times \cancel{3} \times \cancel{4} \times 5}(-3x)^5 + \dots$$

Don't forget to square the -3 as well.

Stop here.

$$= 1 + 4(-3x) + \frac{12}{2}(9x^2) + \frac{4}{1}(-27x^3) + (81x^4) + \frac{0}{5}(-243x^5) + \dots$$

$$= 1 - 12x + 54x^2 - 108x^3 + 81x^4$$

Tip: Make life easier for yourself by cancelling down the fractions before you multiply.

Validity

Some binomial expansions are **only valid for certain values of x**. When you find a binomial expansion, you usually have to state which values of x the expansion is valid for.

> If n is a **positive integer**, the binomial expansion of $(p + qx)^n$ is valid for **all values of x**.

There's more on the validity of other expansions on page 123.

Tip: So far you've only dealt with expansions of $(p + qx)^n$ where p is 1, but there are more complicated examples to come on page 126.

Exercise 1.1

Use the binomial expansion formula to expand each of the following functions in ascending powers of x.

Q1 Expand fully: $(1 + x)^3$

Q2 Expand $(1 + x)^7$ up to and including the term in x^3.

Q3 Expand fully: $(1 - x)^4$

Q3 Hint: Watch out for that minus sign — replace 'x' in the formula with $(-x)$.

Q4 Give the first 3 terms of $(1 + 3x)^6$.

Q5 Give the first 4 terms of $(1 + 2x)^8$.

Q6 Expand $(1 - 5x)^5$ up to and including the term in x^2.

Q7 Expand fully: $(1 - 4x)^3$

Q8 Expand $(1 + 6x)^6$ up to and including the term in x^3.

Expansions where n is negative or a fraction

n is negative

If n is **negative**, the expansion gets more complicated. You can still use the **formula** in the same way, but it will produce an **infinite** number of terms (see the example below). You can just write down the **first few terms** in the series, but this will only be an **approximation** to the whole expansion.

Tip: The question will usually tell you how many terms to give.

This type of expansion can be 'hidden' as a fraction — remember:

$$\frac{1}{(1+x)^n} = (1+x)^{-n}$$

Example 1

Find the binomial expansion of $\frac{1}{(1+x)^2}$ up to and including the term in x^3.

- First, **rewrite** the expression: $\frac{1}{(1+x)^2} = (1+x)^{-2}$.
- Now you can use the formula for $(1+x)^n$. This time $n = -2$:

$n = -2$ $n(n-1)$

$$(1+x)^{-2} = 1 + (-2)x + \frac{(-2)\times(-2-1)}{1\times 2}x^2 + \frac{(-2)\times(-2-1)\times(-2-2)}{1\times 2\times 3}x^3 + \dots$$

$$= 1 + (-2)x + \frac{(-\cancel{2})\times(-3)}{1\times\cancel{2}}x^2 + \frac{(-\cancel{2})\times(-\cancel{3})\times(-4)}{1\times\cancel{2}\times\cancel{3}}x^3 + \dots$$

Tip: Again, you can cancel down before you multiply — but be careful with those minus signs.

$$= 1 + (-2)x + \frac{3}{1}x^2 + \frac{(-4)}{1}x^3 + \dots$$

$$= 1 - 2x + 3x^2 - 4x^3 + \dots$$

With a negative n, you'll never get zero as a coefficient. If the question hadn't told you to stop, the expansion could go on forever.

- We've left out all the terms after $-4x^3$, so the cubic expression you've ended up with is an **approximation** to the original expression.
- You could also write the answer like this:

$$\frac{1}{(1+x)^2} \approx 1 - 2x + 3x^2 - 4x^3$$

n is a fraction

The binomial expansion formula doesn't just work for integer values of n. If n is a **fraction**, you'll have to take care multiplying out the fractions in the coefficients, but otherwise the formula is **exactly the same**.

Remember that **roots** are fractional powers: $\sqrt[n]{1+x} = (1+x)^{\frac{1}{n}}$

Example 2

Find the binomial expansion of $\sqrt[3]{1+2x}$, up to and including the term in x^3.

- First rewrite the expression as a fractional power: $\sqrt[3]{1+2x} = (1+2x)^{\frac{1}{3}}$
- This time $n = \frac{1}{3}$, and you also need to replace x with $2x$:

$n = \frac{1}{3}$ $n(n-1)$

$$(1+2x)^{\frac{1}{3}} = 1 + \frac{1}{3}(2x) + \frac{\frac{1}{3}\times\left(\frac{1}{3}-1\right)}{1\times 2}(2x)^2 + \frac{\frac{1}{3}\times\left(\frac{1}{3}-1\right)\times\left(\frac{1}{3}-2\right)}{1\times 2\times 3}(2x)^3 + \ldots$$

$$= 1 + \frac{2}{3}x + \frac{\frac{1}{3}\times\left(-\frac{2}{3}\right)}{1\times 2}(4x^2) + \frac{\frac{1}{3}\times\left(-\frac{2}{3}\right)\times\left(-\frac{5}{3}\right)}{1\times 2\times 3}(8x^3) + \ldots$$

$$= 1 + \frac{2}{3}x + \frac{\left(-\frac{2}{9}\right)}{2}(4x^2) + \frac{\left(\frac{10}{27}\right)}{6}(8x^3) + \ldots$$

$$= 1 + \frac{2}{3}x + \left(-\frac{1}{9}\right)(4x^2) + \left(\frac{5}{81}\right)(8x^3) + \ldots$$

$$= 1 + \frac{2}{3}x - \frac{4}{9}x^2 + \frac{40}{81}x^3 - \ldots$$

Tip: Cancelling down is much trickier with this type of expansion — it's often safer to multiply everything out fully.

Tip: Exam questions often ask for the coefficients as simplified fractions.

Validity when n is negative or a fraction

Binomial expansions where n is negative or a fraction are **not valid** for **all** values of x. The **rule** to work out the validity of an expansion is as follows:

> If n is a negative integer or a fraction,
> the binomial expansion of $(p+qx)^n$ is **valid** when $\left|\frac{qx}{p}\right| < 1$,
> i.e. when $|x| < \left|\frac{p}{q}\right|$.

This just means that the **absolute value** (or **modulus**) of x (i.e. ignoring any negative signs) must be **smaller** than the absolute value of $\frac{p}{q}$ for the expansion to be valid.

Tip: You might already know the rules $|ab| = |a||b|$ and $\left|\frac{a}{b}\right| = \frac{|a|}{|b|}$.

If you don't, then get to know them — they're handy for rearranging these limits.

Example 3

State the validity of the expansions given in the previous two examples.

a) For the expansion $(1+x)^{-2} = 1 - 2x + 3x^2 - 4x^3 + \ldots$, $p = 1$ and $q = 1$.

So the expansion is valid if $\left|\frac{1x}{1}\right| < 1$, i.e. if $|x| < 1$.

b) For $(1+2x)^{\frac{1}{3}} = 1 + \frac{2}{3}x - \frac{4}{9}x^2 + \frac{40}{81}x^3 - \ldots$, $p = 1$ and $q = 2$.

So the expansion is valid if $|2x| < 1 \Rightarrow 2|x| < 1 \Rightarrow |x| < \frac{1}{2}$.

Combinations of expansions

- You can use the binomial expansion formula for more complicated combinations of expansions — e.g. where different brackets raised to different powers are **multiplied together**.
- Start by dealing with the different expansions **separately**, then **multiply** the expressions together at the end.
- For situations where one bracket is being **divided** by another, change the **sign** of the **power** and **multiply** instead (e.g. $\frac{(1+x)}{(1+2x)^3} = (1+x)(1+2x)^{-3}$).
- For the **combined** expansion to be **valid**, x must be in the valid range for **both** expansions, i.e. where they overlap. In practice this just means sticking to the **narrowest** of the valid ranges for each separate expansion.

Tip: You'll usually be asked to give the expansion to a specified number of terms, so the multiplication shouldn't get too complicated as you can ignore terms in higher powers of x.

Example 4

Write down the first three terms in the expansion of $\frac{(1+2x)^3}{(1-x)^2}$.
State the range of x for which the expansion is valid.

- First re-write the expression as a product of two expansions:

$$\frac{(1+2x)^3}{(1-x)^2} = (1+2x)^3(1-x)^{-2}$$

- Expand each of these separately using the formula:

$$(1+2x)^3 = 1 + 3(2x) + \frac{3\times2}{1\times2}(2x)^2 + \frac{3\times2\times1}{1\times2\times3}(2x)^3$$
$$= 1 + 6x + 3(4x^2) + 8x^3 = 1 + 6x + 12x^2 + 8x^3$$

$$(1-x)^{-2} = 1 + (-2)(-x) + \frac{(-2)\times(-3)}{1\times2}(-x)^2 + \frac{(-2)\times(-3)\times(-4)}{1\times2\times3}(-x)^3 + \ldots$$
$$= 1 + 2x + 3x^2 + 4x^3 + \ldots$$

Tip: You actually only need to go up to the term in x^2 in each expansion (as you're only asked for the first three terms), but make sure you don't cut your expansion too short — it's better to expand too many terms than too few.

- **Multiply** the two expansions together.
  Since you're only asked for the **first three terms**, ignore any terms with **higher powers** of x than x^2.

$$(1+2x)^3(1-x)^{-2} = (1 + 6x + 12x^2 + 8x^3)(1 + 2x + 3x^2 + 4x^3 + \ldots)$$
$$= 1(1 + 2x + 3x^2) + 6x(1 + 2x) + 12x^2(1) + \ldots$$
$$= 1 + 2x + 3x^2 + 6x + 12x^2 + 12x^2 + \ldots$$
$$= 1 + 8x + 27x^2 + \ldots$$

Tip: Notice that you can cut down on the amount of working at this stage by leaving out of the brackets any terms that, when multiplied out, would give you an x^3 term or higher.

- Now find the **validity** of each expansion:

$(1+2x)^3$ is valid for all values of x, since n is a positive integer.

$(1-x)^{-2}$ is valid if $|-x| < 1 \Rightarrow |x| < 1$.

So the combined expression $\frac{(1+2x)^3}{(1-x)^2}$ is only valid if $|x| < 1$, since this is the narrower range of the two separate expansions.

Exercise 1.2

Q1 Use the binomial formula to find the first four terms in the expansion of $(1 + x)^{-4}$.

Q2 a) Find the binomial expansion of $(1 - 6x)^{-3}$, up to and including the term in x^3.

b) For what values of x is this expansion valid?

Q3 Use the binomial formula to find the first three terms in the expansion of:

a) $(1 + 4x)^{\frac{1}{3}}$

b) $(1 + 4x)^{-\frac{1}{2}}$

c) For each of the expansions above, state the range of x for which the expansion is valid.

Q4 Find the binomial expansion of the following functions, up to and including the term in x^3.

a) $\dfrac{1}{(1-4x)^2}$ for $|x| < \dfrac{1}{4}$

b) $\sqrt{1+6x}$ for $|x| < \dfrac{1}{6}$

c) $\dfrac{1}{\sqrt{1-3x}}$ for $|x| < \dfrac{1}{3}$

d) $\sqrt[3]{1+\dfrac{x}{2}}$ for $|x| < 2$

Q5 a) Find the coefficient of the x^3 term in the expansion of $\dfrac{1}{(1+7x)^4}$.

b) For what range of x is this expansion valid?

Q5-6 Hint: You don't need to bother doing the full expansion — just work out the coefficient of the term you need.

Q6 a) What is the coefficient of x^5 in the expansion of $\sqrt[4]{1-4x}$?

b) For what values of x is the binomial expansion of $\sqrt[4]{1-4x}$ valid?

Q7 a) Find the first three terms of the expansion of $(1-5x)^{\frac{1}{6}}$.

b) Hence find the binomial expansion of $(1 + 4x)^4(1-5x)^{\frac{1}{6}}$, up to and including the term in x^2.

c) State the validity of the expansion in b).

Q8 a) Find the first three terms of the binomial expansion of $\dfrac{(1+3x)^4}{(1+x)^3}$.

b) State the range of x for which the expansion is valid.

Q9 a) Write out the expansion of:

(i) $(1 + ax)^4$ (all terms) (ii) $(1 - bx)^{-3}$, up to the term in x^2

b) Hence expand $\dfrac{(1+ax)^4}{(1-bx)^3}$, up to and including the term in x^2.

c) Find the two possible pairs of values of a and b if the first three terms of the expansion of $\dfrac{(1+ax)^4}{(1-bx)^3}$ are $1 + x + 24x^2$.

PROBLEM SOLVING

Expanding $(p + qx)^n$

- You've seen over the last few pages that the binomial expansion of $(1 + x)^n$ works for **any** n, and that you can replace the x with other x-terms.
- However, the 1 at the start **has to be a 1** before you can expand.
- If it's **not a 1**, you'll need to **factorise first** before you can use the formula.
- This means you have to rewrite the expression as follows:

$$(p + qx)^n = p^n\left(1 + \frac{qx}{p}\right)^n$$

Tip: This rearrangement uses the power law $(ab)^k = a^k b^k$ — note that the p outside the brackets is still raised to the power n.

Example 1

Give the binomial expansion of $(3 - x)^4$ and state its validity.

- To use the $(1 + x)^n$ formula, you need the constant term in the brackets to be 1. You can take the 3 outside the brackets by factorising:

$$3 - x = 3\left(1 - \frac{1}{3}x\right)$$

$$\Rightarrow (3 - x)^4 = \left[3\left(1 - \frac{1}{3}x\right)\right]^4$$

$$= 3^4\left(1 - \frac{1}{3}x\right)^4$$

$$= 81\left(1 - \frac{1}{3}x\right)^4$$

The aim here is to get an expression in the form $c(1 + dx)^n$, where c and d are constants.

Tip: Don't forget to raise the 3 to the power 4.

- Now use the $(1 + x)^n$ formula, with $n = 4$, and $-\frac{1}{3}x$ instead of x:

$$\left(1 - \frac{1}{3}x\right)^4 = 1 + 4\left(-\frac{1}{3}x\right) + \frac{4 \times 3}{1 \times 2}\left(-\frac{1}{3}x\right)^2 + \frac{4 \times 3 \times 2}{1 \times 2 \times 3}\left(-\frac{1}{3}x\right)^3 + \frac{4 \times 3 \times 2 \times 1}{1 \times 2 \times 3 \times 4}\left(-\frac{1}{3}x\right)^4$$

$$= 1 - \frac{4}{3}x + 6\left(\frac{1}{9}x^2\right) + 4\left(-\frac{1}{27}x^3\right) + \frac{1}{81}x^4$$

$$= 1 - \frac{4x}{3} + \frac{2x^2}{3} - \frac{4x^3}{27} + \frac{x^4}{81}$$

- Finally, put this back into the original expression:

$$(3 - x)^4 = 81\left(1 - \frac{1}{3}x\right)^4$$

$$= 81\left(1 - \frac{4x}{3} + \frac{2x^2}{3} - \frac{4x^3}{27} + \frac{x^4}{81}\right)$$

$$= 81 - 108x + 54x^2 - 12x^3 + x^4$$

- The expression is valid for all values of x.

  This is because n is a positive integer — so the expansion is finite.

Tip: For this example, you could just pop $a = 3$ and $b = -x$ into the formula for $(a + b)^n$, which will be given to you in the exam. This formula only works when n is a natural number (written $n \in \mathbb{N}$ — these are just positive integers), so you'll need to use the method shown here if the power is negative or a fraction.

Example 2

Give the first 3 terms in the binomial expansion of $(3x + 4)^{\frac{3}{2}}$.

- Again, you need to **factorise** before using the formula, taking care to choose the **right factor**...

$$3x + 4 = 4 + 3x = 4\left(1 + \frac{3}{4}x\right)$$

Make sure you've got the bracket written in the form $(p + qx)^n$ before you factorise.

$$\Rightarrow (3x+4)^{\frac{3}{2}} = \left[4\left(1+\frac{3}{4}x\right)\right]^{\frac{3}{2}}$$
$$= 4^{\frac{3}{2}}\left(1+\frac{3}{4}x\right)^{\frac{3}{2}}$$
$$= 8\left(1+\frac{3}{4}x\right)^{\frac{3}{2}}$$

Tip: It's tempting to take the 3 outside the brackets because of the order the numbers are written in, but don't be fooled.

- Use the $(1 + x)^n$ formula with $n = \frac{3}{2}$, and $\frac{3}{4}x$ instead of x:

$$\left(1+\frac{3}{4}x\right)^{\frac{3}{2}} = 1 + \frac{3}{2}\left(\frac{3}{4}x\right) + \frac{\frac{3}{2}\times\left(\frac{3}{2}-1\right)}{1\times 2}\left(\frac{3}{4}x\right)^2 + \dots$$
$$= 1 + \frac{9}{8}x + \frac{\frac{3}{2}\times\frac{1}{2}}{2}\left(\frac{9}{16}x^2\right) + \dots$$
$$= 1 + \frac{9x}{8} + \frac{27x^2}{128} + \dots$$

- Put this back into the **original expression**:

$$(3x+4)^{\frac{3}{2}} = 8\left(1+\frac{3}{4}x\right)^{\frac{3}{2}} = 8\left(1+\frac{9x}{8}+\frac{27x^2}{128}+\dots\right)$$
$$= 8 + 9x + \frac{27x^2}{16} + \dots$$

Tip: This expansion is only valid for $\left|\frac{qx}{p}\right| < 1$, i.e. $\left|\frac{3x}{4}\right| < 1 \Rightarrow |x| < \frac{4}{3}$.

Example 3

Expand $\frac{1+2x}{(2-x)^2}$ up to the term in x^3.

State the range of x for which the expansion is valid.

- First you need to **rearrange** and **separate** the different expansions:

$$\frac{1+2x}{(2-x)^2} = (1 + 2x)(2 - x)^{-2}$$

$p = 2$ $\quad$ $q = -1$

- The first bracket doesn't need expanding, so deal with the second bracket as usual, **factorising** first...

$$2 - x = 2\left(1 - \frac{1}{2}x\right) \quad \Rightarrow \quad (2 - x)^{-2} = 2^{-2}\left(1-\frac{1}{2}x\right)^{-2} = \frac{1}{4}\left(1-\frac{1}{2}x\right)^{-2}$$

Tip: For expressions where both brackets need expanding, just take your time and set everything out in steps, only combining the expansions at the end.

- ... then using the **formula** with $n = -2$, and $-\frac{1}{2}x$ instead of x:

$$\left(1-\frac{1}{2}x\right)^{-2} = 1 + (-2)\left(-\frac{1}{2}x\right) + \frac{-2\times-3}{1\times 2}\left(-\frac{1}{2}x\right)^2 + \frac{-2\times-3\times-4}{1\times2\times3}\left(-\frac{1}{2}x\right)^3 + \dots$$
$$= 1 + x + \frac{3x^2}{4} + \frac{x^3}{2} + \dots$$

- So this means that:

$$(2-x)^{-2} = \frac{1}{4}\left(1-\frac{1}{2}x\right)^{-2} = \frac{1}{4}\left(1+x+\frac{3x^2}{4}+\frac{x^3}{2}+...\right)$$
$$= \frac{1}{4}+\frac{x}{4}+\frac{3x^2}{16}+\frac{x^3}{8}+...$$

- Putting all this into the **original expression** gives:

$$\frac{1+2x}{(2-x)^2} = (1+2x)(2-x)^{-2} = (1+2x)\left(\frac{1}{4}+\frac{x}{4}+\frac{3x^2}{16}+\frac{x^3}{8}+...\right)$$
$$= \frac{1}{4}+\frac{x}{4}+\frac{3x^2}{16}+\frac{x^3}{8}+2x\left(\frac{1}{4}+\frac{x}{4}+\frac{3x^2}{16}\right)+...$$
$$= \frac{1}{4}+\frac{x}{4}+\frac{3x^2}{16}+\frac{x^3}{8}+\frac{x}{2}+\frac{x^2}{2}+\frac{3x^2}{8}+...$$
$$= \frac{1}{4}+\frac{3x}{4}+\frac{11x^2}{16}+\frac{x^3}{2}+...$$

- The **validity** of the whole expansion will depend on the validity of the expansion of $(2-x)^{-2}$.

This is valid only if $\left|\frac{-x}{2}\right| < 1 \Rightarrow |x| < 2$

Tip: Remember — if both expansions have a limited validity, choose the one with the narrower valid range of x for the combined expansion.

Exercise 1.3

Q1 Find the binomial expansion of the following functions, up to and including the term in x^3:

a) $(2+4x)^3$ b) $(3+4x)^5$ c) $(4+x)^{\frac{1}{2}}$ d) $(8+2x)^{-\frac{1}{3}}$

Q2 If the x^2 coefficient of the binomial expansion of $(a+5x)^5$ is 2000, what is the value of a? PROBLEM SOLVING

Q3 a) Find the binomial expansion of $(2-5x)^7$ up to and including the term in x^2.

b) Hence, or otherwise, find the binomial expansion of $(1+6x)^3(2-5x)^7$, up to and including the term in x^2.

Q4 a) Find the binomial expansion of $\left(1+\frac{6}{5}x\right)^{-\frac{1}{2}}$, up to and including the term in x^3, stating the range of x for which it is valid.

b) Hence, or otherwise, express $\sqrt{\frac{20}{5+6x}}$ in the form $a+bx+cx^2+dx^3+...$ PROBLEM SOLVING

Q5 $f(x) = \frac{1}{\sqrt{5-2x}}$

a) Find the binomial expansion of $f(x)$ in ascending powers of x, up to and including the term in x^2.

b) Hence show that $\frac{3+x}{\sqrt{5-2x}} \approx \frac{3}{\sqrt{5}}+\frac{8x}{5\sqrt{5}}+\frac{19x^2}{50\sqrt{5}}$.

Q6 a) Find the binomial expansion of $(9+4x)^{-\frac{1}{2}}$, up to and including the term in x^2.

b) Hence, or otherwise, find the binomial expansion of $\frac{(1+6x)^4}{\sqrt{9+4x}}$, up to and including the term in x^2.

2. Using the Binomial Expansion as an Approximation

One of the reasons that binomial expansions are so useful is that they can be used to estimate nasty-looking roots, powers and fractions. All you need to work out is the right value of x to use.

Learning Objective:

- Be able to substitute values into a binomial expansion in order to find approximations.

Approximating with binomial expansions

- When you've done an expansion, you can use it to work out the **value** of the original expression for **given values of x**, by **substituting** those values into the **expansion**.
- For most expansions this will only be an **approximate** answer, because you'll have had to limit the expansion to the first few terms.
- Often you'll have to do some **rearranging** of the expression so that you know what value of x to substitute.
- For example, $\sqrt[3]{1.3}$ can be written $(1+0.3)^{\frac{1}{3}}$, which can be approximated by expanding $(1+x)^{\frac{1}{3}}$ and substituting $x = 0.3$ into the expansion.

Tip: You also need to check the validity of the expansion — the approximation will only work for values of x in the valid range.

Example 1

The binomial expansion of $(1 + 3x)^{-1}$ up to the term in x^3 is:
$(1 + 3x)^{-1} \approx 1 - 3x + 9x^2 - 27x^3$. The expansion is valid for $|x| < \frac{1}{3}$.
Use this expansion to approximate $\frac{100}{103}$. Give your answer to 4 d.p.

- For this type of question, you need to find the **right value of x** to make the expression you're expanding equal to the thing you're looking for.
- This means a bit of clever **rearranging**:

$$\frac{100}{103} = \frac{1}{1.03} = \frac{1}{1+0.03} = (1+0.03)^{-1}$$

- This is the same as an expansion of $(1 + 3x)^{-1}$ with $3x = 0.03 \Rightarrow$ $\boldsymbol{x = 0.01}$.
- Check that this value is in the **valid range**:

$0.01 < \frac{1}{3}$, so the expansion is valid for this value of x.

- **Substituting** this value for x into the expansion gives:

$$(1+3(0.01))^{-1} \approx 1 - 3(0.01) + 9(0.01^2) - 27(0.01^3)$$
$$= 1 - 0.03 + 0.0009 - 0.000027$$
$$= 1.0009 - 0.030027$$
$$= 0.970873$$
$$(1+3(0.01))^{-1} \approx 0.9709 \text{ to 4 d.p.}$$

This is the expansion given in the question, with $x = 0.01$.

$\frac{100}{103} = 0.97087...$ so this is a pretty good approximation.

Tip: You need to use a "≈" when you give the answer — it's an approximation because you're only using the first few terms of the expansion.

In some cases you might have to **rearrange the expansion** first to get it into a form that fits with the given expression.

Example 2

The binomial expansion of $(1-5x)^{\frac{1}{2}}$ up to the term in x^2 is

$(1-5x)^{\frac{1}{2}} \approx 1 - \frac{5x}{2} - \frac{25x^2}{8}$. The expansion is valid for $|x| < \frac{1}{5}$.

a) Use $x = \frac{1}{50}$ in this expansion to show that $\sqrt{10} \approx \frac{800}{253}$.

- First, substitute $x = \frac{1}{50}$ into **both sides** of the given expansion:

$$\sqrt{\left(1-5\left(\frac{1}{50}\right)\right)} \approx 1 - \frac{5}{2}\left(\frac{1}{50}\right) - \frac{25}{8}\left(\frac{1}{50}\right)^2$$

$$\sqrt{\left(1-\frac{1}{10}\right)} \approx 1 - \frac{1}{20} - \frac{1}{800}$$

$$\sqrt{\frac{9}{10}} \approx \frac{759}{800}$$

- Now **simplify the square root**...

$$\sqrt{\frac{9}{10}} = \frac{\sqrt{9}}{\sqrt{10}} = \frac{3}{\sqrt{10}} \approx \frac{759}{800}$$

- ...and **rearrange** to find an estimate for $\sqrt{10}$:

$$\frac{3}{\sqrt{10}} \approx \frac{759}{800}$$

$$3 \times 800 \approx 759\sqrt{10}$$

$$\sqrt{10} \approx \frac{3 \times 800}{759}$$

$$\sqrt{10} \approx \frac{800}{253}$$ as required.

b) Find the percentage error in your approximation, to 2 s.f.

Work out the percentage error by finding the difference between your estimate and a calculated 'real' value, and give this as a percentage of the real value:

$$\left|\frac{\text{real value} - \text{estimate}}{\text{real value}}\right| \times 100$$

$$= \left|\frac{\sqrt{10} - \frac{800}{253}}{\sqrt{10}}\right| \times 100$$

$$= 0.0070\% \text{ (2 s.f.)}$$

The % error is really small, which means the approximation is very close to the real answer.

Tip: If you're not quite sure how you'll get the approximation you need from the expansion, the best thing to do is put the numbers in and see what comes out. It's much clearer at this stage in the example where your $\sqrt{10}$ is coming from, but you'd be forgiven for not making the link at the start of the question.

Tip: The modulus sign means you always get a positive answer, whether the estimate is bigger or smaller than the real value. You're only interested in the difference between them.

Exercise 2.1

Q1 a) Find the binomial expansion of $(1 + 6x)^{-1}$, up to and including the term in x^2.

b) What is the validity of the expansion in part a)?

c) Use an appropriate substitution to find an approximation for $\frac{100}{106}$.

d) What is the percentage error of this approximation? Give your answer correct to 1 significant figure.

Q1c) Hint: Always check that the value you've decided to use for x is within the valid range for the expansion.

Q2 a) Use the binomial theorem to expand $(1 + 3x)^{\frac{1}{4}}$ in ascending powers of x, up to and including the term in x^3.

b) For what values of x is this expansion valid?

c) Use this expansion to find an approximate value of $\sqrt[4]{1.9}$ correct to 4 decimal places.

d) Find the percentage error of this approximation, correct to 3 significant figures.

Q3 a) Find the first four terms in the binomial expansion of $(1-2x)^{-\frac{1}{2}}$.

b) For what range of x is this expansion valid?

c) Use $x = \frac{1}{10}$ in this expansion to find an approximate value of $\sqrt{5}$.

d) Find the percentage error of this approximation, correct to 2 significant figures.

Q3c) Hint: Put the value for x into both sides of the expansion, and rearrange the left-hand side until you get $\sqrt{5}$.

Q4 a) Expand $(2 - 5x)^6$ up to and including the x^2 term.

b) By substituting an appropriate value of x into the expansion in a), find an approximate value for 1.95^6.

c) What is the percentage error of this approximation? Give your answer to 2 significant figures.

Q5 a) Find the first three terms in the binomial expansion of $\sqrt{3-4x}$.

b) For what values of x is this expansion valid?

c) Use $x = \frac{3}{40}$ in this expansion to estimate the value of $\frac{3}{\sqrt{10}}$. Leave your answer as a fraction.

d) Find the percentage error of this approximation, correct to 1 significant figure.

3. Binomial Expansion and Partial Fractions

Learning Objective:
- Be able to split rational functions into partial fractions, then find the binomial expansion of the function.

You'll need to deal with some tricky expressions — but you can use partial fractions to make things simpler. You met these back in Chapter 2.

Finding binomial expansions using partial fractions

You can find the binomial expansion of more complicated functions by:

- splitting them into **partial fractions** first,
- expanding **each fraction** using the formula (usually with $n = -1$),
- then **adding** the expansions together.

Example

The function $f(x) = \frac{x-1}{(3+x)(1-5x)}$ **can be expressed as partial fractions in the form:** $\frac{A}{(3+x)} + \frac{B}{(1-5x)}$.

a) Find the values of A and B, and hence express f(x) as partial fractions.

Tip: Look back at pages 39-43 for a recap on this method.

- Start by writing out the problem as an **identity**:

$$\frac{x-1}{(3+x)(1-5x)} \equiv \frac{A}{(3+x)} + \frac{B}{(1-5x)}$$

$$\Rightarrow \frac{x-1}{(3+x)(1-5x)} \equiv \frac{A(1-5x)+B(3+x)}{(3+x)(1-5x)}$$

$$\Rightarrow \quad x - 1 \equiv A(1-5x) + B(3+x)$$

Add the fractions together and cancel the denominators on either side of the identity.

- You can then work out the values of A and B by putting in values of x that make each bracket in turn equal to zero. (This is known as the '**substitution**' method.)

Tip: You could also **equate the coefficients** of x and the constant terms to find A and B.

Let $x = -3$, then: $-3 - 1 = A(1 - (-15))$

$$\Rightarrow -4 = 16A$$

$$\Rightarrow A = -\frac{1}{4}$$

Let $x = \frac{1}{5}$, then: $\frac{1}{5} - 1 = B\left(3 + \frac{1}{5}\right)$

$$\Rightarrow -\frac{4}{5} = \frac{16}{5}B$$

$$\Rightarrow B = -\frac{1}{4}$$

Tip: Remember to put A and B back into the expression to give f(x) as partial fractions.

- So f(x) can also be written as: $-\frac{1}{4(3+x)} - \frac{1}{4(1-5x)}$

b) Use your answer to part a) to find the binomial expansion of f(x), up to and including the term in x^2.

- Start by **rewriting** the **partial fractions** from a) in **$(p + qx)^n$ form**:

$$f(x) = -\frac{1}{4}(3 + x)^{-1} - \frac{1}{4}(1 - 5x)^{-1}$$

- Now do the two binomial expansions **separately**:

$$\begin{aligned}(3+x)^{-1} &= \left(3\left(1+\frac{1}{3}x\right)\right)^{-1} \\ &= \frac{1}{3}\left(1+\frac{1}{3}x\right)^{-1} \\ &= \frac{1}{3}\left(1+(-1)\left(\frac{1}{3}x\right)+\frac{(-1)(-2)}{2}\left(\frac{1}{3}x\right)^2+\dots\right) \\ &= \frac{1}{3}\left(1-\frac{1}{3}x+\frac{1}{9}x^2-\dots\right) \\ &= \frac{1}{3}-\frac{1}{9}x+\frac{1}{27}x^2-\dots\end{aligned}$$

$$\begin{aligned}(1-5x)^{-1} &= 1 + (-1)(-5x) + \frac{(-1)(-2)}{2}(-5x)^2 + \dots \\ &= 1 + 5x + 25x^2 + \dots\end{aligned}$$

- Finally, put everything together by adding the expansions in the rearranged form of f(x):

$$\begin{aligned}f(x) &= -\frac{1}{4}(3 + x)^{-1} - \frac{1}{4}(1 - 5x)^{-1} \\ &\approx -\frac{1}{4}\left(\frac{1}{3}-\frac{1}{9}x+\frac{1}{27}x^2\right) - \frac{1}{4}(1 + 5x + 25x^2) \\ &= -\frac{1}{12}+\frac{x}{36}-\frac{x^2}{108}-\frac{1}{4}-\frac{5x}{4}-\frac{25x^2}{4} \\ &= -\frac{1}{3}-\frac{11x}{9}-\frac{169x^2}{27}\end{aligned}$$

c) Find the range of values of x for which your answer to part b) is valid.

- The two expansions from b) are valid for **different values of x**.

- The **combined** expansion of f(x) is valid where these **two ranges overlap**, i.e. over the **narrower** of the two ranges.

(This is the same as when you combine expansions by multiplying them together, as shown on page 124.)

The expansion of $(3 + x)^{-1}$ is valid when $\left|\frac{x}{3}\right| < 1 \Rightarrow |x| < 3$.

The expansion of $(1 - 5x)^{-1}$ is valid when $|-5x| < 1 \Rightarrow |x| < \frac{1}{5}$.

- The expansion of f(x) is valid for values of x in both ranges, so the expansion of f(x) is valid for $|x| < \frac{1}{5}$.

Tip: There are a lot of different stages to this type of question — which means a lot of places that you could make a mistake, especially with all these negatives and fractions flying around. Set your working out clearly and don't skip stages.

Tip: Remember, the expansion of $(p + qx)^n$ is valid when $\left|\frac{qx}{p}\right| < 1$.

Exercise 3.1

Q1 a) $\frac{5-12x}{(1+6x)(4+3x)} \equiv \frac{A}{(1+6x)} + \frac{B}{(4+3x)}$. Find A and B.

b) (i) Find the binomial expansion of $(1 + 6x)^{-1}$, up to and including the term in x^2.

(ii) Find the binomial expansion of $(4 + 3x)^{-1}$, up to and including the term in x^2.

c) Hence find the binomial expansion of $\frac{5-12x}{(1+6x)(4+3x)}$, up to and including the term in x^2.

d) For what values of x is this expansion valid?

Q2 $f(x) = \frac{6}{(1-x)(1+x)(1+2x)}$

a) Show that $f(x)$ can be expressed as:

$$\frac{1}{(1-x)} - \frac{3}{(1+x)} + \frac{8}{(1+2x)}$$

b) Give the binomial expansion of $f(x)$ in ascending powers of x, up to and including the term in x^2.

c) Find the percentage error when you use this expansion to estimate $f(0.01)$, giving your answer to 2 significant figures.

Q3 a) Factorise fully $2x^3 + 5x^2 - 3x$.

b) Hence express $\frac{5x-6}{2x^3+5x^2-3x}$ as partial fractions.

c) Find the binomial expansion of $\frac{5x-6}{2x^3+5x^2-3x}$, up to and including the term in x^2.

d) For what values of x is this expansion valid?

Q3c) Hint: You'll end up with a term in x^{-1}, which you wouldn't usually get with a binomial expansion — this comes from the partial fractions and you can just leave it as it is.

Q4 $f(x) = \frac{55x+7}{(2x-5)(3x+1)^2}$

a) Express $f(x)$ in the form $\frac{A}{(2x-5)} + \frac{B}{(3x+1)} + \frac{C}{(3x+1)^2}$, where A, B and C are integers to be found.

b) Hence, or otherwise, use the binomial formula to expand $f(x)$ in ascending powers of x, up to and including the term in x^2.

1. Points of Inflection

You've already done some differentiation in Year 1, and seen how you can use it to identify maximum and minimum points of graphs. This section shows you how to use the second derivative to identify other parts of a curve.

Learning Objectives:
- Understand the terms convex and concave in relation to curves.
- Be able to use the second derivative of a function to work out when the graph of that function is concave or convex, and to identify points of inflection.
- Determine the nature of stationary points of a curve where the second derivative is zero.

Convex and concave curves

Continuous curves of the form $y = f(x)$ can be described as **convex** or **concave**.

- **Convex curves** are ones that curve **downwards**. A straight line joining any two points on a convex curve lies **above** the curve between those points.
- **Concave curves** are ones that curve **upwards**. A straight line joining any two points on a concave curve lies **below** the curve between those points.

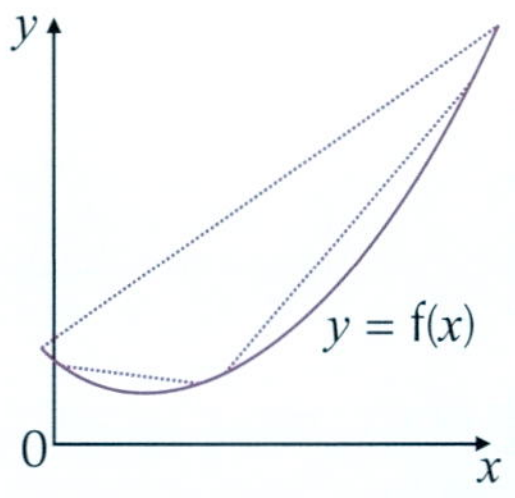

If you join any two points on this curve, the line is above the curve, so $y = f(x)$ is **convex**.

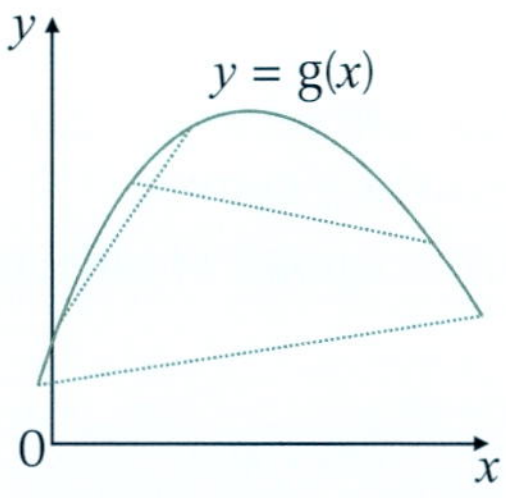

If you join any two points on this curve, the line is below the curve, so $y = g(x)$ is **concave**.

Tip: It can be tricky to remember which one is convex and which is concave. Just remember: concave is the one that looks like the entrance to a cave.

In Year 1, you learnt how to find **first** and **second derivatives** of a function, and how they relate to the **graph** of that function.

$\frac{dy}{dx}$ or $f'(x)$ is the **gradient of the graph** $y = f(x)$ — the rate of change of x.

$\frac{d^2y}{dx^2}$ or $f''(x)$ is the **rate of change of the gradient** of $y = f(x)$.

Tip: Remember, $f'(x)$ is the first derivative and $f''(x)$ is the second derivative of a function $f(x)$ with respect to x.

Convex curves have an **increasing gradient** and **concave curves** have a **decreasing gradient**. This means you can use the **second derivative** to work out if a curve is convex or concave:

A curve $y = f(x)$ is **convex** if $\mathbf{f''(x) > 0}$ for all values of x.
A curve $y = f(x)$ is **concave** if $\mathbf{f''(x) < 0}$ for all values of x.

Most curves aren't **entirely** convex or concave — but the definitions of convex and concave can be applied to **part** of a curve rather than the whole thing. You can use this to divide a curve into concave and convex **sections**.

Tip: If the straight line joining two points on a curve crosses the curve, then that section of the curve isn't entirely convex or concave.

Example

The graph of $y = x^3 + x^2 - x$ has concave and convex sections.
Find the range of values of x where the graph of $y = x^3 + x^2 - x$ is convex.

- Find the second derivative: $\frac{dy}{dx} = 3x^2 + 2x - 1 \Rightarrow \frac{d^2y}{dx^2} = 6x + 2$
- The graph is convex when the second derivative is positive:

$$\frac{d^2y}{dx^2} > 0 \Rightarrow 6x + 2 > 0 \Rightarrow 6x > -2 \Rightarrow x > -\frac{1}{3}$$

Points of inflection

A point where the curve **changes** between concave and convex (i.e. where $f''(x)$ changes between positive and negative) is called a **point of inflection**.

At a point of inflection, $\mathbf{f''(x) = 0}$, but not all points where $f''(x) = 0$ are points of inflection. You need to look what's happening on **either side** of the point to see if the sign of $f''(x)$ is changing. For example:

Tip: You might have seen 'points of inflection' used as the name for 'the other type' of stationary point of a graph (i.e. not a maximum or a minimum). More accurately, these are called **stationary points of inflection**. There's more about this on the next page.

- The graph of $\mathbf{f(x) = x^3 - 6x^2 + 9x + 1}$ has second derivative $f''(x) = 0$ when $x = 2$.
- When $x < 2$, $f''(x) < 0$ so the curve is **concave**, and when $x > 2$, $f''(x) > 0$ so the curve is **convex**.
- So $x = 2$ is a **point of inflection** of the curve.

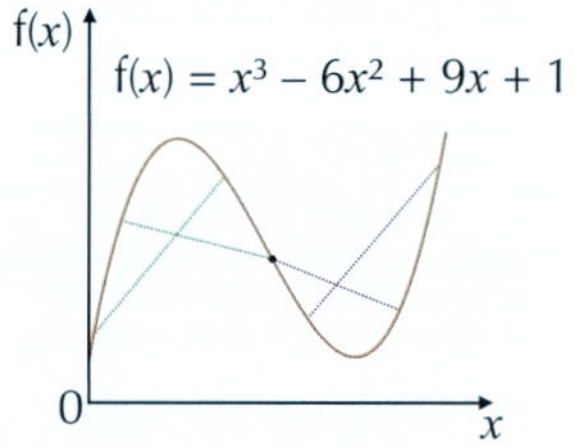

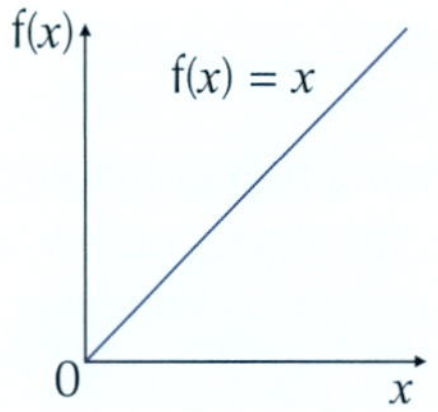

- On the graph of $\mathbf{f(x) = x}$, $f''(x) = 0$ for **all** values of x.
- So at any particular value of x, $f''(x) = 0$, but it's **not** a point of inflection because $f''(x)$ **isn't changing** from positive to negative — it's just constant.

- On the graph of $\mathbf{f(x) = x^4}$, $f''(x) = 12x^2$.
- At the point (0, 0), $f''(x) = 0$, but it's **not** a point of inflection.
- $12x^2$ is positive for **all** non-zero values of x. The whole curve is **convex**, so it doesn't have any points of inflection.

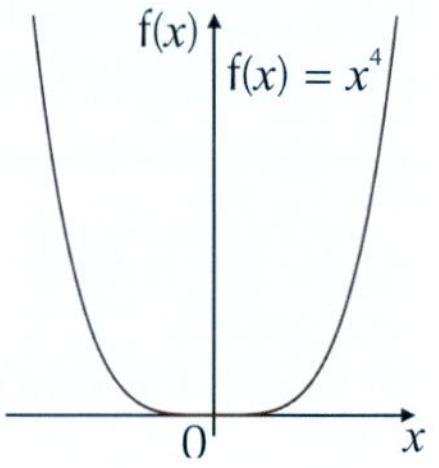

Example 1

Show that the graph of $y = x^3 - 3x^2 + 3$ has a point of inflection at $x = 1$.

- Start by finding $f''(x)$: $f(x) = x^3 - 3x^2 + 3 \Rightarrow f'(x) = 3x^2 - 6x$
  $\Rightarrow f''(x) = 6x - 6$
- Show that $f''(1) = 0$: At $x = 1$, $f''(x) = 6(1) - 6 = 0$
- Look at what happens either side of $x = 1$:
  When $x < 1$, $6x < 6 \Rightarrow 6x - 6 < 0 \Rightarrow f''(x) < 0$
  When $x > 1$, $6x > 6 \Rightarrow 6x - 6 > 0 \Rightarrow f''(x) > 0$

So at $x = 1$, $f''(x) = 0$ and $f''(x)$ changes sign from negative to positive.
So the graph of $y = x^3 - 3x^2 + 3$ has a point of inflection at $x = 1$.

Tip: Notice that when $x = 1$, the gradient $f'(1) \neq 0$, so this is **not** a **stationary** point of inflection.

Example 2

Find the coordinates of the points of inflection of the graph of $y = 2x^4 + 4x^3 - 72x^2$.

- Find the second derivative of $y = 2x^4 + 4x^3 - 72x^2$:

$$y = 2x^4 + 4x^3 - 72x^2 \Rightarrow \frac{dy}{dx} = 8x^3 + 12x^2 - 144x$$

$$\Rightarrow \frac{d^2y}{dx^2} = 24x^2 + 24x - 144$$

- Now find the points where $\frac{d^2y}{dx^2} = 0$. These could be points of inflection.

$$\frac{d^2y}{dx^2} = 0 \Rightarrow 24x^2 + 24x - 144 = 0$$
$$\Rightarrow x^2 + x - 6 = 0$$
$$\Rightarrow (x + 3)(x - 2) = 0$$
$$\Rightarrow x = -3 \text{ and } x = 2$$

- So there could be points of inflection at $x = -3$ and $x = 2$.
  Think about what happens to $\frac{d^2y}{dx^2}$ either side of these points.

  $\frac{d^2y}{dx^2}$ is a quadratic, so you can do this easily with a sketch.

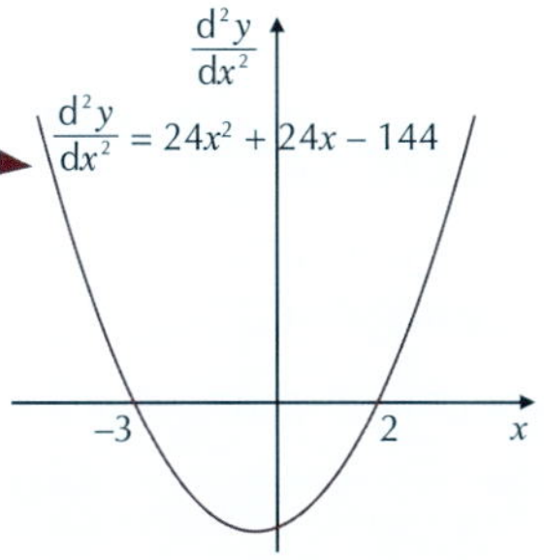

  $\frac{d^2y}{dx^2} > 0$ for $x < -3$,

  $\frac{d^2y}{dx^2} < 0$ for $-3 < x < 2$,

  and $\frac{d^2y}{dx^2} > 0$ for $x > 2$.

  So $\frac{d^2y}{dx^2}$ changes sign at $x = -3$ and $x = 2$, so both are points of inflection.

- The question asks for the coordinates of the points of inflection, so find the y-values:

  At $x = -3$, $y = 2(-3)^4 + 4(-3)^3 - 72(-3)^2 = -594$

  At $x = 2$, $y = 2(2)^4 + 4(2)^3 - 72(2)^2 = -224$

  So the points of inflection are at $(-3, -594)$ and $(2, -224)$.

Tip: If it's not easy to sketch, just work out whether $\frac{d^2y}{dx^2}$ is positive or negative either side of each point.

Tip: Look back at your Year 1 notes if you need a reminder about quadratic inequalities.

Stationary points of inflection

In Year 1, you used the first and second derivatives of a function to find **maximum** and **minimum points** of its graph.

Stationary points are points where $\mathbf{f'(x) = 0}$. A stationary point can be a **maximum** point, a **minimum** point or a **stationary point of inflection**. The value of $f''(x)$ can help you work out which type of stationary point it is:

If $f'(x) = 0$ and $\mathbf{f''(x) > 0}$, it's a **minimum**.
If $f'(x) = 0$ and $\mathbf{f''(x) < 0}$, it's a **maximum**.
If $f'(x) = 0$ and $\mathbf{f''(x) = 0}$, it could be **any one** of the three types — so you have to look at $f''(x)$ on either side of the stationary point.

Tip: If $f''(x) > 0$ on either side of a stationary point, the curve is convex near the point, so it's a minimum. If $f''(x) < 0$ on either side, the curve is concave, so it's a maximum. And if $f''(x)$ changes sign, it's a stationary point of inflection.

Example 3

For $f(x) = x^5 - 60x^3$, find the value of x at each stationary point of the graph of $y = f(x)$, and determine the nature of each one.

- Find the first derivative to locate the stationary points (i.e. where $f'(x) = 0$):
  $f(x) = x^5 - 60x^3 \Rightarrow f'(x) = 5x^4 - 180x^2 = 5x^2(x^2 - 36) = 5x^2(x + 6)(x - 6)$
  So $f'(x) = 0$ when $x = 0$, $x = -6$ and $x = 6$.
- Now find the second derivative and use it to work out the nature of the three stationary points:

  $f'(x) = 5x^4 - 180x^2 \Rightarrow f''(x) = 20x^3 - 360x$
  $f''(-6) = 20(-6)^3 - 360(-6) = -2160$
  $f''(-6) < 0$, so the stationary point at $x = -6$ is a maximum.

  $f''(6) = 20(6)^3 - 360(6) = 2160$
  $f''(6) > 0$, so the stationary point at $x = 6$ is a minimum.

  $f''(0) = 20(0)^3 - 360(0) = 0$, so the stationary point at $x = 0$ could be a maximum, a minimum or a point of inflection.
- Look at the values of $f''(x)$ close to $x = 0$:
  $f''(x) = 20x^3 - 360x = 20x(x^2 - 18)$
  When x is small and negative, $20x$ is negative and $x^2 - 18$ is negative, so $f''(x)$ is positive.
  When x is small and positive, $20x$ is positive and $x^2 - 18$ is negative, so $f''(x)$ is negative.
  The sign of $f''(x)$ changes at $x = 0$, so $x = 0$ is a point of inflection.

Tip: So the graph of $y = x^5 - 60x^3$ looks like this:

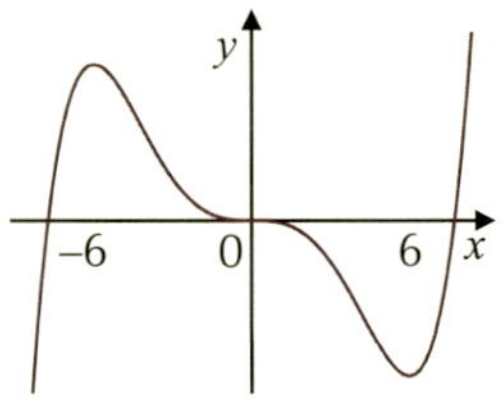

Exercise 1.1

Q1 Find the values of x for which the following graphs are concave:

a) $y = \frac{1}{6}x^3 - \frac{5}{2}x^2 + \frac{1}{4}x + \frac{1}{9}$ b) $y = 4x^2 - x^4$

Q2 The graph of $y = \frac{3}{2}x^4 - x^2 - 3x$ has two points of inflection. Find the x-coordinates of these two points.

Q2 Hint: There are only two points where the second derivative is zero here, and the question tells you there are two points of inflection. So you don't actually need to prove they're points of inflection in this case, you just need to find them.

Q3 If $f(x) = \frac{1}{16}x^4 + \frac{3}{4}x^3 - \frac{21}{8}x^2 - 6x + 20$, identify the ranges of values of x for which the graph of $y = f(x)$ is concave and convex.

Q4 Show that the graph of $y = x^2 - \frac{1}{x}$ has a point of inflection at (1, 0).

Q5 For $f(x) = x^3 + 2x^2 + 3x + 3$, show that:

a) the graph of $y = f(x)$ has one point of inflection,

b) the point of inflection is not a stationary point.

Q6 Find the coordinates of the stationary points of the graph of $y = \frac{1}{10}x^5 - \frac{1}{3}x^3 + \frac{1}{2}x + 4$ and determine their nature.

2. Chain Rule

The differentiation you covered in Year 1 only really works for reasonably straightforward functions — for more complicated functions, you'll need some new rules. The first of the new rules is the chain rule, which you use for differentiating functions of functions, like sin x^2.

Learning Objectives:

- Be able to identify which part of a function to use as u in the chain rule.
- Be able to use the chain rule to differentiate functions of functions.
- Be able to use the chain rule to convert $\frac{dx}{dy}$ into $\frac{dy}{dx}$.

The chain rule

The **chain rule** helps you **differentiate** complicated functions by **splitting them up** into functions that are easier to differentiate. The trick is spotting **how** to split them up, and choosing the right bit to **substitute**.

Once you've worked out how to split up the function, you can differentiate it using this **formula**:

If $y = f(u)$ and $u = g(x)$ then:

$$\frac{dy}{dx} = \frac{dy}{du} \times \frac{du}{dx}$$

Tip: It may help you to think of the derivatives as fractions (they're not, but it's a good way to think of it).

Then the du's cancel:

$$\frac{dy}{\cancel{du}} \times \frac{\cancel{du}}{dx} = \frac{dy}{dx}$$

If you remember this you'll never get the order wrong.

To differentiate a function using the chain rule, just follow these steps:

- Pick a suitable function of x for 'u' and rewrite y in terms of u.
- Differentiate u (with respect to x) to get $\frac{du}{dx}$...
- ...and differentiate y (with respect to u) to get $\frac{dy}{du}$.
- Stick it all in the formula and write everything in terms of x.

The important part is **choosing** which bit to make into u. The aim is to split it into **two separate functions** that you can **easily differentiate**.

- If you have a function inside **brackets**, then the part **inside** the brackets is normally u:

  $y = (x + 1)^2$ can be written as $y = u^2$ where $u = x + 1$.

  Now both u^2 and $x + 1$ are **easy to differentiate**, and the chain rule formula does the hard work for you.

- If there's a **trig function** or **log** involved, it's usually the part **inside** the trig function or log:

  $y = \sin x^2$ can be written as $y = \sin u$ where $u = x^2$

  Again, you end up with 2 functions that are **easy to differentiate**, so you just differentiate each one **separately** then put it all in the **formula.**

Tip: Don't worry if you don't know how to differentiate trig functions yet — it's covered later in this chapter.

Example 1

Find $\frac{dy}{dx}$ if $y = (6x - 3)^5$.

- First of all, decide which part of the function to replace with u. In this case, the bit inside the brackets is easy to differentiate, so that's u.

$$y = (6x - 3)^5, \text{ so let } y = u^5 \text{ where } u = 6x - 3$$

- Next, differentiate the two parts separately:

$$\frac{dy}{du} = 5u^4 \quad \text{and} \quad \frac{du}{dx} = 6$$

- Finally, put everything back into the chain rule formula.

$$\frac{dy}{dx} = \frac{dy}{du} \times \frac{du}{dx} = 5u^4 \times 6 = 30u^4 = 30(6x - 3)^4$$

Tip: Make sure you always substitute the function back in for u. It'll cost you marks if you don't.

Now that you can differentiate **functions of functions** using the chain rule, you can find the equation of a **tangent** or **normal** to a curve that has an equation given by a function of a function.

Example 2

Find the equation of the tangent to the curve $y = \frac{1}{\sqrt{x^2 + 3x}}$ at $(1, \frac{1}{2})$.

- This function's a little more complicated than the previous one, so it will help to first rewrite it in terms of powers:

$$y = \frac{1}{\sqrt{x^2 + 3x}} = (x^2 + 3x)^{-\frac{1}{2}}$$

- Then identify which part to turn into u.

$$y = (x^2 + 3x)^{-\frac{1}{2}}, \text{ so let } y = u^{-\frac{1}{2}} \text{ where } u = (x^2 + 3x)$$

- Once again you now have two functions you can differentiate:

$$y = u^{-\frac{1}{2}} \Rightarrow \frac{dy}{du} = \left(-\frac{1}{2}\right)u^{-\frac{3}{2}} = -\frac{1}{2(\sqrt{x^2 + 3x})^3}$$

$$u = x^2 + 3x \Rightarrow \frac{du}{dx} = 2x + 3$$

- Now put it all back into the chain rule formula:

$$\frac{dy}{dx} = \frac{dy}{du} \times \frac{du}{dx} = \left(-\frac{1}{2(\sqrt{x^2 + 3x})^3}\right) \times (2x + 3) = -\frac{2x + 3}{2(\sqrt{x^2 + 3x})^3}$$

- To find the equation of the tangent, you first need to know the gradient at the point $(1, \frac{1}{2})$, so put the x-value into your equation for $\frac{dy}{dx}$:

$$\frac{dy}{dx} = -\frac{(2 \times 1) + 3}{2(\sqrt{1^2 + (3 \times 1)})^3} = -\frac{5}{16}$$

- Then use your gradient and the values you're given to find c:

$$y = mx + c \Rightarrow \frac{1}{2} = (-\frac{5}{16} \times 1) + c \Rightarrow c = \frac{13}{16}$$

- So the equation of the tangent at $(1, \frac{1}{2})$ is $y = -\frac{5}{16}x + \frac{13}{16}$

Tip: When you get a question with a root or fraction, always rewrite it in terms of powers.

Remember, $\frac{1}{x}$ is the same as x^{-1} and $\sqrt{x}$ is the same as $x^{\frac{1}{2}}$.

Tip: Look back at your Year 1 notes if you want a reminder about finding equations of straight lines, tangents and normals.

Exercise 2.1

Q1 Differentiate with respect to x:

a) $y = (x + 7)^2$ b) $y = (2x - 1)^5$

c) $y = 3(4 - x)^8$ d) $y = (3 - 2x)^7$

e) $y = (x^2 + 3)^5$ f) $y = (5x^2 + 3)^2$

Q2 Find $f'(x)$ for the following:

a) $f(x) = (4x^3 - 9)^8$ b) $f(x) = (6 - 7x^2)^4$

c) $f(x) = (x^2 + 5x + 7)^6$ d) $f(x) = (x + 4)^{-3}$

e) $f(x) = (5 - 3x)^{-2}$ f) $f(x) = \dfrac{1}{(5-3x)^4}$

g) $f(x) = (3x^2 + 4)^{\frac{3}{2}}$ h) $f(x) = \dfrac{1}{\sqrt{5-3x}}$

Q2 Hint: Don't forget: $f'(x)$ is just another way of writing $\frac{dy}{dx}$.

Q3 Find the exact value of $\frac{dy}{dx}$ when $x = 1$ for:

a) $y = \dfrac{1}{\sqrt{5x-3x^2}}$ b) $y = \dfrac{12}{\sqrt[3]{x+6}}$

Q4 Differentiate $\left(\sqrt{x} + \frac{1}{\sqrt{x}}\right)^2$ with respect to x by:

a) Multiplying the brackets out and differentiating term by term.

b) Using the chain rule.

Q5 Find the equation of the tangent to the curve $y = (x - 3)^5$ at $(1, -32)$.

Q6 Find the equation of the normal to the curve $y = \frac{1}{4}(x-7)^4$ when $x = 6$.

Q6 Hint: The gradient of the normal to a curve is just $-1 \div$ gradient of the tangent (this was covered in Year 1 if you need a refresher).

Q7 Find the value of $\frac{dy}{dx}$ when $x = 1$ for $y = (7x^2 - 3)^{-4}$.

Q8 Find $f'(x)$ if $f(x) = \dfrac{7}{\sqrt[3]{3-2x}}$.

Q9 Find the equation of the tangent to the curve $y = \sqrt{5x-1}$ when $x = 2$, in the form $ax + by + c = 0$, $a, b, c \in \mathbb{Z}$.

Q9 Hint: $a, b, c \in \mathbb{Z}$ just means that a, b and c are integers.

Q10 Find the equation of the normal to the curve $y = \sqrt[3]{3x-7}$ when $x = 5$.

Q11 Find the equation of the tangent to the curve $y = (x^4 + x^3 + x^2)^2$ when $x = -1$.

Q12 Show that the curve $y = (2x - 3)^7$ has one point of inflection, and find the coordinates of that point.

Q13 Find the ranges of values of x for which the curve $y = \left(\frac{x}{4} - 2\right)^3$ is convex and concave.

Q12-13 Hint: You'll need to use the chain rule twice for these questions.

Finding $\frac{dy}{dx}$ when $x = f(y)$

The principle of the chain rule can also be used where **x is given in terms of y** (i.e. $x = f(y)$). This comes from a little mathematical rearranging, and you'll find it's often quite useful:

Tip: As with the chain rule, treat the derivatives as fractions to make this result easier to follow (they're not actually fractions though).

$$\frac{dy}{dx} \times \frac{dx}{dy} = \frac{dy}{dy} = 1, \text{ so rearranging gives } \frac{dy}{dx} = \frac{1}{\left(\frac{dx}{dy}\right)}$$

So to differentiate $x = f(y)$, use:

$$\frac{dy}{dx} = \frac{1}{\left(\frac{dx}{dy}\right)}$$

Example

A curve has the equation $x = y^3 + 2y - 7$. Find $\frac{dy}{dx}$ at the point $(-4, 1)$.

- Forget that the x's and y's are in the 'wrong' place and differentiate as usual:

$$x = y^3 + 2y - 7 \Rightarrow \frac{dx}{dy} = 3y^2 + 2$$

- Use $\frac{dy}{dx} = \frac{1}{\left(\frac{dx}{dy}\right)}$ to find $\frac{dy}{dx}$: $\frac{dy}{dx} = \frac{1}{3y^2 + 2}$

- $y = 1$ at the point $(-4, 1)$, so put this in the equation:

$$\frac{dy}{dx} = \frac{1}{3(1)^2 + 2} = \frac{1}{5} = 0.2, \text{ so } \frac{dy}{dx} = 0.2 \text{ at the point } (-4, 1).$$

Exercise 2.2

Q1 Find $\frac{dy}{dx}$ for each of the following functions at the given point. In each case, express $\frac{dy}{dx}$ in terms of y.

a) $x = 3y^2 + 5y + 7$ at $(5, -1)$

b) $x = y^3 - 2y$ at $(-4, -2)$

c) $x = (2y + 1)(y - 2)$ at $(3, -1)$

d) $x = \frac{4 + y^2}{y}$ at $(5, 4)$

Q2 Find $\frac{dy}{dx}$ in terms of y if $x = (2y^3 - 5)^3$.

Q3 Given that $x = \sqrt{4 + y}$, find $\frac{dy}{dx}$ in terms of x by:

PROBLEM SOLVING

a) finding $\frac{dx}{dy}$ first,

b) rearranging into the form $y = f(x)$.

3. Differentiation of e^x, $\ln x$ and a^x

Differentiating exponentials and logarithms is actually a lot easier than you might think because each one follows certain rules.

Learning Objectives:

- Be able to differentiate e^x and $\ln x$.
- Be able to use the rules of differentiation for e^x and $\ln x$ to differentiate more complex functions using the chain rule.
- Be able to use these methods to answer questions on tangents, normals, stationary points, convex and concave curves and points of inflection.
- Be able to differentiate functions of the form $y = a^x$ and $y = a^{f(x)}$.

Differentiating e^x

Remember from Year 1 that 'e' is just a number for which the **gradient of e^x is e^x**, which makes it pretty simple to **differentiate**:

$$y = e^x \quad \Rightarrow \quad \frac{dy}{dx} = e^x$$

You can use the chain rule to show another useful relation involving exponentials. If you replace x with $f(x)$, you have a **function of a function**:

$y = e^{f(x)}$, so let $y = e^u$ where $u = f(x)$

So $\frac{dy}{du} = e^u = e^{f(x)}$ (see above) and $\frac{du}{dx} = f'(x)$

Putting it into the chain rule formula you get:

$$\frac{dy}{dx} = \frac{dy}{du} \times \frac{du}{dx} = e^{f(x)} \times f'(x) = f'(x)e^{f(x)}$$

This works because $e^{f(x)}$ is a special case — the 'e' part stays the same when you differentiate, so you only have to worry about the f(x) part. You can just learn the formula:

$$y = e^{f(x)} \quad \Rightarrow \quad \frac{dy}{dx} = f'(x)e^{f(x)}$$

Tip: In Year 1, you saw that the gradient of Ae^{kx} is kAe^{kx} — this comes from the chain rule.

You can use this formula to differentiate difficult exponentials.

Example 1

Find $\frac{dy}{dx}$ if $y = e^{(3x-2)}$

- Using the formula above, you can see that $y = e^{f(x)}$, where $f(x) = 3x - 2$.
- Differentiating f(x) is very easy: $f'(x) = 3$
- Now just put the right parts back into the formula for $\frac{dy}{dx}$.

$$y = e^{(3x-2)} \Rightarrow \frac{dy}{dx} = f'(x)e^{f(x)} = 3e^{(3x-2)}$$

Example 2

If $f(x) = e^{x^2} + 2e^x$, find $f'(x)$ when $x = 0$.

- The function is in 2 parts, so let's break it down into its two bits and differentiate them separately.
- The second bit's easy: If $f(x) = 2e^x$ then $f'(x) = 2e^x$ too.
- For the first bit, you could just use the formula you used on the previous example, but let's use the chain rule here just to show how it works.

$y = e^{x^2}$, so let $y = e^u$ where $u = x^2$

- Both u and y are now easy to differentiate:

$$\frac{dy}{du} = e^u \text{ and } \frac{du}{dx} = 2x$$

$$\frac{dy}{dx} = \frac{dy}{du} \times \frac{du}{dx} = e^u \times 2x = 2xe^{x^2}$$

- Now put the bits back together:

$$f'(x) = 2xe^{x^2} + 2e^x$$

- And finally, work out the value of $f'(x)$ at $x = 0$

$$f'(0) = (2 \times 0 \times e^{0^2}) + 2e^0$$

$$f'(0) = 0 + 2 = 2$$

Tip: Remember that $e^0 = 1$.

Example 3

The graph of $y = e^{2x} - 6x^2$ has one point of inflection. Find the exact coordinates of this point.

- To find the point of inflection, you need to find where $\frac{d^2y}{dx^2} = 0$.
- Start by finding $\frac{dy}{dx}$. Like in Example 2, split the function up into two parts and differentiate the parts separately.
- The second bit's easy: $y = -6x^2 \Rightarrow \frac{dy}{dx} = -12x$
- For the first bit, use the formula given at the start of this section.

$$y = e^{2x} = e^{f(x)}$$

$$f(x) = 2x \text{ so } f'(x) = 2 \Rightarrow \frac{dy}{dx} = f'(x)e^{f(x)} = 2e^{2x}$$

- Now just put the two parts back together:

$$\frac{dy}{dx} = 2e^{2x} - 12x$$

- Now differentiate again to find $\frac{d^2y}{dx^2}$:

$$\frac{dy}{dx} = 2e^{2x} - 12x = 2e^{f(x)} - 12x$$

$$\frac{d^2y}{dx^2} = 2f'(x)e^{f(x)} - 12 = 2(2 \times e^{2x}) - 12 = 4e^{2x} - 12$$

Tip: You should be familiar with the log rules from Year 1.

- Now to find the point of inflection, set $\frac{d^2y}{dx^2} = 0$:

$$4e^{2x} - 12 = 0 \Rightarrow e^{2x} = 3 \Rightarrow 2x = \ln 3$$

Take logs of both sides to solve for x: $\ln(e^x) = x$

$$\Rightarrow x = \frac{1}{2}\ln 3 = \ln 3^{\frac{1}{2}} = \ln\sqrt{3}$$

$k \log x = \log x^k$

Tip: There's only one point where $\frac{d^2y}{dx^2} = 0$, and the question tells you there's one point of inflection, so that must be the point you're looking for here — there's no need to prove it's a point of inflection.

- Substitute $x = \ln\sqrt{3}$ into $y = e^{2x} - 6x^2$ to find the y-coordinate:

$$y = e^{2\ln\sqrt{3}} - 6(\ln\sqrt{3})^2 \Rightarrow y = e^{\ln 3} - 6(\ln\sqrt{3})^2 \Rightarrow y = 3 - 6(\ln\sqrt{3})^2$$

$2\ln\sqrt{3} = \ln\sqrt{3}^2 = \ln 3$

$e^{\ln x} = x$

- So the point of inflection is at $(\ln\sqrt{3}, 3 - 6(\ln\sqrt{3})^2)$

Exercise 3.1

Q1 Differentiate with respect to x:

a) $y = e^{3x}$ b) $y = e^{2x - 5}$ c) $y = e^{x + 7}$

d) $y = e^{3x + 9}$ e) $y = e^{7 - 2x}$ f) $y = e^{x^3}$

Q2 Find f′(x) if:

a) $f(x) = e^{x^3+3x}$ b) $f(x) = e^{x^3-3x-5}$ c) $f(x) = e^{x(2x + 1)}$

Q3 Find f′(x) if:

a) $f(x) = \frac{1}{2}(e^x - e^{-x})$ b) $f(x) = e^{(x + 3)(x + 4)}$ c) $f(x) = e^{x^4+3x^2} + 2e^{2x}$

Q4 Find the equation of the tangent to the curve $y = e^{2x}$ at the point (0, 1).

Q5 Find the exact value of the x-coordinate of the point of the inflection on the curve $y = \frac{1}{2}x^2 - e^{2x - 6}$.

Q6 Find the equation of the tangent to the curve $y = e^{2x^2}$ when $x = 1$. Leave the numbers in your answer in exact form.

Q7 Show that the curve $y = e^{2x - 4} - x$ is convex for all values of x.

Q8 Find the equation of the normal to the curve $y = e^{3x} + 3$ where it cuts the y-axis.

Q9 Find the exact coordinates of any points of inflection on the curve $y = 2e^{2x} - \frac{1}{2}e^{3 - 4x}$.

Q10 Show that the curve $y = e^{x^3-3x-5}$ has stationary points at $x = \pm 1$.

Q11 Find the x-coordinate of the stationary point for the curve $y = e^{3x} - 6x$ and determine the nature of this point. Leave the numbers in your answer in exact form.

Q3c) Hint: Differentiating each term separately will make this easier.

Q5-6 Hint: Where the questions ask for exact answers, that means they're likely to include e^a or ln a (where a is a number) rather than working out the actual numbers.

Q11 Hint: To determine the nature of the stationary point you need to look at the sign of $\frac{d^2y}{dx^2}$ at the point.

Differentiating ln x

The natural logarithm of a function is the logarithm with base e, written as ln x. Differentiating natural logarithms also uses the chain rule:

- If $y = \ln x$, then $x = e^y$.
- Differentiating gives $\frac{dx}{dy} = e^y$, and $\frac{dy}{dx} = \frac{1}{\left(\frac{dx}{dy}\right)} = \frac{1}{e^y} = \frac{1}{x}$ (since $x = e^y$).
- This gives the result: $y = \ln x \Rightarrow \frac{dy}{dx} = \frac{1}{x}$

Tip: Look back at p.142 for more on $\frac{dx}{dy}$.

Example 1

a) Find $\frac{dy}{dx}$ if $y = \ln(2x + 3)$.

- It's a function of a function, so use the **chain rule**:

$$y = \ln(2x+3), \text{ so let } y = \ln u \text{ where } u = 2x+3$$

$$\Rightarrow \frac{dy}{du} = \frac{1}{u} = \frac{1}{2x+3} \text{ and } \frac{du}{dx} = 2$$

- Now put all the parts into the chain rule formula:

$$\frac{dy}{dx} = \frac{dy}{du} \times \frac{du}{dx} = \frac{1}{2x+3} \times 2 = \frac{2}{2x+3}$$

b) Find $\frac{dy}{dx}$ if $y = \ln(x^2 + 3)$.

Use the **chain rule** again for this one: $y = \ln u$ and $u = x^2 + 3$.

$$\frac{dy}{du} = \frac{1}{u} = \frac{1}{x^2+3} \text{ and } \frac{du}{dx} = 2x.$$

$$\Rightarrow \frac{dy}{dx} = \frac{dy}{du} \times \frac{du}{dx} = \frac{1}{x^2+3} \times 2x = \frac{2x}{x^2+3}$$

- Look at the final answer from those examples. It comes out to $\frac{f'(x)}{f(x)}$.
- This isn't a coincidence — it will always be the case for $y = \ln(f(x))$, so you can just learn the result:

$$y = \ln(f(x)) \Rightarrow \frac{dy}{dx} = \frac{f'(x)}{f(x)}$$

Example 2

Find $f'(x)$ if $f(x) = \ln(x^3 - 4x)$.

- $f(x)$ is in the form $\ln(g(x))$, so use the formula above:

$$f'(x) = \frac{g'(x)}{g(x)}$$

$$g(x) = x^3 - 4x \Rightarrow g'(x) = 3x^2 - 4$$

- Put this into the formula:

$$f(x) = \ln(x^3 - 4x) \Rightarrow f'(x) = \frac{g'(x)}{g(x)} = \frac{3x^2-4}{x^3-4x}$$

Tip: You can check this answer using the chain rule, like in the previous example.

Exercise 3.2

Q1 Differentiate with respect to x:

a) $y = \ln(3x)$

b) $y = \ln(1 + x)$

c) $y = \ln(1 + 5x)$

d) $y = 4\ln(4x - 2)$

Q2 Differentiate with respect to x:

a) $y = \ln(1 + x^2)$

b) $y = \ln(2 + x)^2$

c) $y = 3\ln x^3$

d) $y = \ln(x^3 + x^2)$

Q2 Hint: It might help to simplify the logs before differentiating them.

Q3 Find $f'(x)$ if:

a) $f(x) = \ln \frac{1}{x}$ b) $f(x) = \ln \sqrt{x}$

Q4 Find $f'(x)$ if $f(x) = \ln((2x + 1)^2\sqrt{x-4})$.

Q5 Find $f'(x)$ if $f(x) = \ln(x - \sqrt{x-4})$.

Q6 Find $f'(x)$ if $f(x) = \ln\left(\frac{(3x+1)^2}{\sqrt{2x+1}}\right)$.

Q7 Find the equation of the tangent to the curve $y = \ln(3x)^2$:

a) when $x = -2$ b) when $x = 2$

Q8 Find the equation of the normal to the curve $y = \ln(x + 6)^2$:

a) when $x = -3$ b) when $x = 0$

Q9 Find any stationary points for the curve $y = \ln(x^3 - 3x^2 + 3x)$.

Q4-6 Hint: You'll need to rewrite some of these questions as the sum or difference of two logarithms before differentiating them.

Q7-8 Hint: Rewrite $\ln(f(x))^k$ as $k\ln(f(x))$ to make differentiation simpler.

Differentiating a^x

For any constant a:

$$\frac{d}{dx}(a^x) = a^x \ln a$$

Tip: The proof for this rule uses implicit differentiation — you can see it on p.175.

Example 1

Differentiate the following:

a) $y = 2^x$ $\qquad \frac{dy}{dx} = 2^x \ln 2$

b) $y = \left(\frac{1}{2}\right)^x$ $\qquad$ Use the **log laws** to tidy this up...

$$\frac{dy}{dx} = \left(\frac{1}{2}\right)^x \ln\left(\frac{1}{2}\right) = -2^{-x}\ln 2$$

$\ln\left(\frac{1}{2}\right) = -\ln 2$

$\left(\frac{1}{2}\right)^x = 2^{-x}$

Tip: The rule $\frac{d}{dx}(e^x) = e^x$ is actually just a special case of the rule for a^x:

$\frac{d}{dx}(e^x) = e^x \ln e$
$= e^x \times 1 = e^x$

Differentiating $a^{f(x)}$

Use the **chain rule** (see p.139) to differentiate functions of the form $a^{f(x)}$, by differentiating $y = a^u$ and $u = f(x)$ separately, then using the **formula**:

$$\frac{dy}{dx} = \frac{dy}{du} \times \frac{du}{dx}$$

Example 2

Find the equation of the tangent to the curve $y = 3^{-2x}$ at the point $\left(\frac{1}{2}, \frac{1}{3}\right)$.

- Use the **chain rule** to find $\frac{dy}{dx}$:

Let $u = -2x$ and $y = 3^u$.

Differentiating separately gives: $\frac{du}{dx} = -2$ and $\frac{dy}{du} = 3^u \ln 3$

So $\frac{dy}{dx} = \frac{dy}{du} \times \frac{du}{dx} = 3^u \ln 3 \times -2 = -2(3^{-2x} \ln 3)$

Tip: As with all the other types of differentiation in this chapter, you might be asked to use the gradient function — e.g. to find equations of tangents and normals, or to find where the curve is convex or concave, etc. After the differentiation bit, the methods are the same, whatever the function.

- Now we can find the **gradient** of the tangent:

$$\text{At } \left(\frac{1}{2}, \frac{1}{3}\right), \frac{dy}{dx} = -2(3^{-1} \ln 3) = -\frac{2}{3} \ln 3$$

- So if the equation of the tangent at $\left(\frac{1}{2}, \frac{1}{3}\right)$ has the form $y = mx + c$, then

$$\frac{1}{3} = (-\frac{2}{3} \ln 3)\frac{1}{2} + c \Rightarrow c = \frac{1}{3} + \frac{1}{3} \ln 3$$

- So the equation of the tangent to $y = 3^{-2x}$ at $\left(\frac{1}{2}, \frac{1}{3}\right)$ is:

$$y = -\frac{2x}{3} \ln 3 + \frac{1}{3} + \frac{1}{3} \ln 3 \text{ or } 3y = (1 - 2x)\ln 3 + 1$$

Exercise 3.3

Q1 Differentiate the following:

a) $y = 5^x$ b) $y = 3^{2x}$ c) $y = 10^{-x}$ d) $y = p^{qx}$

Q2 A curve has the equation $y = 2^{4x}$.

a) Show that the gradient of the curve is $\frac{dy}{dx} = 4(2^{4x} \ln 2)$.

b) Find the equation of the tangent to the curve when $x = 2$.

Q2b) Hint: You can leave your answer in terms of ln if you're not asked for a specific degree of accuracy. That way you're giving the exact answer.

Q3 A curve $y = 2^{px}$ passes through the point (1, 32).

a) Find p.

b) Hence find the gradient of the curve at this point.

Q4 A curve has the equation $y = p^{x^3}$.

a) Show that the gradient of the curve is $\frac{dy}{dx} = 3x^2(p^{x^3} \ln p)$.

b) If the curve passes through the point (2, 6561), find p.

c) Hence find the equation of the tangent to the curve when $x = 1$.

Q4 Hint: Leave the ln's in your working and final answer.

Q5 The curve $y = 4^{\sqrt{x}}$ passes through the point (25, a). Show that the equation of the tangent to the curve at (25, a) is $y = 142x - 2520$ (to 3 s.f.). PROBLEM SOLVING

Q6 A curve C has the equation $y = 2^{-3x}$. It passes through the point (2, b).

a) Find the gradient $\frac{dy}{dx}$ of the curve.

b) Find b and the gradient of the curve at (2, b).

c) Hence show that the equation of the tangent to the curve at (2, b) is $64y = 1 + 6 \ln 2 - (3 \ln 2)x$.

4. Differentiating Trig Functions

Trig functions are also pretty easy to differentiate once you learn the rules for sin, cos and tan. In this section you'll see how to differentiate trig functions and then use the chain rule to differentiate the more tricky ones.

Differentiating sin, cos and tan

For **trigonometric functions** where the angle is measured in **radians** the following rules apply:

$y = \sin x$, $\dfrac{dy}{dx} = \cos x$

$y = \cos x$, $\dfrac{dy}{dx} = -\sin x$

$y = \tan x$, $\dfrac{dy}{dx} = \sec^2 x$

You can prove the rules for $\sin x$ and $\cos x$ using **differentiation from first principles**. Recall the diagram from Year 1:

As h gets smaller, the gradient of the line passing through $(x, f(x))$ and $(x + h, f(x + h))$ gets closer to $f'(x)$.

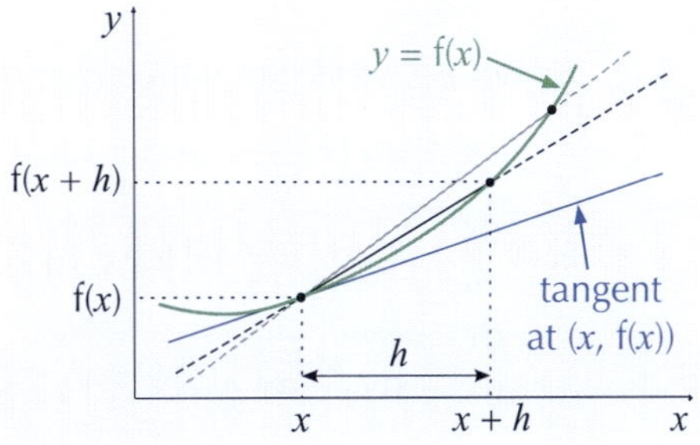

Example 1

Show using differentiation from first principles that if $y = \sin x$, $\dfrac{dy}{dx} = \cos x$.

- Write down the formula for differentiating from first principles. → $\dfrac{dy}{dx} = \lim\limits_{h \to 0}\left[\dfrac{f(x+h) - f(x)}{(x+h) - x}\right]$
- Use the fact that $y = f(x)$ to replace $f(x)$ with $\sin x$. → $= \lim\limits_{h \to 0}\left[\dfrac{\sin(x+h) - \sin x}{(x+h) - x}\right]$
- You can use the sin addition formula (see p.68) to expand $\sin(x + h)$ on top of the fraction. → $= \lim\limits_{h \to 0}\left[\dfrac{\sin x \cos h + \cos x \sin h - \sin x}{(x+h) - x}\right]$

$$= \lim_{h \to 0}\left[\frac{\sin x(\cos h - 1) + \cos x \sin h}{h}\right]$$

- It's helpful to separate the $\sin x$ and $\cos x$ terms into two fractions at this point. → $= \lim\limits_{h \to 0}\left[\dfrac{\sin x(\cos h - 1)}{h} + \dfrac{\cos x \sin h}{h}\right]$
- Now, because you're interested in when h gets really small, you can use the small angle approximations (see p.49) $\sin\theta \approx \theta$ and $\cos\theta \approx 1 - \frac{1}{2}\theta^2$. → $= \lim\limits_{h \to 0}\left[\dfrac{\sin x \times \left(-\frac{1}{2}h^2\right)}{h} + \dfrac{\cos x \times h}{h}\right]$

$$= \lim_{h \to 0}\left[-\frac{h \sin x}{2} + \cos x\right]$$

- As $h \to 0$, the first term $\to 0$, so it disappears.

$$= \cos x$$

Learning Objectives:

- Be able to differentiate sin, cos and tan.
- Be able to differentiate sin and cos from first principles.
- Be able to use the rules for differentiating trig functions in more complicated functions that require the chain rule.

Tip: Remember that $\sec x = \dfrac{1}{\cos x}$ — there's more about this on page 55.

Tip: If you can't remember how to do differentiation from first principles, look back at your Year 1 notes.

Tip: You might see 'δx' used instead of 'h' in differentiation from first principles — they mean the same thing. You can write '$(x + h) - x$' as just 'h' to simplify the denominator.

Tip: The addition formula for sin is $\sin(A + B) \equiv \sin A \cos B + \cos A \sin B$.

The equations for differentiating trig functions can be combined with the **chain rule** to differentiate **more complicated** functions.

Example 2

Differentiate the following with respect to x:

a) $y = \cos(2x)$

- Rewrite the function in 'chain rule notation':

  $y = \cos(2x)$, so let $y = \cos u$ where $u = 2x$

- So $\frac{dy}{du} = -\sin u = -\sin(2x)$ and $\frac{du}{dx} = 2$

  $\frac{dy}{dx} = \frac{dy}{du} \times \frac{du}{dx} = -2\sin(2x)$

b) $y = 4\sin(x^2 + 1)$

- As before, work out which part needs to be u for the chain rule:

  $y = 4\sin(x^2 + 1)$, so let $y = 4\sin u$ where $u = x^2 + 1$

- So $\frac{dy}{du} = 4\cos u = 4\cos(x^2 + 1)$ and $\frac{du}{dx} = 2x$

  $\frac{dy}{dx} = \frac{dy}{du} \times \frac{du}{dx} = 4\cos(x^2 + 1) \times 2x$

  $= 8x\cos(x^2 + 1)$

c) Find $\frac{dy}{dx}$ when $x = \tan(3y)$.

- First find $\frac{dx}{dy}$ with, you guessed it, the chain rule:

  $x = \tan u,\ u = 3y \Rightarrow \frac{dx}{du} = \sec^2 u = \sec^2(3y)$ and $\frac{du}{dy} = 3$

  $\frac{dx}{dy} = \frac{dx}{du} \times \frac{du}{dy} = 3\sec^2(3y)$

- Then use $\frac{dy}{dx} = \frac{1}{\left(\frac{dx}{dy}\right)}$ to get the final answer:

  $\frac{dy}{dx} = \frac{1}{3\sec^2(3y)} = \frac{1}{3}\cos^2(3y)$

Tip: See p.142 for more on $\frac{1}{\left(\frac{dx}{dy}\right)}$.

Once you get the hang of it, you don't need to use the chain rule every time. If it's a **simple function** inside, e.g. $\sin(kx)$, it just differentiates to $k\cos(kx)$. If it's more **complicated** though, like $\sin(x^3)$, it's worth using the **chain rule**.

When differentiating trig functions it's important to know **which part of the function** to turn into u.

- It's **not** always the part **inside the brackets.**
- When you have a trig function multiplied by itself, like $\cos^2 x$, it's often easiest to turn the **trig function itself** into u.

Tip: Remember $\cos^2 x$ is another way of writing $(\cos x)^2$.

Example 3

Find $\frac{dy}{dx}$ if $y = \sin^3 x$.

- Start off by rewriting the function as $y = (\sin x)^3$
- Now you have a function of a function and can carry out the chain rule in exactly the same way as before:

$$y = (\sin x)^3, \text{ so let } y = u^3 \text{ where } u = \sin x$$

- Differentiate y and u:

$$y = u^3 \Rightarrow \frac{dy}{du} = 3u^2 = 3\sin^2 x \quad \text{and} \quad \frac{du}{dx} = \cos x$$

- Then put it all into the chain rule formula:

$$\frac{dy}{dx} = \frac{dy}{du} \times \frac{du}{dx} = 3\sin^2 x \cos x$$

Tip: Don't get $(\sin x)^3$ confused with $\sin x^3$ — for this, you'd take $u = x^3$, so end up with $3x^2 \cos x^3$ when you differentiate.

When you're differentiating trig functions, you'll sometimes be asked to **rearrange** your answer to show it's equal to a **different** trig function. It's worth making sure you're familiar with **trig identities** (see Chapter 3) so you can spot which ones to use and when to use them.

Example 4

For $y = 2\cos^2 x + \sin(2x)$, show that $\frac{dy}{dx} = 2(\cos(2x) - \sin(2x))$.

- First rewrite the equation to make the chain rule easier to use:

$$y = 2\cos^2 x + \sin(2x) \Rightarrow y = 2(\cos x)^2 + \sin(2x)$$

- Then differentiate the parts separately. For the first bit:

$$y = 2u^2 \text{ where } u = \cos x \Rightarrow \frac{dy}{du} = 4u = 4\cos x, \ \frac{du}{dx} = -\sin x$$

- For the second bit:

$$y = \sin u \text{ where } u = 2x \Rightarrow \frac{dy}{du} = \cos u = \cos 2x, \ \frac{du}{dx} = 2$$

- Putting it all back into the chain rule formula:

$$\frac{dy}{dx} = [(4\cos x) \times (-\sin x)] + [(\cos(2x)) \times 2]$$
$$= 2\cos(2x) - 4\sin x \cos x$$

- From the target answer in the question it looks like you need a $\sin(2x)$ from somewhere, so use the double angle formula for sin (page 75):

$$\sin(2x) \equiv 2\sin x \cos x \Rightarrow 4\sin x \cos x \equiv 2\sin(2x)$$

$$\Rightarrow \frac{dy}{dx} = 2\cos(2x) - 2\sin(2x)$$
$$= 2(\cos(2x) - \sin(2x)) \text{ as required.}$$

Exercise 4.1

Q1d) Hint: Don't worry, you don't need to use the sin addition formula — use the chain rule with $u = x + \frac{\pi}{4}$.

Q1 Differentiate with respect to x:

a) $y = \sin(3x)$

b) $y = \cos(-2x)$

c) $y = \cos\frac{x}{2}$

d) $y = \sin\left(x + \frac{\pi}{4}\right)$

e) $y = 6\tan\frac{x}{2}$

f) $y = 3\tan(5x)$

Q2 Find $f'(x)$ if $f(x) = 3\tan(2x - 1)$.

Q3 Find $f'(x)$ if $f(x) = 3\tan x + \tan(3x)$.

Q4 Find $f'(x)$ if $f(x) = \sin\left(x^2 + \frac{\pi}{3}\right)$.

Q5 Find $f'(x)$ if $f(x) = \sin^2 x$.

Q6 Find $f'(x)$ if $f(x) = 2\sin^3 x$.

Q7 a) Find $f'(x)$ if $f(x) = 3\sin x + 2\cos x$.

b) Find the value of x for which $f'(x) = 0$ and $0 \leq x \leq \frac{\pi}{2}$.

Q8 Find $\frac{dy}{dx}$ if $y = \frac{1}{\cos x}$.

Q9 Use differentiation from first principles to show that $\frac{dy}{dx} = -\sin x$ when $y = \cos x$.

PROBLEM SOLVING

Q10-11 Hint: You'll need the double angle formulas here — see page 75.

Q10 Differentiate $y = \cos^2 x$ by:

a) Using the chain rule directly.

b) Expressing y in terms of $\cos(2x)$ and differentiating the result.

Q11 For $y = 6\cos^2 x - 2\sin(2x)$ show that $\frac{dy}{dx} = -6\sin(2x) - 4\cos(2x)$.

Q12 Find the gradient of the curve $y = \sin x$ when $x = \frac{\pi}{4}$.

Q13 Find the equation of the normal to the curve $y = \cos(2x)$ when $x = \frac{\pi}{4}$.

Q14 For the curve $x = \sin(2y)$:

a) Find the equation of the tangent at the point $\left(\frac{\sqrt{3}}{2}, \frac{\pi}{6}\right)$.

b) Find the equation of the normal at the point $\left(\frac{\sqrt{3}}{2}, \frac{\pi}{6}\right)$.

Q15 a) If $y = 2\sin(2x)\cos x$, express y as a difference of two expressions involving $\sin x$ and $\sin^3 x$.

PROBLEM SOLVING

b) Hence find $\frac{dy}{dx}$.

Differentiating by using the chain rule twice

Sometimes you'll have to use the chain rule **twice** when you have a function of a function of a function, like $\sin^3(x^2)$.

Example

Find $\frac{dy}{dx}$ if $y = \sin^2(2x + 1)$

- Start by setting up the first stage of differentiation with the **chain rule**, remembering to rewrite the $\sin^2$ part to make differentiating easier:

$$y = \sin^2(2x+1) = [\sin(2x+1)]^2 \Rightarrow y = u^2,\ u = \sin(2x+1)$$

- Finding $\frac{dy}{du}$ is easy, so start with that:

$$\frac{dy}{du} = 2u = 2\sin(2x+1)$$

- To find $\frac{du}{dx}$ you're going to need the **chain rule again**, so just set it up with u in terms of v instead of y in terms of u.

$$u = \sin(2x+1) \text{ so let } u = \sin v \text{ where } v = 2x+1$$

- Then go through the usual stages:

$$u = \sin v \Rightarrow \frac{du}{dv} = \cos v = \cos(2x+1)$$

$$v = 2x + 1 \Rightarrow \frac{dv}{dx} = 2$$

$$\frac{du}{dx} = \frac{du}{dv} \times \frac{dv}{dx} = 2\cos(2x+1)$$

- Now you have the value of $\frac{du}{dx}$ needed to complete the question:

$$\frac{dy}{dx} = \frac{dy}{du} \times \frac{du}{dx} = [2\sin(2x+1)] \times [2\cos(2x+1)]$$

$$= 4\sin(2x+1)\cos(2x+1)$$

Tip: Calling it v just means you don't end up with a load of u's floating around.

Exercise 4.2

Q1 Find $\frac{dy}{dx}$ if:

a) $y = \sin(\cos(2x))$ b) $y = 2\ln(\cos(3x))$

c) $y = \ln(\tan^2(x))$ d) $y = e^{\tan(2x)}$

Q2 Differentiate the following functions with respect to x:

a) $y = \sin^4(x^2)$ b) $y = e^{\sin^2 x}$

c) $y = \tan^2(3x) + \sin x$ d) $y = e^{2\cos(2x)} + \cos^2(2x)$

Hint: Differentiating e was covered on page 143 and differentiating ln was covered on page 145.

5. Product Rule

Learning Objectives:

- Be able to use the product rule for differentiation and recognise when it's needed.
- Be able to use the product rule together with other rules for differentiation in order to differentiate more complicated functions.

The product rule is a way of differentiating two functions multiplied together. It's fairly simple to use, but can get tricky when put together with other rules.

Differentiating functions multiplied together

To differentiate two functions multiplied together, use the **product rule**:

$$\text{If } y = uv$$
$$\frac{dy}{dx} = u\frac{dv}{dx} + v\frac{du}{dx}$$

Where u and v are functions of x, i.e. $u(x)$ and $v(x)$.

Proving this rule is a lot trickier than anything covered in this section (and you won't need to know how to do it in the exam), but it goes like this:

You need the formula for differentiation from first principles:

$$f'(x) = \lim_{h \to 0} \frac{f(x+h) - f(x)}{h}$$

So when you have a product to differentiate, this can be written as:

$$(fg)'(x) = \lim_{h \to 0} \frac{fg(x+h) - fg(x)}{h} = \lim_{h \to 0} \frac{f(x+h)g(x+h) - f(x)g(x)}{h}$$

The numerator of this fraction can be seen as the area of a **rectangle f(x + h) by g(x + h)** minus the area of a **rectangle f(x) by g(x)**.

It can therefore be rewritten as a sum of the areas of the "extra bits" on the diagram:

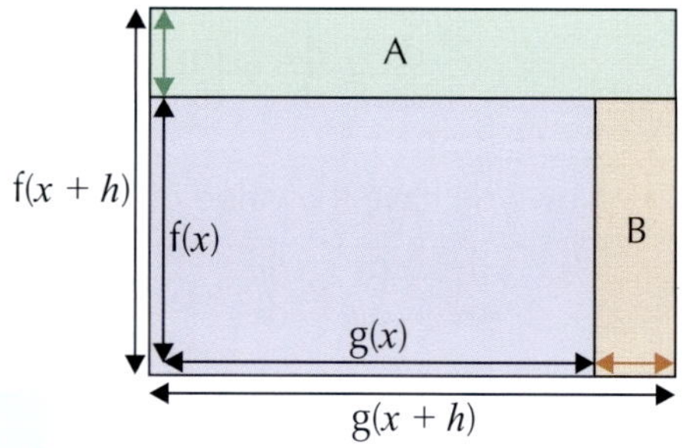

Tip: The length of the green arrow is $f(x+h) - f(x)$ and the length of the orange arrow is $g(x+h) - g(x)$.

$$f(x+h)g(x+h) - f(x)g(x) = \text{Area(A)} + \text{Area(B)}$$

$\text{Area(A)} = g(x+h)[f(x+h) - f(x)]$ $\qquad$ $\text{Area(B)} = f(x)[g(x+h) - g(x)]$

$$(fg)'(x) = \lim_{h \to 0}\left(\frac{g(x+h)[f(x+h) - f(x)] + f(x)[g(x+h) - g(x)]}{h}\right)$$

$$= \lim_{h \to 0}\left(\frac{g(x+h)[f(x+h) - f(x)]}{h}\right) + \lim_{h \to 0}\left(\frac{f(x)[g(x+h) - g(x)]}{h}\right)$$

As $h \to 0$, $x + h \to x$ so this bit is just $g(x)$.

This bit has no h's in it, so the limit as $h \to 0$ is just $f(x)$.

$$= \lim_{h \to 0}(g(x+h)) \lim_{h \to 0}\left(\frac{f(x+h) - f(x)}{h}\right) + \lim_{h \to 0}(f(x)) \lim_{h \to 0}\left(\frac{g(x+h) - g(x)}{h}\right)$$

These are the definitions of $f'(x)$ and $g'(x)$.

$$= g(x)f'(x) + f(x)g'(x)$$

Tip: $g(x)f'(x) + f(x)g'(x)$ is just another way of writing $u\frac{dv}{dx} + v\frac{du}{dx}$.

Example 1

Differentiate $x^3 \tan x$ with respect to x.

- The crucial thing is to write down everything in **steps**. Start by identifying 'u' and 'v':

$$u = x^3 \text{ and } v = \tan x$$

- Now differentiate these two **separately**, with respect to x:

$$\frac{du}{dx} = 3x^2 \text{ and } \frac{dv}{dx} = \sec^2 x$$

- Very **carefully** put all the bits into the formula:

$$\frac{dy}{dx} = u\frac{dv}{dx} + v\frac{du}{dx} = (x^3 \times \sec^2 x) + (\tan x \times 3x^2)$$

- Finally, **rearrange** to make it look nicer:

$$\frac{dy}{dx} = x^3 \sec^2 x + 3x^2 \tan x$$

Tip: See page 149 for how to differentiate $\tan x$.

You might have to differentiate functions using a mixture of the **product rule** and the **chain rule** (as well as the rules for e, ln and trig functions). In a question, you might be told which rules to use, but it's not guaranteed, so make sure you get used to spotting when the different rules are needed.

Example 2

Differentiate $e^{2x}\sqrt{2x-3}$ with respect to x.

- It's a **product** of two functions, so start by identifying 'u' and 'v':

$$u = e^{2x} \text{ and } v = \sqrt{2x-3}$$

- Each of these needs the **chain rule** to differentiate:

$$\frac{du}{dx} = 2e^{2x} \text{ and } \frac{dv}{dx} = \frac{1}{\sqrt{2x-3}}$$

- Put it all into the **product rule formula**:

$$\frac{dy}{dx} = u\frac{dv}{dx} + v\frac{du}{dx} = \left(e^{2x} \times \frac{1}{\sqrt{2x-3}}\right) + (\sqrt{2x-3} \times 2e^{2x})$$

- As before, **rearrange** it and then **simplify**:

$$\frac{dy}{dx} = e^{2x}\left(\frac{1}{\sqrt{2x-3}} + 2(\sqrt{2x-3})\right) = e^{2x}\left(\frac{1+2(2x-3)}{\sqrt{2x-3}}\right)$$

$$= \frac{e^{2x}(4x-5)}{\sqrt{2x-3}}$$

Tip: The chain rule bit has been done all in one go here to save time, but if you need to, do it in steps just to make sure nothing goes wrong.

You might also see questions that ask you to rearrange the final answer to 'show that' it's equal to something, or to find the stationary points of a graph.

Example 3

Show that the derivative of $x^2(2x-1)^3$ is $2x(2x-1)^2(5x-1)$.

- As usual, **identify** u and v then differentiate them **separately**:

$$u = x^2 \text{ and } v = (2x-1)^3$$

$$\Rightarrow \frac{du}{dx} = 2x \text{ and } \frac{dv}{dx} = 2 \times 3(2x-1)^2 \text{ (using the chain rule for } \frac{dv}{dx}\text{)}$$

- Then put it all into the **product rule formula**:

$$\frac{dy}{dx} = u\frac{dv}{dx} + v\frac{du}{dx} = [x^2 \times 6(2x-1)^2] + [(2x-1)^3 \times 2x]$$
$$= 6x^2(2x-1)^2 + 2x(2x-1)^3$$

- This isn't exactly how the question wants the answer, so it needs a little more **rearranging**:

$$6x^2(2x-1)^2 + 2x(2x-1)^3 = 2x(2x-1)^2(3x + (2x-1))$$
$$= 2x(2x-1)^2(5x-1)$$

Tip: In a 'show that' question it's good to look at the final answer so that you know you're on the right track. Here you can see you need to take out $2x(2x-1)^2$ at some point, so it's worth starting with that.

Example 4

The graph of $y = x^3 \ln x$ has two points of inflection. Find the x-coordinates of these points, leaving your answers as exact values.

- There'll be a point of inflection when $\frac{d^2y}{dx^2} = 0$.
- Start by finding $\frac{dy}{dx}$ — identify u and v and use the **product rule**:

$$u = x^3 \text{ and } v = \ln x$$
$$\Rightarrow \frac{du}{dx} = 3x^2 \text{ and } \frac{dv}{dx} = \frac{1}{x}$$
$$\text{So } \frac{dy}{dx} = u\frac{dv}{dx} + v\frac{du}{dx} = \left(x^3 \times \frac{1}{x}\right) + (\ln x \times 3x^2)$$
$$= x^2 + 3x^2 \ln x$$

- Now differentiate again to find $\frac{d^2y}{dx^2}$:

First part: $\frac{d}{dx}(x^2) = 2x$

Second part: $\frac{d}{dx}(3x^2 \ln x)$ — use the **product rule** again:

$$u = 3x^2 \text{ and } v = \ln x$$
$$\Rightarrow \frac{du}{dx} = 6x \text{ and } \frac{dv}{dx} = \frac{1}{x}$$
$$v\frac{du}{dx} + u\frac{dv}{dx} = \left(3x^2 \times \frac{1}{x}\right) + (\ln x \times 6x) = 3x + 6x \ln x$$

Putting these together: $\frac{d^2y}{dx^2} = 2x + 3x + 6x \ln x = x(5 + 6 \ln x)$

- So $\frac{d^2y}{dx^2} = 0$ either when $x = 0$

or when $5 + 6 \ln x = 0 \Rightarrow 6 \ln x = -5 \Rightarrow \ln x = -\frac{5}{6} \Rightarrow x = e^{-\frac{5}{6}}$

Tip: You're told that there are two points of inflection, so these two x-coordinates must be the answers. The question asks for exact values of x, so leave the second answer in terms of e rather than giving a rounded decimal from your calculator.

Exercise 5.1

Q1 Differentiate $y = x(x + 2)$ with respect to x by:

a) Multiplying the brackets out and differentiating directly.

b) Using the product rule.

Q2 Differentiate with respect to x:

a) $y = x^2(x + 6)^3$ b) $y = x^3(5x + 2)^4$ c) $y = x^3e^x$

d) $y = xe^{4x}$ e) $y = xe^{x^2}$ f) $y = e^{2x}\sin x$

Q3 Find $f'(x)$ if:

a) $f(x) = x^3(x + 3)^{\frac{1}{2}}$ b) $f(x) = \dfrac{x^2}{\sqrt{x-7}}$ c) $f(x) = x^4\ln x$

d) $f(x) = 4x\ln x^2$ e) $f(x) = 2x^3\cos x$ f) $f(x) = x^2\cos(2x)$

Q3b) Hint: Remember that $\frac{1}{\sqrt{x}} = x^{-\frac{1}{2}}$.

Q4 For parts a) and b), multiply out the brackets in your answer and simplify.

a) Differentiate $y = (x + 1)^2(x^2 - 1)$.

b) Differentiate $y = (x + 1)^3(x - 1)$.

c) Your answers to part a) and part b) should be the same. Show by rearranging that the expressions for y in parts a) and b) are the same.

Q5 Find the range of values of x for which the curve $y = xe^x$ is concave.

Q5 Hint: You'll have to differentiate twice here.

Q6 Find the equation of the tangent to the curve $y = (\sqrt{x+2})(\sqrt{x+7})$ at the point (2, 6). Write your answer in the form $ax + by + c = 0$, where a, b and c are integers.

Q7 For the curve $y = \dfrac{\sqrt{x-1}}{\sqrt{x+4}}$

a) Find the equation of the tangent to the curve when $x = 5$ in the form $ax + by + c = 0$ where a, b and c are integers.

b) Find the equation of the normal to the curve when $x = 5$ in the form $ax + by + c = 0$ where a, b and c are integers.

Q7 Hint: To use the product rule here you'll need to rewrite the function at the bottom as a negative power.

Q8 Differentiate $y = e^{x^2\sqrt{x+3}}$.

Q9 Find any stationary points for the curve $y = xe^{x-x^2}$.

Q10 a) Find any stationary points of the curve $y = (x - 2)^2(x + 4)^3$.

b) By writing the first derivative of $y = (x - 2)^2(x + 4)^3$ in the form $\frac{dy}{dx} = (Ax^2 + Bx + C)(x + D)^n$, find $\frac{d^2y}{dx^2}$ and hence identify the nature of the stationary points of the curve.

6. Quotient Rule

Learning Objectives:
- Be able to use the quotient rule for differentiation and understand when it's needed.
- Be able to use the quotient rule alongside other methods for differentiating complex functions.

You've seen how to differentiate products with the product rule, and now you'll see how to differentiate quotients (divisions) with the quotient rule.

Differentiating a function divided by a function

In maths a **quotient** is one thing **divided** by another. As with the product rule, there's a rule that lets you differentiate quotients easily — the **quotient rule**:

If $y = \frac{u}{v}$

$$\frac{dy}{dx} = \frac{v\frac{du}{dx} - u\frac{dv}{dx}}{v^2}$$

Where u and v are functions of x, i.e. $u(x)$ and $v(x)$.

Tip: The quotient rule is basically just the product rule on $y = uv^{-1}$ — try it on two simple functions and see for yourself. It's usually quicker to use the quotient rule though.

There's also a proof for the quotient rule — again you won't need to know it for the exam, but you might find it helpful in understanding how it works.

- As before, start with the differentiation from first principles.

$$\frac{d}{dx}f(x) = \lim_{h \to 0} \frac{f(x+h) - f(x)}{h}$$

- And so for the quotient $\frac{f(x)}{g(x)}$, this becomes

$$\frac{d}{dx}\left(\frac{f(x)}{g(x)}\right) = \lim_{h \to 0} \frac{\frac{f(x+h)}{g(x+h)} - \frac{f(x)}{g(x)}}{h}$$

- Put the top of the fraction over a common denominator $(g(x+h)g(x))$ and multiply this common denominator by the h.

$$\frac{d}{dx}\left(\frac{f(x)}{g(x)}\right) = \lim_{h \to 0} \frac{f(x+h)g(x) - f(x)g(x+h)}{g(x+h)g(x)h}$$

- The next stage is to add and subtract $f(x)g(x)$ and then factorise.

$$\frac{d}{dx}\left(\frac{f(x)}{g(x)}\right) = \lim_{h \to 0} \frac{f(x+h)g(x) - f(x)g(x) + f(x)g(x) - f(x)g(x+h)}{g(x+h)g(x)h}$$

$$\frac{d}{dx}\left(\frac{f(x)}{g(x)}\right) = \lim_{h \to 0} \frac{g(x)\,[f(x+h) - f(x)] - f(x)\,[g(x+h) - g(x)]}{g(x+h)g(x)h}$$

Tip: Adding and subtracting the same thing is a classic trick in algebra. It's just like adding zero, and it can get you from algebraic mess to perfectly formed equations.

- You might start to recognise the top row here. To make it a bit clearer, divide both the top and bottom by h, keeping $f(x)$ and $g(x)$ aside.

$$\frac{d}{dx}\left(\frac{f(x)}{g(x)}\right) = \lim_{h \to 0} \frac{g(x)\frac{f(x+h) - f(x)}{h} - f(x)\frac{g(x+h) - g(x)}{h}}{g(x+h)g(x)}$$

- The green bits on top are the definition of $f'(x)$ and $g'(x)$. As h tends to zero, the blue bit at the bottom becomes $g(x)g(x)$, or $(g(x))^2$:

$$\frac{d}{dx}\left(\frac{f(x)}{g(x)}\right) = \frac{g(x)f'(x) - f(x)g'(x)}{(g(x))^2}$$

Tip: The expression $\frac{f'(x)g(x) - f(x)g'(x)}{(g(x))^2}$ is just another way of writing the quotient rule $\frac{v\frac{du}{dx} - u\frac{dv}{dx}}{v^2}$.

It's given in terms of $f(x)$ and $g(x)$ in the formula booklet.

Example 1

Find $\frac{dy}{dx}$ if $y = \frac{\sin x}{2x+1}$.

- You can see that y is a **quotient** in the form of $\frac{u}{v}$.
  First identify u and v and differentiate them **separately**:

$$u = \sin x \quad \Rightarrow \quad \frac{du}{dx} = \cos x$$

and $$v = 2x + 1 \quad \Rightarrow \quad \frac{dv}{dx} = 2$$

- Then just put the correct bits into the quotient rule. It's important that you get things in the right order, so concentrate on what's going where:

$$\frac{dy}{dx} = \frac{v\frac{du}{dx} - u\frac{dv}{dx}}{v^2} = \frac{(2x+1)(\cos x) - (\sin x)(2)}{(2x+1)^2}$$

- Now just neaten it up:

$$\frac{dy}{dx} = \frac{(2x+1)\cos x - 2\sin x}{(2x+1)^2}$$

Tip: This is quite similar to the product rule from the last section — the first stage is to identify u and v and differentiate them separately.

Example 2

Find the gradient of the tangent to the curve with equation $y = \frac{2x^2-1}{3x^2+1}$ at the point (1, 0.25).

- 'Find the gradient of the tangent' means you have to **differentiate**.
- First identify u and v for the **quotient** rule, and differentiate **separately**:

$$u = 2x^2 - 1 \quad \Rightarrow \quad \frac{du}{dx} = 4x$$

and $$v = 3x^2 + 1 \quad \Rightarrow \quad \frac{dv}{dx} = 6x$$

- Then put everything into the quotient rule:

$$\frac{dy}{dx} = \frac{v\frac{du}{dx} - u\frac{dv}{dx}}{v^2} = \frac{(3x^2+1)(4x) - (2x^2-1)(6x)}{(3x^2+1)^2}$$

- To make the expression **easier** to work with, **simplify** it where possible:

$$\frac{dy}{dx} = \frac{2x[2(3x^2+1) - 3(2x^2-1)]}{(3x^2+1)^2} = \frac{2x[6x^2+2-6x^2+3]}{(3x^2+1)^2}$$

$$= \frac{10x}{(3x^2+1)^2}$$

- Finally, put in $x = 1$ to find the gradient at (1, 0.25):

$$\frac{dy}{dx} = \frac{10}{(3+1)^2} = 0.625$$

Tip: Don't try and simplify straight away or you're more likely to get things mixed up.

Tip: If it's a normal rather than a tangent, do –1 ÷ gradient.

Example 3

> **Tip:** To determine the nature of the stationary point you're going to have to differentiate twice.

Determine the nature of the stationary point of the curve $y = \frac{\ln x}{x^2}$ ($x > 0$).

- First use the quotient rule to find $\frac{dy}{dx}$:

$$u = \ln x \Rightarrow \frac{du}{dx} = \frac{1}{x} \quad \text{and} \quad v = x^2 \Rightarrow \frac{dv}{dx} = 2x$$

$$\text{So } \frac{dy}{dx} = \frac{(x^2)\left(\frac{1}{x}\right) - (\ln x)(2x)}{x^4} = \frac{x - 2x\ln x}{x^4} = \frac{1 - 2\ln x}{x^3}$$

- The stationary point occurs where $\frac{dy}{dx} = 0$ (i.e. zero gradient), so this is when:

$$\frac{1 - 2\ln x}{x^3} = 0 \Rightarrow \ln x = \frac{1}{2} \Rightarrow x = e^{\frac{1}{2}}$$

- To find out whether it's a maximum or minimum, differentiate $\frac{1 - 2\ln x}{x^3}$ using the quotient rule to get $\frac{d^2y}{dx^2}$.

$$u = 1 - 2\ln x \Rightarrow \frac{du}{dx} = -\frac{2}{x} \quad \text{and} \quad v = x^3 \Rightarrow \frac{dv}{dx} = 3x^2.$$

$$\text{So } \frac{d^2y}{dx^2} = \frac{(x^3)(-\frac{2}{x}) - (1 - 2\ln x)(3x^2)}{x^6} = \frac{6x^2\ln x - 5x^2}{x^6} = \frac{6\ln x - 5}{x^4}$$

- Now put in the x-value of your stationary point:

$$\frac{d^2y}{dx^2} = \frac{6\ln e^{\frac{1}{2}} - 5}{(e^{\frac{1}{2}})^4} = \frac{3 - 5}{e^2} = -0.27...$$

- $\frac{d^2y}{dx^2}$ is negative, so it's a maximum stationary point.

> **Tip:** A negative second derivative means a maximum and positive means a minimum. If the second derivative is zero, it could be either, or a point of inflection. See p.137-138 for more on this.

As you saw on page 149, the derivative of $\tan x$ is $\sec^2 x$.

Because $\tan x = \frac{\sin x}{\cos x}$, you can prove this using the quotient rule.

Example 4

Prove that the derivative of $\tan x$ with respect to x is $\sec^2 x$. PROBLEM SOLVING

- First write $\tan x$ out as a quotient and set up u and v for the quotient rule:

$$\tan x = \frac{\sin x}{\cos x} = \frac{u}{v}, \text{ so } u = \sin x, \ \frac{du}{dx} = \cos x \ \text{ and } \ v = \cos x, \ \frac{dv}{dx} = -\sin x$$

- Then just put all the right bits into the quotient rule:

$$\frac{d}{dx}(\tan x) = \frac{v\frac{du}{dx} - u\frac{dv}{dx}}{v^2} = \frac{\cos x \cos x - \sin x(-\sin x)}{\cos^2 x} = \frac{\cos^2 x + \sin^2 x}{\cos^2 x}$$

$$= \frac{1}{\cos^2 x} = \sec^2 x \text{ as required}$$

> **Tip:** The identity $\cos^2 x + \sin^2 x \equiv 1$ was used here to simplify.

Exercise 6.1

Q1 Differentiate with respect to x:

a) $y = \frac{x+5}{x-3}$ b) $y = \frac{(x-7)^4}{(5-x)^3}$

c) $y = \frac{e^x}{x^2}$ d) $y = \frac{3x}{(x-1)^2}$

Q2 Find $f'(x)$ for each of the following functions:

a) $f(x) = \frac{x^3}{(x+3)^3}$ b) $f(x) = \frac{x^2}{\sqrt{x-7}}$

c) $f(x) = \frac{e^{2x}}{e^{2x}+e^{-2x}}$ d) $f(x) = \frac{x}{\sin x}$

e) $f(x) = \frac{\sin x}{x}$

Q3 Find $f'(x)$ if $f(x) = \frac{x^2}{\tan x}$,
giving your answer in terms of $\cot x$ and $\operatorname{cosec} x$.

Q4 The graph of $y = \frac{5x-4}{2x^2}$ has one stationary point.
Find the coordinates of this point, and show that it is a maximum.

Q5 Use the quotient rule to find the coordinates of any points of inflection of the graph $y = \frac{x}{e^x}$.

Q5 Hint: Remember from page 136 that if (a, b) is a point of inflection: $\frac{d^2y}{dx^2} = 0$ when $x = a$, and $\frac{d^2y}{dx^2}$ changes sign either side of $x = a$.

Q6 a) Differentiate $y = \frac{x}{\cos(2x)}$ with respect to x.

b) Show that $\frac{dy}{dx} = 0$ when $x = -\frac{1}{2}\cot(2x)$
(you do not need to solve this equation).

Q7 For the curve $y = \frac{1}{1+4\cos x}$:

a) Find the equation of the tangent to the curve when $x = \frac{\pi}{2}$.

b) Find the equation of the normal to the curve when $x = \frac{\pi}{2}$.

Q8 For the curve $y = \frac{2x}{\cos x}$, find the exact value of $\frac{dy}{dx}$ when $x = \frac{\pi}{3}$.

Q9 Show that if $y = \frac{x - \sin x}{1+\cos x}$ then $\frac{dy}{dx} = \frac{x\sin x}{(1+\cos x)^2}$.

Q10 Find any turning points on the curve $y = \frac{\cos x}{4-3\cos x}$ in the range $0 \leq x \leq 2\pi$.

Q11 Differentiate $y = e^{\frac{1+x}{1-x}}$ with respect to x.

Q12 Find the set of values of x for which $\frac{2+3x^2}{3x-1}$ is increasing.

Q12 Hint: You've seen increasing functions in Year 1 — a function $f(x)$ is increasing when its gradient $f'(x) > 0$.

7. More Differentiation

Learning Objectives:

- Be able to differentiate cosec, sec and cot.
- Be able to use these results to differentiate more complicated functions.

In this section you'll see how to differentiate reciprocals of trig functions — sec x, cosec x and cot x. All of these can be derived from the quotient rule.

Differentiating cosec, sec and cot

Remember from page 55 the definitions of these trig functions:

$$\text{cosec}\,x \equiv \frac{1}{\sin x} \qquad \sec x \equiv \frac{1}{\cos x} \qquad \cot x \equiv \frac{1}{\tan x} \equiv \frac{\cos x}{\sin x}$$

- Since **cosec**, **sec** and **cot** are just **reciprocals** of **sin**, **cos** and **tan**, the quotient rule can be used to differentiate them.
- The following results are in the **formula booklet**, but it will help a lot if you understand where they come from.

If $y =$	$\frac{dy}{dx} =$
$\text{cosec}\,x$	$-\text{cosec}\,x\cot x$
$\sec x$	$\sec x\tan x$
$\cot x$	$-\text{cosec}^2\,x$

Tip: If you can't remember which trig functions give a negative result when you differentiate them, just remember it's all the ones that begin with c — cos, cosec and cot.

Example 1

a) Use the quotient rule to differentiate $y = \frac{\cos x}{\sin x}$, and hence show that for $y = \cot x$, $\frac{dy}{dx} = -\text{cosec}^2\,x$.

- Start off by identifying $u = \cos x$ and $v = \sin x$.
- Differentiating separately gives:

$$\frac{du}{dx} = -\sin x \text{ and } \frac{dv}{dx} = \cos x \text{ (see page 150)}$$

$$\frac{dy}{dx} = \frac{(\sin x \times -\sin x) - (\cos x \times \cos x)}{(\sin x)^2} = \frac{-\sin^2 x - \cos^2 x}{\sin^2 x}$$

- Simplify using a trig identity: $\sin^2 x + \cos^2 x \equiv 1$ seems fitting.

$$\frac{dy}{dx} = \frac{-(\sin^2 x + \cos^2 x)}{\sin^2 x} = \frac{-1}{\sin^2 x}$$

- Linking this back to the question, $\tan x \equiv \frac{\sin x}{\cos x}$ and $\cot x \equiv \frac{1}{\tan x}$, so $y = \frac{\cos x}{\sin x} = \cot x$.
- And as $\text{cosec}\,x \equiv \frac{1}{\sin x}$, then:

$$\frac{dy}{dx} = \frac{-1}{\sin^2 x} = -\text{cosec}^2\,x$$

Tip: 'Show that' questions on trig functions often involve using a common identity, so make sure you know them — see Chapter 3.

b) Show that $\frac{d}{dx}\text{cosec } x = -\text{cosec } x \cot x$.

- $\text{cosec } x \equiv \frac{1}{\sin x}$, so use the **quotient rule**:

$$u = 1 \Rightarrow \frac{du}{dx} = 0 \text{ and } v = \sin x \Rightarrow \frac{dv}{dx} = \cos x$$

$$\frac{dy}{dx} = \frac{v\frac{du}{dx} - u\frac{dv}{dx}}{v^2} = \frac{(\sin x \times 0) - (1 \times \cos x)}{\sin^2 x} = -\frac{\cos x}{\sin^2 x}$$

- Since $\cot x \equiv \frac{\cos x}{\sin x}$, and $\text{cosec } x \equiv \frac{1}{\sin x}$:

$$\frac{dy}{dx} = \frac{1}{\sin x} \times \left(-\frac{\cos x}{\sin x}\right) = -\text{cosec } x \cot x$$

c) Show that $\frac{d}{dx}\sec x = \sec x \tan x$.

- Using the **quotient rule** for $\sec x \equiv \frac{1}{\cos x}$:

$$u = 1 \Rightarrow \frac{du}{dx} = 0 \text{ and } v = \cos x \Rightarrow \frac{dv}{dx} = -\sin x$$

$$\frac{dy}{dx} = \frac{v\frac{du}{dx} - u\frac{dv}{dx}}{v^2} = \frac{(\cos x \times 0) - (1 \times -\sin x)}{\cos^2 x} = \frac{\sin x}{\cos^2 x}$$

- Since $\tan x \equiv \frac{\sin x}{\cos x}$, and $\sec x \equiv \frac{1}{\cos x}$,

$$\frac{dy}{dx} = \frac{1}{\cos x} \times \frac{\sin x}{\cos x} = \sec x \tan x$$

Tip: You could also use the chain rule on $\frac{1}{\sin x} = (\sin x)^{-1}$.

Tip: As above, you could write $\sec x$ as $\frac{1}{\cos x} = (\cos x)^{-1}$, and then use the chain rule.

As with other rules covered in this chapter, the rules for $\sec x$, $\text{cosec } x$ and $\cot x$ can be used with the **chain**, **product** and **quotient rules** and in combination with all the other functions you've seen so far.

Example 2

a) Find $\frac{dy}{dx}$ if $y = \cot \frac{x}{2}$.

- This is a function (cot) of a function ($\frac{x}{2}$), so you need the **chain rule**.
- Although $\cot x \equiv \frac{\cos x}{\sin x}$, you don't need the quotient rule as you know that $\cot x$ differentiates to give $-\text{cosec}^2 x$.
- You can go straight to identifying u to use in the chain rule:

$$y = \cot u \Rightarrow \frac{dy}{du} = -\text{cosec}^2 u = -\text{cosec}^2 \frac{x}{2} \text{ and } u = \frac{x}{2} \Rightarrow \frac{du}{dx} = \frac{1}{2}$$

$$\text{So } \frac{dy}{dx} = \frac{dy}{du} \times \frac{du}{dx} = -\frac{1}{2}\text{cosec}^2 \frac{x}{2}$$

Tip: Look back at p.139 for a reminder of the chain rule.

b) Find $\frac{dy}{dx}$ if $y = \sec(2x^2)$.

This is another **function of a function**, so more **chain rule**:

$$y = \sec u \quad \text{and} \quad u = 2x^2$$

$$\frac{dy}{du} = \sec u \tan u = \sec(2x^2) \tan(2x^2) \quad \text{and} \quad \frac{du}{dx} = 4x$$

$$\text{So } \frac{dy}{dx} = \frac{dy}{du} \times \frac{du}{dx} = 4x \sec(2x^2) \tan(2x^2)$$

c) Find $\frac{dy}{dx}$ if $y = e^x \cot x$.

This is a **product** of two functions, so think '**product rule**' (p.154):

$$u = e^x \text{ and } v = \cot x$$

$$\Rightarrow \frac{du}{dx} = e^x \text{ and } \frac{dv}{dx} = -\text{cosec}^2\, x$$

So $\frac{dy}{dx} = u\frac{dv}{dx} + v\frac{du}{dx} = (e^x \times -\text{cosec}^2\, x) + (\cot x \times e^x)$

$= e^x(\cot x - \text{cosec}^2\, x)$

Tip: If it was a more difficult function than x inside the 'cot', you'd do this in exactly the same way but use the chain rule for working out $\frac{dv}{dx}$.

Exercise 7.1

Q1 Differentiate with respect to x:

a) $y = \text{cosec}\,(2x)$ b) $y = \text{cosec}^2\, x$ c) $y = \cot\,(7x)$

d) $y = \cot^7 x$ e) $y = x^4 \cot x$ f) $y = (x + \sec x)^2$

g) $y = \text{cosec}\,(x^2 + 5)$ h) $y = e^{3x} \sec x$ i) $y = (2x + \cot x)^3$

Q2 Find f′(x) if:

a) $f(x) = \frac{\sec x}{x + 3}$ b) $f(x) = \sec \frac{1}{x}$ c) $f(x) = \sec \sqrt{x}$

Q3 Find f′(x) if $f(x) = (\sec x + \text{cosec}\, x)^2$.

Q4 Find f′(x) if $f(x) = \frac{1}{x \cot x}$.

Q5 Find f′(x) if $f(x) = e^x \text{cosec}\, x$.

Q6 Find f′(x) if $f(x) = e^{3x} \cot\,(4x)$.

Q7 Find f′(x) if $f(x) = e^{-2x} \text{cosec}\,(4x)$.

Q8 Find f′(x) if $f(x) = \ln\,(x)\,\text{cosec}\, x$.

Q9 Find f′(x) if $f(x) = \sqrt{\sec x}$.

Q10 Find f′(x) if $f(x) = e^{\sec x}$.

Q11 a) Find f′(x) if $f(x) = \ln\,(\text{cosec}\, x)$.

b) Show that the function in part a) can be written as $-\ln\,(\sin x)$ and differentiate it — you should get the same answer as in part a).

Q11 Hint: Remember the log laws from Year 1.

Q12 Find f′(x) if $f(x) = \ln\,(x + \sec x)$.

Q13 Differentiate $y = \sec\,(\sqrt{x^2 + 5})$.

8. Connected Rates of Change

A derivative like $\frac{dy}{dx}$ *is also called a 'rate of change' — it's the rate of change in* y *with respect to* x*. You can apply different rates of change to a given situation, and if they have variables in common you say they're 'connected'.*

Learning Objective:
- Be able to form differential equations from situations involving connected rates of change.

Connected rates of change

- Some situations have a number of **linked variables**, like length, surface area and volume or distance, speed and acceleration.
- If you know the **rate of change** of one of these linked variables, and the equations that connect the variables, you can use the **chain rule** to help you find the rate of change of the other variables.
- An equation connecting variables with their rates of change (i.e. with a derivative term) is called a **differential equation**.
- When something changes over **time**, the derivative is $\frac{\mathbf{d}}{\mathbf{d}t}$ of that variable.

Tip: There's more on solving differential equations in Chapter 8 (see pages 221-228).

Example 1

a) If $y = 3e^{5x}$ and $\frac{dx}{dt} = 2$, work out $\frac{dy}{dt}$ when $x = -1$.

- Start off by **differentiating** the expression for y, with respect to x:

$$y = 3e^{5x} \Rightarrow \frac{dy}{dx} = 15e^{5x}$$

- Write out the **chain rule** for $\frac{dy}{dt}$, using the information available:

$$\frac{dy}{dt} = \frac{dy}{dx} \times \frac{dx}{dt}$$

- Put in all the things you know to work out $\frac{dy}{dt}$:

$$\frac{dy}{dt} = 15e^{5x} \times 2 = 30e^{5x}$$

- Now find the value of $\frac{dy}{dt}$ at $x = -1$: $\frac{dy}{dt} = 30e^{-5}$

Tip: Sometimes it helps to write out the chain rule first so you know which rate of change is missing and needs to be worked out.

b) y is the surface area of a sphere, and x is its radius. The rate of change of the radius, $\frac{dx}{dt} = -2$. Find $\frac{dy}{dt}$ when $x = 2.5$.

PROBLEM SOLVING

- This is trickier because you're not given the expression for y. But you should know (or be able to look up) the surface area of a sphere:

$$y = 4\pi x^2$$

- Now **differentiate** as before: $\frac{dy}{dx} = 8\pi x$
- Write out the **chain rule** for $\frac{dy}{dt}$: $\frac{dy}{dt} = \frac{dy}{dx} \times \frac{dx}{dt} = 8\pi x \times -2 = -16\pi x$
- Now find $\frac{dy}{dt}$ when $x = 2.5$: $\frac{dy}{dt} = -16\pi \times 2.5 = -40\pi$

Often you'll see much wordier questions involving related rates of change, like the one in the example below, where you have to do a bit more work to figure out where to start.

Tip: If you've forgotten the formulas for areas and volumes of 3D solids, you should brush up on them now. They crop up quite a bit in these questions.

Example 2

MODELLING

A scientist is testing how a new material expands when it is gradually heated. The diagram below shows the sample being tested, which is modelled as a triangular prism.

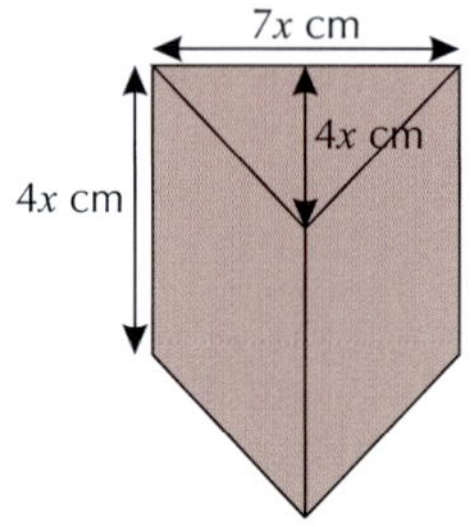

After t minutes, the triangle that forms the base of the prism has base length $7x$ cm and height $4x$ cm, and the length of the prism is also $4x$ cm.

If the sample expands at a constant rate, given by $\frac{dx}{dt} = 0.05$ cm min^{-1}, find an expression in terms of x for $\frac{dV}{dt}$, where V is the volume of the prism.

- The best way to start this kind of question is to write down what you know. There's enough information to write an expression for the **volume of the prism**:

Area of cross-section — Length of prism

$$V = \left(\frac{1}{2} \times 7x \times 4x\right) \times 4x$$

$$\Rightarrow V = 56x^3 \text{ cm}^3$$

- **Differentiate** this expression for the volume with respect to x:

$$\frac{dV}{dx} = 168x^2$$

- You know that $\frac{dx}{dt} = 0.05$ cm min^{-1}.

  So you can use the **chain rule** to find $\frac{dV}{dt}$:

$$\frac{dV}{dt} = \frac{dV}{dx} \times \frac{dx}{dt}$$

$$\Rightarrow \frac{dV}{dt} = 168x^2 \times 0.05 = 8.4x^2 \text{ cm}^3 \text{ min}^{-1}$$

There are a couple of tricks in this type of question that could catch you out if you're not prepared for them. In the example coming up on the next page, you have to spot that there's a hidden derivative described in words.

You also need to remember the rule $\frac{dy}{dx} = \frac{1}{\left(\frac{dx}{dy}\right)}$ (see p.142).

Example 3

A giant metal cube from space is cooling after entering the Earth's atmosphere. As it cools, the surface area of the cube is modelled as decreasing at a constant rate of 0.027 m^2 s^{-1}.

MODELLING

If the side length of the cube after t seconds is x m, find $\frac{dx}{dt}$ at the point when $x = 15$ m.

- Start with what you know:

The cube has side length x **m**.

So the surface area of the cube is: $A = 6x^2 \Rightarrow \frac{dA}{dx} = 12x$

A **decreases** at a constant rate of 0.027 m^2 s^{-1}.

We can write this as $\frac{dA}{dt} = -0.027$.

This value is negative because A is decreasing.

Use $\frac{d}{dt}$ because it's a rate of time.

- Now use the **chain rule** to find $\frac{dx}{dt}$:

$$\frac{dx}{dt} = \frac{dx}{dA} \times \frac{dA}{dt} = \frac{1}{\left(\frac{dA}{dx}\right)} \times \frac{dA}{dt} = \frac{1}{12x} \times -0.027 = -\frac{0.00225}{x}$$

- So when $x = 15$, $\frac{dx}{dt} = -\frac{0.00225}{x} = -\frac{0.00225}{15} = -0.00015 \text{ m s}^{-1}$

Tip: There's a 'hidden' derivative here — the rate of decrease in area. You can see from the units (m^2 s^{-1}) that it's a change in area (m^2) with respect to time (s).

Tip: $\frac{dx}{dt}$ is a change in length over time. Throughout the question the units used for length are metres, and the units for time are seconds, so $\frac{dx}{dt}$ must be in units of m s^{-1}.

Exercise 8.1

MODELLING

Q1 A cube with sides x cm is cooling and the sides are shrinking by 0.1 cm min^{-1}. Find an expression for $\frac{dV}{dt}$, the rate of change of volume with respect to time.

Q2 A cuboid block of sides $2x$ cm by $3x$ cm by $5x$ cm expands when heated such that x increases at a rate of 0.15 cm °C^{-1}. If the volume of the cuboid at temperature θ °C is V cm^3, find $\frac{dV}{d\theta}$ when $x = 3$.

Q3 A snowball of radius r cm is melting. Its radius decreases by 1.6 cm h^{-1}. If the surface area of the snowball at time t hours is A cm^2, find $\frac{dA}{dt}$ when $r = 5.5$ cm. Give your answer to 2 d.p.

Q3 Hint: Model the snowball as a sphere (Surface area $A = 4\pi r^2$).

Q4 A spherical satellite, radius r m, expands as it enters the atmosphere. It grows by 2×10^{-2} mm for every 1 °C rise in temperature. Find an expression $\frac{dV}{d\theta}$ for the rate of change of volume with respect to temperature.

Q4 Hint: Take extra care with the units here.

Q5 Heat, H, is lost from a closed cylindrical tank of radius r cm and height $3r$ cm at a rate of 2 J cm^{-2} of surface area, A. Find $\frac{dH}{dr}$ when $r = 12.3$. Give your answer to 2 d.p.

Q6 Hint: Treat r as a constant as the radius will not change.

Q6 A cylindrical polishing block of radius r cm and length H cm is worn down at one circular end at a rate of 0.5 mm h^{-1}. Find an expression for the rate of change of the volume of the block with respect to time.

Q7 A crystal of a salt is shaped like a prism. Its cross section is an equilateral triangle with sides x mm and the height of the crystal is 20 mm. New material is deposited only on the rectangular faces of the prism (i.e. the height does not change), so that x increases at a rate of 0.6 mm per day.

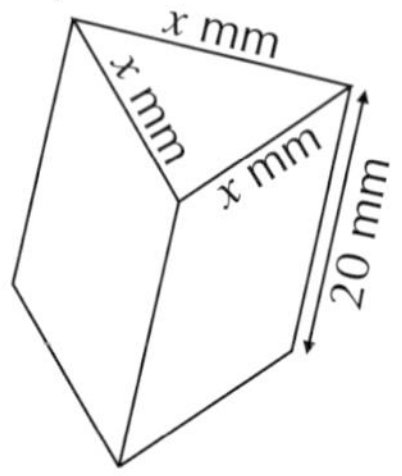

Q7a) Hint: You'll need to use Pythagoras or trig.

a) Find an expression for the area of the end of the prism in terms of x.

b) Find an expression for the rate of change of the volume of the crystal with respect to time.

c) Find the rate of change of the volume of the crystal with respect to time when $x = 0.5$.

Q8 The growth of a population of bacteria in a sample dish is modelled by the equation:

$$D = 1 + 2^{\lambda t}$$

where D is the diameter of the colony in mm, t is time in days, and λ is a constant. The number of bacteria in the colony, n, is directly proportional to the diameter. A biologist counts the bacteria in the colony when its diameter is 2 mm and estimates that there are approximately 208 bacteria.

a) Find an expression for the rate of change of n with respect to time.

b) Find the rate of increase in number of bacteria after 1 day if $\lambda = 5$.

Q9 Hint: The amount of the water remaining is **decreasing** at a rate of 0.3 $cm^3\ s^{-1}$ — so this value should be negative.

Q9 Water is dripping from a hole in the base of a cylinder of radius r cm, where the water height is h cm, at a rate of 0.3 $cm^3\ s^{-1}$.

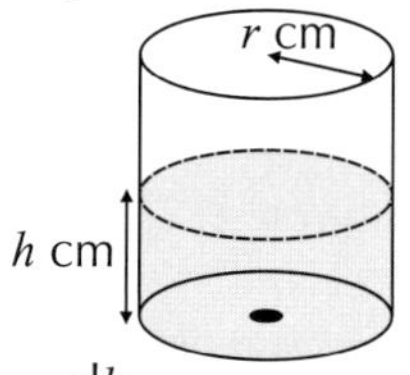

a) Find an expression for $\frac{dh}{dt}$, the rate at which the water level falls in the cylinder.

b) Hence find the rate of change in the water level, per minute, in a cylinder of radius 6 cm when the height of water is 4 cm.

Q10 Hint: A hemisphere is half a sphere.

Q10 The volume, V, of a hemisphere of radius r cm, varies with its temperature, θ, at a rate of k $cm^3\ °C^{-1}$, where k is a constant that depends on the material of which the hemisphere is made.

a) Find an expression for the rate of change of radius with respect to temperature.

b) Hence find $\frac{dr}{d\theta}$ for a material with $k = 1.5$, when $V = 4\ cm^3$.

9. Differentiation with Parametric Equations

You met parametric equations in Chapter 4 — they're equations where x and y are defined separately in terms of a third variable. Differentiating them is simpler than you might expect — but you'll have to combine this with all the other differentiation you've done in this chapter so far.

Learning Objectives:
- Be able to differentiate functions defined parametrically.
- Be able to find the equations of tangents and normals to curves given parametrically.

Differentiating parametric equations

- A curve can be defined by **two parametric equations**, often with the parameter t:

$$y = \mathrm{f}(t) \text{ and } x = \mathrm{g}(t)$$

- To find the gradient, $\frac{dy}{dx}$, you could convert the equations into **Cartesian** form (see pages 93-94), but this isn't always possible or convenient.
- The **chain rule** you met on p.139 can be used to differentiate parametric equations without needing to convert to Cartesian form. It looks like this:

$$\frac{dy}{dx} = \frac{dy}{dt} \div \frac{dx}{dt}$$

- So to find $\frac{dy}{dx}$ from parametric equations, **differentiate** each equation with respect to the parameter t, then put them into the formula.

Tip: On page 139, you saw this given in the form '$\times \frac{dt}{dx}$' rather than '$\div \frac{dx}{dt}$', but it means the same thing.

Example

The curve C is defined by the parametric equations $x = t^2 - 1$ and $y = t^3 - 3t + 4$.

a) Find $\frac{dy}{dx}$ in terms of t.

- Start by **differentiating** the two parametric equations **with respect to t**:

$$x = t^2 - 1 \quad \Rightarrow \frac{dx}{dt} = 2t$$

$$y = t^3 - 3t + 4 \Rightarrow \frac{dy}{dt} = 3t^2 - 3 = 3(t^2 - 1)$$

- Now use the **chain rule** to combine them:

$$\frac{dy}{dx} = \frac{dy}{dt} \div \frac{dx}{dt} = \frac{3(t^2-1)}{2t}$$

b) Find the gradient of C when $t = -2$.

- Use the answer to a) to find the gradient for a specific value of t. So, when $t = -2$:

$$\frac{dy}{dx} = \frac{3((-2)^2-1)}{2(-2)} = \frac{3(3)}{-4} = -\frac{9}{4}$$

c) Find the coordinates of the turning points.

- The turning points occur when $\frac{dy}{dx} = 0$,
  so solve to find the values of t at the turning points:

$$\frac{dy}{dx} = \frac{3(t^2 - 1)}{2t} = 0$$

$$\Rightarrow 3(t^2 - 1) = 0 \Rightarrow t^2 = 1 \Rightarrow t = \pm 1$$

- Now put these values for t into the original parametric equations to find the Cartesian coordinates of the turning points:

  When $t = 1$ $\quad x = (1)^2 - 1 = 0$
  $y = (1)^3 - 3(1) + 4 = 2$

  So there's a turning point at (0, 2)

  When $t = -1$ $\quad x = (-1)^2 - 1 = 0$
  $y = (-1)^3 - 3(-1) + 4 = 6$

  So there's another turning point at (0, 6)

Tip: Remember — Cartesian coordinates are just the (x, y) coordinates.

Exercise 9.1

Q1 For each curve C, defined by the parametric equations given below, find $\frac{dy}{dx}$ in terms of t.

a) $x = t^2$, $y = t^3 - t$ b) $x = t^3 + t$, $y = 2t^2 + 1$

c) $x = t^4$, $y = t^3 - t^2$ d) $x = \cos t$, $y = 4t - t^2$

Q2 The curve C is defined by the parametric equations $x = t^2$, $y = e^{2t}$.

a) Find $\frac{dy}{dx}$ in terms of t. b) Find the gradient of C when $t = 1$.

Q3 The curve C is defined by the parametric equations $x = e^{3t}$, $y = 4t^3 - 2t^2$.

a) Find $\frac{dy}{dx}$ in terms of t. b) Find the gradient of C when $t = 0$.

Q4-5 Hint: You'll need to use the product rule to find $\frac{dy}{dt}$ and $\frac{dx}{dt}$.

Q4 The curve C is defined by the parametric equations $x = t^3$, $y = t^2 \cos t$.

a) Find $\frac{dy}{dx}$ in terms of t. b) Find the gradient of C when $t = \pi$.

Q5 The curve C is defined by the parametric equations $x = t^2 \sin t$, $y = t^3 \sin t + \cos t$.

a) Find $\frac{dy}{dx}$ in terms of t. b) Find the gradient of C when $t = \pi$.

Q6 The curve C is defined by the parametric equations $x = \ln t$, $y = 3t^2 - t^3$.

a) Find $\frac{dy}{dx}$ in terms of t. b) Evaluate $\frac{dy}{dx}$ when $t = -1$.

c) Find the exact coordinates of the turning point of the curve C.

Finding tangents and normals

Once you've found the gradient of a parametric curve at a particular point, you can use this to find the **equation** of the **tangent** or **normal** to the curve at that point. You'll have seen this before, but here's a recap of the method:

- The gradient of the **tangent** is the **same** as the gradient of the curve at that point.
- The gradient of the **normal** at that point is $\frac{-1}{\text{gradient of tangent}}$.
- Put the values for the gradient, m, and the (x, y) coordinates of the point into $y = mx + c$ to find the equation of the line.

Tip: You could also use $y - y_1 = m(x - x_1)$ to get the equation.

Example

The curve C is defined by the following parametric equations:
$x = \sin t$, $y = 2t \cos t$.

a) **Find the gradient of the curve, and the (x, y) coordinates, when $t = \pi$.**

$\frac{dx}{dt} = \cos t$

$\frac{dy}{dt} = 2\cos t - 2t \sin t$

$$\frac{dy}{dx} = \frac{2\cos t - 2t\sin t}{\cos t} = 2 - 2t \tan t$$

When $t = \pi$, $\frac{dy}{dx} = 2 - 2\pi(0) = 2$

When $t = \pi$, $x = 0$, and $y = -2\pi$, so the coordinates are $(0, -2\pi)$.

b) **Hence find the equation of the tangent to C when $t = \pi$.**

- The gradient of the tangent at $t = \pi$ is equal to $\frac{dy}{dx}$ at that point, i.e. 2.
- So substitute $m = 2$, $x = 0$ and $y = -2\pi$ into $y = mx + c$, to find c:

$$y = mx + c$$
$$-2\pi = 2(0) + c \Rightarrow c = -2\pi$$

- Putting c back into the equation gives:

$$y = 2x - 2\pi \text{ or } y = 2(x - \pi)$$

c) **Find the equation of the normal to C when $t = \pi$.**

- The gradient of the normal at $t = \pi$ is $-\frac{1}{2}$.
- So substitute $m = -\frac{1}{2}$, $x = 0$ and $y = -2\pi$ into $y = mx + c$, to find c:

$$y = mx + c$$
$$-2\pi = -\frac{1}{2}(0) + c \Rightarrow c = -2\pi$$

- Putting c back into the equation gives:

$$y = -\frac{1}{2}x - 2\pi \text{ or } x + 2y + 4\pi = 0$$

Exercise 9.2

Q1 A curve is defined by the parametric equations $x = t^2$, $y = t^3 - 6t$. Find the equation of the tangent to the curve at $t = 3$, giving your answer in the form $ax + by + c = 0$.

Q2 A curve C is defined parametrically by $x = t^3 - 2t^2$, $y = t^3 - t^2 + 5t$. Find the equation of the tangent at the point $t = -1$.

Q3-4 Hint: You'll need to use the product rule.

Q3 A curve C is defined by the parametric equations $x = \sin 2t$, $y = t \cos t + 2 \sin t$. Find the equation of the normal to the curve at $t = \pi$.

Q4 The parametric representation of a curve is given by $x = t \ln t$, $y = t^3 - t^2 + 3$. Find the equation of the tangent to the curve at $t = 1$.

Q5 The path of a particle is given parametrically by $x = \theta \sin 2\theta$, $y = \theta^2 + \theta \cos \theta$. Find the equation of the normal to the particle's path at $\theta = \frac{\pi}{2}$.

Q6 The motion of a particle is modelled by the parametric equations $x = t^2 - t$, $y = 3t - t^3$.

a) Find the equation of the tangent to the path of the particle when $t = 2$, giving your answer in a suitable form.

b) Find the Cartesian coordinates of the point at which the normal to the path at $t = 2$ cuts the x-axis.

Q7 A particle moves along a path modelled by the parametric equations $x = \sin 2\theta + 2 \cos \theta$, $y = \theta \sin \theta$.

a) Find the gradient $\frac{dy}{dx}$ of the particle's path in terms of θ.

b) Evaluate $\frac{dy}{dx}$ at $\theta = \frac{\pi}{2}$ and hence obtain equations of the tangent and normal to the path at this point.

Q8 A particle moves along a path given parametrically by $x = s^3 \ln s$, $y = s^3 - s^2 \ln s$.

a) Give the value(s) of s at which the path cuts the y-axis.

b) Hence show that the equation of a tangent to the curve when $x = 0$ is $y = 2x + 1$.

Q9 A curve is given parametrically by $x = \theta^2 \sin \theta$, $y = \frac{\cos \theta}{\theta^3}$.

a) Show that the gradient of the curve when $\theta = \pi$ is $-\frac{3}{\pi^6}$.

b) Hence find the equation of the normal to the curve at this point.

10. Implicit Differentiation

For equations that you can't write in the form $y = f(x)$, you need to use implicit differentiation. It works for equations that contain a mixture of x and y terms, such as xy^2.

Learning Objectives:

- Be able to differentiate functions defined implicitly.
- Be able to find the equations of tangents and normals to curves given implicitly.

Implicit differentiation

An '**implicit relation**' is the mathematical name for any equation in x and y that's written in the form $\mathbf{f(x, y) = g(x, y)}$ instead of $y = f(x)$. For example, $y^2 = xy + x + 2$ is implicit.

Some implicit relations are either awkward or impossible to rewrite in the form $y = f(x)$. This can happen, for example, if the equation contains a number of different powers of y, or terms where x is multiplied by y.

This can make implicit relations tricky to differentiate — the solution is **implicit differentiation**:

Tip: $f(x, y)$ and $g(x, y)$ don't actually both have to include x and y — one of them could even be a constant.

To find $\frac{dy}{dx}$ for an implicit relation between x and y:

- **Step 1:** Differentiate terms in x **only** (and **constant** terms) with respect to x, as normal.
- **Step 2:** Use the **chain rule** to differentiate terms in y **only**:

$$\frac{d}{dx}f(y) = \frac{d}{dy}f(y)\frac{dy}{dx}$$

In practice, this means 'differentiate with respect to y, and stick a $\frac{dy}{dx}$ on the end'.

- **Step 3:** Use the **product rule** to differentiate terms in **both x and y**:

$$\frac{d}{dx}u(x)v(y) = u(x)\frac{d}{dx}v(y) + v(y)\frac{d}{dx}u(x)$$

- **Step 4:** **Rearrange** the resulting equation in x, y and $\frac{dy}{dx}$ to make $\frac{dy}{dx}$ the subject.

Tip: $\frac{d}{dx}f(y)$ just means 'the derivative of $f(y)$ with respect to x'.

Tip: This version of the product rule is slightly different from the one on page 154 — it's got $v(y)$ instead of $v(x)$.

Example 1

Use implicit differentiation to find $\frac{dy}{dx}$ for $y^3 + y^2 = e^x + x^3$.

You need to differentiate each term of the equation with respect to x.

- Start by sticking '$\frac{d}{dx}$' in front of each term:

$$\frac{d}{dx}y^3 + \frac{d}{dx}y^2 = \frac{d}{dx}e^x + \frac{d}{dx}x^3$$

- **Step 1:** Differentiate the terms in x only.

$$\frac{d}{dx}y^3 + \frac{d}{dx}y^2 = e^x + 3x^2$$

- **Step 2:** Use the **chain rule** for the terms in y only.

$$3y^2\frac{dy}{dx} + 2y\frac{dy}{dx} = e^x + 3x^2$$

Tip: Just differentiate with respect to y, but don't forget to put a $\frac{dy}{dx}$ after each term.

- **Step 3:** There are no terms in both x and y to deal with, so...
- **Step 4: Rearrange** to make $\frac{dy}{dx}$ the subject: $(3y^2 + 2y)\frac{dy}{dx} = e^x + 3x^2$

$$\Rightarrow \frac{dy}{dx} = \frac{e^x + 3x^2}{3y^2 + 2y}$$

Example 2

a) Use implicit differentiation to find $\frac{dy}{dx}$ for $2x^2y + y^3 = 6x^2 - 15$.

- Again, start by sticking '$\frac{d}{dx}$' in front of each term:

$$\frac{d}{dx}2x^2y + \frac{d}{dx}y^3 = \frac{d}{dx}6x^2 - \frac{d}{dx}15$$

- First, deal with the **terms in x** and **constant** terms — in this case that's the two terms on the right-hand side:

$$\frac{d}{dx}2x^2y + \frac{d}{dx}y^3 = 12x + 0$$

- Now use the **chain rule** on the term in y.

$$\frac{d}{dx}2x^2y + 3y^2\frac{dy}{dx} = 12x + 0$$

- Use the **product rule** on the term in x and y, where $u(x) = 2x^2$ and $v(y) = y$:

$$2x^2\frac{d}{dx}(y) + y\frac{d}{dx}(2x^2) + 3y^2\frac{dy}{dx} = 12x + 0$$

$$\Rightarrow 2x^2\frac{dy}{dx} + y4x + 3y^2\frac{dy}{dx} = 12x + 0$$

This $\frac{dy}{dx}$ term comes from using the chain rule: $\frac{d}{dx}(y) = \frac{d}{dy}(y)\frac{dy}{dx} = 1\frac{dy}{dx} = \frac{dy}{dx}$

- Finally, **rearrange** to make $\frac{dy}{dx}$ the subject:

$$\frac{dy}{dx}(2x^2 + 3y^2) = 12x - 4xy \Rightarrow \frac{dy}{dx} = \frac{12x - 4xy}{2x^2 + 3y^2}$$

b) Find the gradient of the curve $2x^2y + y^3 = 6x^2 - 15$ at the point (2, 1).

Just put the values for x and y into $\frac{dy}{dx}$: $\frac{dy}{dx} = \frac{12(2) - 4(2)(1)}{2(2)^2 + 3(1)^2} = \frac{16}{11}$

Tip: Once you're happy with this method you might not need to write all the steps out separately, but you can do if you find it easier.

Tip: Take your time using the product rule as it's easy to forget terms if you're not careful. Always start by identifying $u(x)$ and $v(y)$, and do it in steps if you need to.

Exercise 10.1

Q1 Use implicit differentiation to find $\frac{dy}{dx}$ for each of these curves:

a) $y + y^3 = x^2 + 4$

b) $x^2 + y^2 = 2x + 2y$

c) $3x^3 - 4y = y^2 + x$

d) $5x - y^2 = x^5 - 6y$

e) $\cos x + \sin y = x^2 + y^3$

f) $x^3y^2 + \cos x = 4xy$

g) $e^x + e^y = x^3 - y$

h) $3xy^2 + 2x^2y = x^3 + 4x$

Q2 Find the gradient, $\frac{dy}{dx}$, for each of these curves given below:

a) $x^3 + 2xy = y^4$

b) $x^2y + y^2 = x^3$

c) $y^3x + y = \sin x$

d) $y \cos x + x \sin y = xy$

e) $e^x + e^y = xy$

f) $\ln x + x^2 = y^3 + y$

g) $e^{2x} + e^{3y} = 3x^2y^2$

h) $x \ln x + y \ln x = x^5 + y^3$

Q3 a) Show that the curve C, defined implicitly by $e^x + 2 \ln y = y^3$, passes through (0, 1).

b) Find the gradient of the curve at this point.

Q3a) Hint: Evaluate the left-hand and right-hand sides of the equation separately to show they're the same.

Q4 A curve is defined implicitly by $x^3 + y^2 - 2xy = 0$.

a) Find $\frac{dy}{dx}$ for this curve.

b) Show that $y = -2 \pm 2\sqrt{3}$ when $x = -2$.

c) Evaluate the gradient at $(-2, -2 + 2\sqrt{3})$, leaving your answer in surd form.

Q5 The curve $x^3 - xy = 2y^2$ passes through the points (1, –1) and (1, a).

a) Find the value of a.

b) Evaluate the gradient of the curve at each of these points.

Q6 A curve is defined implicitly by $x^2y + y^2x = xy + 4$.

a) At which two values of y does the line $x = 1$ cut the curve?

b) By finding $\frac{dy}{dx}$, evaluate the gradient at each of these points.

Applications of implicit differentiation

Sometimes you can use implicit differentiation when you might not expect it. For some equations of the form $y = f(x)$, the easiest way to differentiate them is to **rearrange** them and use implicit differentiation.

For example, the proof of the rule for **differentiating a^x** from page 147 uses implicit differentiation:

- Take **ln** of both sides of the equation: $y = a^x \Rightarrow \ln y = \ln a^x$
- Use the **log laws** to rearrange the right-hand side: $\ln y = x \ln a$
- Now use **implicit differentiation**: $\frac{d}{dx}(\ln y) = \frac{d}{dx}(x \ln a)$

  Since a is a **constant, ln a** is also a constant.

  $\frac{1}{y}\frac{dy}{dx} = \ln a$ — Use the **chain rule** to deal with $\frac{d}{dx}(\ln y)$

  $\frac{dy}{dx} = y \ln a$
- Use the original equation to get rid of y: $\frac{dy}{dx} = a^x \ln a$

Tip: Remember — $\frac{d}{dy}\ln y = \frac{1}{y}$.

This method of rearranging and using implicit differentiation is also used to differentiate the **inverse trig functions**.

Example 1

Find $\frac{dy}{dx}$ if $y = \arcsin x$ for $-1 \le x \le 1$.

PROBLEM SOLVING

- Rearrange the equation to get rid of the arcsin by taking sin of both sides:

$$\sin y = \sin(\arcsin x)$$
$$\Rightarrow \sin y = x$$

Tip: Sin and arcsin are inverses of each other, so they cancel out.

- Now use implicit differentiation:

$$\frac{d}{dx}(\sin y) = \frac{d}{dx}(x)$$

  - Differentiate the x-term:

$$\frac{d}{dx}(\sin y) = 1$$

  - Use the chain rule on the y-term:

$$\cos y \frac{dy}{dx} = 1$$

  - Rearrange to give an equation for $\frac{dy}{dx}$:

$$\frac{dy}{dx} = \frac{1}{\cos y}$$

- To get $\frac{1}{\cos y}$ in terms of x, first use the identity $\cos^2 x + \sin^2 x \equiv 1$ to write $\cos y$ in terms of $\sin y$:

$$\frac{dy}{dx} = \frac{1}{\sqrt{\cos^2 y}}$$
$$\frac{dy}{dx} = \frac{1}{\sqrt{1 - \sin^2 y}}$$

- Now, use the equation $\sin y = x$ to get $\frac{dy}{dx}$ in terms of x:

$$\frac{dy}{dx} = \frac{1}{\sqrt{1 - x^2}}$$

You can use a similar method to differentiate the other inverse trig functions as well — the derivatives are given below:

If $y =$	**$\frac{dy}{dx} =$**
$\arcsin x$	$\frac{1}{\sqrt{1-x^2}}$
$\arccos x$	$-\frac{1}{\sqrt{1-x^2}}$
$\arctan x$	$\frac{1}{1+x^2}$

Tip: You can have a go at finding the derivatives of $\arccos x$ and $\arctan x$ for yourself in Exercise 10.2 on p.179.

Most implicit differentiation questions aren't really that different from any other differentiation question. Once you've got an expression for the gradient, you'll have to use it to do the sort of things you'd normally expect, like finding **stationary points** of curves and equations of **tangents** and **normals**.

Example 2

Curve A has the equation $x^2 + 2xy - y^2 = 10x + 4y - 21$.

a) Show that when $\frac{dy}{dx} = 0$, $y = 5 - x$.

For starters, we're going to need to find $\frac{dy}{dx}$ by implicit differentiation:

$$\frac{d}{dx}x^2 + \frac{d}{dx}2xy - \frac{d}{dx}y^2 = \frac{d}{dx}10x + \frac{d}{dx}4y - \frac{d}{dx}21$$

$$\Rightarrow 2x + \frac{d}{dx}2xy - \frac{d}{dx}y^2 = 10 + \frac{d}{dx}4y - 0$$

Differentiate x^2, $10x$ and 21 with respect to x.

$$\Rightarrow 2x + \frac{d}{dx}2xy - 2y\frac{dy}{dx} = 10 + 4\frac{dy}{dx}$$

Use the chain rule to differentiate y^2 and $4y$.

$$\Rightarrow 2x + 2x\frac{dy}{dx} + y\frac{d}{dx}2x - 2y\frac{dy}{dx} = 10 + 4\frac{dy}{dx}$$

$$\Rightarrow 2x + 2x\frac{dy}{dx} + 2y - 2y\frac{dy}{dx} = 10 + 4\frac{dy}{dx}$$

Use the product rule to differentiate $2xy$.

$$\Rightarrow 2x\frac{dy}{dx} - 2y\frac{dy}{dx} - 4\frac{dy}{dx} = 10 - 2x - 2y$$

$$\Rightarrow (2x - 2y - 4)\frac{dy}{dx} = 10 - 2x - 2y$$

$$\Rightarrow \frac{dy}{dx} = \frac{10 - 2x - 2y}{2x - 2y - 4} = \frac{5 - x - y}{x - y - 2}$$

So when $\frac{dy}{dx} = 0$: $\frac{5 - x - y}{x - y - 2} = 0 \Rightarrow 5 - x - y = 0 \Rightarrow y = 5 - x$ as required

b) Find the coordinates of the stationary points of A.

Use the answer to part a) to find the points where $\frac{dy}{dx} = 0$.

When $\frac{dy}{dx} = 0$, $y = 5 - x$. So at the stationary points:

$$x^2 + 2xy - y^2 = 10x + 4y - 21$$
$$\Rightarrow x^2 + 2x(5 - x) - (5 - x)^2 = 10x + 4(5 - x) - 21$$
$$\Rightarrow x^2 + 10x - 2x^2 - 25 + 10x - x^2 = 10x + 20 - 4x - 21$$
$$\Rightarrow -2x^2 + 14x - 24 = 0 \Rightarrow x^2 - 7x + 12 = 0$$
$$\Rightarrow (x - 3)(x - 4) = 0 \Rightarrow x = 3 \text{ or } x = 4$$

$x = 3 \Rightarrow y = 5 - 3 = 2$

$x = 4 \Rightarrow y = 5 - 4 = 1$

Put each value of x into $y = 5 - x$ to find the corresponding y-coordinate.

So the stationary points of A are (3, 2) and (4, 1).

Tip: Here you're just putting $y = 5 - x$ into the original equation to find the values of x at the stationary points.

Example 3

A curve defined implicitly by $\sin x - y\cos x = y^2$ passes through two points (π, a) and (π, b), where $a < b$.

a) Find the values of a and b.

- Put $x = \pi$ into the equation and solve for y:

$$\sin \pi - y\cos \pi = y^2 \Rightarrow 0 + y = y^2$$
$$\Rightarrow y^2 - y = 0 \Rightarrow y(y - 1) = 0$$
$$\Rightarrow y = 0 \text{ and } y = 1$$

- So $a = 0$ and $b = 1$.

> **Tip:** You should know $\cos \pi = -1$ and $\sin \pi = 0$ — these are standard trig results that were covered on page 53.

b) Find the equations of the tangents to the curve at each of these points.

- First find $\frac{dy}{dx}$ using implicit differentiation as usual:

$$\cos x + y \sin x - \cos x \frac{dy}{dx} = 2y \frac{dy}{dx}$$

$$\Rightarrow \frac{dy}{dx} = \frac{\cos x + y \sin x}{2y + \cos x}$$

- Now put in $x = \pi$ and $y = 0$ to find the gradient at $(\pi, 0)$.

$$\frac{dy}{dx} = \frac{\cos \pi + 0 \sin \pi}{2(0) + \cos \pi} = 1$$

- So the gradient of the tangent at $(\pi, 0)$ is 1.
  Putting these values into $y = mx + c$ gives:

$$0 = \pi + c \Rightarrow c = -\pi$$

  So the equation of the tangent at $(\pi, 0)$ is $y = x - \pi$.

- Do the same to find the equation of the tangent at $(\pi, 1)$.

$$\frac{dy}{dx} = \frac{\cos \pi + \sin \pi}{2(1) + \cos \pi} = -1$$

$$1 = -\pi + c \Rightarrow c = 1 + \pi$$

  So the equation of the tangent at $(\pi, 1)$ is $y = 1 + \pi - x$.

c) Show that the tangents intersect at the point $\left(\frac{1 + 2\pi}{2}, \frac{1}{2}\right)$.

- The two tangents intersect when:

$$x - \pi = 1 + \pi - x$$

$$\Rightarrow 2x = 1 + 2\pi$$

$$\Rightarrow x = \frac{1 + 2\pi}{2}$$

- Putting this value of x into one of the equations gives:

$$y = \left(\frac{1 + 2\pi}{2}\right) - \pi = \frac{1}{2} + \pi - \pi = \frac{1}{2}$$

- So they intersect at $\left(\frac{1 + 2\pi}{2}, \frac{1}{2}\right)$.

Exercise 10.2

Q1 A curve is defined implicitly by $x^2 + 2x + 3y - y^2 = 0$.

a) Find the coordinates of the stationary points (to 2 decimal places).

b) Show that the curve intersects the y-axis when $y = 0$ and $y = 3$.
Hence find the equation of the tangent at each of these points.

> **Q2 Hint:** You should find 4 stationary points.

Q2 A curve is defined implicitly by $x^3 + x^2 + y = y^2$.

a) Find the coordinates of the stationary points (to 2 decimal places).

b) Show that the curve intersects the line $x = 2$ when $y = 4$ and $y = -3$.
Hence find the equation of the tangent at each of these points.

Q3 A curve is defined implicitly by $x^2y + y^3 = x + 7$.

a) Calculate the x-coordinates of the points on the curve where $y = 1$ and hence find the equations of the normals at these points.

b) Find the coordinates of the point where the normals intersect.

Q4 $e^x + y^2 - xy = 5 - 3y$ is a curve passing through two points (0, a) and (0, b), where a < b.

a) Find the values of a and b and show that one of these points is a stationary point of the curve.

b) Find the equations of the tangent and normal to the curve at the other point.

Q5 Differentiate arccos x with respect to x. PROBLEM SOLVING

Q6 If $y = \arctan x$, show that $\frac{dy}{dx} = \frac{1}{1+x^2}$.

Q6 Hint: You'll need to use the identity $\sec^2\theta \equiv 1 + \tan^2\theta$.

Q7 The curve C is defined by $\ln x + y^2 = x^2y + 6$.

a) Show that C passes through (1, 3) and (1, –2).

b) Find the equations of the normals to the curve at each of these points and explain why these normals cannot intersect.

Q8 A curve is defined implicitly by $e^y + x^2 = y^3 + 4x$. Find the equations of the tangents that touch the curve at (a, 0) and (b, 0). Leave your answer in surd form. PROBLEM SOLVING

Q9 Show that any point on the curve $y \ln x + x^2 = y^2 - y + 1$ which satisfies $y + 2x^2 = 0$ is a stationary point.

Q10 If $f(x) = \arccos(x^2)$ for $-1 \le x \le 1$, find the equation of the tangent to the graph of $y = f(x)$ when $x = \frac{1}{\sqrt{2}}$ and $0 \le y \le \pi$. Give your answer in the form $y = mx + c$, using exact values for m and c. PROBLEM SOLVING

Q11 A curve is defined by $e^{2y} + e^x - e^4 = 2xy + 1$.

a) Find the equation of the tangent to the curve when $y = 0$.

b) Find the equation of the normal to the curve when $y = 0$.

c) Show that these two lines intersect when $x = \frac{4e^8 + 144}{e^8 + 36}$.

Q11 Hint: Remember, $\frac{1}{e^x}$ is the same as e^{-x}.

Q12 $y^2x + 2xy - 3x^3 = x^2 + 2$ passes through two points where $x = 2$. Find the equations of the tangents to the curve at these points, and hence show that they intersect at $\left(-\frac{14}{25}, -1\right)$.

Q13 The curve C is defined implicitly by $\cos y \cos x + \cos y \sin x = \frac{1}{2}$.

a) Find y when $x = \frac{\pi}{2}$ and when $x = \pi$, $0 \le y \le \pi$.

b) Find the equations of the tangents at these points.

Q13 Hint: You won't need your calculator for the trig here but you will need to remember your common angles.

Q14 Find the coordinates of the stationary points of the graph $\frac{1}{3}y^2 = 6x^3 - 2xy$. PROBLEM SOLVING

Chapter 8 Integration

1. Integration of $(ax + b)^n$

Learning Objectives:

- Be able to integrate functions of the form $(ax + b)^n$, $n \neq -1$, where a, b and n are constants.
- Be able to solve integration problems with functions of the form $(ax + b)^n$.

You've already seen how to integrate functions of the form x^n in Year 1. In this section you'll see how to integrate functions which are linear transformations of x^n — functions of the form $(ax + b)^n$.

Integrating $(ax + b)^n$, $n \neq -1$

In Year 1, you learnt to think of **integration** as the **opposite of differentiation**. This means that if you differentiate a function, then **integrating** the result will get you back to the function you started with. Here's an example to show how you can use this technique to integrate functions of the form $(ax + b)^n$.

Example 1

a) Differentiate $(3x + 4)^5$ with respect to x.

Using the **chain rule**, $\frac{d}{dx}(3x + 4)^5 = 5(3x + 4)^4 \times 3 = 15(3x + 4)^4$.

b) Use your answer to a) to find $\int (3x + 4)^4 \, dx$.

- From part a) you know that:

$$(3x + 4)^5 \xrightarrow{\text{Differentiation}} 15(3x + 4)^4$$

- Integration is the **opposite of differentiation** so:

$$(3x + 4)^5 + C \xleftarrow[\text{Integration}]{} 15(3x + 4)^4$$

This means: $\int 15(3x + 4)^4 \, dx = (3x + 4)^5 + c$

$$\Rightarrow 15\int (3x + 4)^4 \, dx = (3x + 4)^5 + c$$

$$\Rightarrow \int (3x + 4)^4 \, dx = \frac{1}{15}(3x + 4)^5 + \frac{c}{15}$$

$$= \frac{1}{15}(3x + 4)^5 + C$$

$\frac{c}{15}$ is just another constant term — you can call it C.

Divide by the constant term to get the integral you're after.

This method gives us a **general result** for integrating all functions of the form $(ax + b)^n$.

Differentiating $(ax + b)^{n+1}$ using the chain rule gives $a(n + 1)(ax + b)^n$.

So $\int a(n+1)(ax + b)^n \, dx = (ax + b)^{n+1} + c$

$$a(n + 1)\int (ax + b)^n \, dx = (ax + b)^{n+1} + c$$

Dividing by $a(n + 1)$ gives the general expression:

$$\int (ax + b)^n \, dx = \frac{1}{a(n+1)}(ax + b)^{n+1} + C \quad \text{for } n \neq -1,\ a \neq 0$$

Tip: The **chain rule** (p.139) is used to differentiate a function of a function f(g(x)): If $y = f(u)$ and $u = g(x)$ then: $\frac{dy}{dx} = \frac{dy}{du} \times \frac{du}{dx}$

Tip: Don't forget that for indefinite integrals (integrals without limits) you need to add a constant of integration, C.

Tip: You can take constant factors outside of integrations and put them at the front to make things easier.

Tip: This doesn't work for $n = -1$ because you'd end up having to divide by $n + 1 = 0$. See page 184 for a method of integrating x^{-1} and $(ax + b)^{-1}$.

Example 2

Find $\int (3-4x)^2\,dx$, using the general expression for $\int (ax+b)^n\,dx$.

Write down the values of a, b and n and then substitute them into the formula. Here $a = -4$, $b = 3$ and $n = 2$.

$$\int (3-4x)^2\,dx = \frac{1}{-4\times 3}(3-4x)^3 + C = -\frac{1}{12}(3-4x)^3 + C$$

$a = -4$ $\qquad$ $n + 1 = 3$

Tip: You can always differentiate your answer to check it — you should get back to what you started with.

In Year 1, you learnt that definite integrals work out the **area** between a curve and the x-axis. To find the area between a curve $y = f(x)$ and the x-axis over an interval, just integrate $f(x)$ with respect to x over that interval.

Example 3

Work out the area enclosed by the curve $y = (x-2)^3$, the x-axis and the lines $x = 2$ and $x = 3$.

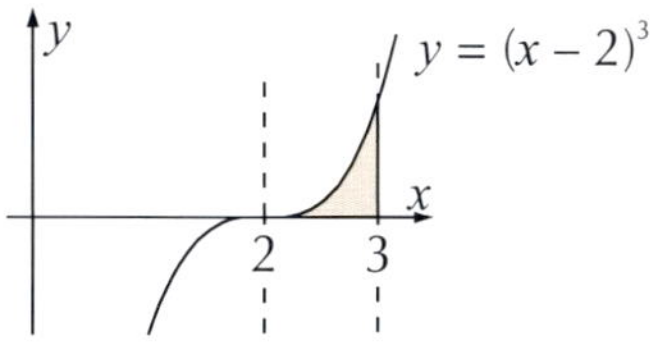

- You just need to integrate the curve $y = (x-2)^3$ between $x = 2$ and $x = 3$, i.e. evaluate $\int_2^3 (x-2)^3\,dx$
- Use the formula with $a = 1$, $b = -2$ and $n = 3$ to work out the integral.

$$\int_2^3 (x-2)^3\,dx = \frac{1}{4}[(x-2)^4]_2^3$$

- Substitute in the limits of integration to find the area.

$x = 3$ $\qquad$ $x = 2$

$$\int_2^3 (x-2)^3\,dx = \frac{1}{4}[(3-2)^4] - \frac{1}{4}[(2-2)^4] = \frac{1}{4}[1^4] - \frac{1}{4}[0^4]$$

$$= \frac{1}{4} - 0 = \frac{1}{4}$$

Tip: There's no constant of integration for definite integrals, as they just cancel out when you do the subtraction.

Tip: Remember from Year 1 that when you integrate a definite integral, you put the function in square brackets and write the limits on the right.

Exercise 1.1

Q1 Integrate with respect to x:

a) $(x+10)^{10}$ $\qquad$ b) $(5x)^7$ $\qquad$ c) $(3-5x)^{-2}$ $\qquad$ d) $(3x-4)^{-\frac{4}{3}}$

Q2 a) By using the general expression for $\int (ax+b)^n\,dx$,

show that the integral $A = \int 8(2x-4)^4\,dx = \dfrac{4(2x-4)^5}{5} + C$.

b) Hence evaluate A between the values $x = \frac{3}{2}$ and $x = \frac{5}{2}$.

Q3 Evaluate $\int_0^1 (6x+1)^{-3}\,dx$.

Q4 The curve $y = f(x)$ goes through the point $\left(1, \frac{3}{35}\right)$ and $f'(x) = (8-7x)^4$. Find $f(x)$.

2. Integration of e^x and $\frac{1}{x}$

Learning Objectives:

- Be able to integrate functions containing e^x and $\frac{1}{x}$ terms.
- Be able to integrate linear transformations of e^x and $\frac{1}{x}$, i.e. functions of the form e^{ax+b} and $\frac{1}{ax+b}$.

The functions e^x and $\frac{1}{x}$ are pretty easy to integrate — it's just the opposite of differentiating e^x and $\ln x$. Simple.

Integrating e^x and e^{ax+b}

e^x differentiates to give e^x, so it makes sense that e^x **integrates** to give $\mathbf{e^x + C}$.

$$\int e^x \, dx = e^x + C$$

Example 1

Integrate the function $6x^2 - 4x + 3e^x$ with respect to x.

- When integrating a function like this, you integrate each term separately, so:
  $$\int 6x^2 - 4x + 3e^x \, dx = \int 6x^2 \, dx - \int 4x \, dx + \int 3e^x \, dx$$
- The first two terms are of the form x^n so they integrate to $2x^3$ and $2x^2$ respectively.
- Using the rule above, the third term is:
  $$\int 3e^x \, dx = 3\int e^x \, dx = 3e^x + c$$
- Putting this all together gives:
  $$\int 6x^2 - 4x + 3e^x \, dx = 2x^3 - 2x^2 + 3e^x + C$$

On p.180 you saw how to integrate **linear transformations** of x^n by differentiating with the **chain rule** and working backwards. You can do the same with functions of the form $\mathbf{e^{ax+b}}$ where a and b are constants. Start by considering what you get when you **differentiate** functions of the form e^{ax+b} using the chain rule and work backwards.

Tip: This method of differentiating $f(ax + b)$ using the chain rule and working backwards to find an integral can be used with any of the functions that you know the derivative of — you'll see it used a lot in the next few sections.

Example 2

a) Differentiate the function e^{4x-1} with respect to x.

Using the chain rule, $\frac{d}{dx}(e^{4x-1}) = 4e^{4x-1}$

b) Using your answer to a), find the integral $\int e^{4x-1} \, dx$.

Reversing the process of differentiation to get integration we have:

$$\frac{d}{dx}(e^{4x-1}) = 4e^{4x-1} \Rightarrow \int 4e^{4x-1} \, dx = e^{4x-1} + c$$

$$\Rightarrow 4\int e^{4x-1} \, dx = e^{4x-1} + c$$

$$\Rightarrow \int e^{4x-1} \, dx = \frac{1}{4}e^{4x-1} + C$$

Take out the factor of 4 and divide by it.

This method gives you a **general rule** for integrating functions of the form $\mathbf{e}^{ax+b}$:

$$\int e^{ax+b}\,dx = \frac{1}{a}e^{ax+b} + C \quad \text{for } a \neq 0$$

This means you just need to **divide** by the **coefficient of x** and add a constant of integration — the e^{ax+b} bit **doesn't change**.

Example 3

Integrate the following:

a) $\mathbf{e^{7x}}$

$$\int e^{7x}\,dx = \frac{1}{7}e^{7x} + C$$

If you differentiated e^{7x} you'd get $7e^{7x}$, so you need to divide by 7 (the coefficient of x) when integrating.

b) $\mathbf{e^{\frac{x}{2}}}$

$$\int e^{\frac{x}{2}}\,dx = \int e^{\frac{1}{2}x}\,dx = 2e^{\frac{x}{2}} + C$$

If you differentiated this one using the **chain rule**, you'd get $\frac{1}{2}e^{\frac{x}{2}}$, so you need to multiply by 2 (divide by $\frac{1}{2}$) to integrate.

c) $\mathbf{2e^{4-3x}}$

$$\int 2e^{4-3x}\,dx = -\frac{2}{3}e^{4-3x} + C$$

Multiplying by 2 doesn't change the integration — the coefficient of x is –3, so divide by –3 and you're done.

Tip: Notice that if you differentiated $2e^{4-3x}$, you'd get $-6e^{4-3x}$, so you need to divide by –3 when you integrate — it's simple really.

Exercise 2.1

Q1 Find the following indefinite integrals:

a) $\int 2e^{x}\,dx$ b) $\int 4x + 7e^{x}\,dx$ c) $\int e^{10x}\,dx$

d) $\int e^{-3x} + x\,dx$ e) $\int e^{\frac{7}{2}x}\,dx$ f) $\int e^{4x-2}\,dx$

g) $\int \frac{1}{2}e^{2-\frac{3}{2}x}\,dx$ h) $\int e^{4\left(\frac{x}{3}+1\right)}\,dx$

Q2 Find the equation of the curve that has the derivative $\frac{dy}{dx} = 10e^{-5x-1}$ and passes through the origin.

Q3 Integrate the function e^{8y+5} with respect to y.

Q4 Evaluate the following definite integrals, giving exact answers:

a) $\int_{2}^{3} e^{2x}\,dx$ b) $\int_{-1}^{0} 12e^{12x+12}\,dx$

c) $\int_{-\frac{\pi}{2}}^{\frac{\pi}{2}} e^{\pi-2x}\,dx$ d) $\int_{3}^{6} \sqrt[6]{e^{x}} + \frac{1}{\sqrt[3]{e^{x}}}\,dx$

Q4 d) Hint: Remember:
$\sqrt[n]{e^{x}} = e^{\frac{x}{n}}$
$\frac{1}{\sqrt[n]{e^{x}}} = e^{-\frac{x}{n}}$

Integrating $\frac{1}{x}$ and $\frac{1}{ax+b}$

The method for integrating x^n and $(ax + b)^n$ on p.180 doesn't work for $n = -1$. For these functions you need to consider the fact that $\frac{d}{dx}(\ln x) = \frac{1}{x}$, which you should remember from Chapter 7.

Example 1

Let $f(x) = \frac{1}{x}$, $x > 0$.

Integrate $f(x)$ with respect to x, given that $\frac{d}{dx}(\ln x) = \frac{1}{x}$.

Given the derivative of $\ln x$, integration is the opposite of differentiation, so:

$\ln x$ —Differentiation→ $\frac{1}{x}$

$\ln x + C$ ←Integration— $\frac{1}{x}$

So $\int \frac{1}{x}\,dx =$ $\ln x + C$

Tip: You'll be working with logs all the time when integrating functions of the form $\frac{1}{x}$ — it'll help to remember the log laws:

$\log(ab) = \log a + \log b$

$\log\left(\frac{a}{b}\right) = \log a - \log b$

$\log(a^b) = b\log a$

So now we have a general result for integrating $\frac{1}{x}$:

$$\int \frac{1}{x}\,dx = \ln|x| + C$$

Notice that this result uses $|x|$ instead of just x. This is because the function $\ln x$ is **not defined** for **negative values** of x. Using the modulus means you'll never end up taking ln of a negative value.

Example 2

Find the following integrals:

a) $\int \frac{5}{x}\,dx$

5 is a **constant coefficient** — you can take it outside the integral so that you're just integrating $\frac{1}{x}$.

$$\int \frac{5}{x}\,dx = 5\int \frac{1}{x}\,dx = 5\ln|x| + C$$

b) $\int_3^9 \frac{1}{3x}\,dx$

- Here $\frac{1}{3}$ **is the coefficient**, so it goes outside the integral.

$$\int_3^9 \frac{1}{3x}\,dx = \frac{1}{3}\int_3^9 \frac{1}{x}\,dx = \frac{1}{3}[\ln|x|]_3^9$$

- Now put in the limits and use log laws to simplify:

$$= \frac{1}{3}(\ln|9| - \ln|3|) = \frac{1}{3}(\ln(\tfrac{9}{3}))$$

← $\ln a - \ln b = \ln \frac{a}{b}$

$$= \frac{1}{3}\ln 3$$

You can integrate **linear transformations** of $\frac{1}{x}$ (i.e. functions of the form $\frac{1}{ax+b}$) by considering the result of differentiating $\ln|ax+b|$.

Example 3

Given that $\frac{d}{dx}(\ln|4x+2|) = \frac{4}{4x+2}$, find $\int \frac{1}{4x+2}\,dx$.

$$\frac{d}{dx}(\ln|4x+2|) = \frac{4}{4x+2} \Rightarrow \int \frac{4}{4x+2}\,dx = \ln|4x+2| + c$$

$$\Rightarrow 4\int \frac{1}{4x+2}\,dx = \ln|4x+2| + c$$

$$\Rightarrow \int \frac{1}{4x+2}\,dx = \frac{1}{4}\ln|4x+2| + C$$

The **general result** for integrating functions of the form $\frac{1}{ax+b}$ is:

$$\int \frac{1}{ax+b}\,dx = \frac{1}{a}\ln|ax+b| + C$$

Example 4

Find $\int \frac{1}{2x+5}\,dx$.

Using the general rule, $a = 2$ and $b = 5$ so the integral is:

$$\int \frac{1}{2x+5}\,dx = \frac{1}{2}\ln|2x+5| + C$$

Tip: Finding $\int \frac{1}{3x}\,dx$ (from the example on page 184) in this way would give a 'different' answer: $\frac{1}{3}\ln|3x| + C$.
But this expands to $\frac{1}{3}\ln|x| + \frac{1}{3}\ln 3 + C$ and the $\frac{1}{3}\ln 3$ becomes part of the constant of integration. So both answers are equivalent.

Exercise 2.2

Q1 Find the following:

a) $\int \frac{19}{x}\,dx$ b) $\int \frac{1}{7x}\,dx$ c) $\int \frac{1}{7x+2}\,dx$ d) $\int \frac{4}{1-3x}\,dx$

Q2 Integrate $y = \frac{1}{8x} - \frac{20}{x}$ with respect to x.

Q3 a) Show that $\int \frac{6}{x} - \frac{3}{x}\,dx = \ln|x^3| + C$.

b) Evaluate $\int_4^5 \frac{6}{x} - \frac{3}{x}\,dx$, giving an exact answer.

Q3-4 Hint: Use the log laws from Year 1 (given on p.184).

Q4 Show that $\int_b^a 15(5+3x)^{-1}\,dx = \ln\left|\frac{5+3a}{5+3b}\right|^5$.

Q5 The graph of the curve $y = f(x)$ passes through the point (1, 2). The derivative of $f(x)$ is given by $f'(x) = \frac{4}{10-9x}$. Find $f(x)$.

Q6 a) Express the area bounded by the curve $y = \frac{-7}{16-2x}$, the x-axis, the y-axis, and the line $x = -3$ as an integral with respect to x.

b) Show that the area is equal to $\ln\left[\left(\frac{8}{11}\right)^{\frac{7}{2}}\right]$.

Q7 Given that $\int_1^A \frac{4}{6x-5}\,dx = 10$ and $A \geq 1$, find A in terms of e.

3. Integration of Trigonometric Functions

Learning Objectives:

- Be able to integrate functions of $\sin x$, $\cos x$ and $\sec^2 x$.
- Be able to integrate other trig functions by considering the derivatives of $\operatorname{cosec} x$, $\sec x$ and $\cot x$.

There are a few trig functions which are really easy to integrate — once you've learnt them, you'll be able to integrate loads of complicated-looking trig functions quickly.

Integration of sin x and cos x

In Chapter 7 you learnt how to differentiate $\sin x$ and $\cos x$. You should remember that $\sin x$ differentiates to $\cos x$, and $\cos x$ differentiates to $-\sin x$.

Working backwards from this, we get:

$$\int \sin x \, dx = -\cos x + C$$
$$\int \cos x \, dx = \sin x + C$$

Tip: You won't be given the integrals for sin and cos in your formula booklet, so make sure you remember them.

Example 1

Find the following integrals:

a) $\int 4\cos x \, dx$

$$\int 4\cos x \, dx = 4\int \cos x \, dx = 4\sin x + C$$

$\int \cos x \, dx = \sin x + C$

b) $\int_0^{\pi} \frac{\sin x}{2} + \frac{1}{\pi} \, dx$

- Integrate each term separately:

$$\int_0^{\pi} \frac{\sin x}{2} + \frac{1}{\pi} \, dx = \int_0^{\pi} \frac{1}{2}\sin x + \frac{1}{\pi} \, dx = \left[\frac{1}{2}(-\cos x) + \frac{1}{\pi}x\right]_0^{\pi}$$
$$= \left[-\frac{1}{2}\cos x + \frac{1}{\pi}x\right]_0^{\pi}$$

Don't forget the minus sign.

- Put in the limits:

$$\left[-\frac{1}{2}\cos x + \frac{1}{\pi}x\right]_0^{\pi} = \left[-\frac{1}{2}\cos\pi + \left(\frac{1}{\pi}\times\pi\right)\right] - \left[-\frac{1}{2}\cos 0 + \left(\frac{1}{\pi}\times 0\right)\right]$$
$$= \left[-\frac{1}{2}(-1) + 1\right] - \left[-\frac{1}{2}(1) + 0\right]$$
$$= \left[\frac{1}{2} + 1\right] - \left[-\frac{1}{2} + 0\right] = 2$$

c) $\int \frac{1}{2}(\cos x + 2\sin x) \, dx$

Multiply out and integrate each term separately:

$$\int \frac{1}{2}(\cos x + 2\sin x) \, dx = \int \frac{1}{2}\cos x + \sin x \, dx$$
$$= \frac{1}{2}\sin x + (-\cos x) + C$$
$$= \frac{1}{2}\sin x - \cos x + C$$

You can integrate **linear transformations** of sin x and cos x of the form sin($ax + b$) and cos($ax + b$).

- Differentiating sin($ax + b$) using the **chain rule** gives:
  $a\cos(ax + b)$
- So when **integrating** cos($ax + b$), you need to divide by a, giving:
  $\frac{1}{a}\sin(ax + b)$.

The same can be done when integrating sin($ax + b$), so we get:

$$\int \sin(ax+b)\,dx = -\frac{1}{a}\cos(ax+b) + C$$
$$\int \cos(ax+b)\,dx = \frac{1}{a}\sin(ax+b) + C$$

Example 2

Find $\int \sin(1-6x)\,dx$.

Using the general formula with $a = -6$ and $b = 1$ gives:

$$\int \sin(1-6x)\,dx = \frac{1}{-6} \times -\cos(1-6x) + C$$
$$= \frac{1}{6}\cos(1-6x) + C$$

Tip: You could also do this by noticing that differentiating $\cos(1-6x)$ with the chain rule gives $6\sin(1-6x)$, so $\int \sin(1-6x)\,dx = \frac{1}{6}\cos(1-6x) + C$

Exercise 3.1

Q1 Integrate the following functions with respect to x.

a) $\frac{1}{7}\cos x$ b) $-3\sin x$ c) $-3\cos x - 3\sin x$

d) $\sin 5x$ e) $\cos\left(\frac{x}{7}\right)$ f) $2\sin(-3x)$

g) $5\cos\left(3x + \frac{\pi}{5}\right)$ h) $-4\sin\left(4x - \frac{\pi}{3}\right)$ i) $\cos(4x + 3) + \sin(3 - 4x)$

Q2 Integrate $\frac{1}{2}\cos 3\theta - \sin\theta$ with respect to θ.

Q3 Evaluate the following definite integrals:

a) $\int_0^{\frac{\pi}{2}} \sin x\,dx$ b) $\int_{\frac{\pi}{6}}^{\frac{\pi}{3}} \sin 3x\,dx$ c) $\int_{-1}^{2} 3\sin(\pi x + \pi)\,dx$

Q4 a) Integrate the function $y = 2\pi\cos\left(\frac{\pi x}{2}\right)$ with respect to x between the limits $x = 1$ and $x = 2$.

b) Given that the function doesn't cross the x-axis between these limits, state whether the area between the curve and the x-axis for $1 \le x \le 2$ lies above or below the x-axis, justifying your answer.

Q5 Show that $\int_{\frac{\pi}{3}}^{\frac{\pi}{2}} \sin(-x) + \cos(-x)\,dx = \frac{1-\sqrt{3}}{2}$.

Q6 Show that the integral of the function $y = 5\cos\left(\frac{x}{6}\right)$ between $x = -2\pi$ and $x = \pi$ is $15(1 + \sqrt{3})$.

Integration of $\sec^2 x$

Another trigonometric function which is easy to integrate is the derivative of $\tan x$, $\mathbf{\sec^2 x}$. Since $\tan x$ differentiates to $\sec^2 x$, you get:

$$\int \sec^2 x \, dx = \tan x + C$$

Tip: You can reach this integral by using the differentiation of $\tan x$ — you just need to integrate both sides.

Example 1

Find $\int 2\sec^2 x + 4x \, dx$.

Integrate each term separately.

$$\int 2\sec^2 x + 4x \, dx = 2\int \sec^2 x \, dx + \int 4x \, dx = 2\tan x + 2x^2 + C$$

Unsurprisingly, you can use the chain rule in reverse again to integrate functions of the form $\sec^2(ax + b)$:

$$\int \sec^2(ax+b) \, dx = \frac{1}{a}\tan(ax+b) + C$$

Example 2

Find $\int \cos 4x - 2\sin 2x + \sec^2\left(\frac{1}{2}x\right) dx$.

Integrate each term separately using the results from above and p.187:

$$\int \sec^2\left(\tfrac{1}{2}x\right) dx = \frac{1}{\left(\frac{1}{2}\right)}\tan\left(\tfrac{1}{2}x\right) = 2\tan\left(\tfrac{1}{2}x\right)$$

$$\int \cos 4x \, dx = \frac{1}{4}\sin 4x$$

$$\int \cos 4x - 2\sin 2x + \sec^2\left(\tfrac{1}{2}x\right) dx = \frac{1}{4}\sin 4x + \cos 2x + 2\tan\left(\tfrac{1}{2}x\right) + C$$

$$\int -2\sin 2x \, dx = -2\left(-\tfrac{1}{2}\cos 2x\right) = \cos 2x$$

Exercise 3.2

Q1 Find the following integrals:

a) $\int 2\sec^2 x + 1 \, dx$ b) $\int \sec^2 9x \, dx$ c) $\int 20\sec^2 3y \, dy$

d) $\int \sec^2 \frac{x}{7} \, dx$ e) $\int_0^{\frac{\pi}{3}} -\frac{1}{\cos^2\theta} \, d\theta$ f) $\int_0^{\frac{\pi}{4}} 3\sec^2(-3x) \, dx$

Q2 Find the value of the integral of the function $y = \sec^2 x$ between the limits $x = \frac{2}{3}\pi$ and $x = \pi$.

Q3 Integrate $\sec^2(x + \alpha) + \sec^2(3x + \beta)$ with respect to x, where α and β are constants.

Q4 Let A be a constant. Integrate $5A\sec^2\left(\frac{\pi}{3} - 2\theta\right)$ with respect to θ between the limits of $\theta = \frac{\pi}{12}$ and $\theta = \frac{\pi}{6}$.

Integration of other trigonometric functions

There are some other more complicated trig functions which are really easy to integrate. They are the **derivatives** of the functions **cosec** x, **sec** x and **cot** x.

You may remember these derivatives from Chapter 7, but here's a recap:

$$\frac{d}{dx}(\operatorname{cosec} x) = -\operatorname{cosec} x \cot x$$

$$\frac{d}{dx}(\sec x) = \sec x \tan x \qquad \frac{d}{dx}(\cot x) = -\operatorname{cosec}^2 x$$

Tip: You'll be given these derivatives in the formula booklet.

Reversing the differentiation gives the following three integrals. They'll be really useful when integrating complicated trig functions.

$$\int \operatorname{cosec} x \cot x \, dx = -\operatorname{cosec} x + C$$
$$\int \sec x \tan x \, dx = \sec x + C$$
$$\int \operatorname{cosec}^2 x \, dx = -\cot x + C$$

As always, you can integrate **linear transformations** of these functions (functions of the form **cosec(**$ax + b$**)cot(**$ax + b$**)**, **sec(**$ax + b$**)tan(**$ax + b$**)** and **cosec²(**$ax + b$**)**) by **dividing** by the coefficient of x.

$$\int \operatorname{cosec}(ax+b)\cot(ax+b) \, dx = -\frac{1}{a}\operatorname{cosec}(ax+b) + C$$
$$\int \sec(ax+b)\tan(ax+b) \, dx = \frac{1}{a}\sec(ax+b) + C$$
$$\int \operatorname{cosec}^2(ax+b) \, dx = -\frac{1}{a}\cot(ax+b) + C$$

Tip: The $ax + b$ bit has to be the same in each trig function — e.g. you couldn't integrate $\sec x \tan 3x$ using these formulas.

Example 1

Find the following:

a) $\int 2 \sec x \tan x \, dx$

Take the constant outside the integral.

$$\int 2\sec x \tan x \, dx = 2\int \sec x \tan x \, dx = 2(\sec x + c) = 2 \sec x + C$$

b) $\int_0^\pi \operatorname{cosec}^2\left(\frac{x}{2} - \frac{\pi}{4}\right) dx$

This is a definite integral, so you need to evaluate between the limits.

$$\int_0^\pi \operatorname{cosec}^2\left(\frac{x}{2} - \frac{\pi}{4}\right) dx = \left[-\frac{1}{\left(\frac{1}{2}\right)}\cot\left(\frac{x}{2} - \frac{\pi}{4}\right)\right]_0^\pi$$

Divide by the coefficient of x.

$$= -2\left[\cot\left(\frac{x}{2} - \frac{\pi}{4}\right)\right]_0^\pi = -2\left[\frac{1}{\tan\left(\frac{x}{2} - \frac{\pi}{4}\right)}\right]_0^\pi$$

$\cot x = \frac{1}{\tan x}$

Put in the limits.

$$= -2\left(\frac{1}{\tan\left(\frac{\pi}{2} - \frac{\pi}{4}\right)} - \frac{1}{\tan\left(0 - \frac{\pi}{4}\right)}\right)$$

$$= -2\left(\frac{1}{\tan\left(\frac{\pi}{4}\right)} - \frac{1}{\tan\left(-\frac{\pi}{4}\right)}\right) = -2\left(\frac{1}{1} - \frac{1}{(-1)}\right) = -4$$

c) $\int 8\,\text{cosec}(2x+1)\cot(2x+1)\,dx$

Take the constant outside the integral.

$$\int 8\,\text{cosec}(2x+1)\cot(2x+1)\,dx = 8\int \text{cosec}(2x+1)\cot(2x+1)\,dx$$
$$= 8\left(-\frac{1}{2}\text{cosec}(2x+1)+c\right)$$
$$= -4\,\text{cosec}(2x+1)+C$$

Don't forget to **divide** by the x coefficient.

Example 2

Find $\int 10\sec 5x\tan 5x + \frac{1}{2}\text{cosec}\,3x\cot 3x - \text{cosec}^2(6x+1)\,dx$.

- Integrate each bit in turn:

$$\int 10\sec 5x\tan 5x\,dx = 10\left(\frac{1}{5}\sec 5x\right)$$
$$= 2\sec 5x$$

$$\int \frac{1}{2}\text{cosec}\,3x\cot 3x\,dx = \frac{1}{2}\left(-\frac{1}{3}\text{cosec}\,3x\right)$$
$$= -\frac{1}{6}\text{cosec}\,3x$$

$$\int -\text{cosec}^2(6x+1)\,dx = -\left(-\frac{1}{6}\cot(6x+1)\right)$$
$$= \frac{1}{6}\cot(6x+1)$$

Don't forget the minus that comes from the integration.

Tip: These three integrals should really all have a constant of integration on the end, but we'll just add a combined constant of integration when we do the final integration.

- Putting these terms together and adding the constant gives:

$$\int 10\sec 5x\tan 5x + \frac{1}{2}\text{cosec}\,3x\cot 3x - \text{cosec}^2(6x+1)\,dx$$
$$= 2\sec 5x - \frac{1}{6}\text{cosec}\,3x + \frac{1}{6}\cot(6x+1)+C$$

Exercise 3.3

Q1 Find the following integrals:

a) $\int \text{cosec}^2 11x\,dx$

b) $\int 5\sec 10\theta\tan 10\theta\,d\theta$

c) $\int -\text{cosec}(x+17)\cot(x+17)\,dx$

d) $\int -3\,\text{cosec}\,3x\cot 3x\,dx$

e) $\int 13\sec\left(\frac{\pi}{4}-x\right)\tan\left(\frac{\pi}{4}-x\right)dx$

Q2 Find $\int 10\,\text{cosec}^2\left(\alpha-\frac{x}{2}\right) - 60\sec(\alpha-6x)\tan(\alpha-6x)\,dx$

Q3 Integrate the function $6\sec 2x\tan 2x + 6\,\text{cosec}\,2x\cot 2x$ with respect to x between the limits of $x = \frac{\pi}{12}$ and $x = \frac{\pi}{8}$.

Q4 Find the area of the region bounded by $y = \text{cosec}^2(3x)$, the x-axis and the lines $x = \frac{\pi}{12}$ and $x = \frac{\pi}{6}$.

4. Integration of $\frac{f'(x)}{f(x)}$

Fractions in which the numerator is the derivative of the denominator are pretty easy to integrate — there is a general formula which comes from the chain rule.

Learning Objectives:

- Be able to integrate functions of the form $\frac{f'(x)}{f(x)}$, including multiples of these functions.
- Be able to integrate functions of tan, cosec, sec and cot.

Integrating $\frac{f'(x)}{f(x)}$

- If you have a fraction that has a function of x as the numerator and a different function of x as the denominator, e.g. $\frac{x-2}{x^3+1}$, you'll probably struggle to integrate it.
- However, if you have a fraction where the **numerator** is the **derivative** of the **denominator**, e.g. $\frac{3x^2}{x^3+1}$, it integrates to give ln of the denominator.
- In general terms, this is written as:

$$\int \frac{f'(x)}{f(x)}\,dx = \ln|f(x)| + C$$

Tip: You sometimes see this rule written without the modulus signs, but if it's possible for f(x) to take a negative value, you need the modulus so you don't end up trying to take ln of a negative number.

- This rule won't surprise you if you remember differentiating ln (f(x)) using the chain rule (see p.146) — the derivative with respect to x of ln (f(x)) is $\frac{f'(x)}{f(x)}$.

The hardest bit about integrations like this is recognising that the denominator differentiates to give the numerator — once you've spotted that you can just use the formula.

Example 1

Integrate the following functions with respect to x.

a) $\frac{2x}{x^2+1}$

- Differentiate the denominator to see what it gives:

$$\frac{d}{dx}(x^2+1) = 2x$$ ← This is the **numerator**.

- The numerator is the derivative of the denominator so use the formula:

$$\int \frac{2x}{x^2+1}\,dx = \ln|x^2+1| + C$$

You could leave out the modulus sign as $x^2 + 1 > 0$

b) $\frac{x(3x-4)}{x^3-2x^2-1}$

- Differentiate the denominator:

$$\frac{d}{dx}(x^3-2x^2-1) = 3x^2 - 4x = x(3x-4)$$ ← This is the **numerator**.

- Use the formula:

$$\int \frac{x(3x-4)}{x^3-2x^2-1}\,dx = \ln|x^3-2x^2-1| + C$$

Tip: You should get used to spotting when the numerator is the derivative of the denominator — you won't have to differentiate the denominator every time. Sometimes you might need to expand out some brackets before you notice it.

You might see questions where the numerator is a **multiple** of the derivative of the denominator just to confuse things. When this happens, just put the multiple **in front** of the ln.

Example 2

Find:

a) $\int \frac{8x^3 - 4}{x^4 - 2x} dx$

- Differentiating: $\frac{d}{dx}(x^4 - 2x) = 4x^3 - 2$
  and $8x^3 - 4 = 2(4x^3 - 2)$
- The numerator is 2 × the derivative of the denominator, so
  $$\int \frac{8x^3 - 4}{x^4 - 2x} dx = 2\int \frac{4x^3 - 2}{x^4 - 2x} dx = 2 \ln|x^4 - 2x| + C$$

b) $\int \frac{3 \sin 3x}{\cos 3x + 2} dx$

- Differentiating: $\frac{d}{dx}(\cos 3x + 2) = -3 \sin 3x$
- The numerator is –1 × the derivative of the denominator, so
  $$\int \frac{3 \sin 3x}{\cos 3x + 2} dx = -\int \frac{-3 \sin 3x}{\cos 3x + 2} dx = -\ln|\cos 3x + 2| + C$$
- You can make the answer a lot neater by combining it all into **one logarithm** — a question might ask you to do this.
  $$= -\ln|\cos 3x + 2| - \ln k = -\ln|k(\cos 3x + 2)|$$

The minus sign is just to avoid fractions in the logarithm.

C is just a constant. We can express C as a logarithm — call it $-\ln k$, where k is a constant.

Tip: Any constant can be expressed as a logarithm because the range of the ln function is $f(x) \in \mathbb{R}$.

You can use this method to integrate **trig functions** by writing them as fractions:

- You can work out the integral of **tan *x*** using this method:
  $$\tan x = \frac{\sin x}{\cos x}, \text{ and } \frac{d}{dx}(\cos x) = -\sin x$$
- The numerator is **–1** × the **derivative** of the **denominator**, so
  $$\int \tan x \, dx = \int \frac{\sin x}{\cos x} dx = -\ln|\cos x| + C$$

This is a useful result — it's given in the formula booklet in the following form:

$$\int \tan x \, dx = \ln|\sec x| + C$$

Tip: The integral of tan, $-\ln|\cos x|$, is the same as $\ln|\sec x|$ by the laws of logs.

There are some other **trig functions** that you can integrate in the same way.

$$\int \text{cosec } x \, dx = -\ln|\text{cosec } x + \cot x| + C$$
$$\int \sec x \, dx = \ln|\sec x + \tan x| + C$$
$$\int \cot x \, dx = \ln|\sin x| + C$$

Tip: As always, if you're integrating a linear transformation of any of these functions, of the form $f(ax + b)$, then divide by a when you integrate.

You can check these results easily by using **differentiation** — differentiate the right-hand side of the results to get the left-hand sides.
Remember that differentiating $\ln|f(x)|$ gives $\frac{f'(x)}{f(x)}$.

Example 3

Find the following integrals:

a) $\int 2 \sec x \, dx$

Use the result for sec x above.
There is a constant of 2 so put that at the front.

$$\int 2 \sec x \, dx = 2\ln|\sec x + \tan x| + C$$

b) $\int \frac{\cot x}{5} \, dx$

Use the result for cot x above.

$\frac{\cot x}{5} = \frac{1}{5}\cot x$ so there is a constant of $\frac{1}{5}$ — put that at the front.

$$\int \frac{\cot x}{5} \, dx = \frac{1}{5}\ln|\sin x| + C$$

c) $\int 2(\text{cosec } x + \sec x) \, dx$

$$\int 2(\text{cosec } x + \sec x) \, dx = \int 2 \text{ cosec } x + 2 \sec x \, dx$$
$$= -2\ln|\text{cosec } x + \cot x| + 2\ln|\sec x + \tan x| + C$$
$$= 2\ln\left|\frac{\sec x + \tan x}{\text{cosec } x + \cot x}\right| + C$$

Use log laws to simplify.

d) $\int \frac{1}{2} \text{cosec } 2x \, dx$

You can just use the result above — so all you have to do is work out what happens to the coefficient of x.
The coefficient of x is 2, so divide by 2 when you integrate:

$$\int \frac{1}{2} \text{cosec } 2x \, dx = -\frac{1}{4}\ln|\text{cosec } 2x + \cot 2x| + C$$

Divide $\frac{1}{2}$ by 2

Tip: Check this by differentiating (using the chain rule with $u = \text{cosec } 2x + \cot 2x$).

Exercise 4.1

Q1 Find the following integrals:

a) $\int \frac{4x^3}{x^4 - 1}\,dx$

b) $\int \frac{2x - 1}{x^2 - x}\,dx$

c) $\int \frac{x^4}{3x^5 + 6}\,dx$

d) $\int \frac{12x^3 + 18x^2 - 3}{x^4 + 2x^3 - x}\,dx$

Q2 Find the indefinite integrals below:

a) $\int \frac{e^x}{e^x + 6}\,dx$

b) $\int \frac{2(e^{2x} + 3e^x)}{e^{2x} + 6e^x}\,dx$

c) $\int \frac{e^x}{3(e^x + 3)}\,dx$

Q3 Find the following integrals:

a) $\int \frac{2\cos 2x}{1 + \sin 2x}\,dx$

b) $\int \frac{\sin 3x}{\cos 3x - 1}\,dx$

c) $\int \frac{3\,\text{cosec}\,x\cot x + 6x}{\text{cosec}\,x - x^2 + 4}\,dx$

d) $\int \frac{\sec^2 x}{\tan x}\,dx$

e) $\int \frac{\sec x\tan x}{\sec x + 5}\,dx$

Q4 Show that $\int \frac{4\cos(2x + 7)}{\sin(2x + 7)}\,dx = 2\ln|k\sin(2x + 7)|$.

Q5 Hint: Try multiplying the bit inside the integral by $\frac{\sec x + \tan x}{\sec x + \tan x}$ in part a) — there's a similar trick for part b) as well.

Q5 Prove that:

a) $\int \sec x\,dx = \ln|\sec x + \tan x| + C$

b) $\int \text{cosec}\,x\,dx = -\ln|\text{cosec}\,x + \cot x| + C$

Q6 Find the following integrals:

a) $\int 2\tan x\,dx$

b) $\int \tan 2x\,dx$

c) $\int 4\,\text{cosec}\,x\,dx$

d) $\int \cot 3x\,dx$

e) $\int \frac{1}{2}\sec 2x\,dx$

f) $\int 3\,\text{cosec}\,6x\,dx$

Q7 Find $\int \frac{\sec^2 x}{2\tan x} - 4\sec 2x\tan 2x + \frac{\text{cosec}\,2x\cot 2x - 1}{\text{cosec}\,2x + 2x}\,dx$.

5. Integrating $\frac{du}{dx} f'(u)$

This section will show you how to integrate certain products of functions. You can use the chain rule in reverse to integrate special products of functions and their derivatives.

Learning Objectives:

- Be able to integrate products of the form $\frac{du}{dx} f'(u)$ using the chain rule in reverse.
- Know and be able to use a result for integrating products of the form $f'(x)[f(x)]^n$.

Integrating using the reverse of the chain rule

In Chapter 7, you saw the chain rule for differentiating a **function of a function**.

- Here it is in the form it was given on page 139:

If $y = f(u)$ and $u = g(x)$ then:

$$\frac{dy}{dx} = \frac{dy}{du} \times \frac{du}{dx}$$

- Since integration is the opposite of differentiation, you have:

$$y \xrightarrow{\text{Differentiation}} \frac{dy}{du} \times \frac{du}{dx}$$

$$y + C \xleftarrow[\text{Integration}]{} \frac{dy}{du} \times \frac{du}{dx}$$

- So $\int \frac{dy}{du} \times \frac{du}{dx}\, dx = y + C$.
- Writing $f(u)$ instead of y and $f'(u)$ instead of $\frac{dy}{du}$ gives:

$$\int \frac{du}{dx} f'(u)\, dx = f(u) + C$$

Tip: To evaluate integrals like this you have to integrate with respect to u (because $f'(u)$ is $\frac{dy}{du}$).

If you're integrating an expression which contains a **function of a function**, $f(u)$, try differentiating the function u. If the **derivative** of u is also part of the expression, you might be able to use the formula above.

This result's quite difficult to grasp — but after a few examples it should make complete sense.

Example 1

a) Differentiate $y = e^{2x^2}$ using the chain rule.

Let $u = 2x^2$, then $y = e^u$. By the chain rule,

$$\frac{dy}{dx} = \frac{dy}{du} \times \frac{du}{dx} = e^u \times 4x = e^{2x^2} \times 4x = 4xe^{2x^2}$$

b) Find $\int 4xe^{2x^2}\,dx$ using your answer to part a).

Look for the bit that would have been u in the chain rule — here it's $2x^2$.

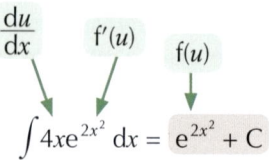

$$\int 4xe^{2x^2}\,dx = e^{2x^2} + C$$

($4x$ is $\frac{du}{dx}$, e^{2x^2} is $f'(u)$, and e^{2x^2} on the right is $f(u)$.)

Example 2

Find the following integrals:

a) $\int 6x^5 e^{x^6}\,dx$

- Here, $u = x^6$ — it appears once differentiated ($6x^5$) and once within a function (e^{x^6}).
- Split the integral into $\frac{du}{dx}$ and $f'(u)$: $\int 6x^5 e^{x^6}\,dx$ ($6x^5$ is $\frac{du}{dx}$, e^{x^6} is $f'(u)$)
- Now use the formula $\int \frac{du}{dx} f'(u)\,dx = f(u) + C$ to write down the result.

$$\int 6x^5 e^{x^6}\,dx = e^{x^6} + C$$

($6x^5$ is $\frac{du}{dx}$, e^{x^6} is $f'(u)$, $f(u) = e^u$)

Tip: You might need to take a constant outside of the integral to get it in the form $\int \frac{du}{dx} f'(u)\,dx$. There's an example of this in part c).

b) $\int e^{\sin x} \cos x\,dx$

- Here, $u = \sin x$.
- Write $\frac{du}{dx}$ and $f'(u)$: $\frac{du}{dx} = \cos x$ and $f'(u) = e^{\sin x}$
- Use the formula to write down the result.

$$\int e^{\sin x} \cos x\,dx = e^{\sin x} + C$$

($e^{\sin x}$ is $f'(u)$, $\cos x$ is $\frac{du}{dx}$, $f(u) = e^u$)

c) $\int x^4 \sin(x^5)\,dx$

- Here, $u = x^5$.
- You'll need to take out a constant to get the integral you want.

$$\int x^4 \sin(x^5)\,dx = \frac{1}{5}\int 5x^4 \sin(x^5)\,dx$$

- Now split up the integral: $\frac{du}{dx} = 5x^4$ and $f'(u) = \sin(x^5)$
- Use the formula.

$$\frac{1}{5}\int 5x^4 \sin(x^5)\,dx = \frac{1}{5}(-\cos(x^5) + c) = -\frac{1}{5}\cos(x^5) + C$$

($5x^4$ is $\frac{du}{dx}$, $\sin(x^5)$ is $f'(u)$, $f(u) = -\cos u$)

Exercise 5.1

Find the following integrals:

Q1 $\int 2xe^{x^2}\,dx$

Q2 $\int 6x^2e^{2x^3}\,dx$

Q3 $\int \frac{1}{2\sqrt{x}}e^{\sqrt{x}}\,dx$

Q4 $\int x^3e^{x^4}\,dx$

Q5 $\int (4x-1)e^{(x^2-\frac{1}{2}x)}\,dx$

Q6 $\int 2x\sin(x^2+1)\,dx$

Q7 $\int x^3\cos(x^4)\,dx$

Q8 $\int x\sec^2(x^2)\,dx$

Q9 $\int e^{\cos x}\sin x\,dx$

Q10 $\int \cos 2x\, e^{\sin 2x}\,dx$

Q11 $\int \sec^2 x\, e^{\tan x}\,dx$

Q12 $\int \sec x\tan x\, e^{\sec x}\,dx$

Integrating $f'(x) \times [f(x)]^n$

Some products are made up of a **function** and its **derivative**:

This bracket is the **derivative**...

e.g. $3(3x^2+4)(x^3+4x)^2$

...of this bracket.

If you spot that part of a product is the **derivative** of the other part of it (which is raised to a **power**), you can integrate it using this rule (which is just a special case of the 'reverse chain rule' on p.195):

$$\int (n+1)f'(x)[f(x)]^n\,dx = [f(x)]^{n+1} + C$$

This function is the **derivative**... ...of this function.

Remember that this result needs you to have a multiple of $\boldsymbol{n+1}$ (not n) — you can check this by **differentiating** the right-hand side using the **chain rule**.

Watch out for any other multiples too — you might have to **multiply** or **divide** by a **constant**.

Tip: To get this formula, rewrite the 'reverse chain rule' on p.195, replacing the function 'u' with '$f(x)$' and the function '$f(u)$' with '$[f(x)]^{n+1}$'. Then $\frac{du}{dx}$ will be replaced with $f'(x)$ and $f'(u)$ will be replaced with $(n+1)[f(x)]^n$.

This will probably make more sense if you have a look at some examples:

Example 1

Evaluate $\int 12x^3(2x^4 - 5)^2 \, dx$.

- Here, **$f(x) = 2x^4 - 5$**, so $f'(x) = 8x^3$. $n = 2$, so $n + 1 = 3$.

  So $\int (n + 1)f'(x)[f(x)]^n \, dx = [f(x)]^{n+1} + C$

  $\Rightarrow \int 3(8x^3)(2x^4 - 5)^2 = \int 24x^3(2x^4 - 5)^2 \, dx = (2x^4 - 5)^3 + C$

- Divide everything by **2** to match the original integral:

  $\int 12x^3(2x^4 - 5)^2 \, dx = \frac{1}{2}(2x^4 - 5)^3 + C$

Example 2

Tip: This one looks pretty horrific, but it isn't too bad once you spot that $-\text{cosec}^2 x$ is the derivative of $\cot x$.

Find $\int 8 \,\text{cosec}^2 x \cot^3 x \, dx$.

- For this one, **$f(x) = \cot x$**, so $f'(x) = -\text{cosec}^2 x$. $n = 3$, so $n + 1 = 4$.

  So $\int (n + 1)f'(x)[f(x)]^n \, dx = [f(x)]^{n+1} + C$

  $\Rightarrow \int -4 \,\text{cosec}^2 x \cot^3 x \, dx = \cot^4 x + C$

- Multiply everything by **–2** to match the original integral:

  $\int 8 \,\text{cosec}^2 x \cot^3 x \, dx = -2 \cot^4 x + C$

Example 3

Find $\int (x - 2)\sqrt{x^2 - 4x + 5} \, dx$.

- You can write the square root as a **fractional power**:

  $\int (x - 2)\sqrt{x^2 - 4x + 5} \, dx = \int (x - 2)(x^2 - 4x + 5)^{\frac{1}{2}} \, dx$

- Now, **$f(x) = x^2 - 4x + 5$**, so $f'(x) = 2x - 4$. $n = \frac{1}{2}$, so $n + 1 = \frac{3}{2}$.

  So $\int (n + 1)f'(x)[f(x)]^n \, dx = [f(x)]^{n+1} + C$

  $\Rightarrow \int \frac{3}{2}(2x - 4)(x^2 - 4x + 5)^{\frac{1}{2}} \, dx = (x^2 - 4x + 5)^{\frac{3}{2}} + C$

- $\frac{3}{2}(2x - 4) = 3(x - 2)$, so you need to divide everything by **3** to match the original integral:

  $\int (x - 2)\sqrt{x^2 - 4x + 5} \, dx = \frac{1}{3}(x^2 - 4x + 5)^{\frac{3}{2}} + C$

Example 4

Evaluate $\int \frac{\cos x}{\sin^4 x}\,dx$.

- Write $\frac{1}{\sin^4 x}$ as a negative power.

$$\int \frac{\cos x}{\sin^4 x}\,dx = \int \frac{\cos x}{(\sin x)^4}\,dx = \int \cos x(\sin x)^{-4}\,dx$$

- Now, $\mathbf{f(x) = \sin x}$, so $f'(x) = \cos x$. $n = -4$, so $n + 1 = -3$.

$$\text{So } \int (n+1)f'(x)[f(x)]^n\,dx = [f(x)]^{n+1} + C$$

$$\Rightarrow \int -3\cos x(\sin x)^{-4}\,dx = (\sin x)^{-3} + C$$

$$\Rightarrow \int \frac{-3\cos x}{\sin^4 x}\,dx = \frac{1}{\sin^3 x} + C$$

- Divide everything by **–3** to match the original integral:

$$\int \frac{\cos x}{\sin^4 x} = -\frac{1}{3\sin^3 x} + C$$

Exercise 5.2

Q1 Find the following indefinite integrals:

a) $\int 6x(x^2 + 5)^2\,dx$ b) $\int (2x + 7)(x^2 + 7x)^4\,dx$

c) $\int (x^3 + 2x)(x^4 + 4x^2)^3\,dx$ d) $\int \frac{2x}{(x^2 - 1)^3}\,dx$

e) $\int \frac{6e^{3x}}{(e^{3x} - 5)^2}\,dx$ f) $\int \sin x\cos^5 x\,dx$

g) $\int 2\sec^2 x\tan^3 x\,dx$ h) $\int 3e^x(e^x + 4)^2\,dx$

i) $\int 32(2e^{4x} - 3x)(e^{4x} - 3x^2)^7\,dx$ j) $\int \frac{\cos x}{(2 + \sin x)^4}\,dx$

k) $\int 5\,\text{cosec}\,x\cot x\,\text{cosec}^4 x\,dx$ l) $\int 2\,\text{cosec}^2 x\cot^3 x\,dx$

Q1-2 Hint: You'll need the derivatives of cosec, sec and cot:
$\frac{d}{dx}(\text{cosec}\,x) = -\text{cosec}\,x\cot x$
$\frac{d}{dx}(\sec x) = \sec x\tan x$
$\frac{d}{dx}(\cot x) = -\text{cosec}^2 x$

Q2 Find the following integrals:

a) $\int 6\tan x\sec^6 x\,dx$ b) $\int \cot x\,\text{cosec}^3 x\,dx$

Q3 Integrate the following functions with respect to x:

a) $4\cos x\,e^{\sin x}(e^{\sin x} - 5)^3$ b) $(\sin x\,e^{\cos x} - 4)(e^{\cos x} + 4x)^6$

Q4 Integrate:

a) $\int \frac{\sec^2 x}{\tan^4 x}\,dx$ b) $\int \cot x\,\text{cosec}\,x\sqrt{\text{cosec}\,x}\,dx$

6. Using Trigonometric Identities in Integration

Learning Objective:

- Be able to use the double angle formulas for sin, cos and tan alongside the methods learnt throughout this chapter to simplify difficult trig integrals.

You can sometimes use the trig identities that you learnt in Chapter 3 to manipulate nasty trig integrations to give functions you know how to integrate.

Integrating using trig identities

Integrating using the double angle formulas

If you're given a tricky **trig function** to integrate, you might be able to simplify it using one of the **double angle formulas**. They're especially useful for things like $\cos^2 x$, $\sin^2 x$ and $\sin x \cos x$. Here are the double angle formulas again:

$$\sin 2x \equiv 2 \sin x \cos x \qquad \cos 2x \equiv \cos^2 x - \sin^2 x$$

$$\tan 2x \equiv \frac{2 \tan x}{1 - \tan^2 x}$$

You came across two other ways of writing the double angle formula for cos, which come from using the identity $\cos^2 x + \sin^2 x \equiv 1$:

$$\cos 2x \equiv 2 \cos^2 x - 1 \qquad \cos 2x \equiv 1 - 2 \sin^2 x$$

Once you've rearranged the original function using one of the **double angle formulas**, the function you're left with should be easier to integrate using the rules you've seen in this chapter.

Example 1

Find the following:

a) $\int \sin^2 x \, dx$

- Rearranging the **cos** double angle formula: $\cos 2x \equiv 1 - 2 \sin^2 x$

  gives $\sin^2 x \equiv \frac{1}{2}(1 - \cos 2x)$.

- So rewrite the integration:

$$\int \sin^2 x \, dx = \int \frac{1}{2}(1 - \cos 2x) dx = \frac{1}{2} \int (1 - \cos 2x) dx$$

$$= \frac{1}{2}\left(x - \frac{1}{2}\sin 2x\right) + C = \frac{1}{2}x - \frac{1}{4}\sin 2x + C$$

Tip: Use one of the cos double angle formulas when you've got a $\cos^2 x$ or a $\sin^2 x$ to integrate.

b) $\int \cos^2 5x \, dx$

- Rearranging the **cos** double angle formula: $\cos 2x \equiv 2 \cos^2 x - 1$

  gives $\cos^2 x \equiv \frac{1}{2}(\cos 2x + 1)$.

- So rewrite the integration:

Don't forget to double the x coefficient and to divide by 10 when you integrate.

$$\int \cos^2 5x \, dx = \int \frac{1}{2}(\cos 10x + 1) dx = \frac{1}{2} \int (\cos 10x + 1) dx$$

$$= \frac{1}{2}\left(\frac{1}{10}\sin 10x + x\right) + C = \frac{1}{20}\sin 10x + \frac{1}{2}x + C$$

Tip: For a reminder of integrating $\sin x$ and $\cos x$ (and linear transformations of them) see pages 186-187.

Example 2

Find the following integrals:

a) $\int \sin x \cos x \, dx$

- Rearranging the **sin** double angle formula: $\sin 2x \equiv 2 \sin x \cos x$

  gives $\sin x \cos x \equiv \frac{1}{2}\sin 2x$.

- So rewrite the integration:

  $\int \sin x \cos x \, dx = \int \frac{1}{2}\sin 2x \, dx = \frac{1}{2}\left(-\frac{1}{2}\cos 2x\right) + C = -\frac{1}{4}\cos 2x + C$

b) $\int_0^{\frac{\pi}{4}} \sin 2x \cos 2x \, dx$

- Rearranging the **sin** double angle formula with x replaced with $2x$: $\sin 4x \equiv 2 \sin 2x \cos 2x$

  gives $\sin 2x \cos 2x \equiv \frac{1}{2}\sin 4x$.

- So rewrite the integration:

$$\int_0^{\frac{\pi}{4}} \sin 2x \cos 2x \, dx = \int_0^{\frac{\pi}{4}} \frac{1}{2}\sin 4x \, dx = \left[\frac{1}{2}\left(-\frac{1}{4}\cos 4x\right)\right]_0^{\frac{\pi}{4}} = -\frac{1}{8}[\cos 4x]_0^{\frac{\pi}{4}}$$

$$= -\frac{1}{8}\left(\left[\cos\frac{4\pi}{4}\right] - [\cos 0]\right) = \frac{1}{8}(\cos 0 - \cos \pi)$$

$$= \frac{1}{8}(1-(-1)) = \frac{2}{8} = \frac{1}{4}$$

c) $\int \frac{4\tan\frac{x}{2}}{1-\tan^2\frac{x}{2}} \, dx$

- Rearrange, then use the double angle formula for **tan**:

$$\frac{4\tan\frac{x}{2}}{1-\tan^2\frac{x}{2}} = 2\left(\frac{2\tan\frac{x}{2}}{1-\tan^2\frac{x}{2}}\right) = 2\left(\tan\left(2\times\frac{x}{2}\right)\right) = 2\tan x$$

- Rewrite the integration:

$$\int \frac{4\tan\frac{x}{2}}{1-\tan^2\frac{x}{2}} \, dx = \int 2\tan x \, dx = -2\ln|\cos x| + C$$

Tip: If you need to integrate a function of the form $\sin x \cos x$, use the double angle formula for sin.

Tip: Don't forget to 'double the angle' when using these formulas.

Tip: Remember the integral of $\tan x$ is $-\ln|\cos x| + C$ (which is the same as $\ln|\sec x| + C$ using the laws of logs). It comes from writing $\tan x = \frac{\sin x}{\cos x}$ (as shown on p.192).

Integrating using other trigonometric identities

- There are a couple of other **identities** you can use to simplify trig functions:

$$\sec^2 x \equiv 1 + \tan^2 x \qquad \text{cosec}^2 x \equiv 1 + \cot^2 x$$

- These identities are really useful if you have to integrate $\mathbf{tan^2} x$ or $\mathbf{cot^2} x$, as you already know how to integrate $\sec^2 x$ and $\text{cosec}^2 x$ (see pages 188-189).

$$\int \sec^2 x \, dx = \tan x + C \qquad \int \text{cosec}^2 x \, dx = -\cot x + C$$

Tip: If you use one of these identities to get rid of a $\cot^2 x$ or a $\tan^2 x$, don't forget the stray 1s flying around — they'll just integrate to x.

Example 3

a) Find $\int \tan^2 x - 1\,dx$

- Rewrite the function in terms of $\sec^2 x$:

$\sec^2 x \equiv 1 + \tan^2 x$

$$\tan^2 x - 1 = (\sec^2 x - 1) - 1 = \sec^2 x - 2$$

- Now integrate: $\int \tan^2 x - 1\,dx = \int \sec^2 x - 2\,dx = \tan x - 2x + C$

b) Find $\int \cot^2 3x\,dx$

- Get the function in terms of $\operatorname{cosec}^2 x$:

$\operatorname{cosec}^2 x \equiv 1 + \cot^2 x$

$$\cot^2 3x = \operatorname{cosec}^2 3x - 1$$

$$\int \cot^2 3x\,dx = \int \operatorname{cosec}^2 3x - 1\,dx = -\frac{1}{3}\cot 3x - x + C$$

c) $\int \cos^3 x\,dx$

- You don't know how to integrate $\cos^3 x$ but you can split it into $\cos^2 x$ and $\cos x$ and use identities.

$\cos^2 x + \sin^2 x \equiv 1$

$$\cos^3 x = \cos^2 x \cos x = (1 - \sin^2 x)\cos x$$
$$= \cos x - \cos x \sin^2 x = \cos x - \cos x(\sin x)^2$$

- Now write out the integral:

$$\int \cos^3 x\,dx = \int \cos x\,dx - \int \cos x(\sin x)^2\,dx$$

- If $f(x) = \sin x$, then $\cos x(\sin x)^2 = f'(x) \times [f(x)]^2$. So:

$$\int \cos x \sin^2 x\,dx = \frac{1}{3}\sin^3 x + c$$

Tip: See p.197 for more on how to do this type of integration.

- So the whole integral is:

$$\int \cos^3 x\,dx = \int \cos x\,dx - \int \cos x(\sin x)^2\,dx$$
$$= \sin x - \frac{1}{3}\sin^3 x + C$$

d) Evaluate $\int_0^{\frac{\pi}{3}} 6\sin 3x\cos 3x + \tan^2 \frac{1}{2}x + 1\,dx$.

- Using the **sin** double angle formula:

$$6\sin 3x\cos 3x \equiv 3\sin 6x$$

and using the identity for $\tan^2 x$:

$$\tan^2 \tfrac{1}{2}x + 1 \equiv \sec^2 \tfrac{1}{2}x$$

- Now integrate:

$$\int_0^{\frac{\pi}{3}} 6\sin 3x\cos 3x + \tan^2 \frac{1}{2}x + 1\,dx$$
$$= \int_0^{\frac{\pi}{3}} 3\sin 6x + \sec^2 \frac{1}{2}x\,dx = \left[-\frac{3}{6}\cos 6x + 2\tan \frac{1}{2}x\right]_0^{\frac{\pi}{3}}$$
$$= \left[-\frac{1}{2}\cos(2\pi) + 2\tan\left(\frac{\pi}{6}\right)\right] - \left[-\frac{1}{2}\cos(0) + 2\tan(0)\right]$$
$$= \left[-\frac{1}{2}(1) + 2\left(\frac{1}{\sqrt{3}}\right)\right] - \left[-\frac{1}{2}(1) + 2(0)\right] = -\frac{1}{2} + \frac{2}{\sqrt{3}} + \frac{1}{2} = \frac{2}{\sqrt{3}} = \frac{2\sqrt{3}}{3}$$

Put in the limits.

Exercise 6.1

Q1 Find the following indefinite integrals:

a) $\int \cos^2 x \, dx$

b) $\int 6\sin x \cos x \, dx$

c) $\int \sin^2 6x \, dx$

d) $\int \frac{2\tan 2x}{1-\tan^2 2x} \, dx$

e) $\int 2\sin 4x \cos 4x \, dx$

f) $\int 2\cos^2 4x \, dx$

g) $\int \cos x \sin x \, dx$

h) $\int \sin 3x \cos 3x \, dx$

i) $\int \frac{6\tan 3x}{1-\tan^2 3x} \, dx$

j) $\int 5\sin 2x \cos 2x \, dx$

k) $\int (\sin x + \cos x)^2 \, dx$

l) $\int 4\sin x \cos x \cos 2x \, dx$

m) $\int (\cos x + \sin x)(\cos x - \sin x) \, dx$

n) $\int \sin^2 x \cot x \, dx$

Q1 l) Hint: Use the sin double angle formula twice.

Q2 Evaluate the following definite integrals, giving exact answers:

a) $\int_0^{\frac{\pi}{4}} \sin^2 x \, dx$

b) $\int_0^{\pi} \cos^2 2x \, dx$

c) $\int_0^{\pi} \sin\frac{x}{2}\cos\frac{x}{2} \, dx$

d) $\int_{\frac{\pi}{4}}^{\frac{\pi}{2}} \sin^2 2x \, dx$

e) $\int_0^{\frac{\pi}{4}} \cos 2x \sin 2x \, dx$

f) $\int_{\frac{\pi}{4}}^{\frac{\pi}{2}} \sin^2 x - \cos^2 x \, dx$

Q3 Find the following integrals:

a) $\int \cot^2 x - 4 \, dx$

b) $\int \tan^2 x \, dx$

c) $\int 3\cot^2 x \, dx$

d) $\int \tan^2 4x \, dx$

Q4 Find the exact value of $\int_0^{\frac{\pi}{4}} \tan^2 x + \cos^2 x - \sin^2 x \, dx$.

Find the integrals in Q5-8:

Q5 $\int (\sec x + \tan x)^2 \, dx$

Q6 $\int (\cot x + \operatorname{cosec} x)^2 \, dx$

Q7 $\int 4 + \cot^2 3x \, dx$

Q8 $\int \cos^2 4x + \cot^2 4x \, dx$

Q9 Integrate the following functions with respect to x:

a) $\tan^3 x + \tan^5 x$

b) $\cot^5 x + \cot^3 x$

c) $\sin^3 x$

Q10 Use the identity $\sin A + \sin B \equiv 2\sin\left(\frac{A+B}{2}\right)\cos\left(\frac{A-B}{2}\right)$ to find $\int 2\sin 4x \cos x \, dx$.

PROBLEM SOLVING

7. Finding Area using Integration

Learning Objectives:

- Be able to use integration to find the area between two curves, or the area between a line and a curve.
- Be able to use integration to find the area under a curve given its parametric equations.

In Year 1 you saw how to find the area between a curve and the x-axis using integration. But you can also use integration to find the area between a curve and a line, or even two curves.

Finding enclosed areas

To find the area between a curve, a line and the x-axis, you'll either have to **add** or **subtract** integrals to find the area you're after — it's always best to **draw a diagram** of the area.

Example 1

Find the area enclosed by the curve $y = x^2$, the line $y = 2 - x$ and the x-axis.

- Draw a diagram of the curve and the line.

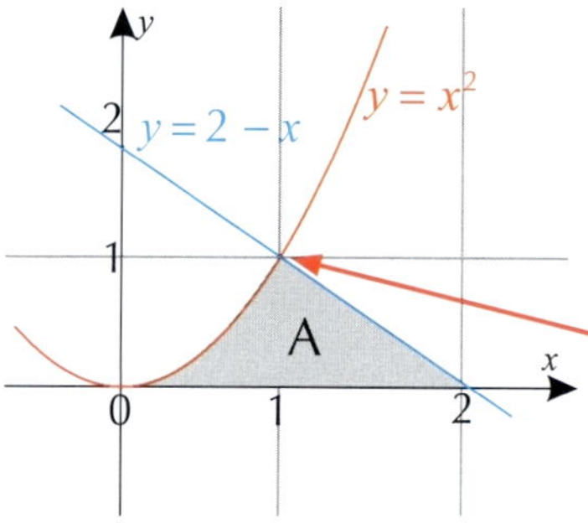

You have to find **area A** — but you'll need to **split** it into two smaller bits.

Find out where the curve and line meet by **solving** $x^2 = 2 - x$. They meet at $x = 1$ (they also meet at $x = -2$, but this isn't in A).

Tip: You need to find the x-coordinate of the point of intersection between the line and the curve, as well as the x-intercept of each function, so that you know the limits to integrate between.

- The area A is the area under the **red** curve between 0 and 1 **added to** the area under the **blue** line between 1 and 2.

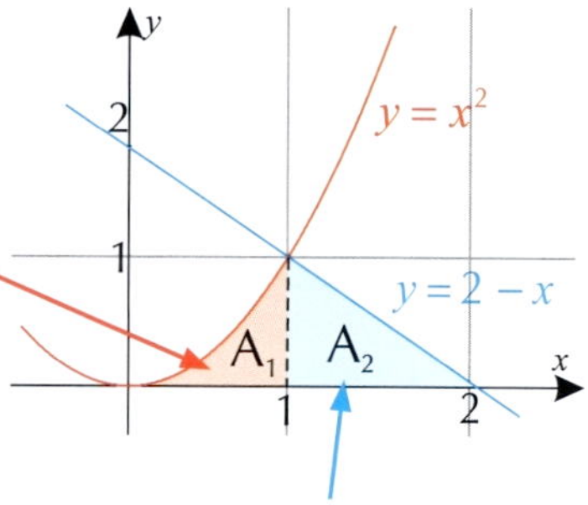

A_1 is the area under the curve $y = x^2$ between 0 and 1, so integrate between these limits to find the area:

$$\int_0^1 x^2\,dx = \left[\frac{x^3}{3}\right]_0^1 = \frac{1}{3} - 0 = \frac{1}{3}$$

A_2 is the area under the line $y = 2 - x$ between 1 and 2, so integrate between these limits to find the area:

$$\int_1^2 (2 - x)\,dx = \left[2x - \frac{x^2}{2}\right]_1^2$$

$$= \left(2(2) - \frac{2^2}{2}\right) - \left(2(1) - \frac{1^2}{2}\right)$$

$$= 2 - \frac{3}{2} = \frac{1}{2}$$

Tip: A_2 is just a triangle with base 1 and height 1, so you could also calculate its area using the formula for the area of a triangle.

- **Add** the areas together to find the area A:

$$A = A_1 + A_2 = \frac{1}{3} + \frac{1}{2} = \frac{5}{6}$$

Sometimes you'll need to find the area **enclosed** by the graphs of two functions — this usually means **subtracting** some area from another. Here is an example with two curves:

Example 2

The diagram below shows the curves $y = \sin x + 1$ and $y = \cos x + 1$. Find the area of the shaded grey region.

PROBLEM SOLVING

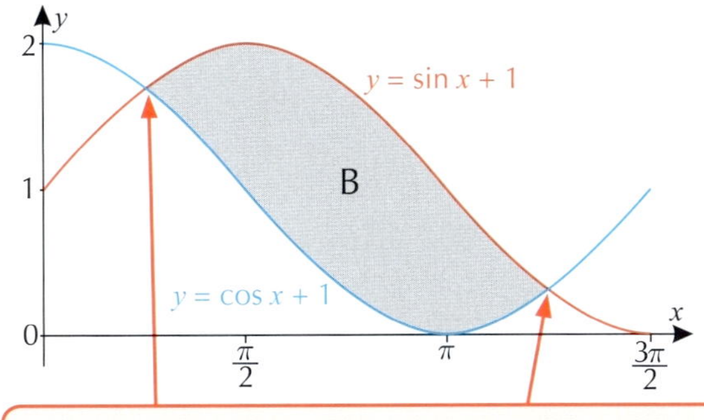

Find out where the graphs meet by **solving** $\sin x + 1 = \cos x + 1$

$\Rightarrow \sin x = \cos x \Rightarrow \frac{\sin x}{\cos x} = 1 \Rightarrow \tan x = 1$

$\Rightarrow x = \tan^{-1}(1) = \frac{\pi}{4}, \frac{5\pi}{4}$. They meet at $\boldsymbol{x = \frac{\pi}{4}}$ and $\boldsymbol{x = \frac{5\pi}{4}}$.

- You need to find two different integrals to work out the area of B:

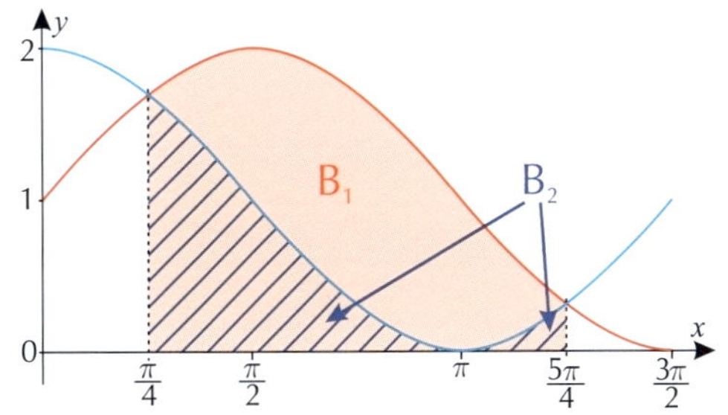

- The area B is the area under the **red** curve between $\frac{\pi}{4}$ and $\frac{5\pi}{4}$ (B_1) **minus** the area under the **blue** curve between $\frac{\pi}{4}$ and $\frac{5\pi}{4}$ (B_2).

$$B_1 = \int_{\frac{\pi}{4}}^{\frac{5\pi}{4}} \sin x + 1\,dx = [-\cos x + x]_{\frac{\pi}{4}}^{\frac{5\pi}{4}} = \left(-\cos\left(\frac{5\pi}{4}\right) + \frac{5\pi}{4}\right) - \left(-\cos\left(\frac{\pi}{4}\right) + \frac{\pi}{4}\right)$$
$$= \left(-\left(-\frac{\sqrt{2}}{2}\right) + \frac{5\pi}{4}\right) - \left(-\frac{\sqrt{2}}{2} + \frac{\pi}{4}\right) = \sqrt{2} + \pi$$

$$B_2 = \int_{\frac{\pi}{4}}^{\frac{5\pi}{4}} \cos x + 1\,dx = [\sin x + x]_{\frac{\pi}{4}}^{\frac{5\pi}{4}} = \left(\sin\left(\frac{5\pi}{4}\right) + \frac{5\pi}{4}\right) - \left(\sin\left(\frac{\pi}{4}\right) + \frac{\pi}{4}\right)$$
$$= \left(-\frac{\sqrt{2}}{2} + \frac{5\pi}{4}\right) - \left(\frac{\sqrt{2}}{2} + \frac{\pi}{4}\right) = -\sqrt{2} + \pi$$

- So the area $B = B_1 - B_2 = (\sqrt{2} + \pi) - (-\sqrt{2} + \pi) = 2\sqrt{2}$

Tip: In the table on page 54, $\cos\frac{\pi}{4}$ is given as $\frac{1}{\sqrt{2}}$ — here it's written as $\frac{\sqrt{2}}{2}$ but it's the same thing (you can rationalise the denominator to check).

Sometimes you might need to add **and** subtract integrals to find the right area. You'll often need to do this when the curve goes **below** the x-axis. The integrations you need to do should be obvious if you draw a **picture**.

Exercise 7.1

> **Q1 Hint:** If it doesn't look like you've been given enough information, look for any hints on the graph that may tell you what line to use.

Q1 Find the shaded area in the following diagrams:

a) 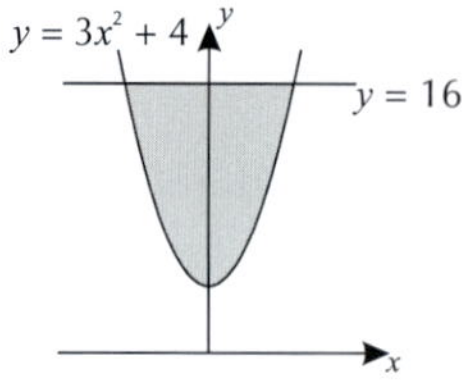

b) 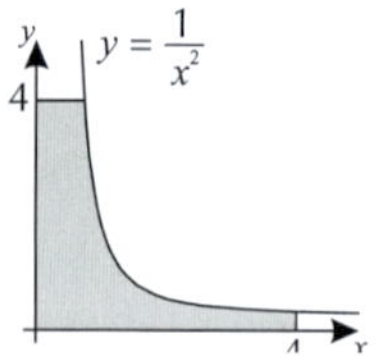

c) 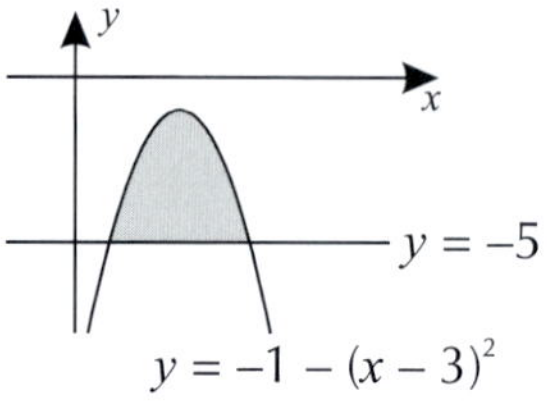

d) 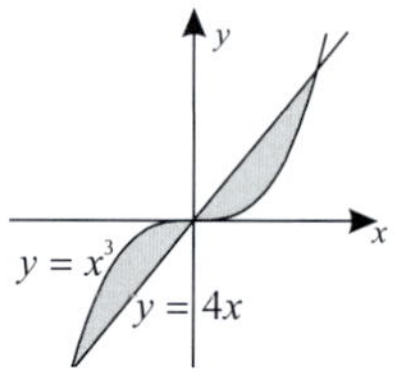

> **Q2b) Hint:** This one's a bit tricky — the best way to do it is to consider the bits above and below the x-axis separately.

Q2 For each part, find the area enclosed by the curve and line:

a) $y = x^2 + 4$ and $y = x + 4$ b) $y = x^2 + 2x - 3$ and $y = 4x$

Q3 Find the shaded area shown to the right:

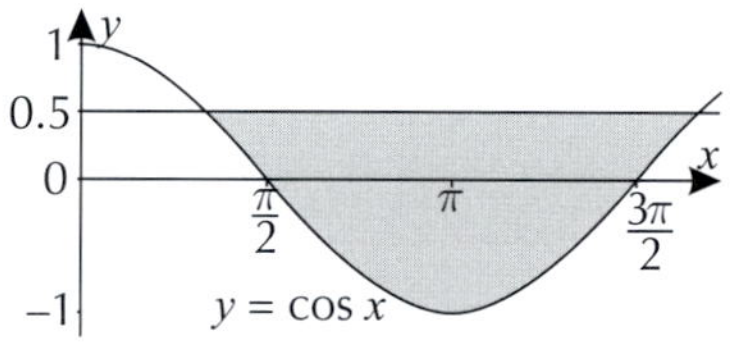

Q4 A company has designed a logo based on multiples of the sine curve, shown to the right. Calculate the total area of the grey sections of the logo.

PROBLEM SOLVING

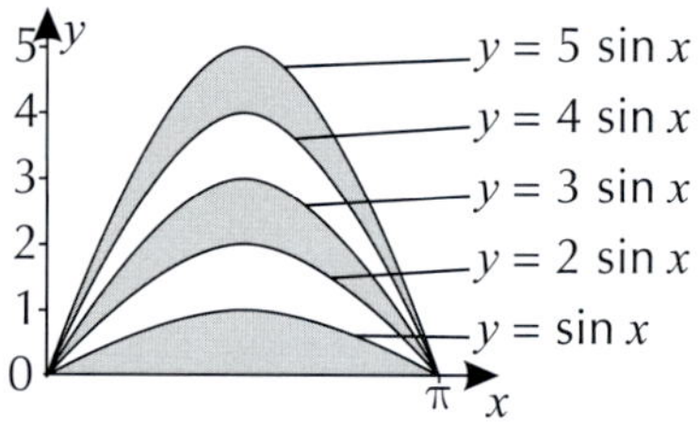

Parametric integration

- Normally, to find the **area** under a graph, you can do a **simple integration**. But if you've got **parametric equations**, things are more difficult — you can't find $\int y\,dx$ if y isn't written in terms of x.
- There's a sneaky way to get around this. Suppose your parameter is t. Then:

$$\int y\,dx = \int y\frac{dx}{dt}\,dt$$

> **Tip:** This comes from the chain rule for differentiation (p.139). If you think of dx as $\frac{dx}{1}$, then $\frac{dx}{1} = \frac{dx}{dt} \times \frac{dt}{1}$.

- Both y and $\frac{dx}{dt}$ are written **in terms of t**, so you can **multiply** them together to get an expression you can **integrate with respect to t**.

With a **definite integral**, you need to **alter the limits** as well. So if you have x-values as limits, work out the corresponding values of t before you integrate.

> **Tip:** There's more about altering the limits of definite integrals on pages 210-211.

Example

The shaded region marked A on the sketch is bounded by the x-axis, the line $x = 2$, and by the curve with parametric equations $x = t^2 - 2$ and $y = t^2 - 9t + 20$, $t \geq 0$, which crosses the x-axis at $x = 14$.

Find the area of A.

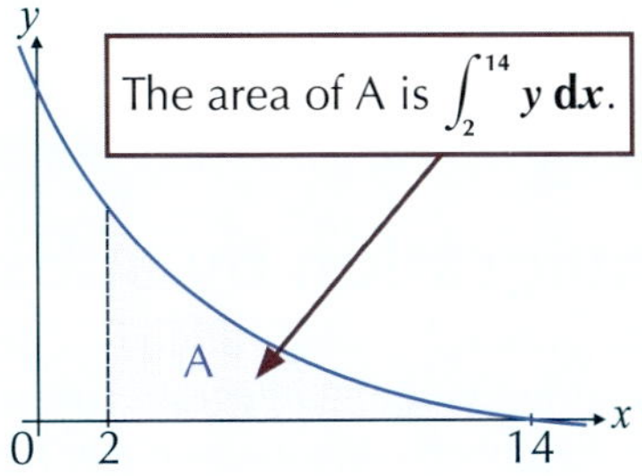

- You need to use $\int y\,dx = \int y \frac{dx}{dt}\,dt$ to integrate, so find $\frac{dx}{dt}$:

$$\frac{dx}{dt} = \frac{d}{dt}(t^2 - 2) = 2t$$

- Now sort out the **limits**. 14 and 2 are the limits for integrating **with respect to x**. You need to find the **corresponding values of t**:

$$x = 2 \Rightarrow t^2 - 2 = 2 \Rightarrow t^2 = 4 \Rightarrow t = 2$$
$$x = 14 \Rightarrow t^2 - 2 = 14 \Rightarrow t^2 = 16 \Rightarrow t = 4$$

Tip: $t \geq 0$ so ignore the negative square roots.

- Now **integrate** to find the area of A:

$$A = \int_2^{14} y\,dx = \int_2^4 y\frac{dx}{dt}\,dt$$
$$= \int_2^4 (t^2 - 9t + 20)(2t)\,dt = \int_2^4 2t^3 - 18t^2 + 40t\,dt$$
$$= \left[\frac{1}{2}t^4 - 6t^3 + 20t^2\right]_2^4$$
$$= \left(\frac{1}{2}(4)^4 - 6(4)^3 + 20(4)^2\right) - \left(\frac{1}{2}(2)^4 - 6(2)^3 + 20(2)^2\right)$$
$$= 64 - 40 = 24$$

Exercise 7.2

Q1 For each of the following curves, find an expression in parametric form that is equivalent to the indefinite integral $\int y\,dx$.

a) $x = \frac{3}{t}$, $y = 4t^2$ b) $x = \tan 5\theta$, $y = \sec^2 5\theta$

Q1 Hint: You don't need to integrate these, but you will need to differentiate to get $\frac{dx}{dt}$ or $\frac{dx}{d\theta}$.

Q2 For each of the following curves, find an expression equivalent to $\int y\,dx$ and integrate it.

a) $x = (4t - 5)^2$, $y = t^2 - 3t$ b) $x = t^2 + 3$, $y = 4t - 1$

Q3 A curve has parametric equations $x = 3t^2$, $y = \frac{5}{t}$, where $t > 0$.

Find an expression for $y\frac{dx}{dt}$, and hence evaluate $\int_3^{75} y\,dx$.

Q3 Hint: Don't forget to change the limits — they're given as values for x but you need the corresponding values of t.

Q4 The curve shown here has parametric equations $x = 4t(t + 1)$, $y = 3t^3$.

a) Find the values t_1 and t_2 that correspond to $x = 8$ and $x = 120$, given that $t > 0$.

b) Hence find a parametric integral corresponding to $\int_8^{120} y\,dx$, and evaluate this to find the area A.

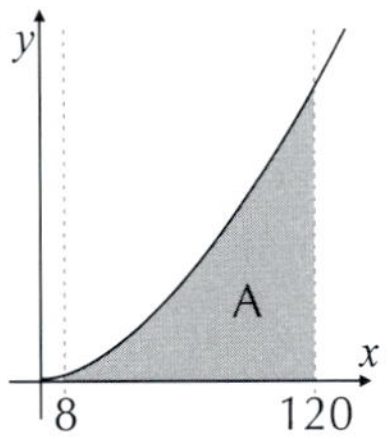

8. Integration by Substitution

Learning Objectives:

- Understand that integration by substitution is the reverse of differentiating using the chain rule.
- Use integration by substitution to integrate functions, including calculating definite integrals.
- Be able to integrate using substitutions that require trig identities.

Earlier, you saw how to use the chain rule in reverse to integrate a function of the form $\frac{du}{dx}f'(u)$. Integration by substitution is a more general version of this technique. You'll often find there's more than one way to get to the answer.

Integration by substitution

Integration by substitution is a way of integrating a **function of a function** by simplifying the integral. Like differentiating with the chain rule, to integrate by substitution you have to write part of the function in terms of u, where u is some **function** of x.

Here's the method:

- You'll be given an integral that's made up of **two functions of x**.
- **Substitute** u for one of the functions of x to give a function that's **easier to integrate**.
- Next, find $\frac{du}{dx}$, and **rewrite** it so that **dx** is on its own.
- **Rewrite** the original integral in terms of u and **du**.
- You should now be left with something that's **easier** to integrate — just **integrate** as normal, then at the last step **replace** u with the **original substitution**.

Tip: You might have to do a bit of rearranging to get u in terms of x.

Example 1

Use the substitution $u = x^2 - 2$ to find $\int 4x(x^2 - 2)^4\, dx$.

- Start by differentiating u with respect to x:

$$u = x^2 - 2 \Rightarrow \frac{du}{dx} = 2x$$

- Now rearrange the equation for $\frac{du}{dx}$ to get dx on its own:

$$\frac{du}{dx} = 2x \Rightarrow du = 2x\, dx$$

$$\Rightarrow dx = \frac{1}{2x}du$$

$\frac{du}{dx}$ isn't really a fraction, but you can treat it like one for this bit.

- Substitute what you've got so far back into the original expression:

$$\int 4x(x^2 - 2)^4\, dx = \int 4xu^4 \frac{1}{2x}\, du = \int 2u^4\, du$$

The remaining x's cancel.

- Now you've got a much simpler expression to integrate with respect to u:

$$\int 2u^4\, du = \frac{2}{5}u^5 + C$$

- And finally, substitute $u = x^2 - 2$ back in:

$$= \frac{2}{5}(x^2 - 2)^5 + C$$

Tip: You won't always be told what substitution to use, so you might have to take an educated guess. Functions inside brackets, denominators of fractions and square roots are often good places to start. There might be multiple different substitutions that you could use to get the right answer.

Tip: This integral is of the form $f'(x) \times [f(x)]^n$ (p.197). You could use the rule you learnt back there for this kind of integral, but you've been asked to do it by substitution.

That first example worked out nicely, because the x's **cancelled out** when you substituted in the expressions for u and dx. It isn't always quite so straightforward — sometimes you need to get rid of some x's by **rearranging** the equation to get x in terms of u.

Example 2

Find $\int x(3x + 2)^3\,dx$, using the substitution $u = 3x + 2$.

- Start by finding $\frac{du}{dx}$ and then rearrange to get dx on its own:

$$u = 3x + 2 \Rightarrow \frac{du}{dx} = 3 \Rightarrow dx = \frac{1}{3}\,du$$

- If you substitute for u and dx, you end up with an x still in the integral:

$$\int x(3x + 2)^3\,dx = \int xu^3 \frac{1}{3}\,du$$

- To get rid of that x, you have to rearrange the equation for u:

$$u = 3x + 2 \Rightarrow x = \frac{u-2}{3}$$

- So

$$\begin{aligned}\int x(3x + 2)^3\,dx &= \int \left(\frac{u-2}{3}\right)u^3 \frac{1}{3}\,du \\ &= \int \frac{u^4 - 2u^3}{9}\,du \\ &= \frac{1}{9}\left(\frac{u^5}{5} - \frac{u^4}{2}\right) + C \\ &= \frac{u^5}{45} - \frac{u^4}{18} + C \\ &= \frac{(3x+2)^5}{45} - \frac{(3x+2)^4}{18} + C\end{aligned}$$

Tip: The expression in brackets is often the thing you substitute.

Tip: Note that this integral cannot be done using methods from previous sections.

Tip: Don't forget to rewrite your answer in terms of x again at the end.

Some integrations look really tricky, but with a clever substitution they can be made a lot simpler.

Example 3

Find $\int 3x\sqrt{2 - x^2}\,dx$, using a suitable substitution.

- $\sqrt{2 - x^2}$ looks like the most awkward bit to integrate, so that might be a good choice for the substitution. Let $u = \sqrt{2 - x^2}$.

- Now differentiate to find $\frac{du}{dx}$:

$$\frac{du}{dx} = -\frac{x}{\sqrt{2 - x^2}} = -\frac{x}{u}$$

$$\Rightarrow u\,du = -x\,dx \Rightarrow -\frac{u}{x}\,du = dx$$

- Now substitute what you've got into the original integral:

$$\begin{aligned}\int 3x\sqrt{2 - x^2}\,dx &= \int 3x \times u \times -\frac{u}{x}\,du \\ &= \int -3u^2\,du \\ &= -u^3 + C \\ &= -(\sqrt{2 - x^2})^3 + C\end{aligned}$$

Tip: The chain rule is used here to differentiate.

Tip: You could also use the substitution $u = 2 - x^2$ to solve this — the differentiation is easier but you get a tougher integral with fractional powers. Give it a try on your own.

Exercise 8.1

Q1 Find the following integrals using the given substitutions:

a) $\int 12(x+3)^5\,dx,\ u = x + 3$ b) $\int (11-x)^4\,dx,\ u = 11 - x$

c) $\int 24x(x^2+4)^3\,dx,\ u = x^2 + 4$ d) $\int \sin^5 x \cos x\,dx,\ u = \sin x$

e) $\int x(x-1)^5\,dx,\ u = x - 1$

Q2 Use an appropriate substitution to find:

a) $\int 21(x+2)^6\,dx$ b) $\int (5x+4)^3\,dx$

c) $\int x(2x+3)^3\,dx$ d) $\int 24x(x^2-5)^7\,dx$

Q3 Use the given substitutions to find the following integrals:

a) $\int 6x\sqrt{x+1}\,dx,\ u = \sqrt{x+1}$ b) $\int \frac{x}{\sqrt{4-x}}\,dx,\ u = \sqrt{4-x}$

c) $\int \frac{15(\ln x)^4}{x}\,dx,\ u = \ln x$

Q4 Find the following integrals by substitution:

a) $\int \frac{4x}{\sqrt{(2x-1)}}\,dx$ b) $\int \frac{1}{4-\sqrt{x}}\,dx$ c) $\int \frac{e^{2x}}{1+e^x}\,dx$

Q5 Use integration by substitution to prove that

$$\int (n+1)f'(x)[f(x)]^n\,dx = [f(x)]^{n+1} + C$$

PROBLEM SOLVING

Q5 Hint: This is the formula from p.197.

Definite integrals

If you're given a **definite integral** to find using a substitution, it's important that you remember to **change the limits** to u. To do this, put the x-limits into the equation for u to find the corresponding values of u.

Doing it this way means you **don't** have to **put x back in** at the last step — just put the values of u into the integration for u.

Tip: You could also solve this one using the method on p.197.

Examples

a) Use the substitution $u = \cos x$ to find $\int_{\frac{\pi}{2}}^{2\pi} -12\sin x \cos^3 x\,dx$.

- As with indefinite integrals, start by differentiating u, and rearranging to get dx on its own:

$$u = \cos x \Rightarrow \frac{du}{dx} = -\sin x \Rightarrow dx = -\frac{1}{\sin x}\,du$$

- Now use the substitution to change the limits of the integral from x-values to u-values:

$$x = \frac{\pi}{2} \Rightarrow u = \cos\frac{\pi}{2} = 0$$

$$x = 2\pi \Rightarrow u = \cos 2\pi = 1$$

- Substitute all that back into the original integral, and solve:

$$\int_{\frac{\pi}{2}}^{2\pi} -12\sin x\cos^3 x\, dx = \int_0^1 -12\sin x\, u^3 \frac{-1}{\sin x}\, du$$
$$= \int_0^1 12u^3\, du$$
$$= [3u^4]_0^1$$
$$= [3(1)^4] - [3(0)^4] = 3 - 0 = 3$$

b) Use a suitable substitution to find $\int_2^{\frac{7}{2}} x\sqrt{2x-3}\, dx$.

- The square root looks like the most awkward part of the integral, so try the substitution $u = \sqrt{2x-3}$.
- Differentiate the substitution, and rearrange to get dx on its own:

$$u = \sqrt{2x-3} \Rightarrow \frac{du}{dx} = \frac{1}{\sqrt{2x-3}} = \frac{1}{u} \Rightarrow dx = u\, du$$

- Rearrange the substitution to get an expression for x:

$$u = \sqrt{2x-3} \Rightarrow x = \frac{u^2+3}{2}$$

- Convert the limits from x-values to u-values:

$$x = 2 \Rightarrow u = \sqrt{2(2)-3} = \sqrt{1} = 1$$
$$x = \frac{7}{2} \Rightarrow u = \sqrt{2\left(\frac{7}{2}\right)-3} = \sqrt{4} = 2$$

- Substituting everything back into the original integral gives:

$$\int_2^{\frac{7}{2}} x\sqrt{2x-3}\, dx = \int_1^2 \frac{u^2+3}{2} \times u \times u\, du$$
$$= \frac{1}{2}\int_1^2 u^4 + 3u^2\, du$$
$$= \frac{1}{2}\left[\frac{u^5}{5} + u^3\right]_1^2$$
$$= \left[\frac{2^5}{10} + \frac{2^3}{2}\right] - \left[\frac{1^5}{10} + \frac{1^3}{2}\right] = \frac{36}{5} - \frac{3}{5} = \frac{33}{5}$$

Tip: Like Example 3 on p.209, there are other substitutions you could use. Try solving this on your own with the substitution $u = 2x - 3$.

Sometimes when you convert the limits of a definite integral, the **upper limit** converts to a **lower number** than the **lower limit** does.

You can either keep the converted limits in the **same places** as the corresponding original limits and carry on as normal or **swap them** so the higher value is the upper limit and stick a **minus sign** in front of the whole integral.

Tip: Swapping the limits and putting a minus in front of the integral might seem more complicated, but it often cancels with another minus, making the whole integration easier.

Exercise 8.2

Q1 Find the exact values of the following using the given substitutions:

a) $\int_{\frac{2}{3}}^{1}(3x-2)^4\,dx,\ u = 3x-2$

b) $\int_{-2}^{1} 2x(x+3)^4\,dx,\ u = x+3$

c) $\int_{0}^{\frac{\pi}{6}} 8\sin^3 x\cos x\,dx,\ u = \sin x$

d) $\int_{0}^{3} x\sqrt{x+1}\,dx,\ u = \sqrt{x+1}$

Q2 Use an appropriate substitution to find the exact value of each of the following:

a) $\int_{2}^{\sqrt{5}} x(x^2-3)^4\,dx$

b) $\int_{1}^{2} x(3x-4)^3\,dx$

c) $\int_{2}^{10} \frac{x}{\sqrt{x-1}}\,dx$

Q3 Integrate the function $y = \frac{1}{3-\sqrt{x}}$ between $x = 1$ and $x = 4$, using a suitable substitution.
Give your answer in the form $a + b\ln 2$, where a and b are integers.

Q4 Find $\int_{0}^{1} 2e^x(1+e^x)^3\,dx$, using the substitution $u = 1 + e^x$.
Give your answer to 1 decimal place.

Q5 Use integration by substitution to find the integral of the function $y = \frac{x}{\sqrt{3x+1}}$ between the limits $x = 1$ and $x = 5$.

Trig identities

As you know by now, there's a vast range of **trig identities** and **formulas** to deal with in A-level maths. This can make for some pretty tricky **integration questions** involving trig functions. Here are a couple of examples:

Tip: If you need a reminder of the trig identities, they're given in Chapter 3.

Examples

a) Use the substitution $u = \tan x$ to find $\int \frac{\sec^4 x}{\sqrt{\tan x}}\,dx$.

- First, work out what all the substitutions will be. Start by finding dx:

$$u = \tan x \Rightarrow \frac{du}{dx} = \sec^2 x \Rightarrow dx = \frac{1}{\sec^2 x}\,du$$

- This substitution for dx will leave $\sec^2 x$ on the numerator — so now you need to find $\sec^2 x$ in terms of u:

From the identity $\sec^2 x \equiv 1 + \tan^2 x$

$$u = \tan x \Rightarrow \sec^2 x \equiv 1 + u^2$$

- Then substitute all these bits into the integral:

$$\int \frac{\sec^4 x}{\sqrt{\tan x}}\,dx = \int \frac{(1+u^2)\times\sec^2 x}{\sqrt{u}} \times \frac{1}{\sec^2 x}\,du$$

$$= \int \frac{1}{\sqrt{u}} + \frac{u^2}{\sqrt{u}}\,du = \int u^{-\frac{1}{2}} + u^{\frac{3}{2}}\,du$$

$$= 2u^{\frac{1}{2}} + \frac{2}{5}u^{\frac{5}{2}} + C = 2\sqrt{\tan x} + \frac{2}{5}\sqrt{\tan^5 x} + C$$

b) Calculate $\int_{\frac{1}{2}}^{\frac{\sqrt{3}}{2}} \frac{4}{\sqrt{1-x^2}}\, dx$, using the substitution $x = \sin\theta$, where $-\frac{\pi}{2} \le \theta \le \frac{\pi}{2}$.

- Start by differentiating x with respect to θ, and use the result to find dx:

  $x = \sin\theta \Rightarrow \frac{dx}{d\theta} = \cos\theta \Rightarrow dx = \cos\theta\, d\theta$

- Use the substitution to convert the limits from x to θ:

  $x = \sin\theta \Rightarrow \theta = \sin^{-1} x$

  So $x = \frac{\sqrt{3}}{2} \Rightarrow \theta = \frac{\pi}{3}$ and $x = \frac{1}{2} \Rightarrow \theta = \frac{\pi}{6}$

- Now solve the integral:

$$\int_{\frac{1}{2}}^{\frac{\sqrt{3}}{2}} \frac{4}{\sqrt{1-x^2}}\, dx = \int_{\frac{\pi}{6}}^{\frac{\pi}{3}} \frac{4}{\sqrt{1-\sin^2\theta}} \cos\theta\, d\theta$$

Use the identity $\sin^2\theta + \cos^2\theta \equiv 1$

$$= \int_{\frac{\pi}{6}}^{\frac{\pi}{3}} \frac{4\cos\theta}{\sqrt{\cos^2\theta}}\, d\theta$$

$$= \int_{\frac{\pi}{6}}^{\frac{\pi}{3}} 4\, d\theta = [4\theta]_{\frac{\pi}{6}}^{\frac{\pi}{3}} = \frac{4\pi}{3} - \frac{2\pi}{3} = \frac{2\pi}{3}$$

Tip: Notice that $\sin\theta$ has an inverse because θ is restricted to between $-\frac{\pi}{2}$ and $\frac{\pi}{2}$.

Exercise 8.3

Q1 Find the exact value of $\int_0^1 \frac{1}{1+x^2}\, dx$ using the substitution $x = \tan\theta$ where $-\frac{\pi}{2} < \theta < \frac{\pi}{2}$.

Q2 Find the exact value of $\int_0^{\frac{\pi}{6}} 3\sin x \sin 2x\, dx$ using the substitution $u = \sin x$.

Q3 Use the substitution $x = 2\sin\theta$, where $-\frac{\pi}{2} \le \theta \le \frac{\pi}{2}$, to find the exact value of $\int_1^{\sqrt{3}} \frac{1}{(4-x^2)^{\frac{3}{2}}}\, dx$.

Q4 Find the exact value of $\int_{\frac{1}{2}}^{1} \frac{1}{x^2\sqrt{1-x^2}}\, dx$.
Use the substitution $x = \cos\theta$ where $0 \le \theta \le \pi$.

Q5 Find $\int 2\tan^3 x\, dx$ using the substitution $u = \sec^2 x$.

Q1-4 Hint: Remember, the phrase 'exact value' is usually a clue that the answer will include a surd or π.

9. Integration by Parts

Learning Objectives:

- Understand that integration by parts comes from the product rule.
- Be able to use integration by parts to integrate functions, including $\ln x$.
- Be able to integrate functions where integration by parts has to be applied more than once.

Sadly, not every integration problem can be solved with a nifty substitution or a clever trick. Integration by parts is another way to deal with integrating a product of two functions — it involves both differentiation and integration.

Integration by parts

If you have a **product** to integrate but you can't use any of the methods you've learnt so far, you might be able to use **integration by parts**.

The **formula** for integrating by parts is:

$$\int u\frac{dv}{dx}\,dx = uv - \int v\frac{du}{dx}\,dx$$

where u and v are both functions of x.

Here's the **proof** of this formula. You're **not** expected to know it for the exam, but you might find it useful.

Start with the **product rule**:

- If u and v are both functions of x, then

$$\frac{d}{dx}uv = u\frac{dv}{dx} + v\frac{du}{dx}$$

- Integrate both sides of the product rule with respect to x:

$$\int \frac{d}{dx}uv\,dx = \int u\frac{dv}{dx}\,dx + \int v\frac{du}{dx}\,dx$$

- On the left-hand side, uv is differentiated, then integrated — so you end up back at uv:

$$uv = \int u\frac{dv}{dx}\,dx + \int v\frac{du}{dx}\,dx$$

- Now just rearrange to get:

$$\int u\frac{dv}{dx}\,dx = uv - \int v\frac{du}{dx}\,dx$$

Tip: Take a look back at p.154 for more about the product rule.

- The hardest thing about integration by parts is **deciding** which bit of your product should be $\boldsymbol{u}$ and which bit should be $\boldsymbol{\frac{dv}{dx}}$.
- There's no set rule for this — you just have to look at both parts, see which one **differentiates** to give something **nice**, then set that one as u.
- For example, if you have a product that has a **single** $\boldsymbol{x}$ as one part of it, choose this to be u. It differentiates to **1**, which makes **integrating** $v\frac{du}{dx}$ very easy.

Tip: The integration by parts formula is sometimes written $\int uv'\,dx = uv - \int vu'\,dx$ — you might find this version easier to use.

Example 1

Find $\int 2xe^x\, dx$.

- Start by working out what should be u and what should be $\frac{dv}{dx}$ — choose them so that $v\frac{du}{dx}$ is easier to integrate than $2xe^x$.

- The two factors are $2x$ and e^x, so try them both ways round:

If $u = 2x$ and $\frac{dv}{dx} = e^x$, then $v\frac{du}{dx} = 2e^x$ ← Easier to integrate than $2xe^x$.

If $u = e^x$ and $\frac{dv}{dx} = 2x$, then $v\frac{du}{dx} = x^2e^x$

- So let $u = 2x$ and $\frac{dv}{dx} = e^x$

- Put u, v, $\frac{du}{dx}$ and $\frac{dv}{dx}$ into the integration by parts formula:

$$u = 2x \Rightarrow \frac{du}{dx} = 2 \quad \frac{dv}{dx} = e^x \Rightarrow v = e^x$$

$$\int 2xe^x\, dx = \int u\frac{dv}{dx}\, dx = uv - \int v\frac{du}{dx}\, dx$$

$$= 2xe^x - \int 2e^x\, dx$$

$$= 2xe^x - 2e^x + C$$

Don't forget the constant of integration.

Tip: You don't always need to work out both possible versions. Here, e^x won't change whether you integrate it or differentiate it, so you just need to think about whether the integration would be made easier by differentiating $2x$ or by integrating it.

Example 2

Find $\int x^3 \ln x\, dx$.

- Choose u and $\frac{dv}{dx}$:

Let $u = \ln x$ and $\frac{dv}{dx} = x^3$

$$u = \ln x \Rightarrow \frac{du}{dx} = \frac{1}{x} \quad \frac{dv}{dx} = x^3 \Rightarrow v = \frac{x^4}{4}$$

- Put u, v, $\frac{du}{dx}$ and $\frac{dv}{dx}$ into the integration by parts formula:

$$\int x^3 \ln x\, dx = \ln x \times \frac{x^4}{4} - \int \frac{x^4}{4} \times \frac{1}{x}\, dx$$

$$= \frac{x^4 \ln x}{4} - \frac{1}{4}\int x^3\, dx$$

$$= \frac{x^4 \ln x}{4} - \frac{x^4}{16} + C$$

Tip: If you have a product that has $\ln x$ as one of its factors, let $u = \ln x$, as $\ln x$ is easy to differentiate but quite tricky to integrate (see next page).

Until now, you haven't been able to integrate **ln x**, but **integration by parts** gives you a way to get around this. The trick is to write ln x as (1 × ln x).

- You can write ln x as (ln x × 1). So let $u = \ln x$ and let $\frac{dv}{dx} = 1$.

  $u = \ln x \Rightarrow \frac{du}{dx} = \frac{1}{x}$

  $\frac{dv}{dx} = 1 \Rightarrow v = x$

- Putting these into the formula gives:

$$\int \ln x \, dx = \int (\ln x \times 1) \, dx = \ln x \times x - \int x \frac{1}{x} \, dx$$
$$= x \ln x - \int 1 \, dx$$
$$= \mathbf{x \ln x - x + C}$$

You can use **integration by parts** on definite integrals too. The only change from the method for indefinite integrals is that you have to **apply the limits** of the integral to the ***uv*** bit.

The integration by parts formula for definite integrals can be written like this:

$$\int_a^b u \frac{dv}{dx} \, dx = [uv]_a^b - \int_a^b v \frac{du}{dx} \, dx$$

Example 3

Find the exact value of $\int_0^{\frac{\pi}{2}} 4x \sin\left(\frac{x}{2}\right) dx$.

- Choose u and $\frac{dv}{dx}$.

  $\sin\left(\frac{x}{2}\right)$ will give a cos function whether you integrate or differentiate it, so the only way to get a simpler $\int v \frac{du}{dx} \, dx$ is to make $u = 4x$.

- Let $u = 4x$ and $\frac{dv}{dx} = \sin\left(\frac{x}{2}\right)$

  $u = 4x \Rightarrow \frac{du}{dx} = 4 \quad \frac{dv}{dx} = \sin\left(\frac{x}{2}\right) \Rightarrow v = -2\cos\left(\frac{x}{2}\right)$

- Substitute everything into the formula and complete the integration:

$$\int_0^{\frac{\pi}{2}} 4x \sin\left(\frac{x}{2}\right) dx = \left[-8x\cos\left(\frac{x}{2}\right)\right]_0^{\frac{\pi}{2}} - \int_0^{\frac{\pi}{2}} -8\cos\left(\frac{x}{2}\right) dx$$
$$= -8\left[x\cos\left(\frac{x}{2}\right)\right]_0^{\frac{\pi}{2}} + 16\left[\sin\left(\frac{x}{2}\right)\right]_0^{\frac{\pi}{2}}$$
$$= -8\left[\frac{\pi}{2}\cos\left(\frac{\pi}{4}\right) - 0\cos(0)\right] + 16\left[\sin\left(\frac{\pi}{4}\right) - \sin(0)\right]$$
$$= -8\left[\frac{\pi}{2}\frac{1}{\sqrt{2}}\right] + 16\left[\frac{1}{\sqrt{2}}\right]$$
$$= -\frac{4\pi}{\sqrt{2}} + \frac{16}{\sqrt{2}}$$
$$= 8\sqrt{2} - 2\pi\sqrt{2}$$

Tip: Go back to p.186 if you want a reminder about integrating trig functions.

Exercise 9.1

Q1 Use integration by parts to find:

a) $\int xe^x \, dx$ b) $\int xe^{-x} \, dx$ c) $\int xe^{-\frac{x}{3}} \, dx$ d) $\int x(e^x + 1) \, dx$

Q2 Use integration by parts to find:

a) $\int_0^{\pi} x \sin x \, dx$ b) $\int 2x \cos x \, dx$

c) $\int 3x \cos\left(\frac{1}{2}x\right) dx$ d) $\int_{-\frac{\pi}{2}}^{\frac{\pi}{2}} 2x(1 - \sin x) \, dx$

Q3 Use integration by parts to find:

a) $\int 2 \ln x \, dx$ b) $\int x^4 \ln x \, dx$ c) $\int \ln 4x \, dx$ d) $\int \ln x^3 \, dx$

Q4 Use integration by parts to find:

a) $\int_{-1}^{1} 20x(x + 1)^3$ b) $\int_0^{1.5} 30x\sqrt{2x + 1} \, dx$

Q5 Use integration by parts to find the exact values of the following:

a) $\int_0^1 12xe^{2x} \, dx$ b) $\int_0^{\frac{\pi}{3}} 18x \sin 3x \, dx$ c) $\int_1^2 \frac{1}{x^2} \ln x \, dx$

Q6 Find $\int \frac{x}{e^{2x}} \, dx$.

Q7 Find $\int (x + 1)\sqrt{x + 2} \, dx$.

Q8 Find $\int \ln(x + 1) \, dx$.

Repeated use of integration by parts

Sometimes **integration by parts** leaves you with a function for $v\frac{du}{dx}$ which is **simpler** than the function you started with, but still **tricky to integrate**.

You might have to carry out integration by parts **again** to find $\int v\frac{du}{dx} \, dx$.

Example 1

Find $\int x^2 \sin x \, dx$.

- Let $u = x^2$ and let $\frac{dv}{dx} = \sin x$. Then $\frac{du}{dx} = 2x$ and $v = -\cos x$.
- Putting these into the formula gives:

  $\int x^2 \sin x \, dx = -x^2\cos x - \int -2x \cos x \, dx = -x^2\cos x + \int 2x \cos x \, dx$
- $2x\cos x$ isn't very easy to integrate, but you can integrate by parts again:

  Let $u_1 = 2x$ and let $\frac{dv_1}{dx} = \cos x$. Then $\frac{du_1}{dx} = 2$ and $v_1 = \sin x$.
- Putting these into the formula gives:

  $\int 2x \cos x \, dx = 2x \sin x - \int 2 \sin x \, dx = 2x \sin x + 2\cos x + C$
- So $\int x^2 \sin x \, dx = -x^2\cos x + \int 2x \cos x \, dx$

  $= -x^2\cos x + 2x \sin x + 2\cos x + C$

Tip: Calling the different parts u_1 and v_1 just means you don't get confused with the u and v used in the first integration.

Tip: The formula for integrating $(ax + b)^n$ is used a few times in this example — go back to p.180 if you've forgotten how it works.

Example 2

Use integration by parts to find $\int_2^3 x^2(x-1)^{-4}\,dx$.

- Let $u = x^2$ and let $\frac{dv}{dx} = (x-1)^{-4}$. Then $\frac{du}{dx} = 2x$ and $v = -\frac{1}{3}(x-1)^{-3}$.
- Putting these into the formula gives:

$$\int_2^3 x^2(x-1)^{-4}\,dx = \left[-\frac{x^2}{3}(x-1)^{-3}\right]_2^3 - \int_2^3 -\frac{2x}{3}(x-1)^{-3}\,dx$$
$$= \left[-\frac{x^2}{3}(x-1)^{-3}\right]_2^3 + \frac{2}{3}\int_2^3 x(x-1)^{-3}\,dx$$

- $\int_2^3 x(x-1)^{-3}\,dx$ is still tricky to integrate. Use integration by parts again:

  Let $u_1 = x$ and let $\frac{dv_1}{dx} = (x-1)^{-3}$. Then $\frac{du_1}{dx} = 1$ and $v_1 = -\frac{1}{2}(x-1)^{-2}$.
- Put these into the formula:

$$\int_2^3 x(x-1)^{-3}\,dx = \left[-\frac{x}{2}(x-1)^{-2}\right]_2^3 - \int_2^3 -\frac{1}{2}(x-1)^{-2}\,dx$$
$$= \left[-\frac{x}{2}(x-1)^{-2}\right]_2^3 - \frac{1}{2}[(x-1)^{-1}]_2^3$$
$$= \left[-\frac{3}{2}(2)^{-2} + \frac{2}{2}(1)^{-2}\right] - \frac{1}{2}[2^{-1} - 1^{-1}]$$
$$= \left[-\frac{3}{8} + 1\right] - \frac{1}{2}\left[\frac{1}{2} - 1\right]$$
$$= \frac{5}{8} + \frac{1}{4} = \frac{7}{8}$$

- Now you can evaluate the original integral:

$$\int_2^3 x^2(x-1)^{-4}\,dx = \left[-\frac{x^2}{3}(x-1)^{-3}\right]_2^3 + \frac{2}{3}\int_2^3 x(x-1)^{-3}\,dx$$
$$= \left[-\frac{x^2}{3}(x-1)^{-3}\right]_2^3 + \frac{2}{3}\left(\frac{7}{8}\right)$$
$$= \left[\left(-\frac{9}{3}(2)^{-3}\right) - \left(-\frac{4}{3}(1)^{-3}\right)\right] + \frac{7}{12}$$
$$= \left[-\frac{9}{24} + \frac{4}{3}\right] + \frac{7}{12}$$
$$= \frac{23}{24} + \frac{7}{12} = \frac{37}{24}$$

Exercise 9.2

Q1 Use integration by parts twice to find:

a) $\int x^2 e^x\,dx$ b) $\int x^2 \cos x\,dx$

c) $\int 4x^2 \sin 2x\,dx$ d) $\int 40x^2(2x-1)^4\,dx$

Q2 Find $\int_{-1}^0 x^2(x+1)^4\,dx$ using integration by parts.

Q3 Use integration by parts to find the area enclosed by the curve $y = x^2e^{-2x}$, the x-axis and the lines $x = 0$ and $x = 1$.

Q3 Hint: Since $x^2 \geq 0$ and $e^{-2x} > 0$ for all x, the area will be entirely above the x-axis.

10. Integration Using Partial Fractions

By rewriting algebraic expressions as partial fractions, you can turn difficult-looking integrations into ones you're more familiar with.

Learning Objective:

- Be able to integrate rational expressions in which the denominator can be written as a product of linear factors, by splitting them into partial fractions.

Use of partial fractions

You can integrate algebraic fractions where the denominator can be written as a product of **linear factors** by splitting them up into **partial fractions**. Each fraction can then be **integrated separately** using the method on pages 184-185.

Example 1

Find $\int \frac{12x+6}{4x^2-9}\,dx$ where $x > 2$.

- The first step is to write the function as **partial fractions** as follows:
  - **Factorise** the denominator:
  $$\frac{12x+6}{4x^2-9} \equiv \frac{12x+6}{(2x+3)(2x-3)}$$
  - Write as an **identity** with partial fractions:
  $$\frac{12x+6}{(2x+3)(2x-3)} \equiv \frac{A}{2x+3} + \frac{B}{2x-3}$$
  - **Add** the partial fractions and **cancel** the denominators:
  $$\frac{12x+6}{(2x+3)(2x-3)} \equiv \frac{A(2x-3)+B(2x+3)}{(2x+3)(2x-3)}$$
  $$12x + 6 \equiv A(2x-3) + B(2x+3)$$
  - Use the **substitution method** to find A and B:

    Substituting $x = \frac{3}{2}$ into the identity gives
    $$18 + 6 = 0A + 6B \Rightarrow 24 = 6B \Rightarrow B = 4$$
    Substituting $x = -\frac{3}{2}$ into the identity gives
    $$-18 + 6 = -6A + 0B \Rightarrow -12 = -6A \Rightarrow A = 2$$
  - Replace A and B in the **original identity**:
  $$\frac{12x+6}{4x^2-9} \equiv \frac{2}{2x+3} + \frac{4}{2x-3}$$
- So $\int \frac{12x+6}{4x^2-9}\,dx = \int \frac{2}{2x+3} + \frac{4}{2x-3}\,dx$.
- Integrate each term **separately**:
  $$= 2 \times \frac{1}{2}\ln|2x+3| + 4 \times \frac{1}{2}\ln|2x-3| + C$$
  $$= \ln|2x+3| + 2\ln|2x-3| + C$$
- $x > 2$ so $2x + 3 > 7$ and $2x - 3 > 1$ so remove the modulus signs:
  $$= \ln(2x+3) + 2\ln(2x-3) + C$$
  $$= \ln(2x+3)(2x-3)^2 + C$$

Tip: Writing expressions as partial fractions is covered in Chapter 2 — but it's recapped here to remind you what they're all about.

Tip: You could also use the 'equating coefficients' method to find A and B.

Tip: It's a good idea to tidy up your answers using the log laws.

Example 2

Find the exact value of $\int_3^4 \frac{2}{x(x-2)}\,dx$, writing it as a single logarithm.

- Start by writing $\frac{2}{x(x-2)}$ as **partial fractions**.

  - Write as an **identity** with partial fractions:
    $$\frac{2}{x(x-2)} \equiv \frac{A}{x} + \frac{B}{x-2} \equiv \frac{A(x-2)+Bx}{x(x-2)}$$
    $$\Rightarrow 2 \equiv A(x-2) + Bx$$
  - Use the **equating coefficients method** to find A and B:
    Equating constant terms: $2 = -2A \Rightarrow \mathbf{A = -1}$
    Equating x coefficients: $0 = A + B \Rightarrow 0 = -1 + B \Rightarrow \mathbf{B = 1}$
  - Replace A and B in the **original identity**:
    $$\frac{2}{x(x-2)} \equiv \frac{-1}{x} + \frac{1}{x-2} \equiv \frac{1}{x-2} - \frac{1}{x}$$

- So $\int_3^4 \frac{2}{x(x-2)}\,dx = \int_3^4 \frac{1}{(x-2)} - \frac{1}{x}\,dx$
- Integrate each term separately:
  $$= [\ln|x-2| - \ln|x|]_3^4$$
  $$= \left[\ln\left|\frac{x-2}{x}\right|\right]_3^4 = \ln\left|\frac{4-2}{4}\right| - \ln\left|\frac{3-2}{3}\right|$$
  $$= \ln\left(\frac{1}{2}\right) - \ln\left(\frac{1}{3}\right) = \ln\left(\frac{3}{2}\right)$$

$\log a - \log b = \log\left(\frac{a}{b}\right)$

Tip: When the constant term only contains one of the unknowns, it'll often be easier to compare coefficients than to substitute.

Tip: The question asks for the answer as a single logarithm so make sure you fully simplify your answer.

Exercise 10.1

Q1 Integrate the following functions by writing them as partial fractions:

a) $\int \frac{24(x-1)}{9-4x^2}\,dx$ b) $\int \frac{21x-82}{(x-2)(x-3)(x-4)}\,dx$

Q2 Find $\int_0^1 \frac{x}{(x-2)(x-3)}\,dx$ by expressing as partial fractions. Give your answer as a single logarithm.

Q3 a) Express $\frac{6}{2x^2-5x+2}$ in partial fractions.

b) Hence find $\int \frac{6}{2x^2-5x+2}\,dx$ where $x > 2$.

c) Evaluate $\int_3^5 \frac{6}{2x^2-5x+2}\,dx$, expressing your answer as a single logarithm.

Q4 Given that $f(x) = 3x + 5$ and $g(x) = x(x + 10)$, find $\int_1^2 \frac{f(x)}{g(x)}\,dx$ to 3 d.p. by expressing $\frac{f(x)}{g(x)}$ as partial fractions.

Q5 Show that $\int_0^{\frac{2}{3}} \frac{-(t+3)}{(3t+2)(t+1)}\,dt = 2\ln\frac{5}{3} - \frac{7}{3}\ln 2$.

PROBLEM SOLVING

Q6 Use the substitution $u = \sqrt{x}$ to find the exact value of $\int_9^{16} \frac{4}{\sqrt{x}(9x-4)}\,dx$.

11. Differential Equations

Differential equations involve differentiation as well as integration. The differentiation comes in because they always include a derivative term, usually to describe a rate of change, and integration is used to solve them.

Differential equations

- A **differential equation** is an equation that includes a **derivative term** such as $\frac{dy}{dx}$ (or $\frac{dP}{dt}$, $\frac{ds}{dt}$, $\frac{dV}{dr}$ etc, depending on the variables), as well as **other variables** (like x **and** y).
- Before you even think about **solving** them, you have to be able to **set up** ('**formulate**') differential equations.
- Differential equations tend to involve a **rate of change** (giving a derivative term) and a **proportion relation**, where the rate of change will be directly or inversely proportional to some function of the variables.
- It'll help to think about what the derivative **actually means**. $\frac{dy}{dx}$ is defined as 'the **rate of change** of y with respect to x'. In other words, it tells you how y changes as x changes.

Learning Objectives:

- Be able to formulate a differential equation for a given situation.
- Be able to find general and particular solutions to differential equations.
- Be able to formulate and solve differential equations that model real-life situations and interpret the results in context.
- Be able to identify the limitations of a model and suggest refinements to address them.

Example 1

The number of bacteria in a petri dish, b, is increasing over time, t, at a rate directly proportional to the number of bacteria. Formulate a differential equation that shows this information. MODELLING

- The question tells you that you need to write a differential equation, so you know there'll be a **derivative term**. Work out what that is first.
- You're told that the number of bacteria (b) increases as time (t) increases — so that's the rate of change of b with respect to t, or $\frac{db}{dt}$.
- The rate of change, $\frac{db}{dt}$, is proportional to b, so $\frac{db}{dt} \propto b$
- We're looking for an equation, not a proportion relation, so rewrite it:

So $\frac{db}{dt} = kb$ for some constant k, $k > 0$.

Tip: One of the variables in a differential equation will often be time, t, — so the question will be about how something changes over time.

Tip: Remember — if $a \propto b$, then $a = kb$ for some constant k.

Example 2

The volume of interdimensional space jelly, V, in a container is decreasing over time, t, at a rate inversely proportional to the square of its volume. Show this as a differential equation. MODELLING

- This time the question tells you how V decreases as t increases — the derivative term is the rate of change of V with respect to t, or $\frac{dV}{dt}$.
- $\frac{dV}{dt}$ is **inversely proportional** to the **square** of V, so $\frac{dV}{dt} \propto \frac{1}{V^2}$
- The equation needs a minus sign, because V is **decreasing** as t increases.

$\frac{dV}{dt} = -\frac{k}{V^2}$ for some constant k, $k > 0$.

Tip: 'x is inversely proportional to y' means x is directly proportional to $\frac{1}{y}$.

Tip: If the variables aren't given in the question, you'll have to come up with them for yourself.

Example 3

The rate of cooling of a hot liquid is proportional to the difference between the temperature of the liquid and the temperature of the room. Formulate a differential equation to represent this situation.

MODELLING

- Let L = temperature of the liquid, R = room temperature and t = time.
- The derivative term is the rate of change of L with respect to t, or $\frac{dL}{dt}$.
- The rate of change is proportional to the difference between L and R, so $\frac{dL}{dt} \propto (L - R)$
- $\frac{dL}{dt}$ is the rate of cooling — so the temperature is **decreasing** and you need a minus sign in the equation again.

$$\frac{dL}{dt} = -k(L - R) \text{ for some constant } k, k > 0.$$

Exercise 11.1

MODELLING

Q1 The number of fleas (N) on a cat is increasing over time, t, at a rate directly proportional to the number of fleas. Show this as a differential equation.

Q2 The value, x, of a house is increasing over time, t, at a rate inversely proportional to the square of x. Formulate a differential equation to show this.

Q3 The rate of depreciation of the amount (£A) a car is worth is directly proportional to the square root of A. Show this as a differential equation.

Q4 The rate of decrease of a population, y, with respect to time is directly proportional to the difference between y and λ where λ is a constant. Formulate a differential equation to show this.

Q5 Hint: The overall rate of change of V is the difference between the rate at which water is flowing in and the rate at which it's flowing out.

Q5 The volume of water which is being poured into a container is directly proportional to the volume of water (V) in the container. The container has a hole in it from which water flows out at a rate of 20 cm^3s^{-1}. Show this as a differential equation.

Solving differential equations

Tip: Remember — your equation might be in terms of variables other than x and y.

- **Solving** a differential equation means using it to find an **equation** in terms of the two variables, **without** a derivative term. To do this, you need to use **integration**.
- The only differential equations containing x and y terms that you'll have to solve in A-level Maths are ones with **separable variables** — where x and y can be separated into functions **f(x)** and **g(y)**.

Solving differential equations

- **Step 1:** Write the differential equation in the form $\frac{dy}{dx} = f(x)g(y)$.
- **Step 2:** **Rearrange** the equation into the form: $\frac{1}{g(y)}dy = f(x)\,dx$.
  To do this, get all the terms containing y on the **left-hand side**, and all the terms containing x on the **right-hand side** and split up the $\frac{dy}{dx}$.
- **Step 3:** Now **integrate both sides**: $\int \frac{1}{g(y)}\,dy = \int f(x)\,dx$.
  Don't forget the **constant of integration** (you only need one — not one on each side). It might be useful to write the constant as **ln k** rather than **C** (see p.224).
- **Step 4:** **Rearrange** your answer to get it in a **nice form** — you might be asked to find it in the form $y = h(x)$.
- **Step 5:** If you're asked for a **general solution**, leave C (or k) in your answer. If they want a **particular solution**, they'll give you x and y values for a certain point. All you do is put these values into your equation and use them to **find C** (or k).

Tip: Like in integration by substitution, you can treat $\frac{dy}{dx}$ as if it's a fraction here.

Tip: Sometimes there'll already be another k in the differential equation (e.g. if you've formulated the equation yourself). Obviously, in that case you can't use ln k as the constant of integration, so pick another letter — you might see examples that use ln A instead.

Example 1

Find the general solution of the differential equation $\frac{ds}{dt} = -6t^2$.

- **Step 1** is already done: $f(t) = -6t^2$, $g(s) = 1$.
- **Step 2** — rearrange the equation: $ds = -6t^2\,dt$
- **Step 3** — integrate both sides: $\int 1\,ds = \int -6t^2\,dt$

$$\Rightarrow s = -2t^3 + C$$

Steps 4 and 5 aren't needed here — the equation doesn't need rearranging, and you're only looking for the general solution, so you're done.

Tip: There's no 's' term in this differential equation. When this is the case, you can just integrate 'normally' without separating the variables.

$$s = \int \frac{ds}{dt}\,dt$$

Example 2

Find the particular solution of $\frac{dy}{dx} = 2y(1 + x)^2$ when $x = -1$ and $y = 4$.

- Identify f(x) and g(y): $f(x) = 2(1 + x)^2$ and $g(y) = y$.
- Separate the variables: $\frac{1}{y}\,dy = 2(1 + x)^2\,dx$
- And integrate: $\int \frac{1}{y}\,dy = \int 2(1 + x)^2\,dx$

$$\Rightarrow \ln|y| = \frac{2}{3}(1 + x)^3 + C$$

- Now to find the particular solution, work out the value of C for the given values of x and y:

$$\ln 4 = \frac{2}{3}(1 + (-1))^3 + C$$

$$\Rightarrow \ln 4 = C$$

$$\text{so } \ln|y| = \frac{2}{3}(1 + x)^3 + \ln 4$$

Example 3

Find the general solution of $(x-2)(2x+3)\frac{dy}{dx} = xy + 5y$, where $x > 2$.

Give your answer in the form $y = f(x)$.

- First, separate the variables: $\frac{dy}{dx} = \frac{x+5}{(x-2)(2x+3)} \times y$

$$\frac{1}{y}\,dy = \frac{x+5}{(x-2)(2x+3)}\,dx$$

- To make the right-hand side easier to integrate, write it as partial fractions (see p.219):

$$\frac{x+5}{(x-2)(2x+3)} \equiv \frac{A}{x-2} + \frac{B}{2x+3}$$

$$\Rightarrow x + 5 \equiv A(2x+3) + B(x-2)$$

- Solving for A and B gives $A = 1$, $B = -1$, so $\frac{1}{y}\,dy = \frac{1}{x-2} - \frac{1}{2x+3}\,dx$

- Now you can integrate: $\int \frac{1}{y}\,dy = \int \frac{1}{x-2} - \frac{1}{2x+3}\,dx$

$$\Rightarrow \ln|y| = \ln|x-2| - \frac{1}{2}\ln|2x+3| + \ln k$$

$$\Rightarrow \ln|y| = \ln\left|\frac{k(x-2)}{\sqrt{2x+3}}\right|$$

You know $x > 2$, so $x - 2$ is positive and the modulus can be removed.

$$\Rightarrow y = \frac{k(x-2)}{\sqrt{2x+3}}$$

Tip: Since all the other terms are ln (something), it makes sense to use ln k as the constant of integration here, then you can use the log laws to simplify.

Tip: You might be asked to sketch a graph for various values of C (or ln k etc.) — check your Year 1 notes on graph sketching and translations if you need a reminder of this.

Example 4

Find the particular solution to the differential equation $\frac{db}{dt} = 4\sqrt{b}$, given that when $t = 12$, $b = 900$.

Give your answer in the form $b = f(t)$.

- Rearranging $\frac{db}{dt} = 4\sqrt{b}$ gives: $\frac{1}{\sqrt{b}}\,db = 4\,dt$

- Integrate both sides: $\int b^{-\frac{1}{2}}\,db = \int 4\,dt$

$$\Rightarrow 2b^{\frac{1}{2}} = 4t + C$$

- In this case, it's easier to find C for the given values of b and t before you rearrange the equation:

$$2\sqrt{b} = 4t + C \Rightarrow 2\sqrt{900} = 4(12) + C$$

$$\Rightarrow 60 = 48 + C$$

$$\Rightarrow C = 12$$

- Now rearrange to get the form $b = f(t)$:

$$2\sqrt{b} = 4t + 12 \Rightarrow \sqrt{b} = 2t + 6$$

$$\Rightarrow b = 4t^2 + 24t + 36$$

Tip: Rearranging the equation first, then finding C would give you a quadratic in C to solve, so it's easier to leave the rearrangement till last here.

Exercise 11.2

Q1 Find the general solutions of the following differential equations where $x \geq 0$. Give your answers in the form $y = f(x)$.

a) $\frac{dy}{dx} = 8x^3$ b) $\frac{dy}{dx} = 5y$ c) $\frac{dy}{dx} = 6x^2y$

d) $\frac{dy}{dx} = \frac{y}{x}$ e) $\frac{dy}{dx} = (y + 1)\cos x$ f) $\frac{dy}{dx} = \frac{3xy - 6y}{(x-4)(2x-5)}$

Q1f) Hint: You'll need to do some work before you can integrate with respect to x.

Q2 Find the particular solutions of the following differential equations at the given conditions:

a) $\frac{dy}{dx} = -\frac{x}{y}$ $x = 0, y = 2$

b) $\frac{dx}{dt} = \frac{2}{\sqrt{x}}$ $t = 5, x = 9$

c) $\frac{dV}{dt} = 3(V - 1)$ $t = 0, V = 5$

d) $\frac{dy}{dx} = \frac{\tan y}{x}$ $x = 2, y = \frac{\pi}{2}$

e) $\frac{dx}{dt} = 10x(x + 1)$ $t = 0, x = 1$

Q3 The rate of increase of the variable V at time t satisfies the differential equation $\frac{dV}{dt} = a - bV$, where a and b are positive constants.

a) Show that $V = \frac{a}{b} - Ae^{-bt}$, where A is a positive constant.

b) Given that $V = \frac{a}{4b}$ when $t = 0$, find A in terms of a and b.

c) Find the value V approaches as t gets very large.

Applying differential equations to real-life problems

- Some questions involve taking **real-life problems** and using differential equations to **model** them.
- **Population** questions come up quite often — the population might be **increasing** or **decreasing**, and you have to find and solve differential equations to show it. In cases like this, one variable will usually be t, **time**.
- You might be given a **starting condition** — e.g. the **initial population**. The important thing to remember is that:

the starting condition occurs when $\boldsymbol{t = 0}$

Tip: This might seem pretty obvious, but it's really important.

- Once you've solved the differential equation you can use it to **answer questions** about the model. For example, if the equation is for population you might be asked to find the **population** after a certain number of years, or the **number of years** it takes to reach a certain population. Don't forget to relate the answer back to the situation given in the question.

You may also have to identify **limitations** of the model, as well as suggest possible **changes** that would **improve** it. Common things that you should think about are:

- Is there any information **missing** from the model?
- What happens to the model when the variables get really **big/small**?
- Is the model appropriate? Is a **continuous** function used for a **discrete** variable? Does the function allow **negative** values that don't make sense?
- Are there **other factors** that have not been accounted for in the model? Some examples might be **natural immunity** to a disease, **immigration/emigration** of a population or **seasonal variation** in weather.

Tip: Continuous and discrete variables were covered back in Year 1 of the course — discrete variables only take certain values (e.g. integers, names etc.) while continuous variables can take infinitely many values (e.g. real numbers between 0 and 1).

When suggesting some refinements to the model, you don't have to make up a whole new model — just identify what the changes would be and how you could make them.

Questions like the following examples can seem a bit **overwhelming** at first, but follow things through **step by step** and they shouldn't be too bad.

Example 1

The population of rabbits in a park is decreasing as winter approaches. The rate of decrease is directly proportional to the current number of rabbits (*P*).

MODELLING

a) Explain why this situation can be modelled by the differential equation $\frac{dP}{dt} = -kP$, where *t* is the time in days and *k* is a positive constant.

- The model states that the rate of decrease in the rabbit population (i.e. $\frac{dP}{dt}$) is **proportional** to P. This means $\frac{dP}{dt} \propto P$.
- By introducing a **constant** of proportionality, the model becomes:

$$\frac{dP}{dt} = -kP$$

- The minus sign shows that the population is decreasing.

b) If the initial population is P_0, solve your differential equation to find *P* in terms of P_0, *k* and *t*.

- First, solve the differential equation to find the general solution:

$$\frac{dP}{dt} = -kP \Rightarrow \frac{1}{P}\,dP = -k\,dt$$
$$\Rightarrow \int \frac{1}{P}\,dP = \int -k\,dt$$
$$\Rightarrow \ln P = -kt + C$$

- At $t = 0$, $P = P_0$. Putting these values into the equation gives:

$$\ln P_0 = -k(0) + C \Rightarrow \ln P_0 = C$$

- So the equation becomes:

$$\ln P = -kt + \ln P_0 \Rightarrow P = e^{(-kt + \ln P_0)} = e^{-kt}e^{\ln P_0}$$
$$\Rightarrow P = P_0 e^{-kt}$$

Tip: You don't need modulus signs when you integrate to get $\ln P$ here. $P \geq 0$ as you can't have a negative population.

c) Given that $k = 0.1$, find the time at which the population of rabbits will have halved, to the nearest day.

- When the population of rabbits has halved, $P = \frac{1}{2}P_0$.
  You've been told that $k = 0.1$, so substitute these values into the equation above and solve for t:

$$\frac{1}{2}P_0 = P_0e^{-0.1t} \Rightarrow \frac{1}{2} = e^{-0.1t}$$
$$\Rightarrow \ln\frac{1}{2} = -0.1t$$
$$\Rightarrow -0.6931 = -0.1t$$
$$\Rightarrow t = 6.931$$

- So to the nearest day, $t = 7$.
  This means that it will take 7 days for the population to halve.

Tip: Make sure you always link the numbers back to the situation.

d) Give a limitation of the model and suggest a possible improvement that could be made to address it.

There are several ways to answer this:

- The value of P_0 is not given, so the model could be improved by finding a suitable value of P_0.
- As t becomes very large, the population becomes increasingly small but never reaches 0 — it may be more realistic to choose a model where the population can reach 0.
- The population of rabbits is a discrete variable while the model is continuous — choosing a function that limits the possible values of P to integers would solve this problem.
- The situation being modelled is the approach of winter — the model could give a limit on t to show at which point the model stops being appropriate.

Tip: You may be able to think of other limitations and improvements for part d) — these aren't the only possible answers.

Example 2

Water is leaking from the bottom of a water tank shaped like a vertical cylinder, so that at time t seconds the depth, D, of water in the tank is decreasing at a rate proportional to $\frac{1}{D^2}$.

MODELLING

a) Explain why the depth of water satisfies the differential equation $\frac{dD}{dt} = -\frac{k}{D^2}$ for some constant $k > 0$.

The question tells you that the rate at which D decreases (i.e. $\frac{dD}{dt}$) is inversely proportional to D^2. This can be written as:

$$\frac{dD}{dt} \propto \frac{1}{D^2} \quad \Rightarrow \quad \frac{dD}{dt} = -\frac{k}{D^2} \text{ for some } k > 0$$

The minus sign indicates that the depth of the water is decreasing.

b) Given that D is decreasing at a rate of 2 cm s^{-1} when $D = 40$ cm, find k.

Use the differential equation for D: $\frac{dD}{dt} = -\frac{k}{D^2}$

So $\frac{dD}{dt} = -2$ when $D = 40 \Rightarrow -\frac{k}{40^2} = -2$

$$\Rightarrow k = 2 \times 40^2 = 3200$$

Tip: There are a few steps to part c) — first you have to find the general solution, then sub in the values given to find the particular solution, then use this solution to answer the question.

c) Given that D = 60 cm at t = 0 s , find a particular solution to the differential equation for D, and hence calculate how long it takes for the tank to empty.

- Solve the differential equation, using the value of k from part b), to find the general solution:

$$\frac{dD}{dt} = -\frac{3200}{D^2} \Rightarrow D^2\,dD = -3200\,dt$$
$$\Rightarrow \int D^2\,dD = \int -3200\,dt$$
$$\Rightarrow \frac{1}{3}D^3 = -3200t + C$$

- When $t = 0$, $D = 60$. Putting these values into the equation gives:

$$\frac{1}{3}60^3 = -3200(0) + C \Rightarrow C = 72\,000$$
$$\Rightarrow \frac{1}{3}D^3 = 72\,000 - 3200t$$

- The tank is empty when $D = 0 \Rightarrow 72\,000 = 3200t$

$$\Rightarrow t = 22.5 \text{ s}$$

Tip: Part d) is a 'connected rates of change' question — see p.165 if you've forgotten how to tackle them.

d) Given that the radius of the cylinder is 20 cm, calculate the rate at which the volume of water in the tank is decreasing when t = 10 s.

- The volume of water in the tank is $V = \pi r^2 h = \pi(20)^2 D = 400\pi D$.

$$\Rightarrow \frac{dV}{dD} = 400\pi$$

- So using the chain rule:

$$\frac{dV}{dt} = \frac{dV}{dD} \times \frac{dD}{dt} = 400\pi \times -\frac{3200}{D^2} = -\frac{1280000\pi}{D^2}$$

- When $t = 10$, $\frac{1}{3}D^3 = 72000 - 3200(10) = 40000$

$$\Rightarrow D = \sqrt[3]{3 \times 40000} = 49.324\ldots \text{ cm}$$

$$\text{So } \frac{dV}{dt} = \frac{-1280000\pi}{49.324^2} = -1652.87\ldots \text{ cm}^3\text{ s}^{-1}$$

- So the volume is decreasing at a rate of 1650 cm^3s^{-1} (3 s.f.)

Exercise 11.3

MODELLING

Q1 Hint: Don't be put off by the amount of information given in the question. Just go through it step by step.

Q1 A virus spreads so that t hours after infection, the rate of increase of the number of germs (N) in the body of an infected person is directly proportional to the number of germs in the body.

a) Given that this can be represented by the differential equation $\frac{dN}{dt} = kN$, show that the general solution of this equation is $N = Ae^{kt}$, where A and k are positive constants.

b) Given that a person catching the virus will initially be infected with 200 germs and that this will double to 400 germs in 8 hours, find the number of germs an infected person has after 24 hours.

c) Give one possible limitation of this model.

Q2 The rate of depreciation of the value (V) of a car at time t after it is first purchased is directly proportional to V.

a) If the initial value of the car is V_0, show that $V = V_0e^{-kt}$, where k is a positive constant.

b) If the car drops to one half of its initial value in the first year after purchase, how long (to the nearest month) will it take to be worth 5% of its initial value?

Q3 It is thought that the rate of increase of the number of field mice (N) in a given area is directly proportional to N.

a) Formulate a differential equation for N.

b) Given that in 4 weeks the number of mice in a particular field has risen from 20 to 30, find the length of time, to the nearest week, before the field is over-run with 1000 mice.

A biologist believes that the rate of increase of the number field mice is actually directly proportional to the square root of N when natural factors such as predators and disease are taken into account.

c) Repeat parts a) and b) using this new model.

d) Suggest another refinement that could be made to improve this model.

Q4 A cube has side length x. At time t seconds, the side length is increasing at a rate of $\frac{1}{x^2(t+1)}$ cm s^{-1}.

a) Show that the volume (V) is increasing at a rate which satisfies the differential equation $\frac{dV}{dt} = \frac{3}{t+1}$.

b) Given that the volume of the cube is initially 15 cm^3, find the length of time, to 3 s.f., for it to reach a volume of 18 cm^3.

Q5 A local activist is trying to get lots of signatures on his petition, and has just launched a new online campaign.
The rate of increase of the number of signatures (y) he's gathered can be represented by the differential equation $\frac{dy}{dt} = k(p - y)$, where p is the population of his town and t is the time in days since the new campaign was launched.

a) Find the general solution of this equation.

b) Given that the population of his town is 30 000, he initially has 10 000 signatures on his petition and it takes 5 days for him to reach 12 000 signatures, how long, to the nearest day, will it take for him to reach 25 000 signatures?

c) Draw a graph to show this particular solution.

d) The activist wants 28 000 signatures within 92 days of launching his new campaign. According to the model, will he achieve this?

e) What is a likely limitation of this model and how could it be addressed?

Q5c) Hint: You just need to sketch the graph of the equation you found in part b).

Chapter 9 Numerical Methods

1. Location of Roots

Learning Objectives:

- Be able to locate the roots of $f(x) = 0$ by finding changes in sign of $f(x)$ between two values of x.
- Be able to choose upper and lower bounds to show that a root is accurate to a certain number of decimal places.
- Be able to sketch functions and use the sketches to find approximate locations of roots.

Sometimes finding the solutions of an equation algebraically is quite difficult. In these situations, it's often helpful to find roughly where the roots are (the points where $f(x) = 0$) by looking at the graph of $y = f(x)$.

Locating roots by changes of sign

'Solving' or 'finding the roots of' an equation (where $f(x) = 0$) is the same as finding the values of x where the graph crosses the x-axis.

- The graph of the function gives you a rough idea of **how many** roots there are and **where** they are. E.g. the function $f(x) = 3x^2 - x^3 - 2$ (below) has 3 roots, since it crosses the x-axis three times (i.e. there are 3 solutions to the equation $3x^2 - x^3 - 2 = 0$). From the graph you can see there's a root at $x = 1$ and two others near $x = -1$ and $x = 3$.

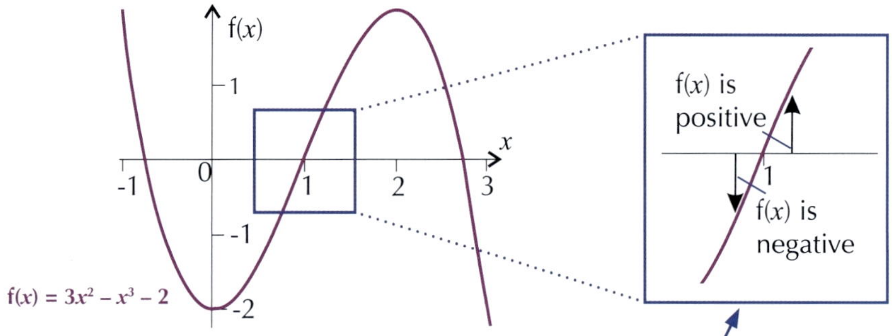

For each root in the graph above, $f(x)$ goes from positive to negative or vice versa — **$f(x)$ changes sign as it passes through a root.** So to find if there's a root between two values 'a' and 'b', work out f(a) and f(b). If the signs are different, there's a root somewhere between them.

- Be careful though — this only applies when the **range of x** you're interested in is **continuous** (no break or jump in the line of the graph). Some graphs, like $\tan x$, have an **asymptote** where the line jumps from positive to negative without actually crossing the x-axis, so you might think there's a root when there isn't.
- This method might **fail** to find a particular root. If the function **touches** the x-axis but doesn't cross it, then there won't be a change of sign. An accurate sketch of the graph will help to avoid this.
- Often you'll be given an **approximation** to a root and be asked to show that it's correct to a certain accuracy. To do this, choose the right **upper and lower bounds** and work out if there's a sign change between them.

The **lower bound** is the **lowest** value a number could have and still be **rounded up** to the correct answer. The **upper bound** is the **upper limit** of the values which will be **rounded down** to the correct answer.

Tip: You need to be careful that the interval you are looking at is not **too large**. If it's too big, there could be **multiple** roots within the interval. If there were an **even** number of roots, the sign would appear not to change and you would miss those roots — e.g. on the graph above, $f(-1)$ and $f(2)$ are both positive, but there are two roots between them.

Example

Show that one root of the equation $x^3 - x^2 - 9 = 0$ is $x = 2.472$ correct to 3 d.p.

- If $x = 2.472$ is a root rounded to 3 decimal places, the exact root must lie between the **upper and lower bounds** of this value — **2.4715** and **2.4725**. Any value in this interval would be rounded to 2.472 to 3 d.p.

- The function $f(x) = x^3 - x^2 - 9$ is **continuous**, so a root lies in the interval $2.4715 \le x < 2.4725$ if f(2.4715) and f(2.4725) have **different signs**.
- $f(2.4715) = 2.4715^3 - 2.4715^2 - 9 = -0.0116...$

  $f(2.4725) = 2.4725^3 - 2.4725^2 - 9 = 0.0017...$
- f(2.4715) and f(2.4725) have different signs, so a root must lie between them. Since any value between them would be rounded to 2.472 to 3 d.p. this answer must be correct.

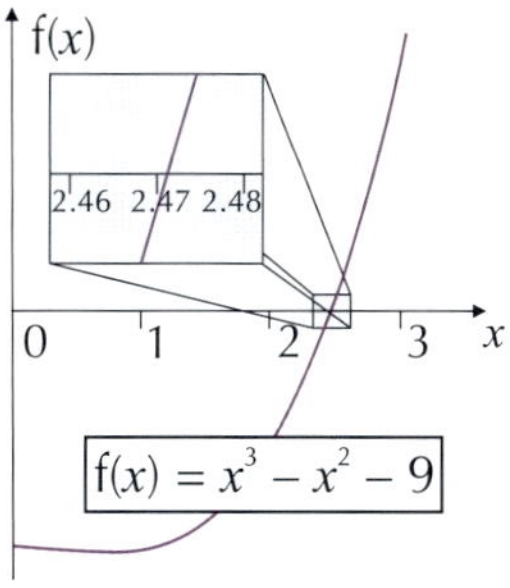

Tip: Although 2.4725 would actually be rounded up to 2.473, everything below it would be rounded down to 2.472. It's just a bound.

Tip: The function is continuous because it's just a cubic curve:

It has no breaks or jumps.

Exercise 1.1

Q1 $f(x) = x^3 - 5x + 1$
Show that there is a root of $f(x) = 0$ in the interval $2 < x < 3$.

Q2 $f(x) = \sin 2x - x$ (x is in radians).
Show that there is a root of $f(x) = 0$ between $x = 0.9$ and $x = 1.0$.

Q2 Hint: Remember to set your calculator to radians before doing this question.

Q3 $f(x) = x^3 + \ln x - 2, x > 0$
Show that there is a root of $f(x) = 0$ in the interval $[1.2, 1.3]$.

Q3 Hint: You might see brackets describing intervals — x lies in the interval [a, b] means $a \le x \le b$ and x lies in the interval (a, b) means $a < x < b$.

Q4 A bird is observed diving into the sea. Its height above the water after x seconds is modelled by the equation $f(x) = 2x^2 - 8x + 7$, where $f(x)$ is the height in metres. Show that the bird hits the water in the interval $1.2 < x < 1.3$ seconds.

MODELLING

Q5 Show that there are 2 solutions, α and β, to the equation $3x - x^4 + 3 = 0$, such that $1.6 < \alpha < 1.7$ and $-1 < \beta < 0$.

Q6 Show that there are 2 solutions, α and β, to the equation $e^{x-2} - \sqrt{x} = 0$, such that $0.01 < \alpha < 0.02$ and $2.4 < \beta < 2.5$.

Q7 Show that $x = 2.8$ is a solution to the equation $x^3 - 7x - 2 = 0$ to 1 d.p.

Q8 Show that $x = 0.7$ is a solution to the equation $2x - \frac{1}{x} = 0$ to 1 d.p.

Q8 Hint: The function $f(x) = 2x - \frac{1}{x}$ is not continuous for all x, but it is for $x > 0$.

Q9 $f(x) = e^x - x^3 - 5x$
Verify that a root of the equation $f(x) = 0$ is $x = 0.25$ correct to 2 d.p.

Q10 Hint: Roots of f(x) are given when f(x) = 0, so you might need to rearrange the equation before doing any calculations.

Q10 Show that a solution to the equation $4x - 2x^3 = 15$ lies between –2.3 and –2.2.

Q11 Show that a solution to the equation $\ln (x + 3) = 5x$ lies between 0.23 and 0.24.

Q12 Show that a solution to the equation $e^{3x}\sin x = 5$ lies between $x = 0$ and $x = 1$ (x is in radians).

Sketching graphs to find approximate roots

Sometimes it's easier to find the number of roots and roughly where they are if you **sketch** the graphs first.

Tip: Setting the equations equal to each other and then rearranging them to get f(x) = 0 gets you to where you were in the previous section.

- In some questions you might be asked to **sketch** the graphs of **two equations** on the same set of axes. Sketching graphs was covered in Year 1 if you need a reminder.
- At the points where they **cross** each other, the two equations are equal. So for $y = x + 3$ and $y = x^2$, at the points of intersection you know that $x + 3 = x^2$, which you can rearrange to get $x^2 - x - 3 = 0$.

> The **number of roots** of this 'combined' equation is the same as the number of **points of intersection** of the original two graphs. The sketch you made will also show roughly **where** the roots are (it's the same x-value for both), so locating them is a bit easier.

Tip: Don't worry about trying to make your sketch perfect, the important things are that the graphs are the correct shape and that they cross the axes in the right places.

Example

a) On the same set of axes, sketch the graphs $y = \ln x$ and $y = (x - 3)^2$.

Just sketch the two graphs on a set of axes.

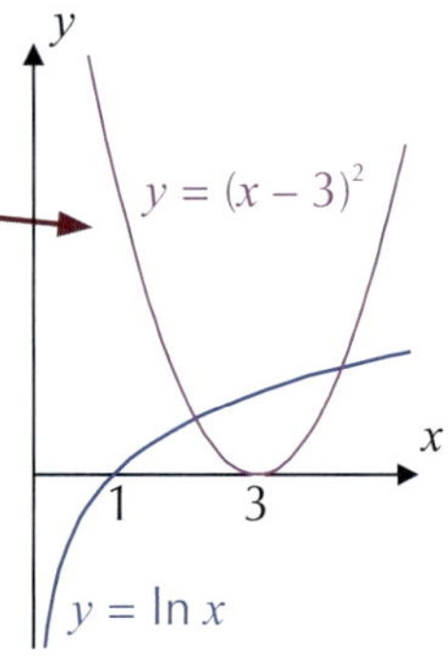

b) Hence work out the number of roots of the equation $\ln x - (x - 3)^2 = 0$.

This equation is a combination of the previous two. The graphs cross twice, so the equation has two roots.

c) Show that there is a solution between 2 and 3, and find this solution to 1 decimal place.

- Put 2 and 3 into the equation and check for a sign change.

$$\ln 2 - (2 - 3)^2 = -0.306...$$
$$\ln 3 - (3 - 3)^2 = 1.098...$$

- The sign has changed, so there is a root between 2 and 3.
- On the sketch, it looks like the root is much closer to 2 than 3, so try again using $x = 2$ and $x = 2.2$

$$\ln 2.2 - (2.2 - 3)^2 = 0.148...$$

Tip: You might need to choose a wider interval depending on how accurate your sketch is.

- f(2.2) is positive, so the root is between 2 and 2.2.
  Now try again with $x = 2.1$, as it's halfway between the two and will tell you which it's closer to.

  $\ln 2.1 - (2.1 - 3)^2 = -0.068...$ $\Rightarrow$ root is between 2.1 and 2.2.

- Now you just need to check if it rounds up to 2.2 or down to 2.1 — 2.15 is the upper bound for 2.1 and lower bound for 2.2, so try that.

  $\ln 2.15 - (2.15 - 3)^2 = 0.0429...$ $\Rightarrow$ root is between 2.1 and 2.15.

- So the answer rounds down, and the value of the root to 1 d.p. is 2.1.

Exercise 1.2

Sketch all graphs in this exercise for $-5 < x < 5$ unless otherwise stated.

Q1 a) On the same axes, sketch the graphs of $y = \frac{1}{x}$ and $y = x - 2$.

b) Using your graph from part a) write down the number of solutions of the equation $\frac{1}{x} = x - 2$ in this interval.

c) Show that one solution of the equation $\frac{1}{x} = x - 2$ lies in the interval $2.4 < x < 2.5$.

Q1 Hint: You might find it useful to rearrange the equation into the form $f(x) = 0$.

Q2 a) On the same axes, sketch the graphs of $y = 2x^3 - 7x$ and $y = x^2$.

b) Using your graph from part a) write down the number of solutions of the equation $2x^3 - x^2 - 7x = 0$ in this interval.

c) Show that the equation $2x^3 - x^2 - 7x = 0$ has a root between $x = -2$ and $x = -1$.

Q3 a) Sketch the graphs of $y = 2^x - 3$ and $y = \ln x$ on the same axes.

b) $f(x) = \ln x - 2^x + 3$. Using your graph write down the number of roots of the equation $f(x) = 0$.

c) Show that the equation $f(x) = 0$ has a root between 1.8 and 2.2 and find this root to 1 decimal place.

Q4 a) Sketch the graphs of $y = \sqrt{x+1}$ and $y = 2x$ on the same axes.

b) Write down the number of solutions of the equation $\sqrt{x+1} = 2x$.

c) Show that the equation $\sqrt{x+1} = 2x$ has a solution in the interval (0.6, 0.7).

d) By rearranging the equation $\sqrt{x+1} = 2x$, use the quadratic formula to find the solution of the equation from part c) to 3 s.f.

Q5 a) Sketch the graphs of $y = e^{2x}$ and $y = 3 - x^2$ on the same axes.

b) Using your graph from part a), explain how you know that the equation $e^{2x} + x^2 = 3$ has two solutions.

c) Show that the negative solution of the equation $e^{2x} + x^2 = 3$ lies between $x = -2$ and $x = -1$, and find this solution to 1 d.p.

2. Iterative Methods

Learning Objectives:

- Be able to use iteration formulas to find a solution of an equation to a given level of accuracy.
- Be able to find and rearrange iteration formulas and understand that some don't converge to a solution of an equation.

Another way of finding roots of an equation is with iteration formulas. They can seem fiddly to work with but are actually pretty simple to use.

Using iteration formulas

Some equations are too difficult to solve algebraically, so you need to find **approximations** to the roots to a certain level of accuracy.
In exam questions you'll usually be told the value of x that a root is close to, and then **iteration** does the rest.

Iteration is a numerical method for **solving equations**, like **trial and improvement**.

- You put an approximate value of a root x into an iteration formula, and it gives you a **slightly more accurate** value.
- You then put the new value into the iteration formula, and keep going until your answers are the same when rounded to the level of accuracy needed.

Example 1

Use the iteration formula $x_{n+1} = \sqrt[3]{x_n + 4}$ to solve $x^3 - 4 - x = 0$, to 2 d.p. Start with $x_0 = 2$.

- The notation x_n just means the approximation of x at the nth iteration. Putting x_0 in the formula for x_n gives you x_1 — the first iteration.
- $x_0 = 2$, so $x_1 = \sqrt[3]{x_0 + 4} = \sqrt[3]{2 + 4} = 1.8171...$
- This value now gets put back into the formula to find x_2:
  $x_1 = 1.8171...$, so $x_2 = \sqrt[3]{x_1 + 4} = \sqrt[3]{1.8171... + 4} = 1.7984...$
- Carry on until you get answers that are the same when rounded to 2 d.p:
  $x_2 = 1.7984...$, so $x_3 = \sqrt[3]{x_2 + 4} = \sqrt[3]{1.7984... + 4} = 1.7965...$
- x_2, x_3 and all further iterations are the same when rounded to 2 d.p., so the root is $x = 1.80$ to 2 d.p.

Tip: Here you have been given the iteration formula, but sometimes you have to find it yourself (see p.236).

- Sometimes an iteration formula will just **not find a root**. In these cases, no matter how close to the root you have x_0, the iteration sequence **diverges** — the numbers get further and further apart.
- The iteration might also **stop working**, like if you have to take the square root of a negative number.
- However, you'll nearly always be given a formula that converges to a certain root, otherwise there's not much point in using it.
- If your formula diverges when it shouldn't, chances are you went wrong somewhere, so go back and double check every stage.

Tip: This doesn't mean you'll never see a diverging formula in an exam, but if you do it will usually be followed by a question like 'what do you notice about the iterations?'. If the iterations seem to bounce up and down, or do something else unexpected, then it probably diverges.

Example 2

The equation $x^3 - x^2 - 9 = 0$ has a root close to $x = 2.5$.
What is the result of using $x_{n+1} = \sqrt{x_n^3 - 9}$ with $x_0 = 2.5$ to find this root?

- Start with $x_1 = \sqrt{2.5^3 - 9} = 2.5739...$ (seems okay so far).
- Subsequent iterations give: $x_2 = 2.8376...$, $x_3 = 3.7214...$, $x_4 = 6.5221...$
- The results are getting further and further apart with each iteration.
- So the sequence diverges.

Tip: The list of results $x_1, x_2, x_3...$ is called the iteration sequence.

Exercise 2.1

Q1 a) Show that the equation $x^3 + 3x^2 - 7 = 0$ has a root in the interval (1, 2).

b) Use the iterative formula $x_{n+1} = \sqrt{\frac{7 - x_n^3}{3}}$ with $x_0 = 1$ to find values for x_1, x_2, x_3 and x_4 to 3 decimal places.

Q2 An intersection of the curves $y = \ln x$ and $y = x - 2$ is at the point $x = \alpha$, where α is 3.1 to 1 decimal place.

a) Starting with $x_0 = 3.1$, use the iterative formula $x_{n+1} = 2 + \ln x_n$ to find the first 5 iterations, giving your answers to 4 decimal places.

b) Write down an estimate of the value of α to 3 decimal places.

Q3 a) Show that the equation $x^4 - 5x + 3 = 0$ has a root between $x = 1.4$ and $x = 1.5$.

b) Use the iterative formula $x_{n+1} = \sqrt[3]{5 - \frac{3}{x_n}}$ and $x_0 = 1.4$ to find iterations x_1 to x_6 to 3 decimal places.

c) Hence write down an approximation of the root from part a) to 2 decimal places.

Q4 a) Show that the function $f(x) = x^2 - 5x - 2$ has a root that lies between $x = 5$ and $x = 6$.

b) The root in part a) can be estimated using the iterative formula $x_{n+1} = \frac{2}{x_n} + 5$. Using a starting value of $x_0 = 5$ find the values of x_1, x_2, x_3 and x_4, giving your answers to 4 significant figures.

Q5 Use the iterative formula $x_{n+1} = 2 - \ln x_n$ with $x_0 = 1.5$ to find the root of the equation $\ln x = 2 - x$ to 2 decimal places.

Q6 a) Show that the equation $e^x - 10x = 0$ has a root in the interval (3, 4).

b) Using the iterative formula $x_{n+1} = \ln(10x_n)$ with an appropriate starting value, find values for x_1, x_2, x_3 and x_4 to 3 d.p.

c) Verify that the value of the root from part a) is $x = 3.577$ to 3 d.p.

d) Describe what happens when you use the alternative formula $x_{n+1} = \frac{e^{x_n}}{10}$ with $x_0 = 3$.

Hint: Leave the numbers from each iteration in your calculator so that you don't lose any accuracy with each step. You can use the ANS button on your calculator to speed things up. Enter the starting value, and then type the iteration formula replacing x_n with 'ANS' — each time you press enter you'll get another iteration.

Q7 The iterative formula $x_{n+1} = \frac{x_n^2 - 3x_n}{2} - 5$ is used to try and find approximations to a root of $f(x) = x^2 - 5x - 10$.

a) Find the values of x_1, x_2, x_3 and x_4, starting with $x_0 = -1$ and describe what is happening to the sequence x_1, x_2, x_3, x_4...

b) Using the alternative iterative formula $x_{n+1} = \sqrt{5x_n + 10}$ with starting value $x_0 = 6$, find a root to the equation $f(x) = 0$ to 3 significant figures. Verify your answer is correct to this level of accuracy.

Finding iteration formulas

- The **iteration formula** is just a **rearrangement** of the equation, leaving a single 'x' on **one side.**
- There are often lots of **different ways** to rearrange the equation, so in the exam you might be asked to '**show that**' it can be rearranged in a **certain way**, rather than starting from scratch.
- Sometimes a rearrangement of the equation leads to a **divergent** iteration when you come to working out the steps.
- This is the reason you probably **won't** be asked to **both** rearrange **and** use a formula to find a root without **prompting**.

Example

a) Show that $x^3 - x^2 - 9 = 0$ can be rearranged into $x = \sqrt{\frac{9}{x-1}}$. Use this to make an iteration formula and find the value of a root to 2 d.p. with starting value $x_0 = 2.5$.

- The '9' is on its own in the fraction so try:
  $x^3 - x^2 - 9 = 0 \Rightarrow x^3 - x^2 = 9$
- The LHS can be factorised now: $x^2(x - 1) = 9$
- Get the x^2 on its own by dividing by $x - 1$: $x^2 = \frac{9}{x-1}$
- Finally take the square root of both sides: $x = \sqrt{\frac{9}{x-1}}$ as required
- You can now use the iteration formula $x_{n+1} = \sqrt{\frac{9}{x_n - 1}}$ to find approximations of the roots.

  $x_1 = \sqrt{\frac{9}{2.5-1}} = 2.449...$ $\quad x_2 = \sqrt{\frac{9}{2.449-1}} = 2.491...$

  and so on, until you find after 16 iterations that the value of the root, to 2 d.p., is:

  $x = 2.47$

b) Show that $x^3 - x^2 - 9 = 0$ can also be rearranged into $x = \sqrt{x^3 - 9}$ and use this to make an iteration formula.

- Start by isolating the x^2 term: $x^2 = x^3 - 9$
- Now just take the square root of both sides: $x = \sqrt{x^3 - 9}$ as required
- This makes the iteration formula: $x_{n+1} = \sqrt{x_n^3 - 9}$
- This is the iteration formula used in the example on page 235, so if you tried to use it to find a root you'd end up with a diverging sequence and it wouldn't find a root.

Tip: This shows why you'll never just be given an equation and told to find a root by first making an iteration formula.

Exercise 2.2

Q1 Show that the equation $x^4 + 7x - 3 = 0$ can be written in the form:

a) $x = \sqrt[4]{3 - 7x}$

b) $x = \dfrac{3 - 5x - x^4}{2}$

c) $x = \dfrac{\sqrt{3 - 7x}}{x}$

Q1 Hint: Think about which parts you need to get on their own before starting to rearrange the equation. In part b), for example, turn $7x$ into $5x + 2x$ to get where you want.

Q2 a) Show that the equation $x^3 - 2x^2 - 5 = 0$ can be rewritten as $x = 2 + \dfrac{5}{x^2}$.

b) Use the iterative formula $x_{n+1} = 2 + \dfrac{5}{x_n^2}$ with starting value $x_0 = 2$ to find x_5 to 1 decimal place.

c) Verify that the value found in part b) is a root of the equation $x^3 - 2x^2 - 5 = 0$ to 1 decimal place.

Q3 a) Rearrange the equation $x^2 + 3x - 8 = 0$ into the form $x = \dfrac{a}{x} + b$ where a and b are values to be found.

b) Verify that a root of the equation $x^2 + 3x - 8 = 0$ lies in the interval $(-5, -4)$.

c) Use the iterative formula $x_{n+1} = \dfrac{a}{x_n} + b$ with $x_0 = -5$ to find the values for $x_1 - x_6$, giving your answers to 3 d.p. Hence find a value of the root of the equation $x^2 + 3x - 8 = 0$ to 2 d.p.

Q4 a) Show that the equation $2^{x-1} = 4\sqrt{x}$ can be written as $x = 2^{2x-6}$.

b) Use the iterative formula $x_{n+1} = 2^{2x_n - 6}$ starting with $x_0 = 1$ to find the values of x_1, x_2, x_3 and x_4, giving your answers to 4 d.p.

c) Verify that the value for x_4 is a correct approximation to 4 d.p. for the root of the equation $2^{x-1} = 4\sqrt{x}$.

Q4 Hint: Start by rewriting everything as powers of 2 or x if you're struggling. You're going to need the rules for multiplying powers for this question.

Q5 $f(x) = \ln 2x + x^3$

a) Show that $f(x) = 0$ has a solution in the interval $0.4 < x < 0.5$.

b) Show that $f(x) = 0$ can be rewritten in the form $x = \dfrac{e^{-x^3}}{2}$.

c) Using an iterative formula based on part b) and an appropriate value for x_0 find an approximation of the root of the equation $f(x) = 0$ to 3 decimal places.

Q5 Hint: Remember that $\ln e^x = x$.

Q6 $f(x) = x^2 - 9x - 20$

a) Find an iterative formula for $f(x) = 0$ in the form $x_{n+1} = \sqrt{px_n + q}$ where p and q are constants to be found.

b) By using the formula in part a) and a starting value of $x_0 = 10$, find an approximation to a root of the equation $f(x) = 0$. Give your answer to 3 significant figures.

c) Show that an alternative iterative formula is $x_{n+1} = \dfrac{x_n^2 - 4x_n}{5} - 4$.

d) By using the iterative formula in part c) with starting value $x_0 = 1$ find the value of x_1, x_2... x_8.

e) Describe the behaviour of this sequence.

3. Sketching Iterations

Learning Objectives:

- Be able to sketch cobweb and staircase diagrams.
- Be able to identify diagrams of iterations that show convergence or divergence.

Once you've calculated a sequence of iterations, you can show on a diagram whether your sequence converges or diverges.

Cobweb and staircase diagrams

The instructions below show you how to sketch an iteration diagram.

- First, sketch the graphs of $\boldsymbol{y = x}$ and $\boldsymbol{y = f(x)}$ (where $f(x)$ is the iterative formula). The point where the two graphs **meet** is the **root** you're aiming for.
- Draw a **vertical line** from the x-value of your starting point ($\boldsymbol{x_0}$) until it meets the curve $\boldsymbol{y = f(x)}$.
- Now draw a **horizontal line** from this point to the line $\boldsymbol{y = x}$. At this point, the x-value is $\boldsymbol{x_1}$, the value of your first iteration. This is one **step**.
- Draw another step — a **vertical line** from this point to the curve, and a **horizontal line** joining it to the line $y = x$. Repeat this for each of your iterations.
- If your steps are getting **closer and closer** to the root, the sequence of iterations is **converging**.
- If the steps are moving **further and further away** from the root, the sequence is **diverging**.

Tip: In the exam you'll usually be given the graphs of $y = x$ and $y = f(x)$ and you'll just have to draw in the iteration steps between the two.

The method above produces two different types of diagrams — **cobweb** diagrams and **staircase** diagrams.

Cobweb diagrams

Cobweb diagrams look like they're **spiralling** in to the root (or away from it). The example below shows a **convergent cobweb diagram**.

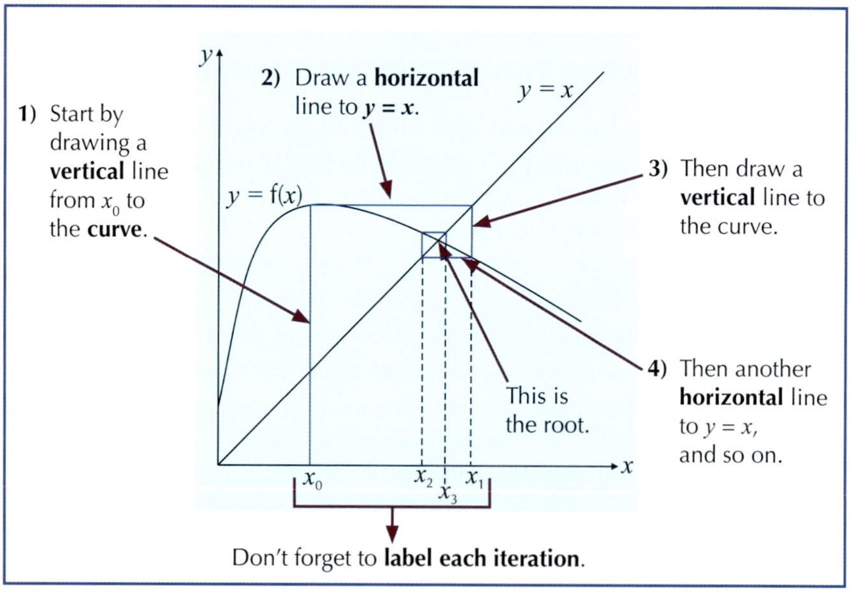

y
y = x
y = f(x)
x
1) Start by drawing a vertical line from x_0 to the curve.
2) Draw a horizontal line to $y = x$.
3) Then draw a vertical line to the curve.
4) Then another horizontal line to $y = x$, and so on.
This is the root.
x_0
x_2
x_3
x_1
Don't forget to label each iteration.

Tip: In this case, the iterations alternate between being below the root and above the root, but are getting closer each time.

A **divergent** cobweb diagram would have a similar shape, but each iteration would **spiral away from the root** rather than towards it.

Staircase diagrams

Staircase diagrams look like a set of **steps** leading to (or away from) the root. The examples below show a **convergent** and a **divergent** staircase diagram.

This is an example of a **convergent staircase diagram**:

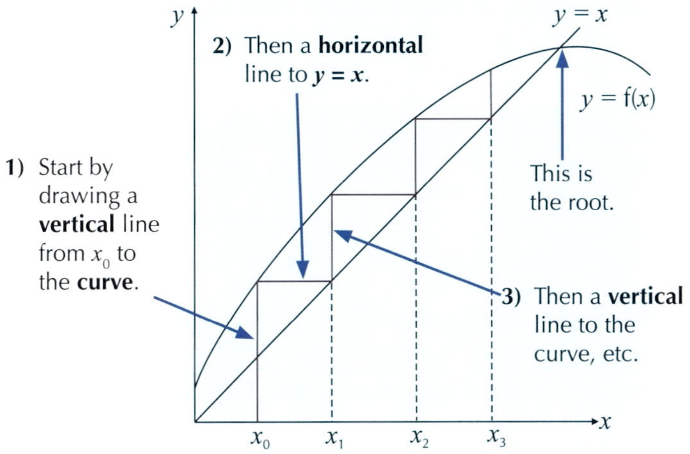

Starting at x_0, the next iterations x_1, x_2 and x_3 are getting **closer** to the point where the two graphs intersect (the root).

This is an example of a **divergent staircase diagram**:

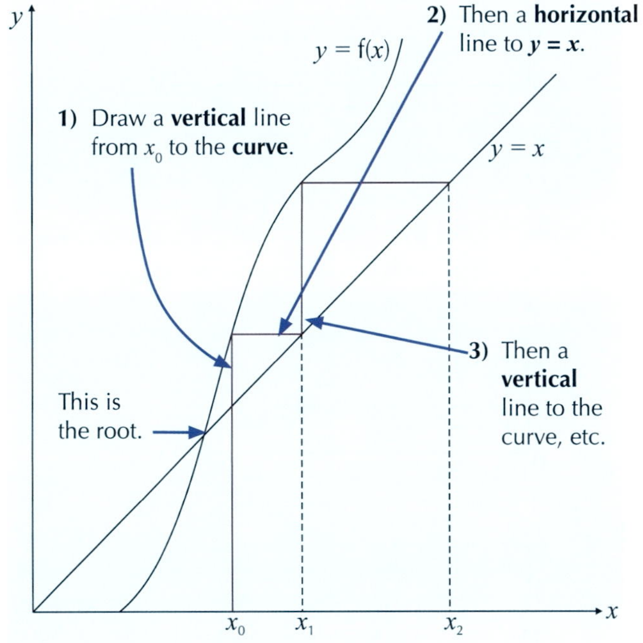

Starting at x_0, the next iterations x_1 and x_2 are getting **further away** from the root.

In general, these diagrams will only converge if your starting value of x_0 is **close enough** to the root, and if the graph of f(x) **isn't too steep**.
For a root, a, of a function f(x), the iterations **will converge** if the gradient of f(x) at a is **between –1 and 1** (i.e. $|f'(a)| < 1$) for a suitable choice of x_0.

Exercise 3.1

Q1 Hint: You'll need to draw the line $y = x$ on each diagram.

Q1 For each graph below, draw a diagram to show the convergence or divergence of the iterative sequence for the given value of x_0, and say whether it is a convergent or divergent staircase or cobweb diagram. Label x_0, x_1 and x_2 on each diagram where possible.

a) $x_0 = 3.5$

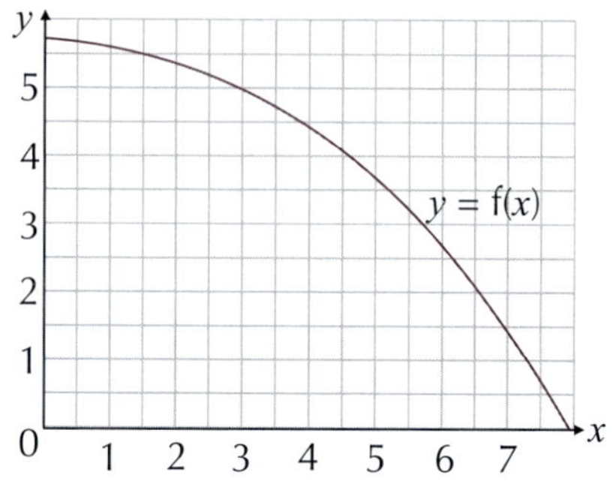

b) $x_0 = 3$

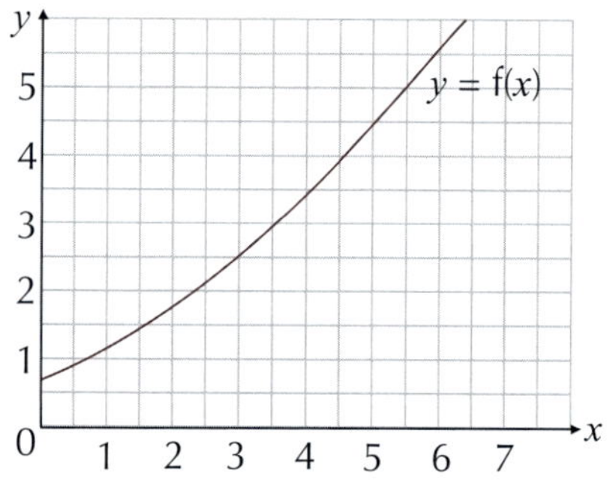

Q1c) Hint: Take care with the scale on this diagram.

c) $x_0 = 1.75$

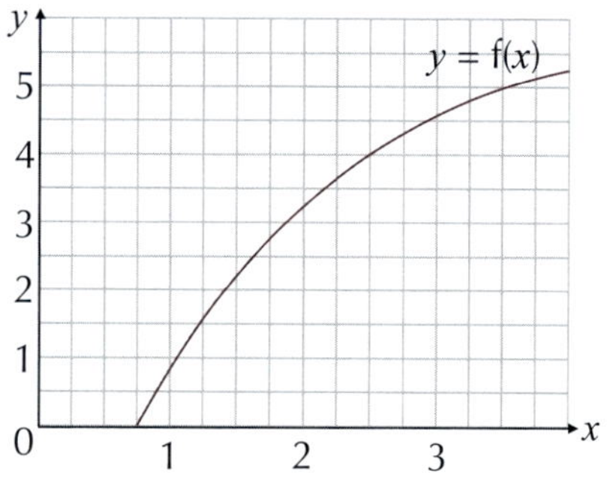

d) $x_0 = 4$

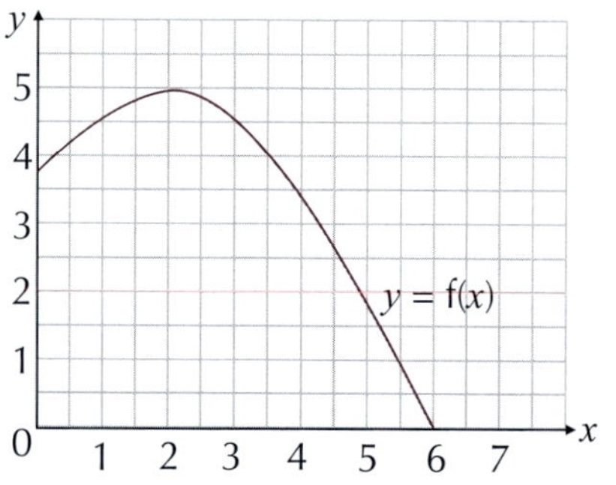

4. The Newton-Raphson Method

The next numerical method that you need to know for finding the root of an equation is the Newton-Raphson method, which uses differentiation.

Learning Objectives:

- Be able to use the Newton-Raphson method to find the root of an equation.
- Be able to answer questions that combine multiple numerical methods.

The Newton-Raphson method

The **Newton-Raphson method** works by finding the **tangent** to a function at a point x_0, and using its ***x*-intercept** for the next iteration, x_1. Repeating the process **iteratively** to find x_2, x_3, etc. gets you **closer** to the root, as shown:

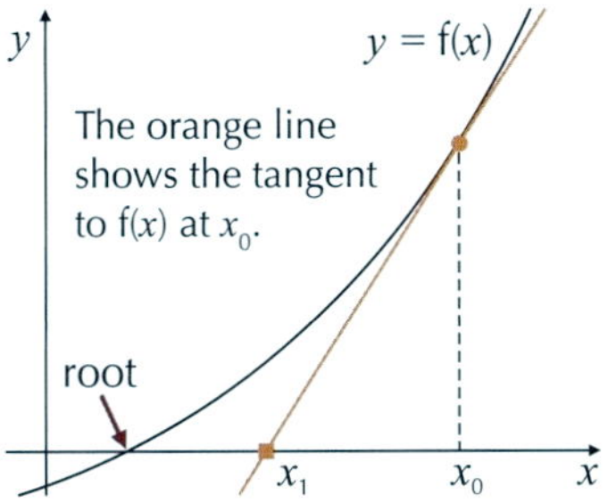

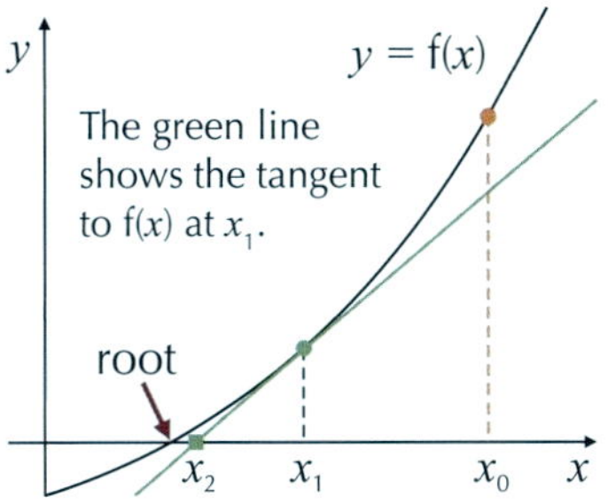

The iteration formula for this method can be derived as follows:

- The tangent to f(*x*) at the point x_n is a straight line that can be written in the form: $y - y_n = m(x - x_n)$

  $y - f(x_n) = f'(x_n)\,(x - x_n)$ ← gradient at x_n

- The next iteration, x_{n+1}, should be at the *x*-intercept of this tangent line, i.e. $x = x_{n+1}$ when $y = 0$. So substitute $(x_{n+1}, 0)$ into the equation:

$$0 - f(x_n) = f'(x_n)\,(x_{n+1} - x_n)$$

$$x_{n+1} - x_n = \frac{-f(x_n)}{f'(x_n)}$$

$$x_{n+1} = x_n - \frac{f(x_n)}{f'(x_n)}$$

Tip: Remember that differentiating an equation gives you the gradient which you can use to find the tangent at a given point. Here, the line is a tangent to f(*x*) at x_n.

Tip: This formula is in the formula booklet. Don't be put off by the differentiated term — use it the same way as any other iteration formula.

Example

Use the Newton-Raphson method with a starting point of $x_0 = -2$ to find a root of $f(x) = x^3 - 3x + 5$ correct to 5 decimal places.

- First, differentiate f(*x*) to find $f'(x) = 3x^2 - 3$.
- This means the iteration formula is $x_{n+1} = x_n - \dfrac{x_n^3 - 3x_n + 5}{3x_n^2 - 3}$ (numerator: $f(x_n)$; denominator: $f'(x_n)$)
- From the question we take $x_0 = -2$, so

$$x_1 = -2 - \frac{(-2)^3 - 3(-2) + 5}{3(-2)^2 - 3} = -\frac{7}{3}$$

$$x_2 = -\frac{7}{3} - \frac{\left(-\frac{7}{3}\right)^3 - 3\left(-\frac{7}{3}\right) + 5}{3\left(-\frac{7}{3}\right)^2 - 3} = -2.280555556...$$

$x_3 = -2.279020068...$, $x_4 = -2.279018786...$, $x_5 = -2.279018786...$

- So the root is −2.27902 to 5 d.p.

Tip: Make sure to use the ANS button on your calculator when you're doing these iterations, so you avoid rounding errors.

There are times when the Newton-Raphson method can't be used to find a root.

- If the function f(x) cannot be differentiated, then you won't be able to form an iteration formula to use.
- Like other iterative methods, if you choose a start point too far away from the root, the sequence might diverge.
- If the tangent to f(x) is **horizontal** at the point x_n (i.e. if x_n is a **stationary point**) the method will fail. This is shown on the diagram below:

Tip: You know from the definition of a stationary point that f′(x) = 0 at that point. So you'd have to divide by zero in the iteration formula, which will cause the method to fail.

The tangent does not meet the x-axis, so there is no x_{n+1} value to continue the iterations with.

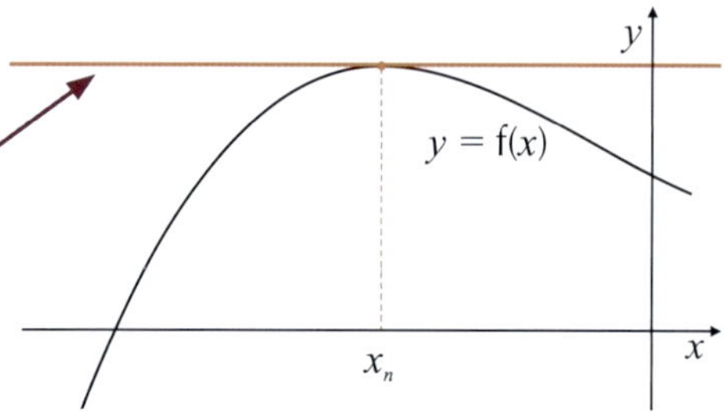

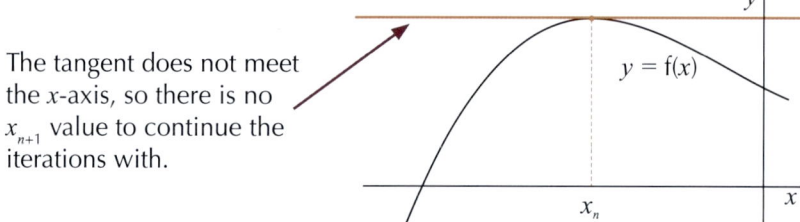

- Similarly, if you try a point where f(x) has a **shallow gradient**, the tangent meets the x-axis a really long way away from the root, which could cause the iterations to diverge:

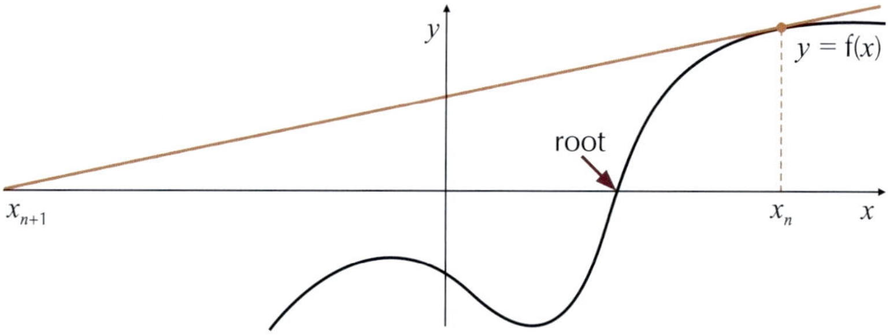

Exercise 4.1

Hint: Look back at Chapter 7 if you need help differentiating any of these functions.

Q1 Using the Newton-Raphson method, give an iteration formula for finding the roots of each of these functions:

a) $f(x) = 5x^2 - 6$

b) $g(x) = e^{3x} - 4x^2 - 1$

c) $h(x) = \sin x + x^3 - 1$

Q2 Use the Newton-Raphson method to find a root of $x^4 - 2x^3 - 5 = 0$ to 5 s.f., starting with $x_0 = 2.5$.

Q3 Use the Newton-Raphson method to find the negative root of $f(x) = x^2 - 5x - 12$ to 5 s.f. Use $x_0 = -1$ as your start value.

Q4 Find a root to the equation $x^2 \ln x = 5$ to 5 s.f., using $x_0 = 2$. PROBLEM SOLVING

Q5 Show that the Newton-Raphson method will fail to find a root of the equation $2x^3 - 15x^2 + 109 = 0$ if $x_0 = 1$ is used as the starting value.

Combining the methods

You might see a question that **combines** all (or at least most) of the methods covered so far in this chapter into one long question.

Here you can see an **exam-style question** worked from start to finish, just how they'd want you to do it in the real thing.

Tip: Remember to show **all** of your working. That way, even if you get your final answer wrong, you can still get some marks for showing that you understood what the question was asking.

Example

The graph below shows both roots of the continuous function $f(x) = 6x - x^2 + 13$.

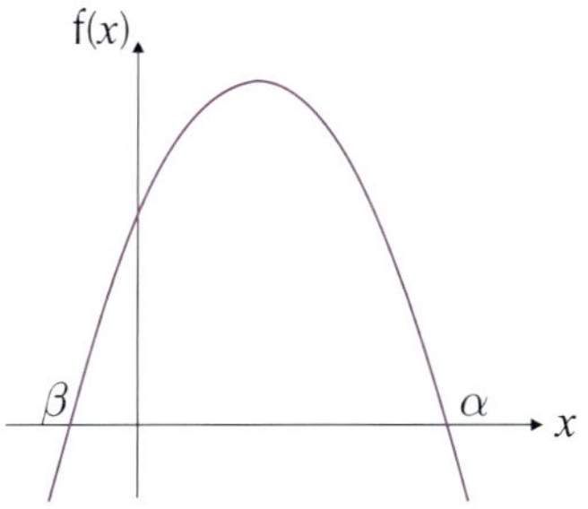

a) **Show that the positive root, α, lies in the interval $7 < x < 8$.**

f(x) is a **continuous function**, so if f(7) and f(8) have **different signs** then there is a root in the interval $7 < x < 8$:

$$f(7) = (6 \times 7) - 7^2 + 13 = 6.$$
$$f(8) = (6 \times 8) - 8^2 + 13 = -3.$$

There is a **change of sign** so $7 < \alpha < 8$.

b) **Show that $6x - x^2 + 13 = 0$ can be rearranged into the formula: $x = \sqrt{6x + 13}$**

Get the x^2 **on its own** to make: $6x + 13 = x^2$

Now take the (positive) square root to leave: $x = \sqrt{6x + 13}$.

Tip: The key with the rearranging questions is being able to pick up hints that show how it needs to be rearranged. Here the $\sqrt{\ }$ sign shows you'll need to get the x^2 on its own and square root both sides.

c) **Use the iteration formula $x_{n+1} = \sqrt{6x_n + 13}$ and $x_0 = 7$ to find α to 1 d.p.**

Using $x_{n+1} = \sqrt{6x_n + 13}$ with $x_0 = 7$, gives $x_1 = \sqrt{6 \times 7 + 13}$
$= 7.4161...$

Continuing the iterations:

$$x_2 = \sqrt{6 \times 7.4161... + 13} = 7.5826...$$
$$x_3 = \sqrt{6 \times 7.5826... + 13} = 7.6482...$$
$$x_4 = \sqrt{6 \times 7.6482... + 13} = 7.6739...$$
$$x_5 = \sqrt{6 \times 7.6739... + 13} = 7.6839...$$

x_4 and x_5 both round to 7.7 to 1 d.p., so $\alpha = 7.7$ (1 d.p.)

d) **Sketch a diagram to show the convergence of the sequence for x_1, x_2 and x_3.**

In an exam question, $y = \sqrt{6x + 13}$ and $y = x$ would usually be drawn on a graph for you, and the position of x_0 would be marked.

All you have to do is draw on the lines and label the values of x_1, x_2 and x_3.

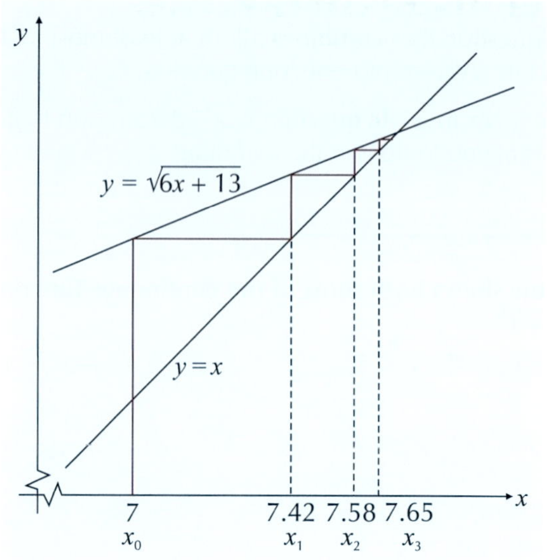

Tip: Here the values of x_1, x_2 and x_3 have been rounded to 2 d.p. to make it easier to label them on the graph.

e) Use the Newton-Raphson formula to find the negative root, β, to 5 s.f. Start with $x_0 = -1$.

- Differentiate f(x): $f'(x) = 6 - 2x$.
- Putting this into the Newton-Raphson formula gives:

$$x_{n+1} = x_n - \frac{f(x_n)}{f'(x_n)} = x_n - \frac{6x_n - x_n^2 + 13}{6 - 2x_n}$$

- Starting with $x_0 = -1$, this gives $x_1 = -1 - \frac{6(-1) - (-1)^2 + 13}{6 - 2(-1)} = -1.75$
- Further iterations give: $x_2 = -1.690789...$
  $x_3 = -1.690415...$
  $x_4 = -1.690415...$
- So, $\beta = -1.6904$ (5 s.f.)

Tip: Notice how much quicker this Newton-Raphson formula converges to the root — in part c) it took 5 iterations to find a root to 1 d.p.

Exercise 4.2

Q1 Let $f(x) = x^4 + 2x^3 - 4x^2 - 7x + 2$ and $g(x) = x^3 - 4x + 1$

a) Show that $f(x) = (x + 2)g(x)$. Hence give an integer root of f(x).

b) Show that a root of g(x), α, lies between 1 and 2.

c) Show that $g(x) = 0$ can be rearranged into the formula $x = \sqrt[3]{4x - 1}$.

d) Use the iteration formula $x_{n+1} = \sqrt[3]{4x_n - 1}$ with $x_0 = 2$ to find α to 3 s.f.

e) Use the Newton-Raphson method with $x_0 = -2$ to find another root of g(x), β, to 4 s.f.

f) Explain why the Newton-Raphson method for g(x) fails when $x_n = \frac{2\sqrt{3}}{3}$.

5. The Trapezium Rule

It's not always possible to integrate a function using the methods you learn at A-level, and some functions can't be integrated at all. When this happens, all is not lost — you can approximate the integral using the trapezium rule.

Learning Objectives:

- Be able to use the trapezium rule to approximate the value of definite integrals.
- Be able to calculate an upper and lower bound of the area beneath a curve.
- Be able to explain and use the fact that integration is the limit of a sum of rectangles.

The trapezium rule

When you find yourself with a function which is too difficult to integrate, you can **approximate** the area under the curve using lots of **trapeziums**, which gives an approximate value of the integral.

- The **area** under this curve between a and b can be approximated by the green **trapezium** shown.
- It has height $(b - a)$ and parallel sides of length $f(a)$ and $f(b)$.
- The area of the trapezium is an **approximation** of the integral $\int_a^b f(x)\,dx$.

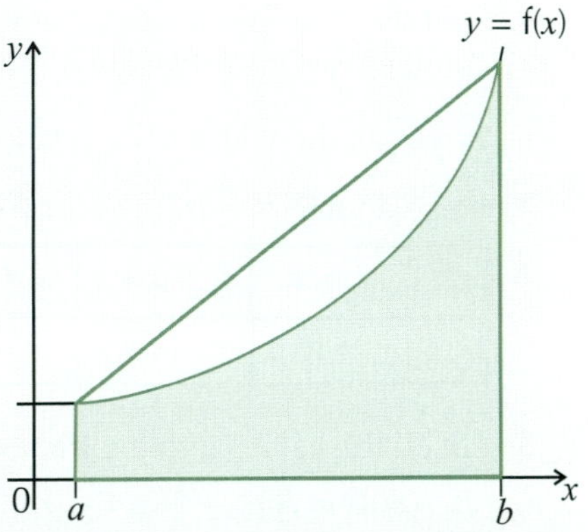

- It's not a very good approximation, but if you split the area up into **more** trapeziums of equal width, the approximation will get more and more **accurate** because the **difference** between the trapeziums and the curve will get **smaller**.

Tip: The area of a trapezium is given by the formula $\frac{h}{2}(a + b)$.

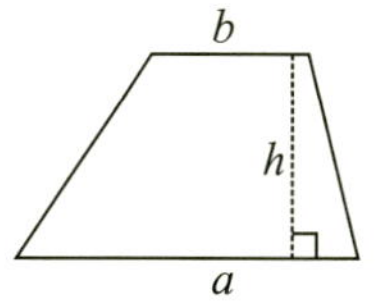

The **trapezium rule** for approximating $\int_a^b$ **f(x) dx** works like this:

- n is the **number** of strips i.e. trapeziums.
- h is the **width** of each strip — it's equal to $\frac{(b-a)}{n}$.
- The **x-values** go up in steps of h, starting with $x_0 = a$.
- The **y-values** are found by putting the x-values into the equation of the curve — so $y_1 = f(x_1)$. They give the **heights** of the sides of the trapeziums.

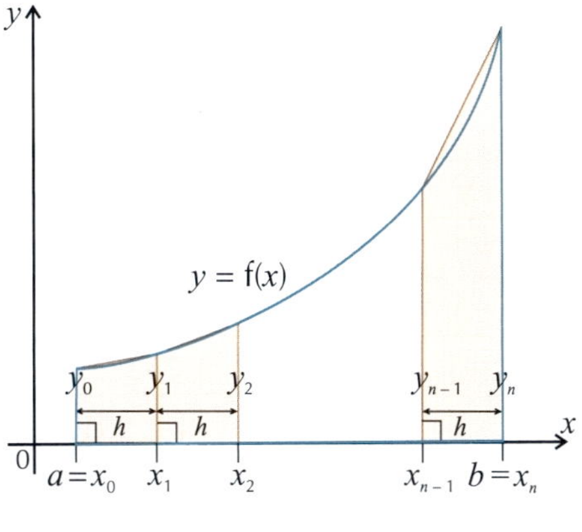

- The **area** of each trapezium is $A = \frac{h}{2}(y_r + y_{r+1})$.

Then an **approximation** for $\int_a^b f(x)\,dx$ is found by **adding** the **areas** of all the trapeziums:

$$\int_a^b f(x)\,dx \approx \frac{h}{2}(y_0 + y_1) + \frac{h}{2}(y_1 + y_2) + \ldots + \frac{h}{2}(y_{n-1} + y_n)$$

$$= \frac{h}{2}[y_0 + 2(y_1 + y_2 + \ldots + y_{n-1}) + y_n]$$

Tip: If the number of strips is n, the number of y-values is $n + 1$. The y-values are sometimes called **ordinates**.

Tip: This just says 'Add the first and last heights $(y_0 + y_n)$ and add this to twice all the other heights added up — then multiply by $\frac{h}{2}$.'

Tip: The trapezium rule is in the formula booklet — always look it up. It's actually given as $\int_a^b y\,dx$, but this is the same since $y = f(x)$.

So the **trapezium rule** says:

$$\int_a^b f(x)\,dx \approx \frac{h}{2}[y_0 + 2(y_1 + y_2 + ... + y_{n-1}) + y_n]$$

This may seem like a lot of information, but it's simple if you follow this **step by step** method:

To approximate the integral $\int_a^b f(x)\,dx$:

- **Split** the interval up into a number of equal sized strips, n. You'll always be told what n is (it could be 4, 5 or even 6).
- Work out the **width** of each strip: $h = \frac{(b-a)}{n}$
- Make a **table** of x and y values:

x	$x_0 = a$	$x_1 = a + h$	$x_2 = a + 2h$	...	$x_n = b$
y	$y_0 = f(x_0)$	$y_1 = f(x_1)$	$y_2 = f(x_2)$	...	$y_n = f(x_n)$

- Put all the values into the **trapezium rule**:
$\int_a^b f(x)\,dx \approx \frac{h}{2}[y_0 + 2(y_1 + y_2 + ... + y_{n-1}) + y_n]$

Example 1

Use the trapezium rule with 3 strips to find an approximate value for $\int_0^{1.5} \sqrt{x^2 + 2x}\,dx$.

- $n = 3$, $a = 0$ and $b = 1.5$, so the width of each strip is $h = \frac{1.5 - 0}{3} = 0.5$.
- This gives x-values of $x_0 = 0$, $x_1 = 0.5$, $x_2 = 1$ and $x_3 = 1.5$.
- Calculate the value of y for each of x_0, x_1, x_2, x_3:

x	$y = \sqrt{x^2 + 2x}$
$x_0 = 0$	$y_0 = 0$
$x_1 = 0.5$	$y_1 = \sqrt{1.25} = 1.118...$
$x_2 = 1$	$y_2 = \sqrt{3} = 1.732...$
$x_3 = 1.5$	$y_3 = \sqrt{5.25} = 2.291...$

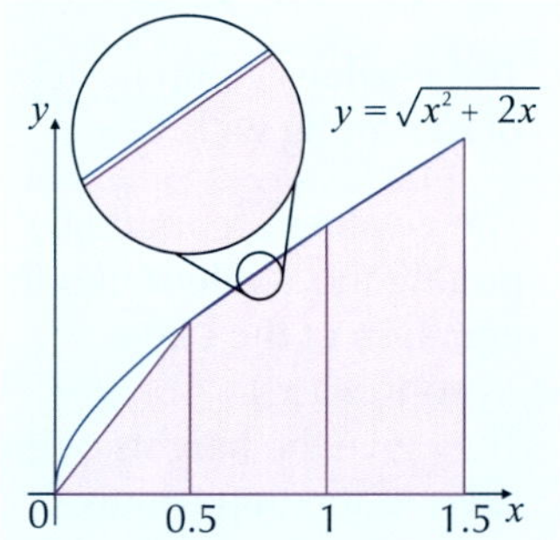

Tip: The graph shows that the estimate here is less than the actual value of the integral — there's a gap between the curve and the top of each strip. Even where it looks like the curve and the top of the trapezium are the same, if you zoom in far enough there will always be a gap.

- Now use the formula to find the approximate value of the integral:

$$\int_0^{1.5} \sqrt{x^2 + 2x}\,dx \approx \frac{h}{2}[y_0 + 2(y_1 + y_2) + y_3]$$
$$= \frac{0.5}{2}[0 + 2(1.118... + 1.732...) + 2.291...]$$
$$= \frac{1}{4}[7.991...]$$
$$= 2.00 \text{ (3 s.f.)}$$

The **approximation** that the trapezium rule gives will either be an **overestimate** (too big) or an **underestimate** (too small).

This will depend on the **shape** of the graph — a sketch can show whether the tops of the trapeziums lie **above** the curve or stay **below** it.

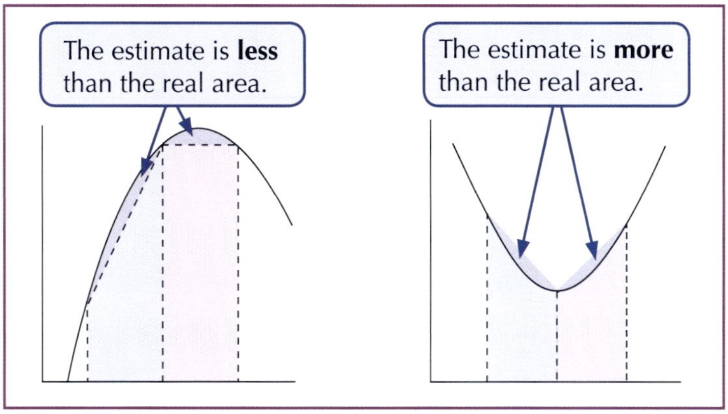

The estimate is **less** than the real area.
The estimate is **more** than the real area.

Tip: An exam question might ask you to draw the trapeziums on a given graph.

Tip: Another way to say this is that the trapezium rule gives an underestimate if the curve is concave and an overestimate if the curve is convex — see p.135 for more on convex and concave curves.

Using **more strips** (i.e. **increasing** n) gives you a **more accurate** approximation.

Example 2

Use the trapezium rule to approximate $\int_0^4 \frac{6x^2}{x^3+2}\,dx$ to 3 d.p. using:

a) $n = 2$

- For 2 strips, the width of each strip is $h = \frac{4-0}{2} = 2$, so the x-values are 0, 2 and 4.
- Calculate the corresponding y-values:

x	$y = \frac{6x^2}{x^3+2}$
$x_0 = 0$	$y_0 = 0$
$x_1 = 2$	$y_1 = 2.4$
$x_2 = 4$	$y_2 = 1.4545$ (4 d.p.)

- Putting these values into the formula gives:

$$\int_0^4 \frac{6x^2}{x^3+2}\,dx \approx \frac{h}{2}[y_0 + 2y_1 + y_2]$$
$$= \frac{2}{2}[0 + 2(2.4) + 1.4545]$$
$$= 6.255 \text{ (3 d.p.)}$$

b) $n = 4$

- For 4 strips, the width of each strip is $h = \frac{4-0}{4} = 1$, so the x-values are 0, 1, 2, 3 and 4.

Tip: If you have to round the y-values, make sure you leave enough decimal places. For example, if your final answer has to be to 3 d.p., find the y-values to at least 4 d.p.

Tip: Increasing the number of strips increases the accuracy because the gaps between curve and line are smaller:

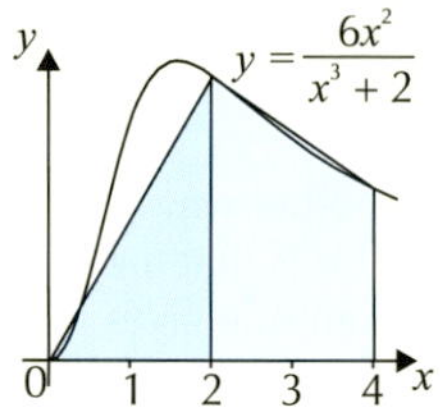

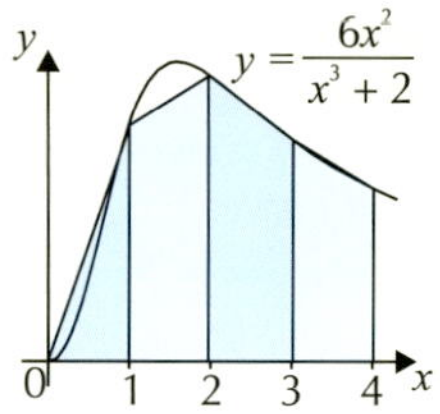

- Calculate the corresponding y-values:

x	$y = \frac{6x^2}{x^3+2}$
$x_0 = 0$	$y_0 = 0$
$x_1 = 1$	$y_1 = 2$
$x_2 = 2$	$y_2 = 2.4$
$x_3 = 3$	$y_3 = 1.8621$ (4 d.p.)
$x_4 = 4$	$y_4 = 1.4545$ (4 d.p.)

- Putting these values into the formula gives:

$$\int_0^4 \frac{6x^2}{x^3+2}\,dx \approx \frac{h}{2}[y_0 + 2(y_1 + y_2 + y_3) + y_4]$$
$$= \frac{1}{2}[0 + 2(2 + 2.4 + 1.8621) + 1.4545]$$
$$= \frac{1}{2}[13.9787] = 6.989 \text{ (3 d.p.)}$$

Upper and lower bounds

You can calculate an **upper** and **lower bound** for the area under a curve using a simplified version of the trapezium rule that uses rectangles:

$$\int_a^b f(x)\,dx \approx h[y_0 + y_1 + y_2 + \dots + y_{n-1}]$$

This formula sums the areas of the rectangles which meet f(x) with their **left hand corner**.

In this example, the rectangles are **below** the curve, so they calculate a **lower bound**.

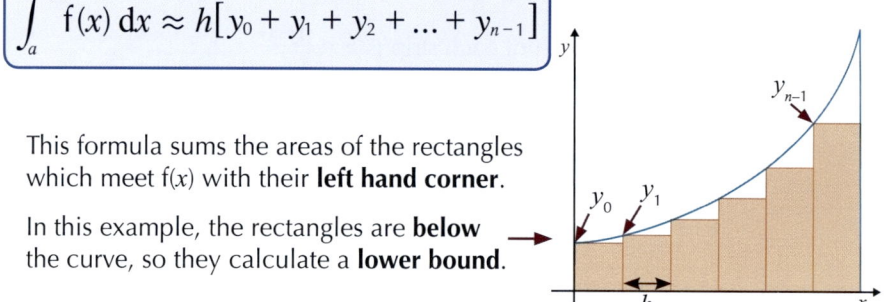

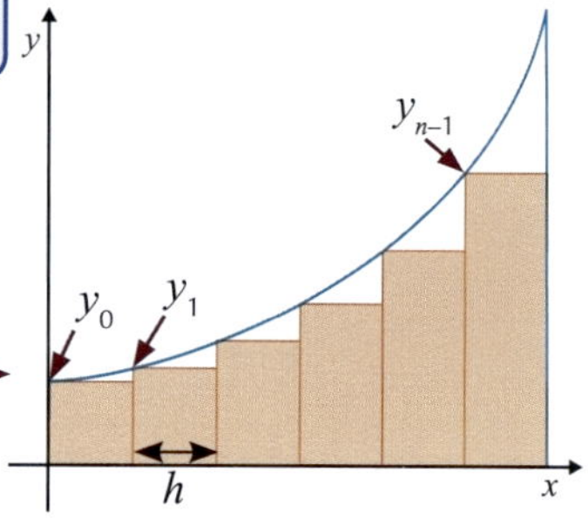

Tip: For an increasing function like the one shown here, calculating the bound using the 'left hand corner' will give the lower bound, and using the 'right hand corner' will give the upper bound. But for a **decreasing** function, it's the other way around. So to calculate a bound of a curve across a turning point, do a separate calculation either side of the turning point — one will use the right hand corner, and the other will use the left.

$$\int_a^b f(x)\,dx \approx h[y_1 + y_2 + \dots + y_{n-1} + y_n]$$

This formula sums the areas of the rectangles which meet f(x) with their **right hand corner**.

In this example, the rectangles are **above** curve, so they calculate an **upper bound**.

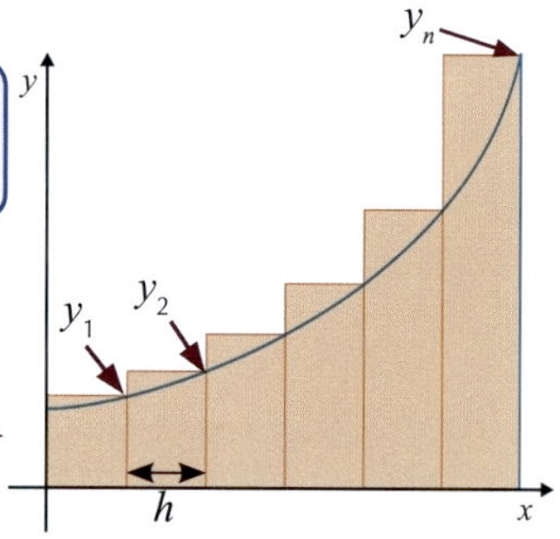

When you approximate the area under a curve using the **trapezium rule**, that value will lie **between** these two bounds.

Integration as the limit of a sum

When **differentiating** a function, you're finding the **gradient** of the curve. You saw in Year 1 that you do this by finding the gradient of a straight line over a smaller and smaller interval, until the interval is virtually nothing. So, for a function f(x):

$$f'(x) = \lim_{\delta x \to 0} \left[\frac{f(x + \delta x) - f(x)}{(x + \delta x) - x} \right]$$

Similarly, you can define the **integral** of a curve f(x) between two points a and b using limits:

$$\int_a^b f(x)\,dx = \lim_{\delta x \to 0} \sum_{x=a}^{b} f(x) \times \delta x$$

You saw on the previous page that you can find upper and lower bounds for the area under a curve by adding up the area of rectangles. These are **approximations** of the actual area under the curve.

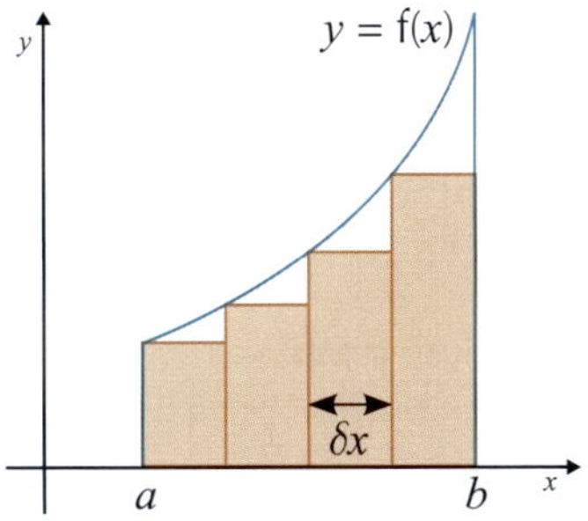

The area of each rectangle is height × width.
Its height is the y-value, f(x), and its width is just δx.

So the area of each rectangle is $R = f(x) \times \delta x$.

Then the **sum** of all the rectangles between a and $b - \delta x$ is:

$$\sum_{x=a}^{b-\delta x} R = \sum_{x=a}^{b-\delta x} f(x)\,\delta x$$

As δx gets smaller and smaller, then this sum of rectangles comes closer and closer to the actual area under the curve.
This is how we **define** an integral — as the **limit of a sum** of areas.
So as the width δx approaches 0, the sum of areas is indistinguishable from the actual area. And as $\delta x \to 0$, $b - \delta x \to b$, so the upper value of the sum can be replaced with b:

$$\lim_{\delta x \to 0} \sum_{x=a}^{b-\delta x} f(x)\,\delta x = \int_a^b f(x)\,dx$$

This is really just a **change in notation**, so that you don't have to write out the whole limit of a sum each time you want to write an integral.

When you're writing the **exact** integral, you use **d** instead of δ, and you replace $\sum$ with $\int$, which are both just ways of writing 'S' (for 'sum').

Tip: In Chapter 7, you saw this with h instead, but it's the same formula. δ is the Greek letter delta, and δx just means "change in x".

Tip: You should have seen the formula for differentiating from first principles back in Year 1. This formula uses the same ideas to integrate, although you don't need to use it for the integration you came across in Chapter 8.

Tip: It doesn't matter if you use rectangles above or below the curve — the result will be the same.

Tip: You're summing the areas of the rectangles where the left hand corner meets the curve. So, to avoid adding on an extra rectangle beyond the point b, you need to sum only up to the point $b - \delta x$.

If you use rectangles **above** the curve, then you would sum the 'right hand corner' rectangles from $a + \delta x$ to b.

Exercise 5.1

Q1 Use the trapezium rule to find approximations of each of the following integrals. Use the given number of intervals in each case. Give your answers to 3 significant figures.

a) $\int_0^2 \sqrt{x+2}\, dx$, 2 intervals

b) $\int_1^3 2(\ln x)^2\, dx$, 4 intervals

c) $\int_0^{0.4} e^{x^2}\, dx$, 2 intervals

d) $\int_{-\frac{\pi}{4}}^{\frac{\pi}{4}} 4x\tan x\, dx$, 4 intervals

e) $\int_0^{0.3} \sqrt{e^x+1}\, dx$, 6 intervals

f) $\int_0^{\pi} \ln(2+\sin x)\, dx$, 6 intervals

Q2 For questions 1a) and b) above, find an upper and lower bound for the area beneath the curve to 4 s.f.

Q3 Use the trapezium rule with 3 intervals to find an estimate to $\int_0^{\frac{\pi}{2}} \sin^3\theta\, d\theta$. Give your answer to 3 d.p.

Q4 The shape of an aeroplane's wing is modelled by the curve $y = \sqrt{\ln x}$, where x and y are measured in metres. Use the trapezium rule with 5 intervals to estimate the area of a cross-section of the wing enclosed by the curve $y = \sqrt{\ln x}$, the x-axis and the lines $x = 2$ and $x = 7$. Give your answer to 3 d.p.

MODELLING

Q5 a) Complete the following table of values to 3 d.p. for $y = e^{\sin x}$.

x	0	$\frac{\pi}{8}$	$\frac{\pi}{4}$	$\frac{3\pi}{8}$	$\frac{\pi}{2}$
y	1	1.466			2.718

b) (i) Using the trapezium rule with 2 intervals, estimate $\int_0^{\frac{\pi}{2}} e^{\sin x}\, dx$ to 2 d.p.

(ii) Repeat the calculation using 4 intervals.

c) Which is the better estimate? Explain your answer.

Q6 The diagram below shows part of the curve $y = \frac{3}{\ln x}$.

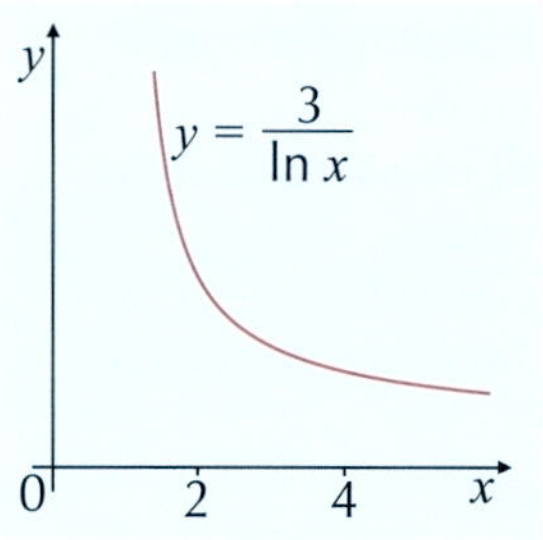

a) Using the trapezium rule with 4 intervals, find an estimate to 2 d.p. for $\int_2^4 \frac{3}{\ln x}\, dx$.

b) Without further calculation, state whether your answer to part a) is an over-estimate or under-estimate of the true area. Explain your answer.

Q7 a) Using the trapezium rule with $h = \frac{\pi}{6}$, show that $\int_{-\frac{\pi}{2}}^{\frac{\pi}{2}} \cos x\, dx$ can be approximated as $\frac{\pi(2+\sqrt{3})}{6}$.

b) Without further calculation, state whether this approximation is an under- or over-estimation. Explain your answer.

Chapter 10 Vectors

1. Vectors in Three Dimensions

In Year 1 you learnt how to represent, calculate with, and model situations using vectors in two dimensions. Now you're going to find out how to do all of that in three dimensions. You might want to look back at your notes on 2D vectors first, to refresh your memory.

Learning Objectives:
- Understand how to represent vectors in three dimensions, using both unit vectors and column vectors.
- Be able to add and subtract three-dimensional vectors and multiply them by scalars.
- Be able to show that two 3D vectors are parallel, and that three points are collinear.

Three-dimensional position vectors

- **Three-dimensional vectors** have components in the direction of the ***x*-, *y*-** and ***z*-axes**. Imagine that the *x*- and *y*-axes lie flat on the page. Then imagine a **third axis** sticking straight through the page at right angles to it — this is the ***z*-axis**.
- The points in three dimensions are given **(*x*, *y*, *z*) coordinates**.
- The **unit vector** in the direction of the *z*-axis is **k**, so three-dimensional vectors can be written like this: $x\mathbf{i} + y\mathbf{j} + z\mathbf{k}$ or $\begin{pmatrix} x \\ y \\ z \end{pmatrix}$
- Three-dimensional vectors are used to describe things in **three-dimensional space**, e.g. an aeroplane moving through the sky.
- Calculating with 3D vectors is just the same as with 2D vectors, as the next example shows.

Example 1

The diagram below shows the position of the points P and Q.

a) Write the position vectors $\overrightarrow{OP}$ and $\overrightarrow{OQ}$ as column vectors.

$\overrightarrow{OP} = 4\mathbf{i} + 3\mathbf{j} + 0\mathbf{k} = \begin{pmatrix} 4 \\ 3 \\ 0 \end{pmatrix}$

$\overrightarrow{OQ} = 2\mathbf{i} + 5\mathbf{j} + 4\mathbf{k} = \begin{pmatrix} 2 \\ 5 \\ 4 \end{pmatrix}$

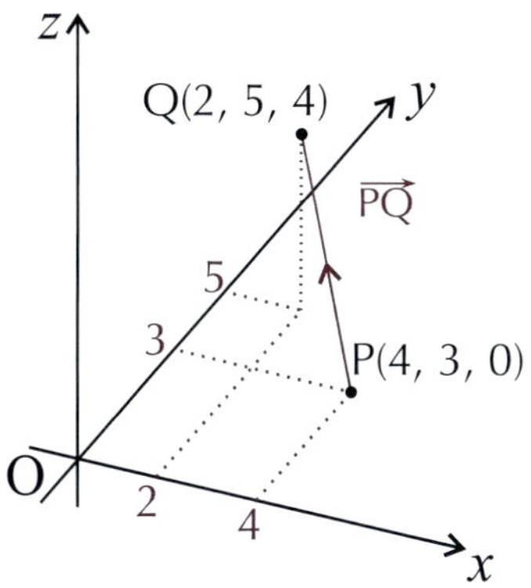

b) Hence find $\overrightarrow{PQ}$ as a column vector.

$\overrightarrow{PQ} = -\overrightarrow{OP} + \overrightarrow{OQ}$

$= -\begin{pmatrix} 4 \\ 3 \\ 0 \end{pmatrix} + \begin{pmatrix} 2 \\ 5 \\ 4 \end{pmatrix} = \begin{pmatrix} 2-4 \\ 5-3 \\ 4-0 \end{pmatrix} = \begin{pmatrix} -2 \\ 2 \\ 4 \end{pmatrix}$

Tip: Don't get confused between a point and its position vector. Points are given by (*x*, *y*, *z*) coordinates and position vectors are the movement needed to get to the point from the origin, given in the form $x\mathbf{i} + y\mathbf{j} + z\mathbf{k}$. You'll need to be able to swap between the forms.

Tip: Just add along each row of the column as you do with 2D column vectors — don't forget that you've multiplied $\overrightarrow{OP}$ by the scalar '–1' so all its entries will change signs.

Tip: Remember that to multiply by a scalar, each number in the column vector should be multiplied by that scalar.

Tip: Imagine a vector triangle: to get from O to R via P you go along $\overrightarrow{OP}$ and then along the arrow on $\overrightarrow{PR}$, hence $\overrightarrow{OP} + \overrightarrow{PR}$.

The point R lies on the line PQ such that PR:RQ = 3:1.

c) Find the position vector of point R, in unit vector form.

- The ratio tells you that point R is $\frac{3}{4}$ of the way along the line PQ from point P, so $\overrightarrow{PR} = \frac{3}{4}\overrightarrow{PQ}$:

$$\overrightarrow{PR} = \frac{3}{4} \times \begin{pmatrix} -2 \\ 2 \\ 4 \end{pmatrix} = \begin{pmatrix} -1.5 \\ 1.5 \\ 3 \end{pmatrix}$$

- Position vector $\overrightarrow{OR} = \overrightarrow{OP} + \overrightarrow{PR}$

$$\overrightarrow{OR} = \begin{pmatrix} 4 \\ 3 \\ 0 \end{pmatrix} + \begin{pmatrix} -1.5 \\ 1.5 \\ 3 \end{pmatrix} = \begin{pmatrix} 4-1.5 \\ 3+1.5 \\ 0+3 \end{pmatrix} = \begin{pmatrix} 2.5 \\ 4.5 \\ 3 \end{pmatrix} = 2.5\mathbf{i} + 4.5\mathbf{j} + 3\mathbf{k}$$

Parallel lines and collinear points

- As with two-dimensional vectors, **parallel** vectors in three dimensions are **scalar multiples** of each other.
- To determine whether two vectors are parallel, check whether one vector can be produced by multiplying the other by a scalar.
- If two vectors are **parallel**, and also have a **point in common**, then they must lie on the same straight line — their points are **collinear**.

Tip: Make sure you're comfortable working with vectors in all forms — as column vectors, using **i**, **j**, **k** notation, and as the movement between points given as Cartesian coordinates.

Example 2

PROBLEM SOLVING

Show that points A, B and C are collinear, if $\overrightarrow{OA} = 2\mathbf{i} - \mathbf{j} + \mathbf{k}$, $\overrightarrow{OB} = \mathbf{i} + 2\mathbf{j} + 3\mathbf{k}$ and $\overrightarrow{OC} = -\mathbf{i} + 8\mathbf{j} + 7\mathbf{k}$.

- First, find vectors $\overrightarrow{AB}$ and $\overrightarrow{BC}$:

$$\begin{aligned}\overrightarrow{AB} &= -\overrightarrow{OA} + \overrightarrow{OB} \\ &= (-2 + 1)\mathbf{i} + (1 + 2)\mathbf{j} + (-1 + 3)\mathbf{k} \\ &= -\mathbf{i} + 3\mathbf{j} + 2\mathbf{k}\end{aligned}$$

$$\begin{aligned}\overrightarrow{BC} &= -\overrightarrow{OB} + \overrightarrow{OC} \\ &= (-1 - 1)\mathbf{i} + (-2 + 8)\mathbf{j} + (-3 + 7)\mathbf{k} \\ &= -2\mathbf{i} + 6\mathbf{j} + 4\mathbf{k}\end{aligned}$$

- Show that $\overrightarrow{AB}$ and $\overrightarrow{BC}$ are parallel by finding a **scalar multiple**:

$$\begin{aligned}\overrightarrow{BC} &= -2\mathbf{i} + 6\mathbf{j} + 4\mathbf{k} \\ &= 2(-\mathbf{i} + 3\mathbf{j} + 2\mathbf{k}) \\ &= 2\overrightarrow{AB}\end{aligned}$$

- $\overrightarrow{BC}$ is a scalar multiple of $\overrightarrow{AB}$, so they are **parallel**, and also have a **point in common** (B).

  So A, B and C must all lie on the same line — they are **collinear**.

Exercise 1.1

Q1 R is the point (4, –5, 1) and S is the point (–3, 0, –1). Write down the position vectors of R and S, giving your answers:

a) as column vectors, and

b) in unit vector form.

Q2 Give $\overrightarrow{GH}$ and $\overrightarrow{HG}$ as column vectors, where

$\overrightarrow{OG} = \begin{pmatrix} 2 \\ -3 \\ 4 \end{pmatrix}$ and $\overrightarrow{OH} = \begin{pmatrix} -1 \\ 4 \\ 9 \end{pmatrix}$.

Q3 Triangle JKL has vertices at the points J (4, 0, –3), K (–1, 3, 0) and L (2, 2, 7). Find the vectors $\overrightarrow{JK}$, $\overrightarrow{KL}$ and $\overrightarrow{LJ}$.

Q3 Hint: The question doesn't mention **i** and **j** components or column vectors so you can answer it using either.

Q4 A 3D printer is being used to make the plastic toy sketched below.

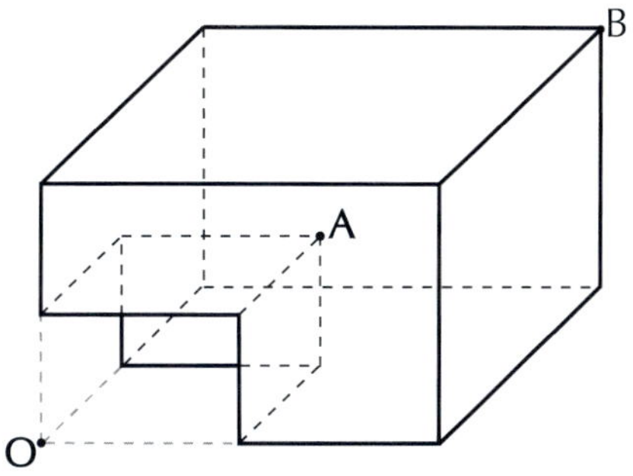

The toy can be modelled as a large cuboid with a smaller cuboid section missing. The large cuboid section is an enlargement of the smaller, 'missing' cuboid, by a scale factor of 2, centred at point O.

a) Prove that $\overrightarrow{AB} = \overrightarrow{OA}$.

b) Find the position vector of point B, if $\overrightarrow{OA} = 3\mathbf{i} + \mathbf{j} + 2\mathbf{k}$.

Q5 M is a point on the line CD, where C has coordinates (–1, 3, –5), M has coordinates (1, 1, –2) and $\overrightarrow{CD} = \begin{pmatrix} 4 \\ -4 \\ 6 \end{pmatrix}$.

Show that M is the midpoint of CD.

Q6 Show that vectors $\mathbf{a} = \frac{3}{4}\mathbf{i} + \frac{1}{3}\mathbf{j} - 2\mathbf{k}$ and $\mathbf{b} = \frac{1}{4}\mathbf{i} + \mathbf{j} - \frac{2}{3}\mathbf{k}$ are <u>not</u> parallel.

Q7 In a 3D board game, players take turns to position counters at points inside a cuboid grid. Players score by forming a straight line with three of their counters, unblocked by their opponent.

Show that counters placed at coordinates (1, 0, 3), (3, 1, 2) and (7, 3, 0) lie on a straight line.

2. Calculating with Vectors

Learning Objectives:

- Be able to find the magnitude of any vector in three dimensions.
- Be able to find the unit vector in the direction of any vector in three dimensions.
- Be able to find the distance between any two three-dimensional points using vectors.
- Be able to calculate the angle between any two vectors in three dimensions using trigonometry.

Calculating with 2D vectors should already be familiar to you, and 3D vector calculations are pretty similar. Magnitude is a scalar quantity that tells you a vector's length. You can use Pythagoras' theorem and trigonometry to find the magnitude of any vector.

Calculating with 3D vectors

Magnitude of a 3D vector

- The **magnitude** (sometimes called **modulus**) of a vector **v** is written as |**v**|, and for a vector $\overrightarrow{AB}$ it is written $|\overrightarrow{AB}|$.
- Magnitude is always a **positive**, **scalar** quantity.

You will have used **Pythagoras' theorem** to find the length of a 2D vector, and it works in **three dimensions** too:

The **distance** of point ***(a, b, c)*** from the origin is: $\sqrt{a^2+b^2+c^2}$

You can **derive** this formula from the **two-dimensional** Pythagoras' theorem:

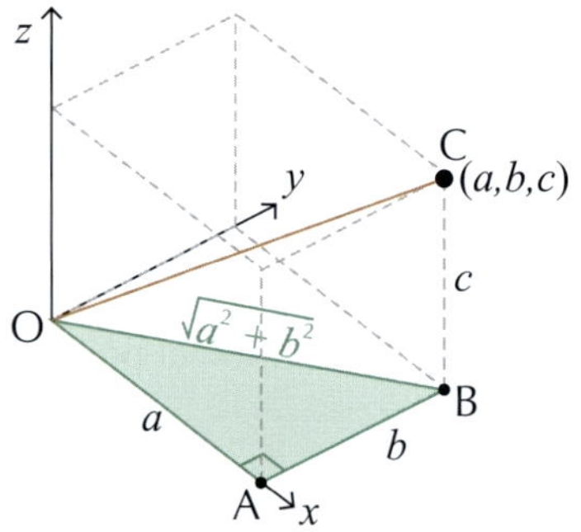

- The **orange** line shows the distance of the point (a, b, c) from the origin.
- First use Pythagoras on the **green triangle**.
- OB is $\sqrt{a^2+b^2}$.

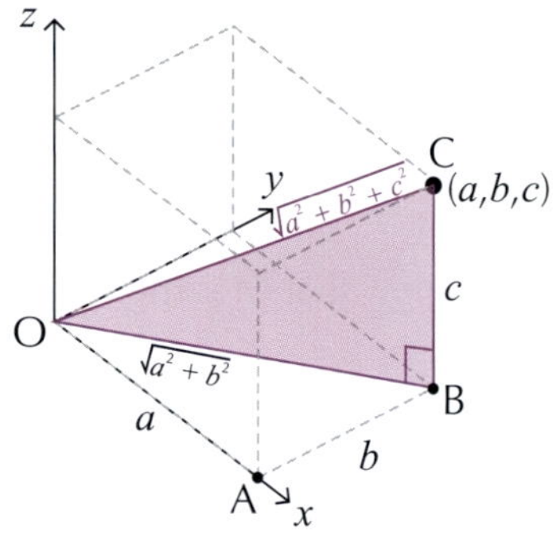

- Now use Pythagoras on the **purple triangle**.
- OC is the distance from the origin to (a, b, c).
- So the distance from the origin to (a, b, c) is: $\sqrt{(\sqrt{a^2+b^2})^2+c^2} = \sqrt{a^2+b^2+c^2}$

Tip: Remember the position vector of the point (a, b, c) is $a\mathbf{i} + b\mathbf{j} + c\mathbf{k}$, or as a column vector: $\begin{pmatrix} a \\ b \\ c \end{pmatrix}$

So the **magnitude** of any vector $\mathbf{v} = a\mathbf{i} + b\mathbf{j} + c\mathbf{k}$ is the same as the distance of point (a, b, c) from the origin:

$$|\mathbf{v}| = |a\mathbf{i} + b\mathbf{j} + c\mathbf{k}| = \sqrt{a^2+b^2+c^2}$$

The **unit vector** in the direction of **v** is $\frac{\mathbf{v}}{|\mathbf{v}|}$, as it is for 2D vectors.

Tip: A unit vector has a magnitude of 1. Magnitude is always a positive scalar, so the unit vector is always parallel to the vector, and has the same direction.

Example 1

The diagram below shows the position of point Q.

a) Find $|\overrightarrow{OQ}|$, giving your answer in reduced surd form.

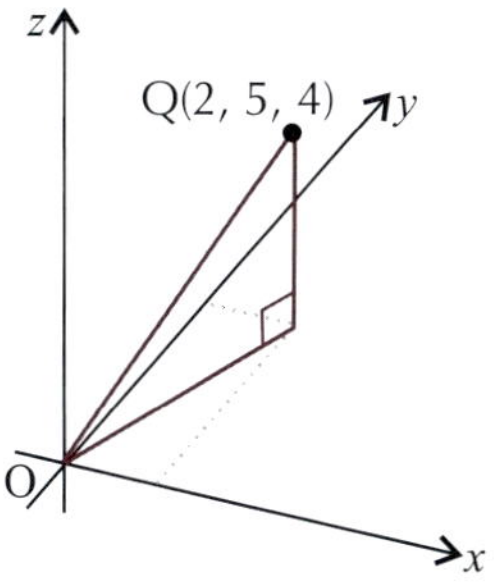

$\overrightarrow{OQ} = 2\mathbf{i} + 5\mathbf{j} + 4\mathbf{k}$

Plug the coordinates, which are the **i**, **j** and **k** coefficients in $\overrightarrow{OQ}$, into the formula:

$$|\overrightarrow{OQ}| = \sqrt{x^2 + y^2 + z^2} = \sqrt{2^2 + 5^2 + 4^2} = \sqrt{45} = \mathbf{3\sqrt{5}}$$

b) Find the unit vector in the direction of $\overrightarrow{OQ}$.

To find a **unit vector**, divide the vector by its magnitude:

$$\frac{\overrightarrow{OQ}}{|\overrightarrow{OQ}|} = \frac{1}{3\sqrt{5}}(2\mathbf{i} + 5\mathbf{j} + 4\mathbf{k})$$

$$= \frac{2}{3\sqrt{5}}\mathbf{i} + \frac{5}{3\sqrt{5}}\mathbf{j} + \frac{4}{3\sqrt{5}}\mathbf{k}$$

$$= \frac{2\sqrt{5}}{15}\mathbf{i} + \frac{\sqrt{5}}{3}\mathbf{j} + \frac{4\sqrt{5}}{15}\mathbf{k}$$

Tip: There's no need to draw a triangle out to use Pythagoras — you can just plug the coordinates into the formula.

Distance between two points in three dimensions

- The **distance** between any two points $P(x_1, y_1, z_1)$ and $Q(x_2, y_2, z_2)$ is the **magnitude** of the vector $\overrightarrow{PQ}$.
- $\overrightarrow{PQ} = \overrightarrow{OQ} - \overrightarrow{OP} = (x_2\mathbf{i} + y_2\mathbf{j} + z_2\mathbf{k}) - (x_1\mathbf{i} + y_1\mathbf{j} + z_1\mathbf{k})$
  $= (x_2 - x_1)\mathbf{i} + (y_2 - y_1)\mathbf{j} + (z_2 - z_1)\mathbf{k}$

So the magnitude of the vector $\overrightarrow{PQ} = (x_2 - x_1)\mathbf{i} + (y_2 - y_1)\mathbf{j} + (z_2 - z_1)\mathbf{k}$ is $\sqrt{(x_2 - x_1)^2 + (y_2 - y_1)^2 + (z_2 - z_1)^2}$, by Pythagoras' theorem.

Tip: To see visually why this formula works, draw a diagram with two right-angled triangles like the one on the previous page. The triangles will have sides whose lengths are the distances between the two points in the x, y and z directions.

Example 2

The position vector of point A is 3i + 2j + 4k, and the position vector of point B is 2i + 6j – 5k.

Find $|\overrightarrow{AB}|$ to 1 decimal place.

- A has the coordinates (3, 2, 4), B has the coordinates (2, 6, –5).
- $\overrightarrow{AB} = (x_2 - x_1)\mathbf{i} + (y_2 - y_1)\mathbf{j} + (z_2 - z_1)\mathbf{k}$
  $= (2 - 3)\mathbf{i} + (6 - 2)\mathbf{j} + (-5 - 4)\mathbf{k}$
- $|\overrightarrow{AB}| = \sqrt{(x_2 - x_1)^2 + (y_2 - y_1)^2 + (z_2 - z_1)^2}$
  $= \sqrt{(2-3)^2 + (6-2)^2 + (-5-4)^2}$
  $= \sqrt{1 + 16 + 81} = \sqrt{98} = \mathbf{9.9}$ (1 d.p.)

Example 3

A = (–3, –6, 4), B = (2*t*, 1, –*t*). $|\overrightarrow{AB}| = 3\sqrt{11}$.
Find the possible values of *t*.

$|\overrightarrow{AB}| = \sqrt{(2t+3)^2 + (1+6)^2 + (-t-4)^2}$

So: $\sqrt{(2t+3)^2 + (1+6)^2 + (-t-4)^2} = 3\sqrt{11}$

$\Rightarrow 4t^2 + 12t + 9 + 49 + t^2 + 8t + 16 = 99$

$\Rightarrow 5t^2 + 20t + 74 = 99$

$\Rightarrow 5t^2 + 20t - 25 = 0$

$\Rightarrow t^2 + 4t - 5 = 0$

$\Rightarrow (t+5)(t-1) = 0$

$\Rightarrow$ **$t = -5$ or $t = 1$**

The angle between two vectors

To find the angle between two vectors:

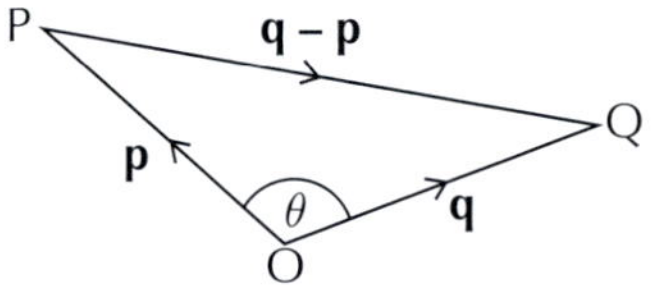

- Create a **triangle** with the vectors as two sides, and angle θ between them.
- Find the **magnitude** (i.e. the length) of each side of the triangle.
- Use the **cosine rule** to find the angle θ from these lengths.

Tip: The cosine rule is usually given as: $a^2 = b^2 + c^2 - 2bc\cos A$. But, if you're finding angles, it's best to rearrange it:
$\cos A = \frac{b^2 + c^2 - a^2}{2bc}$

See your Year 1 notes if you need a reminder.

Example 4

The flight path of two different aeroplanes taking off from a runway are modelled as position vectors a = 5i –2j + k, and b = –3i + 3j + k. Find the angle between the two flight paths, in degrees, to 1 d.p.

MODELLING

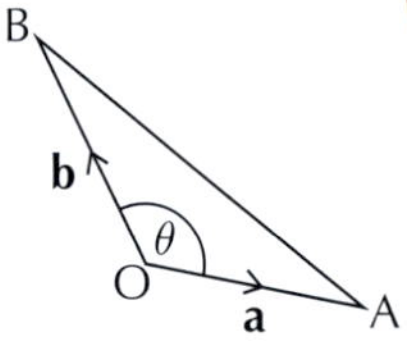

- **a** and **b** form two sides of the triangle AOB. The lengths of these sides are:

  $|\mathbf{a}| = \sqrt{5^2 + (-2)^2 + 1^2} = \sqrt{30}$

  $|\mathbf{b}| = \sqrt{(-3)^2 + 3^2 + 1^2} = \sqrt{19}$

- $\overrightarrow{AB} = \mathbf{b} - \mathbf{a} = -8\mathbf{i} + 5\mathbf{j} + 0\mathbf{k}$

- $|\overrightarrow{AB}| = \sqrt{(-8)^2 + 5^2 + 0^2} = \sqrt{89}$

- Using the cosine rule:

  $\cos A = \frac{b^2 + c^2 - a^2}{2bc}$

  $\cos\theta = \frac{(\sqrt{30})^2 + (\sqrt{19})^2 - (\sqrt{89})^2}{2 \times \sqrt{30} \times \sqrt{19}} = \frac{-40}{2\sqrt{570}}$

- So $\theta = \cos^{-1}\left(\frac{-40}{2\sqrt{570}}\right) =$ **146.9°** (1 d.p.)

Exercise 2.1

Unless specified, give each answer in this exercise as an integer or as a simplified surd.

Q1 Find the magnitude of each of the following vectors:

a) $\mathbf{i} + 4\mathbf{j} + 8\mathbf{k}$ b) $\begin{pmatrix} 4 \\ 2 \\ 4 \end{pmatrix}$ c) $\begin{pmatrix} -4 \\ -5 \\ 20 \end{pmatrix}$ d) $7\mathbf{i} + \mathbf{j} - 7\mathbf{k}$ e) $\begin{pmatrix} -2 \\ 4 \\ -6 \end{pmatrix}$

Q2 Find the magnitude of the resultant of each pair of vectors.

a) $\mathbf{i} + \mathbf{j} + 2\mathbf{k}$ and $\mathbf{i} + 2\mathbf{j} + 4\mathbf{k}$ b) $2\mathbf{i} + 11\mathbf{j} + 25\mathbf{k}$ and $3\mathbf{j} - 2\mathbf{k}$

c) $\begin{pmatrix} 4 \\ 2 \\ 8 \end{pmatrix}$ and $\begin{pmatrix} -2 \\ 4 \\ 1 \end{pmatrix}$ d) $\begin{pmatrix} 3 \\ 0 \\ 10 \end{pmatrix}$ and $\begin{pmatrix} -1 \\ 5 \\ 4 \end{pmatrix}$ e) $\begin{pmatrix} 8 \\ 4 \\ 10 \end{pmatrix}$ and $\begin{pmatrix} 2 \\ -2 \\ 4 \end{pmatrix}$

Q3 Find the distances between each of the following pairs of points:

a) (3, 4, 5), (5, 6, 6) b) (7, 2, 9), (–11, 1, 15)

c) (10, –2, –1), (6, 10, –4) d) (0, –4, 10), (7, 0, 14)

e) (–4, 7, 10), (2, 4, –12) f) (7, –1, 4), (30, 9, –6)

Q4 Hint: The distance is the length of $2\mathbf{m} - \mathbf{n}$, i.e. its magnitude.

Q4 The flight path of a toy aeroplane is modelled by the vector $2\mathbf{m} - \mathbf{n}$, where $\mathbf{m} = \begin{pmatrix} -5 \\ -2 \\ 6 \end{pmatrix}$ metres, and $\mathbf{n} = \begin{pmatrix} -4 \\ 1 \\ 2 \end{pmatrix}$ metres.

MODELLING

Find the distance of the plane's destination from its starting position. Give your answer in metres to 1 decimal place.

Q5 $\overrightarrow{OA} = \mathbf{i} - 4\mathbf{j} + 3\mathbf{k}$ and $\overrightarrow{OB} = -\mathbf{i} - 3\mathbf{j} + 5\mathbf{k}$. Find $|\overrightarrow{AO}|$, $|\overrightarrow{BO}|$ and $|\overrightarrow{BA}|$. Show that triangle AOB is right-angled.

PROBLEM SOLVING

Q6 Find the unit vector in the direction of $\mathbf{v}$, where $\mathbf{v} = 4\mathbf{i} - 4\mathbf{j} - 7\mathbf{k}$.

Q7 P is the point (2, –1, 4) and Q is the point $(q - 2, 5, 2q + 1)$. Given that the length of the line PQ is 11, find the possible coordinates of the point Q.

Q8 Find the angle, in degrees to 1 d.p., that the vector $\begin{pmatrix} 1 \\ 3 \\ 2 \end{pmatrix}$ makes with:

a) vector $\begin{pmatrix} 3 \\ 2 \\ 1 \end{pmatrix}$,

b) the unit vector $\mathbf{j}$.

Q8b) Hint: Write this as a column vector first.

Q9 A toy rocket is launched at an angle of 60° to the unit vector in the $\mathbf{i}$ direction, with a velocity $\mathbf{v} = (\mathbf{i} + a\mathbf{j} + \mathbf{k})$ ms^{-1}. Find the possible values of a, in surd form, and hence the launch speed of the rocket.

PROBLEM SOLVING

Chapter 11 Probability

1. Conditional Probability

Learning Objectives:
- Be able to use set notation to represent probabilities.
- Be able to use the conditional probability formula.
- Be able to use Venn diagrams, tree diagrams and two-way tables to find conditional probabilities.

A lot of the probability in this chapter is stuff you've seen before, just using different notation. What's new in Year 2 is conditional probability, which is the probability of an event happening, given that another event happens.

Set notation

In Year 1, you saw some of the laws of probability, such as the **addition law**: **P(A or B) = P(A) + P(B) – P(A and B)**.
To avoid having to write "and", "or" and "not" all the time, you can use **set notation** to describe probabilities:

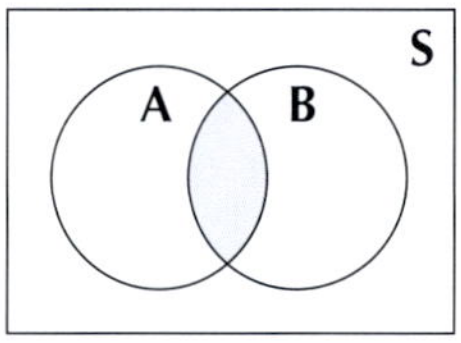

P(A and B) = P(A ∩ B)

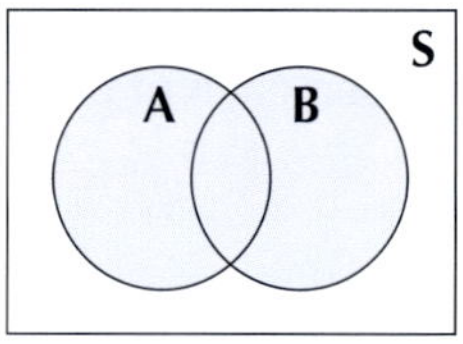

P(A or B) = P(A ∪ B)

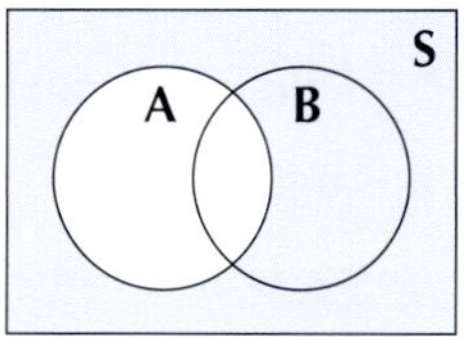

P(not A) = P(A′)

So, in this notation, the **addition law** would be written:

P(A ∪ B) = P(A) + P(B) – P(A ∩ B)

You should also know that:

P(A′) = 1 – P(A)
P(A) = P(A ∩ B) + P(A ∩ B′)

For **mutually exclusive** events,

P(A ∩ B) = 0 ⇒ P(A ∪ B) = P(A) + P(B)

and for **independent** events,

P(A ∩ B) = P(A) × P(B)

Tip: ∪ means the **union** of two sets (everything that is in either set — i.e. the outcomes corresponding to either event A or B or both happening), and ∩ means the **intersection** (only the things that are in both — i.e. the outcomes corresponding to A and B both happening).

Tip: Mutually exclusive means that the two events can't both happen at the same time (like rolling a 3 and rolling a 4 at the same time on the same dice). Events are **independent** if their probabilities are not affected by whether or not the other event happens (like rolling a 3 then rolling a 4 on a dice — the result of the first roll has no effect on the second).

Example

For two independent events A and B, P(A) = 0.5 and P(A ∪ B) = 0.6. Find P(B) and P(A ∩ B).

- Use the addition law:

  P(A ∪ B) = P(A) + P(B) – P(A ∩ B)
  0.6 = 0.5 + P(B) – P(A ∩ B)
  0.1 = P(B) – P(A ∩ B)

- Since the events are independent, you know that P(A ∩ B) = P(A)P(B).

  0.1 = P(B) – P(A)P(B)
  0.1 = P(B) – 0.5P(B) ⇒ 0.1 = 0.5P(B) ⇒ P(B) = 0.2

- Finally, you can use the formula for independent events to find P(A ∩ B):

  P(A ∩ B) = P(A)P(B) = 0.5 × 0.2 = 0.1

Conditional probability

A probability is **conditional** if it **depends** on what has already happened. The probability that an event B happens, **given** that an event A has already happened, is called the **conditional probability** of 'B given A', written **P(B | A)**.

For example, if you picked two cards from a deck **without replacement**, the probability of the second card being a heart would be **conditional** on whether or not the first card was a heart.

You can work out the probability of **B given A**, using this formula:

$$\mathbf{P(B|A) = \frac{P(A \cap B)}{P(A)}}$$

Tip: The probability of A given B is:

$$P(A|B) = \frac{P(A \cap B)}{P(B)}$$

Here's an explanation of where this formula comes from...

Event A has happened and for B|A to happen, B will also happen.

- If you know that A has already happened, then the only remaining possible outcomes must be the ones corresponding to A.

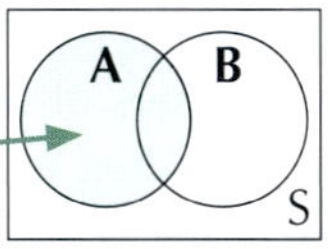

- And the only remaining possible outcomes corresponding to B also happening must be the ones in A ∩ B.

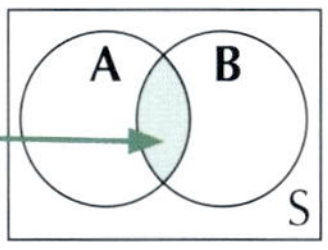

- Using the **probability formula** (and assuming event **A** has **already happened**):

$$P(B|A) = \frac{\text{number of possible outcomes corresponding to B}}{\text{total number of possible outcomes}}$$

$$= \frac{\text{number of outcomes in } A \cap B}{\text{number of outcomes in A}}$$

- So if $n(A \cap B)$ = number of outcomes in A ∩ B, and $n(A)$ = number of outcomes in A:

$$P(B|A) = \frac{n(A \cap B)}{n(A)}$$

- Now, if you divide the top and bottom of the fraction by $n(S)$ (the total number of outcomes in S), its value doesn't change, but you can write it in a different way:

$$\mathbf{P(B|A)} = \frac{n(A \cap B)}{n(A)} = \frac{n(A \cap B)/n(S)}{n(A)/n(S)} = \frac{\mathbf{P(A \cap B)}}{\mathbf{P(A)}}$$

← the formula in the box above

And if you rearrange this formula for conditional probability, you get a formula that's known as the **product law**. For events A and B:

$$\mathbf{P(A \cap B) = P(A)P(B|A)}$$

Tip: The product law is given in this form in the formula booklet.

- So to find the probability that A **and** B **both** happen, you multiply the probability of A by the probability of B given that A has happened.
- Or you can write this the other way around, by swapping A and B:

$$\mathbf{P(A \cap B) = P(B)P(A|B)}$$

Example 1

For events A and B: $P(A) = 0.6$, $P(B) = 0.5$, $P(A \cap B) = 0.3$, $P(B' | A) = 0.5$.

a) Find $P(A | B)$.

Using the formula for conditional probability:

$$P(A|B) = \frac{P(A \cap B)}{P(B)} = \frac{0.3}{0.5} = 0.6$$

b) Find $P(A \cap B')$.

Using the product law: $P(A \cap B') = P(A)P(B'|A)$
$= 0.6 \times 0.5 = 0.3$

Tip: Just replace B with B′ in the product law.

Example 2

The Venn diagram represents two events, C and D. The numbers of equally likely outcomes corresponding to the events are shown.

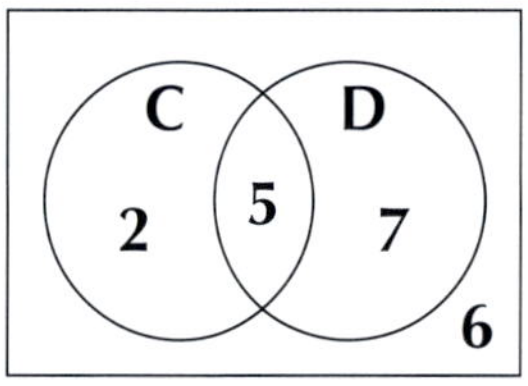

a) Find $P(C | D)$.

- Here you're given the numbers of outcomes, so you have to find the probability of the events you're interested in.
- Once event **D has happened**, there are **5** possible outcomes corresponding to event **C** happening and **12** possible remaining outcomes in **total**. (5 outcomes in $C \cap D$)
- So, $P(C|D) = \frac{\text{number of outcomes corresponding to C}}{\text{total number of possible outcomes}} = \frac{5}{12}$

b) Find $P(D | C')$.

- Once event **C hasn't happened**, there are **7** possible outcomes corresponding to event **D** happening and **13** possible remaining outcomes in **total**. (7 outcomes in $D \cap C'$)
- So, $P(D|C') = \frac{\text{number of outcomes corresponding to D}}{\text{total number of possible outcomes}} = \frac{7}{13}$

Tip: To find the probability of an event, divide the number of equally likely outcomes corresponding to that event by the total number of possible outcomes.

Tip: Doing it this way is a bit simpler than using the conditional probability formula. But if you did use the formula, you'd work out:

$P(C \cap D) = \frac{5}{20}$ and $P(D) = \frac{12}{20}$

And so, $P(C|D) = \frac{P(C \cap D)}{P(D)} = \frac{\frac{5}{20}}{\frac{12}{20}} = \frac{5}{12}$

Example 3

MODELLING

Vikram either walks or runs to the bus stop. The probability that he walks is 0.4. The probability that he catches the bus is 0.54. If he walks to the bus stop, the probability that he catches the bus is 0.3.

a) Draw a Venn diagram representing the events W, 'Vikram walks to the bus stop', and C, 'Vikram catches the bus'.

- Okay, this one looks a bit scary, but it'll make more sense when you write down the probabilities you know:
  $P(W) = 0.4$, $P(C) = 0.54$ and $P(C|W) = 0.3$

Tip: The probability that Vikram catches the bus, given that he walks to the bus stop, is 0.3.

- To draw the Venn diagram, you need to find the intersection $C \cap W$. And you can do that using the **product law**:

  $P(C \cap W) = P(C|W)P(W) = 0.3 \times 0.4 = 0.12$

- Now you can draw the Venn diagram:

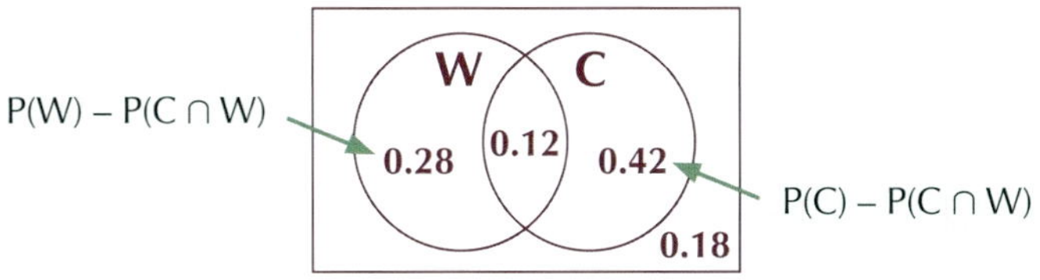

b) Find the probability that Vikram catches the bus, given that he runs to the bus stop.

- This is the probability $P(C|W')$.
- Using the formula for conditional probability: $P(C|W') = \frac{P(C \cap W')}{P(W')}$
- $P(C \cap W') = 0.42$ and $P(W') = 1 - 0.4 = 0.6$
- So, $P(C|W') = \frac{P(C \cap W')}{P(W')} = \frac{0.42}{0.6} = 0.7$
- So P(Vikram catches the bus, given that he runs to the bus stop) = 0.7

Tip: He either walks or runs to the bus stop, so P(runs) = $P(W')$.

Exercise 1.1

Q1 Given that $P(A) = 0.3$, $P(B) = 0.5$ and $P(A \cup B) = 0.7$, find:

a) $P(A')$ b) $P(A \cap B)$ c) $P(A \cap B')$

Q2 The events X and Y are independent, where $P(X) = 0.8$ and $P(Y) = 0.15$. Find the probability that neither X nor Y occurs.

Q2 Hint: You can use $P(A \cup B) + P(A' \cap B') = 1$ — you can see this from the Venn diagram below:

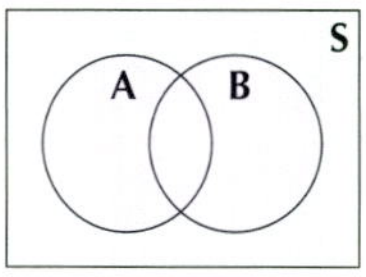

Q3 If $P(G) = 0.7$, $P(H) = 0.63$ and $P(G \cap H) = 0.24$, find:

a) $P(G|H)$ b) $P(H|G)$

Q4 $P(A) = 0.68$, $P(B') = 0.44$, $P(C) = 0.44$, $P(A \cap B) = 0.34$, $P(A \cap C) = 0.16$ and $P(B \cap C') = 0.49$. Find:

a) $P(B|A)$ b) $P(A|C)$ c) $P(C'|B)$

Q5 Events J and K are such that $P(J) = 0.4$, $P(J|K) = 0.64$ and $P(K|J) = 0.2$.

a) Find $P(J \cup K)$. b) Find $P(J' \cap K')$.

Q6 In a group of eleven footballers, five are over 6 feet tall. Two of the three players who can play in goal are over 6 feet tall. One of the players is selected at random.

a) If the player is over 6 feet tall, what is the probability that they can play in goal?

b) If the player can play in goal, what is the probability that they are over 6 feet tall?

Q7 The Venn diagram below shows the numbers of students studying Maths, English and Art, from a group of 100 students. One of the students is selected at random.

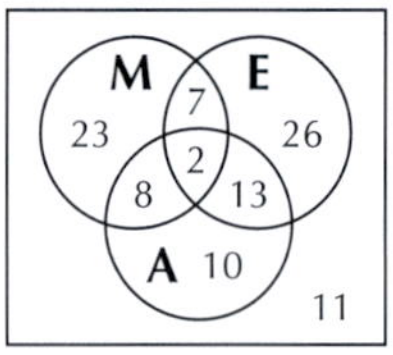

a) If the student is studying Art, what is the probability that they are also studying Maths?

b) If the student is studying English and Maths, what is the probability that they are also studying Art?

c) If the student is not studying Maths, what is the probability that they are studying English?

d) Find $P(A|E')$.

e) Find $P(M|A \cap E)$.

Q8 Given that $P(X) = 0.44$, $P(Y') = 0.72$, $P(Z) = 0.61$, $P(X|Y) = 0.75$, $P(Z|X) = 0.25$, $P(Y \cap Z') = 0.2$ and $P(X \cap Y \mid Z) = \frac{7}{61}$, find:

a) $P(Y)$

b) $P(X \cap Y)$

c) $P(X \cap Z)$

d) $P(Y|Z')$

e) $P(X \cap Y \cap Z)$

Using conditional probability

Independent events

If A and B are **independent**, then P(B) is the same, whether A happens or not. So the probability of B, given that A happens, is just the probability of B. This means you have the following results:

- $\mathbf{P(B \mid A) = P(B \mid A') = P(B)}$. Similarly, $\mathbf{P(A \mid B) = P(A \mid B') = P(A)}$.
- The **conditional probability** formula becomes: $\mathbf{P(B \mid A) = P(B) = \frac{P(A \cap B)}{P(A)}}$
- The **product law** becomes: $\mathbf{P(A \cap B)} = P(A)P(B|A) = \mathbf{P(A)P(B)}$ (this is the formula for independent events that you've already met).

To show that events A and B are independent, you just need to show that **one** of the following statements is true:

- $P(B|A) = P(B)$ [or $P(A|B) = P(A)$]
- $P(A) \times P(B) = P(A \cap B)$

Example 1

For events A and B, $P(A) = 0.25$, $P(B \mid A) = 0.8$ and $P(A' \cap B) = 0.4$.

a) Find: (i) $P(A \cap B)$, (ii) $P(A')$, (iii) $P(B' \mid A)$, (iv) $(B \mid A')$, (v) $P(B)$, (vi) $P(A \mid B)$.

(i) Using the **product law**: $\mathbf{P(A \cap B)} = P(A)P(B|A) = 0.25 \times 0.8 = 0.2$

(ii) $\mathbf{P(A')} = 1 - P(A) = 1 - 0.25 = 0.75$

Now at this stage, things are starting to get a bit trickier. It'll help a lot if you draw a **Venn diagram** showing what you know so far.

- You know from the question that:
  $P(A) = 0.25$, $P(B|A) = 0.8$ and $P(A' \cap B) = 0.4$
- And you've found:
  $P(A \cap B) = 0.2$ and $P(A') = 0.75$

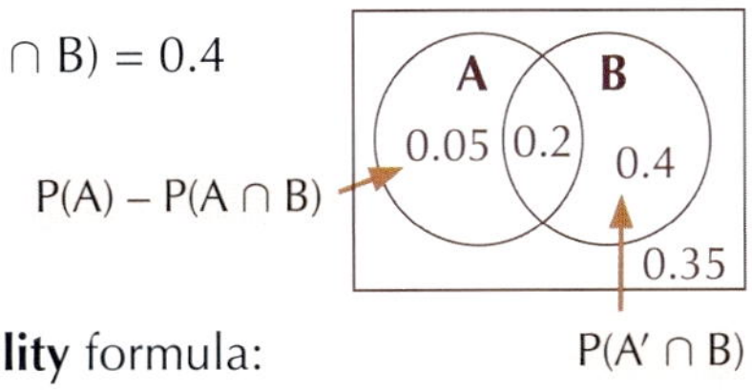

(iii) Using the **conditional probability** formula:

$$P(B'|A) = \frac{P(B' \cap A)}{P(A)}$$

You can see from the Venn diagram that $P(B' \cap A) = 0.05$, so:

$$\mathbf{P(B' | A)} = \frac{P(B' \cap A)}{P(A)} = \frac{0.05}{0.25} = 0.2$$

(iv) Using the **conditional probability** formula:

$$\mathbf{P(B | A')} = \frac{P(B \cap A')}{P(A')} = \frac{0.4}{0.75} = \frac{40}{75} = \frac{8}{15}$$

(v) You can see from the Venn diagram that:

$$\mathbf{P(B)} = P(B \cap A) + P(B \cap A') = 0.2 + 0.4 = 0.6$$

And finally...

(vi) $\mathbf{P(A | B)} = \frac{P(A \cap B)}{P(B)} = \frac{0.2}{0.6} = \frac{1}{3}$

b) Say whether or not A and B are independent.

There are different ways you can do this. For example, if you compare the values of $P(B|A)$ and $P(B)$, you can see that:

$P(B|A) = 0.8 \neq P(B) = 0.6$, so A and B are not independent.

Tip: Another way of getting $P(B'|A)$ is to do $1 - P(B|A)$, since $B|A$ is the complement of $B'|A$.

Tip: You can also find $P(B)$ by doing:
$P(B|A)P(A) + P(B|A')P(A')$
That's because
$P(B|A)P(A) = P(B \cap A)$,
$P(B|A')P(A') = P(B \cap A')$
And
$P(B \cap A) + P(B \cap A') = P(B)$

Tip: Or you can show:
$P(A|B) \neq P(A)$, or
$P(A) \times P(B) \neq P(A \cap B)$.

Tree diagrams

You may not have realised it when you saw them in Year 1, but **tree diagrams** are actually all about conditional probability. You can label the branches of a tree diagram using **set notation** like this:

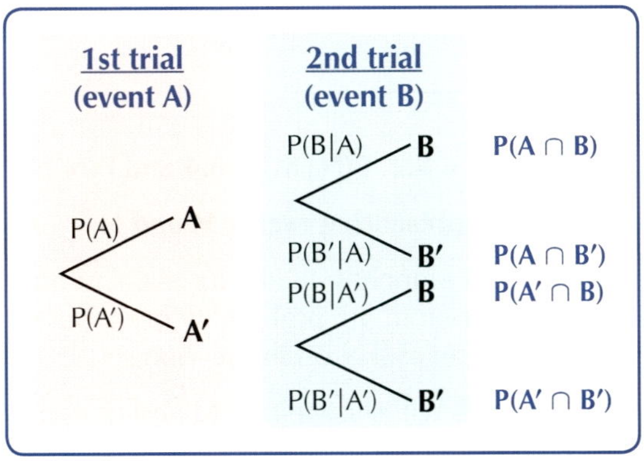

Tree diagrams are really useful for seeing how the **result** of one event **affects** the probability of the other event happening.

Tip: If the events are independent, then the probabilities for the second trial are just $P(B)$ and $P(B')$ — it doesn't matter whether or not A has happened.

Example 2

Horace is either late for school or on time for school, and when he gets to school he is either shouted at or not shouted at. The probability that he's late for school is 0.4. If he's late, the probability that he's shouted at is 0.7. If he's on time, the probability that he's shouted at is 0.2.

MODELLING

Given that Horace is shouted at, what is the probability that he was late?

- It's best to take complicated questions like this step by step. You're given information about two events: Let **L** = 'Horace is late' and let **S** = 'Horace is shouted at'.
- Start by writing down the probability you want to find: that's **P(L | S)**.

  Using the conditional probability formula: $P(L|S) = \frac{P(L \cap S)}{P(S)}$

- So you need to find $P(L \cap S)$ and $P(S)$ — and the easiest way is by drawing a **tree diagram** using the information in the question:

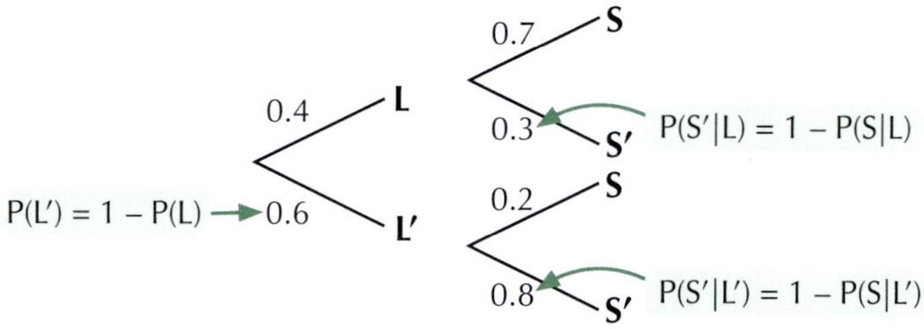

- Using the tree diagram:

  $\mathbf{P(L \cap S)} = P(L)P(S|L) = 0.4 \times 0.7 = \mathbf{0.28}$

  $\mathbf{P(S)} = P(L \cap S) + P(L' \cap S) = 0.28 + P(L')(S|L')$
  $= 0.28 + 0.6 \times 0.2$
  $= \mathbf{0.4}$

- So $P(L|S) = \frac{P(L \cap S)}{P(S)} = \frac{0.28}{0.4} = 0.7$
- This means P(Horace was late, given he is shouted at) = 0.7

Tip: Be careful with questions like this — the question tells you the probability of S conditional on L (and L′). But you need to think of the situation the 'other way round' — with L conditional on S. So don't just rush in.

Tip: In general, if B depends on A, then A depends on B. Here, S depends on L, so L depends on S.

Example 3

For events M and N: P(M) = 0.2, P(N | M) = 0.4 and P(N′ | M′) = 0.7.

a) Draw a tree diagram representing events M and N.

- You need **two** sets of branches — one for event M and one for event N. Since you're told the probability of M, but not that of N, show **M** on the **first** set and N on the second set.
- Each pair of branches will show either **M** and its complement **M′**, or **N** and its complement **N′**.

- Draw the tree diagram and write on as much as you know so far:

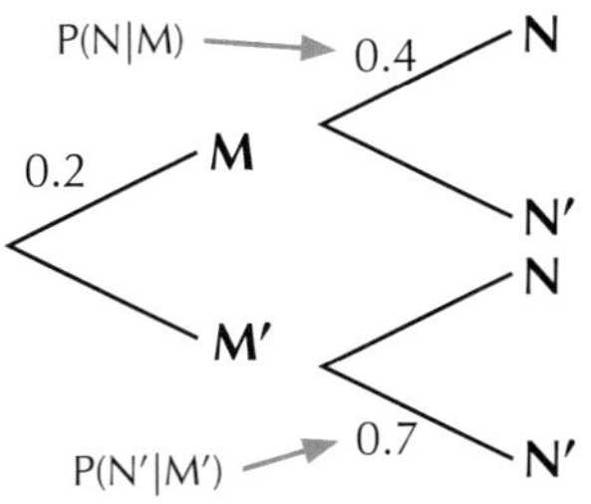

- You can find the probabilities of the remaining branches by doing 1 – the probability on the other branch:

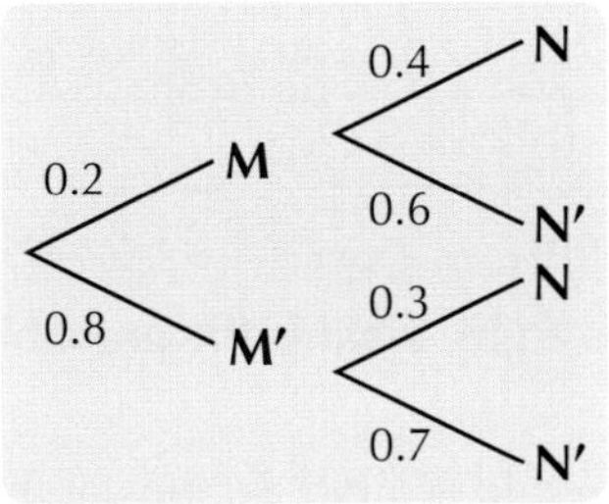

b) Find P(N).

$P(N) = P(M \cap N) + P(M' \cap N) = P(M)P(N|M) + P(M')P(N|M')$

$= 0.2 \times 0.4 + 0.8 \times 0.3$

$= 0.32$

Tip: There are 2 'paths' giving N, so find the probability of each, then add them together.

c) Find P(M′ | N′).

- Remember, if N′ is conditional on M′, then M′ is conditional on N′.
- Using the conditional probability formula: $P(M'|N') = \frac{P(M' \cap N')}{P(N')}$
- Using the probabilities you found above:

  P(N′) = 1 – P(N) = **0.68**

  P(M′ ∩ N′) = P(M′)P(N′|M′) = 0.8 × 0.7 = **0.56**

- So, $P(M'|N') = \frac{P(M' \cap N')}{P(N')} = \frac{0.56}{0.68} = \frac{56}{68} = \frac{14}{17}$

Two-way tables

Probabilities can also be shown in **two-way tables**, like this:

	A	A′	Total
B	P(A ∩ B)	P(A′ ∩ B)	P(B)
B′	P(A ∩ B′)	P(A′ ∩ B′)	P(B′)
Total	P(A)	P(A′)	1

Tip: You might also see tables with numbers of objects or equally-likely outcomes instead of probabilities (like in the example on the next page).

Two-way tables can make finding conditional probabilities quite easy — if you're trying to find a probability "given A", simply look in the A column (or row) of the table. **Dividing** the **relevant entry** by the **total** for that row or column will give you the conditional probability.

Example 4

MODELLING

22 children are at a Halloween party. The two-way table below shows the costumes that they are wearing:

	Werewolf	Zombie	Ghost	Total
Boys	4	3	3	10
Girls	2	4	6	12
Total	6	7	9	22

A child is picked at random.

a) Given that the child is a girl, what is the probability that they're dressed as a zombie?

- You're told that the child is a girl, so look at the Girls row of the table — there are 12 girls in total.
- The Zombie entry in this row is 4, so:

$$\text{P(Zombie|Girl)} = \frac{4}{12} = \frac{1}{3}$$

b) Two girls dressed as ghosts go home. Draw an updated two-way table showing the <u>probabilities</u> of selecting a child from each category.

To find the probabilities, divide every entry by the new total of 20. Don't forget to change the Ghost ∩ Girls entry to 6 – 2 = 4 and update the Girl, Ghost and overall totals before dividing.

2 ÷ 20 = 0.1

3 ÷ 20 = 0.15

	Werewolf	Zombie	Ghost	Total
Boys	0.2	0.15	0.15	0.5
Girls	0.1	0.2	0.2	0.5
Total	0.3	0.35	0.35	1

7 ÷ 20 = 0.35

20 ÷ 20 = 1

c) Another child is picked at random. Given that the child is dressed as a werewolf, what is the probability that they're a boy?

Here you're told that the child is dressed as a werewolf, so look in the Werewolf column — the Boys entry is 0.2, and the total is 0.3, so:

$$\text{P(Boy|Werewolf)} = \frac{0.2}{0.3} = \frac{2}{3}$$

Tip: It doesn't matter whether the table gives the numbers or the probabilities — you can always find the conditional probability by dividing the entry by the total.

Exercise 1.2

Q1 Events A and B have the following probabilities:
$P(A) = 0.75$, $P(B|A') = 0.4$, $P(A \cap B') = 0.45$
Are A and B independent?

Q1 Hint: It might help to draw a suitable diagram, even if the question doesn't ask for it.

Q2 Joe records the ages of people going to a cinema on a Sunday evening. He obtains the following data:

- 26 of the 40 people were under 20 years old.
- 18 people went to see a comedy film.
- Of the people who watched an action film, 6 of them were 20 years old or above.

a) Copy the two-way table below and fill it in using Joe's data.

	Action film	Comedy film	Total
Under 20			
20 and over			
Total			

b) One of the cinema-goers is picked at random. Given that they saw a comedy film, what is the probability that they are under 20?

Q3 The Venn diagram below shows the probabilities for events R, S and T:

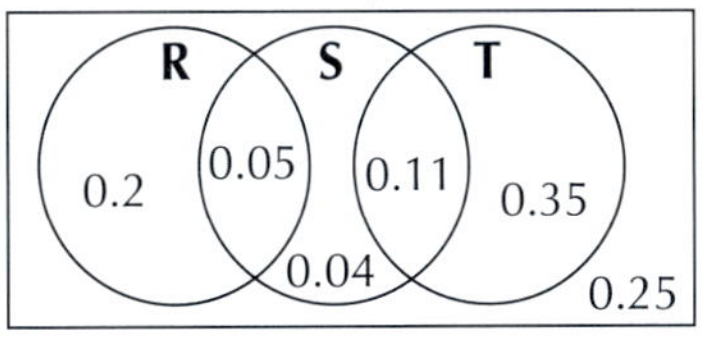

a) Which events are: (i) mutually exclusive? (ii) independent?

b) Find: (i) $P(R|S)$ (ii) $P(R|S \cap T')$

Q4 X, Y and Z are independent events, with $P(X) = 0.84$, $P(Y) = 0.68$ and $P(Z) = 0.48$. Find the following probabilities:

a) $P(X \cap Y)$ b) $P(Y' \cap Z')$ c) $P(Y|Z)$ d) $P(Z'|Y')$ e) $P(Y|X')$

Q5 Amena sometimes goes to the park, depending on the weather. If it's a sunny day, the probability that she goes to the park is 0.6. If it isn't sunny, she goes to the park with probability 0.2. The probability that it is sunny is 0.45.

a) Draw a Venn diagram to show the event that it is sunny (S) and the event that she goes to the park (P).

b) What is the probability that it's sunny, given that she goes to the park?

Q6 A group of people were asked about their mobile phones. 62% of them own a smartphone and 53.9% of them have a contract costing over £25 a month. Of the people with a smartphone, 29% have a contract costing £25 a month or less.

Use a tree diagram to find the probability that a person from the group owns a smartphone, given that their contract costs over £25 a month.

2. Modelling with Probability

Learning Objective:
- Be able to evaluate assumptions made when modelling the probability of events.

Doing calculations based on probabilities is all very well and good, but in real-life situations, the hardest part is often finding the probabilities in the first place. You usually need to make a few assumptions, which will affect the accuracy of your model.

MODELLING

Modelling assumptions and probability

In most models, the probability of an event is based on some **assumptions**. For example, when flipping a coin, you assume that each side is **equally likely** to come up, so the probability is 0.5 for both. However, it's possible that the coin is **biased**, giving the two outcomes **different probabilities**.

Evaluating and **criticising** the assumptions being made is an important part of the modelling process.

Some common issues to think about are:

- Have you assumed that two (or more) events are **equally likely**? Is this true? Could the probabilities be **biased** in some way?
- Is the probability based on **past data**? Is the data **appropriate**? How **reliable** is the data? How was the data **sampled**?
- Is the experiment itself **truly random**? Is there anything about the way that the experiment is being **carried out** that could affect the outcome?

Tip: There's more on criticising modelling assumptions on p.226 — even though it's a different topic, some of the ideas are the same.

Example

Sanaa wants to know the probability that it will rain tomorrow. She looks up the weather data for the previous 30 days, and finds that it has rained on 12 of them. She concludes that the probability that it will rain is $\frac{12}{30} = 0.4$.

Give a reason why this model might be inaccurate.

There are lots of answers you could give. For example:

- She has only taken data from the past 30 days, which might not be a large enough sample to give an accurate estimate, or might not take seasonal variations into account.
- She has assumed that the probability that it rains on one day is not affected by whether or not it rained the day before (i.e. that they are independent events) but this might not be true.

Tip: She has used the relative frequency of the event to estimate the probability — you might have seen this at GCSE.

Exercise 2.1

MODELLING

For the following models, give an assumption that has been made and explain how it may lead to inaccuracy.

Q1 Ella throws a wooden cuboid into the air. She says, "The probability that it lands on any given face is $\frac{1}{6}$."

Q2 Harriet shows a selection of 10 playing cards to a volunteer. The volunteer chooses one, replaces it and chooses another. Harriet says, "The probability of a particular card being chosen twice is 0.01."

Q3 In a class of primary school children, 56% walked to school one day. Their teacher says, "The probability that a child from any primary school walks to school on a given day is 0.56."

Chapter 12 The Normal Distribution

1. The Normal Distribution

In this section you'll be introduced to the normal distribution and use it to find different probabilities. Many variables can be modelled by normal distributions, which can be very useful in real-life situations.

Learning Objectives:

- Know the shape and properties of the normal distribution.
- Be able to find probabilities for normal distributions.
- Be able to find the range of values within which a normally distributed random variable will fall with a given probability.
- Be able to find the mean and standard deviation of a normal distribution given some probabilities for the distribution.
- Be able to apply the normal distribution to real-life situations.

The normal distribution

The shape of a normal distribution

In real-life situations, the distribution of a set of data often follows a **particular pattern** — with most of the data values falling **somewhere in the middle**, and only a small proportion taking much higher or lower values.

- For example, this histogram shows the distribution of the weights of some hedgehogs.
- Most of the weights lie close to the mean weight — with similar numbers distributed symmetrically above and below.

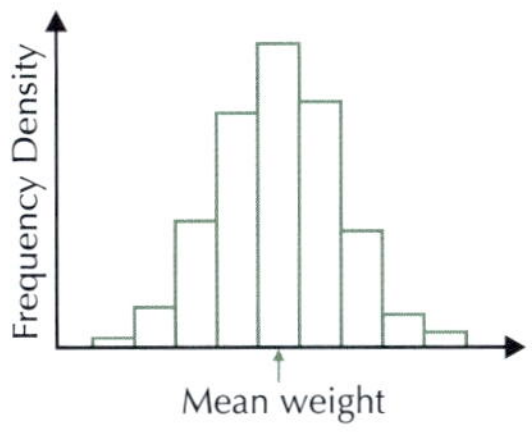

- A variable like this can often be **modelled** by a **normal distribution**.
- A normal distribution is **continuous**, so can easily model continuous variables — such as height, weight, length, etc.

If X is a **continuous random variable** that follows a **normal** distribution, you can describe the probability distribution of X using just two measures — its **mean**, μ and **variance**, σ^2.

Tip: Remember, **continuous** variables can take any value within a certain range. It's called a **random** variable because it takes different values with different probabilities.

- Whatever the values of μ and σ^2, the **graph** of a normal distribution always looks like the **curve** below.
  - The curve is **'bell-shaped'**.
  - There's a **peak** at the **mean**, μ.
  - It's **symmetrical** about the mean — so values the same distance above and below the mean are equally likely.

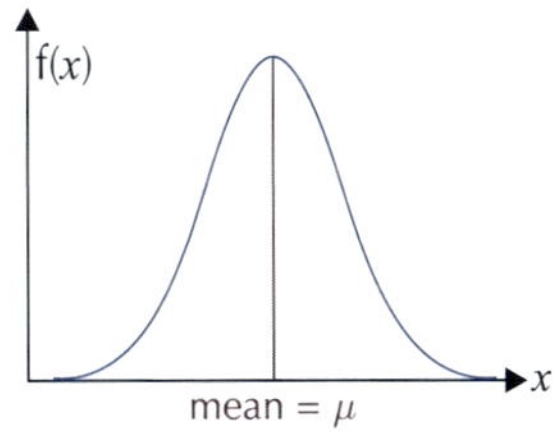

Tip: The vertical axis is labelled f(x) because the equation of the curve is a function of x.

- You can see that the shape of the normal curve **approximately fits** the shape of the histogram for hedgehog weights above. The peak at the mean reflects the fact that values close to the mean are most likely.
- As the distribution is symmetrical (i.e. not skewed), this means that **mean = median = mode**.
- The width and height of the curve depend on the **variance** of the normal distribution. The three graphs on the next page all show normal distributions with the **same mean** (μ), but **different variances** (σ^2).

Tip: As x gets further away from the mean, f(x) gets closer to 0 but never reaches it (the x-axis is an asymptote).

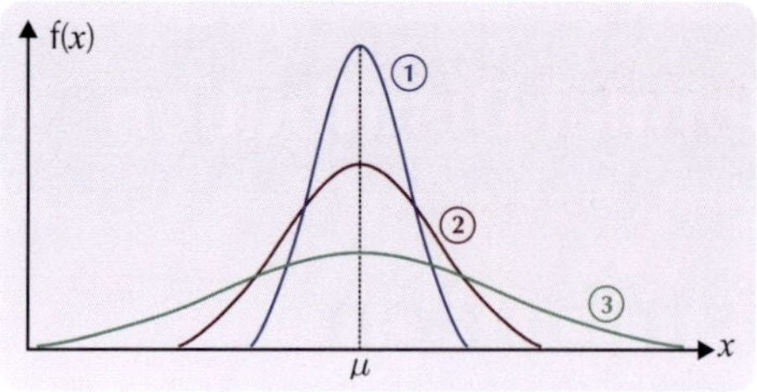

- The **larger** the variance, the **wider** the curve, so **graph 3** has the **largest** variance and **graph 1** has the **smallest** variance.
- The **total area** under the curve is always the **same**, so a wider curve needs to have a lower height.

The area under a normal curve

The **area** under a normal curve shows **probabilities**.

- The **total area** under the curve represents the **total probability** of the random variable taking one of its possible values. And since the total probability is 1, the **total area under the curve** must also be **1**.
- The **probability** of the variable taking a value **between two limits** is the **area under the curve** between those limits.

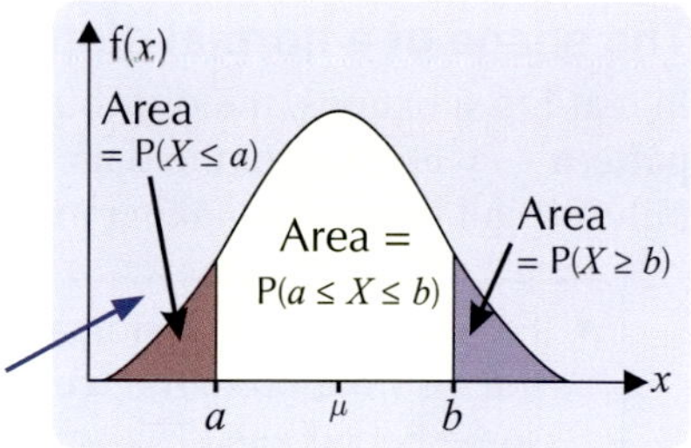

- Values of the **cumulative distribution function** (cdf) are the **areas** under the curve to the **left of x** (the **probability** that $X \leq x$) for different values of x.
- Since X is a **continuous** random variable, $P(X = x) = 0$, which means that **$P(X \leq x)$** and **$P(X < x)$** are the **same thing**. So you can **interchange** the ≤ and < signs.

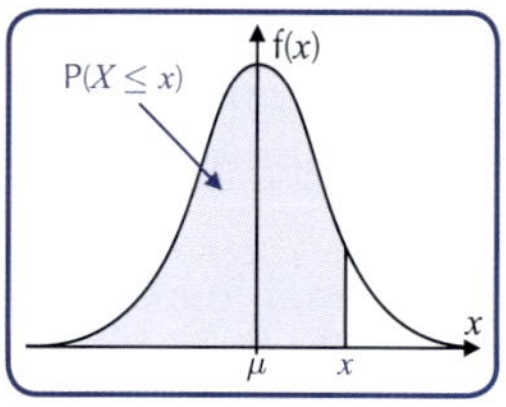

Tip: You'll see how to find and use areas under the normal curve over the next few pages.

Tip: For any continuous random variable X, $P(X = a) = 0$ for any value of a.

That's because the area under a graph at a single point is zero.

There are some **facts** about the **area** under the curve that apply to **all** normal distributions.

- There are **points of inflection** (where the curve changes between concave and convex — see p.135-136) at $x = \mu + \sigma$ and $x = \mu - \sigma$.

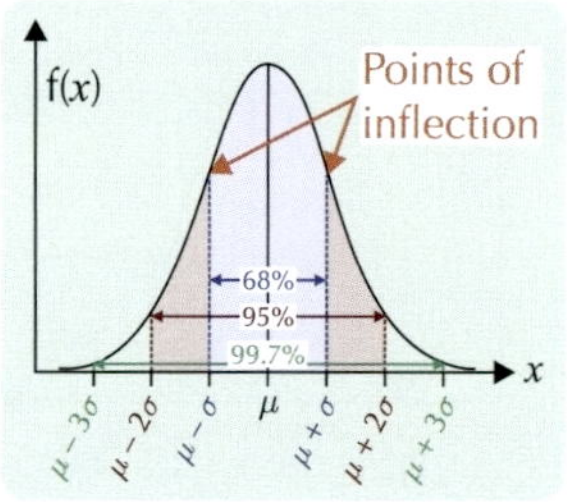

- **68%** of the total **area** lies within **±1** standard deviation ($\pm\sigma$) of the mean.
- **95%** of the total **area** lies within **±2** standard deviations ($\pm 2\sigma$) of the mean.
- **99.7%** of the total **area** lies within **±3** standard deviations ($\pm 3\sigma$) of the mean.

So **68%** of **observations** are within $\pm\sigma$ of the mean, **95%** of **observations** are within $\pm 2\sigma$ of the mean and **99.7%** of **observations** are within $\pm 3\sigma$ of the mean.

You can use these facts to check if a normal distribution is **suitable** for the situation you're modelling — see page 288.

Tip: These facts show how values close to the mean are **much** more likely than those further away.

Tip: Remember, standard deviation is the square root of the variance — so it's a measure of **dispersion** (how spread out values are from the mean).

Describing a normal distribution

If a continuous random variable X is **normally** distributed with mean μ and variance σ^2, it is written like this: → $X \sim N(\mu, \sigma^2)$

'**N**' stands for '**normal**' and '**~**' is short for '**is distributed**'.

- So going back to the hedgehog weights on page 269, you could define a random variable, $W \sim N(\mu, \sigma^2)$, where W represents hedgehog weight.
- Here, μ would represent the mean weight of the hedgehogs and σ would represent the standard deviation of hedgehog weights.

Calculating probabilities

- Before calculating any probabilities, **draw a sketch** to show the probability you want to find. This is useful as it shows the area you're looking for and also helps check that your answer seems **sensible**.
- You can use a **calculator** to find probabilities for **any** normal distribution. The method can **vary** depending on the model of your calculator, but you usually enter the **distribution** mode and choose the **normal cdf** function. Then input the required **x-value(s)**, **mean** and **standard deviation**.
- If you're given only **one x-value** (i.e. if you need to find $P(X \geq x)$ or $P(X \leq x)$ for some x) you might need to choose your own **lower** or **upper bound** — just pick a really **large negative** or **positive** number (see Example 1 below).

Tip: Make sure you know how to calculate normal probabilities on your calculator. If your calculator can only do the standard normal distribution (see p.273) you'll have to read p.274 first, then look at Example 1.

Example 1

a) If $X \sim N(100, 16)$, find $P(X \leq 105)$ to 3 s.f.

- Draw a sketch showing the area you need to find.

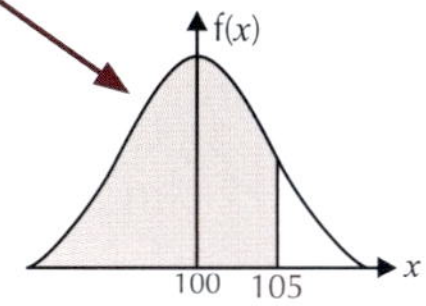

- You just input the upper bound $x = 105$, a lower bound (e.g. –9999), $\mu = 100$, and $\sigma = \sqrt{16} = 4$ into the normal cdf function on your calculator.
- So $P(X \leq 105) = 0.894350... = 0.894$ (3 s.f.)

Tip: Take the square root of the variance to get the standard deviation.

b) If $X \sim N(13, 5.76)$, find $P(X > 17)$ to 3 s.f.

- X is a continuous random variable, so $P(X > 17)$ is just the same as $P(X \geq 17)$. Draw a sketch showing the area you need to find.

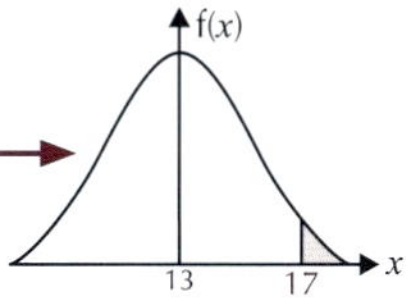

- Here, input the lower bound $x = 17$, an upper bound (e.g. 9999), $\mu = 13$ and $\sigma = \sqrt{5.76} = 2.4$ into the normal cdf function.
- So $P(X > 17) = P(X \geq 17) = 0.047790... = 0.0478$ (3 s.f.)

Tip: Almost all the possible values of x are within 3 standard deviations of the mean (see p.270), so choosing –9999 or 9999 means that the area you're calculating will cover the whole region you want.

Sometimes you might need to use the following trick to find the probability that X takes a value **greater than** (or greater than or equal to) x.

Using the fact that the **total area** under the curve is **1**, we get this definition. → $P(X > x) = 1 - P(X \leq x)$

Example 2

a) If $X \sim N(102, 144)$, find $P(X > 78)$ to 3 s.f.

- Draw a sketch showing the area you need to find.

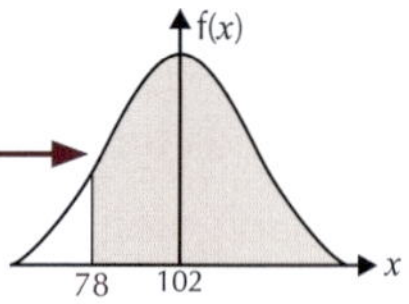

- As the total area under the curve is 1:
  $P(X > 78) = 1 - P(X \le 78)$
  You might find it useful to add this to the sketch.

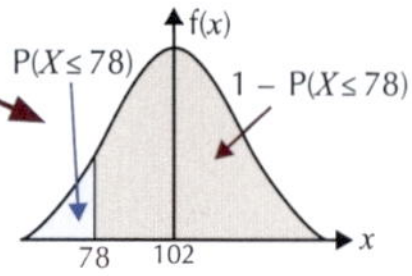

- Input the upper bound $x = 78$, a lower bound, $\mu = 102$ and $\sigma = \sqrt{144} = 12$ into the normal cdf function on your calculator.
  $P(X \le 78) = 0.022750...$
  So $P(X > 78) = 1 - P(X \le 78) = 1 - 0.022750... =$ **0.977** (3 s.f.)

Tip: You might be able to do this straight away by using 78 as a lower bound and choosing a large upper bound. However, this method can be useful when the mean and variance are unknown (see p.277-279).

b) If $X \sim N(102, 144)$, find $P(84 \le X \le 106)$ to 3 s.f.

- Draw a sketch showing the area you need to find.

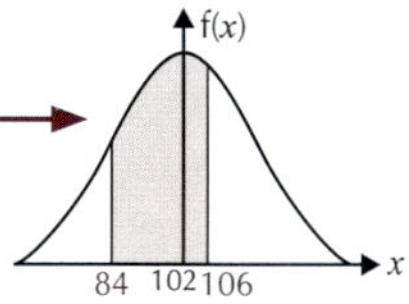

- For this example you can either split up the probability or calculate it directly.

Method 1 — Split up

- You need to find the area to the left of $x = 106$ and subtract the area to the left of $x = 84$.
- So, $P(84 \le X \le 106) = P(X \le 106) - P(X < 84)$
  $= 0.630558... - 0.066807...$
  $= 0.564$ (3 s.f.)

Using a calculator

Method 2 — Direct

- Use 84 as the lower bound and 106 as the upper bound in the normal cdf function on your calculator.
- This gives $P(84 \le X \le 106) = 0.563751... = 0.564$ (3 s.f.)

- Both methods give the answer: $P(84 \le X \le 106) =$ **0.564** (3 s.f.)

Tip: The direct method is much quicker here but for some questions you might need to split the probability up — see page 275.

Exercise 1.1

Q1 If $X \sim N(40, 25)$, find: a) $P(X < 50)$ b) $P(X \le 43)$

Q2 If $X \sim N(24, 6)$, find: a) $P(X \ge 28)$ b) $P(X > 25)$

Q3 If $X \sim N(120, 40)$, find: a) $P(X > 107)$ b) $P(X > 115)$

Q4 If $X \sim N(17, 3^2)$, find: a) $P(X \le 15)$ b) $P(X < 12)$

Q4 Hint: You'll often see the variance written as a number squared.

Q5 If $X \sim N(50, 5^2)$, find:
a) $P(52 < X < 63)$ b) $P(57 \le X < 66)$

Q6 If $X \sim N(0.6, 0.04)$, find:

a) $P(0.45 \le X \le 0.55)$ b) $P(0.53 < X < 0.58)$

Q7 If $X \sim N(260, 15^2)$, find:

a) $P(240 < X \le 280)$ b) $P(232 < X < 288)$

The standard normal distribution, Z

The **standard normal distribution**, Z, is a normal distribution that has mean $\mu = 0$ and variance $\sigma^2 = 1$.

$$Z \sim N(0, 1)$$

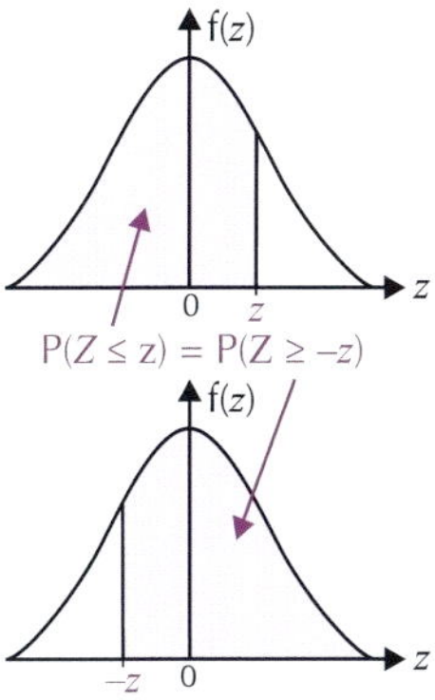

- The curve is **symmetrical** about the mean 0.
- $P(Z \le 0) = P(Z \ge 0) = \frac{1}{2}$, so you'd expect $P(Z \le z)$ to be **$< \frac{1}{2}$** for **negative values** of z and **$> \frac{1}{2}$** for **positive values** of z.
- Because of the symmetry of the curve, **$P(Z \le z) = P(Z \ge -z)$**.
- The **facts** about the area under the curve on p.270 still apply but with $\mu = 0$ and $\sigma = 1$ — e.g. points of inflection are at $z = 1$ and $z = -1$.

- Values of the **cumulative distribution function** (cdf) are usually written $\Phi(z)$ — i.e. $\Phi(z) = P(Z \le z)$.
- Since $P(Z = z) = 0$, you can **interchange** the $\le$ and $<$ signs — i.e. $\Phi(z) = P(Z < z)$ as well.

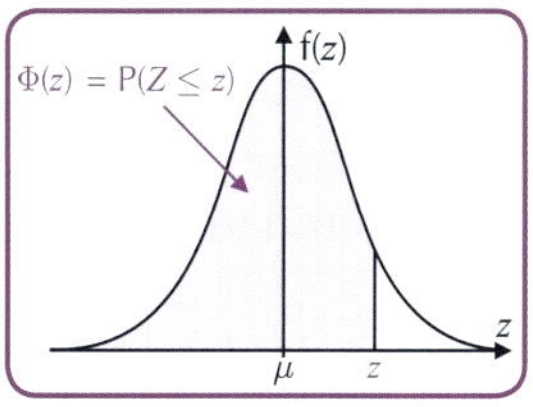

As before, you can use a **calculator** to find probabilities.

Examples

Given that $Z \sim N(0, 1)$, find the following probabilities to 3 s.f.

a) $P(Z \le 0.64)$

- Draw a sketch of the area you need to find.
- Input $z = 0.64$, a lower bound, $\mu = 0$ and $\sigma = 1$ into the normal cdf function.

So $\Phi(0.64) = P(Z \le 0.64) = 0.738913...$

$= 0.739$ (3 s.f.)

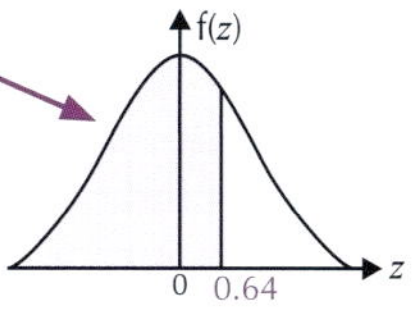

b) $P(Z > -0.42)$

- Draw a sketch of the area you need to find.
- Z is a continuous random variable, so $P(Z > -0.42)$ is just the same as $P(Z \ge -0.42)$.

Your calculator still works for negative z-values. This gives $P(Z > -0.42) = 0.662757...$

$= 0.663$ (3 s.f.)

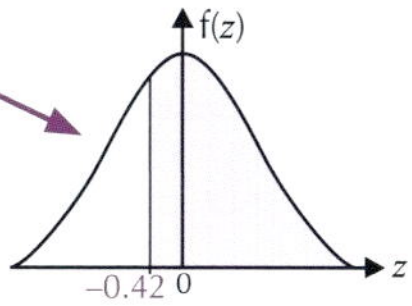

Tip: Remember to use a lower bound (e.g. –9999) for 'less than' probabilities and an upper bound (e.g. 9999) for 'greater than' probabilities.

Tip: You could use symmetry here — i.e. $P(Z > -0.42) = P(Z < 0.42) = \Phi(0.42)$.

Tip: You could split this up like on p.272:
$\Phi(0.82) - \Phi(0.12)$
$= 0.79389... - 0.54775...$
$= 0.246$ (3 s.f.)

c) P(0.12 < Z ≤ 0.82)

- Draw a sketch of the area you need to find.

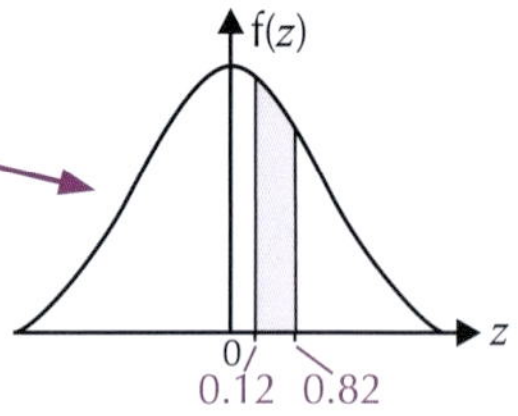

- So $P(0.12 < Z \leq 0.82) = 0.246133...$
$= 0.246$ (3 s.f.)

Exercise 1.2

Q1 Use a calculator to find the following probabilities to 3 s.f.

a) $P(Z \leq 1.87)$ b) $P(Z < 0.99)$ c) $P(Z > 2.48)$
d) $P(Z \geq 0.14)$ e) $P(Z > -0.24)$ f) $P(Z > -1.21)$
g) $P(Z < -0.62)$ h) $P(Z \leq -2.06)$ i) $P(Z \geq 1.23)$

Q2 Use a calculator to find the following probabilities to 3 s.f.

a) $P(1.34 < Z < 2.18)$ b) $P(0.76 \leq Z < 1.92)$
c) $P(-1.45 \leq Z \leq 0.17)$ d) $P(-2.14 < Z < 1.65)$
e) $P(-1.66 < Z \leq 1.66)$ f) $P(-0.34 \leq Z < 0.34)$
g) $P(-3.25 \leq Z \leq -2.48)$ h) $P(-1.11 < Z < -0.17)$

Converting to the Z distribution

Any continuous random variable, X, where $X \sim N(\mu, \sigma^2)$, can be **transformed** to the **standard normal variable**, Z, by:

- **subtracting the mean** (μ), and then
- **dividing by the standard deviation** (σ).

Tip: This method comes in handy when the mean and/or standard deviation are unknown (see pages 277-279), or if your calculator only works out standard normal probabilities.

$$\text{If } X \sim N(\mu, \sigma^2), \text{ then } \frac{X-\mu}{\sigma} = Z, \text{ where } Z \sim N(0, 1)$$

Examples

a) If $X \sim N(5, 16)$, find $P(X < 7)$ by transforming X to the standard normal variable Z.

- Start by transforming X to Z.

$\mu = 5, \sigma = \sqrt{16} = 4$

$P(X < 7) = P\left(Z < \frac{7-5}{4}\right) = P(Z < 0.5)$

Tip: Remember, you take the square root of the variance to get the standard deviation.

- Draw a sketch showing the area you need to find.

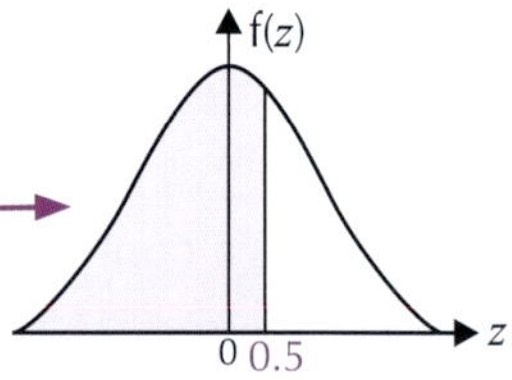

- Using a calculator, $P(Z < 0.5) = 0.691$ (3 s.f.)

b) If $X \sim N(5, 16)$, find $P(5 < X < 11)$.

- Start by **transforming** X to Z.

  $P(5 < X < 11) = P\left(\frac{5-5}{4} < Z < \frac{11-5}{4}\right) = P(0 < Z < 1.5)$

- Draw a **sketch** showing the area you need to find.

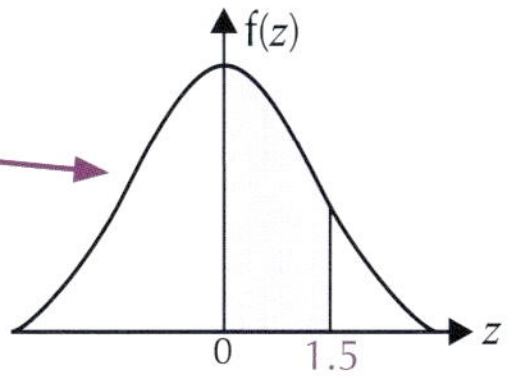

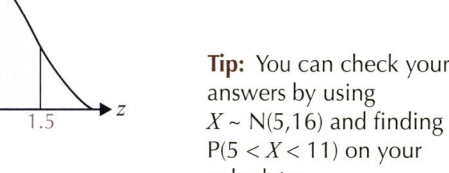

- You need to find the area to the left of $z = 1.5$, then **subtract** the area to the left of $z = 0$.
- So, $P(0 < Z < 1.5) = P(Z < 1.5) - P(Z \leq 0)$
  $= 0.93319... - 0.5 =$ **0.433** (3 s.f.)

Tip: You can check your answers by using $X \sim N(5,16)$ and finding $P(5 < X < 11)$ on your calculator.

Exercise 1.3

Q1 If $X \sim N(106, 100)$, transform X to the standard normal variable to find:
a) $P(X \leq 100)$ b) $P(X > 122)$

Q2 If $X \sim N(11, 2^2)$, transform X to the standard normal variable to find:
a) $P(X \geq 9)$ b) $P(10 \leq X \leq 12)$

Q3 If $X \sim N(260, 15^2)$, transform X to the standard normal variable to find:
a) $P(240 < X < 280)$ b) $P(232 < X < 288)$

Finding x-values

You might be given a **probability**, p, and be asked to find a specific **range of x-values** where the probability of X falling in this range is equal to p.

For any probability involving **less than** or **less than or equal to** (e.g. find x where $P(X \leq x) = p$) these questions can be done **directly** on your calculator. To do this, choose the **inverse normal** function on your calculator. Then input the **probability**, **mean** and **standard deviation**.

Tip: The inverse normal function on your calculator might be labelled 'InvNorm' or 'InvN'.

Example 1

$X \sim N(85, 25)$. If $P(X < a) = 0.9192$, find the value of a to 2 s.f.

- Draw a sketch of the area you need to find.

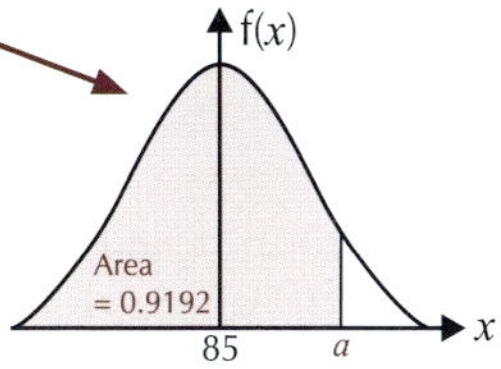

- Input probability = 0.9192, $\mu = 85$ and $\sigma = \sqrt{25} = 5$ into the inverse normal function on your calculator.

  So $P(X < x) = 0.9192$ for $x =$ **92** (2 s.f.)

More complicated problems involving two x-values need splitting up (like in the example on page 272) with the formula below:

$$P(a \leq X \leq b) = P(X \leq b) - P(X < a)$$

Tip: Remember, for a continuous random variable X, $P(X \leq a) = P(X < a)$. So for a continuous variable you could write this as: $P(a \leq X \leq b) = P(X \leq b) - P(X \leq a)$.

Example 2

$X \sim N(24, 9)$. If $P(x < X < 24) = 0.2$, find the value of x to 3 s.f.

- Draw a sketch of the area you need to find.

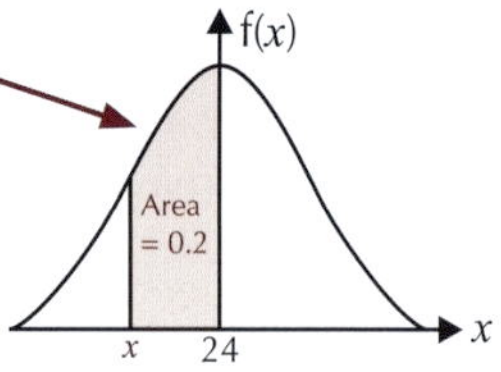

- Splitting up the probability gives:
  $P(x < X < 24) = P(X < 24) - P(X < x) = 0.2$
  So $P(X < x) = P(X < 24) - 0.2$
- Since the mean is 24, $P(X < 24) = 0.5$.
  So $P(X < x) = 0.5 - 0.2 = 0.3$
- Use a calculator to find the value of x.
  $P(X < x) = 0.3$ for $x = 22.426798...$, so $x =$ 22.4 (3 s.f.)

To find z for a 'greater than' probability, you have to use $P(Z \le z) = 1 - P(Z > z)$ before you can use your calculator.

Example 3

$Z \sim N(0, 1)$. If $P(Z > z) = 0.15$, find the value of z.

- Draw a sketch of the area you need to find.

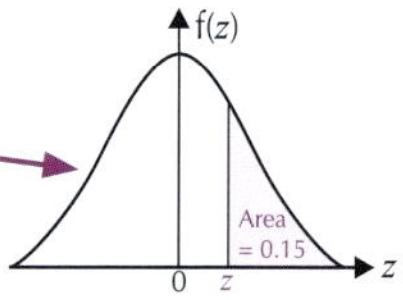

- The area under the curve is 1, so subtract 0.15 from 1 to get the area to the **left** of z: $1 - 0.15 = 0.85$
- Draw another sketch showing the new area.

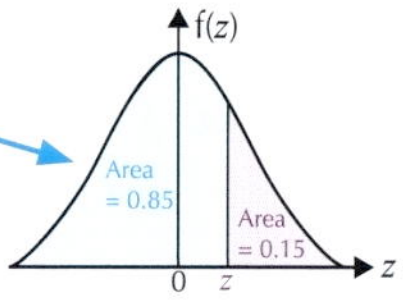

- Use a calculator to find the value of z.
  $\Phi(z) = 0.85$ for $z = 1.036433...$
- So $P(Z > z) = 0.15$ for $z =$ 1.04 (3 s.f.)

Tip: Rearrange the formula $P(Z > z) = 1 - \Phi(z)$ to give $\Phi(z) = 1 - P(Z > z)$.

Exercise 1.4

Q1 If $X \sim N(12, 0.64)$, find the value of x such that:

a) $P(X < x) = 0.8944$ b) $P(X \le x) = 0.0304$

c) $P(X \ge x) = 0.2660$ d) $P(X > x) = 0.7917$

Q2 If $X \sim N(25, 4^2)$, find the value of x such that:

a) $P(X < x) = 0.975$ b) $P(X > x) = 0.4$

Q3 If $X \sim N(48, 5^2)$, find the value of x such that:

a) $P(53 < X < x) = 0.05$ b) $P(x < X < 49) = 0.4312$

Q4 For the standard normal distribution, find the value of z to 3 s.f. that gives each of the following probabilities:

a) $P(Z < z) = 0.7$ b) $P(Z < z) = 0.999$

c) $P(Z > z) = 0.005$ d) $P(Z > z) = 0.2$

Q5 Find the value of z such that:

a) $P(Z < z) = 0.004$ b) $P(Z > z) = 0.0951$

c) $P(-z < Z < z) = 0.9426$ d) $P(z < Z < -1.25) = 0.0949$

Q5c) Hint: Draw a graph of the area you're given, then use the symmetry of the graph to find the values of z.

Finding the mean and standard deviation of a normal distribution

On page 274 you saw how any normally distributed variable, $X \sim N(\mu, \sigma^2)$, can be **transformed** to the standard normal variable, Z, by using $Z = \frac{X - \mu}{\sigma}$.

So far you've used this relationship to find probabilities when the mean and standard deviation have been **known**. But you can use it to find the **mean** and **standard deviation** when they're **unknown** and you know some **probabilities** for the distribution.

Example 1

If the random variable $X \sim N(\mu, 4)$ and $P(X < 23) = 0.9015$, find μ.

- Start by **transforming the probability** you're given for X into a probability for Z. The mean is unknown, so just leave it as μ for now.

  $P(X < 23) = P\left(Z < \frac{23 - \mu}{2}\right) = 0.9015$

- Draw a **sketch** to show the information.

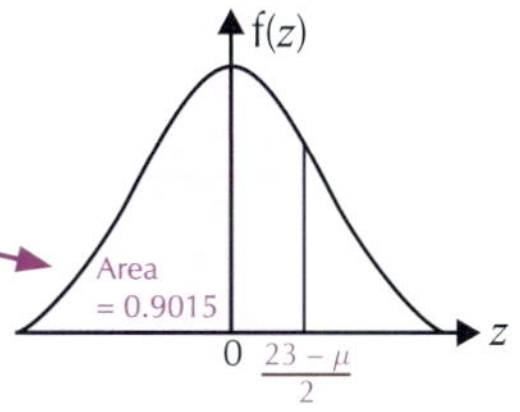

- Now, if you use a calculator to find z for which $\Phi(z) = 0.9015$, you can form an **equation** in μ.

  From the calculator, $\Phi(z) = 0.9015$ for $z = 1.2901...$

  So $\frac{23 - \mu}{2} = 1.2901...$ ← Equation in μ.

- Now **solve** this equation for μ.

  $\frac{23 - \mu}{2} = 1.2901... \Rightarrow 23 - \mu = 2.5802... \Rightarrow \mu = 20.42$ (2 d.p.)

Tip: $\sigma = \sqrt{4} = 2$

Tip: So $X \sim N(20.42, 4)$.

Example 2

If the random variable $X \sim N(\mu, 4^2)$ and $P(X > 19.84) = 0.025$, find μ.

- Transform the probability you're given for X into a probability for Z.

  $P(X > 19.84) = P\left(Z > \frac{19.84 - \mu}{4}\right) = 0.025$

- Draw a **sketch** to show the information.

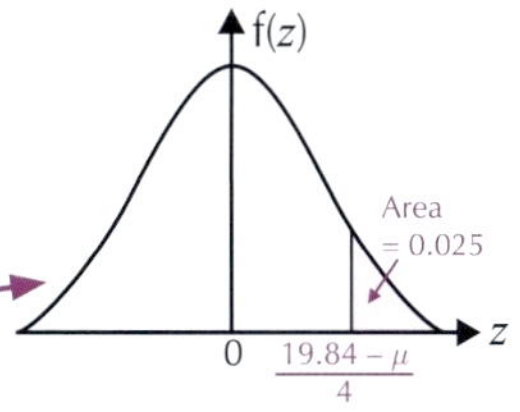

- Remember, your calculator finds the z-value for $P(Z \le z)$.

  The area under the curve is 1, so $1 - 0.025 = 0.975$.
  From the calculator, $\Phi(z) = 0.975$ for $z = 1.96$ to 2 d.p.

  So $\frac{19.84 - \mu}{4} = 1.96$ ← Equation in μ.

- Now **solve** this equation for μ.

  $\frac{19.84 - \mu}{4} = 1.96 \Rightarrow 19.84 - \mu = 7.84 \Rightarrow \mu = 12$

Tip: So $X \sim N(12, 4^2)$.

In the first two examples you found the mean, but you can find the **standard deviation** (s.d.) in exactly the same way.

Example 3

If the random variable $X \sim N(53, \sigma^2)$ and $P(X > 56) = 0.2$, find σ.

- Again, start by transforming the probability you're given for X into a probability for Z. The s.d. is unknown, so just leave it as σ for now.

  $P(X > 56) = P\left(Z > \frac{56-53}{\sigma}\right) = P\left(Z > \frac{3}{\sigma}\right) = 0.2$

- Draw a **sketch** to show the information.

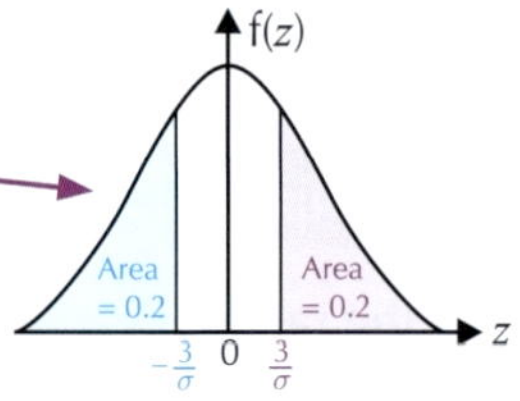

- Using symmetry, $P\left(Z < -\frac{3}{\sigma}\right) = 0.2$

  From the calculator, $\Phi(z) = 0.2$ for $z = -0.84162...$

  So $-\frac{3}{\sigma} = -0.84162...$ ← Equation in σ.

- Now **solve** the equation for σ:

  $\sigma = 3 \div 0.84162... \Rightarrow \sigma = 3.56$ (3 s.f.)

Tip: Z is symmetrical about the mean 0 so $P(Z < -z) = P(Z > z)$ (see page 273).

Tip: So $X \sim N(53, 3.56^2)$.

When you're asked to find the mean **and** the standard deviation, the method is a little bit **more complicated**.

You start off as usual, but instead of getting one equation in one unknown to solve, you end up with **two equations** in **two unknowns**, μ and σ. In other words, you have **simultaneous equations**, which you **solve** to find μ and σ.

Tip: Remember, you can solve simultaneous equations by adding or subtracting them to get rid of one unknown.

Example 4

The random variable $X \sim N(\mu, \sigma^2)$.
If $P(X < 9) = 0.5596$ and $P(X > 14) = 0.0322$, find μ and σ.

PROBLEM SOLVING

- **Transform** the first probability for X into a probability for Z.

  $P(X < 9) = P\left(Z < \frac{9-\mu}{\sigma}\right) = 0.5596$

- Draw a **sketch** to show the information.

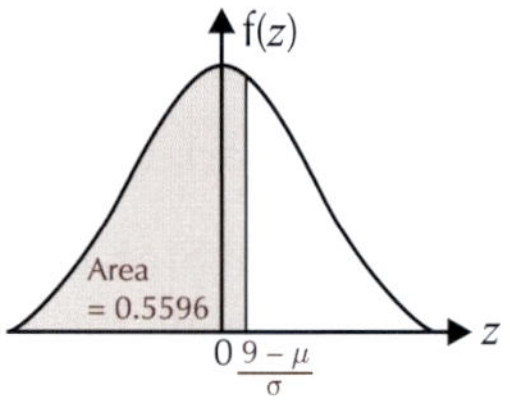

- Now, if you use a calculator to find z for which $\Phi(z) = 0.5596$, you can form an **equation** in μ and σ.

  From the calculator, $\Phi(z) = 0.5596$ for $z = 0.14995...$

  So $\frac{9-\mu}{\sigma} = 0.14995... \Rightarrow 9 - \mu = 0.14995...\sigma$ ← 1st equation in μ and σ

- Now do the same thing for the second probability for X. **Transforming** to Z, you get:

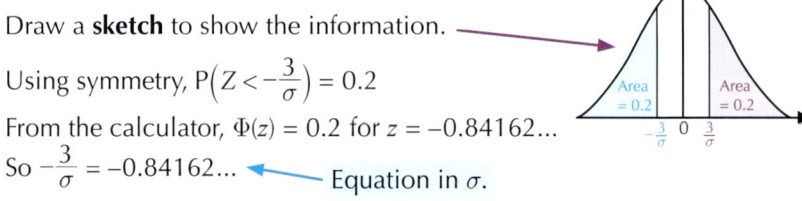

  $P(X > 14) = P\left(Z > \frac{14-\mu}{\sigma}\right) = 0.0322$

- Draw a **sketch** to show the information.

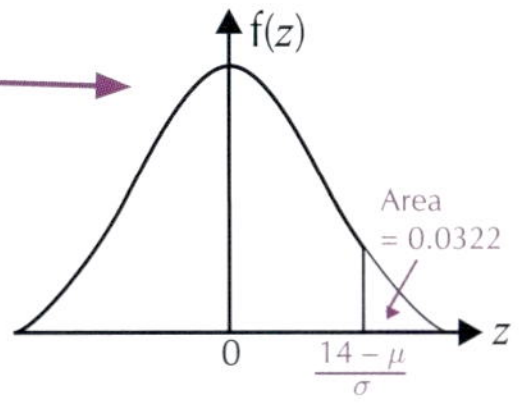

- You can see from the graph that:

$$P\left(Z \leq \frac{14-\mu}{\sigma}\right) = 1 - P\left(Z > \frac{14-\mu}{\sigma}\right)$$
$$= 1 - 0.0322 = 0.9678$$

- Now use a calculator to find z for which $\Phi(z) = 0.9678$.

  From the calculator, $\Phi(z) = 0.9678$ for $z = 1.8494...$

  So $\frac{14-\mu}{\sigma} = 1.8494... \Rightarrow 14 - \mu = 1.8494...\sigma$

  2nd equation in μ and σ

- So now you have your **simultaneous** equations:

  $9 - \mu = 0.14995...\sigma$ ①

  $14 - \mu = 1.8494...\sigma$ ②

  It helps to number the equations, so you can refer to them later.

Tip: Remember to use the unrounded z-values in your calculations.

- Each equation has one 'μ', so you can **subtract** them to get rid of μ, which will leave you with an **equation in σ** to solve.

  ② – ①: $14 - 9 - \mu - (-\mu) = 1.8494...\sigma - 0.14995...\sigma$

  $\Rightarrow 5 = 1.6994...\sigma$

  $\Rightarrow \sigma = 2.9421... \Rightarrow \sigma = 2.94$ (3 s.f.)

- Finally, find μ by **substituting** $\sigma = 2.94...$ back into one of the equations.

  Using equation ①,

  $9 - \mu = 0.14995...\sigma$

  $\Rightarrow \mu = 9 - 0.14995... \times 2.9421... \Rightarrow \mu = 8.5588... = 8.56$ (3 s.f.)

Tip: $X \sim N(8.56, 2.94^2)$.

Exercise 1.5

Q1 For each of the following, use the information to find μ.

a) $X \sim N(\mu, 6^2)$ and $P(X < 23) = 0.9332$

b) $X \sim N(\mu, 8^2)$ and $P(X < 57) = 0.9970$

c) $X \sim N(\mu, 100^2)$ and $P(X > 528) = 0.1292$

d) $X \sim N(\mu, 0.4^2)$ and $P(X < 11.06) = 0.0322$

e) $X \sim N(\mu, 0.02^2)$ and $P(X > 1.52) = 0.9938$

Q2 $X \sim N(\mu, 3.5^2)$. If the middle 95% of the distribution lies between 6.45 and 20.17, find the value of μ. PROBLEM SOLVING

Q3 For each of the following, use the information to find σ.

a) $X \sim N(48, \sigma^2)$ and $P(X < 53) = 0.8944$

b) $X \sim N(510, \sigma^2)$ and $P(X < 528) = 0.7734$

c) $X \sim N(17, \sigma^2)$ and $P(X > 24) = 0.0367$

d) $X \sim N(0.98, \sigma^2)$ and $P(X < 0.95) = 0.3085$

e) $X \sim N(5.6, \sigma^2)$ and $P(X > 4.85) = 0.8365$

Q4 $X \sim N(68, \sigma^2)$. If the middle 70% of the distribution lies between 61 and 75, find the value of σ. PROBLEM SOLVING

Q5 For each of the following, find μ and σ. PROBLEM SOLVING

a) $X \sim N(\mu, \sigma^2)$, $P(X < 30) = 0.9192$ and $P(X < 36) = 0.9953$

b) $X \sim N(\mu, \sigma^2)$, $P(X < 4) = 0.9332$ and $P(X < 4.3) = 0.9987$

c) $X \sim N(\mu, \sigma^2)$, $P(X < 20) = 0.7881$ and $P(X < 14) = 0.0548$

d) $X \sim N(\mu, \sigma^2)$, $P(X < 696) = 0.9713$ and $P(X < 592) = 0.2420$

e) $X \sim N(\mu, \sigma^2)$, $P(X > 33) = 0.1056$ and $P(X > 21) = 0.9599$

f) $X \sim N(\mu, \sigma^2)$, $P(X > 66) = 0.3632$ and $P(X < 48) = 0.3446$

The normal distribution in real-life situations

MODELLING

Now it's time to use everything you've learnt about normal distributions to answer questions in **real-life** contexts. These are the kind of questions that usually come up in **exams**.

You always start by **defining** a **normally-distributed random variable** to represent the information you're given. Then you use the methods you've seen on pages 269-279 to find out what you need to know.

Tip: The first step is always to define the variable. So the variable here is the mass of cereal in a box.

Tip: Always check the answer seems sensible. Using the rules on p.270, you know that 97.5% of values are less than $2 \times \sigma$ above the mean (and $2 \times 8 + 766 = 782$). So this answer seems reasonable.

Example 1

A machine which fills boxes of cereal is set so that the mass of cereal going into the boxes follows a normal distribution with mean 766 g and standard deviation 8 g.

a) Find the probability that a randomly selected box of cereal contains less than 780 g of cereal.

- First, define a random variable to represent the mass of cereal in a box. If X represents the mass of cereal in g, then $X \sim N(766, 64)$. ← 8^2
- Next, turn the question into a probability for X. So you want to find $P(X < 780)$.
- Draw a sketch showing the area you need to find. →
- Using a calculator, $P(X < 780) =$ 0.960 (3 s.f.)

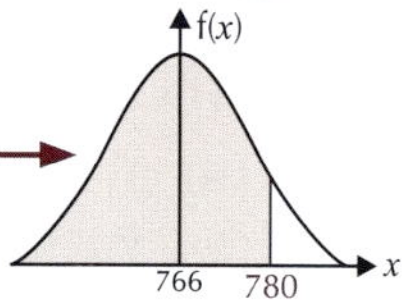

b) The machine fills 2138 boxes of cereal in an hour. Find the number of boxes that you would expect to contain less than 780 g.

Multiply the total number of boxes by the probability calculated in a):
$2138 \times P(X < 780) = 2138 \times 0.960 \approx$ 2052 boxes

c) Find the probability that a randomly selected box of cereal contains between 780 g and 790 g of cereal.

- Again, turn the question into a probability for X. So you want to find $P(780 < X < 790)$.
- Draw a sketch to help you picture this area. →
- Using a calculator, $P(780 < X < 790) = 0.038709... =$ 0.0387 (3 s.f.)

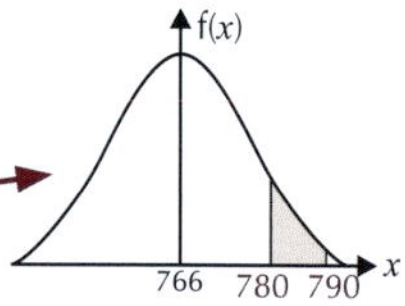

Example 2

The times taken by 666 people to complete an assault course are normally distributed with a mean of 600 seconds and a variance of 105 seconds.

a) Find the probability that a randomly selected person took more than 620 seconds.

- Start by defining a random variable to represent the time taken. If X represents the time taken in seconds, then $X \sim N(600, 105)$.
- Next, turn the question into a probability for X. So you want to find $P(X > 620)$.
- Draw a sketch to show the information.

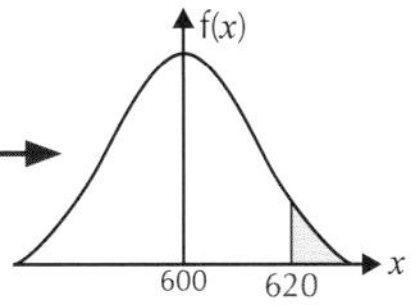

- $P(X > 620) = 0.025480... = 0.0255$ (3 s.f.)

b) Find the number of people you would expect to complete the assault course in over 620 seconds.

$666 \times P(X > 620) = 666 \times 0.0255$

$= 17$ people (to the nearest whole number)

Real-life normal distribution questions also ask you to **find values**, when you're **given probabilities** (see page 275).

Example 3

The forces needed to snap lengths of a certain type of elastic are normally distributed with $\mu = 13$ N and $\sigma = 1.8$ N.

a) The probability that a randomly selected length of elastic is snapped by a force of less than a N is 0.7580. Find the value of a.

- Start by **defining a random variable** to represent the force needed. If F represents the force needed in N, then $F \sim N(13, 1.8^2)$.
- Next, turn the question into a **probability for F**. So you know that $P(F < a) = 0.7580$.
- Next, **transform** F to Z.

  $P(F < a) = P\left(Z < \frac{a-13}{1.8}\right) = 0.7580$
- Draw a **sketch** to show the information.

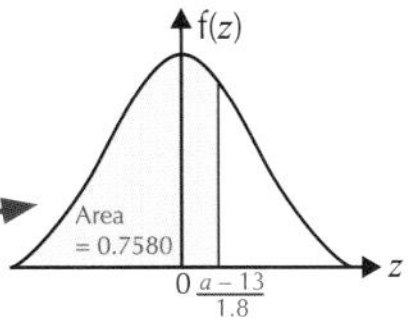

- Use a calculator to find z for which $\Phi(z) = 0.7580$.

  $\Phi(z) = 0.7580$ for $z = 0.70$, so $\frac{a-13}{1.8} = 0.70$
- Now, just **solve** for a:

  $\frac{a-13}{1.8} = 0.7 \Rightarrow a - 13 = 1.26 \Rightarrow a = 14.26$

Tip: It's neater to write the variance as 1.8^2, rather than 3.24.

Tip: Use a sensible letter for the variable — it doesn't have to be X.

Tip: You can find the value of a directly using the inverse normal function on some calculators, without having to transform to Z.

b) Find the range of values that includes the middle 80% of forces needed to snap the length of elastic.

PROBLEM SOLVING

- It's difficult to know where to start with this one. So it's a good idea to **sketch the distribution of F**, to show the range you need to find.

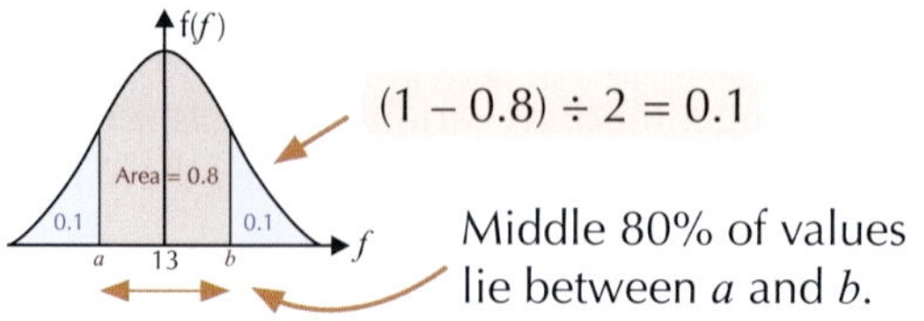

- Now you can write **two probability statements** — one for a and one for b: $P(F < a) = 0.1$ and $P(F > b) = 0.1$.
- Next, **transform** F to Z.
  $P(F < a) = P\left(Z < \frac{a-13}{1.8}\right) = 0.1$ and $P(F > b) = P\left(Z > \frac{b-13}{1.8}\right) = 0.1$
- Draw another **sketch** to show this information.

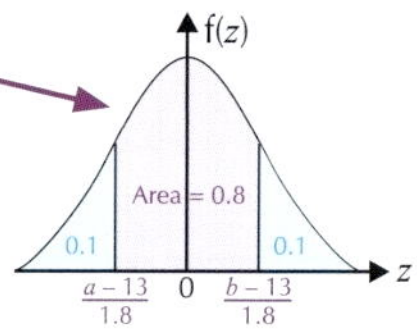

- Now, use your calculator to find a and b.
  Starting with a, $\Phi(z) = 0.1$ for $z = -1.281...$
  So, $\frac{a-13}{1.8} = -1.281... \Rightarrow a = 10.7$ (3 s.f.)
- And using symmetry, $\frac{b-13}{1.8} = 1.281... \Rightarrow b = 15.3$ (3 s.f.)
- So the range of values is 10.7 N to 15.3 N

Tip: Using Z is useful because the range is symmetrical about $z = 0$. So $\frac{b-13}{1.8} = -\frac{a-13}{1.8}$, making a and b easier to find.

Exercise 1.6

MODELLING

Q1 The lengths of time taken by a group of 56 blood donors to replace their red blood cells are modelled by a normal distribution with a mean of 36 days and a standard deviation of 6 days.

a) It takes Edward 28 days to replace his red blood cells. Find the probability that a randomly selected donor from the group takes less time than Edward to replace their red blood cells.

b) Find the number of blood donors that this model would predict to take less time than Edward to replace their blood cells.

c) 6.3% of the group take longer than Bella to replace their red blood cells. How long does it take Bella?

Q2 The personal best times taken by athletes at a sports club to run 400 m are known to follow a normal distribution with a mean of 51 seconds and a standard deviation of 2.1 seconds.

a) Gary's personal best time is 49.3 seconds. What percentage of the athletes have a slower personal best time than Gary?

b) The athletes with personal bests in the top 20% of times are selected for a special training programme. What time do they have to beat to be selected for the programme?

Q3 The volume of vinegar contained in bottles is modelled by a normal distribution with a standard deviation of 5 ml. It is found that 71.9% of bottles contain less than 506 ml of vinegar.

a) Find the mean volume of vinegar contained in the bottles.

b) The label on each bottle says it contains 500 ml of vinegar. Find the probability that a random bottle contains less than 500 ml.

c) A shop's store room has 1303 bottles of vinegar. Find the number of bottles that this model would predict to contain at least 500 ml.

Q3 Hint: See p.277 for a reminder of how to find the mean of a normal distribution.

Q4 A particular type of toy car uses two identical batteries. The lifetimes of individual batteries can be modelled by a normal distribution with a mean of 300 hours and a standard deviation of 50 hours.

a) Find the probability that a battery lasts less than 200 hours.

b) Find the probability that a battery lasts at least 380 hours.

c) Stating any assumptions you make, find the probability that both of the batteries in a car last at least 380 hours.

d) The probability that a randomly selected battery lasts more than 160 hours, but less than h hours, is 0.9746. Find the value of h.

Q4 Hint: See p.258 for a reminder of finding probabilities of multiple events.

Q5 The heights of a population of 17-year-old boys are assumed to follow a normal distribution with a mean of 175 cm. 80% of this population of 17-year-old boys are taller than 170 cm.

a) Find the standard deviation of the heights of the 17-year-old boys in this population.

b) One 17-year-old boy is selected from the population at random. Find the probability that his height is within 4 cm of the mean height.

Q6 The masses of the eggs laid by the hens on farmer Elizabeth's farm are assumed to follow a normal distribution with mean 60 g and standard deviation 3 g.

a) The probability that a randomly selected egg has a mass of at least $60 - m$ grams is 0.9525. Find the value of m to the nearest gram.

b) Farmer Elizabeth keeps the lightest 10% of eggs for herself and uses them to make sponge cakes. Find the maximum mass of an egg that could end up in one of farmer Elizabeth's sponge cakes.

Q7 In a particularly wet village, it rains almost continuously. The daily rainfall, in cm, is modelled by a normal distribution. The daily rainfall is less than 4 cm on only 10.2% of days, and it's greater than 7 cm on 64.8% of days. PROBLEM SOLVING

Find the mean and standard deviation of the daily rainfall.

2. Normal Approximation to a Binomial Distribution

Learning Objectives:

- Be able to apply a continuity correction when using a normal distribution to approximate a discrete random variable.
- Be able to use a normal approximation to a binomial random variable.

In Year 1 you saw the binomial distribution B(n, p). Under certain conditions you can use a normal distribution N(μ, σ^2) to approximate the binomial distribution — this can simplify calculations and is usually pretty accurate.

Continuity corrections

The **binomial distribution** is **discrete** and occurs in situations with a **fixed number** (n) of **independent** trials, where the outcome of each trial can be either a **success** or **failure**. If the probability of success is p and X is the number of successes, then:

$$X \sim \text{B}(n, p)$$

The normal distribution can be used to **approximate** the binomial distribution under particular circumstances.

But using the normal distribution (which is continuous) to approximate a discrete distribution is slightly awkward.

- The binomial distribution is **discrete**, so if the random variable X follows this distribution, you can work out P(X = 0), P(X = 1), etc.
- But a normally-distributed variable is **continuous**, and so if $Y \sim \text{N}(\mu, \sigma^2)$, P($Y$ = 0) = P(Y = 1) = 0, etc. (see page 270).

To allow for this, you might have to use a **continuity correction**.

- You assume that the discrete value $X = 1$ is '**spread out**' over the interval $0.5 < Y < 1.5$.

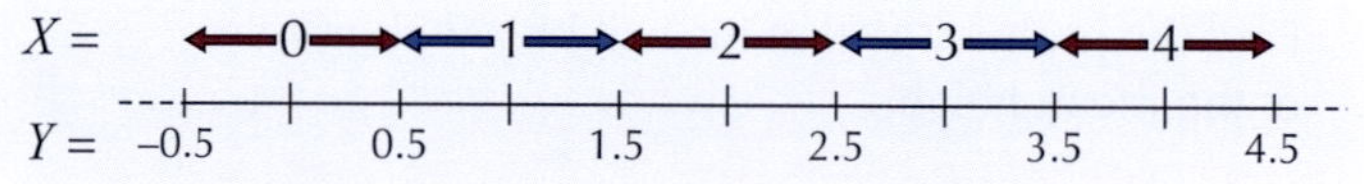

- Then to **approximate** the discrete probability **P(X = 1)**, you find the continuous probability **P(0.5 < Y < 1.5)**.
- Similarly, the discrete value $X = 2$ is **spread out** over the interval $1.5 < Y < 2.5$, so P(X = 2) is **approximated** by P($1.5 < Y < 2.5$), and so on.

The **interval** you need to use with the normal distribution depends on the discrete **probability** you're trying to find.

The general idea is always the same, though — each **discrete value b** covers the **continuous interval** from $b - \frac{1}{2}$ up to $b + \frac{1}{2}$.

Discrete	Normal	
$P(X = b)$	$P(b - \frac{1}{2} < Y < b + \frac{1}{2})$	
$P(X \le b)$	$P(Y < b + \frac{1}{2})$	...to include b
$P(X < b)$	$P(Y < b - \frac{1}{2})$	...to exclude b
$P(X \ge b)$	$P(Y > b - \frac{1}{2})$	...to include b
$P(X > b)$	$P(Y > b + \frac{1}{2})$	...to exclude b

Exercise 2.1

Q1 A discrete random variable X has possible values 0, 1, 2, 3...
X is to be approximated by the normal variable Y.
What interval for Y would you use to approximate:

a) $P(X = 5)$? b) $P(12 \leq X \leq 15)$? c) $P(X \leq 10)$?

Q2 The random variable X follows a binomial distribution.
The normal random variable Y is to be used to approximate probabilities for X.
Write down the probability for Y that would approximate:

a) $P(X = 50)$ b) $P(X < 300)$ c) $P(X \geq 99)$

Q3 An unfair coin is to be tossed 1000 times and the number of heads (X) recorded. A random variable Y following a normal distribution is to be used to approximate the probabilities below.
Write down the probability for Y that would approximate:

a) the probability of getting exactly 200 heads
b) the probability of getting at least 650 heads
c) the probability of getting less than 300 heads
d) the probability of getting exactly 499, 500 or 501 heads
e) the probability of getting between 250 and 750 heads (inclusive)
f) the probability of getting at least 100 heads but less than 900 heads

Normal approximation to a binomial distribution

Certain **binomial** distributions can be approximated by a normal distribution, and if n is **large** (and you're not trying to find $P(X = x)$), you may not have to use a continuity correction.

For the normal approximation to a binomial distribution to work well, you need the following conditions to be true:

Suppose the random variable X follows a **binomial distribution**, i.e. $X \sim B(n, p)$.

If (i) $p \approx \frac{1}{2}$,

and (ii) n is large,

then X can be approximated by the normal random variable $Y \sim N(np, np(1-p))$

Tip: Remember... the symbol '≈' means 'approximately equal to'.

Tip: This can also be written $Y \sim N(np, npq)$, where $q = 1 - p$.

- This means that as long as p isn't too far from $\frac{1}{2}$ and n is quite **large**, then you don't need to use $B(n, p)$ to work out probabilities for X.
- Instead you can get a **good approximation** to the probabilities for X using a normal distribution.
- In fact, even if p isn't all that close to 0.5, this approximation usually works well as long as np and $n(1-p)$ are both **bigger than 5**.

Example 1

The random variable $X \sim B(90, 0.47)$.
Find a suitable normal approximation for X, giving μ and σ^2.

- $X \sim B(n, p)$ where $n = 90$ and $p = 0.47$.
- Approximate X with $Y \sim N(\mu, \sigma^2)$, where:

  i) $\mu = np = 90 \times 0.47 =$ 42.3

  ii) $\sigma^2 = np(1 - p) = 90 \times 0.47 \times (1 - 0.47) =$ 22.419
- So X can be approximated by $Y \sim N(42.3, 22.419)$.

Tip: This approximation is appropriate because $n = 90$ is large and $p = 0.47$ is close to 0.5.

Example 2

The random variable $X \sim B(800, 0.4)$ is to be approximated using the normally distributed random variable $Y \sim N(\mu, \sigma^2)$.

a) Verify that a normal approximation is appropriate, and specify the distribution of Y.

- n is large, and p is close to $\frac{1}{2}$, so a normal approximation is suitable.
- $\mu = np = 800 \times 0.4 = 320$ and $\sigma^2 = np(1 - p) = 800 \times 0.4 \times 0.6 = 192$
- So use the approximation $Y \sim N(320, 192)$

Tip: So the standard deviation of Y is $\sigma = \sqrt{192}$.

b) Find an approximate value for $P(320 < X \leq 350)$.

n is large, so you don't need to use a continuity correction and you can find the probability directly using your calculator.

So $P(320 < Y \leq 350) =$ 0.485 (3 s.f.)

Tip: If you used a continuity correction here, you'd work out $P(320.5 < Y < 350.5)$ instead, which is 0.472 (3 s.f.).

This is one of the reasons why the normal distribution is so useful — because it can be used to approximate other distributions.
It means you can use it in all sorts of real-life situations.

Example 3

Each piglet born on a farm is equally likely to be male or female.

MODELLING

a) 250 piglets are born. Use a suitable normal approximation to model the number of male piglets born.

- $n = 250$, and p (the probability that the piglet is male) is 0.5. So if X represents the number of male piglets born, then $\boldsymbol{X \sim B(250, 0.5)}$.
- Since n is large and p is 0.5, a normal approximation is appropriate — X can be approximated by a normal random variable $Y \sim N(\mu, \sigma^2)$:

  $\mu = np = 250 \times 0.5 = 125$ and $\sigma^2 = np(1 - p) = 250 \times 0.5 \times 0.5 = 62.5$

  So $Y \sim N(125, 62.5)$

b) Use your approximation to estimate the probability that there will be more than 130 male piglets born.

- Using your approximation, $P(X > 130) \approx P(Y > 130)$.
- From your calculator, $P(Y > 130) =$ 0.264 (3 s.f.)

Tip: If you used a continuity correction here, you'd work out $P(Y > 130.5)$ instead, which is 0.243 (3 s.f.).

Example 4

a) Only 23% of robin chicks survive to adulthood. If 200 robin chicks are randomly sampled from across the country, use a suitable normal approximation to estimate the probability that at least 25% of them survive to adulthood.

MODELLING

- Start by defining the random variable, and stating how it is distributed. If X represents the number of survivors, then $X \sim B(200, 0.23)$.
- p **isn't** particularly close to 0.5, but n is large, so find np and $n(1-p)$:

  $np = 200 \times 0.23 = \mathbf{46}$ and $n(1-p) = 200 \times (1 - 0.23) = \mathbf{154}$.

  Both np and $n(1-p)$ are much **greater than 5** (see p.285), so X can be approximated by $Y \sim N(\mu, \sigma^2)$.
- Work out μ and σ^2:

  $\mu = np = 46$ and $\sigma^2 = np(1-p) = 200 \times 0.23 \times (1 - 0.23) = 35.42$

  So approximate X with the random variable $\mathbf{Y \sim N(46, 35.42)}$.
- 25% of 200 = 50, which means you need to find $P(X \geq 50) \approx P(Y \geq 50)$. Using your calculator, $P(Y \geq 50) =$ 0.251 (3 s.f.)

Tip: If you used a continuity correction here, you'd work out $P(Y > 49.5)$ instead, which is 0.278 (3 s.f.).

b) A student takes a sample of 11 robin chicks from her garden. Explain whether or not a normal approximation would still be appropriate to estimate the probability of at least 25% surviving to adulthood.

- This time, if X is the number of survivors, then $X \sim B(11, 0.23)$ — so $np = 11 \times 0.23 = \mathbf{2.53}$, which is less than 5.
- n is small and p isn't close to 0.5, so a normal approximation is not appropriate.

Exercise 2.2

Q1 Which of the binomial distributions described below would a normal approximation be suitable for? Give reasons for your answers.

a) $X \sim B(600, 0.51)$ b) $X \sim B(100, 0.98)$

c) $X \sim B(100, 0.85)$ d) $X \sim B(6, 0.5)$

Q2 The normal random variable $Y \sim N(\mu, \sigma^2)$ is to be used to approximate these binomial distributions. Find μ and σ^2 in each case.

a) $X \sim B(350, 0.45)$ b) $X \sim B(250, 0.35)$ c) $X \sim B(70, 0.501)$

Q3 It is estimated that 5% of people are carriers of a certain disease. A health authority tests a sample of 1000 people to see if they carry the disease. If more than 75 people test positive they will offer a vaccination to the whole population.

MODELLING

a) Explain why the normal distribution would be a suitable approximation for the number of people in the sample (X) carrying the disease.

b) Use an appropriate normal distribution to estimate the probability that the whole population will be offered a vaccination.

3. Choosing Probability Distributions

MODELLING

Learning Objectives:

- Be able to choose a suitable probability distribution for a given context.
- Be able to explain why a probability distribution is or isn't suitable for a given context.

This section shows you how to choose a suitable probability distribution for a given context and explain why it is suitable for that context.

Choosing probability distributions

An exam question might ask you to choose a suitable probability distribution or explain why one is appropriate.

This can get quite confusing — so here's a handy summary of it all.

Normal distribution: $X \sim N(\mu, \sigma^2)$

Conditions:

- The values are **continuous**.
- The values are **symmetrically distributed**, with a **peak** in the **middle** — this peak represents the **mean** of the values.
- The values **'tails off'** either side of the mean — i.e. values become **less likely** as you move **further away** from the mean.

Then the situation can be modelled by a **normal distribution**, $N(\mu, \sigma^2)$.

For the continuous random variable $X \sim N(\mu, \sigma^2)$:

- X is a **continuous** random variable with **mean** μ and **standard deviation** σ. The mean of X (μ) will correspond to the mean of the values.
- The graph of X has **points of inflection** at $x = \mu - \sigma$ and $x = \mu + \sigma$.
- **68%** of the area under the graph of X is between $x = \mu + \sigma$ and $x = \mu - \sigma$. This means that 68% of the values should fall between $x = \mu \pm \sigma$. Similarly **95%** of the values should fall between $x = \mu \pm 2\sigma$ and **nearly all** of the values (99.7%) should fall between $x = \mu \pm 3\sigma$.
- You can estimate the **probability** that a value will fall within a range by using the normal distribution functions on your **calculator**.

Tip: You first saw the normal distribution on page 269.

Tip: Points of inflection are where the curve changes between concave and convex — see pages 135-136.

Binomial distribution: $X \sim B(n, p)$

Conditions:

- The values represent the **number of 'successes'** in a **fixed number** (n) of trials, where each trial involves either **'success'** or **'failure'**.
- Trials are **independent**, and the probability of 'success' (p) is **constant**.

Then the situation can be modelled by a **binomial distribution**, $B(n, p)$.

For the discrete random variable $X \sim B(n, p)$:

- $P(X = x) = \binom{n}{x} \times p^x \times (1 - p)^{n-x}$
- Probabilities can be found using the binomial distribution functions on your **calculator**.
- If n is large and $p \approx \frac{1}{2}$, then the values can also be approximately modelled using a **normal distribution**, $N(np, np(1 - p))$. In this case, you might need to use a **continuity correction**.

Tip: In Year 1 you were introduced to the binomial distribution — look back at your notes if you need a reminder.

Tip: On pages 284-287 you saw how a normal distribution can be used to approximate a binomial distribution.

The **conditions** for the normal distribution are useful when you're given a **diagram** and asked to find μ and σ.

Example 1

The times taken by runners to finish a 10 km race are normally distributed. Use the diagram to estimate the mean and standard deviation of the times.

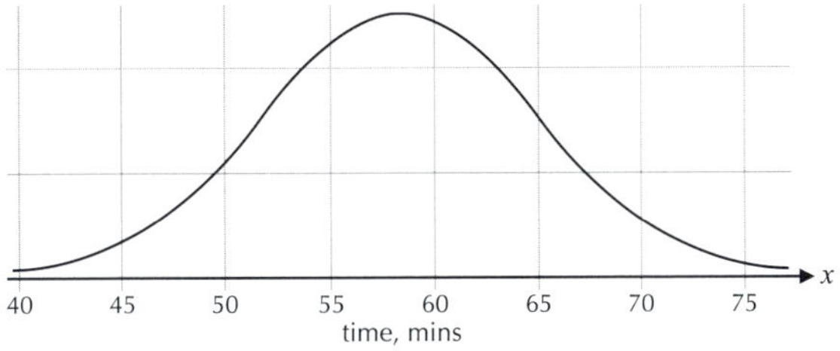

- The mean is in the middle, so $\mu \approx 58$ minutes.
- For a normal distribution there is a point of inflection at $x = \mu + \sigma$.
- Use the diagram to find the point of inflection. This is where the line changes from concave to convex — i.e. at $x = 65$.
- Use your values for x and μ to find σ.
  $65 = 58 + \sigma \Rightarrow \sigma = 7$ minutes

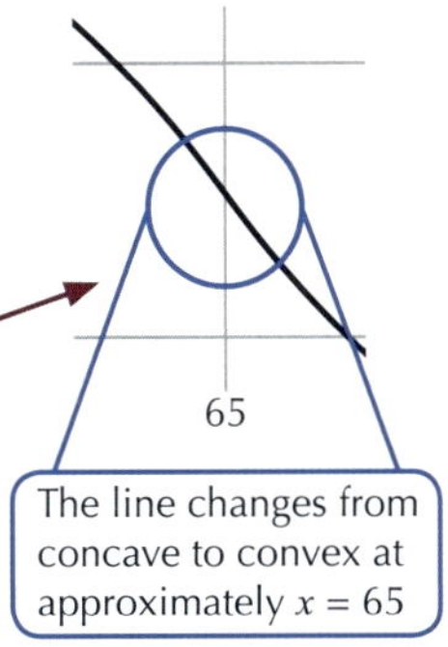

Tip: You could have used any of the normal conditions involving μ and σ here. These would give slightly different estimates so a range of answers would be marked correct. Using the point of inflection is probably the simplest method though.

You might be asked to **explain** why a distribution **is** or **isn't suitable** so try to spot the conditions from the previous page.

Example 2

A restaurant has several vegetarian meal options on its menu. The probability of any person ordering a vegetarian meal is 0.15. One lunch time, 20 people order a meal.

a) Suggest a suitable model to describe the number of people ordering a vegetarian meal and give the values of any parameters.

There are a fixed number of trials (20 meals), with probability of success (i.e. vegetarian meal) 0.15. If X is the number of people ordering a vegetarian meal, then $X \sim B(20, 0.15)$.

b) Explain why it wouldn't be suitable to use a normal approximation here.

$n = 20$ is not particularly large and $p = 0.15$ is quite far away from 0.5.

Tip: For b) you could have calculated $np = 3$ — which is less than 5.

c) Find the probability that at least 5 people order a vegetarian meal.

- Use the binomial cdf on your calculator, with $n = 20$ and $p = 0.15$.
- Your calculator will probably give probabilities for $P(X \leq x)$ so you might need to do a bit of rearranging.

$$P(X \geq 5) = 1 - P(X < 5) = 1 - P(X \leq 4)$$
$$= 1 - 0.8298...$$
$$= 0.170 \text{ (3 s.f.)}$$

Example 3

The heights of 1000 sunflowers are measured and the mean height is calculated to be 9.8 ft. The distribution of the sunflowers' heights is symmetrical about the mean, with the shortest sunflower measuring 5.8 ft and the tallest measuring 13.7 ft. The standard deviation of the sunflowers' heights is 1.3 ft.

a) Explain why the distribution of the sunflowers' heights might reasonably be modelled using a normal distribution.

- The data collected is **continuous** and the distribution of the heights is **symmetrical** about the **mean**. This is also true for a normally distributed random variable X.
- **Almost all** the data is within **3 standard deviations** of the mean: $9.8 - (3 \times 1.3) = \mathbf{5.9}$ and $9.8 + (3 \times 1.3) = \mathbf{13.7}$
- So the random variable $\boldsymbol{X \sim \mathrm{N}(9.8, 1.3^2)}$ seems a reasonable model.

b) Sunflowers that measure 7.5 ft or taller are harvested. Estimate the number of sunflowers that will be harvested.

- Using a calculator: $\mathrm{P}(X \geq 7.5) = 0.961572...$
- Multiply the total number of sunflowers by this probability: $1000 \times 0.961572... = 962$ (to the nearest whole number)

Exercise 3.1

MODELLING

Q1 The speeds of 100 randomly selected cars on a stretch of road are recorded by a speed camera. The table below shows the results.

Speed s, mph	$s < 35$	$35 \leq s < 40$	$40 \leq s < 45$	$45 \leq s < 50$	$50 \leq s < 55$	$55 \leq s < 60$	$s \geq 60$
Freq.	0	3	28	40	27	2	0

a) Explain why it would be reasonable to use a normal distribution to model the cars' speeds.

b) The distribution of cars' speeds on a different stretch of road is shown in the diagram below. Estimate the standard deviation of the speeds.

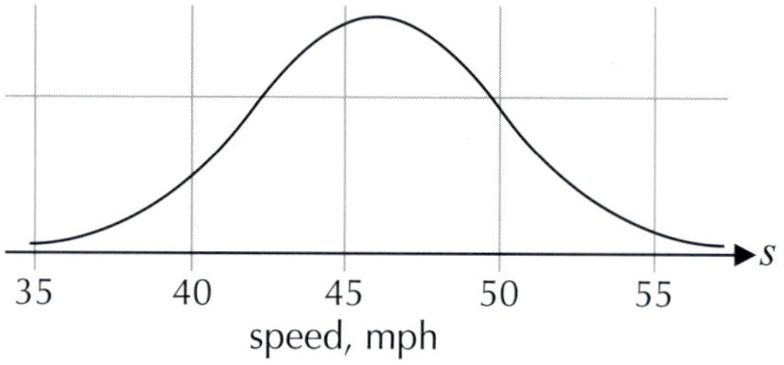

Q2 A tennis player successfully completes a first-serve 65% of the time. Give a suitable distribution to model the number of successful first-serves in each of the situations below. State any parameters.

a) The number of successful first-serves in one game, given that the player attempts 10 first-serves.

b) The number of successful first-serves in one match, given that the player attempts 100 first-serves.

1. The Product Moment Correlation Coefficient

The product moment correlation coefficient (PMCC) measures correlation. You also need to be able to carry out hypothesis tests on the PMCC.

Learning Objectives:

- Be able to interpret different values of the product moment correlation coefficient.
- Be able to carry out hypothesis tests on the product moment correlation coefficient.

Understanding the PMCC

The **product moment correlation coefficient** (r for short) measures the **strength** of the linear correlation between two variables. It basically tells you how close to a straight line the points on a scatter diagram lie.

The PMCC is always between +1 and –1.

- If all your points lie **exactly on a straight line** with a **positive** gradient (perfect positive correlation), $r = +1$.
- If all your points lie **exactly on a straight line** with a **negative** gradient (perfect negative correlation), $r = -1$.
- If $r = 0$ (or more likely, pretty close to 0), that would mean the variables **aren't correlated**.

These graphs give you an idea of what different values of r look like.

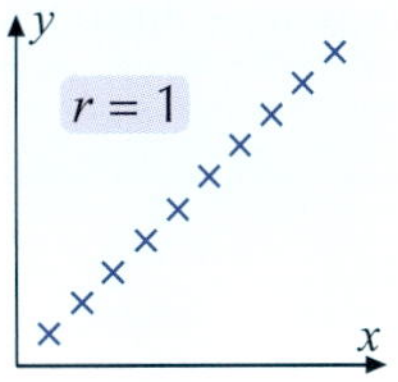

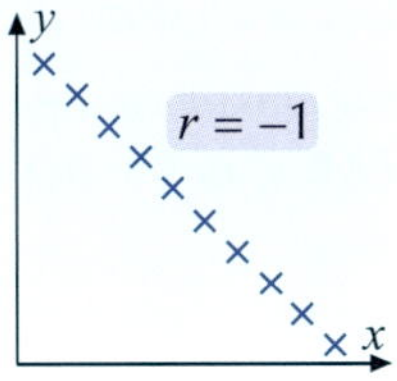

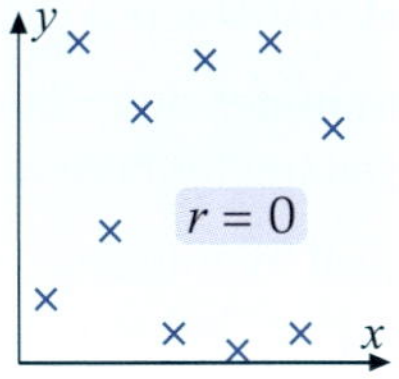

Tip: Correlation was covered in Year 1 — take a look back at your notes if you need a reminder.

Tip: The closer the value of r to +1 or –1, the stronger the correlation between the two variables (values close to +1 mean a strong positive correlation, and values close to –1 mean a strong negative correlation).

Values of r close to zero mean there is only a weak correlation.

Example 1

MODELLING

Marcus records the amount of exercise (in hours) that 30 people do in a month, and the length of time (in mins) it takes them to complete a lap of a cycle route. He calculates the product moment correlation coefficient of his data to be $r = -0.866$.

Interpret this result in context.

- r is close to –1, so this means the data has a **strong negative correlation**.
- This shows that as one variable increases, the other decreases.
- In context, this suggests that the **more** hours of exercise a person does, the **less** time it takes them to cycle the route (i.e. they can cycle quicker).
- However, you need to be careful with your conclusion. There isn't enough data to conclude that one variable necessarily **causes** the change — there might be **another** factor that has an effect on **both** variables, such as age.

Example 2

For the following sets of data, explain why the PMCC given might be misleading.

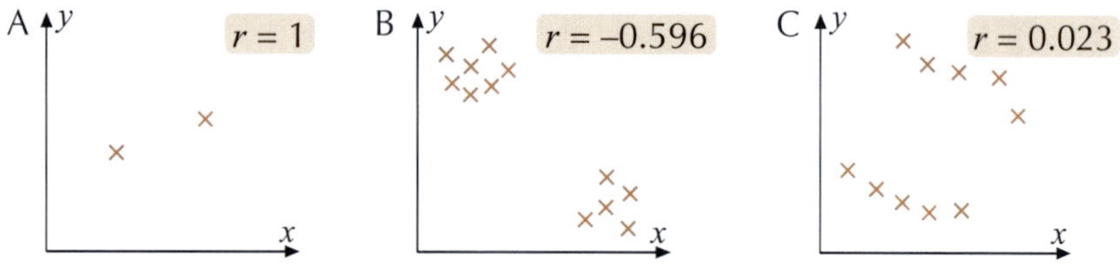

- In A, there are only two data points, so r would always be +1 or –1 (you can always draw a straight line between two points). This doesn't tell you anything about the relationship between the variables.
- The scatter graph of B shows two clusters of data — there appears to be negative correlation overall, but none within the clusters. Similarly, C contains two data clusters, each with negative correlation, despite r suggesting virtually no correlation.
  Establishing what has caused the two clusters in each case and separating the data would lead to better conclusions.

Tip: Scatter graph B could show age on the x-axis and time spent playing outside on the y-axis — then the clusters could be children and adults.

PMCC hypothesis testing

To test whether a value of r is likely to mean that the two variables are actually correlated, you need to do a hypothesis test. The method used is very similar to the hypothesis testing you did in Year 1.

- A **test statistic** is a statistic calculated from the sample data — here it's r.
- A **parameter** describes a characteristic of a population. The parameter here is ρ, the PMCC of the population.
- The **null hypothesis** (**H_0**) is a statement about the value of the population parameter, ρ.
- The null hypothesis here is always that ρ is **zero** — i.e. that there's **no correlation** between the two variables.

$H_0: \rho = 0$

Tip: ρ is the Greek letter rho.

The **alternative hypothesis** (**H_1**) is what you're going to conclude if you **reject** the null hypothesis — there are two kinds of alternative hypothesis:

- A **one-tailed** alternative hypothesis specifies whether the parameter you're investigating is **greater than** or **less than** the value you used in H_0.
- A **two-tailed** alternative hypothesis says that the parameter you're investigating is **not equal** to the value in H_0.

For a **one-tailed** test: $H_1: \rho > 0$ or $\rho < 0$
For a **two-tailed** test: $H_1: \rho \neq 0$

The **significance level** of a test (α) determines **how unlikely** your data needs to be under the null hypothesis (H_0) before you reject H_0.

- Significance levels can be given as a **decimal** or **percentage**.
- You're usually told what significance level to use, but the most common values are $\alpha = 0.1$ (or 10%), $\alpha = 0.05$ (or 5%) and $\alpha = 0.01$ (or 1%).

Tip: This should all be familiar to you — have a look back at your Year 1 notes if you need a reminder.

On page 369 there's a table of critical values. For a given sample size and significance level (for either a one-tailed or a two-tailed test) you can find the **critical value** — the point at which your test statistic r would be **significant**.

The examples below show how to use the table of critical values to test for significance.

Tip: You'll be given the critical values as part of the question in the exam, so you won't need to hunt through pages of tables to find what you need.

Example 1

A teacher thinks that test scores are related to the number of hours spent revising. She samples 10 students and finds that the PMCC is 0.76. Carry out a hypothesis test at the 10% significance level to investigate whether the evidence suggests that test scores and hours spent revising are correlated.

MODELLING

- You're given the test statistic for the sample: $r = 0.76$.
- The null hypothesis is that there is no correlation, so $H_0: \rho = 0$
- This is a two-tailed test, so the alternative hypothesis is $H_1: \rho \neq 0$
- Test for significance using a two-tailed significance level of $\alpha = 10\%$ and a sample size of $n = 10$.

Using the table, the critical value is 0.5494.

So you would reject H_0 if $r \geq 0.5494$ or if $r \leq -0.5494$.

Since $0.76 > 0.5494$, the result is significant.

one-tailed	10%	5%	2.5%	1%	0.5%
two-tailed	20%	10%	5%	2%	1%
n					
4	0.8000	0.9000	0.9500	0.9800	0.9900
⋮	⋮	⋮	⋮	⋮	⋮
9	0.4716	0.5822	0.6664	0.7498	0.7977
10	0.4428	0.5494	0.6319	0.7155	0.7646
11	0.4187	0.5214	0.6021	0.6851	0.7348

- Write your **conclusion** — you will either reject the null hypothesis H_0 or have insufficient evidence to do so.

There is evidence at the 10% level of significance to reject H_0 and to support the alternative hypothesis that test scores and hours spent revising are correlated.

Tip: Make sure you look up the 10% significance level for a two-tailed rather than a one-tailed test — you want to have 5% in each 'tail', so it's the same column you'd use for a one-tailed test at the 5% significance level.

Tip: To get maximum marks you need to include all the steps above and write a conclusion that links the answer to the context of the question.

Finding the critical region is another way to do a hypothesis test.

- The **critical region** is the set of **all** values of the test statistic that would cause you to **reject** $\mathbf{H_0}$.
- First, find all the values that would make you reject H_0, then work out your test statistic and check if it's in the critical region.

Example 2

Mary believes that the more a person likes romance movies, the less they tend to like horror movies (and vice versa). She asks 20 people at a cinema to give a score out of 100 for each type of movie, and calculates the product moment correlation coefficient of the data.

MODELLING

a) Find the critical region for a test of Mary's claim at the 5% level.

- This is a one-tailed test — Mary thinks there is negative correlation, i.e. the PMCC is less than zero. So the hypotheses are:

$$H_0: \rho = 0 \quad \text{and} \quad H_1: \rho < 0$$

Tip: The alternative hypothesis is $\rho < 0$, so you need to make the critical value from the table negative, since it's negative correlation you're expecting to see. If the alternative hypothesis was $\rho > 0$, you would use $r \geq 0.3783$ as your critical region instead.

- Find the critical value for a sample size of $n = 20$ and a one-tailed significance level of 5%.

Using the table, the critical value is 0.3783.

You would reject H_0 if $r \leq -0.3783$.

one-tailed	10%	5%	2.5%	1%	0.5%
two-tailed	20%	10%	5%	2%	1%
n					
4	0.8000	0.9000	0.9500	0.9800	0.9900
⋮	⋮	⋮	⋮	⋮	⋮
18	0.3170	0.4000	0.4683	0.5425	0.5897
19	0.3077	0.3887	0.4555	0.5285	0.5751
20	0.2992	0.3783	0.4438	0.5155	0.5614

- So the critical region is $r \leq -0.3783$

b) Mary calculates a PMCC of $r = -0.28$. State the conclusion of the hypothesis test.

- Compare this value of r to the critical region — $r = -0.28$ is greater than the critical value of -0.3783, so r is not in the critical region and the result is not significant.
- Write your conclusion.

$r = -0.28$ is outside the critical region so there is insufficient evidence to reject H_0 at the 5% significance level. The data does not provide significant evidence to support the claim that the more a person likes romance movies, the less they like horror movies.

Exercise 1.1

MODELLING

Q1 Hint: Use the PMCC table on p.369.

Q1 The lengths and widths (in cm) of 8 leaves from a tree were measured.

a) Find the critical region for a hypothesis test, at the 2.5% significance level, of the claim that the PMCC for the population, ρ, is greater than zero.

b) The PMCC of the data set, r, is found to be 0.994. Interpret this result in context and state the conclusion of the hypothesis test.

Q2 The results for a random sample of 13 students in a maths test and a science test were recorded. The product moment correlation coefficient was found to be $r = 0.4655$. Carry out a hypothesis test at the 5% significance level to test the claim that maths and science test results are positively correlated.

Q3 Hint: The doctor doesn't know whether the correlation is positive or negative — she just thinks they are linked in some way.

Q3 A doctor checked the kidney function of 9 of her patients to see if it was related to their weight. She thinks kidney function and weight are correlated.

a) Find the critical region for a test of the doctor's claim at the 1% significance level.

b) The sample data gave a product moment correlation coefficient of $r = -0.8925$. Interpret this value in context and say whether this result provides sufficient evidence to back the doctor's claim.

2. Hypothesis Tests of the Mean of a Population

You can also carry out a hypothesis test to check claims about the mean of a population that has a normal distribution. There's not too much to learn here — it's mostly just applying what you already know.

MODELLING

Learning Objectives:

- Be able to conduct a hypothesis test about the population mean of a normal distribution when the variance is known.

Hypothesis tests of a population mean

You'll be carrying out hypothesis tests of the **mean** of a normal distribution, based on a **random sample** of ***n*** **observations** from the population.

In the tests you need to carry out, the population will be **normally distributed** with **known variance** (i.e. $X \sim N(\mu, \sigma^2)$). Here's how the hypothesis test works:

- The **population parameter** you're testing will always be μ, the mean of the population.
- The **null** hypothesis will be: $H_0: \mu = a$ for some constant a.
- The **alternative** hypothesis, H_1, will either be

  $H_1: \mu < a$ or $H_1: \mu > a$ (one-tailed test)

  or $H_1: \mu \neq a$ (two-tailed test)
- State the **significance level**, α — you'll usually be given this.
- To find the value of the **test statistic**:

  — Calculate the **sample mean**, $\bar{x}$.

  — If $X \sim N(\mu, \sigma^2)$, then $\bar{X} \sim N\left(\mu, \frac{\sigma^2}{n}\right) \Rightarrow Z = \frac{\bar{X}-\mu}{\sigma/\sqrt{n}} \sim N(0, 1)$

  — Then the value of your **test statistic** will be: $z = \frac{\bar{x}-\mu}{\sigma/\sqrt{n}}$
- Use a **calculator** to test for significance, either by:

  — finding the **probability** of your test statistic taking a value **at least as extreme** as your observed value (this probability is called the ***p*-value**) and comparing it to the significance level α.

  — finding the **critical value(s)** of the test statistic and seeing if your observed value lies in the **critical region**.

Tip: $\bar{X}$ is a random variable representing the mean of any sample. $\bar{x}$ is the mean for a particular sample.

Tip: Remember, the critical value(s) depend on the significance level α, and whether it's a one-tailed or a two-tailed test.

Example 1

The times, X, in minutes, taken by the athletes in a running club to complete a certain run follow a normal distribution $X \sim N(12, 4)$. The coach increases the number of training sessions per week, and a random sample of 20 times run since the increase has a mean of 11.2 minutes. Assuming that the variance has remained unchanged, test at the 5% level whether there is evidence that the mean time has decreased.

- Let μ = mean time (in minutes) since the increase in training sessions.
- Then the **hypotheses** are: $H_0: \mu = 12$, $H_1: \mu < 12$

  So this is a **one-tailed** test.

Tip: You're testing to see if the mean time has decreased.

Tip: $\sigma^2 = 4$, so $\sigma = 2$.

Tip: Under H_0, $X \sim N(12, 4)$, so $\overline{X} \sim N(12, \frac{4}{20})$, $\overline{X} \sim N(12, 0.2)$.
So under H_0, $Z = \frac{\overline{X} - 12}{\sqrt{0.2}} \sim N(0,1)$

- The **significance level** is 5%, so $\alpha = 0.05$.
- Now find the value of your test statistic:
  $\bar{x} = 11.2$, so $z = \frac{\bar{x} - \mu}{\sigma/\sqrt{n}} = \frac{11.2 - 12}{2/\sqrt{20}} = -1.7888...$
- All that's left is to test for significance. Here are the **two methods**...

Method 1: Finding the *p*-value

- Work out the p-value — this is the probability of the test statistic (Z) being **at least as extreme** as the observed value of $z = -1.7888...$
- This is a **one-tailed test** and values more likely to occur under H_1 are at the **lower end** of the distribution. So 'at least as extreme as the observed value' means '$-1.7888...$ or lower'.
- The p-value is $P(Z \leq -1.7888...) = 0.0368 < 0.05$ ← $\alpha = 0.05$
- So the result is **significant**.

Method 2: Finding the critical region

- This is a **one-tailed test** and the critical region will be at the **lower end** of the distribution. So the critical value is z such that $\mathbf{P(Z < z) = 0.05}$.
- Using your calculator, the critical value is $z = -1.645$, meaning the critical region is $Z < -1.645$.

$P(Z < -1.645) = 0.05$ →

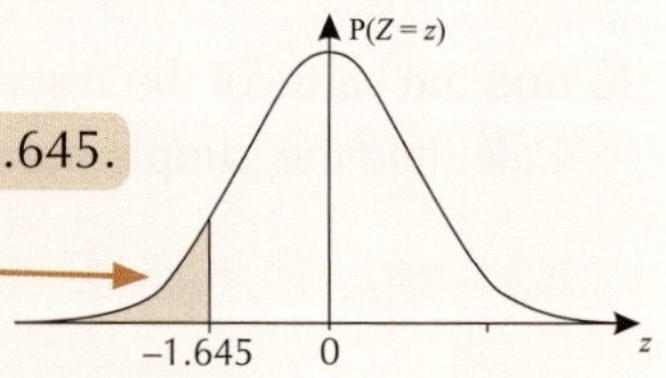

- Since $z = -1.788... < -1.645$, the observed value of the test statistic lies in the critical region.
- So the result is **significant**.

- All you need to do now is write your conclusion.

  There is evidence at the 5% level of significance to **reject H_0** and to suggest that the mean time has decreased.

Example 2

The volume (in ml) of a cleaning fluid dispensed in each operation of a machine is normally distributed with mean μ and standard deviation 3. Out of a random sample of 20 measured volumes, the mean volume dispensed was 30.9 ml.

Does this data provide evidence at the 1% level of significance that the machine is dispensing a mean volume that is different from 30 ml?

Tip: Under H_0, $X \sim N(30, 3^2)$, so $\overline{X} \sim N(30, \frac{3^2}{20})$.
So under H_0, $Z = \frac{\overline{X} - 30}{3/\sqrt{20}} \sim N(0,1)$

- Let μ = mean volume (in ml) dispensed in all possible operations of the machine (i.e. μ is the mean volume of the 'population').
- Your hypotheses will be: $H_0: \mu = 30$ and $H_1: \mu \neq 30$
- The **significance level** is 1%, so $\alpha = 0.01$.

- Now find the value of your test statistic:

  $\bar{x} = 30.9$, so $z = \dfrac{\bar{x}-\mu}{\sigma/\sqrt{n}} = \dfrac{30.9-30}{3/\sqrt{20}} = 1.3416...$

- This is a **two-tailed** test, so you need to check whether the p-value (the probability of the test statistic being at least as extreme as the observed value) is less than $\frac{\alpha}{2} = 0.005$.

  $P(Z \geq 1.3416...) = 0.0899 > \frac{\alpha}{2}$.

  This means this result is **not significant** at this level.

This data does **not** provide evidence at the 1% level to support the claim that the machine is dispensing a mean volume different from 30 ml.

Tip: This is a **two-tailed** test. The critical region is given by $P(Z > z) = \frac{\alpha}{2} = 0.005$ or $P(Z < -z) = \frac{\alpha}{2} = 0.005$.

Using your calculator, this gives a value for z of 2.576.

So the critical region is $Z > 2.576$ or $Z < -2.576$.

Exercise 2.1

MODELLING

Q1 The weight of plums (in grams) from a tree follows the distribution N(42, 16). It is suggested that the weight of plums has increased. A random sample of 25 plums gives a mean weight of 43.5 grams. Assuming the variance has remained unchanged, test this claim at:

a) the 5% significance level b) the 1% significance level

Q2 Bree plays cards with her friends, and the amount of money she wins each week is normally distributed with a mean, μ, of £24 and standard deviation, σ, of £2. A new friend has joined the game and Bree thinks this will change the amount she wins. In a random sample of 12 weeks the sample mean, $\bar{x}$, was found to be £22.50. Test Bree's claim at the 1% level.

Q3 The marks obtained in a language test by students in a school have followed a normal distribution with mean 70 and variance 64. A teacher suspects that this year the mean score is lower. A sample of 40 students is taken and the mean score is 67.7. Assuming the variance remains unchanged, test the teacher's claim at the 5% level of significance.

Q4 The weights (in grams) of pigeons in a city centre are normally distributed with mean 300 and standard deviation 45. The council claims that this year the city's pigeons are, on average, heavier. A random sample of 25 pigeons gives a mean of 314 g.

a) State the null hypothesis and the alternative hypothesis.

b) State whether it is a one- or two-tailed test.

c) Assuming the variance remains unchanged, use a 1% level of significance to test the hypothesis.

Q5 The number of hours that Kyle spends studying each week follows a normal distribution with mean 32 and standard deviation 5. Kyle believes that the mean amount of time he spends studying each week has increased. In a random sample of 20 weeks over the past year, the mean number of hours he spent studying was recorded as 34. Use a 5% and a 1% level to test Kyle's belief. Comment on your answers.

Chapter 14 Kinematics

1. Projectiles

Learning Objectives:

- Be able to resolve a body's velocity into its horizontal and vertical components.
- Be able to perform calculations involving projectile motion, including the use of **i-j** vectors.
- Be able to model projectile motion under gravity, and understand the modelling assumptions.
- Be able to derive general equations that describe the path of a projectile.

Tip: You can split the projectile's velocity at any point along its path into horizontal and vertical components — as long as you know its speed and direction at that particular point.

A projectile is an object which has been projected (e.g. thrown or fired) through the air. When you're doing projectile questions, you'll have to consider the motion of the projectile in two dimensions. That means you'll have to split the projectile's velocity into its horizontal and vertical components, then use the constant acceleration equations to find out more information about the projectile's motion.

The two components of velocity

When you project a body through the air with **initial speed *u***, at an **angle of *θ*** to the horizontal, it will move along a **curved path**:

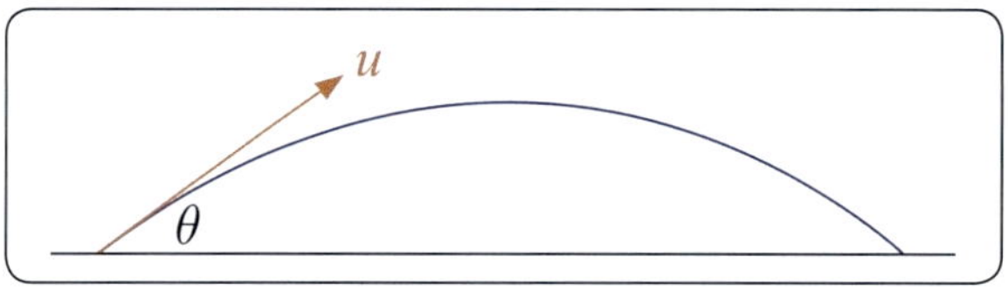

You can use **trigonometry** to resolve the body's initial velocity into its **horizontal** and **vertical components**:

- Horizontal component (x):

  $\cos\theta = \dfrac{\text{adjacent}}{\text{hypotenuse}} = \dfrac{x}{u}$, so $\boldsymbol{x = u\cos\theta}$

- Vertical component (y):

  $\sin\theta = \dfrac{\text{opposite}}{\text{hypotenuse}} = \dfrac{y}{u}$, so $\boldsymbol{y = u\sin\theta}$

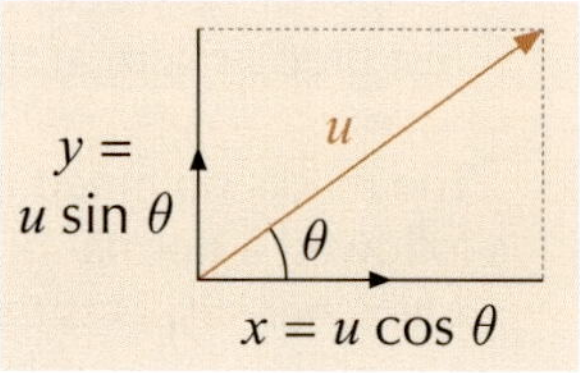

Example 1

A ball is thrown with initial speed 9 ms⁻¹ at an angle of 40° above the horizontal. Find the horizontal and vertical components of the ball's initial velocity.

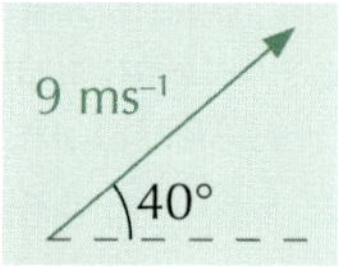

- Resolving horizontally:

  Horizontal component = 9 cos 40° = 6.89 ms^{-1} (3 s.f.)

- Resolving vertically:

  Vertical component = 9 sin 40° = 5.79 ms^{-1} (3 s.f.)

Given the horizontal and vertical components of a projectile's velocity, you can find its **speed** using **Pythagoras' theorem**.

You can find its **direction of motion** using **trigonometry**.

Example 2

At a particular point on its trajectory, a particle has velocity v, with horizontal component 12 ms^{-1} and vertical component –5 ms^{-1}.

Find the particle's speed, v, and direction of motion at this point.

- Draw a diagram to show the particle's motion.

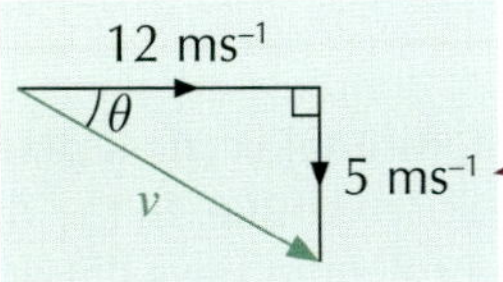

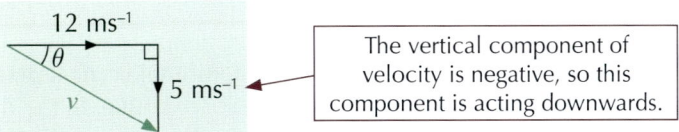

The vertical component of velocity is negative, so this component is acting downwards.

- Using Pythagoras' theorem:

$$v = \sqrt{12^2 + (-5)^2} = 13 \text{ ms}^{-1}$$

- Using trigonometry:

$$\theta = \tan^{-1}\left(\frac{5}{12}\right) = 22.6° \text{ (3 s.f.) below the horizontal}$$

Tip: A projectile's trajectory is just the path that it moves along.

Exercise 1.1

Q1 Each of the following diagrams shows the speed of a projectile and the angle its velocity makes with the horizontal. In each case, find the horizontal and vertical components of the projectile's velocity.

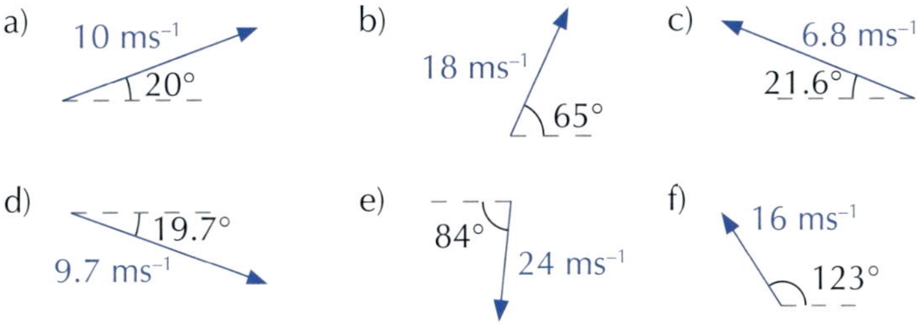

Q1 Hint: If you're not asked to give answers to a specific accuracy, rounding to 3 significant figures is fine for most situations.

Q2 A particle is moving with speed 8 ms^{-1} at an angle of 35° to the horizontal. Find the horizontal and vertical components of its velocity.

Q3 A rocket is fired vertically upwards with speed 45 ms^{-1}. Find the horizontal and vertical components of its initial velocity.

Q4 A body is fired at an angle α to the horizontal with speed 22 kmh^{-1}. Find the horizontal and vertical components of its initial velocity, giving your answer in metres per second.

Q5 A ball is thrown with velocity **v**, with horizontal component 6 ms^{-1} and vertical component 8 ms^{-1}.

Find the speed and direction of projection of the ball.

Q6 A particle moves with velocity **u**. The horizontal component of **u** is 17 ms^{-1} and the vertical component is –2.5 ms^{-1}.

Find the magnitude and direction of **u**.

The constant acceleration equations

To solve problems involving projectile motion, you will need to make certain **modelling assumptions**:

> **Tip:** An external force is any force acting on a body other than the body's weight — e.g. air resistance, friction, tension etc.

- The projectile is moving only under the influence of **gravity** (i.e. there'll be **no external forces** acting on it).
- The projectile is a **particle** — i.e. its weight acts from a single point, so its dimensions don't matter.
- The projectile moves in a **two-dimensional vertical plane** — i.e. it doesn't swerve from side to side. For example, in the diagram of the projectile's path on page 298, the projectile stays flat on the page — it doesn't move towards you (out of the page) or away from you (into the page).

Because the projectile is modelled as moving only under the influence of gravity, the only acceleration the projectile will experience will be **acceleration due to gravity** (g **= 9.8 ms**$^{-2}$).

> **Tip:** You might be asked to use different values for g (e.g. 10 ms^{-2} or 9.81 ms^{-2}), so read the question carefully. The examples and questions in this chapter use g = 9.8 ms^{-2}, unless stated otherwise.

Acceleration due to gravity acts **vertically downwards**, so will only affect the **vertical component** of the projectile's velocity. The **horizontal component** of the velocity will remain **constant** throughout the motion.

When **modelling real situations**, consider which of these assumptions would not be **valid**.

For example, a falling parachutist would encounter significant **air resistance**, so the model should ideally allow for this.

To answer a projectiles question, split all the vector quantities you know — **velocity**, **acceleration** and **displacement** — into their **horizontal** and **vertical components**. Then you can deal with the two components **separately** using the ***suvat*** **equations** below.

The thing that connects the two components is **time** — this will be the same no matter what direction you're resolving in.

> **Tip:** The *suvat* equations only work for motion in a straight line. Using them for projectiles means that you consider the horizontal and vertical components of the motion separately.

> **Tip:** Speed, distance and time must always be positive. But velocity, acceleration and displacement can be positive or negative.

The Constant Acceleration Equations

$$v = u + at$$

$$s = ut + \frac{1}{2}at^2$$

$$s = \left(\frac{u+v}{2}\right)t$$

$$v^2 = u^2 + 2as$$

$$s = vt - \frac{1}{2}at^2$$

s = displacement in m

u = initial speed (or velocity) in ms^{-1}

v = final speed (or velocity) in ms^{-1}

a = acceleration in ms^{-2}

t = time that passes in s (seconds)

Remember — these equations only work if the acceleration is **constant**.

Example 1

A particle is projected with initial speed 18 ms^{-1} from a point on horizontal ground, at an angle of 40° above the horizontal. Find the distance of the particle from its point of projection 1.8 seconds after it is projected.

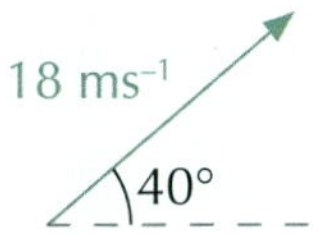

- Consider the horizontal and vertical motion separately.

- Resolving horizontally, taking right as positive:

  $s = s_x,\quad u = 18 \cos 40°,\quad a = 0,\quad t = 1.8$

  Using $s = ut + \frac{1}{2}at^2$:

  $s_x = (18 \cos 40° \times 1.8) + 0$

  $= 24.8198...$

- Resolving vertically, taking up as positive:

  $s = s_y,\quad u = 18 \sin 40°,\quad a = -9.8,\quad t = 1.8$

  Using $s = ut + \frac{1}{2}at^2$:

  $s_y = (18 \sin 40° \times 1.8) + (\frac{1}{2} \times -9.8 \times 1.8^2)$

  $= 4.9503...$

- So the particle's horizontal displacement from its starting point is 24.8198... metres and its vertical displacement is 4.9503... metres. Use Pythagoras' theorem to find its distance from the starting point:

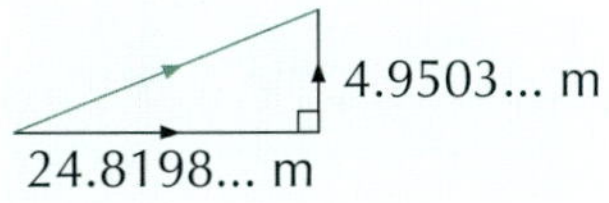

  distance $= \sqrt{(24.8198...)^2 + (4.9503...)^2}$

  $=$ 25.3 m (3 s.f.)

Tip: Subscripts $_x$ and $_y$ are often used to denote the horizontal and vertical components of a vector respectively. So here s_x is the particle's horizontal displacement and s_y is the particle's vertical displacement.

Tip: Distance must be positive, so ignore the negative square root.

Example 2

A ball is kicked from a point on horizontal ground. The initial velocity of the ball is 23 ms^{-1}, at an angle α to the horizontal. The ball reaches a maximum vertical height above the ground of 4.8 m.

Find the value of α.

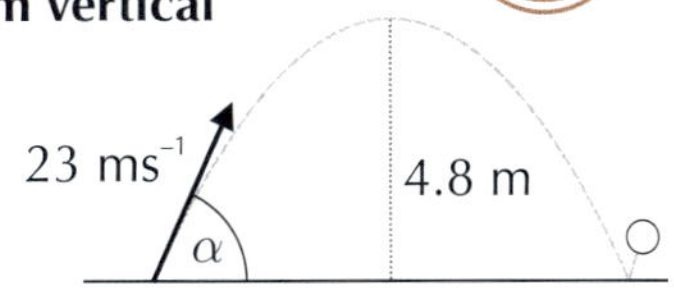

- Resolving vertically, taking up as positive:

  $s = 4.8,\quad u = 23 \sin \alpha,\quad v = 0,\quad a = -9.8$

  Using $v^2 = u^2 + 2as$:

  $0 = (23 \sin \alpha)^2 + (2 \times -9.8 \times 4.8)$

  $0 = 529 \sin^2 \alpha - 94.08$

  $\sin^2 \alpha = 94.08 \div 529$

  $= 0.1778...$

  $\sin \alpha = 0.4217...$

  $\Rightarrow \alpha =$ 24.9° (3 s.f.)

Tip: When the ball reaches its maximum height, the vertical component of its velocity will momentarily be zero.

Example 3

MODELLING

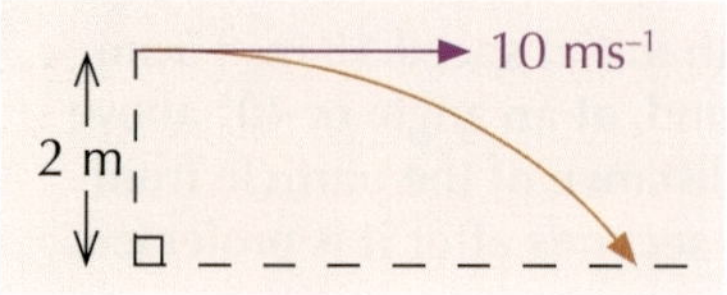

A stone is thrown horizontally with speed 10 ms^{-1} from a height of 2 m above horizontal ground.

a) Find the speed and direction of motion of the stone after 0.5 seconds.

- Consider the vertical and horizontal motion separately.
- Resolving vertically, taking down as positive:

  $u = 0, \quad v = v_y, \quad a = 9.8, \quad t = 0.5$

  Using $v = u + at$:

  $v_y = 9.8 \times 0.5 = 4.9$ ms^{-1}

- The horizontal component of velocity is constant, as there is no acceleration horizontally, so $v_x = u_x = 10$ ms^{-1}.
- You can now find the speed and direction of motion of the stone at this time.

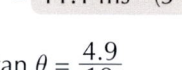

  $v = \sqrt{10^2 + (-4.9)^2}$

  $= 11.1$ ms^{-1} (3 s.f.)

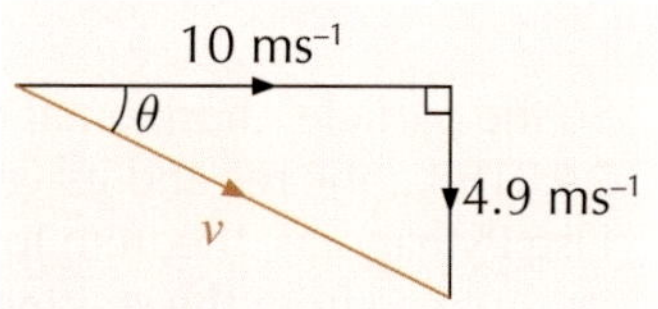

  $\tan\theta = \dfrac{4.9}{10}$

  $\Rightarrow \theta = 26.1°$ (3 s.f.) below the horizontal

Tip: The stone is thrown horizontally, so the vertical component of its initial velocity is zero.

b) Find the stone's horizontal displacement when it lands on the ground.

- First, you need to find the length of time that the stone is in the air before it lands.
- Resolving vertically, taking down as positive:

  $s = 2, \quad u = 0, \quad a = 9.8, \quad t = t$

  Using $s = ut + \frac{1}{2}at^2$:

  $2 = \frac{1}{2} \times 9.8 \times t^2$

  $t^2 = 0.4081...$

  $\Rightarrow t = 0.6388...$ s

- Now you know how long the stone is in the air, you can find its horizontal displacement in this time.
- Resolving horizontally, taking right as positive:

  $s = s, \quad u = 10, \quad a = 0, \quad t = 0.6388...$

  Using $s = ut + \frac{1}{2}at^2$:

  $s = 10 \times 0.6388... = 6.39$ m (3 s.f.)

Tip: You may see similar questions asked as 'find the horizontal range of the stone.'

Tip: The length of time the projectile is in the air is usually referred to as the 'time of flight'.

Example 4

A particle is projected from a point y m above flat horizontal ground. The initial velocity of the particle is U ms^{-1} at an angle θ to the horizontal:

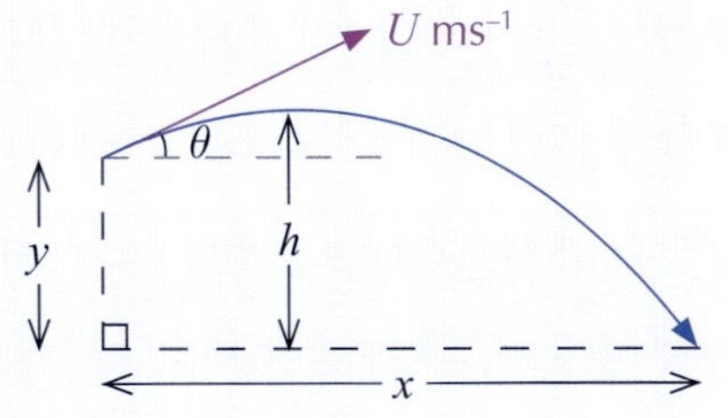

a) Find the maximum height the particle reaches above the ground, h, in terms of y, U and θ.

- Resolving vertically, taking up as positive:

$s = h - y, \quad u = U\sin\theta, \quad v = 0, \quad a = -g,$

Using $v^2 = u^2 + 2as$:

$$0 = (U\sin\theta)^2 + (2 \times -g \times (h - y))$$

$$0 = U^2\sin^2\theta - 2gh + 2gy$$

$$2gh = U^2\sin^2\theta + 2gy$$

$$h = \frac{U^2\sin^2\theta}{2g} + y$$

b) Find the horizontal range of the particle, x, in terms of y, U and θ.

- Resolving vertically, taking up as positive:

$s = -y, \quad u = U\sin\theta, \quad a = -g, \quad t = t$

Using $s = ut + \frac{1}{2}at^2$:

$$-y = (U\sin\theta \times t) + (\frac{1}{2} \times -g \times t^2) = (U\sin\theta)t - \frac{g}{2}t^2$$

$$\frac{g}{2}t^2 - (U\sin\theta)t - y = 0$$

Using the quadratic formula:

$$t = \frac{U\sin\theta \pm \sqrt{U^2\sin^2\theta - \left(4 \times \frac{g}{2} \times -y\right)}}{g}$$

$$= \frac{U\sin\theta \pm \sqrt{U^2\sin^2\theta + 2gy}}{g}$$

- Time should be positive, so $t = \frac{U\sin\theta + \sqrt{U^2\sin^2\theta + 2gy}}{g}$
- Now, resolving horizontally:

$s = x, \quad u = U\cos\theta, \quad a = 0, \quad t = \frac{U\sin\theta + \sqrt{U^2\sin^2\theta + 2gy}}{g}$

Using $s = ut + \frac{1}{2}at^2$:

$$x = \frac{U^2\sin\theta\cos\theta + \left(\sqrt{U^2\sin^2\theta + 2gy}\right)U\cos\theta}{g}$$

Tip: Since g and y are positive, $\sqrt{U^2\sin^2\theta + 2gy}$ is greater than $\sqrt{U^2\sin^2\theta}$ (i.e. $U\sin\theta$). So when the ± is a –, t is negative.

Tip: For projections at ground level ($y = 0$) the formulae for time of flight, t, and horizontal range, x, simplify as follows:

$t = \frac{2U\sin\theta}{g}$

$x = \frac{2U^2\sin\theta\cos\theta}{g}$

The particle is used to model a ball thrown at a velocity of 30 ms^{-1} at an angle of 25° above the horizontal, from a vertical height of 1.5 m.

c) Find the length of time the ball is at least 5 m above the ground.

Tip: Because the ball is thrown from 1.5 m above the ground, it only has to travel 3.5 m to reach a height of 5 m above ground level.

- Resolving vertically, taking up as positive:

  $s = 3.5, \quad u = 30 \sin 25°, \quad a = -9.8, \quad t = t$

  Using $s = ut + \frac{1}{2}at^2$:

  $3.5 = (30 \sin 25° \times t) + (\frac{1}{2} \times -9.8 \times t^2)$

  $4.9t^2 - (12.67...)t + 3.5 = 0$

  Using the quadratic formula:

  $$t = \frac{12.67... \pm \sqrt{(-12.67...)^2 - (4 \times 4.9 \times 3.5)}}{9.8}$$

  $\Rightarrow t = 0.314...$ or $t = 2.273...$

- So the ball is exactly 5 m above the ground 0.314... seconds after being thrown, and again 2.273... seconds after being thrown:

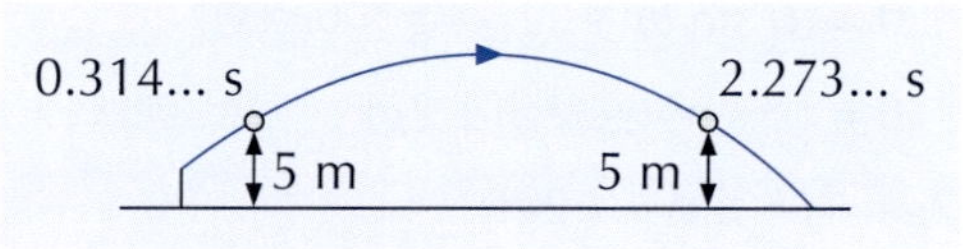

- So the length of time that the ball is at least 5 m above the ground is:

  2.273... – 0.314... = 1.96 s (3 s.f.)

Example 5

A golf ball is struck from a point *A* on a horizontal plane. When the ball has moved a horizontal distance *x*, its height above the plane is *y*. The ball is modelled as a particle projected with initial speed *u* ms^{-1} at an angle α above the horizontal.

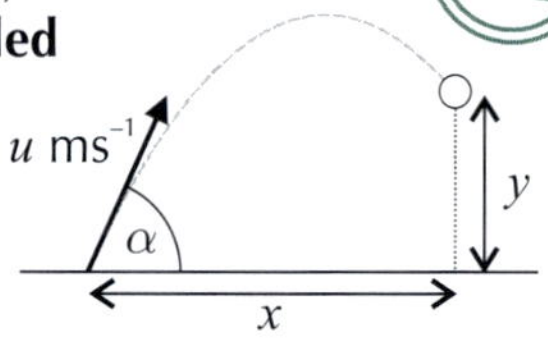

a) Show that $y = x \tan \alpha - \frac{gx^2}{2u^2 \cos^2 \alpha}$.

Tip: This is the general equation of the path of a particle projected from a point on a horizontal plane. The path of a particle projected from a point above the horizontal (as in Example 4) could be derived in the same way.

- Resolving horizontally, taking right as positive:

  $s = x, \quad u_x = u \cos \alpha, \quad a = 0, \quad t = t$

  Using $s = ut + \frac{1}{2}at^2$:

  $x = u \cos \alpha \times t$

  Rearrange to make t the subject:

  $t = \frac{x}{u \cos \alpha}$ — call this equation ①

- Resolving vertically, taking up as positive:

  $s = y, \quad u_y = u \sin \alpha, \quad a = -g, \quad t = t$

  Using $s = ut + \frac{1}{2}at^2$:

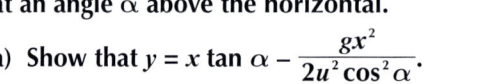

  $y = (u \sin \alpha \times t) - \frac{1}{2}gt^2$ — call this equation ②

Tip: It would be a massive pain to rearrange equation 2 to make t the subject, so you're better off doing it to equation 1.

- t is the same horizontally and vertically, so you can substitute equation ① into equation ② and eliminate t:

$$y = \left(u\sin\alpha \times \frac{x}{u\cos\alpha}\right) - \frac{1}{2}g\left(\frac{x}{u\cos\alpha}\right)^2$$

$$= x\frac{\sin\alpha}{\cos\alpha} - \frac{1}{2}g\left(\frac{x^2}{u^2\cos^2\alpha}\right)$$

$$= x\tan\alpha - \frac{gx^2}{2u^2\cos^2\alpha}$$ — as required.

Tip: t doesn't appear in the formula you're given in the question, so you know that's what you should be trying to eliminate.

b) The ball just passes over the top of a 10 m tall tree, which is 45 m from *A*. Given that $\alpha = 45°$, find the speed of the ball as it passes over the tree.

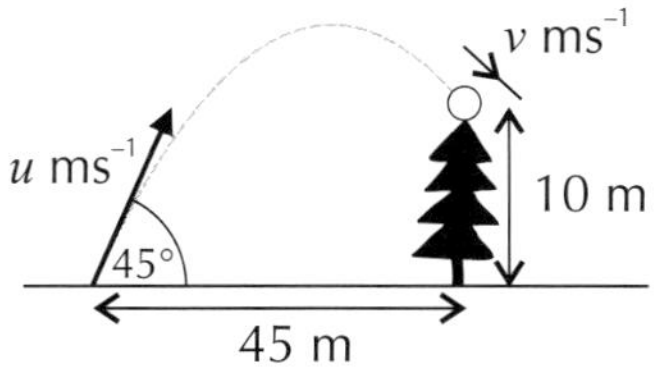

- First of all, you need to find the ball's initial speed, u.
- Using the result from part a), and substituting $x = 45$ m, $y = 10$ m and $\alpha = 45°$:

$$y = x\tan\alpha - \frac{gx^2}{2u^2\cos^2\alpha}$$

$$10 = 45\tan 45° - \frac{9.8 \times 45^2}{2u^2 \times \cos^2 45°}$$

$$10 = 45 - \frac{19\,845}{u^2}$$

$$\frac{19\,845}{u^2} = 35$$

$$\Rightarrow u = 23.811...\ \text{ms}^{-1}$$

Tip: Once you've found the ball's initial velocity, you can resolve horizontally and vertically and use the *suvat* equations as usual.

- So the ball was struck with initial speed 23.811... ms^{-1}, at an angle of 45° above the horizontal.
- Resolving horizontally, taking right as positive:

$$v_x = u_x = (23.811...)\cos 45°$$
$$= 16.837...\ \text{ms}^{-1}$$

Tip: Remember that the horizontal component of velocity stays constant throughout the motion — so v_x always equals u_x.

- Resolving vertically, taking up as positive:

$s = 10$, $u = (23.811...)\sin 45° = 16.837...$, $v = v_y$, $a = -9.8$

Using $v^2 = u^2 + 2as$:

$$v_y^2 = (16.837...)^2 + (2 \times -9.8 \times 10)$$
$$= 87.5\ \text{m}^2\text{s}^{-2}$$

- Now you can find the speed using Pythagoras' theorem:

$$\text{speed} = \sqrt{(16.837...)^2 + 87.5}$$
$$= 19.3\ \text{ms}^{-1}\ \text{(3 s.f.)}$$

Tip: Don't bother square rooting to find v_y — you'll need to use v_y^2 in the next step. And don't worry about the weird m^{2}s^{-2} units — it's just because the speed it squared, so the units should be squared too.

> **Hint:** Drawing a diagram might help you in some questions if you're struggling.

Exercise 1.2

Q1

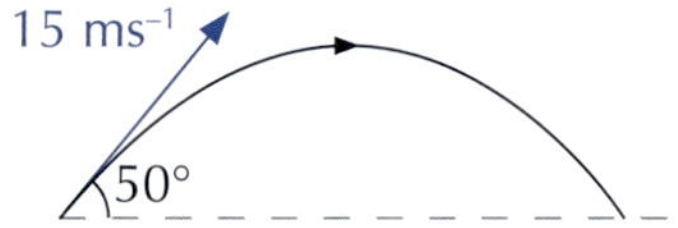

A projectile is launched from a point on horizontal ground with speed 15 ms^{-1} at an angle of 50° above the horizontal. Find:

a) the time taken for the projectile to reach its maximum height,

b) the maximum height the projectile reaches above the ground.

Q2 A stone is catapulted with speed 12 ms^{-1} at an angle of 37° above the horizontal. It hits a target 0.5 seconds after being fired. Find:

a) the stone's horizontal displacement 0.5 seconds after being fired,

b) the stone's vertical displacement 0.5 seconds after being fired,

c) the straight-line distance from the stone's point of projection to the target.

Q3 A ball is kicked from a point on a flat, horizontal field. It has initial speed 8 ms^{-1} and leaves the ground at an angle of 59°. Find:

a) the ball's time of flight,

b) the horizontal range of the ball.

> **Q4 Hint:** When you're given a value for tan θ, it often means you can use it to get sin θ and cos θ without needing a calculator. Watch out for familiar right-angled triangles such as
>
> 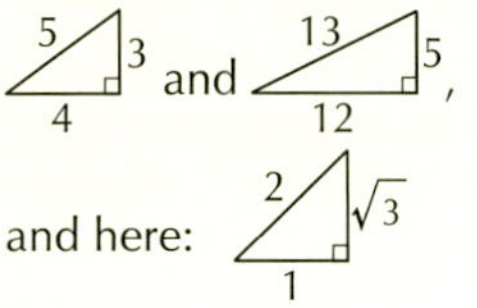
>

Q4 A particle is projected at an angle θ above the horizontal with speed 18 ms^{-1}. Given that $\tan\theta = \sqrt{3}$, find:

a) the particle's speed 2 seconds after being projected,

b) the direction of motion of the particle at this time.

Q5

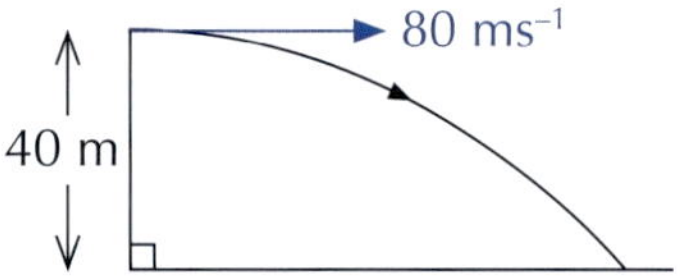

A bullet is fired horizontally at a speed of 80 ms^{-1} from the edge of a vertical cliff 40 m above sea level, as shown.

a) How long after being fired does the bullet hit the sea?

b) Find the horizontal distance from the bottom of the cliff to the point that the bullet hits the sea.

c) Suggest how the model used to calculate the bullet's motion in air should be refined to calculate its motion in the water.

Q6

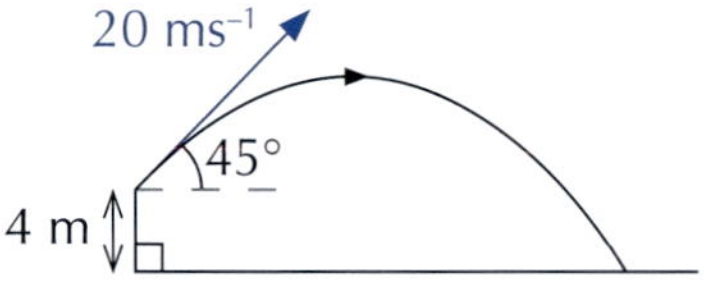

An object is fired from a point 4 m directly above flat, horizontal ground. It has initial speed 20 ms^{-1} at an angle of 45° above the horizontal. For how long is it higher than 11 m above the ground?

Q7

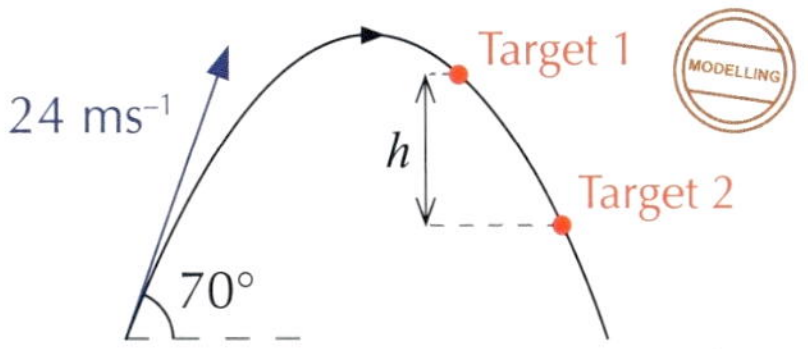

A ball is fired from a machine with speed 24 ms^{-1} at an angle of 70° above the horizontal. The ball passes through a target 3 seconds after being fired, and then passes through a second target 4 seconds after being fired. Find h, the difference in height between the two targets.

Q8 A particle is projected from a point 0.6 m above flat, horizontal ground. The particle's initial speed is 7.5 ms^{-1} at an angle of α above the horizontal. It reaches a maximum height of 2.8 m above the ground. Find the horizontal distance travelled by the particle in the first 1.2 seconds of its motion.

Q9 A ball is hit by a bat from a point on flat, horizontal ground. The ball leaves the ground with speed V ms^{-1}, at an angle of θ above the horizontal, where $\tan\theta = \frac{3}{4}$. The ball lands a horizontal distance of 50 m away. Find the value of V.

MODELLING

Q9 Hint: You should be able to write down the values of $\sin\theta$ and $\cos\theta$ without a calculator.

Q10 A particle is fired from a point 16 m above flat, horizontal ground. The particle's initial speed is 3 ms^{-1} at an angle of 7° below the horizontal. Find:

a) the horizontal range of the particle,

b) the speed and direction of the particle when it lands.

Q10 Hint: Be careful with this one — the particle is fired at an angle below the horizontal.

Q11 A golf ball is struck by a golf club from a point on flat, horizontal ground. It leaves the ground at an angle of 40° to the horizontal. 2 seconds after being struck, the ball is travelling upwards at an angle of 10° to the horizontal. Find:

MODELLING

a) the ball's speed of projection,

b) the height of the ball above the ground 2 seconds after being struck by the golf club.

Q12

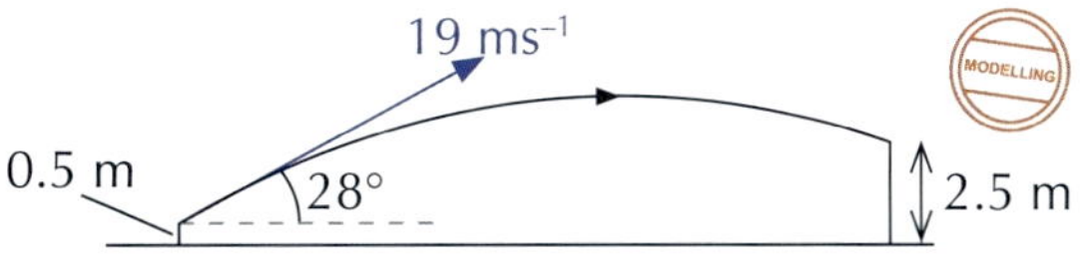

A cricket ball is hit from a point 0.5 m above flat, horizontal ground. The ball's initial speed is 19 ms^{-1}, at an angle of 28° above the horizontal. It is caught on its descent by a fielder when it is 2.5 m above the ground. Find:

a) the length of time that the ball is in the air,

b) the fielder's horizontal distance from where the ball was struck,

c) the speed of the ball at the point where it is caught.

Q13 A body is projected from a point on flat, horizontal ground with speed U ms^{-1} at an angle θ above the horizontal. Show that:

PROBLEM SOLVING

a) the body reaches a maximum height of $\frac{U^2 \sin^2\theta}{2g}$ metres above the ground,

b) the body reaches its maximum height $\frac{U\sin\theta}{g}$ seconds after it is projected.

Q14

MODELLING

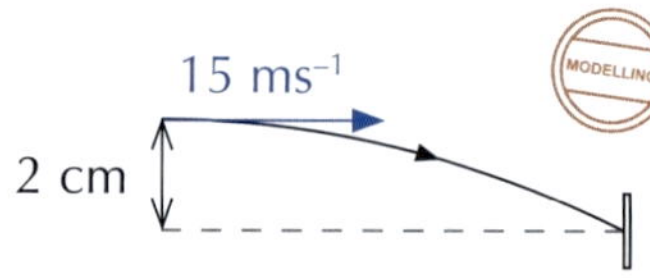

A dart is thrown horizontally towards a dartboard with speed 15 ms^{-1}. It hits the dartboard 2 cm vertically below the level at which it was thrown.

a) Find the horizontal distance from the point the dart is released to the dartboard.

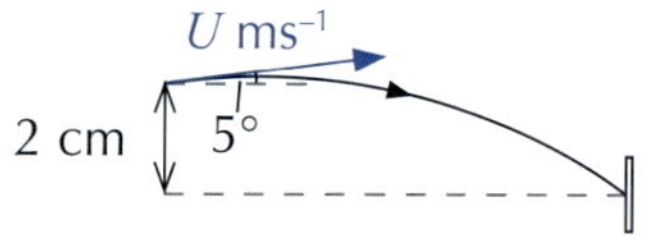

A second dart is thrown from the same point, but at an angle of 5° above the horizontal. It hits the dartboard at the same point as the first dart.

b) Find U, the speed that the second dart is thrown at.

Q15 A projectile is moving relative to the x- and y-coordinate axes. The projectile is fired from the origin with speed 3 ms^{-1}, at an angle α above the x-axis. The projectile moves freely under gravity and passes through the point (3, 1) m. Show that:

PROBLEM SOLVING

$$1 = 3\tan\alpha - \frac{g}{2\cos^2\alpha}$$

Q16

PROBLEM SOLVING

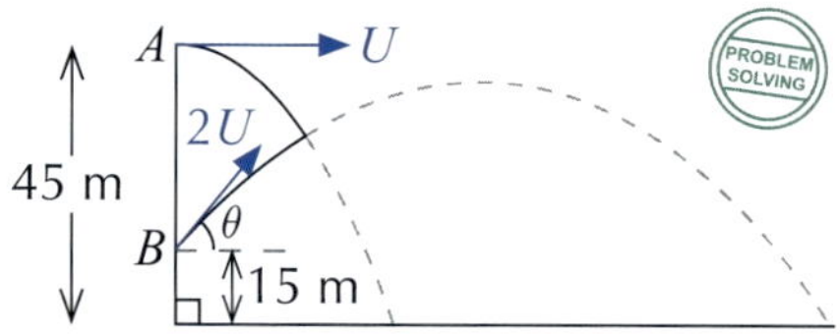

Particle A is projected horizontally from a point 45 m above ground level. At the same time, particle B is projected at an angle θ above the horizontal, from a point 15 m above ground level, directly below the point of projection of particle A.
The particles collide 2 seconds after they are projected.
Given that the speed of projection of B is twice the speed of projection of A, find:

a) the value of θ,

b) the speed of projection of each of the two particles.

i and j vectors

You can describe projectile motion in terms of the **unit vectors i** and **j**.

This isn't really all that different from any of the stuff from the last few pages — the **i**-component of the vector describes the **horizontal** motion of the projectile, and the **j**-component of the vector describes the **vertical** motion of the projectile.

When you're using **i** and **j** vectors to describe projectile motion, you can either consider horizontal and vertical motion **separately**, as before, or you can deal with both components in one go using the ***suvat* equations** in **vector form**:

$$\mathbf{v} = \mathbf{u} + \mathbf{a}t \qquad \mathbf{s} = \mathbf{u}t + \frac{1}{2}\mathbf{a}t^2$$
$$\mathbf{s} = \frac{1}{2}(\mathbf{u} + \mathbf{v})t \qquad \mathbf{s} = \mathbf{v}t - \frac{1}{2}\mathbf{a}t^2$$

If you're using the *suvat* equations in vector form, then acceleration due to gravity will be $\mathbf{a} = -g\mathbf{j}$ ($= -9.8\mathbf{j}$ ms^{-2}). It has no horizontal component as acceleration due to gravity always acts vertically downwards.

Tip: Remember, **i** and **j** are a pair of perpendicular vectors, each of magnitude one unit. **i** and **j** are directed horizontally and vertically respectively.

Tip: $v^2 = u^2 + 2as$ doesn't have a vector equivalent, as you can't really square a vector.

Example 1

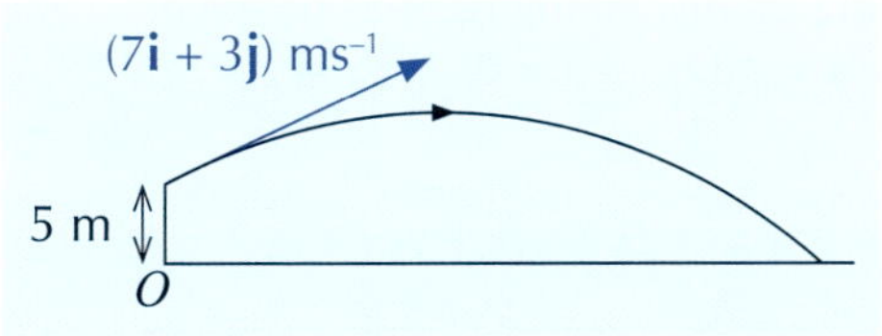

A particle is projected from the point 5j m, relative to a fixed origin *O*. The particle's initial velocity is (7i + 3j) ms^{-1}, and it moves freely under gravity.

a) Find the position vector of the particle 0.8 seconds after it is projected.

- Write down the *suvat* values:
  $\mathbf{s} = \mathbf{s}$, $\quad \mathbf{u} = (7\mathbf{i} + 3\mathbf{j})$, $\quad \mathbf{a} = -9.8\mathbf{j}$, $\quad t = 0.8$
- Using $\mathbf{s} = \mathbf{u}t + \frac{1}{2}\mathbf{a}t^2$:
  $\mathbf{s} = 0.8(7\mathbf{i} + 3\mathbf{j}) - \frac{1}{2} \times (9.8\mathbf{j}) \times 0.8^2$
  $= 5.6\mathbf{i} + 2.4\mathbf{j} - 3.136\mathbf{j}$
  $= (5.6\mathbf{i} - 0.736\mathbf{j})$ m
- This is the particle's displacement from the point of projection 0.8 seconds after it is projected. Add this to its initial position vector to find its new position vector:
  $5\mathbf{j} + 5.6\mathbf{i} - 0.736\mathbf{j} = (5.6\mathbf{i} + 4.264\mathbf{j})$ m

b) Find the velocity of the particle at this time.

- Using $\mathbf{v} = \mathbf{u} + \mathbf{a}t$:
  $\mathbf{v} = (7\mathbf{i} + 3\mathbf{j}) + 0.8(-9.8\mathbf{j})$
  $= 7\mathbf{i} + 3\mathbf{j} - 7.84\mathbf{j}$
  $= (7\mathbf{i} - 4.84\mathbf{j})$ ms^{-1}

Example 2

A stone is launched from a point 1.2 metres vertically above the point O, which is on flat, horizontal ground. The stone is launched with velocity $\begin{pmatrix} 2q \\ q \end{pmatrix}$ ms^{-1}, where q is a constant. It travels freely under gravity for 4 seconds, before landing on the ground.

MODELLING

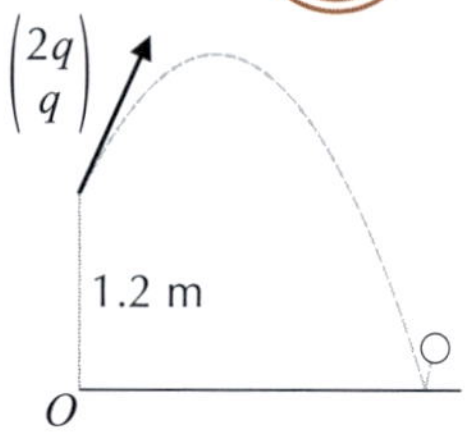

a) Find the value of q and hence the initial speed of the stone.

- Resolving vertically, taking up as positive:

  $s = -1.2, \quad u = q, \quad a = -9.8, \quad t = 4$

  Using $s = ut + \frac{1}{2}at^2$:

  $-1.2 = 4q + (-4.9 \times 4^2) \Rightarrow 4q = 77.2 \Rightarrow q = 19.3$

- So the stone's initial velocity is $\begin{pmatrix} 38.6 \\ 19.3 \end{pmatrix}$ ms^{-1}.

  Now you can find its initial speed using Pythagoras' theorem:

  speed $= \sqrt{38.6^2 + 19.3^2} =$ 43.2 ms^{-1} (3 s.f.)

b) Find the horizontal range of the stone.

- Resolving horizontally, taking right as positive:

  $s = s, \quad u = 38.6, \quad a = 0, \quad t = 4$

  Using $s = ut + \frac{1}{2}at^2$:

  $s = 38.6 \times 4 =$ 154.4 m

c) Find the position vector of the stone relative to O when it is at its highest point.

- First consider only the vertical motion to find the time taken for the stone to reach its maximum height. Resolving vertically, taking up as positive:

  $u = 19.3, \quad v = 0, \quad a = -9.8, \quad t = t$

  Using $v = u + at$:

  $0 = 19.3 - 9.8t \Rightarrow t = 1.969...$ seconds

- Now consider the horizontal and vertical components of motion together using the *suvat* equations in column vector form:

  $\mathbf{s} = \mathbf{s}, \quad \mathbf{u} = \begin{pmatrix} 38.6 \\ 19.3 \end{pmatrix}, \quad \mathbf{a} = \begin{pmatrix} 0 \\ -9.8 \end{pmatrix}, \quad t = 1.969...$

  Using $\mathbf{s} = \mathbf{u}t + \frac{1}{2}\mathbf{a}t^2$:

  $\mathbf{s} = (1.969...) \times \begin{pmatrix} 38.6 \\ 19.3 \end{pmatrix} + \frac{1}{2} \times (1.969...^2) \times \begin{pmatrix} 0 \\ -9.8 \end{pmatrix}$

  $\mathbf{s} = \begin{pmatrix} 76.01... \\ 38.00... \end{pmatrix} + \begin{pmatrix} 0 \\ -19.00... \end{pmatrix} = \begin{pmatrix} 76.01... \\ 19.00... \end{pmatrix}$ m

- This is the stone's displacement from its starting point when it reaches its highest point. Add it to the stone's initial position vector to find its new position vector:

  $\begin{pmatrix} 0 \\ 1.2 \end{pmatrix} + \begin{pmatrix} 76.01... \\ 19.00... \end{pmatrix} = \begin{pmatrix} 76.01... \\ 20.20... \end{pmatrix} = \begin{pmatrix} 76.0 \\ 20.2 \end{pmatrix}$ m (3 s.f.)

Tip: An initial velocity of $\begin{pmatrix} 2q \\ q \end{pmatrix}$ ms^{-1} means that the horizontal component of velocity is $2q$ ms^{-1} and the vertical component is q ms^{-1}. You can use column vectors in the *suvat* equations in the same way that you would use **i** and **j** vectors.

Tip: You could just find the stone's horizontal and vertical displacements separately, then write it in vector form at the end.

Example 3

An ice hockey puck is modelled as a particle moving with constant acceleration, a, across a horizontal surface. It is projected across the surface at a point (0.4i + 1.6j) m from the origin at time t = 0 with velocity (0.5i – 1.2j) ms^{-1}. It comes to rest after 4 seconds.

a) Find the acceleration of the puck.

- Write down the *suvat* values as usual:
  $\mathbf{u} = 0.5\mathbf{i} - 1.2\mathbf{j}$, $\mathbf{v} = 0$, $\mathbf{a} = \mathbf{a}$, $t = 4$
- Using $\mathbf{v} = \mathbf{u} + \mathbf{a}t$:
  $0 = 0.5\mathbf{i} - 1.2\mathbf{j} + 4\mathbf{a} \Rightarrow \mathbf{a} = (-0.125\mathbf{i} + 0.3\mathbf{j})\ \text{ms}^{-2}$

b) Find the position vector, r, at the point that the puck comes to rest.

- First find the displacement, **s**:
  $\mathbf{s} = \mathbf{s}$, $\mathbf{u} = 0.5\mathbf{i} - 1.2\mathbf{j}$, $\mathbf{v} = 0$, $t = 4$

  Using $\mathbf{s} = \frac{1}{2}(\mathbf{u} + \mathbf{v})t$:

  $\mathbf{s} = \frac{1}{2}(0.5\mathbf{i} - 1.2\mathbf{j} + 0) \times 4$

  $\mathbf{s} = (\mathbf{i} - 2.4\mathbf{j})$ m
- Add this displacement to the initial position vector to find **r**:
  $\mathbf{r} = 0.4\mathbf{i} + 1.6\mathbf{j} + \mathbf{i} - 2.4\mathbf{j} = (1.4\mathbf{i} - 0.8\mathbf{j})$ m

Tip: Not all constant acceleration problems involve gravity — you can use the *suvat* equations whenever you're told to assume that acceleration is constant.

Exercise 1.3

Hint: Again, you might find it useful to draw diagrams for these questions.

Q1 A particle is projected from a point on flat, horizontal ground with velocity (12**i** + 16**j**) ms^{-1}. Find the particle's velocity:

a) 2 seconds after projection,

b) when it reaches its maximum height,

c) when it hits the ground.

Q2 A projectile is fired from a height of 5 m directly above a fixed point O, which is on flat, horizontal ground. The particle's initial velocity is $\begin{pmatrix}17\\10\end{pmatrix}$ ms^{-1}. Find:

a) the particle's maximum height above the ground,

b) the speed of the particle as it hits the ground,

c) the direction of motion of the particle when it hits the ground.

Q3 A stone is thrown with velocity (6**i** + 9**j**) ms^{-1} from a window 2.5 m vertically above flat, horizontal ground. It is thrown towards a target on the ground, a horizontal distance of 20 m from the window. Find:

a) the length of time that the stone is at least 6 m above the ground,

b) the distance by which the stone falls short of the target.

Q4 A golf ball is hit off the edge of a 40 m high vertical cliff with velocity $(a\mathbf{i} + b\mathbf{j})$ ms^{-1}, where a and b are constants. It takes 5 seconds to land on the ground below, level with the foot of the cliff, a horizontal distance of 200 m away. Find:

a) the values of a and b,

b) the velocity of the golf ball when it hits the ground.

Q5

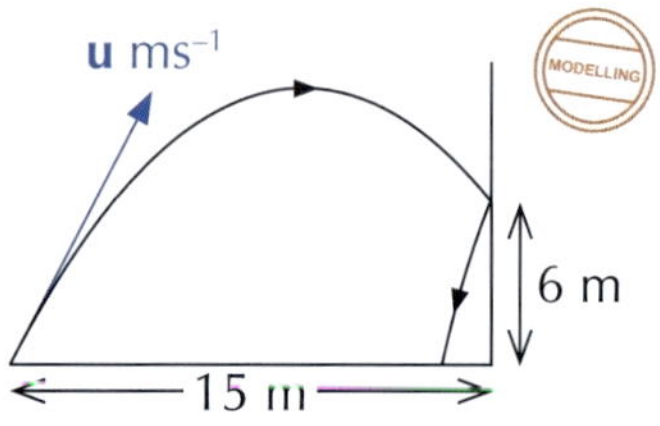

A football is kicked from a point on flat, horizontal ground towards a vertical wall, a horizontal distance of 15 m away. The ball's initial velocity is **u** ms^{-1}. The ball hits the wall 6 m above the ground, 3 seconds after being kicked.

Find, as vectors in **i** + **j** notation:

a) the initial velocity of the ball,

b) the velocity of the ball as it hits the wall.

The ball rebounds from the wall and lands on the ground.
As a result of the impact with the wall, the horizontal component of the ball's velocity is reversed and halved in magnitude.
The vertical component of its velocity is unaffected by the impact.

c) Find the horizontal distance between the wall and the point on the ground where the ball lands.

d) Suggest how the model you have used could be improved.

Q6 A toy submarine is moving underwater at a constant depth, with an initial velocity at time $t = 0$ of $(-0.6\mathbf{i} + 3\mathbf{j})$ ms^{-1}, where the **i** direction is due east and the **j** direction due north. The toy is modelled as a particle moving with constant acceleration of $(0.1\mathbf{i} + 0.5\mathbf{j})$ ms^{-2}.

a) At what value of t will the toy be moving due north?

b) Find the approximate bearing on which the toy is moving at $t = 2.5$ s.

c) Find the distance of the toy from its starting point after 1 minute.

d) Suggest two adaptations that would allow the model to better represent the movement of an object underwater.

Q7 A particle moves across a horizontal surface with a constant, non-zero acceleration of $(p\mathbf{i} + q\mathbf{j})$ ms^{-2}, where the **i** direction is due east and the **j** direction due north.
At $t = 0$, the particle is moving north-west with a speed of 5 ms^{-1}.
After 10 seconds, it is moving due south with a speed of $\sqrt{2}$ ms^{-1}.
Find p and q, in simplified surd form.

2. Non-Uniform Acceleration in 2 Dimensions

The constant acceleration equations can only be used in situations where the acceleration in a particular direction isn't changing — e.g. vertical motion due to a constant gravitational force with no resistance. In real life, the forces acting on an object change as a result of its motion, so its acceleration will vary.

Learning Objectives:

- Be able to differentiate and integrate vectors with respect to time.
- Be able to find a body's position vector, velocity or acceleration at a particular time given an expression for one of the other vectors.

Using vectors

- For a body moving in **two dimensions** (i.e. in a **plane**), you can describe its **position**, **velocity** and **acceleration** using the **unit vectors i** and **j**.
- The **i**-component describes the body's **horizontal** motion, and the **j**-component usually describes the body's **vertical** motion.
- You've already seen how to use the *suvat* equations in vector form on pages 309-311, but if the body is moving with **variable acceleration**, you'll have to use **calculus** instead.
- The relationships between **displacement** (or **position**), **velocity** and **acceleration** are shown below:

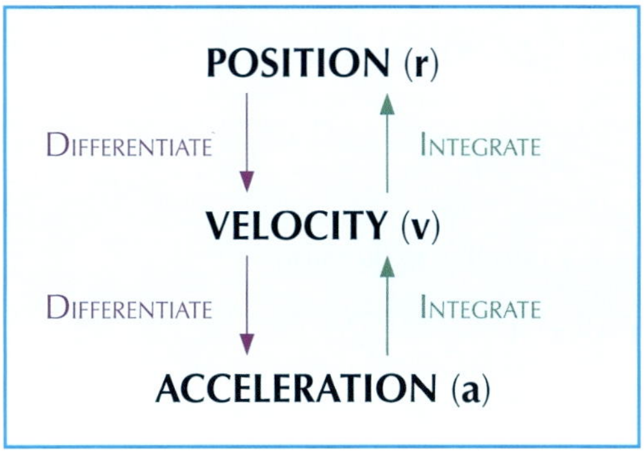

Tip: In two dimensions, questions will usually refer to a body's **position vector**. This is just the vector describing the body's displacement from the origin.

Tip: This should be familiar from Year 1, where you used these relationships for motion in one dimension.

- To differentiate or integrate vectors written in **i** and **j** notation, just differentiate or integrate each component **separately**:

Differentiating Vectors

If $\mathbf{r} = x\mathbf{i} + y\mathbf{j}$ is the position vector of a body, then:

the body's velocity, $\mathbf{v} = \frac{d\mathbf{r}}{dt} = \frac{dx}{dt}\mathbf{i} + \frac{dy}{dt}\mathbf{j}$

and the body's acceleration, $\mathbf{a} = \frac{d\mathbf{v}}{dt} = \frac{d^2\mathbf{r}}{dt^2}$

$$= \frac{d^2x}{dt^2}\mathbf{i} + \frac{d^2y}{dt^2}\mathbf{j}$$

Tip: The shorthand for $\frac{d\mathbf{r}}{dt}$ is $\dot{\mathbf{r}}$ (the single dot means differentiate **r** once with respect to time). The shorthand for $\frac{d^2\mathbf{r}}{dt^2}$ is $\ddot{\mathbf{r}}$ (the double dots mean differentiate **r** twice with respect to time).

Integrating Vectors

If $\mathbf{a} = p\mathbf{i} + q\mathbf{j}$ is the acceleration of a body, then:

the body's velocity, $\mathbf{v} = \int \mathbf{a}\,\mathrm{d}t = \int (p\mathbf{i} + q\mathbf{j})\,\mathrm{d}t$
$= \left(\int p\,\mathrm{d}t\right)\mathbf{i} + \left(\int q\,\mathrm{d}t\right)\mathbf{j}$

If $\mathbf{v} = w\mathbf{i} + z\mathbf{j}$ is the velocity of a body, then:

the body's position vector, $\mathbf{r} = \int \mathbf{v}\,\mathrm{d}t = \int (w\mathbf{i} + z\mathbf{j})\,\mathrm{d}t$
$= \left(\int w\,\mathrm{d}t\right)\mathbf{i} + \left(\int z\,\mathrm{d}t\right)\mathbf{j}$

Tip: When you're integrating, you'll still need to add a constant of integration, **C**, but it will be a vector with **i** and **j** components.

Example 1

A particle is moving in a plane. At time t seconds, its position in the plane is given by r = [(t^2 – 1)i + (2 + 5t)j] m relative to a fixed origin O.

Find the particle's speed and direction of motion at time t = 7.5 seconds.

- Differentiate the expression for the particle's position vector to find its velocity. Remember to treat the **i** and **j** components separately:

$$\mathbf{v} = \frac{\mathrm{d}\mathbf{r}}{\mathrm{d}t} = \left[\frac{\mathrm{d}}{\mathrm{d}t}(t^2 - 1)\right]\mathbf{i} + \left[\frac{\mathrm{d}}{\mathrm{d}t}(2 + 5t)\right]\mathbf{j}$$

$$= [2t\mathbf{i} + 5\mathbf{j}]\ \mathrm{ms}^{-1}$$

- Substitute t = 7.5 into this expression:

$$\mathbf{v} = 2(7.5)\mathbf{i} + 5\mathbf{j}$$

$$= (15\mathbf{i} + 5\mathbf{j})\ \mathrm{m}$$

- Use the **i** and **j** components to draw a right-angled triangle:

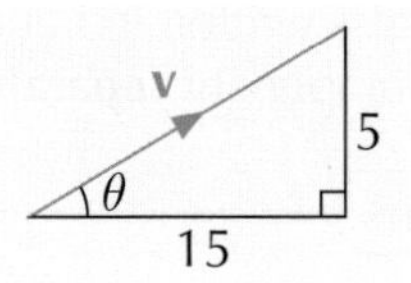

- Now use Pythagoras' theorem to find the particle's speed, and trigonometry to find its direction of motion:

$$\text{speed} = |\mathbf{v}| = \sqrt{15^2 + 5^2}$$

$$= 15.8\ \mathrm{ms}^{-1}\ (3\ \text{s.f.})$$

$$\theta = \tan^{-1}\left(\frac{5}{15}\right)$$

$= 18.4°$ (3 s.f.) from the **i**-direction

Example 2

A particle is moving in a vertical plane so that at time t seconds it has velocity v ms^{-1}, where $v = (8 + 2t)\mathbf{i} + (t^3 - 6t)\mathbf{j}$. When $t = 2$, the particle has position vector $(10\mathbf{i} + 3\mathbf{j})$ m with respect to a fixed origin O.

a) Find the acceleration of the particle at time t.

- Differentiate the expression for velocity with respect to time:

$$\mathbf{a} = \frac{d\mathbf{v}}{dt} = \left[\frac{d}{dt}(8 + 2t)\right]\mathbf{i} + \left[\frac{d}{dt}(t^3 - 6t)\right]\mathbf{j}$$

$$= [2\mathbf{i} + (3t^2 - 6)\mathbf{j}] \text{ ms}^{-2}$$

b) Show that the position vector of the particle when $t = 4$ is $\mathbf{r} = (38\mathbf{i} + 27\mathbf{j})$ m.

- Integrate the expression for velocity with respect to time:

$$\mathbf{r} = \int \mathbf{v}\,dt = \left(\int (8 + 2t)\,dt\right)\mathbf{i} + \left(\int (t^3 - 6t)\,dt\right)\mathbf{j}$$

$$= (8t + t^2)\mathbf{i} + \left(\frac{t^4}{4} - 3t^2\right)\mathbf{j} + \mathbf{C}$$

- When $t = 2$, $\mathbf{r} = (10\mathbf{i} + 3\mathbf{j})$. Use these values to find $\mathbf{C}$:

$$10\mathbf{i} + 3\mathbf{j} = (8(2) + 2^2)\mathbf{i} + \left(\frac{2^4}{4} - 3(2)^2\right)\mathbf{j} + \mathbf{C}$$

$$10\mathbf{i} + 3\mathbf{j} = 20\mathbf{i} - 8\mathbf{j} + \mathbf{C}$$

$$\Rightarrow \quad \mathbf{C} = (10 - 20)\mathbf{i} + (3 - -8)\mathbf{j}$$

$$= -10\mathbf{i} + 11\mathbf{j}$$

$$\Rightarrow \quad \mathbf{r} = \left[(8t + t^2 - 10)\mathbf{i} + \left(\frac{t^4}{4} - 3t^2 + 11\right)\mathbf{j}\right] \text{ m}$$

- Substitute $t = 4$ into the equation:

$$\mathbf{r} = (8(4) + 4^2 - 10)\mathbf{i} + \left(\frac{4^4}{4} - 3(4)^2 + 11\right)\mathbf{j}$$

$$= (32 + 16 - 10)\mathbf{i} + (64 - 48 + 11)\mathbf{j}$$

$$= (38\mathbf{i} + 27\mathbf{j}) \text{ m} \quad \text{— as required.}$$

c) Find the value of t for which the particle is directly above O.

- When the particle is directly above O, the **i**-component of its position vector will be zero:

$$8t + t^2 - 10 = 0$$

- Solve for t using the quadratic formula:

$$t = \frac{-8 \pm \sqrt{8^2 - (4 \times 1 \times -10)}}{2}$$

$$\Rightarrow t = 1.099... \text{ or } t = -9.099...$$

Ignore — t can't be negative

- Check that the **j**-component of **r** is greater than zero for $t = 1.099...$

$$\frac{(1.099...)^4}{4} - 3(1.099...)^2 + 11 = 7.741... > 0$$

- So the particle is directly above O when $t = 1.10$ s (3 s.f.)

Tip: Remember that **C** is a vector, so you need to collect **i** and **j** terms, and add or subtract to simplify.

Tip: You need to check that the **j**-component of **r** is greater than 0 to make sure that the particle is above O (rather than at it or below it).

Example 3

A body is moving on a plane. At time *t* seconds, the body's acceleration is given by a = [(4 – 18*t*)i + 72*t*j] ms^{-2}, relative to a fixed origin *O*. When *t* = 1, the body has position vector r = 2i m, relative to *O*. When *t* = 2, the body is moving with velocity v = (–20i + 135j) ms^{-1}.

PROBLEM SOLVING

Find the body's position vector when it is moving parallel to i.

- First, integrate the expression for acceleration with respect to time to find an expression for velocity:

$$\mathbf{v} = \int \mathbf{a}\,dt = \left(\int (4-18t)\,dt\right)\mathbf{i} + \left(\int 72t\,dt\right)\mathbf{j}$$
$$= (4t - 9t^2)\mathbf{i} + 36t^2\mathbf{j} + \mathbf{C}$$

- When $t = 2$, $\mathbf{v} = -20\mathbf{i} + 135\mathbf{j}$. Use these values to find **C**:

$$-20\mathbf{i} + 135\mathbf{j} = (4(2) - 9(2)^2)\mathbf{i} + 36(2)^2\mathbf{j} + \mathbf{C}$$
$$-20\mathbf{i} + 135\mathbf{j} = -28\mathbf{i} + 144\mathbf{j} + \mathbf{C}$$
$$\Rightarrow \quad \mathbf{C} = 8\mathbf{i} - 9\mathbf{j}$$

- So the particle's velocity at time *t* is given by:

$$\mathbf{v} = [(4t - 9t^2 + 8)\mathbf{i} + (36t^2 - 9)\mathbf{j}]\ \text{ms}^{-1}$$

- When the particle is moving parallel to **i**, the **j**-component of its velocity is zero...

$$36t^2 - 9 = 0$$
$$t^2 = 9 \div 36 = 0.25$$
$$\Rightarrow \quad t = 0.5 \text{ seconds}$$

Tip: As always, ignore the negative value of *t*.

- ...and the **i**-component of its velocity is non-zero:

$$4(0.5) - 9(0.5)^2 + 8 = 7.75 \neq 0$$

Tip: You need to check that the **i** component of velocity is non-zero for the value of *t* you've found. If it was zero, then the particle wouldn't be moving parallel to **i** at this time — it would be stationary.

- So the particle is moving parallel to **i** at time $t = 0.5$ seconds. Now integrate the expression for velocity to find an expression for the particle's position vector:

$$\mathbf{r} = \int \mathbf{v}\,dt = \left(\int (4t - 9t^2 + 8)\,dt\right)\mathbf{i} + \left(\int (36t^2 - 9)\,dt\right)\mathbf{j}$$
$$= (2t^2 - 3t^3 + 8t)\mathbf{i} + (12t^3 - 9t)\mathbf{j} + \mathbf{D}$$

- When $t = 1$, $\mathbf{r} = 2\mathbf{i}$. Use these values to find **D**:

$$2\mathbf{i} = (2(1)^2 - 3(1)^3 + 8(1))\mathbf{i} + (12(1)^3 - 9(1))\mathbf{j} + \mathbf{D}$$
$$2\mathbf{i} = 7\mathbf{i} + 3\mathbf{j} + \mathbf{D}$$
$$\Rightarrow \mathbf{D} = -5\mathbf{i} - 3\mathbf{j}$$

- So the particle's position vector relative to *O* at time *t* is given by:

$$\mathbf{r} = [(2t^2 - 3t^3 + 8t - 5)\mathbf{i} + (12t^3 - 9t - 3)\mathbf{j}]\ \text{m}$$

- Substitute $t = 0.5$ into this equation to find the particle's position vector when it is travelling parallel to **i**:

$$\mathbf{r} = (2(0.5)^2 - 3(0.5)^3 + 8(0.5) - 5)\mathbf{i} + (12(0.5)^3 - 9(0.5) - 3)\mathbf{j}$$
$$= (-0.875\mathbf{i} - 6\mathbf{j})\ \text{m}$$

Example 4

A boat is modelled as a particle moving across a horizontal surface with acceleration a = ($e^{0.2t}$ i + sin t j) ms^{-2} at time t seconds (for $0 \le t \le 5$). At time $t = 0$, the boat is at 2j m from the origin, moving in the positive j direction at a speed of 10 ms^{-1}. Find an expression for the boat's position vector, r, at time t.

MODELLING

- Integrate the expression for acceleration with respect to time to find an expression for velocity:

$$\mathbf{v} = \int \mathbf{a}\,dt = \left(\int e^{0.2t}\,dt\right)\mathbf{i} + \left(\int \sin t\,dt\right)\mathbf{j}$$
$$= \left(\frac{e^{0.2t}}{0.2}\right)\mathbf{i} - (\cos t)\mathbf{j} + \mathbf{C} = 5e^{0.2t}\,\mathbf{i} - (\cos t)\mathbf{j} + \mathbf{C}$$

- When $t = 0$, $\mathbf{v} = 0\mathbf{i} + 10\mathbf{j}$. Use this information to find **C**, and hence **v**:

$$0\mathbf{i} + 10\mathbf{j} = 5e^0\,\mathbf{i} - (\cos 0)\mathbf{j} + \mathbf{C}$$
$$0\mathbf{i} + 10\mathbf{j} = 5\mathbf{i} - 1\mathbf{j} + \mathbf{C}$$
$$\Rightarrow \quad \mathbf{C} = -5\mathbf{i} + 11\mathbf{j}$$
$$\Rightarrow \quad \mathbf{v} = (5e^{0.2t} - 5)\mathbf{i} + (-\cos t + 11)\mathbf{j}$$
$$= 5(e^{0.2t} - 1)\mathbf{i} + (11 - \cos t)\mathbf{j}$$

- Integrate again to find an expression for the position vector **r** at time t:

$$\mathbf{r} = \int \mathbf{v}\,dt = \left(\int 5(e^{0.2t} - 1)\,dt\right)\mathbf{i} + \left(\int (11 - \cos t)\,dt\right)\mathbf{j}$$
$$= 5\left(\frac{e^{0.2t}}{0.2} - t\right)\mathbf{i} + (11t - \sin t)\mathbf{j} + \mathbf{D}$$
$$= 5(5e^{0.2t} - t)\mathbf{i} + (11t - \sin t)\mathbf{j} + \mathbf{D}$$

- When $t = 0$, the boat is at $\mathbf{r} = 2\mathbf{j}$.
Use this information to find **D**, and hence **r**:

$$0\mathbf{i} + 2\mathbf{j} = 5(5e^0 - 0)\mathbf{i} + (0 - \sin 0)\mathbf{j} + \mathbf{D}$$
$$0\mathbf{i} + 2\mathbf{j} = 25\mathbf{i} - 0\mathbf{j} + \mathbf{D}$$
$$\Rightarrow \quad \mathbf{D} = -25\mathbf{i} + 2\mathbf{j}$$
$$\Rightarrow \quad \mathbf{r} = (5(5e^{0.2t} - t) - 25)\mathbf{i} + (11t - \sin t + 2)\mathbf{j}$$
$$= 5(5e^{0.2t} - t - 5)\mathbf{i} + (2 + 11t - \sin t)\mathbf{j}$$

Tip: You'll be expected to know how to differentiate and integrate any of the types of expressions that were covered in Chapters 7 and 8 — so any calculus that comes up elsewhere in the course can also come up in Mechanics.

Example 5

A particle's acceleration is connected to its velocity, v = xi + yj, such that $\frac{dx}{dt} = \frac{9}{x^2}$ and $\frac{dy}{dt} = \frac{2}{y}$. The particle is stationary at $t = 0$. Given that $x \ge 0$ and $y \ge 0$, show that the velocity vector, v, at time t, is given by v = $3t^{\frac{1}{3}}$i + $2t^{\frac{1}{2}}$j.

PROBLEM SOLVING

The question gives each component of acceleration as a differential equation. Solve these separately and then combine to give **v** in terms of t.

- In the **i** direction:

$$\frac{dx}{dt} = \frac{9}{x^2} \quad \Rightarrow \quad \int x^2\,dx = \int 9\,dt \quad \Rightarrow \quad \frac{x^3}{3} = 9t + C$$

Tip: See p.222–224 for a recap on solving differential equations.

- $x = 0$ when $t = 0$, so $C = 0$.

  This means that, in the **i** direction:

  $$\frac{x^3}{3} = 9t \Rightarrow x = \sqrt[3]{27t} = 3\sqrt[3]{t} = 3t^{\frac{1}{3}}$$

- In the **j** direction:

  $$\frac{dy}{dt} = \frac{2}{y} \Rightarrow \int y\,dy = \int 2\,dt \Rightarrow \frac{y^2}{2} = 2t + D$$

- $y = 0$ when $t = 0$, so $D = 0$.

  This means that, in the **j** direction:

  $$\frac{y^2}{2} = 2t \Rightarrow y = \sqrt{4t} = 2\sqrt{t} = 2t^{\frac{1}{2}}$$

  $y \geq 0$, so ignore the negative root here

- Substituting x and y into velocity $\mathbf{v} = x\mathbf{i} + y\mathbf{j}$ gives:

  $\mathbf{v} = 3t^{\frac{1}{3}}\mathbf{i} + 2t^{\frac{1}{2}}\mathbf{j}$ as required

Exercise 2.1

Q1 A particle is moving in a plane. At time t seconds, the particle's position relative to a fixed origin O is given by $\mathbf{r} = [(t^3 - 3t)\mathbf{i} + (t^2 + 2)\mathbf{j}]$ m. Find:

a) (i) an expression for the particle's velocity in terms of t,

(ii) the particle's speed and direction of motion at time $t = 3$ s,

b) (i) an expression for the particle's acceleration in terms of t,

(ii) the magnitude of the particle's acceleration at time $t = 4$ s.

Q2 A car is modelled as a particle travelling in a plane. Its velocity at time t seconds is given by the expression:
$\mathbf{v} = [(\frac{1}{3}t^3 + 2t^2 - 12t)\mathbf{i} + 14\mathbf{j}]$ ms^{-1}, where $0 \leq t \leq 4$.
At a certain time, the car reaches its maximum velocity when the acceleration falls to zero. Find the car's velocity at this time.

MODELLING

Q3 Hint: Column vectors are just another way of showing the horizontal and vertical components:
$\begin{pmatrix} a \\ b \end{pmatrix} = a\mathbf{i} + b\mathbf{j}$.

Q3 A body is moving in a plane.
At time t seconds, the body's acceleration is given by $\mathbf{a} = \begin{pmatrix} 3e^{3t} \\ \frac{6}{\sqrt{t}} \end{pmatrix}$ ms^{-2}.
At $t = 0$, the body has position vector $\mathbf{r} = \begin{pmatrix} 2 \\ -9 \end{pmatrix}$ m relative to a fixed origin O, and is travelling at $\begin{pmatrix} 4 \\ 6 \end{pmatrix}$ ms^{-1}. Find:

a) a column vector expression for the body's velocity in terms of t,

b) the body's position vector, as a column vector, in terms of t.

Q4 Hint: Remember that you need to work in radians when differentiating and integrating trig functions.

Q4 An object is travelling in a plane with non-uniform acceleration. At time t seconds, the particle's position relative to a fixed origin O is given by $\mathbf{r} = [(\sin t)\mathbf{i} + (\cos t)\mathbf{j}]$ m. **i** and **j** are the unit vectors directed due east and due north respectively.

a) What is the first non-zero value of t for which the object is moving in an easterly direction?

b) Describe the object's position and distance from O at this time.

Q5 The velocity of a particle at time t seconds ($t \geq 0$) is given by $\mathbf{v} = [(t^2 - 6t)\mathbf{i} + (4t + 5)\mathbf{j}]$ ms^{-1}. The particle is moving in a plane, and at time $t = 0$ the particle passes through the origin, O. The particle passes through the point with position vector $(b\mathbf{i} + 12\mathbf{j})$ m, where b is a constant. Find the value of b.

PROBLEM SOLVING

Q6 Particle A is moving in a vertical plane. At time t seconds, the particle's position relative to a fixed origin O is given by $\mathbf{r}_A = [(t^3 - t^2 - 4t + 3)\mathbf{i} + (t^3 - 2t^2 + 3t - 7)\mathbf{j}]$ m. Find:

a) the value of t for which the particle's velocity is $(-3\mathbf{i} + 2\mathbf{j})$ ms^{-1},

b) the value of t for which the direction of motion of the particle is 45° above $\mathbf{i}$.

A second particle, B, is moving in the same plane as particle A. At time t seconds, the acceleration of B is given by $\mathbf{a}_B = (6t\mathbf{i} + 6t\mathbf{j})$ ms^{-2}. At time $t = 1$ second, B passes the point $(2\mathbf{i} + 3\mathbf{j})$ m relative to O, with velocity $(4\mathbf{i} - \mathbf{j})$ ms^{-1}. Find:

c) an expression for the position vector of particle B relative to O in terms of t,

d) an expression for the position vector of particle B relative to particle A in terms of t,

e) the distance between particles A and B at time $t = 4$ seconds.

Q6b) Hint: Think about the relationship between the horizontal and vertical components of the particle's velocity when it is moving in this direction.

Q7 A particle moves so that it traces a curve C, as defined by the parametric equations $y = \ln t$ and $x = \frac{1}{t^3}$.

x and y represent the displacement of the particle from the origin in the $\mathbf{i}$ and $\mathbf{j}$ directions respectively, at time t seconds.

a) Find the gradient of C, $\frac{dy}{dx}$, when $t = \frac{1}{2}$.

b) Show that the acceleration of the particle at this time is $384\mathbf{i} - 4\mathbf{j}$.

Q7 Hint: To differentiate parametric equations, use $\frac{dy}{dx} = \frac{dy}{dt} \div \frac{dx}{dt}$ (see p.169).

Q8 The velocity of a particle at time t seconds is given by $\mathbf{v} = [(2\sin^2 t)\mathbf{i} + (4\sin 2t\cos 2t)\mathbf{j}]$ ms^{-1}. At time $t = \frac{\pi}{4}$, the particle has position vector $\mathbf{r} = \left[\frac{\pi}{4}\mathbf{i} + \mathbf{j}\right]$ m, relative to the origin.

What was the particle's initial distance from the origin at time $t = 0$?

Q8 Hint: Trig identities will help here — see Chapter 3 for a recap.

Q9 A particle is moving with a velocity of $\mathbf{v} = [m\mathbf{i} + n\mathbf{j}]$ ms^{-1}. It has a constant acceleration of 0.1 ms^{-2} in the $\mathbf{i}$ direction, and a variable acceleration in the $\mathbf{j}$ direction, such that $\frac{dn}{dt} = (kn - 10)$ ms^{-2} and $\frac{dn}{dt} < 0$ (k is a constant).

PROBLEM SOLVING

The particle is stationary at $t = 0$, and when its acceleration is -2 ms^{-2} in the $\mathbf{j}$ direction, its velocity in the $\mathbf{j}$ direction is -4 ms^{-1}.

a) Show that $n + 5 > 0$.

b) Hence show that the velocity of the particle can expressed as $\mathbf{v} = \left[0.1t\mathbf{i} + 5(e^{-2t} - 1)\mathbf{j}\right]$ ms^{-1}.

Chapter 15 Dynamics

1. Resolving Forces

Learning Objectives:

- Be able to resolve force vectors into horizontal and vertical components.
- Be able to calculate the resultant force and solve problems for objects in equilibrium.

You've seen problems with more than one force acting on an object before, but things get more complicated when these forces are acting at different angles. You'll often need to use trigonometry to resolve the forces into their horizontal and vertical components.

Components of forces

You've already seen resolving vectors (p.298), so some of this should be familiar:

- A **component of a force** is the **magnitude** of the force in a **particular direction**.
- You can use trigonometry to **resolve** forces and find their components in a particular direction.
- Often it's very useful to resolve forces into their **horizontal** and **vertical** components. Together with the force vector, these components form a **right-angled triangle**.

You can find the horizontal and vertical components of a force, F, using **trigonometry**.

- Horizontal component, F_x:
  $\cos\theta = \frac{\text{adjacent}}{\text{hypotenuse}} = \frac{F_x}{F}$, so $\boldsymbol{F_x = F\cos\theta}$
- Vertical component, F_y:
  $\sin\theta = \frac{\text{opposite}}{\text{hypotenuse}} = \frac{F_y}{F}$, so $\boldsymbol{F_y = F\sin\theta}$

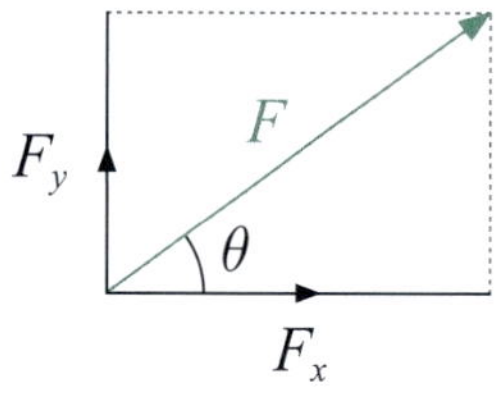

Example 1

A particle is acted on by a force of 15 N at 30° above the positive horizontal. Find the horizontal and vertical components of the force.

- Draw a vector triangle and use trigonometry to calculate the horizontal and vertical components.

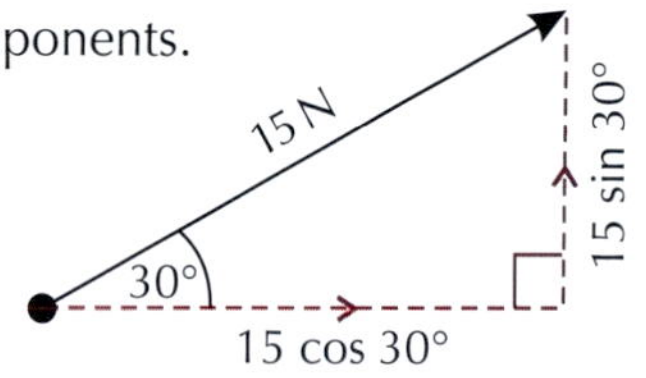

- Vertical component $= 15\sin 30°$
  $= 7.5$ N upwards
- Horizontal component $= 15\cos 30°$
  $= 13.0$ N (to 3 s.f.) to the right

- If you're resolving in a direction which is **perpendicular** to the line of action of a force, then the component of the force in that direction will be **zero**.
- When you're resolving a force, you should state which direction you're taking as being positive. Then any components acting in the **opposite direction** will be **negative**.

Tip: For example, a force which acts horizontally has zero vertical component, and vice versa.

Example 2

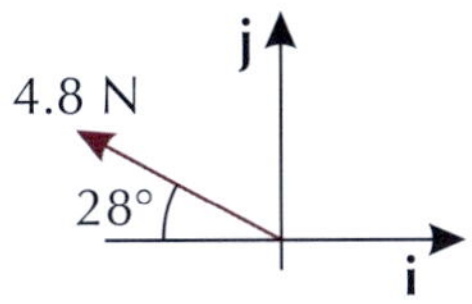

Find the i and j components of the force shown.

- Resolving in the **i**-direction (→):

  $-4.8 \cos 28° = -4.24$ N (to 3 s.f.)

- Resolving in the **j**-direction (↑):

  $4.8 \sin 28° = 2.25$ N (to 3 s.f.)

Tip: Draw an arrow to show which direction you are taking as positive. Here, 'right' is positive, so the **i**-component of the force is negative, because it acts to the left.

- You can resolve forces in **any two perpendicular directions**, not just horizontally and vertically.
- For example, for an object on an **inclined plane**, you might want to resolve the forces in directions **parallel** and **perpendicular** to the plane.

Example 3

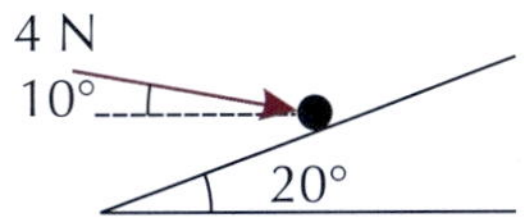

The diagram shows a particle at rest on a smooth inclined plane. A force of magnitude 4 N acts on the particle at an angle of 10° to the horizontal. Find the components of the force that act parallel and perpendicular to the plane.

- By the alternate angles theorem, the angle between the direction of the force and the slope of the plane is 20° + 10° = 30°.

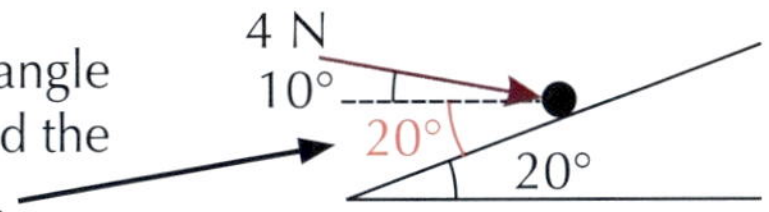

- Resolving parallel to the plane (↗):

  $F_{\text{parallel}} = 4 \cos 30° = 3.46$ N (to 3 s.f.)

- Resolving perpendicular to the plane (↖):

  $F_{\text{perpendicular}} = -4 \sin 30° = -2$ N

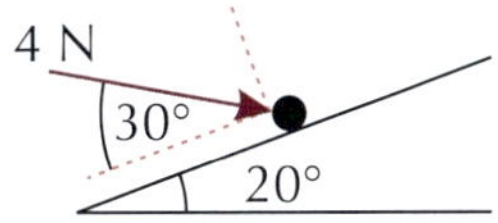

Exercise 1.1

Q1 Find the horizontal and vertical components of each of the forces shown:

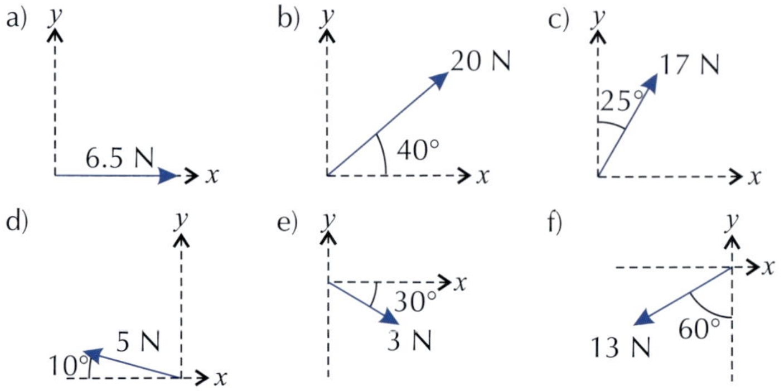

Q2 Each of the following diagrams shows a force and the angle it makes with the horizontal. Write each force in the form $(a\mathbf{i} + b\mathbf{j})$ N.

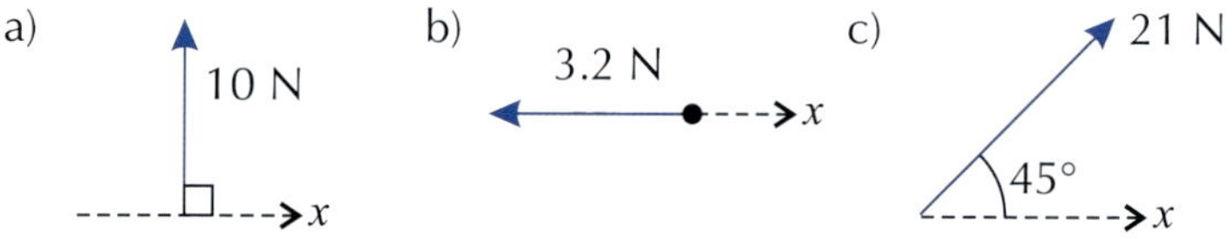

Q3-4 Hint: Finding the direction and magnitude of a vector given in components was covered in Year 1.

Q3 The force $(c\mathbf{i} + 7\mathbf{j})$ N acts at 25° to $\mathbf{i}$, where c is a positive constant.

a) Find the value of c.

b) Find the magnitude of this force.

Q4

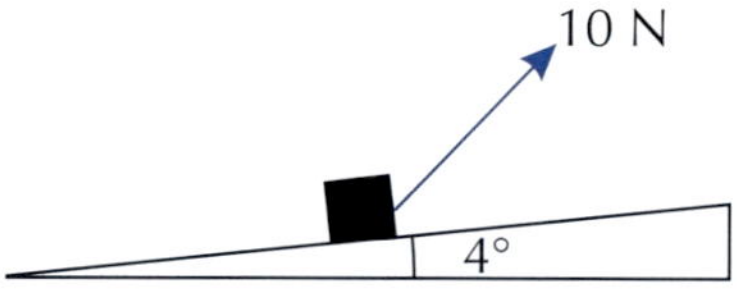

The diagram shows a force of magnitude 10 N acting on an object on an inclined plane. The component of the force in the direction parallel to the plane is 7.1 N.

a) Find the angle between the direction of the force and the incline of the plane.

b) Find the horizontal component of the force.

Resultant forces and equilibrium

In Year 1, you learned about **resultant forces** and what it means for an object to be in **equilibrium**:

- The resultant force is the sum of all the forces acting on an object. You can calculate the resultant force by **adding the corresponding components** of the forces.
- An object is in equilibrium if the resultant force on it is **zero**.

You will often have to resolve one or more forces to find the resultant force on an object, or use the fact that it is in equilibrium to find a missing force.

Example 1

The diagram on the right shows a force of 15 N and a horizontal force of 20 N applied to a particle. Find the magnitude and direction of the resultant force on the particle.

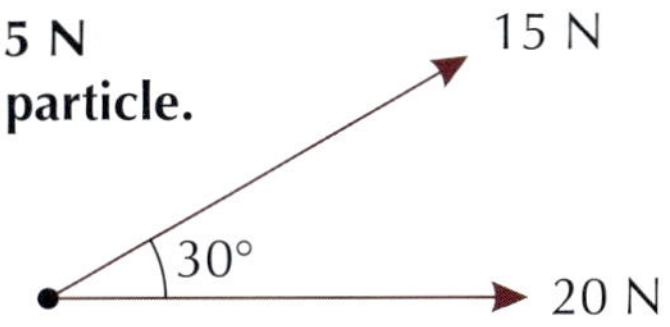

- Resolve to find the horizontal and vertical components of the 15 N force.

  Horizontal component = 15 cos 30° = 12.99... N
  Vertical component = 15 sin 30° = 7.5 N

- So the resultant force has a horizontal component of 20 + 12.99... = 32.99... N and a vertical component of 7.5 N
- Use Pythagoras' theorem to find the magnitude of the force.

$$R^2 = (20 + 12.99...)^2 + (7.5)^2$$
$$R = \sqrt{32.99...^2 + 7.5^2}$$
$$= 33.8 \text{ N (to 3 s.f.)}$$

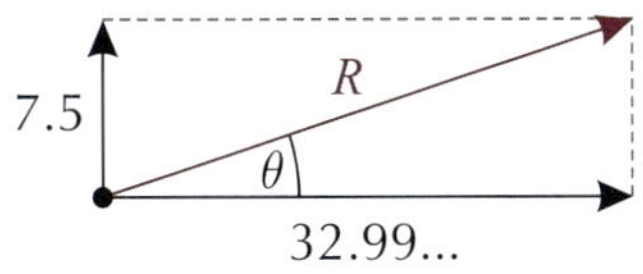

- Use trigonometry to find its direction, θ

$$\tan\theta = \frac{7.5}{32.99...}$$
$$\theta = \tan^{-1}\left(\frac{7.5}{32.99...}\right)$$
$$= 12.8° \text{ (to 3 s.f.) above the horizontal}$$

Example 2

Three forces of magnitude 9 N, 12 N and 13 N act on a particle *P* in the directions shown in the diagram. Find the magnitude and direction of the resultant of the three forces.

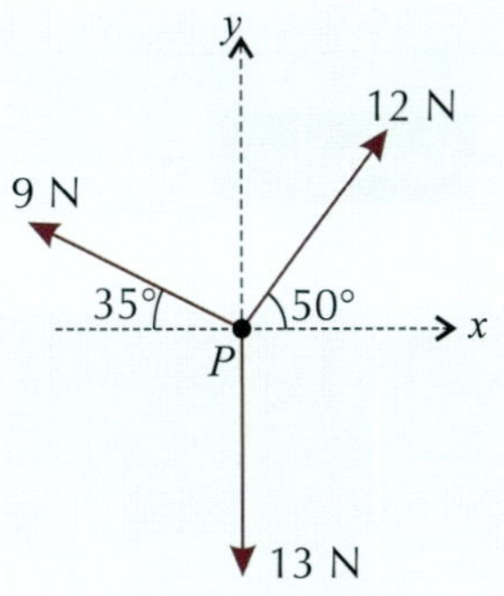

- Resolving in the y-direction (↑):

$$9 \sin 35° + 12 \sin 50° - 13$$
$$= 1.354... = 1.35 \text{ N (to 3 s.f.)}$$

- Resolving in the x-direction (→):

$$12 \cos 50° - 9 \cos 35°$$
$$= 0.3410... = 0.341 \text{ N (to 3 s.f.)}$$

- Use the components to form a right-angled triangle. The resultant is shown by the hypotenuse of the triangle.

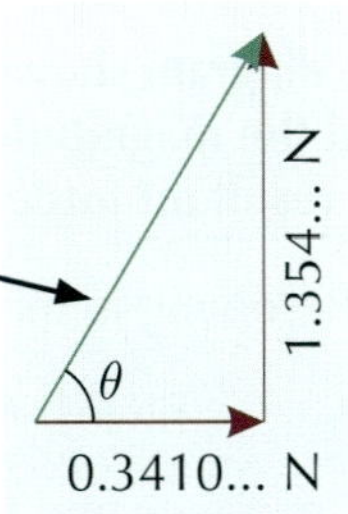

- Use Pythagoras' theorem to find the magnitude:

$$\sqrt{1.354...^2 + 0.3410...^2} = 1.40 \text{ N (to 3 s.f.)}$$

- Use trigonometry to find the required direction:

$$\theta = \tan^{-1}\left(\frac{1.354...}{0.3410...}\right) = 75.9° \text{ (to 3 s.f.) above the positive } x\text{-axis}$$

Example 3

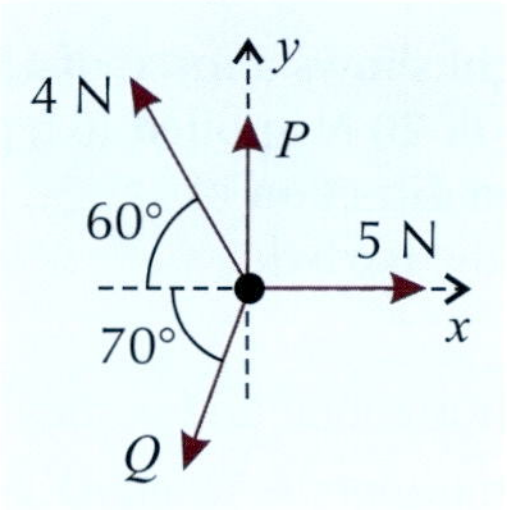

The diagram shows all the forces acting on a particle. Given that the particle is in equilibrium, find the magnitudes of the missing forces *P* and *Q*.

- The particle is in equilibrium, so the resultant force on the particle is zero.
- Resolve to find two perpendicular components of the resultant force, then make each component equal to zero.
- Resolving in the x-direction ($\rightarrow$):

$$5 - Q\cos 70° - 4\cos 60° = 0$$
$$Q\cos 70° = 3$$
$$\Rightarrow Q = \frac{3}{\cos 70°} = 8.77 \text{ N (to 3 s.f.)}$$

- Resolving in the y-direction ($\uparrow$):

$$P + 4\sin 60° - Q\sin 70° = 0$$
$$P = \frac{3}{\cos 70°}\sin 70° - 4\sin 60°$$
$$\Rightarrow P = 4.78 \text{ N (to 3 s.f.)}$$

Example 4

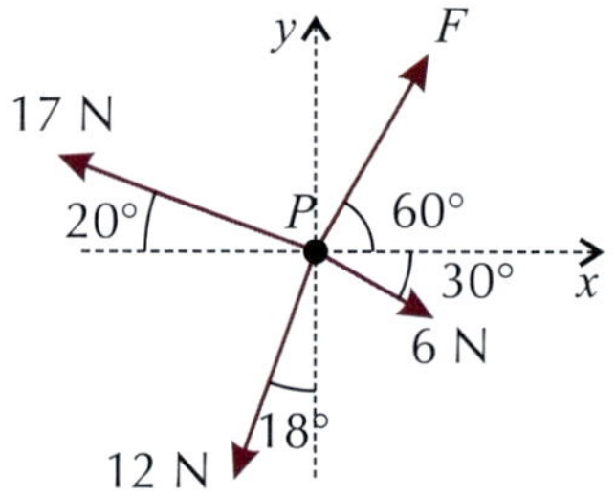

The diagram shows the forces acting on particle *P*. Find the magnitude of force *F*, given that the resultant force on the particle acts in the *x*-direction.

- Resolving in the y-direction ($\uparrow$):

$$F\sin 60° + 17\sin 20° - 12\cos 18° - 6\sin 30° = 0$$

- Rearrange and solve this equation to find the magnitude of F:

$$F = \frac{-17\sin 20° + 12\cos 18° + 6\sin 30°}{\sin 60°}$$
$$= 9.93 \text{ N (to 3 s.f.)}$$

Tip: You're told that the resultant force acts in the x-direction. This means that the vertical component of the resultant must be zero.

Exercise 1.2

Q1 Each of the diagrams below shows the forces acting on a particle. Find the magnitude and direction of the resultant force on each particle.

a)

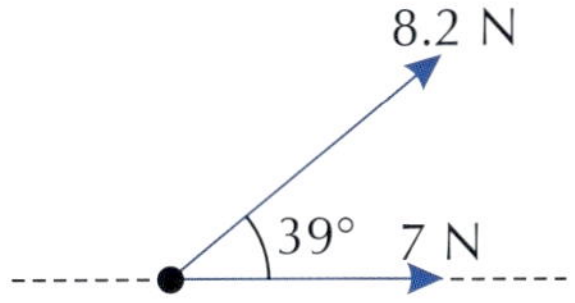

b)

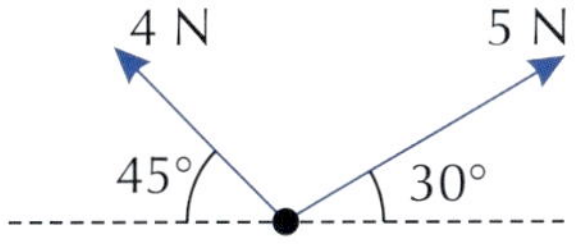

c)

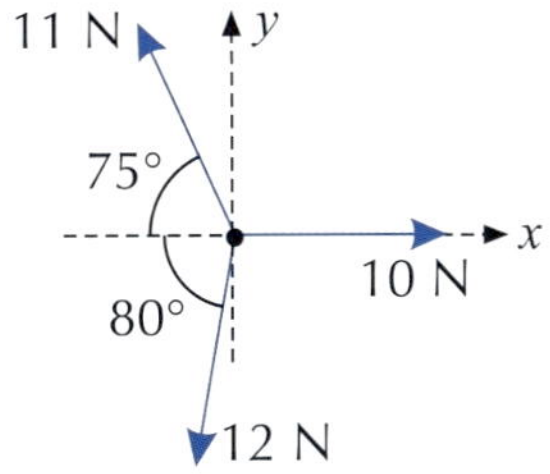

d)

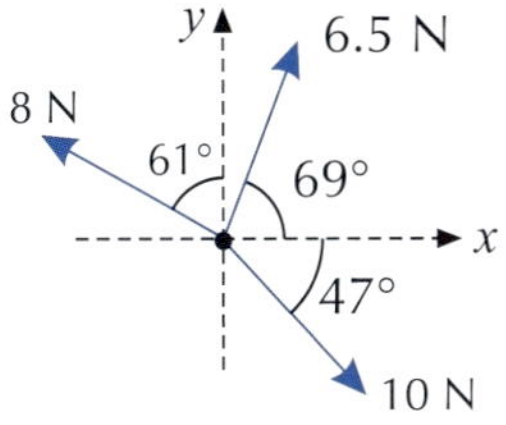

Hint: You can answer these questions either by resolving the forces, or drawing a vector triangle and using trigonometry and Pythagoras' theorem.

Q2 A force of magnitude 18 N acts on a particle at 60° above the positive horizontal. A force of magnitude 16 N also acts vertically upwards on the particle.

a) Draw a diagram to illustrate the forces acting on the particle.

b) Find the magnitude and direction of the resultant of the two forces.

Q3 The diagram shows all the forces acting on a particle in equilibrium. Find the magnitude of the force F.

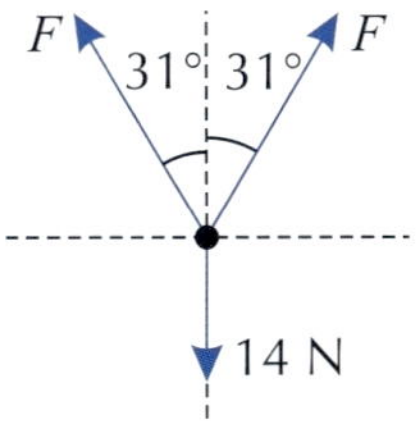

Q4 A particle is acted on by a force of magnitude 9 N in the positive horizontal direction and a force of magnitude 10 N at 30° above the positive horizontal. A third force of magnitude P N acts on the particle at 60° above the positive horizontal. Find P, given that the resultant of the three forces acts at 45° above the positive horizontal.

Q5 Forces of magnitude 1 N and 2 N act at angles of 10° and 35° above **i** respectively. A third force of magnitude 3 N acts at an angle of θ above **i**, where $180° \leq \theta \leq 360°$. The resultant of the three forces has no horizontal component.

a) Find θ.

b) Calculate the magnitude of the resultant of the three forces.

Q5 Hint: Be careful — make sure the angle you calculate is in the range given for θ. Remember $\cos\theta = \cos(360° - \theta)$.

Q6 Forces of magnitude P N and $2P$ N act on a particle vertically upwards and horizontally, as shown in the diagram. A third force of 20 N acts at an angle θ to the downward vertical. No other forces act on the particle. Given that the particle is in equilibrium, find the values of P and θ.

PROBLEM SOLVING

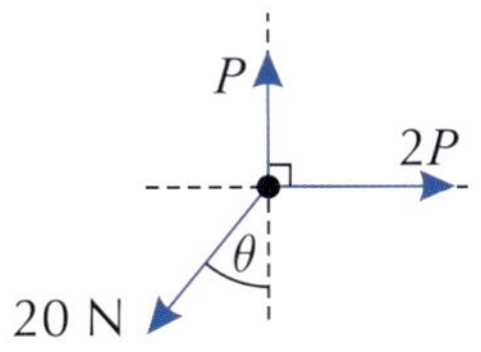

Harder equilibrium problems

You can use everything from the last few pages to answer questions about all sorts of **real-life situations**. Objects hanging from **strings** and objects on **inclined planes** are some common examples of where you might see forces in equilibrium.

> **Tip:** The $8g$ force acting downwards is the body's weight. T_1 and T_2 are the tensions in the two strings. Look back at your Year 1 notes if you want to recap weight and tension.

Example 1

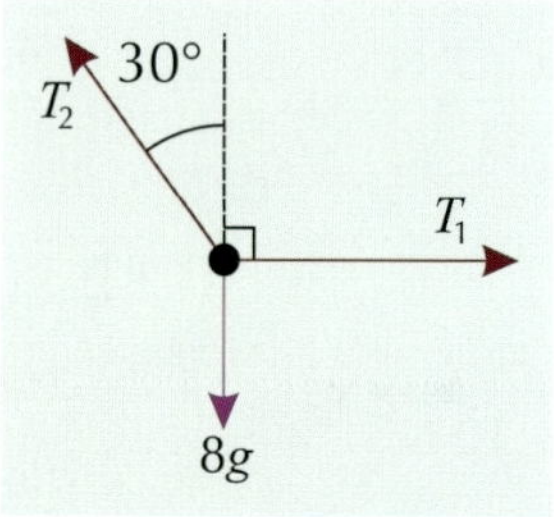

The diagram shows a body of mass 8 kg held in equilibrium by two light, inextensible strings. One string is horizontal, and the other string makes an angle of 30° with the vertical.

Find the magnitude of the tension in each string.

> **Tip:** The components in each direction must sum to zero because the body is in equilibrium.

- Resolving vertically (↑):

  $T_2 \cos 30° - 8g = 0$

  $T_2 \cos 30° = 8g$

  $T_2 = 8g \div \cos 30° =$ 90.5 N (to 3 s.f.)

  Remember to use $g = 9.8$ ms^{-2} unless you're told otherwise.

- Resolving horizontally (→):

  $T_1 - T_2 \sin 30° = 0$

  $T_1 = T_2 \sin 30°$

  $= 90.5 \sin 30°$

  $= 45.264... =$ 45.3 N (to 3 s.f.)

Inclined planes

When you're resolving forces on an object that is on an **inclined plane**, it's usually best to resolve **parallel** and **perpendicular to the slope**.

> **Tip:** Example 3 on page 321 involved resolving forces parallel and perpendicular to a plane, so this shouldn't be completely new.

Mostly, you'll be resolving an object's **weight**, because weight always acts **vertically downwards**, and you'll need to use its **components parallel** and **perpendicular to the slope**.

These are:

- Component of weight **parallel** to the plane: $mg \sin \theta$
- Component of weight **perpendicular** to the plane: $mg \cos \theta$

Where mg is the weight of the object and θ is the angle of the incline of the slope.

Resolving an Object's Weight

- You want to find the **components** of an object's weight **parallel** and **perpendicular** to the plane.

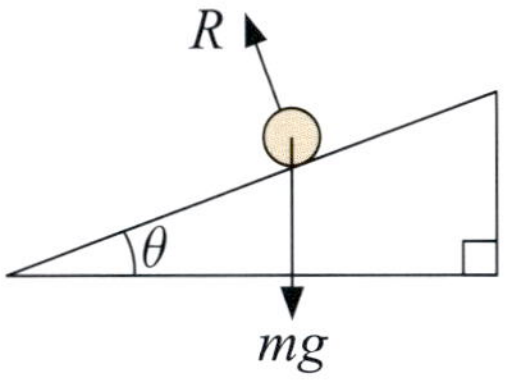

- Using **right-angled triangles**, the angle between the object's weight and its component perpendicular to the plane is equal to θ.

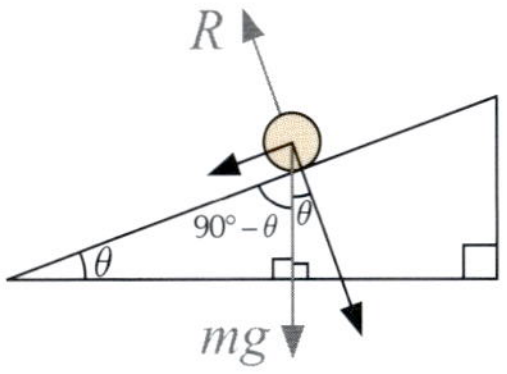

- You can **resolve** as usual to find the components of the weight using **trigonometry**.

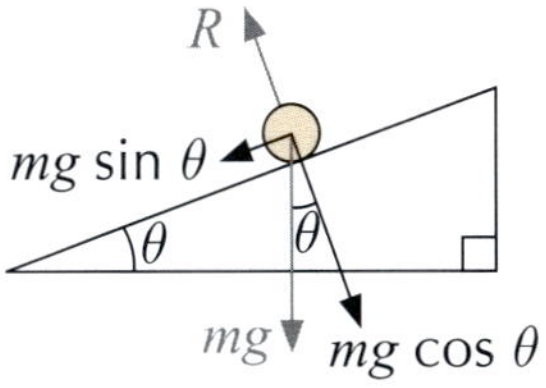

- You may have to resolve **other forces** as well as the weight, depending on their **direction**.

Example 2

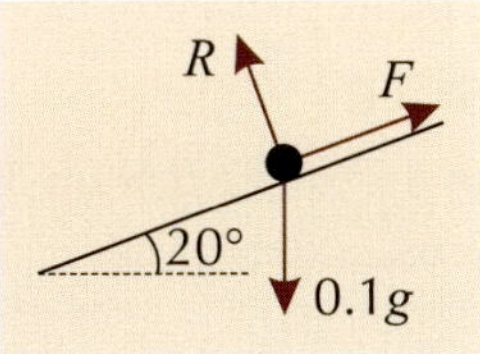

A stone of mass 0.1 kg rests on a smooth plane inclined at 20° to the horizontal. It is held in equilibrium by a force, *F*, acting parallel to the plane, as shown.

Find the magnitude of force *F* and the normal reaction, *R*, of the plane on the stone.

- Resolving parallel to the plane (↗):

$$F - 0.1g \sin 20° = 0$$

$$F = 0.98 \sin 20°$$

$$= 0.335 \text{ N (to 3 s.f.)}$$

This is the component of the stone's weight acting down the plane.

- Resolving perpendicular to the plane (↖):

$$R - 0.1g \cos 20° = 0$$

$$R = 0.98 \cos 20°$$

$$= 0.921 \text{ N (to 3 s.f.)}$$

Tip: The force F acts parallel to the plane already, so you don't need to resolve it into components.

Example 3

MODELLING

Two brothers are fighting over a sledge of weight 100 N. The sledge lies on a smooth slope inclined at an angle of 35° to the horizontal. One brother tries to pull the sledge up the slope with a force of magnitude 70 N, while the other tries to pull the sledge down the slope with a force of magnitude *F*.

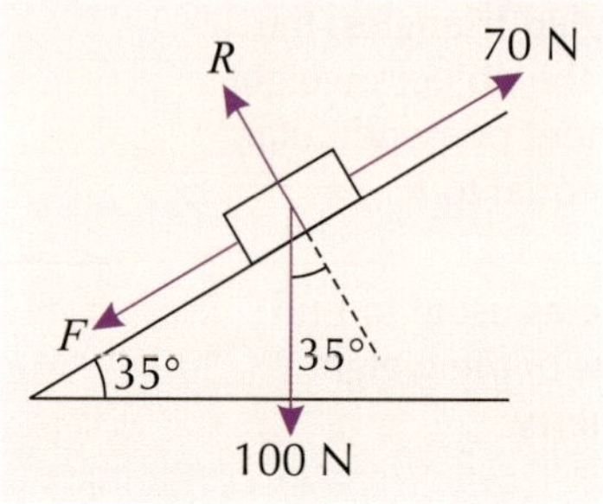

a) Find the magnitude of the normal reaction force, *R*, on the sledge.

- Resolving perpendicular to the plane (↖):

$R - 100 \cos 35° = 0$

$\Rightarrow R = 100 \cos 35° =$ 81.9 N (to 3 s.f.)

b) The sledge remains stationary. Find the magnitude of the force *F*.

- Resolving parallel to the plane (↗):

$70 - F - 100 \sin 35° = 0$

$\Rightarrow F = 70 - 100 \sin 35° =$ 12.6 N (to 3 s.f.)

Example 4

A rock of mass 9 kg rests on a smooth plane inclined at 35° to the horizontal. The rock is held in equilibrium by a horizontal force of magnitude 10 N directed towards the plane and the tension in a string parallel to the plane, as shown in the diagram.

a) Find the normal reaction force, *R*, between the plane and rock.

- Draw a diagram showing all the forces acting on the rock.
- Resolving perpendicular to the plane (↖):

$R - 10 \sin 35° - 9g \cos 35° = 0$

$R = 10 \sin 35° + 9g \cos 35°$

$=$ 78.0 N (to 3 s.f.)

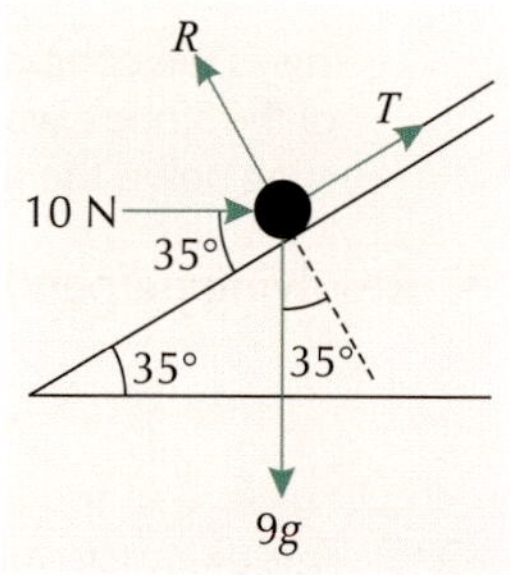

Tip: You need to resolve the 10 N force into components parallel and perpendicular to the plane. Find the angle between the 10 N force and the slope of the plane using alternate angles.

b) Find the magnitude of the tension in the string.

- Resolving parallel to the plane (↗):

$T + 10 \cos 35° - 9g \sin 35° = 0$

$T = 9g \sin 35° - 10 \cos 35°$

$=$ 42.4 N (to 3 s.f.)

Example 5

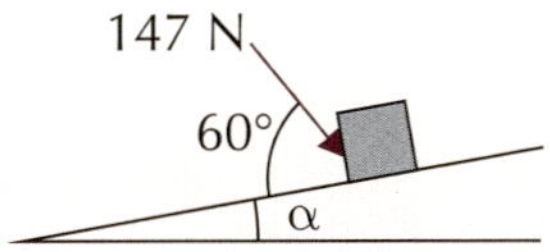

The diagram shows a block of mass 30 kg at rest on a smooth plane inclined at an angle α to the horizontal. The block is held in equilibrium by a force of 147 N, acting at an angle of 60° to the plane.

a) Show that $\sin\alpha = 0.25$.

- Draw a diagram showing all of the forces acting on the block:

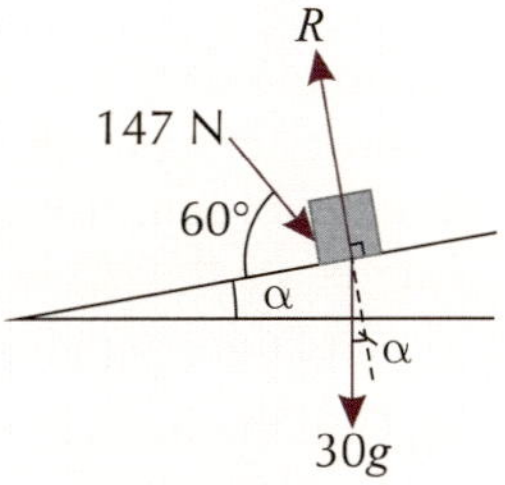

- Resolving parallel to the plane (↗):

$147\cos 60° - 30g\sin\alpha = 0$

$(30 \times 9.8)\sin\alpha = 147 \times 0.5$

$\sin\alpha = 73.5 \div 294$

$= 0.25$ as required

b) Find the normal reaction, R, of the plane on the block.

- First you need to find the value of α:

$\alpha = \sin^{-1}(0.25) = 14.4775...°$

- Now resolving perpendicular to the plane (↖):

$R - 147\sin 60° - 30g\cos\alpha = 0$

$R = 147\sin 60° + 30g\cos(14.4775...°)$

$= 412$ N (to 3 s.f.)

Tip: This time you've got two different angles to deal with — make sure you use the correct one when resolving the different forces.

Exercise 1.3

Q1 Each of the following diagrams shows a particle held in equilibrium by two light, inextensible strings. In each case, find the magnitude of the missing tension, T, and the mass of the particle, m.

a)

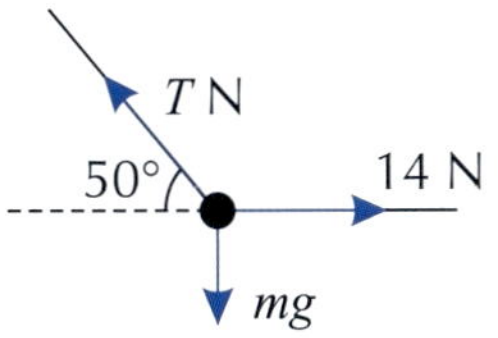

b)

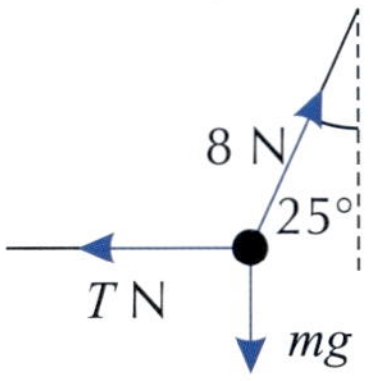

Q2 The diagram shows a particle suspended from two light, inextensible strings. The particle is in equilibrium. Find the weight of the particle, W, and the angle x.

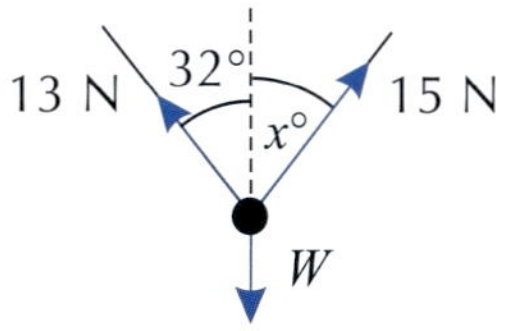

Q3 Hint: The 'line of greatest slope of the plane' just means straight down (or up) the slope of the plane.

Q3 A particle of weight 10 N rests on a smooth inclined plane.
The plane is inclined at an angle $x°$ to the horizontal.
A force of magnitude 8 N acting along the line of greatest slope of the plane holds the particle in equilibrium.

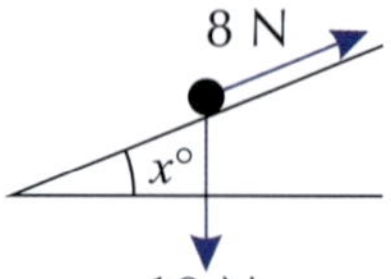

a) Calculate the size of the angle x.

b) Calculate the magnitude of the normal reaction force exerted by the plane on the particle.

Q4 Hint: Use the fact that cos 45° = sin 45°.

Q4 A particle rests on a smooth plane inclined at 45° to the horizontal. A force acting up the plane, parallel to the slope, holds the particle in equilibrium. The magnitude of the normal reaction force exerted by the plane on the particle is 25 N.

a) Find the magnitude of the force which holds the particle at rest.

b) Find the mass of the particle.

Q5 A particle with a mass of 5 kg rests on a smooth plane inclined at 10° to the horizontal, as shown. The particle is held in place by a light, inextensible string inclined at an angle of 30° to the plane.

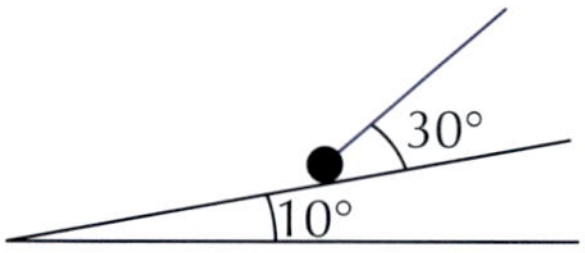

a) Calculate the magnitude of the tension in the string.

b) Calculate the magnitude of the normal reaction of the plane on the particle.

Q6 The diagram shows a 2 kg block held at rest on a smooth inclined plane by a horizontal force of 14.7 N. The plane is inclined at an angle of θ to the horizontal.
Show that $\tan \theta = 0.75$.

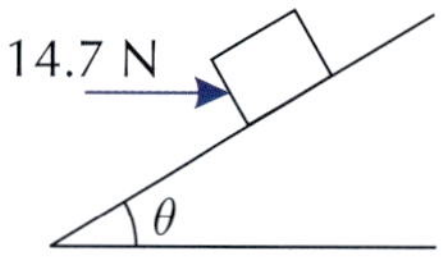

Q7 A particle of mass m is held at rest on a smooth plane inclined at 12° to the horizontal by a light, inextensible string. The string is inclined at 45° to the horizontal. The magnitude of the normal reaction force from the plane on the particle is 15 N. Find:

a) the magnitude of the tension, T, in the string,

b) the mass, m, of the particle in kg.

Q8 A particle is held in equilibrium on a smooth plane by a light, inextensible string. The plane is inclined at 50° to the horizontal and the string is inclined at 70° to the horizontal. The tension in the string has magnitude 5 N. A force, F, also acts on the particle, down the plane, parallel to the slope. Given that the magnitude of the normal reaction of the plane on the particle is four times the magnitude of F, find the weight of the particle.

2. Friction

If a body is in contact with a rough surface, and there is a force acting that could cause the body to move, then friction will act between the body and the surface to try and prevent motion. Pretty much any time you see the word 'rough', you know you're going to be dealing with it.

Learning Objectives:
- Understand and use the relationship $F \leq \mu R$.
- Understand the term 'limiting friction'.
- Use the formula for limiting friction to solve problems for a body in equilibrium.

Friction

- If an object is in contact with a **rough** surface, then **friction** can act to try and **prevent it moving**.
- Friction will **only** act if there is a **force** acting on the object which could cause it to **move** along the surface.

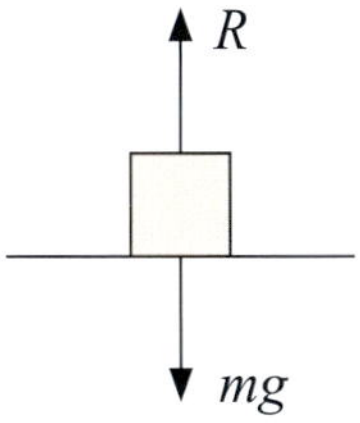

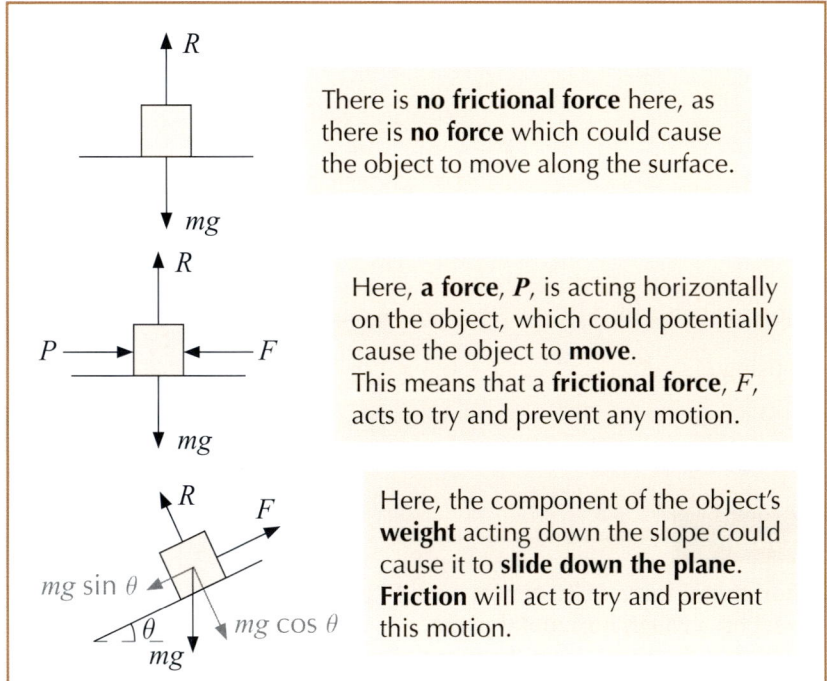

There is **no frictional force** here, as there is **no force** which could cause the object to move along the surface.

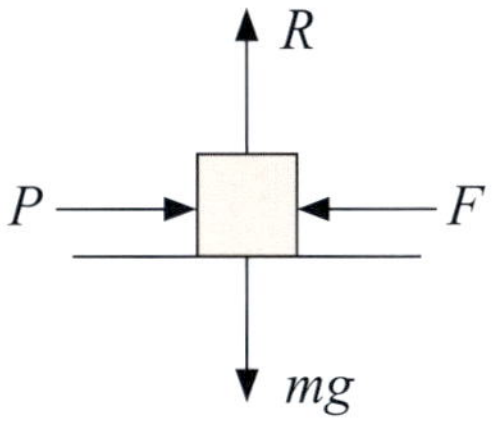

Here, **a force**, ***P***, is acting horizontally on the object, which could potentially cause the object to **move**.
This means that a **frictional force**, F, acts to try and prevent any motion.

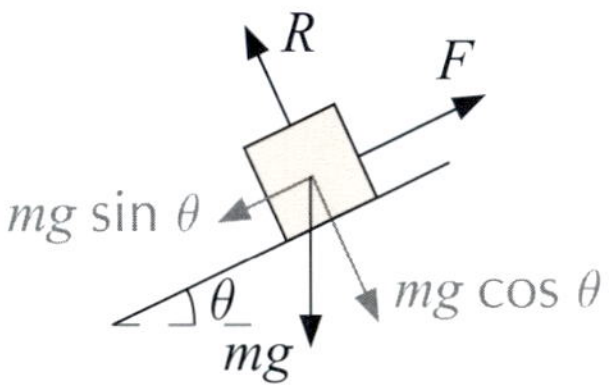

Here, the component of the object's **weight** acting down the slope could cause it to **slide down the plane**. **Friction** will act to try and prevent this motion.

Tip: Remember — there will only be friction if the surface is rough. If the surface is smooth, no frictional force will act on the object.

- Friction always acts in the **opposite direction to motion** (or potential motion).
- The magnitude of the frictional force can take a **range of possible values**. As the magnitude of the force which is trying to move the object **increases**, the magnitude of the frictional force will **also increase**.
- Eventually, the frictional force will reach its **maximum possible value**. At this point, friction is said to be **limiting**.
- If the force which is trying to move the object is **greater** than the maximum possible frictional force, the frictional force won't be large enough to prevent motion, and the object will **move**.
- The maximum possible magnitude of the frictional force depends on the **coefficient of friction**, μ, between the surface and the object.
- μ is a **number greater than or equal to 0**, and is a measure of the effect of friction between the object and the surface — the higher the number, the greater the effect of friction.

Tip: The magnitude of the frictional force will only be as big as is necessary to prevent motion.

- The frictional force, F, acting on an object is given by:

$$F \le \mu R$$

where R is the **normal reaction** of the surface on the object.

- When the object starts to **move**, or is on the point of moving (i.e. friction is **limiting**), the frictional force takes its **maximum value**:

$$F_{max} = \mu R$$

Tip: An object which is on the point of moving is said to be in 'limiting equilibrium'. For friction problems involving a moving object, see p.339-341.

Example 1

A horizontal force, P, acts on a particle of mass 12 kg, which is at rest on a rough horizontal plane.

Find the range of values that the frictional force, F, can take, given that the coefficient of friction between the particle and the plane is 0.4.

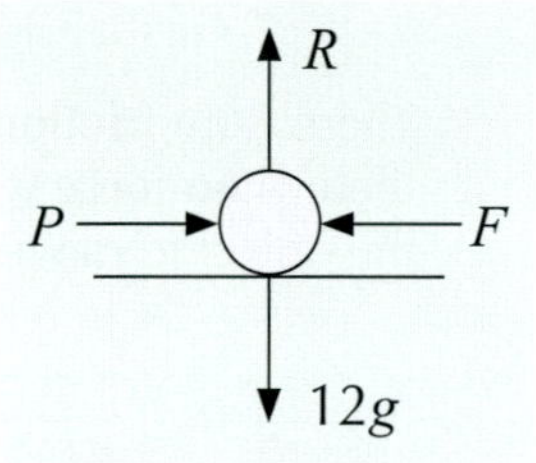

Resolving vertically (↑): $R = 12g$

This is just $R - 12g = 0$ rearranged.

So, using $F \le \mu R$: $F \le 0.4 \times 12g$

$F \le 47.04$ N

Tip: So F can take any value up to 47.04 N, depending on the magnitude of P.
If $P \le 47.04$ N, then the particle will remain at rest (and will be on the point of moving if $P = 47.04$ N).
If $P > 47.04$ N, then the particle will move and F will remain constant at 47.04 N.

Example 2

A horizontal force of 30 N acts on a 4 kg block which is at rest on a rough horizontal plane.
Given that the block is on the point of moving, find the coefficient of friction between the block and the plane.

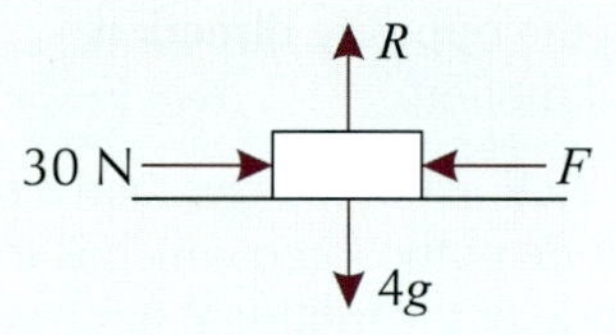

- Resolving horizontally: $F = 30$ N
- Resolving vertically: $R = 4g$
- The block is on the point of moving, so friction is limiting:

$$F = \mu R$$
$$30 = \mu \times 4g$$
$$\mu = 30 \div 4g$$
$$= 0.77 \text{ (to 2 d.p.)}$$

Tip: μ is usually given correct to 2 decimal places.

Example 3

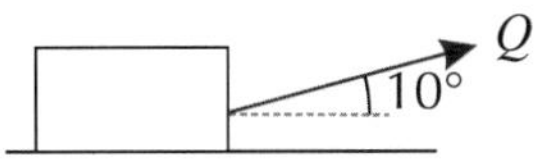

A box of mass 10 kg is on the point of slipping across a rough horizontal plane. A force, Q, acts on the box at an angle of 10° to the horizontal, as shown. Given that the coefficient of friction between the box and the plane is 0.7, find the magnitude of the normal reaction force from the plane and the magnitude of Q.

- Draw a diagram to show all the forces acting on the box:

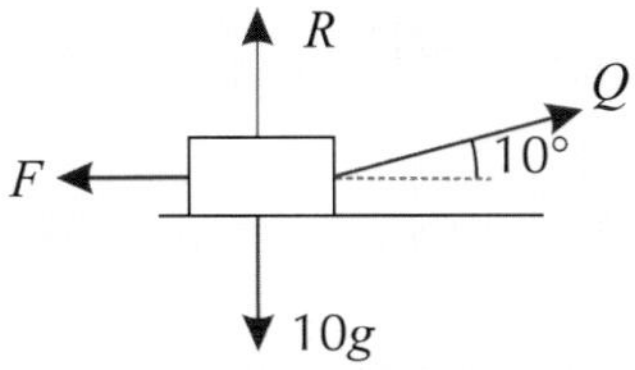

- Resolving vertically: $R + Q \sin 10° = 10g$

  $Q \sin 10° = 10g - R$ — call this equation ①

- Resolving horizontally: $Q \cos 10° = F$

  $Q \cos 10° = \mu R$

  $Q \cos 10° = 0.7R$ — call this equation ②

- Dividing equation ① by equation ②:

$$\frac{Q\sin 10°}{Q\cos 10°} = \frac{10g - R}{0.7R}$$

$$\tan 10° = \frac{10g - R}{0.7R}$$

$$0.7R \tan 10° = 10g - R$$

$$R = \frac{10g}{(1 + 0.7\tan 10°)}$$

$$= 87.232...$$

$$= 87.2 \text{ N (3 s.f.)}$$

- Substituting R into equation ②:

$$Q \cos 10° = 0.7 \times 87.232...$$

$$Q = 61.063... \div \cos 10°$$

$$= 62.0 \text{ N (3 s.f.)}$$

Tip: F always acts parallel to the surface, no matter what angle other forces act at.

Tip: You can assume that the box is about to slide horizontally in the direction of Q.

Tip: The box is on the point of sliding (i.e. in limiting equilibrium), so $F = \mu R$.

Example 4

A block of mass 3.2 kg lies at rest on a rough plane, inclined at an angle of θ to the horizontal. The block is on the point of sliding.

Given that $\mu = 0.75$, find the value of θ.

- Draw a diagram.

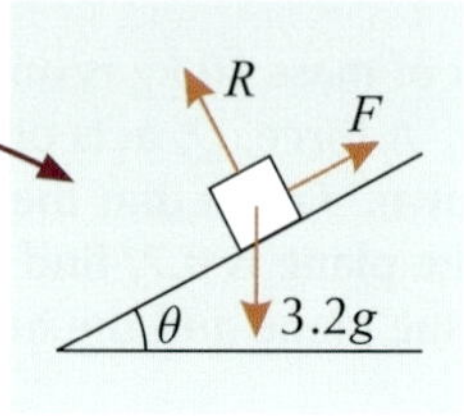

- Resolving parallel to the plane:

  $F = 3.2g \sin \theta$

- Resolving perpendicular to the plane:

  $R = 3.2g \cos \theta$

- The block is on the point of moving, so, using $F = \mu R$:

$$3.2g \sin \theta = 0.75 \times 3.2g \cos \theta$$

$$\frac{3.2g \sin \theta}{3.2g \cos \theta} = 0.75$$

$$\tan \theta = 0.75$$

$$\theta = 36.9° \text{ (to 3 s.f.)}$$

Tip: Make sure you know in which direction friction is acting. Here, the block is on the point of sliding down the slope, because of its weight. Friction will therefore act up the slope, to oppose motion.

Exercise 2.1

Q1 A particle of weight W rests on a rough horizontal surface with coefficient of friction μ between them. A horizontal force is applied to the particle such that the particle is on the point of slipping.

Find the magnitude of the frictional force, F, on the particle if:

a) $W = 5$ N, $\mu = 0.25$

b) $W = 8$ N, $\mu = 0.3$

c) $W = 15$ N, $\mu = 0.75$

Q2

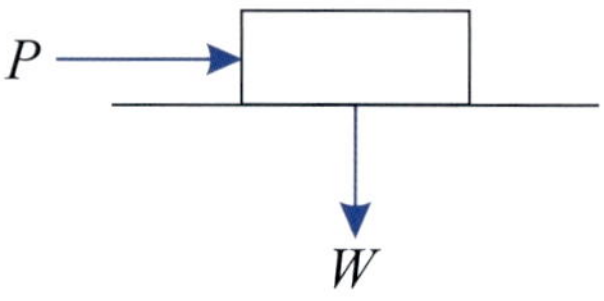

A particle of weight W is at rest on a rough horizontal surface. A horizontal force P acts on the particle. Given that the particle is on the point of moving, calculate the coefficient of friction if:

a) $W = 20$ N, $P = 5$ N b) $W = 15$ N, $P = 8$ N

Q3 A particle of mass 12 kg is placed on a rough horizontal surface. A horizontal force of magnitude 50 N is applied to the particle. The coefficient of friction between the particle and the surface is 0.5.

a) Will the 50 N force cause the particle to move?

b) What is the maximum possible magnitude of force that can be applied horizontally to the particle, such that the particle will remain in equilibrium?

Q4 A pull-along toy of weight 8 N rests on a rough horizontal surface with coefficient of friction $\mu = 0.61$ between the toy and the surface. A light, inextensible string is attached to the toy, at an angle of α to the horizontal, as shown.

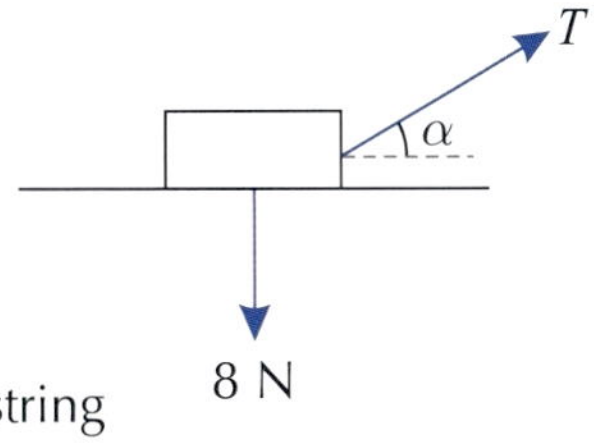

Find the maximum possible tension in the string given that the toy remains stationary when:

a) $\alpha = 0°$ b) $\alpha = 35°$

Q4 Hint: You're looking for the tension in the string when the toy is on the point of sliding.

Q5

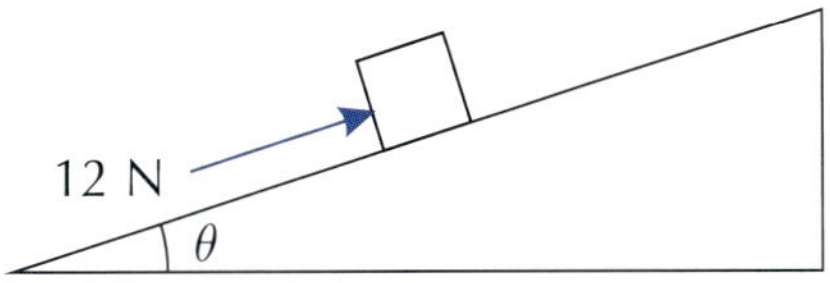

A box of mass 2 kg is at rest on a rough plane inclined at an angle of θ to the horizontal. A force of 12 N acts up the plane on the box, which is on the point of moving up the slope. Given that $\cos\theta = \frac{4}{5}$, find the coefficient of friction between the box and the plane.

Q5 Hint: To find $\sin\theta$ given $\cos\theta$, you can draw a right-angled triangle and use trigonometry and Pythagoras' theorem. Alternatively, you can use $\theta = \cos^{-1}\left(\frac{4}{5}\right)$, as long as you don't round it.

Q6

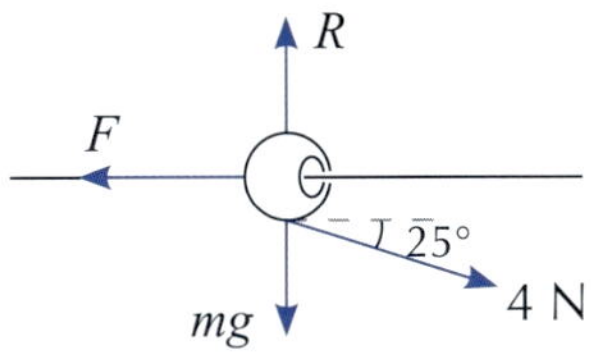

A horizontal wire is threaded through a bead of mass m kg. A force of magnitude 4 N is applied to the bead, at an angle of 25° to the horizontal, as shown. The bead is on the point of sliding along the wire. Given that the coefficient of friction between the wire and the bead is 0.15, find:

a) the magnitude of R, the normal reaction of the wire on the bead,

b) the value of m.

Q6 Hint: In mechanics, a bead is just a particle with a hole in it which a wire can pass through.

Q7 An object is placed on a rough plane inclined at an angle α to the horizontal. The coefficient of friction between the object and the plane is μ. Show that the object will be at rest on the plane if $\tan\alpha \leq \mu$.

PROBLEM SOLVING

Q8 A sledge of mass 20 kg is at rest on a rough slope inclined at 29° to the horizontal. The coefficient of friction between the slope and the sledge is 0.1. A force, T, acts on the sledge at an angle of 15° to the slope, as shown.

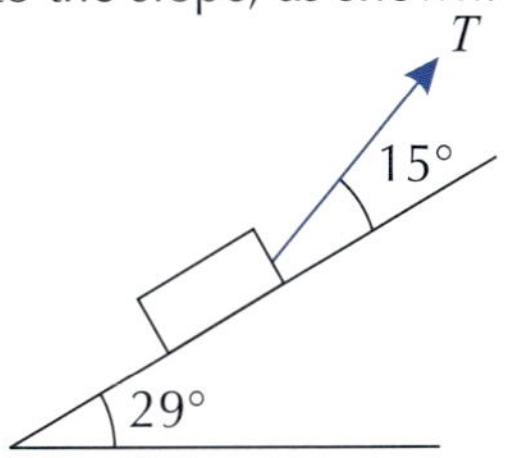

Calculate the magnitude of T, given that this force is only just large enough to stop the sledge sliding down the slope.

Q8 Hint: Think carefully about which direction the frictional force will act in.

3. Newton's Laws of Motion

Learning Objectives:

- Be able to use Newton's laws of motion in situations that require resolving forces.
- Be able to solve problems involving friction when a particle is moving.
- Be able to solve problems involving connected particles, friction forces and inclined planes.

You've seen Newton's three laws before, so there's nothing new here. You need to be able to combine what you learned in Year 1 with what you've seen so far in this chapter.

Newton's laws of motion

Here's a quick recap of Newton's three laws of motion:

- **Newton's first law:**

A body will stay at **rest** or maintain a **constant velocity** unless a **resultant force** acts on the body.

- **Newton's second law:**

The **overall resultant force** (F_{net}) acting on a body is equal to the **mass** of the body multiplied by the body's **acceleration**.

i.e. $F_{net} = ma$

Tip: The formula for weight ($W = mg$) is derived from Newton's second law, where g is acceleration due to gravity.

- **Newton's third law:**

For **two bodies**, A and B, **in contact** with each other, the force exerted by A on B is **equal in magnitude** but **opposite in direction** to the force exerted by B on A.

Example 1

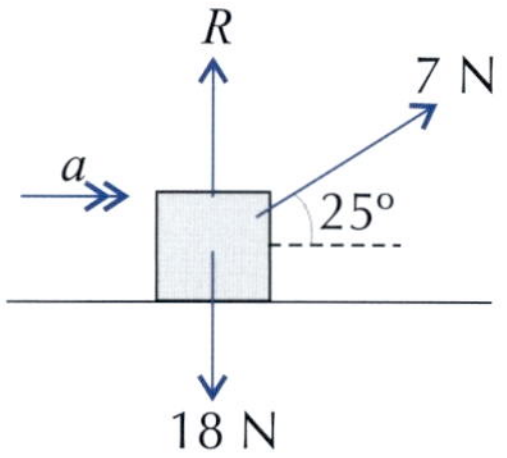

A particle of weight 18 N is being pulled along a smooth horizontal surface by a force of 7 N acting at an angle of 25° to the horizontal, as shown. Find the speed of the particle after 3 seconds, given that it starts from rest.

- Resolving forces horizontally (→), $F_{net} = 7 \cos 25°$
  Using Newton's second law to find the acceleration, a:

$$F_{net} = ma$$
$$7 \cos 25° = \frac{18}{g}a$$
$$6.344... = 1.836... \times a$$
$$a = 3.454... \text{ ms}^{-2}$$

Tip: You're told the weight, so use the formula $W = mg$ to find the mass.

- List the variables:

$u = 0, \quad v = v, \quad a = 3.454..., \quad t = 3$

$v = u + at$
$= 0 + 3.454... \times 3$
$= 10.4 \text{ ms}^{-1}$ (to 3 s.f.)

Tip: It's usually easiest to list the variables you know from the question, and the one you want to find, and pick the *suvat* equation that has them all in it.

Example 2

Two forces, given by the vectors (12i – 4j) N and (–3i + 16j) N, act on a particle of mass 3 kg, causing it to accelerate from rest.

a) Find the acceleration of the particle as a vector.

- The resultant force on the particle

$$\boldsymbol{F}_{net} = (12\mathbf{i} - 4\mathbf{j}) + (-3\mathbf{i} + 16\mathbf{j}) = 9\mathbf{i} + 12\mathbf{j}$$

- Now, using Newton's second law (with vectors):

$$\boldsymbol{F}_{net} = m\boldsymbol{a}$$

$$9\mathbf{i} + 12\mathbf{j} = 3\boldsymbol{a}$$

$$\boldsymbol{a} = (3\mathbf{i} + 4\mathbf{j})\text{ ms}^{-2}$$

b) Find the velocity vector of the particle after 7 seconds.

- List the variables:

$\boldsymbol{u} = 0$ $\boldsymbol{v} = \boldsymbol{v}$ $\boldsymbol{a} = 3\mathbf{i} + 4\mathbf{j}$ $t = 7$

- So use the equation $\boldsymbol{v} = \boldsymbol{u} + \boldsymbol{a}t$:

$$\boldsymbol{v} = \boldsymbol{u} + \boldsymbol{a}t$$

$$\boldsymbol{v} = 0 + (3\mathbf{i} + 4\mathbf{j}) \times 7$$

$$\boldsymbol{v} = (21\mathbf{i} + 28\mathbf{j})\text{ ms}^{-1}$$

c) How far does the particle travel in the first 4 seconds?

- Find the magnitude of the acceleration:

$$|\boldsymbol{a}| = \sqrt{3^2 + 4^2} = 5\text{ ms}^{-2}$$

- Now list the variables

$s = s$ $u = 0$ $a = 5$ $t = 4$

$$s = ut + \frac{1}{2}at^2$$

$$s = (0 \times 4) + \frac{1}{2} \times 5 \times 4^2 = 40\text{ m}$$

d) A third force *K* is applied to the particle, which now moves with acceleration (7i – 2j) ms⁻². Find the force *K* in terms of i and j.

- Use Newton's second law to find the new resultant force:

$$\boldsymbol{F}_{net} = m\boldsymbol{a}$$

$$\boldsymbol{F}_{net} = 3(7\mathbf{i} - 2\mathbf{j}) = 21\mathbf{i} - 6\mathbf{j}$$

- Finally subtract the two original forces from $\boldsymbol{F}_{net}$ to find $\boldsymbol{K}$:

$$(12\mathbf{i} - 4\mathbf{j}) + (-3\mathbf{i} + 16\mathbf{j}) + \boldsymbol{K} = (21\mathbf{i} - 6\mathbf{j})$$

$$\boldsymbol{K} = (21\mathbf{i} - 6\mathbf{j}) - (12\mathbf{i} - 4\mathbf{j}) - (-3\mathbf{i} + 16\mathbf{j})$$

$$= (12\mathbf{i} - 18\mathbf{j})\text{ N}$$

Tip: You could solve part c) by finding the displacement as a vector, and then working out its magnitude — you would get the same answer either way.

Exercise 3.1

Q1

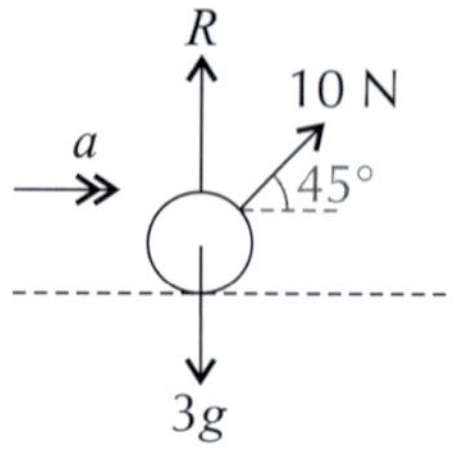

A particle of mass 3 kg is being accelerated across a horizontal plane by a force of 10 N acting at an angle of 45° to the horizontal, as shown above. Find the magnitude of:

a) the acceleration of the particle,

b) the normal reaction with the plane.

Q2

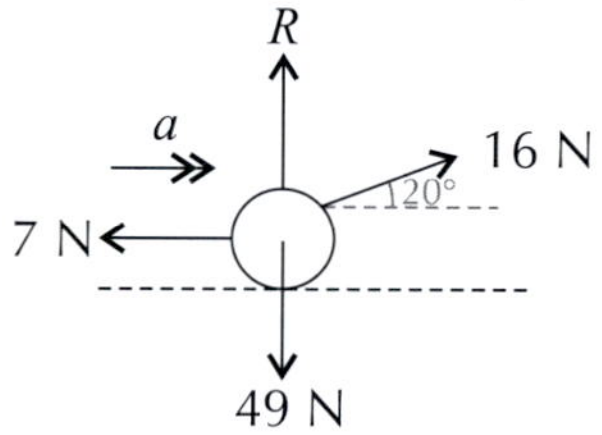

A particle of weight 49 N initially at rest on a horizontal plane is acted on by a constant horizontal resistive force of 7 N and a force of 16 N acting at an angle of 20° to the horizontal, as shown. Find:

a) the acceleration of the particle,

b) the speed of the particle after 7 seconds,

c) the magnitude of the normal reaction with the plane.

Q3 A particle of mass 5 kg is acted on by two forces, $(6\mathbf{i} - 10\mathbf{j})$ N and $(4\mathbf{i} + 5\mathbf{j})$ N.

a) Find the magnitude of the acceleration of the particle.

b) A third force, of $(-6\mathbf{i} + 2\mathbf{j})$ N, begins to act on the particle. Describe the effect this has on the magnitude of the acceleration of the particle.

Q4 Three forces are acting on a particle of mass 10 kg.

a) Two of the forces are given by $(6\mathbf{i} - 2\mathbf{j})$ N and $(5\mathbf{i} + 5\mathbf{j})$ N and the resultant force is $(7\mathbf{i} + \mathbf{j})$ N. Find the third force in vector form.

b) Find the particle's velocity vector 8 seconds after it begins accelerating from rest.

c) A fourth force, of $(-7\mathbf{i} - \mathbf{j})$ N, begins to act on the particle. Describe the effect this has on the particle's motion.

Friction and inclined planes

Inclined planes

- For objects on an inclined plane, resolve forces **parallel** and **perpendicular** to the plane, rather than horizontally and vertically.
- Remember that weight always acts **vertically downwards**, so you need to find the **components** of the weight in these directions.

Example 1

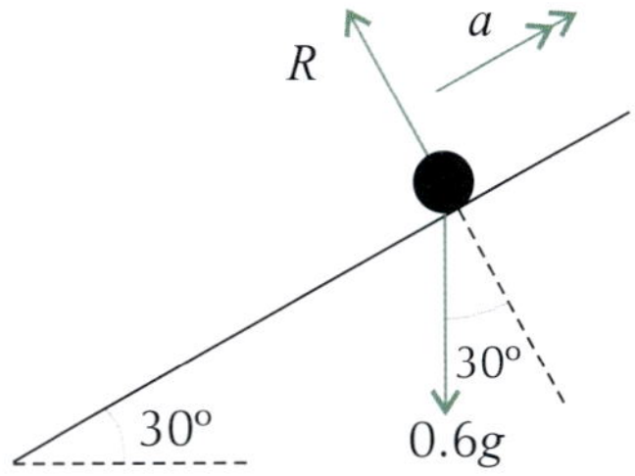

A particle of mass 600 g is propelled up the line of greatest slope of a smooth plane inclined at 30° to the horizontal. Immediately after the propelling force has stopped, the particle has speed 3 ms⁻¹. Find:

a) the distance the particle travels before coming instantaneously to rest.

- First, resolve parallel to the plane (↗) to find a:

$$F_{net} = ma$$
$$-0.6g \sin 30° = 0.6a$$
$$a = -4.9 \text{ ms}^{-2}$$

- Next, use one of the constant acceleration equations:

$$v^2 = u^2 + 2as$$
$$0 = 3^2 + 2(-4.9)s$$
$$s = 0.918 \text{ m (3 s.f.)}$$

b) the magnitude of the normal reaction from the plane.

- Now resolve perpendicular to the plane (↖):

$$F_{net} = ma$$
$$R - 0.6g \cos 30° = 0.6 \times 0$$
$$R = 5.09 \text{ N (3 s.f.)}$$

Tip: Remember — the normal reaction force is always perpendicular to the surface.

Tip: The particle is slowing down, so its acceleration is negative (i.e. it's decelerating).

Tip: The particle is moving parallel to the plane, so the acceleration perpendicular to the plane is zero.

Frictional forces

- You've already covered friction on pages 331-334, but back then you were only concerned with objects which were being held in **equilibrium**.
- When a **moving** object is acted on by a frictional force, friction is **limiting**, and the frictional force (F) is at its maximum value: $F = \mu R$. (Where μ is the coefficient of friction and R is the normal reaction between the object and the surface.)
- Friction **opposes** motion, so it will always act in the **opposite direction** to the direction an object is moving in.
- You'll need to use Newton's laws and $F = \mu R$ to solve problems involving objects which are accelerating and being acted on by friction.

Example 2

A block of mass 12 kg is being accelerated across a rough, horizontal plane by a force of magnitude 60 N, as shown. Given that the magnitude of the acceleration is 1.5 ms⁻², find the coefficient of friction between the block and the plane.

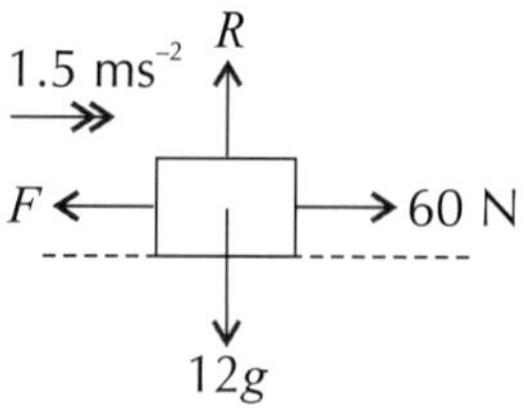

- First, resolve forces horizontally (→), to find the frictional force, F:

$$F_{net} = ma$$
$$60 - F = 12 \times 1.5$$
$$F = 42 \text{ N}$$

- Next, resolve forces vertically (↑) to find the normal reaction, R:

$$F_{net} = ma$$
$$R - 12g = 12 \times 0$$
$$R = 12g = 117.6 \text{ N}$$

- The block is moving, so F takes its maximum possible value:

$$F = \mu R$$
$$42 = \mu \times 117.6$$
$$\mu = 0.36 \text{ (2 d.p.)}$$

Tip: Don't get confused between F and F_{net}. F is the frictional force and F_{net} is the overall resultant force.

Example 3

A small body of weight 20 N is released from rest on a rough plane angled at 15° to the horizontal.

Given that the body accelerates down the plane at a rate of 0.44 ms⁻², find the coefficient of friction between the body and the plane.

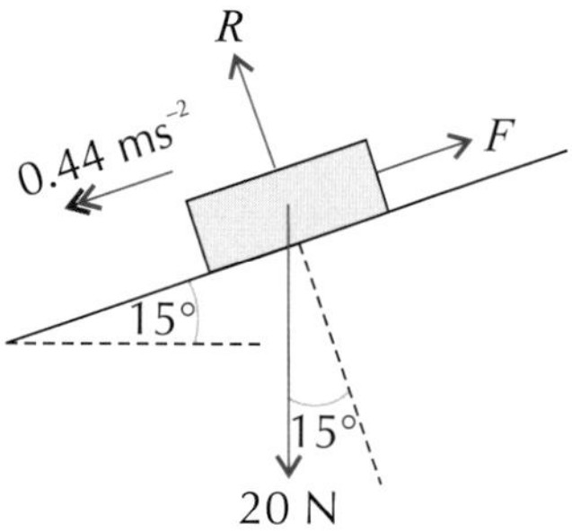

Tip: The body is accelerating down the plane, so the frictional force acts up the slope.

- Resolve parallel to the plane (↙):

$$F_{net} = ma$$
$$20 \sin 15° - F = \frac{20}{g} \times 0.44$$
$$F = 4.278... \text{ N}$$

- Resolve perpendicular to the plane (↖):

$$F_{net} = ma$$
$$R - 20 \cos 15° = \frac{20}{g} \times 0$$
$$R = 20 \cos 15° = 19.318... \text{ N}$$

- It's sliding, so:

$$F = \mu R$$
$$4.278... = \mu \times 19.318...$$
$$\mu = 0.22 \text{ (to 2 d.p.)}$$

Example 4

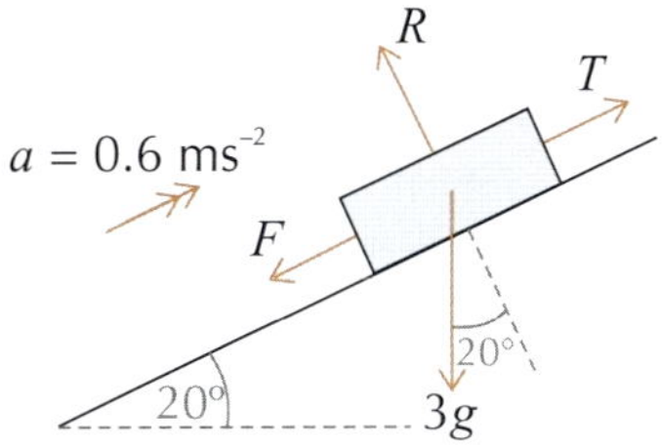

A mass of 3 kg is being pulled up a plane inclined at 20° to the horizontal by a rope parallel to the surface. Given that the mass is accelerating at 0.6 ms^{-2} and the coefficient of friction between the mass and the plane is 0.4, find *T*, the tension in the rope.

- Resolve perpendicular to the plane (↖):

$$F_{net} = ma$$
$$R - 3g\cos 20° = 3 \times 0$$
$$R = 3g\cos 20° = 27.626... \text{ N}$$

- The mass is sliding, so $F = \mu R$:

$$F = 0.4 \times 27.626... = 11.050... \text{ N}$$

- Resolve parallel to the plane (↗):

$$F_{net} = ma$$
$$T - F - 3g\sin 20° = 3 \times 0.6$$
$$T = 1.8 + 11.050... + 3g\sin 20°$$
$$= 22.9 \text{ N (3 s.f.)}$$

Exercise 3.2

Q1 A wooden block of mass 1.5 kg is placed on a smooth plane which is inclined at 25° to the horizontal, as shown in the diagram. The block is released from rest and begins to accelerate down the slope.

a) What is the magnitude of the block's acceleration?

b) Assuming that the block's acceleration is constant, find the speed of the block 3 seconds after it is released.

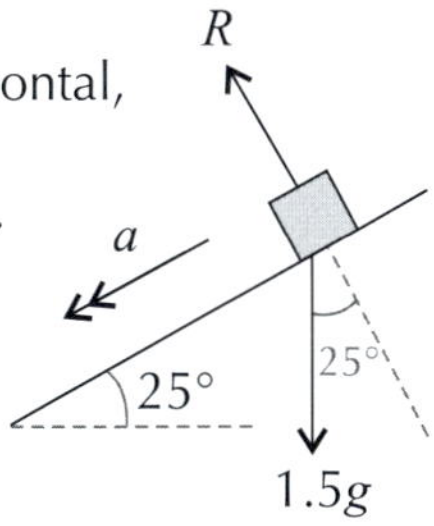

Q1 Hint: If a surface is smooth, it means there's no frictional force.

Q2 A horizontal force of magnitude 26 N is acting on a box of mass 4 kg, causing the box to accelerate across a rough, horizontal plane at a rate of 1.2 ms^{-2}.

a) Draw a labelled diagram to show the forces acting on the box.

b) Calculate the normal reaction, *R*, of the plane on the box.

c) Use Newton's second law to find the magnitude of the frictional force acting on the box.

d) Find the coefficient of friction, μ, between the box and the plane.

Q3 A box of mass 7 kg is accelerated across a rough horizontal plane by a horizontal force, P. The coefficient of friction, μ, between the box and the plane is 0.2 and the box accelerates at a rate of 0.8 ms^{-2}.

a) Draw a labelled diagram to show the forces acting on the box.

b) Calculate the normal reaction, R, of the plane on the box.

c) Calculate the magnitude of P.

Q4 Hint: Remember, taut means that the rope is tight and straight, and has tension in it.

Q4 A taut rope is pulling a block up a rough inclined plane angled at 40° to the horizontal. The rope is parallel to the plane.
The coefficient of friction between the block and the surface is 0.35 and the tension in the rope is 70 N.

a) Draw a diagram to show the forces acting on the block.

b) Given that the block is accelerating at a rate of 3.2 ms^{-2}, find the mass of the block.

Q5 A block of mass 1 kg is sliding down a rough plane of length 2 m, which is inclined at an angle of 50° to the horizontal.
The coefficient of friction between the block and the surface is 0.4.

a) Draw a diagram to show the forces acting on the block.

b) If the block is released from rest at the top of the plane, find how long it will take to reach the bottom of the plane.

Q6

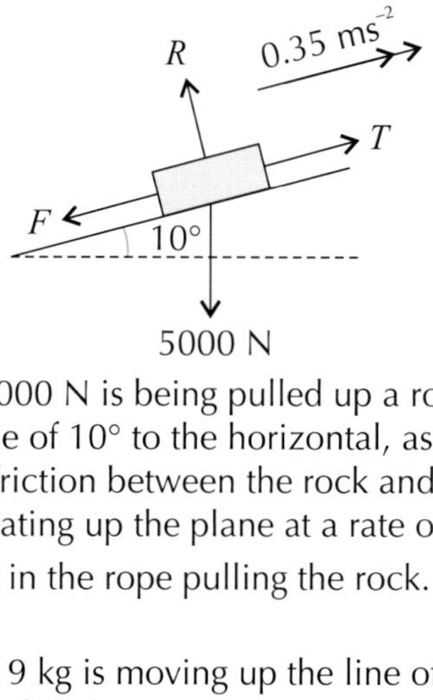

A rock weighing 5000 N is being pulled up a rough inclined plane which is at an angle of 10° to the horizontal, as shown.
The coefficient of friction between the rock and the plane is 0.1.
The rock is accelerating up the plane at a rate of 0.35 ms^{-2}.

Find T, the tension in the rope pulling the rock.

Q7 A package of mass 9 kg is moving up the line of greatest slope of a rough plane inclined at an angle of 22° to the horizontal. The package is being pushed up the slope by a force of magnitude 110 N acting horizontally. Given that the package is accelerating up the slope at a rate of 1.3 ms^{-2}, find μ, the coefficient of friction between the package and the plane. PROBLEM SOLVING

Q8 A block of mass 5 kg is at rest on a rough plane inclined at 35° to the horizontal. The coefficient of friction between the block and the plane is 0.4. Given that the block is being held in equilibrium by a force D acting parallel to the plane, find the range of possible values of the magnitude of D. PROBLEM SOLVING

Connected particles, pegs and pulleys

You also need to be able to resolve forces and calculate friction in situations with **connected particles**. You should remember from Year 1 that when particles are connected by a **taut**, **inextensible string** or a **rigid rod**, it will exert an **equal tension** force or **thrust** force respectively on both particles. The particles will also have the same **magnitude** of **acceleration**.

Normally, the way to approach these problems is to resolve forces on each particle **separately** and form a pair of **simultaneous equations**, although sometimes it will be easier to treat the **whole system** as **one object**.

Example 1

Two particles, A and B, of mass 6 kg and 8 kg respectively, are connected by a light, inextensible string. The particles are accelerating up a rough slope which is inclined at an angle of 20° to the horizontal. The force causing the acceleration has magnitude 200 N and acts on A in a direction parallel to the slope. The coefficient of friction between each particle and the slope is 0.6.

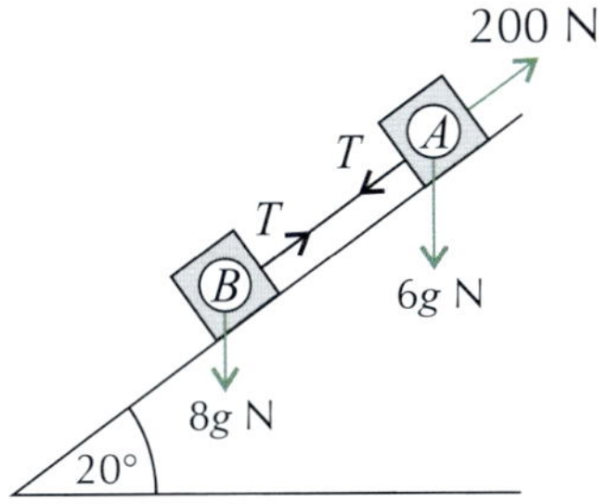

Find the acceleration of A.

- Treat the two particles as one combined particle with mass 6 kg + 8 kg = 14 kg, and resolve perpendicular to the plane (↖):

$$F_{net} = ma$$

$$R - 14g\cos 20° = 0$$

$$R = 14g\cos 20° \text{ N}$$

- The particles are moving, so the total frictional force acting on the two particles is:

$$F = \mu R$$

$$= 0.6 \times 14g\cos 20°$$

$$= 8.4g\cos 20°$$

- Again, treat the two particles as one combined particle, but now resolve parallel to the plane (↗):

$$F_{net} = ma$$

$$200 - 14g\sin 20° - F = 14a$$

$$200 - 14g\sin 20° - 8.4g\cos 20° = 14a$$

$$a = 5.41 \text{ ms}^{-2} \text{ (3 s.f.)}$$

Tip: The acceleration of A will be the same as the acceleration of the two particles combined.

Tip: The total frictional force acting on the two particles is the sum of the individual frictional forces: $F = F_A + F_B$

Example 2

Particles A and B, of mass 4 kg and 10 kg respectively, are connected by a light inextensible string which passes over a fixed smooth pulley. A force of 15 N acts on A at an angle of 25° to a rough horizontal plane, as shown. The coefficient of friction between A and the plane is 0.7. B is released from rest and falls d m vertically to the ground in 2 seconds.

a) Find d.

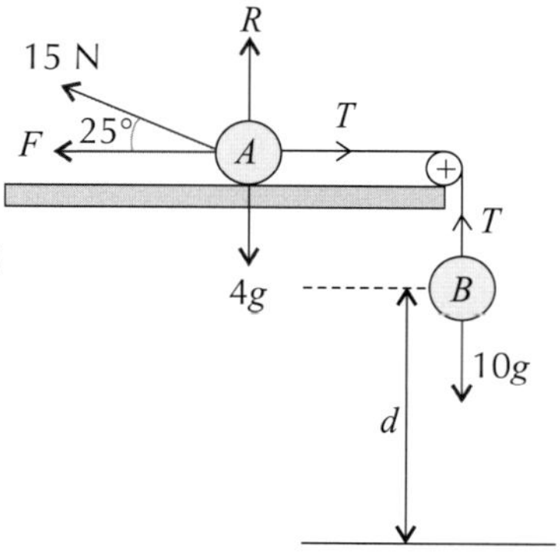

- Resolve vertically for A (↑) to find the normal reaction, R:

  $R + 15 \sin 25° - 4g = 0$

  $R = 32.860...$ N

- The particles are moving, so F takes its maximum value:

  $F = \mu R$

  $= 0.7 \times 32.860... = 23.002...$ N

- Resolve horizontally for A (→):

  $T - 23.002... - 15 \cos 25° = 4a$

  $T = 4a + 36.597...$ — equation ①

- Resolve vertically for B (↓):

  $10g - T = 10a$

  $T = 98 - 10a$ — equation ②

- Substitute equation ① into equation ②:

  $4a + 36.597... = 98 - 10a$

  $a = 4.385...$ ms^{-2}

- Using $s = ut + \frac{1}{2}at^2$: $s = d,\ u = 0,\ t = 2,\ a = 4.385...$

  $d = (0 \times 2) + \frac{1}{2}(4.385... \times 2^2) =$ 8.77 m (to 3 s.f.)

b) When B hits the ground, A carries on moving along the plane. Given that the resistive forces acting on A remain the same as in part a), and that A does not hit the pulley, how long does it take A to stop after B hits the ground?

- First, find the speed of A when B hits the ground:

  $v = u + at$

  $v = 0 + (4.385... \times 2) = 8.771...$ ms^{-1}

- Next, resolve horizontally for A (→) to find its deceleration:

  $-23.002... - 15 \cos 25° = 4a$

  $-36.597... = 4a$

  $a = -9.149...$ ms^{-2}

- Now find the time taken for A to stop:

  $t = \frac{(v - u)}{a}$

  $t = \frac{0 - 8.771...}{-9.149...} =$ 0.959 s (to 3 s.f.)

Tip: Try not to round the numbers in any of the intermediate steps of a calculation, otherwise your final answer might not be correct.

Tip: Remember that A and B will have the same acceleration up until the point when B hits the ground. When it does, the string will go slack and there will be no tension force acting on A.

Tip: Don't confuse the *suvat* variables from before B hits the ground with those from after.

Example 3

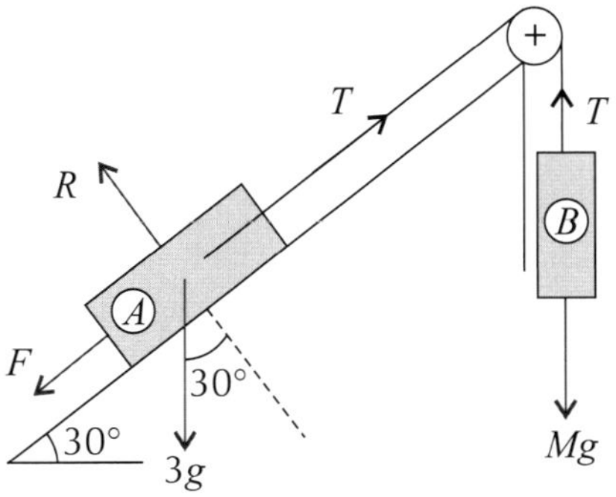

A block of mass 3 kg is held in equilibrium on a rough plane inclined at 30° to the horizontal. It is attached via a light, inextensible string to a mass of M kg hanging vertically beneath a fixed smooth pulley, as shown. The coefficient of friction between A and the plane is 0.4.

a) Find M, given that A is on the point of sliding up the plane.

- Resolve perpendicular to the plane (↖) for A:

$$R - 3g\cos 30° = 3 \times 0$$
$$R = 3g\cos 30°$$

- The friction is limiting, so:

$$F = \mu R$$
$$= 0.4 \times 3g\cos 30°$$
$$= 10.184...\text{ N}$$

- Resolve vertically (↓) for B:

$$Mg - T = M \times 0$$
$$T = Mg$$

- Resolve parallel to the plane (↗) for A:

$$T - F - 3g\sin 30° = 3 \times 0$$
$$Mg - 10.184... - 3g\sin 30° = 0$$
$$M = 2.54\text{ kg (to 3 s.f.)}$$

b) What would M be if A were on the point of sliding down the plane?

- Resolve parallel to the plane (↗) for A:

$$T + F - 3g\sin 30° = 3 \times 0$$
$$Mg + 10.184... - 3g\sin 30° = 0$$
$$M = 0.461\text{ kg (to 3 s.f.)}$$

Tip: A is "on the point of sliding" — this is another way of describing limiting equilibrium, which means that the frictional force takes its maximum possible value: $F = \mu R$. This is really a statics question — the particles aren't moving, so acceleration is zero in all directions.

Tip: You've already done all the hard work in part a), so in b), you only need to repeat the very last step and resolve parallel to the plane, with the frictional force acting in the opposite direction.

Example 4

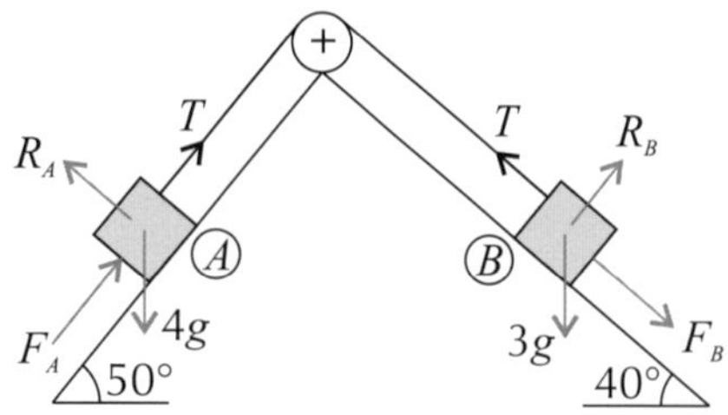

Two particles, A and B, of masses 4 kg and 3 kg respectively, are joined by a light inextensible string which passes over a smooth fixed pulley. The particles are positioned as shown in the diagram above. The system is released from rest and A begins to move down the slope and B begins to move up the slope. Given that the coefficient of friction between each particle and the surface is 0.2, find the magnitude of the acceleration of A.

- The particles are moving, so friction is limiting:

$$F_A = \mu R_A$$
$$= 0.2 \times 4g \cos 50° = 5.039... \text{ N}$$
$$F_B = \mu R_B$$
$$= 0.2 \times 3g \cos 40° = 4.504... \text{ N}$$

- Resolve parallel to the plane (↙) for A:

$$4g \sin 50° - F_A - T = 4a$$
$$24.989... - T = 4a \quad \text{— call this equation (1)}$$

- Resolve parallel to the plane (↖) for B:

$$T - 3g \sin 40° - F_B = 3a$$
$$T - 23.402... = 3a$$
$$T = 3a + 23.402... \quad \text{— call this equation (2)}$$

- Substitute equation (2) into equation (1):

$$24.989... - 3a - 23.402... = 4a$$
$$a = 0.227 \text{ ms}^{-2} \text{ (to 3 s.f.)}$$

Tip: Find R_A and R_B by resolving perpendicular to each plane.

Exercise 3.3

Q1 A light, inextensible string passing over a smooth pulley connects boxes A and B, of mass 10 kg and 8 kg respectively, as shown in the diagram. The system is released from rest and the boxes begin to accelerate at a rate of 0.5 ms^{-2}. Find:

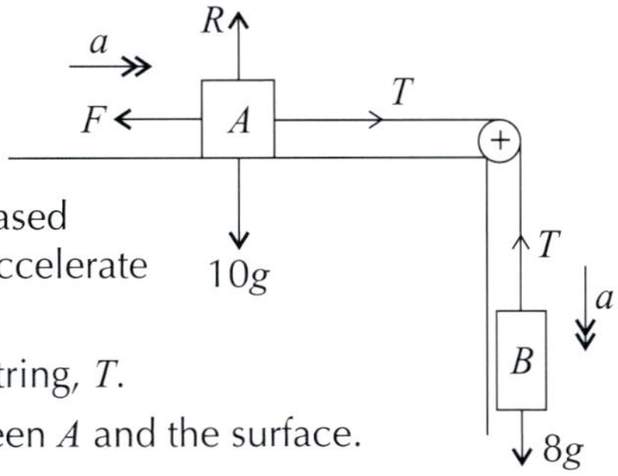

a) the tension in the connecting string, T.

b) the coefficient of friction between A and the surface.

Q2

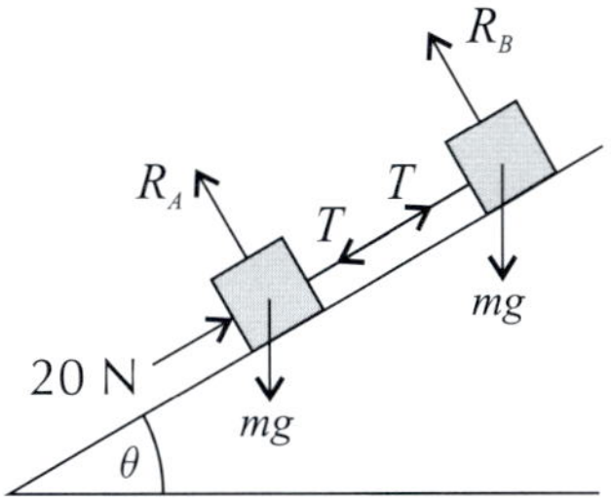

Two identical objects are connected by a rigid rod. The objects are being pushed up a rough slope inclined at an angle θ to the horizontal, by a force of 20 N acting parallel to the slope. Calculate the thrust force in the rod.

Q2 Hint: The thrust force acts just like a tension force, but it pushes against the blocks rather than pulling on them.

Q3

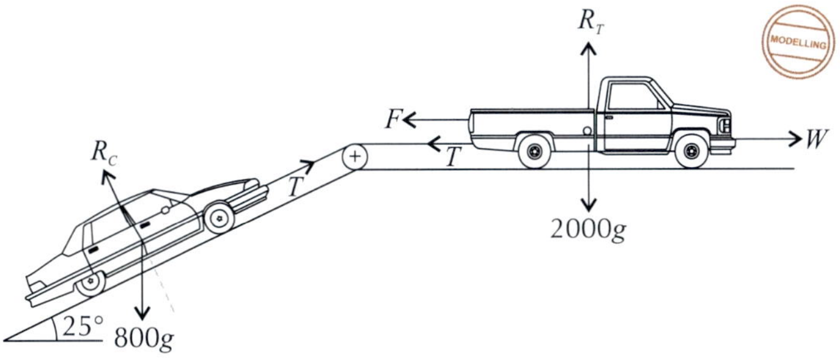

A truck of mass 2 tonnes is moving along a rough, horizontal plane (μ = 0.8), and towing a car of mass 0.8 tonnes up a smooth plane, angled at 25° to the horizontal, as shown in the diagram above.

The chain between the truck and the car is modelled as a light, inextensible string passing over a fixed, smooth pulley.

The car is initially 25 m from the pulley and is towed 10 m towards the pulley in the first 30 seconds after starting from rest. The truck and the car move with constant acceleration.

You may assume that the frictional force between the truck and the horizontal plane takes its maximum value.

a) Find the tension in the chain.

b) Calculate W, the driving force of the truck.

c) The chain connecting the truck to the car snaps. Assuming that all other forces remain constant, calculate the instantaneous acceleration of the truck.

Q3 Hint: Start by finding the acceleration of the truck and the car.

Q4 A light, inextensible string attaches a block P of mass 60 kg to a block Q of mass 20 kg over a fixed, smooth pulley. P lies on a rough plane ($\mu = 0.1$) inclined at an angle of 50° to the horizontal, and Q hangs vertically below the pulley, as shown. The system is held in limiting equilibrium with P on the point of sliding down the plane by a second light, inextensible string attached to Q at one end and the horizontal ground at the other.

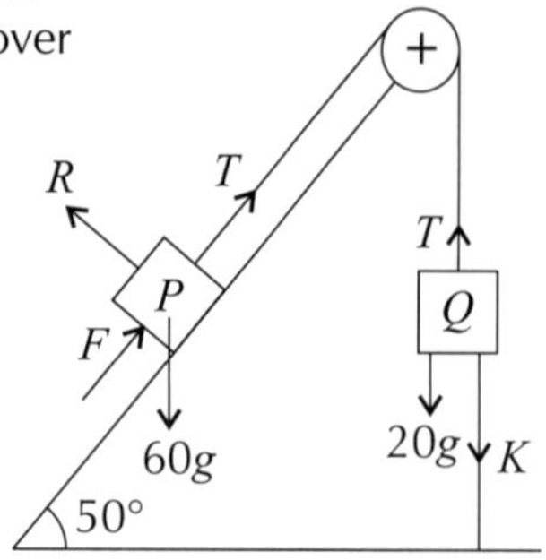

a) Find the tension in the second string, K.

b) The string fixing Q to the ground is cut. Given that Q does not hit the pulley, find the speed of P 5 seconds after it starts moving.

Q4b) Hint: When the string is cut, the tension force K will disappear.

Q5 Particles A and B, of mass 8 kg and 9 kg respectively, are connected by a light, inextensible string which passes over a fixed, smooth pulley. A force of 60 N acts on A at an angle of 40° to the rough, horizontal plane that A rests on, away from the pulley. Particle A is held 20 cm from the pulley and, when released from rest, accelerates towards the pulley at a rate of 0.4 ms^{-2}.

a) Find μ, the coefficient of friction between A and the plane.

b) Find the speed of A when it hits the pulley.

c) When A hits the pulley, the string breaks and B falls freely. Given that B was initially released from a position 60 cm above the horizontal ground, find the speed of B when it hits the ground.

Q5c) Hint: Remember — while the particles are connected, they'll have the same speed.

Q6

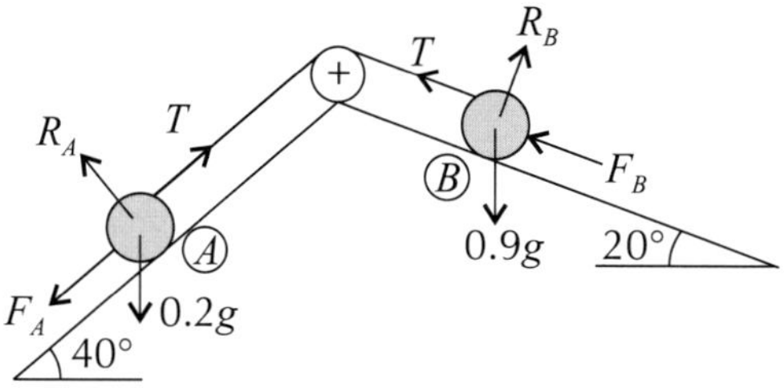

Particles A and B, of mass 0.2 kg and 0.9 kg respectively, are connected by a light inextensible string. The string passes over a smooth pulley fixed to the apex of a wedge.

A rests on a rough plane ($\mu_A = 0.7$) inclined at an angle of 40° to the horizontal and B rests on a different rough plane ($\mu_B = 0.05$) inclined at an angle of 20° to the horizontal, as shown above.

The particles are held at rest with A 35 cm from the pulley.

a) Given that A begins to accelerate up the slope when the particles are released from rest, find the tension in the string.

b) Find the speed of A when it reaches the pulley.

Chapter 16 Moments

1. Moments

So far you've seen that a force applied to a particle can cause the particle to accelerate in a straight line in the direction of the force. However, a force can also have a turning effect, usually when a body is pivoted at a point.

Learning Objectives:
- Be able to find the moment of a force about a point.
- Be able to find the sum of moments about a point.
- Be able to find missing distances and forces in calculations involving rigid bodies in equilibrium.
- Be able to find the centre of mass of a non-uniform rod or beam.

Moments

- A **moment** is the **turning effect** a force has **around a point**.
- The **larger the magnitude** of the force, and the **greater the distance** between the force and the pivot, the **greater the moment**.
- Moments are either **clockwise** or **anticlockwise**.
- You can use the following formula to find the moment of a force about a point:

$$\text{Moment} = \begin{array}{c}\text{Magnitude}\\\text{of Force}\end{array} \times \begin{array}{c}\text{Perpendicular Distance}\\\text{from the line of action of}\\\text{the force to the pivot}\end{array}$$

- Or, more concisely: **Moment =** $\boldsymbol{Fd}$.
- If the force is measured in newtons and the distance is measured in metres, then the moment is measured in newton-metres, **Nm**.

Tip: The 'line of action' of the force is just the direction that the force acts in.

Example 1

A 2 m long plank is attached to a ship at one end, *O*. The plank is horizontal, and a bird lands on the other end, applying a downward force of magnitude 15 N, as shown. Model the plank as a light rod and find the turning effect of the bird on the plank.

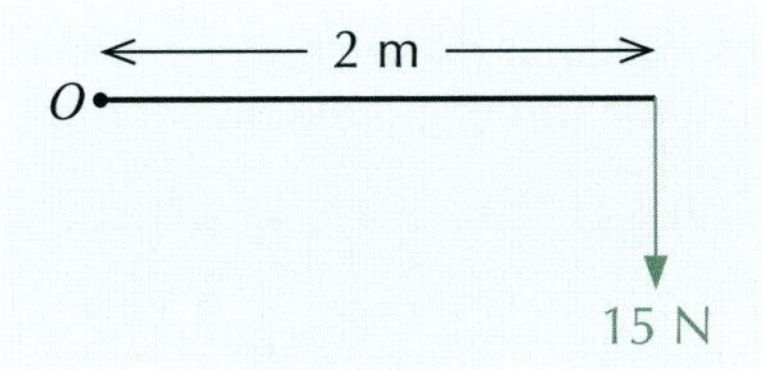

Use the formula:

$$\begin{aligned}\text{Moment} &= Fd\\ &= 15 \times 2\\ &= 30 \text{ Nm clockwise}\end{aligned}$$

Tip: Modelling the plank as a light rod means that you don't have to worry about its weight.

Tip: The point O is the pivot point.

Tip: Make sure you give a direction with your answer — i.e. which way the force will cause the rod to turn.

Example 2

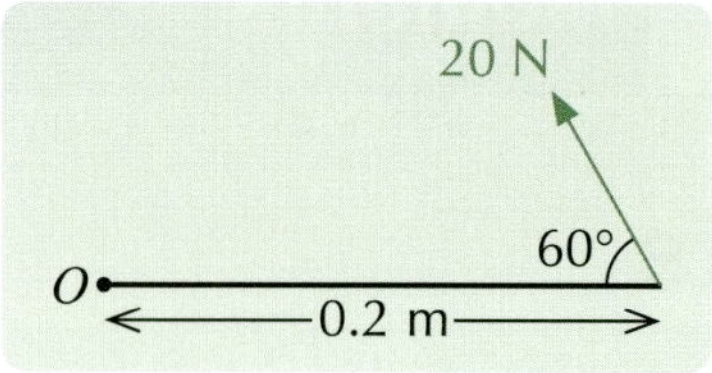

A spanner is attached to a bolt at a point, *O*. A force of 20 N is applied at an angle of 60° to the other end of the spanner, as shown. Find the turning effect of the force upon the bolt.

- This time the force is acting at an angle — resolve the force to find its component acting perpendicular to the spanner:

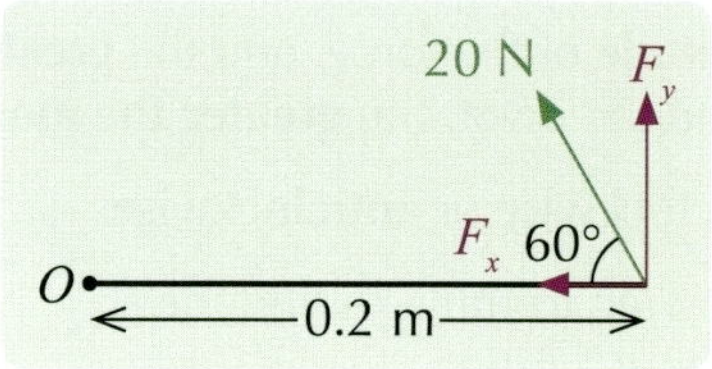

- F_x acts through O, so its moment is zero (it has no turning effect).
- Resolving perpendicular to the spanner, $F_y = 20 \sin 60°$.
- F_y is a perpendicular distance of 0.2 m from O, so, using the formula:

  Moment = 20 sin 60° × 0.2

  = **3.46 Nm (3 s.f.) anticlockwise**

Tip: 'Find the turning effect' means the same as 'find the moment'.

Tip: You can either find the component of the force acting perpendicular to the spanner (as done in the example), or you can find the perpendicular distance to the line of action of the force:

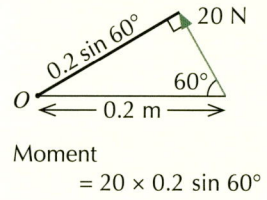

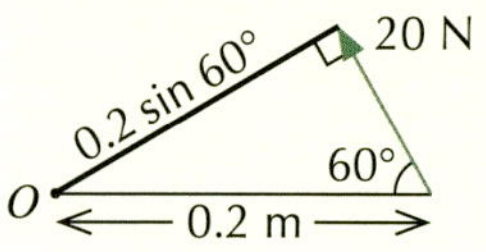

Moment
= 20 × 0.2 sin 60°

Example 3

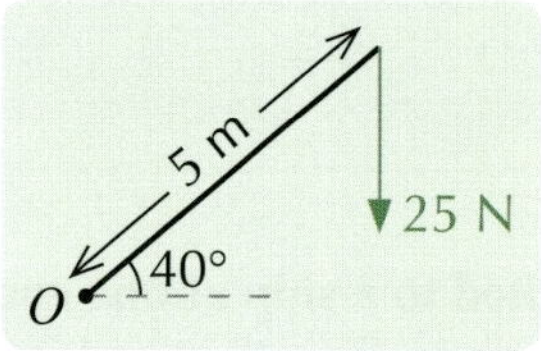

A force of 25 N acts upon a light rod of length 5 m, pivoted at *O*, as shown. The 25 N force acts vertically downwards, and the rod makes an angle of 40° with the horizontal. What is the turning effect about *O*?

- You need the perpendicular distance, d, between the line of action of the force and the pivot point:

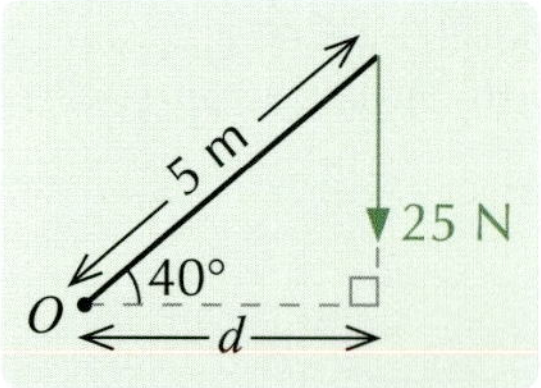

- $d = 5 \cos 40°$, so:

  Moment = 25 × 5 cos 40° = **95.8 Nm (3 s.f.) clockwise**

Tip: You could've resolved the force to find the component acting perpendicular to the rod, as in Example 2:

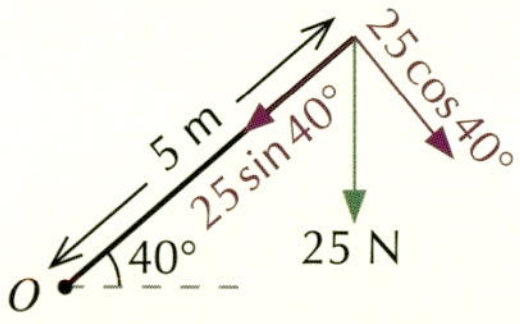

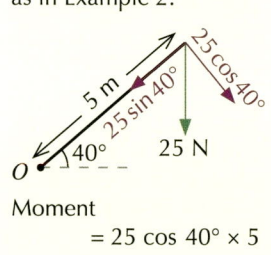

Moment
= 25 cos 40° × 5

Finding the sum of moments about a point

- If there are **two or more** forces acting on a rod or beam, then you can find the **sum of the moments** about a particular point.
- If the sum of the moments about a particular point is **not zero**, the rod will **turn** clockwise or anticlockwise about **that point**.
- The direction that the rod turns (i.e. clockwise or anticlockwise) is called the **sense** of rotation.
- If the sum of the moments about a particular point is **zero**, then the rod **won't rotate** about that point.
- If you know that the rod is **in equilibrium**, then the sum of the moments about **any point** must be **zero**.

Example 4

The diagram shows the light rod AB. A force of magnitude 5 N acts vertically downwards at point C. A force of magnitude 4 N acts at point B, making an angle of 30° with the rod, as shown:

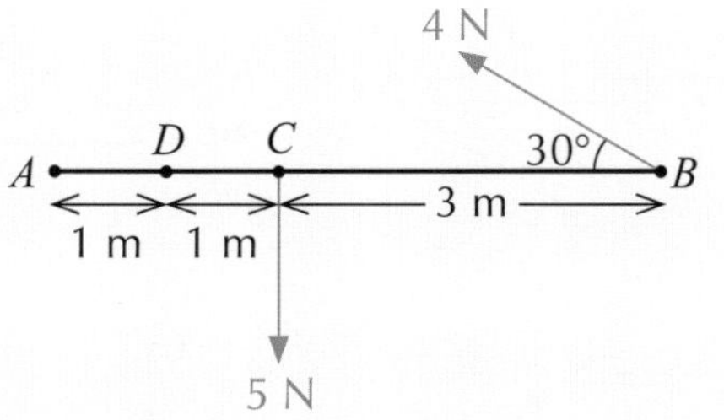

a) The rod is pivoted at point A. Show that there is no overall turning effect about this point.

- Taking moments about A (with clockwise being the positive direction), and finding their sum:

$$\text{Sum of moments} = (5 \times 2) - (4 \sin 30° \times 5)$$
$$= 10 - 10$$
$$= 0 \text{ Nm}$$

This term is negative because the moment is anticlockwise.

- Total moment is zero, so there is no overall turning effect about A.

b) The rod is now pivoted at point D. Given that the forces acting at B and C are the same as in part a), will the rod rotate clockwise or anticlockwise about this point?

- Taking moments about D (with clockwise being the positive direction), and finding their sum:

$$\text{Sum of moments} = (5 \times 1) - (4 \sin 30° \times 4)$$
$$= 5 - 8$$
$$= -3 \text{ Nm}$$

- The sum of moments is negative, so the rod will rotate anticlockwise.

Tip: When you're finding the sum of moments, you have to say which direction is positive — clockwise or anticlockwise.

Tip: Don't forget to resolve so that the line of action of the force and its distance from the pivot are perpendicular to each other.

Exercise 1.1

Hint: Remember to state the sense of rotation when giving the moment of a force.

Q1 For each of the following light rods, find the moment of the force about A.

a)

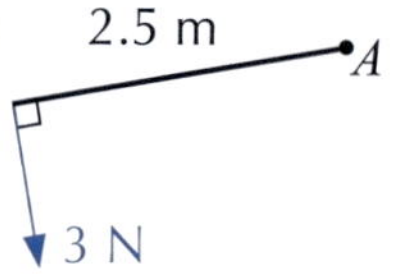

b)

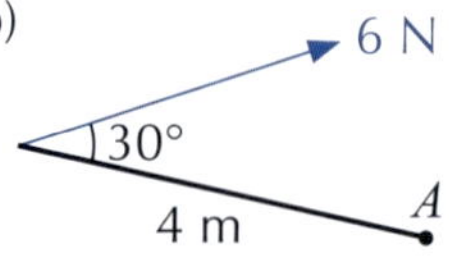

c)

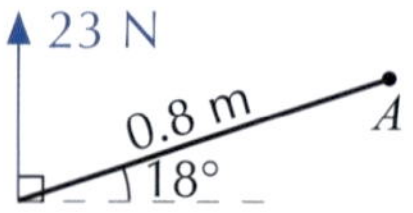

Q2 For each of the following light rods, find the sum of the moments about X.

Hint: Remember to choose either anticlockwise or clockwise as the positive direction each time.

a)

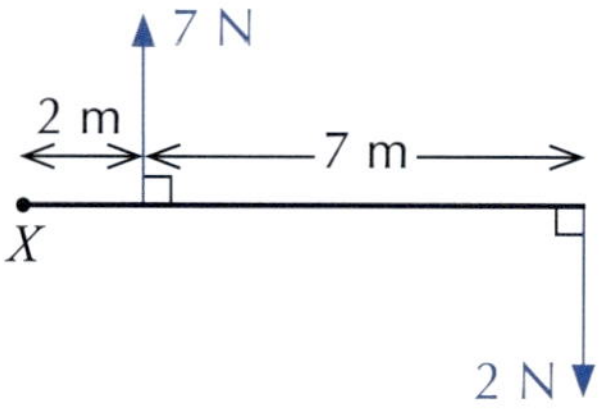

b)

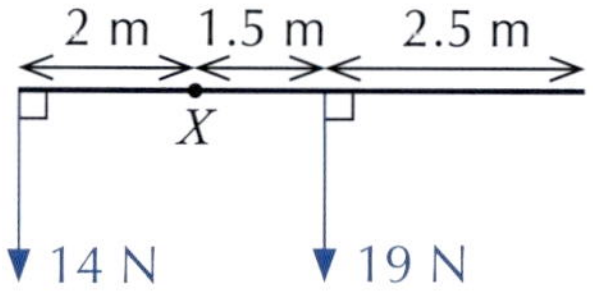

c)

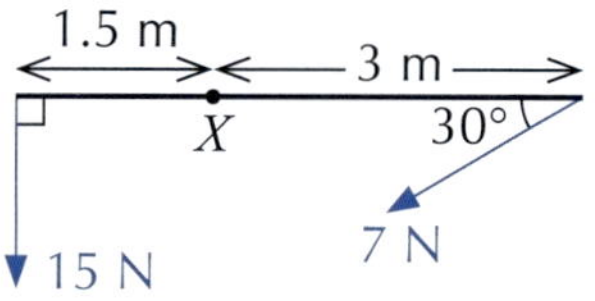

d)

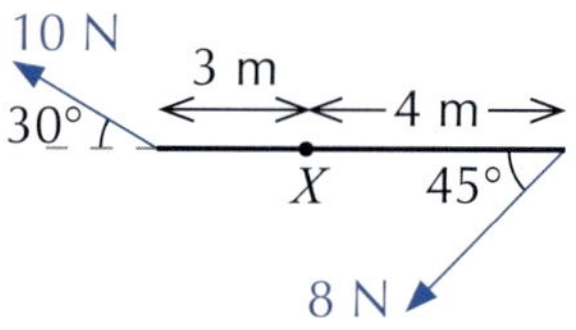

e)

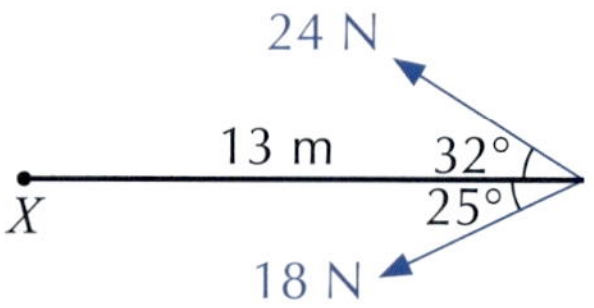

f)

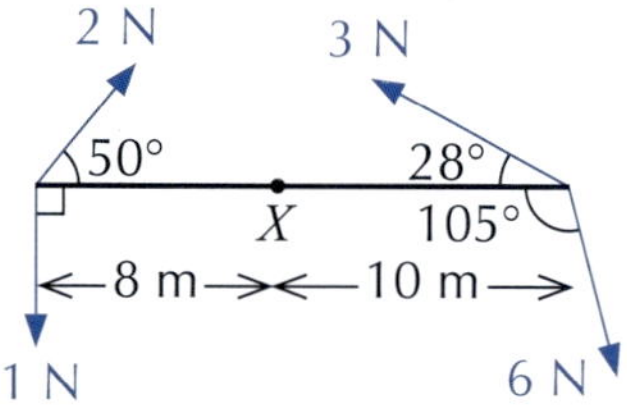

Q3

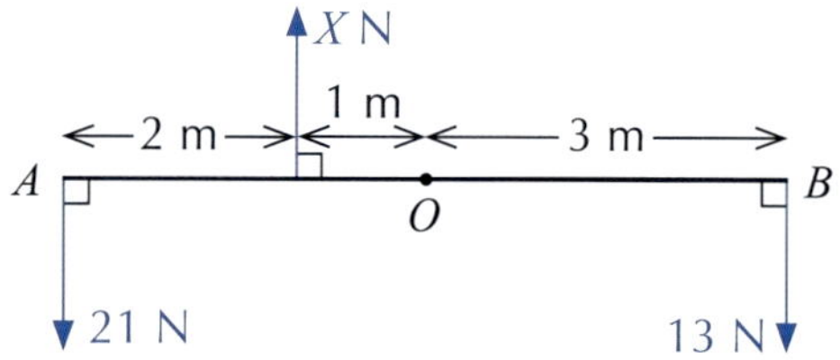

The diagram shows the light rod AB. Find the range of possible values for X, given that AB rotates clockwise about the point O.

Q4

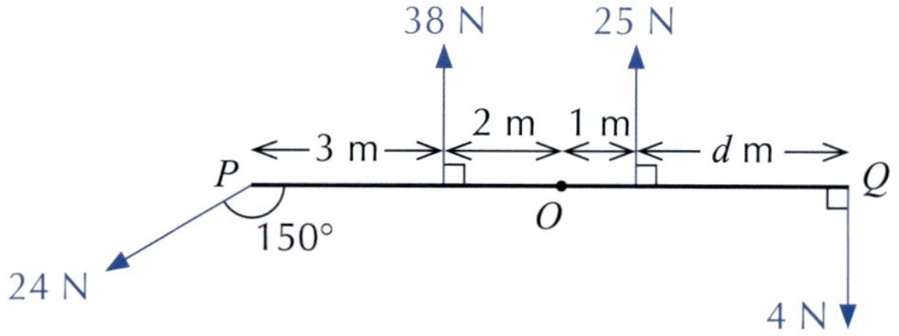

The diagram shows the light rod PQ. Find the range of possible values for d which will cause PQ to rotate anticlockwise about the point O.

Moments in equilibrium

- A rigid body which is in **static equilibrium** will **not move**.
- This means that there is **no resultant force** in **any direction** — any forces acting on the body will cancel each other out.
- It also means that the **sum of the moments** on the body **about any point** is **zero**.
- So, for a body in equilibrium:

Total Clockwise Moment	=	Total Anticlockwise Moment

- By **resolving forces** and **equating clockwise and anticlockwise moments**, you can solve problems involving bodies in equilibrium.

Example 1

Two weights of 30 N and 45 N are placed on a light 8 m beam. The 30 N weight is at one end of the beam, as shown, whilst the other weight is a distance *d* from the midpoint, *M*. The beam is held in equilibrium by a light, inextensible wire with tension *T* attached at *M*.

Find *T* and the distance *d*.

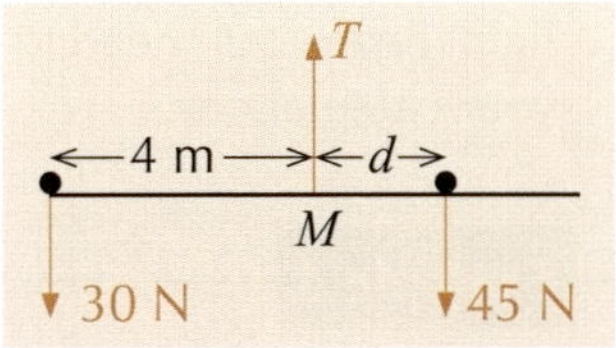

- First resolve forces vertically (forces 'up' = forces 'down'):

$T = 30 + 45 = 75 \text{ N}$

- Then take moments about M (moments clockwise = moments anticlockwise):

$45d = 30 \times 4$

$d = 120 \div 45 = 2.67 \text{ m (3 s.f.)}$

Tip: The line of action of T passes through M, so the moment of T about M is zero.

Example 2

The diagram below shows a light rod *AB*, of length 10 m. Particles of mass M_A kg and M_B kg are placed at *A* and *B* respectively. The rod is supported in equilibrium by two vertical reaction forces of magnitude 145 N and 90 N, as shown in the diagram below.

Find the values of M_A and M_B.

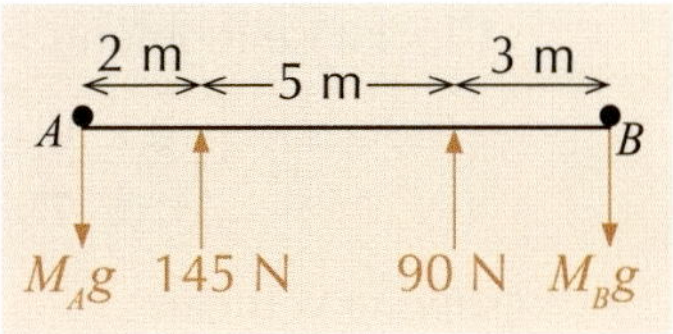

- As the rod is in equilibrium, you know that clockwise moments = anticlockwise moments. So, taking moments about A:

$M_B g \times 10 = (145 \times 2) + (90 \times 7)$

$\Rightarrow M_B = 920 \div 98 = 9.387... = 9.39 \text{ kg (3 s.f.)}$

Tip: It's a good idea to take moments about a point where an unknown force is acting, as the moment of that force about the point will be zero, and it won't appear in the equation.

- You also know that there is no resultant force acting on the rod. So, resolving forces vertically ('forces up' = 'forces down'):

$$145 + 90 = M_A g + M_B g$$
$$\Rightarrow M_A + M_B = 235 \div 9.8 = 23.979...$$
$$\Rightarrow M_A = 23.979... - 9.387... = 14.6 \text{ kg (3 s.f.)}$$

Centres of mass

- The **centre of mass** (**COM**) of an object is the point where the object's **weight** can be considered to act.
- The mass of a **uniform beam** is spread evenly along the length of the beam, and so the centre of mass is at its **midpoint**:

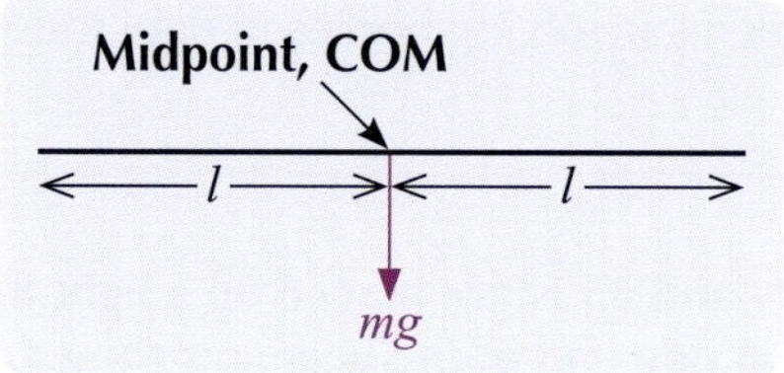

Tip: Up until now, all the beams have been light, so you haven't needed to worry about their weights.

- The centre of mass of a **non-uniform beam** could be at **any point** along the beam.
- When you're **taking moments** and **resolving forces** for a heavy (i.e. not light) beam, you need to remember to include the **weight** of the beam in your calculations.

Example 3

A 6 m long uniform beam *AB* of weight 40 N is supported at *A* by a vertical reaction *R*. *AB* is held horizontally by a vertical wire attached 1 m from the other end. A particle of weight 30 N is placed 2 m from the support *R*.

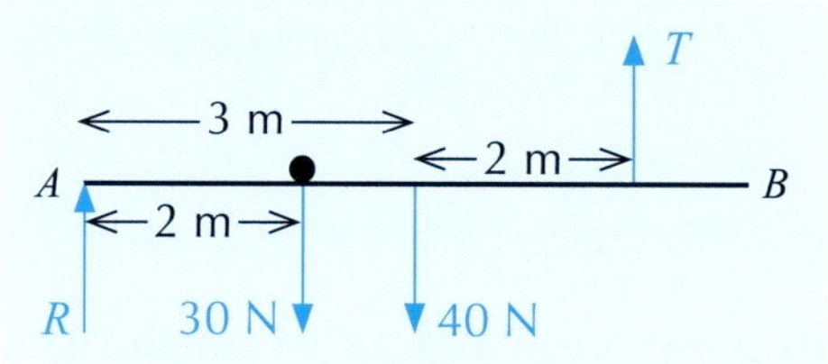

Find the tension *T* in the wire and the force *R*.

- Taking moments about A:

Clockwise moments = Anticlockwise moments

$$(30 \times 2) + (40 \times 3) = 5T$$
$$\Rightarrow T = 180 \div 5 = 36 \text{ N}$$

This is the weight of the beam, acting at its centre.

- Resolving vertically:

$$T + R = 30 + 40$$
$$R = 70 - T = 70 - 36 = 34 \text{ N}$$

The point of tilting

If a rod is '**about to tilt**' about a particular point of support, then any **normal reactions** acting at any other supports along the rod will be **zero**. The **tension** in any strings supporting the rod at any other point will also be **zero**.

Tip: You can usually assume that supports and strings are fixed, and won't move with the rod.

Example 4

A non-uniform wooden plank of mass M kg rests horizontally on supports at A and B, as shown. When a bucket of water of mass 18 kg is placed at point C, the plank is in equilibrium, and is on the point of tilting about B.

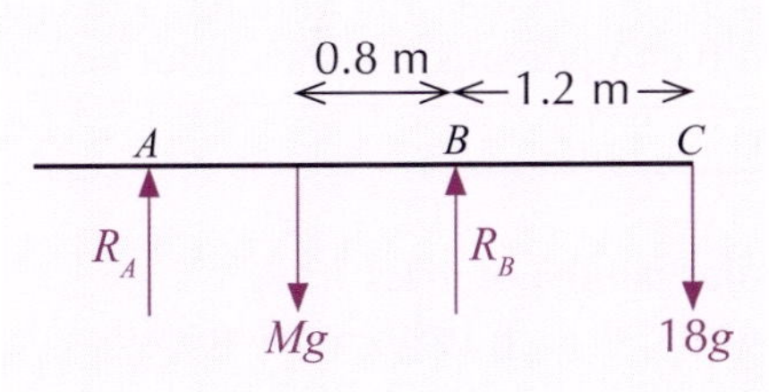

Find the value of M and the magnitude of the reaction at B.

- Taking moments about B:

  $(18g \times 1.2) + 0 = Mg \times 0.8$

  $R_A = 0$, so the moment of the force is also zero.

  $\Rightarrow M = (18 \times 9.8 \times 1.2) \div (9.8 \times 0.8) =$ **27 kg**

- Resolving vertically:

  $R_A + R_B = Mg + 18g$

  $0 + R_B = 27g + 18g \Rightarrow R_B =$ **441 N**

Tip: The plank is on the point of tilting about B, so the reaction at A, R_A, is zero.

Laminas

A **lamina** is a flat 2D object (its thickness can be ignored). The centre of mass of a **uniform rectangular** lamina is at the symmetrical centre of the rectangle.

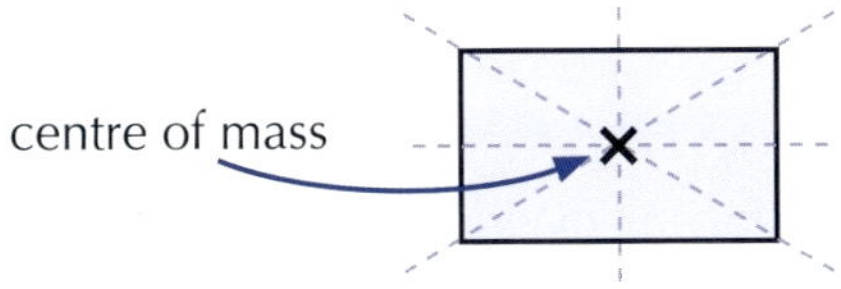

Example 5

A uniform lamina, $ABCD$, of weight 8 N, is pivoted at point A. The lamina is held in equilibrium by a vertical force F acting at point C, as shown.

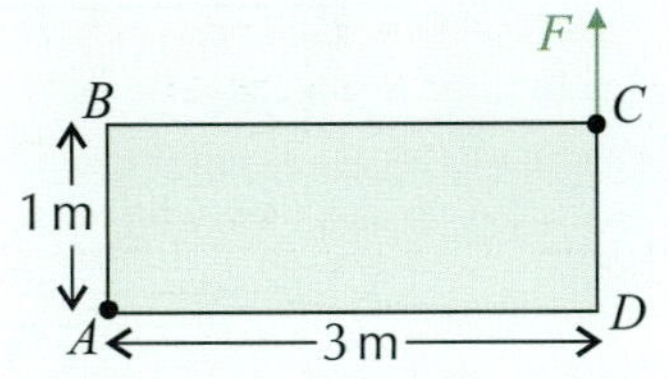

a) **Find the horizontal and vertical distances of the centre of mass of the lamina from A.**

The centre of mass is at the centre of the lamina, so the horizontal distance is $3 \div 2 =$ **1.5 m** and the vertical distance is $1 \div 2 =$ **0.5 m**

Tip: There will also be a reaction force in the pivot at A (you can resolve forces to find it), but since you're taking moments about A, it doesn't affect the answer.

b) Find the magnitude of the force F.

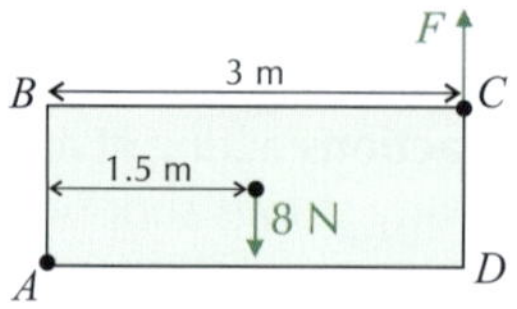

- Both the lamina's weight and F act vertically, so the perpendicular distance from each force to the pivot is just the horizontal distance.
- Taking moments about A: $8 \times 1.5 = F \times 3$
  $\Rightarrow F = 12 \div 3 =$ **4 N**

If there are forces acting **at angles** to the sides of the lamina, it's usually easier to resolve them **perpendicular** and **parallel** to the lamina's sides. But remember that **both components** will have a turning effect on the lamina.

Example 6

A uniform rectangular lamina of mass 12 kg is pivoted at one corner at A. A light, inextensible string attached at the opposite corner applies a tension force T to the lamina as shown.

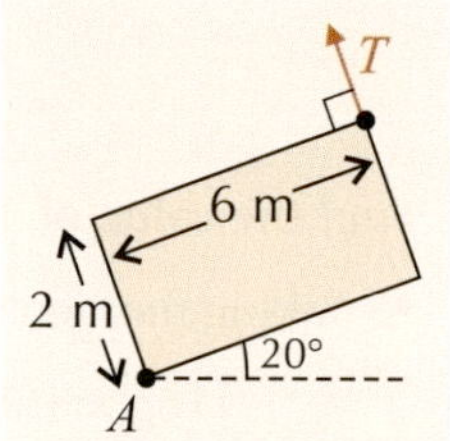

Given that the string holds the lamina in equilibrium at an angle of 20° to the horizontal, find the tension in the string.

Tip: You could do this example without breaking the weight into components — instead, you'd find the horizontal distance from A to the COM, which is the perpendicular distance from A to the line of action of the weight.

- Here, the lamina's weight acts at 20° to its sides. Find the components of its weight parallel and perpendicular to the sides of the lamina.
- Taking moments about A:
  $(mg \cos 20° \times 3) = (mg \sin 20° \times 1) + (T \times 6)$
  $\Rightarrow 6T = mg(3 \cos 20° - \sin 20°)$
  $\Rightarrow T = 19.6(2.4770...) =$ **48.6 N** (3 s.f.)

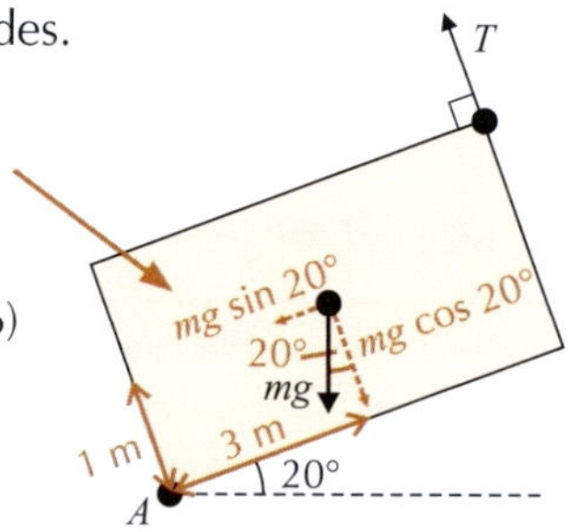

Exercise 1.2

Q1 For each of the following diagrams, find the values of x and y, given that the light rod AB is in equilibrium:

a)

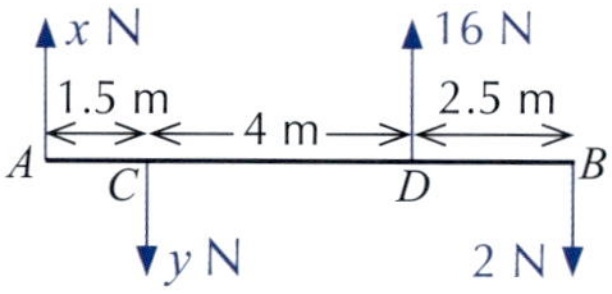

b)

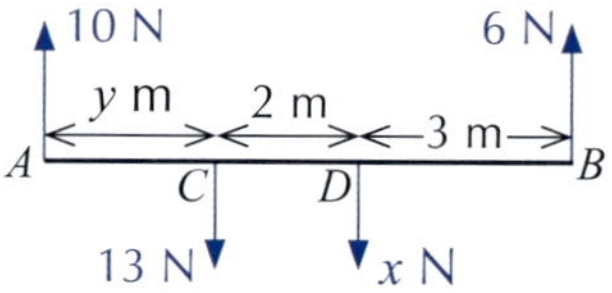

Q2

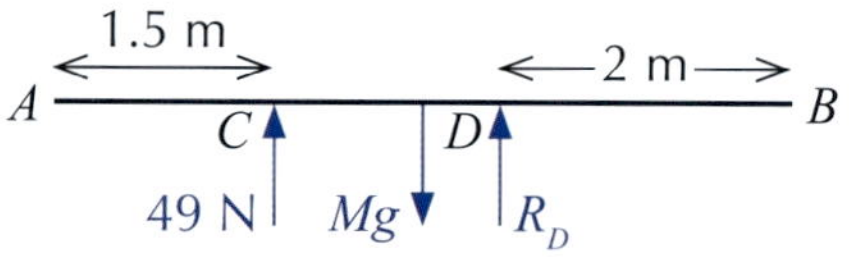

Q2 Hint: The rod is uniform, so its weight acts halfway along its length.

A uniform rod, AB, of length 5 m, rests horizontally in equilibrium on supports at C and D, as shown. If the magnitude of the normal reaction at C is 49 N, find:

a) the magnitude of the normal reaction at D,

b) the mass of the rod.

Q3

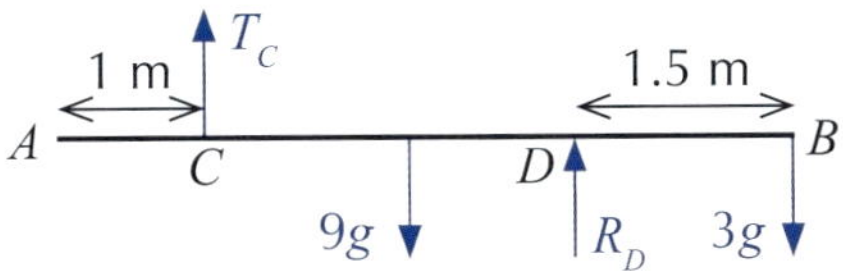

A uniform rod, AB, of mass 9 kg, is held in equilibrium by a fixed vertical wire at C and a support at D, as shown. When an object of mass 3 kg is placed at B, the rod is still horizontal and in equilibrium, but is on the point of tilting clockwise about D.
Find the length of the rod.

Q3 Hint: When the rod is on the point of tilting clockwise about D, the tension in the wire at C will be zero.

Q4 A uniform beam, AB, has length 5 m and mass 8 kg. It is held horizontally in equilibrium by vertical ropes at B and the point C, 1 m from A.

a) Find the tension in each rope.

b) When a particle of mass 16 kg is placed on the beam, the beam remains horizontal and in equilibrium, but is on the point of tilting anticlockwise about C. Find the distance of the particle from A.

Q5 A painter of mass 80 kg stands on a horizontal non-uniform 4 m plank, AB, of mass 20 kg. The plank rests on supports 1 m from each end, at C and D. The painter places paint pots, of mass 2.5 kg, 0.2 m from each end of the plank.

a) He stands at the centre of mass of the plank and finds that the reaction forces at C and D are in the ratio 4 : 1.
Find the distance of the centre of mass of the plank from A.

b) He uses up all the paint in the pot near A, and discards the pot. He then stands on the plank at a point between D and B, and the plank is on the point of tilting about D. How far is he from B?

Q6 A uniform rectangular lamina $ABCD$ of weight 6 N is pivoted at point A and acted on by a force F at point C. F has magnitude 10 N, and acts at an angle of θ to the horizontal, as shown.

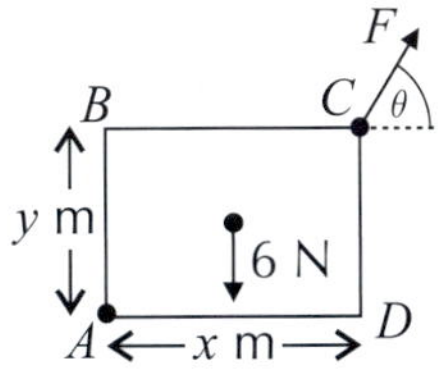

Find the sum of the moments about A if:

a) $x = 10,\ y = 6,\ \theta = 90°$
b) $x = 1,\ y = 5,\ \theta = 0°$
c) $x = 4,\ y = 8,\ \theta = 270°$
d) $x = 7,\ y = 6,\ \theta = 180°$
e) $x = 5,\ y = 5,\ \theta = 30°$
f) $x = 1,\ y = 11,\ \theta = 120°$

Q7 A uniform rectangular lamina of mass m has corners at the following coordinates: A (0, 0), B (–2, 8), C (2, 9), D (4, 1). The lamina is pivoted at A. A horizontal force of 8 N acts on the lamina in the negative x-direction at B, and a vertical force of 11 N pulls upwards on the lamina at C.

a) Draw a diagram to show this information, given that gravity acts in the negative y-direction.

b) Given that the system is in equilibrium, find m.

2. Reaction Forces and Friction

Learning Objectives:

- Be able to solve problems involving rigid bodies that are freely hinged to a surface.
- Be able to solve problems involving rigid bodies under the influence of a frictional force.

When an object is resting against a surface, the surface will exert a normal reaction on the object perpendicular (or normal) to the surface and, if the surface is rough, friction will oppose any motion. However, if the object is fixed by a hinge, the reaction force might not act perpendicular to the surface.

Reaction forces

If a **rigid body** is fixed to a **surface** by a **hinge** or **pivot**, you can deal with the reaction force on the body from the hinge by **splitting it up** into two **components**, **parallel** and **perpendicular** to the **surface**.

Tip: When a body is freely hinged, the reaction force is the force in the hinge that helps keep the body in equilibrium. It's not the same as the normal reaction, which acts when a body is in contact with a surface, and which always acts perpendicular to the surface.

Example 1

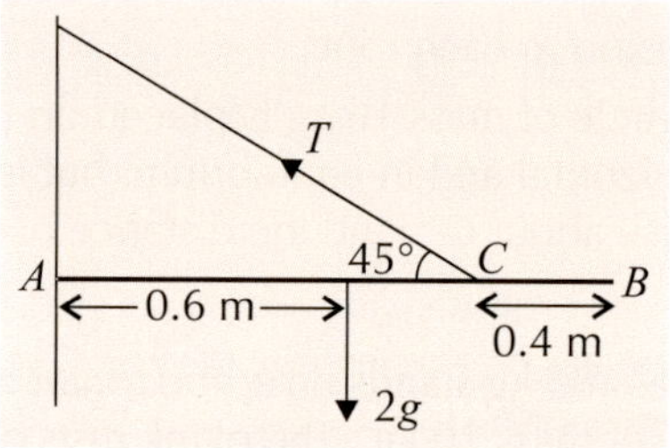

A uniform rod, *AB*, of mass 2 kg and length 1.2 m, is freely hinged on a vertical wall. The rod is held horizontally in equilibrium by a light, inextensible string attached at *C*, 0.4 m from *B*, which makes an angle of 45° with the rod, as shown. Find the tension in the string and the magnitude of the reaction force on the rod at *A*.

- Taking moments about A:

  $T\sin 45° \times (1.2 - 0.4) = 2g \times 0.6$

  $T\sin 45° = 14.7 \Rightarrow T = 20.788... =$ 20.8 N (3 s.f.)

- Split the reaction into horizontal and vertical components, R_H and R_V, and draw a diagram.
- When you're drawing your diagram, you can work out in which directions the components act by considering the other forces — because the rod is in equilibrium, the horizontal components and vertical components of all the forces acting on the rod must sum to zero.
- The horizontal component of the tension in the string acts on the rod to the left, so the horizontal component of the reaction will act to the **right** to balance this out.
- The downward force of the rod's weight ($2g = 19.6$ N) is greater than the upward vertical component of the tension in the string ($T\sin 45° = 14.7$ N), so the reaction will act **upwards** to balance this out:

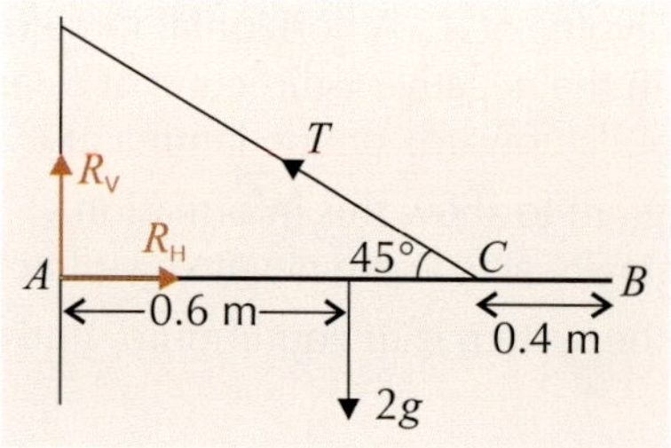

Tip: Taking moments about A means you don't have to worry about the reaction force when finding T.

Tip: Don't worry if you draw the components acting in the wrong direction — the numbers in your calculations will just come out negative.

- Resolving horizontally:

  $R_H = T\cos 45° = (20.788...)\cos 45°$

  $R_H = 14.7$ N

- Resolving vertically:

  $R_V + T\sin 45° = 2g$

  $R_V = 4.9$ N

- Now use Pythagoras' theorem to find R, the magnitude of the reaction at A:

  $|R| = \sqrt{R_H^2 + R_V^2} = \sqrt{14.7^2 + 4.9^2}$

  $= 15.5$ N (3 s.f.)

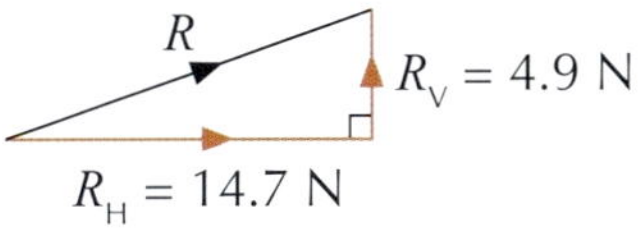

Tip: The rod is in equilibrium, so: 'forces left' = 'forces right' and 'forces up' = 'forces down'.

Example 2

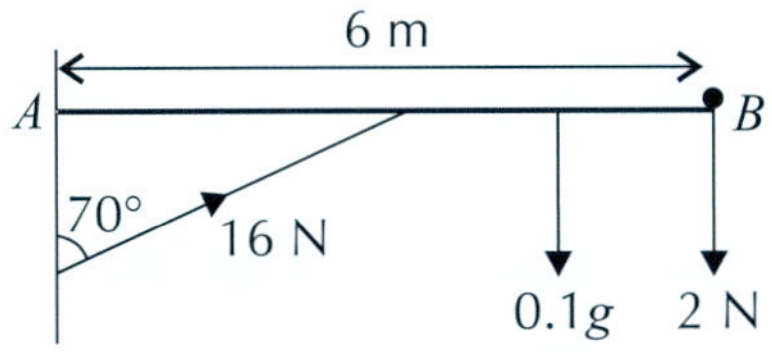

A non-uniform rod, *AB*, of length 6 m and mass 0.1 kg, is freely hinged on a vertical wall. The rod is supported by a light strut at an angle of 70° to the wall, as shown. The strut exerts a thrust of 16 N at the midpoint of the rod. A particle of weight 2 N rests at *B*, and the rod is horizontal and in equilibrium.

Find the distance of the centre of mass of the rod from *A*, and the magnitude and direction of the reaction force on the rod at *A*.

- Let x be the distance of the centre of mass of the rod from A. Taking moments about A:

  $(0.1g \times x) + (2 \times 6) = 16\cos 70° \times 3$

  $0.1gx = 4.416...$

  $\Rightarrow x = 4.507... = 4.51$ m (3 s.f.)

Tip: Taking moments about A means you don't have to worry about the reaction force when finding x.

- As in the previous example, split the reaction into its horizontal and vertical components, R_H and R_V, and draw a diagram.
- The horizontal component of the thrust in the strut acts on the rod to the right, so the horizontal component of the reaction will act to the **left** to balance this out.
- The upward vertical component of the thrust in the strut ($16\cos 70° = 5.472...$ N) is greater than the downward force of the two weights ($0.1g + 2 = 2.98$ N), so the reaction will act **downwards** to balance this out.

Tip: The strut supports the rod at its midpoint — so the vertical component of the thrust acts 3 m from A.

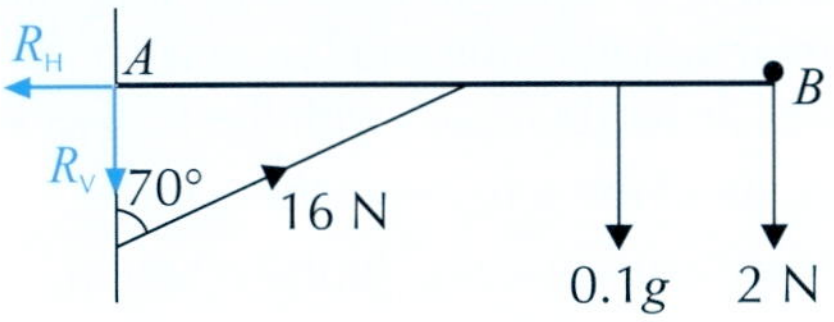

Tip: Remember — the reaction force is in the hinge, not the wall, so that's why R can act to the left.

- Resolving horizontally:

  $R_H = 16 \sin 70° = 15.035...$ N

- Resolving vertically:

  $R_V + 0.1g + 2 = 16 \cos 70°$

  $\Rightarrow R_V = 2.492...$ N

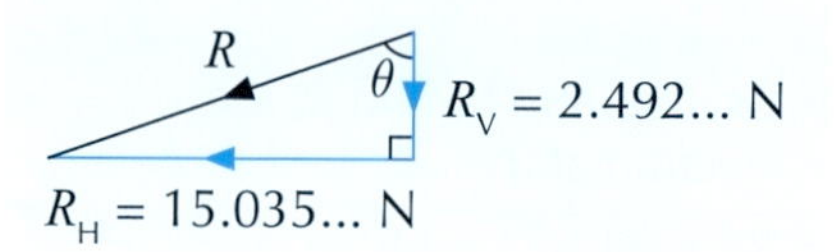

- Using Pythagoras' theorem to find the magnitude:

  $|R| = \sqrt{R_H^2 + R_V^2} = \sqrt{(15.035...)^2 + (2.492...)^2}$

  $= 15.2$ N (3 s.f.)

- And using trigonometry to find the direction:

  $\theta = \tan^{-1}\left(\frac{15.035...}{2.492...}\right) = 80.58...°$ (clockwise from the downwards vertical)

  So direction = 270° – 80.58...° = 189° (3 s.f.)

Tip: Remember that the direction of a force is usually measured anticlockwise from the horizontal.

Exercise 2.1

Q1

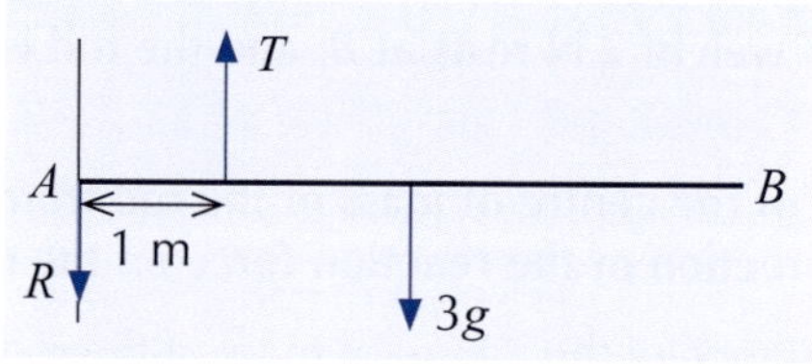

A uniform rod, AB, of mass 3 kg and length 4 m, is freely hinged to a vertical wall at A. The rod is held horizontal in equilibrium by a light vertical wire attached to the rod 1 m from A, as shown.
The reaction force, R, at the hinge, acts vertically downwards. Find:

a) the magnitude of the tension in the wire.

b) the magnitude of the reaction force, R, acting on the rod at A.

Q1 Hint: The only other forces acting on the rod are vertical, so R must also be vertical.

Q2

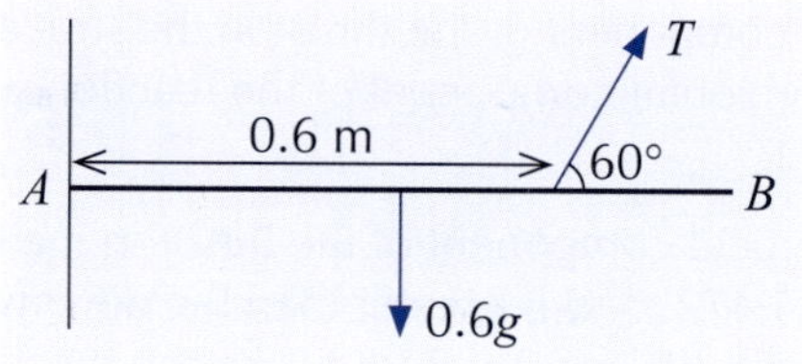

A uniform rod, AB, of mass 0.6 kg and length 0.8 m, is freely hinged to a vertical wall at A. The rod is held horizontally and in equilibrium by a light wire attached to the rod 0.6 m from A.
The wire makes an angle of 60° with the rod, as shown. Find:

a) the magnitude of the tension in the wire.

b) the magnitude and direction of the reaction force acting on the rod at A.

Q3

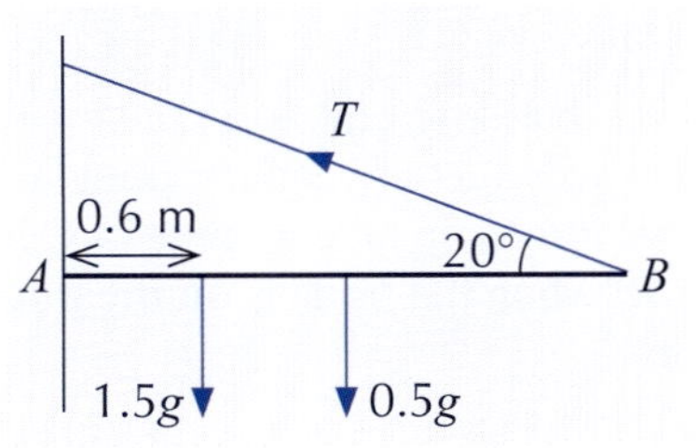

A non-uniform beam, AB, of mass 1.5 kg and length 2.4 m, is freely hinged to a vertical wall at A. The centre of mass of the beam is 0.6 m from A. One end of a light, inextensible string is attached to B, and the other end is fixed to the wall, directly above A, as shown. The string makes an angle of 20° with the beam. A body of mass 0.5 kg is placed at the beam's midpoint. The beam is horizontal and in equilibrium. Find:

a) the tension in the string.

b) the magnitude and direction of the reaction force acting on the beam at A.

Q4

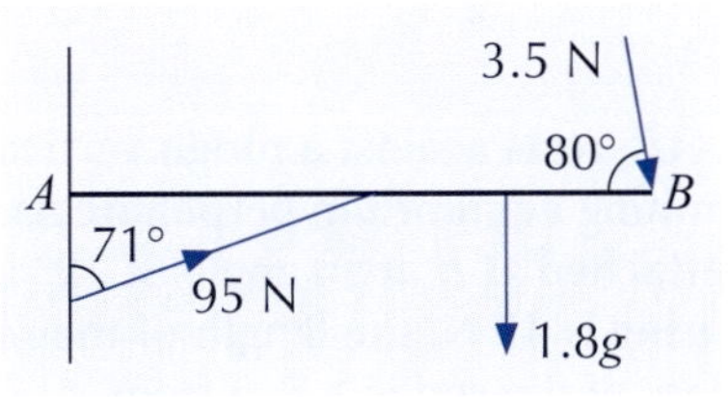

A non-uniform rod, AB, of mass 1.8 kg and length 7 m, is freely hinged to a vertical wall at A. A light strut, fixed to the wall, supports the rod at its midpoint, so the rod is kept horizontal and in equilibrium. The strut makes an angle of 71° with the wall, and the thrust in the strut has magnitude 95 N. A force of magnitude 3.5 N is applied to the rod at B, at an angle of 80° to the rod, as shown. Find:

a) the distance of the centre of mass of the rod from A.

b) the magnitude and direction of the reaction force acting on the rod at A.

Q5

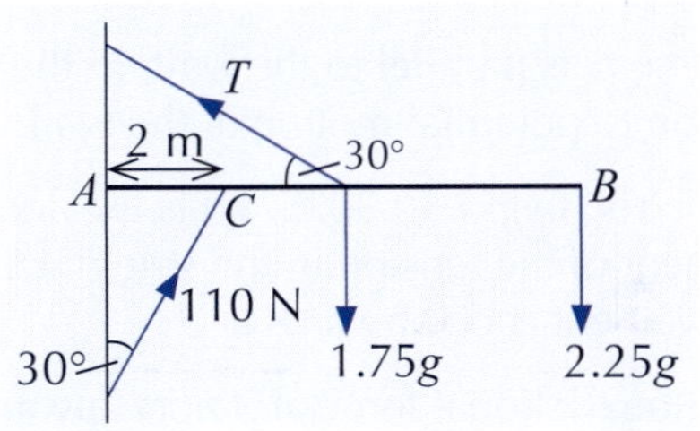

A uniform plank, AB, of mass 1.75 kg and length 8 m, is freely hinged to a vertical wall at A. An object of mass 2.25 kg is placed on the plank at B. The plank is kept horizontal and in equilibrium by a light wire fixed at its midpoint, at an angle of 30° above the plank, and a light strut fixed at C, 2 m along the plank from A, at an angle of 30° to the wall, as shown. The thrust in the strut has magnitude 110 N.

a) Find T, the magnitude of the tension in the wire.

b) Calculate the magnitude and direction of the reaction force acting on the plank at A.

Friction

- You saw in Chapter 15 that if a body is in contact with a **rough** surface, then a **frictional force** will act between the body and the surface to oppose motion. The frictional force can take a range of values, and will reach its maximum ($F = \mu R$) when the body is **on the point of moving**, or in '**limiting equilibrium**'.
- So if you have a rigid body resting on a rough surface, you will have a **normal reaction** force acting **perpendicular** to the surface, and a **frictional force** acting **parallel** to it, in the **opposite** direction to any potential motion.

Example 1

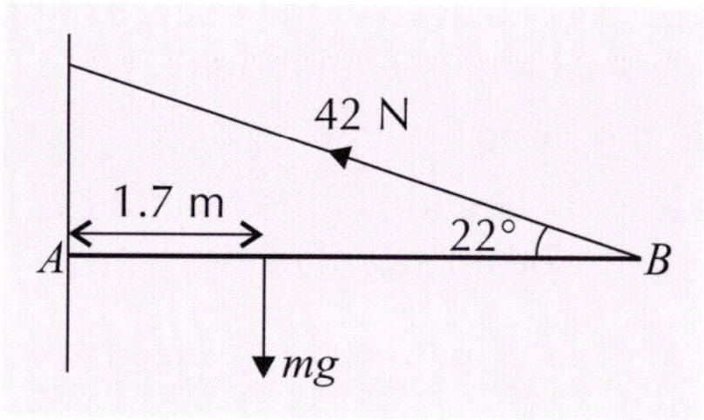

A non-uniform rod, AB, rests against a rough vertical wall at A. The rod is held in limiting equilibrium perpendicular to the wall by a light, inextensible string attached at B at an angle of 22° above the rod. The tension in the string is 42 N, the length of the rod is 5.5 m and the centre of mass of the rod is 1.7 m from A.

Find the mass of the rod, m, and the coefficient of friction, μ, between the wall and the rod.

Tip: Taking moments about A means you don't need to worry about the frictional and reaction forces acting there.

- Taking moments about A:

$$mg \times 1.7 = 42 \sin 22° \times 5.5$$

$$\Rightarrow m = 86.534... \div 16.66 = 5.194...$$

$$= 5.19 \text{ kg (3 s.f.)}$$

- Now draw a diagram showing the forces at A.
- The normal reaction of the wall on the rod acts **perpendicular** to the wall.
- The frictional force acts **parallel** to the wall, in the opposite direction to potential motion at the wall.
- The weight of the rod (mg = 50.902... N) is greater than the upwards vertical component of the tension in the string (42 sin 22° = 15.733... N), so potential movement is downwards at A.
- This means that the frictional force at A acts **upwards**.
- So the diagram looks like this:

Tip: R is the normal reaction of the wall on the rod — don't confuse it with the reaction in the hinge in the examples in the previous section. There's no hinge here — it's friction which keeps the rod from slipping on the wall.

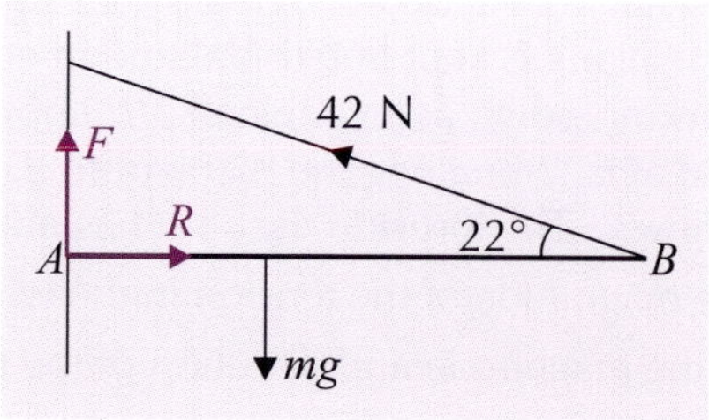

- Resolving horizontally:

  $R = 42 \cos 22° = 38.941...$ N

- Resolving vertically:

  $F + 42 \sin 22° = mg$

  $\Rightarrow F = 50.902... - 15.733... = 35.168...$ N

- The rod is in limiting equilibrium, so friction is at its maximum:

  $F = \mu R$

  $35.168... = \mu \times 38.941...$

  $\Rightarrow \mu = 0.90$ (2 d.p.)

Tip: You could also take moments about B (or another point) to find F.

'Ladder' questions

You might see a question where a rigid body rests at an angle against the ground and a wall — a simple example of a situation like this is a **ladder**.

In these questions, you'll need to consider the **normal reaction** of the **ground**, the **normal reaction** of the **wall**, and any **frictional forces** which may be acting. The question will tell you whether the ground and wall are **rough** or **smooth** — this lets you know whether you need to take friction into account in your calculations.

There are four possible combinations of surfaces for ladder-style questions:

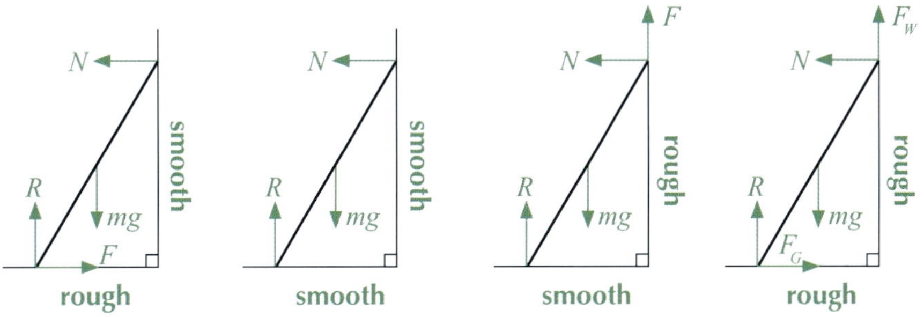

Tip: Friction acts to prevent motion. In each case, think about which way the ladder would slip — the frictional force will act in the opposite direction.

Example 2

A ladder rests against a smooth vertical wall at an angle of 65° to rough horizontal ground. The ladder has mass 4.5 kg and length 5*x* m. A cat of mass 1.3 kg sits on the ladder at *C*, 4*x* m from the base. The ladder is in limiting equilibrium.

Modelling the ladder as a uniform rod and the cat as a particle, find the coefficient of friction between the ground and the ladder.

- Draw a diagram to show the forces acting between the ladder and the ground, and the ladder and the wall:

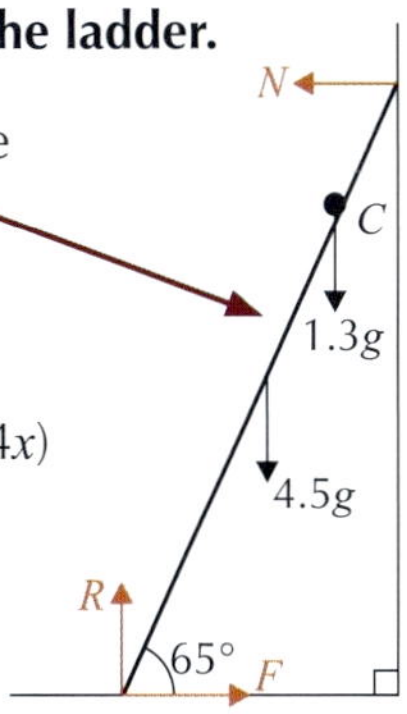

- Taking moments about the base of the ladder:

  $N \sin 65° \times 5x = (4.5g \cos 65° \times 2.5x)$
  $+ (1.3g \cos 65° \times 4x)$

  $(4.531...)xN = (46.593..)x + (21.536...)x$

  $\Rightarrow N = (68.130...)x \div (4.531...)x = 15.034...$ N

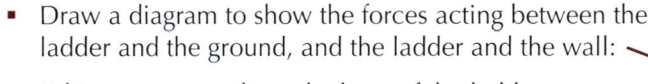

Tip: N is the normal reaction at the wall, R is the normal reaction at the ground. The wall is smooth, so there is no frictional force. The ground is rough, so there is a frictional force between it and the ladder.

- Resolving vertically:

  $R = 1.3g + 4.5g = 56.84$ N
- Resolving horizontally:

  $F = N$
- The ladder is in limiting equilibrium, so $F = \mu R$:

  $F = \mu R \Rightarrow N = \mu R$

  $15.034... = \mu \times 56.84$

  $\Rightarrow \mu = 15.034... \div 56.84 =$ 0.26 (2 d.p.)

Example 3

A uniform ladder of mass 11 kg and length 4 m rests against a smooth vertical wall, at an angle of 58° to rough horizontal ground. The coefficient of friction between the ladder and the ground is 0.45. A bucket of water is hung 3 m from the base of the ladder.

What is the maximum possible mass of the bucket of water for which the ladder remains in equilibrium?

- Draw a diagram to show what's going on:

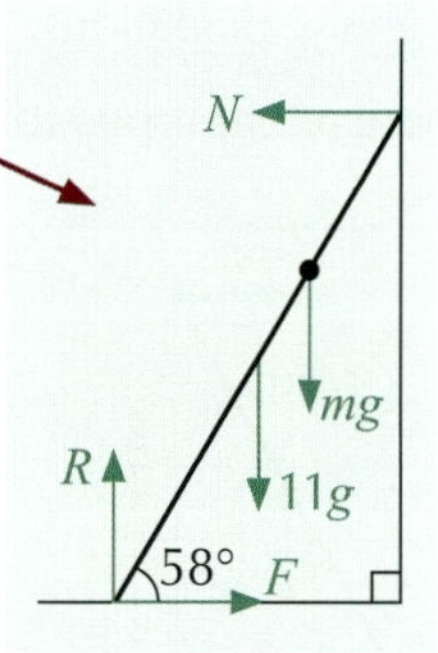

- Resolving vertically:

  $R = 11g + mg = g(11 + m)$
- Resolving horizontally:

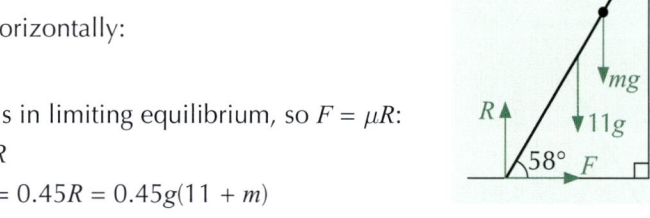

  $F = N$
- The ladder is in limiting equilibrium, so $F = \mu R$:

  $F = \mu R$

  $\Rightarrow N = 0.45R = 0.45g(11 + m)$
- Taking moments about the base of the ladder:

  $N \sin 58° \times 4 = (11g \cos 58° \times 2) + (mg \cos 58° \times 3)$

  $1.8g(11 + m) \sin 58° = 22g \cos 58° + 3mg \cos 58°$

  $19.8 \sin 58° - 22 \cos 58° = m(3 \cos 58° - 1.8 \sin 58°)$

  $\Rightarrow m = 5.133... \div 0.06327... =$ 81.1 kg (3 s.f.)

Tip: When the bucket is at its maximum mass, the ladder is on the point of slipping — i.e. in limiting equilibrium.

Example 4

A uniform ladder of length 6 m and mass 16 kg is held against a rough vertical wall by a horizontal force of magnitude 50 N, being applied at a distance of x m from the base of the ladder.

Given that the ladder is inclined at an angle of 70° to the smooth horizontal ground, and that the coefficient of friction between the ladder and the wall is 0.6, find the range of possible values for x so that the ladder remains in equilibrium.

- To solve this, you need to find the value of x when the ladder is about to slide towards the wall, and when it is on the point of sliding towards the floor.

- First consider the situation where the ladder is about to slide towards the ground.
- Draw a diagram to show the forces on the ladder:

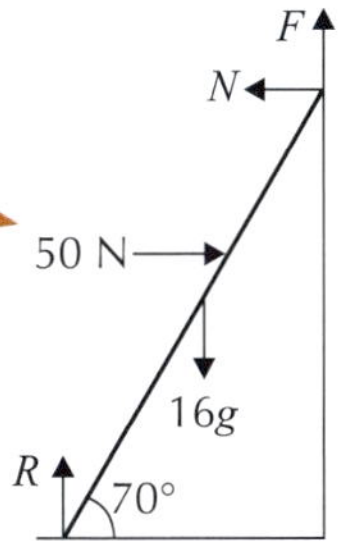

- Resolving forces horizontally:

  $N = 50$ N
- The ladder is in limiting equilibrium, so:

  $F = \mu N \Rightarrow F = 0.6 \times 50 = 30$ N
- Taking moments about the base of the ladder:

  $(50 \sin 70° \times x) + (16g \cos 70° \times 3)$
  $= (N \sin 70° \times 6) + (F \cos 70° \times 6)$

  $(50 \sin 70°)x + 470.4 \cos 70° = 300 \sin 70° + 180 \cos 70°$

  $(50 \sin 70°)x = 300 \sin 70° + 180 \cos 70° - 470.4 \cos 70°$

  $x = 182.58... \div 46.984... =$ 3.886... m

- Now consider the situation where the ladder is on the point of sliding towards the wall — draw a new diagram showing the forces on the ladder:

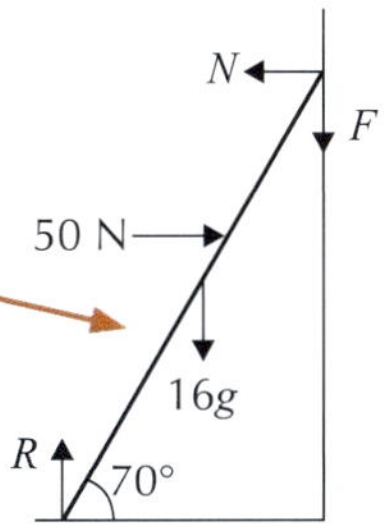

- The only change is that the friction force F acts in the opposite direction.
- Resolving forces horizontally:

  $N = 50$ N $\Rightarrow F = 30$ N
- Taking moments about the base of the ladder:

  $(50 \sin 70° \times x) + (16g \cos 70° \times 3) + (F \cos 70° \times 6)$
  $= (N \sin 70° \times 6)$

  $(50 \sin 70°)x + 470.4 \cos 70° + 180 \cos 70° = 300 \sin 70°$

  $(50 \sin 70°)x = 300 \sin 70° - 180 \cos 70° - 470.4 \cos 70°$

  $x = 59.457... \div 46.984... =$ 1.265... m

So the range of values of x for which the ladder is in equilibrium is

1.265... m $\leq x \leq$ 3.886... m

Tip: The calculation here is almost identical to the one above, except that the friction term has changed sign.

Bodies supported along their lengths

Rather than leaning against a wall, a rigid body may be held in equilibrium by resting on **supports** at points along its length. You can solve problems like this just as before, by **resolving forces** and **taking moments**.

You need to know whether the **ground** and **supports** are **rough** or **smooth**, and you should also remember that the **normal reaction** at a **support** will always act **perpendicular** to the body.

Example 5

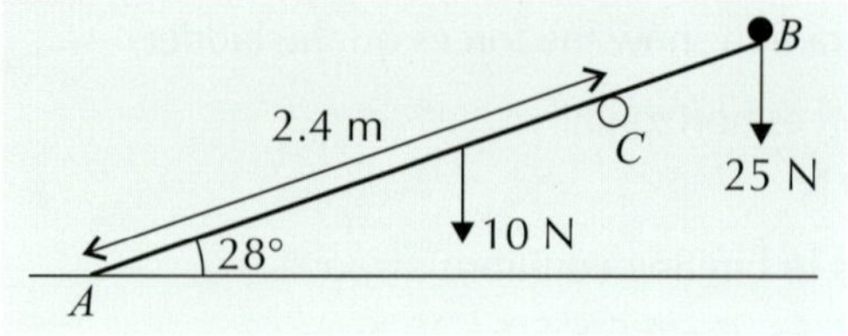

A uniform rod, *AB*, of length 3.3 m and weight 10 N, rests with *A* on rough horizontal ground. The rod is supported by a smooth peg at *C*, where *AC* = 2.4 m, in such a way that the rod makes an angle of 28° with the ground. A particle of weight 25 N is placed at *B*.

Given that the rod is in limiting equilibrium, find the magnitude of the normal reaction, *N*, at the peg and the magnitude of the frictional force, *F*, between the rod and the ground.

- Draw a diagram to show the forces acting on the rod:

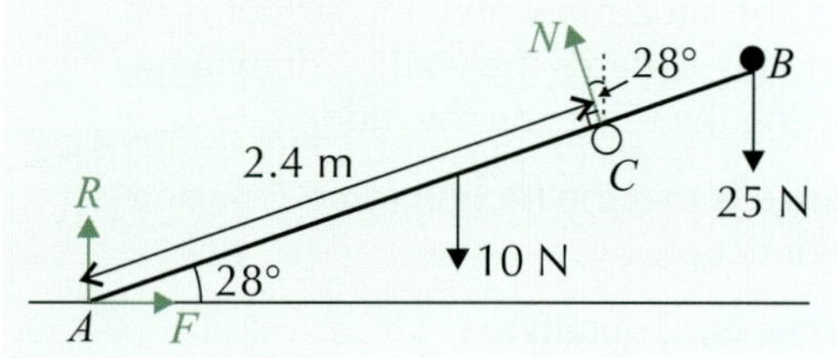

Tip: The peg is smooth, so there is no frictional force at *C*. The normal reaction at *C*, *N*, acts perpendicular to the rod. The rod is on the point of slipping to the left, as viewed in the diagram, so the frictional force at *A* acts to the right.

- Taking moments about *A*:

$2.4N = (10 \cos 28° \times 1.65) + (25 \cos 28° \times 3.3)$

$\Rightarrow N = 36.421...$

$= 36.4 \text{ N (3 s.f.)}$

The rod's weight acts at its midpoint, 1.65 m from *A*.

- Resolving horizontally:

$F = N \sin 28° = (36.421...) \sin 28°$

$= 17.1 \text{ N (3 s.f.)}$

Exercise 2.2

Q1

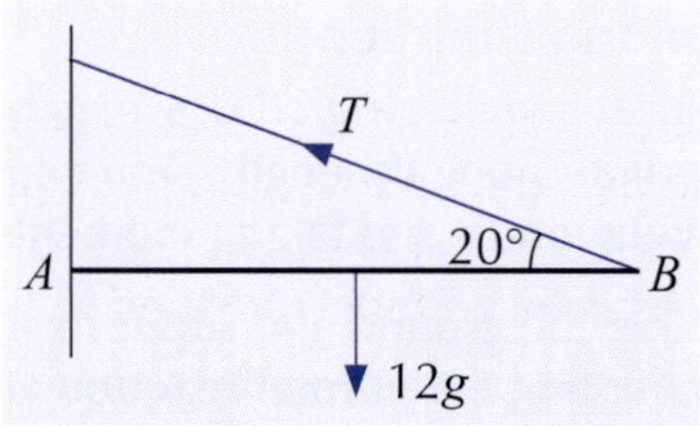

A uniform rod, *AB*, of mass 12 kg and length 16 m, rests against a rough, vertical wall at *A*. The rod is held horizontally in limiting equilibrium by a light wire attached at *B*, at an angle of 20° to the rod, as shown. Find:

a) the magnitude of the tension in the wire,

b) the coefficient of friction between the wall and the rod.

Q2 A uniform ladder of mass 11 kg and length 7 m rests against a rough vertical wall, at an angle of 60° to smooth, horizontal ground, as shown. A horizontal force of magnitude 35 N is applied to the base of the ladder, keeping it in limiting equilibrium, with the ladder on the point of sliding up the wall. Find:

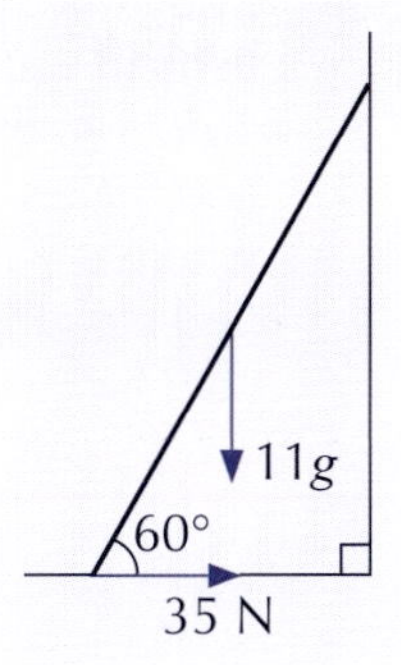

a) the magnitude of the normal reaction of the wall on the ladder,

b) the frictional force between the wall and the ladder,

c) the coefficient of friction between the wall and the ladder.

Q3

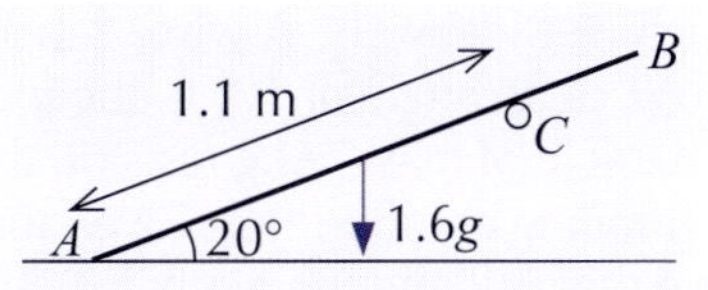

A uniform beam, AB, of mass 1.6 kg and length 1.5 m, rests with A on rough, horizontal ground. The beam is supported by a smooth peg at C, where $AC = 1.1$ m, so that it makes an angle of 20° with the horizontal, as shown. The beam is on the point of slipping. Find:

a) the magnitude of the normal reaction of the peg on the beam,

b) the magnitude of the normal reaction of the ground on the beam,

c) the magnitude of the frictional force between the ground and the beam,

d) the coefficient of friction between the ground and the beam.

Q4 A uniform ladder of mass 10 kg and length 6 m rests with one end on rough, horizontal ground and the other end against a smooth, vertical wall. The coefficient of friction between the ground and the ladder is 0.3, and the ladder makes an angle of 65° with the ground.
A girl of mass 50 kg begins to climb the ladder.
How far up the ladder can she climb before the ladder slips?

Q5 A uniform ladder of mass 9 kg and length 4.8 m rests in limiting equilibrium with one end on rough, horizontal ground and the other end against a rough, vertical wall. The normal reactions at the wall and the ground have magnitude 22 N and 75 N respectively. Find:

a) the angle that the ladder makes with the ground,

b) the coefficient of friction between the wall and the ladder,

c) the coefficient of friction between the ground and the ladder.

Q6 Robert holds an 8 m uniform ladder in place against a smooth vertical wall by applying a horizontal force of K N to it, 1 m from the base of the ladder. The ladder weighs 100 N, and makes an angle of 75° with the rough horizontal floor, where $\mu = 0.1$. Given that the ladder remains at rest, find the range of possible values for the magnitude of the force K.

Q7

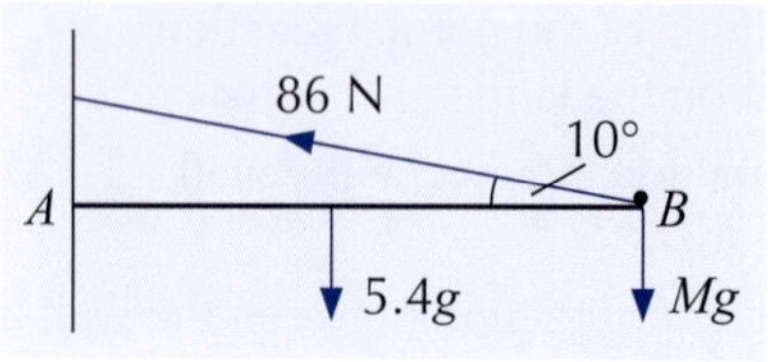

A non-uniform plank, AB, of mass 5.4 kg and length 11 m, rests with A against a rough, vertical wall. A wire with tension of magnitude 86 N is fixed to B, at an angle of 10° to the plank, keeping the plank horizontal. A particle of mass M kg rests on the plank at B, as shown. The plank is in limiting equilibrium. Given that the coefficient of friction between the wall and the plank is 0.55, find:

a) the value of M,

b) the distance of the centre of mass of the plank from A.

Q8

Q8 Hint: You're told that the reaction forces at A and C both act perpendicular to the beam. So it's a good idea to resolve forces parallel and perpendicular to the beam, rather than horizontally and vertically.

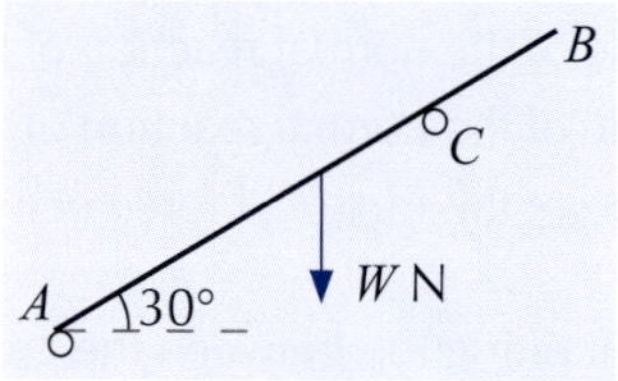

A uniform beam, AB, of weight W N rests in limiting equilibrium at an angle of 30° to the horizontal on a rough peg at A and a smooth peg at C, where $AC = 0.75AB$. The reaction forces at A and C are both perpendicular to the beam. Find the coefficient of friction between the peg and the beam at A.

Q9

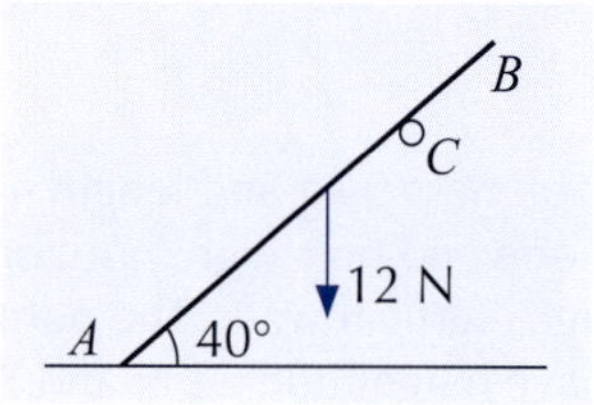

PROBLEM SOLVING

A uniform beam AB of weight 12 N and length 5 m rests on a smooth surface at A and a rough peg at C, 4 m from A. The beam makes an angle of 40° with the horizontal, as shown. Given that the beam is in equilibrium, find the range of possible values of the coefficient of friction between the beam and the peg.

Statistical Tables

Critical values for the Product Moment Correlation Coefficient, *r*

The r-values in the table are the minimum values of the Product Moment Correlation Coefficient for which a one- or two-tailed hypothesis test at the given significance level on a data set of size n is statistically significant.

one-tailed two-tailed	10% 20%	5% 10%	2.5% 5%	1% 2%	0.5% 1%
n					
4	0.8000	0.9000	0.9500	0.9800	0.9900
5	0.6870	0.8054	0.8783	0.9343	0.9587
6	0.6084	0.7293	0.8114	0.8822	0.9172
7	0.5509	0.6694	0.7545	0.8329	0.8745
8	0.5067	0.6215	0.7067	0.7887	0.8343
9	0.4716	0.5822	0.6664	0.7498	0.7977
10	0.4428	0.5494	0.6319	0.7155	0.7646
11	0.4187	0.5214	0.6021	0.6851	0.7348
12	0.3981	0.4973	0.5760	0.6581	0.7079
13	0.3802	0.4762	0.5529	0.6339	0.6835
14	0.3646	0.4575	0.5324	0.6120	0.6614
15	0.3507	0.4409	0.5140	0.5923	0.6411
16	0.3383	0.4259	0.4973	0.5742	0.6226
17	0.3271	0.4124	0.4821	0.5577	0.6055
18	0.3170	0.4000	0.4683	0.5425	0.5897
19	0.3077	0.3887	0.4555	0.5285	0.5751
20	0.2992	0.3783	0.4438	0.5155	0.5614

Answers

Chapter 1: Proof

1. Proof by Contradiction

Exercise 1.1 — Proof by contradiction

Q1 Suppose that there is a number x that is the largest multiple of 3, so it can be written as $x = 3k$ for some integer k.
Then $x + 3 = 3k + 3 = 3(k + 1)$ is also a multiple of 3 and is larger than x, which contradicts the initial assumption.
So there cannot be a largest multiple of 3.

Q2 Suppose that there is some even number x for which x^2 is odd. Since x is even, it can be written as $x = 2n$ for some integer n.
Then $x^2 = (2n)^2 = 4n^2 = 2(2n^2)$ which is even, which contradicts the assumption that x^2 is odd.
So if x^2 is odd, then x must also be odd.

Q3 a) Suppose that there is a rational number $x \neq 0$ and an irrational number y such that xy is rational.
This means that x can be written as $x = \frac{a}{b}$ and xy can be written as $xy = \frac{c}{d}$, where a, b, c and d are all non-zero integers.
So $xy = \frac{a}{b}y = \frac{c}{d} \Rightarrow y = \frac{bc}{ad}$
Since bc and ad are both integers, this means that y is a rational number, which contradicts the assumption that y is irrational. So the statement must be true.

b) To disprove the statement, find a counter-example.
$\sqrt{2}$ is an irrational number, but $\sqrt{2} \times \sqrt{2} = 2$ which is rational, so the statement is false.

Q4 Suppose that there is a smallest positive rational number, and call it x. Since x is rational, it can be written as $x = \frac{a}{b}$, and since it is positive, a and b are both positive integers (or both negative, in which case you can simplify the fraction by dividing top and bottom by –1 to get a and b positive).
Then $\frac{a}{b+1}$ is also a positive rational number, and is smaller than x, which contradicts the assumption that x is the smallest positive rational number.
So there cannot be a smallest positive rational number.

Q5 Assume that $1 + \sqrt{2}$ is rational, so it can be written as $\frac{a}{b}$ where a and b are non-zero integers.
So $1 + \sqrt{2} = \frac{a}{b} \Rightarrow \sqrt{2} = \frac{a}{b} - 1 \Rightarrow \sqrt{2} = \frac{a-b}{b}$
Since a and b are integers, $(a - b)$ is also an integer, which means that $\sqrt{2}$ is rational, which is not true.
So $1 + \sqrt{2}$ must be irrational.

Q6 a) Suppose that there is an integer x such that x^2 is a multiple of 3 but x is not.
If x is not a multiple of 3, then there are two cases to consider: $x = 3k + 1$ and $x = 3k + 2$ for some integer k.
If $x = 3k + 1$, then $x^2 = (3k + 1)^2 = 9k^2 + 6k + 1$
$= 3(3k^2 + 2k) + 1$
So x^2 is not a multiple of 3.
If $x = 3k + 2$, then $x^2 = (3k + 2)^2 = 9k^2 + 12k + 4$
$= 3(3k^2 + 4k + 1) + 1$
So x^2 is not a multiple of 3.
Therefore, by exhaustion, x^2 cannot be a multiple of 3, which contradicts the initial assumption.
So if x^2 is a multiple of 3, then x must also be a multiple of 3.

b) Suppose that $\sqrt{3}$ is rational, so $\sqrt{3} = \frac{a}{b}$ for some non-zero integers a and b that share no common factors.
So $b\sqrt{3} = a \Rightarrow 3b^2 = a^2 \Rightarrow a^2$ is a multiple of 3.
From part a), this means that a is also a multiple of 3, so write $a = 3k$ for some integer k.
$3b^2 = (3k)^2 \Rightarrow 3b^2 = 9k^2 \Rightarrow b^2 = 3k^2$
$\Rightarrow b^2$ is a multiple of 3.
Again, from part a), this means b is a multiple of 3.
But it was assumed at the start that a and b had no common factors.
So $\sqrt{3}$ cannot be written as an integer fraction, so it is irrational.

Chapter 2: Algebra and Functions

1. Simplifying Expressions

Exercise 1.1 — Simplifying algebraic fractions

Q1 $\frac{4}{2x+10} = \frac{4}{2(x+5)} = \frac{2}{x+5}$

Q2 $\frac{5x}{x^2+2x} = \frac{5x}{x(x+2)} = \frac{5}{x+2}$

Q3 $\frac{6x^2-3x}{3x^2} = \frac{3x(2x-1)}{3x^2} = \frac{2x-1}{x}$

Q4 $\frac{4x^3}{x^3+3x^2} = \frac{4x^3}{x^2(x+3)} = \frac{4x}{x+3}$

Q5 $\frac{3x+6}{x^2+3x+2} = \frac{3(x+2)}{(x+1)(x+2)} = \frac{3}{x+1}$

Q6 $\frac{x^2+3x}{x^2+x-6} = \frac{x(x+3)}{(x-2)(x+3)} = \frac{x}{x-2}$

Q7 $\frac{2x-6}{x^2-9} = \frac{2(x-3)}{(x-3)(x+3)} = \frac{2}{x+3}$

Q8 $\frac{5x^2-20x}{2x^2-5x-12} = \frac{5x(x-4)}{(2x+3)(x-4)} = \frac{5x}{2x+3}$

Q9 $\frac{3x^2-7x-6}{2x^2-x-15} = \frac{(3x+2)(x-3)}{(2x+5)(x-3)} = \frac{3x+2}{2x+5}$

Q10 $\frac{x^3-4x^2-19x-14}{x^2-6x-7} = \frac{(x+1)(x+2)(x-7)}{(x+1)(x-7)} = x+2$

To factorise the cubic, try a few different values for x — once you've spotted that f(–1) = 0 (and so (x + 1) is a factor) you can take that out and see what's left to factorise.

Q11 $\frac{x^3-2x^2}{x^3-4x} = \frac{x^2(x-2)}{x(x^2-4)} = \frac{x^2(x-2)}{x(x-2)(x+2)} = \frac{x}{x+2}$

Q12 $\frac{1+\frac{1}{x}}{x+1} = \frac{\left(1+\frac{1}{x}\right)x}{(x+1)x} = \frac{x+1}{x(x+1)} = \frac{1}{x}$

Q13 $\frac{3+\frac{1}{x}}{2+\frac{1}{x}} = \frac{\left(3+\frac{1}{x}\right)x}{\left(2+\frac{1}{x}\right)x} = \frac{3x+1}{2x+1}$

Q14 $\dfrac{1+\frac{1}{2x}}{2+\frac{1}{x}} = \dfrac{\left(1+\frac{1}{2x}\right)2x}{\left(2+\frac{1}{x}\right)2x} = \dfrac{2x+1}{4x+2} = \dfrac{2x+1}{2(2x+1)} = \dfrac{1}{2}$

Q15 $\dfrac{\frac{1}{3x}-1}{3x^2-x} = \dfrac{\left(\frac{1}{3x}-1\right)3x}{(3x^2-x)3x} = \dfrac{-(3x-1)}{3x^2(3x-1)} = -\dfrac{1}{3x^2}$

Q16 $\dfrac{2+\frac{1}{x}}{6x^2+3x} = \dfrac{\left(2+\frac{1}{x}\right)x}{(6x^2+3x)x} = \dfrac{2x+1}{3x^2(2x+1)} = \dfrac{1}{3x^2}$

Q17 $\dfrac{\frac{3x}{x+2}}{\frac{x}{x+2}+\frac{1}{x+2}} = \dfrac{\left(\frac{3x}{x+2}\right)(x+2)}{\left(\frac{x}{x+2}+\frac{1}{x+2}\right)(x+2)} = \dfrac{3x}{x+1}$

Q18 $\dfrac{2+\frac{1}{x+1}}{3+\frac{1}{x+1}} = \dfrac{\left(2+\frac{1}{x+1}\right)(x+1)}{\left(3+\frac{1}{x+1}\right)(x+1)} = \dfrac{2(x+1)+1}{3(x+1)+1} = \dfrac{2x+3}{3x+4}$

Q19 $\dfrac{1-\frac{2}{x+3}}{x+2} = \dfrac{\left(1-\frac{2}{x+3}\right)(x+3)}{(x+2)(x+3)} = \dfrac{x+3-2}{(x+2)(x+3)}$

$= \dfrac{x+1}{(x+2)(x+3)}$

Q20 $\dfrac{4-\frac{1}{x^2}}{2-\frac{1}{x}-\frac{1}{x^2}} = \dfrac{\left(4-\frac{1}{x^2}\right)x^2}{\left(2-\frac{1}{x}-\frac{1}{x^2}\right)x^2} = \dfrac{4x^2-1}{2x^2-x-1}$

$= \dfrac{(2x+1)(2x-1)}{(x-1)(2x+1)} = \dfrac{2x-1}{x-1}$

Exercise 1.2 — Adding and subtracting algebraic fractions

Q1 $\dfrac{2x}{3}+\dfrac{x}{5} = \dfrac{10x}{15}+\dfrac{3x}{15} = \dfrac{13x}{15}$

Q2 $\dfrac{2}{3x}-\dfrac{1}{5x} = \dfrac{10}{15x}-\dfrac{3}{15x} = \dfrac{7}{15x}$

Q3 $\dfrac{3}{x^2}+\dfrac{2}{x} = \dfrac{3}{x^2}+\dfrac{2x}{x^2} = \dfrac{3+2x}{x^2}$

Q4 $\dfrac{x+1}{3}+\dfrac{x+2}{4} = \dfrac{4(x+1)}{12}+\dfrac{3(x+2)}{12}$

$= \dfrac{4x+4+3x+6}{12} = \dfrac{7x+10}{12}$

Q5 $\dfrac{2x}{3}+\dfrac{x-1}{7x} = \dfrac{14x^2}{21x}+\dfrac{3(x-1)}{21x} = \dfrac{14x^2+3x-3}{21x}$

Q6 $\dfrac{3x}{4}-\dfrac{2x-1}{5x} = \dfrac{15x^2}{20x}-\dfrac{4(2x-1)}{20x} = \dfrac{15x^2-8x+4}{20x}$

Q7 $\dfrac{2}{x-1}+\dfrac{3}{x} = \dfrac{2x}{x(x-1)}+\dfrac{3(x-1)}{x(x-1)} = \dfrac{2x+3x-3}{x(x-1)} = \dfrac{5x-3}{x(x-1)}$

Q8 $\dfrac{3}{x+1}+\dfrac{2}{x+2} = \dfrac{3(x+2)}{(x+1)(x+2)}+\dfrac{2(x+1)}{(x+1)(x+2)}$

$= \dfrac{3x+6+2x+2}{(x+1)(x+2)} = \dfrac{5x+8}{(x+1)(x+2)}$

Q9 $\dfrac{4}{x-3}-\dfrac{1}{x+4} = \dfrac{4(x+4)}{(x-3)(x+4)}-\dfrac{x-3}{(x-3)(x+4)}$

$= \dfrac{4x+16-x+3}{(x-3)(x+4)} = \dfrac{3x+19}{(x-3)(x+4)}$

Q10 $\dfrac{6}{x+2}+\dfrac{6}{x-2} = \dfrac{6(x-2)}{(x+2)(x-2)}+\dfrac{6(x+2)}{(x+2)(x-2)}$

$= \dfrac{6x-12+6x+12}{(x+2)(x-2)} = \dfrac{12x}{(x+2)(x-2)}$

Q11 $\dfrac{3}{x-2}-\dfrac{5}{2x+3} = \dfrac{3(2x+3)}{(x-2)(2x+3)}-\dfrac{5(x-2)}{(x-2)(2x+3)}$

$= \dfrac{6x+9-5x+10}{(x-2)(2x+3)} = \dfrac{x+19}{(x-2)(2x+3)}$

Q12 $\dfrac{3}{x+2}+\dfrac{x}{x+1} = \dfrac{3(x+1)}{(x+2)(x+1)}+\dfrac{x(x+2)}{(x+2)(x+1)}$

$= \dfrac{3x+3+x^2+2x}{(x+2)(x+1)} = \dfrac{x^2+5x+3}{(x+2)(x+1)}$

Q13 $\dfrac{5x}{(x+1)^2}-\dfrac{3}{x+1} = \dfrac{5x}{(x+1)^2}-\dfrac{3(x+1)}{(x+1)^2}$

$= \dfrac{5x-3x-3}{(x+1)^2} = \dfrac{2x-3}{(x+1)^2}$

Q14 $\dfrac{5}{x(x+3)}+\dfrac{3}{x+2} = \dfrac{5(x+2)}{x(x+3)(x+2)}+\dfrac{3x(x+3)}{x(x+3)(x+2)}$

$= \dfrac{5x+10+3x^2+9x}{x(x+3)(x+2)} = \dfrac{3x^2+14x+10}{x(x+3)(x+2)}$

Q15 $\dfrac{x}{x^2-4}-\dfrac{1}{x+2} = \dfrac{x}{(x+2)(x-2)}-\dfrac{1}{x+2}$

$= \dfrac{x}{(x+2)(x-2)}-\dfrac{x-2}{(x+2)(x-2)}$

$= \dfrac{x-x+2}{(x+2)(x-2)} = \dfrac{2}{(x+2)(x-2)}$

Q16 $\dfrac{3}{x+1}+\dfrac{6}{2x^2+x-1} = \dfrac{3}{x+1}+\dfrac{6}{(x+1)(2x-1)}$

$= \dfrac{3(2x-1)}{(x+1)(2x-1)}+\dfrac{6}{(x+1)(2x-1)}$

$= \dfrac{6x-3+6}{(x+1)(2x-1)} = \dfrac{3(2x+1)}{(x+1)(2x-1)}$

Q17 $\dfrac{2}{x}+\dfrac{3}{x+1}+\dfrac{4}{x+2}$

$= \dfrac{2(x+1)(x+2)}{x(x+1)(x+2)}+\dfrac{3x(x+2)}{x(x+1)(x+2)}+\dfrac{4x(x+1)}{x(x+1)(x+2)}$

$= \dfrac{2x^2+6x+4+3x^2+6x+4x^2+4x}{x(x+1)(x+2)} = \dfrac{9x^2+16x+4}{x(x+1)(x+2)}$

Q18 $\dfrac{3}{x+4}-\dfrac{2}{x+1}+\dfrac{1}{x-2}$

$= \dfrac{3(x+1)(x-2)}{(x+4)(x+1)(x-2)}-\dfrac{2(x+4)(x-2)}{(x+4)(x+1)(x-2)}$

$+\dfrac{(x+4)(x+1)}{(x+4)(x+1)(x-2)}$

$= \dfrac{3x^2-3x-6-2x^2-4x+16+x^2+5x+4}{(x+4)(x+1)(x-2)}$

$= \dfrac{2(x^2-x+7)}{(x+4)(x+1)(x-2)}$

Q19 $2-\dfrac{3}{x+1}+\dfrac{4}{(x+1)^2} = \dfrac{2(x+1)^2}{(x+1)^2}-\dfrac{3(x+1)}{(x+1)^2}+\dfrac{4}{(x+1)^2}$

$= \dfrac{2x^2+4x+2-3x-3+4}{(x+1)^2} = \dfrac{2x^2+x+3}{(x+1)^2}$

Q20 $\dfrac{2x^2-x-3}{x^2-1}+\dfrac{1}{x(x-1)} = \dfrac{(x+1)(2x-3)}{(x+1)(x-1)}+\dfrac{1}{x(x-1)}$

$= \dfrac{2x-3}{x-1}+\dfrac{1}{x(x-1)} = \dfrac{x(2x-3)}{x(x-1)}+\dfrac{1}{x(x-1)}$

$= \dfrac{2x^2-3x+1}{x(x-1)} = \dfrac{(x-1)(2x-1)}{x(x-1)} = \dfrac{2x-1}{x}$

Exercise 1.3 — Multiplying and dividing algebraic fractions

Q1 a) $\dfrac{2x}{3}\times\dfrac{5x}{4} = \dfrac{x}{3}\times\dfrac{5x}{2} = \dfrac{x\times5x}{3\times2} = \dfrac{5x^2}{6}$

b) $\dfrac{6x^3}{7}\times\dfrac{2}{x^2} = \dfrac{6x}{7}\times\dfrac{2}{1} = \dfrac{6x\times2}{7\times1} = \dfrac{12x}{7}$

c) $\dfrac{8x^2}{3y^2}\times\dfrac{x^3}{4y} = \dfrac{2x^2}{3y^2}\times\dfrac{x^3}{y} = \dfrac{2x^2\times x^3}{3y^2\times y} = \dfrac{2x^5}{3y^3}$

d) $\dfrac{8x^4}{3y}\times\dfrac{6y^2}{5x} = \dfrac{8x^3}{1}\times\dfrac{2y}{5} = \dfrac{8x^3\times2y}{1\times5} = \dfrac{16x^3y}{5}$

Q2 **a)** $\frac{x}{3} \div \frac{3}{x} = \frac{x}{3} \times \frac{x}{3} = \frac{x \times x}{3 \times 3} = \frac{x^2}{9}$

b) $\frac{4x^3}{3} \div \frac{x}{2} = \frac{4x^3}{3} \times \frac{2}{x} = \frac{4x^2}{3} \times \frac{2}{1} = \frac{4x^2 \times 2}{3 \times 1} = \frac{8x^2}{3}$

c) $\frac{3}{2x} \div \frac{6}{x^3} = \frac{3}{2x} \times \frac{x^3}{6} = \frac{1}{2} \times \frac{x^2}{2} = \frac{1 \times x^2}{2 \times 2} = \frac{x^2}{4}$

d) $\frac{2x^3}{3y} \div \frac{4x}{y^2} = \frac{2x^3}{3y} \times \frac{y^2}{4x} = \frac{x^2}{3} \times \frac{y}{2} = \frac{x^2 \times y}{3 \times 2} = \frac{x^2 y}{6}$

Q3 $\frac{x+2}{4} \times \frac{x}{3x+6} = \frac{x+2}{4} \times \frac{x}{3(x+2)} = \frac{1}{4} \times \frac{x}{3} = \frac{1 \times x}{4 \times 3} = \frac{x}{12}$

Q4 $\frac{4x}{5} \div \frac{4x^2+8x}{15} = \frac{4x}{5} \times \frac{15}{4x(x+2)} = \frac{1}{1} \times \frac{3}{(x+2)} = \frac{3}{(x+2)}$

Q5 $\frac{2x^2-2}{x} \times \frac{5x}{3x-3} = \frac{2(x-1)(x+1)}{x} \times \frac{5x}{3(x-1)}$

$= \frac{2(x+1)}{1} \times \frac{5}{3} = \frac{2(x+1) \times 5}{1 \times 3} = \frac{10(x+1)}{3}$

Q6 $\frac{2x^2+8x}{x^2-2x} \times \frac{x-1}{x+4} = \frac{2x(x+4)}{x(x-2)} \times \frac{x-1}{x+4}$

$= \frac{2}{x-2} \times \frac{x-1}{1} = \frac{2(x-1)}{x-2}$

Q7 $\frac{x^2-4}{9} \div \frac{x-2}{3} = \frac{(x+2)(x-2)}{9} \times \frac{3}{x-2}$

$= \frac{x+2}{3} \times \frac{1}{1} = \frac{x+2}{3}$

Q8 $\frac{2}{x^2+4x} \div \frac{1}{x+4} = \frac{2}{x(x+4)} \times \frac{x+4}{1} = \frac{2}{x} \times \frac{1}{1} = \frac{2}{x}$

Q9 $\frac{x^2+4x+3}{x^2+5x+6} \times \frac{x^2+2x}{x+1} = \frac{(x+1)(x+3)}{(x+2)(x+3)} \times \frac{x(x+2)}{x+1}$

$= \frac{1}{1} \times \frac{x}{1} = x$

Q10 $\frac{x^2+5x+6}{x^2-2x-3} \times \frac{3x+3}{x^2+2x} = \frac{(x+2)(x+3)}{(x-3)(x+1)} \times \frac{3(x+1)}{x(x+2)}$

$= \frac{x+3}{x-3} \times \frac{3}{x} = \frac{3(x+3)}{x(x-3)}$

Q11 $\frac{x^2-4}{6x-3} \times \frac{2x^2+5x-3}{x^2+2x} = \frac{(x+2)(x-2)}{3(2x-1)} \times \frac{(2x-1)(x+3)}{x(x+2)}$

$= \frac{x-2}{3} \times \frac{x+3}{x} = \frac{(x-2)(x+3)}{3x}$

Q12 $\frac{x^2+7x+6}{4x-4} \div \frac{x^2+8x+12}{x^2-x} = \frac{(x+1)(x+6)}{4(x-1)} \times \frac{x(x-1)}{(x+2)(x+6)}$

$= \frac{x+1}{4} \times \frac{x}{x+2} = \frac{x(x+1)}{4(x+2)}$

Q13 $\frac{x^2+4x+4}{x^2-4x+3} \times \frac{x^2-2x-3}{2x^2-2x} \times \frac{4x-4}{x^2+2x}$

$= \frac{(x+2)^2}{(x-3)(x-1)} \times \frac{(x-3)(x+1)}{2x(x-1)} \times \frac{4(x-1)}{x(x+2)}$

$= \frac{x+2}{x-1} \times \frac{x+1}{x} \times \frac{2}{x} = \frac{2(x+2)(x+1)}{x^2(x-1)}$

Q14 $\frac{x}{6x+12} \div \frac{x^2-x}{x+2} \times \frac{3x-3}{x+1} = \frac{x}{6(x+2)} \times \frac{x+2}{x(x-1)} \times \frac{3(x-1)}{x+1}$

$= \frac{1}{2} \times \frac{1}{1} \times \frac{1}{x+1} = \frac{1}{2(x+1)}$

Q15 $\frac{x^2+5x}{2x^2+7x+3} \times \frac{2x+1}{x^3-x^2} \div \frac{x+5}{x^2+x-6}$

$= \frac{x(x+5)}{(2x+1)(x+3)} \times \frac{2x+1}{x^2(x-1)} \times \frac{(x+3)(x-2)}{x+5}$

$= \frac{1}{1} \times \frac{1}{x(x-1)} \times \frac{x-2}{1} = \frac{x-2}{x(x-1)}$

Exercise 1.4 — Algebraic division

Q1 **a)** $x^3 - 14x^2 + 6x + 11 \equiv (Ax^2 + Bx + C)(x + 1) + D$

The degree of the quotient is the difference between the degrees of the polynomial and the divisor, in this case $3 - 1 = 2$. The degree of the remainder must be less than the degree of the divisor (1) so it must be 0.

Set $x = -1$: $-1 - 14 - 6 + 11 = D$, so $D = -10$.
Set $x = 0$: $11 = C + D$, so $C = 21$.
Equating the coefficients of x^3 gives $1 = A$.
Equating the coefficients of x^2 gives:
$-14 = A + B$, so $B = -15$.
So $x^3 - 14x^2 + 6x + 11 \equiv (x^2 - 15x + 21)(x + 1) - 10$.
Quotient: $x^2 - 15x + 21$
Remainder: -10

b) $2x^3 + 5x^2 - 8x - 17 \equiv (Ax^2 + Bx + C)(x - 2) + D$
Set $x = 2$: $16 + 20 - 16 - 17 = D$, so $D = 3$.
Set $x = 0$: $-17 = -2C + D$, so $C = 10$.
Equating the coefficients of x^3 gives $2 = A$.
Equating the coefficients of x^2 gives:
$5 = -2A + B$, so $B = 9$.
So $2x^3 + 5x^2 - 8x - 17 \equiv (2x^2 + 9x + 10)(x - 2) + 3$.
Quotient: $2x^2 + 9x + 10$
Remainder: 3

c) $6x^3 + x^2 - 11x - 5 \equiv (Ax^2 + Bx + C)(2x + 1) + D$
Set $x = -\frac{1}{2}$: $-\frac{6}{8} + \frac{1}{4} + \frac{11}{2} - 5 = D$, so $D = 0$.
Set $x = 0$: $-5 = C + D$, so $C = -5$.
Equating the coefficients of x^3 gives $6 = 2A$, so $A = 3$.
Equating the coefficients of x^2 gives:
$1 = A + 2B$, so $B = -1$.
So $6x^3 + x^2 - 11x - 5 \equiv (3x^2 - x - 5)(2x + 1)$.
Quotient: $3x^2 - x - 5$
Remainder: 0

Q2 $3x^4 - 8x^3 - 6x - 4 \equiv (Ax^3 + Bx^2 + Cx + D)(x - 3) + E$
Set $x = 3$: $243 - 216 - 18 - 4 = E$, so $E = 5$.
Set $x = 0$: $-4 = -3D + E$, so $D = 3$.
Equating the coefficients:
x^4 terms $\Rightarrow 3 = A$.
x^3 terms $\Rightarrow -8 = -3A + B$, so $B = 1$.
x terms $\Rightarrow -6 = -3C + D$, so $C = 3$.
So $3x^4 - 8x^3 - 6x - 4 \equiv (3x^3 + x^2 + 3x + 3)(x - 3) + 5$
or $(3x^4 - 8x^3 - 6x - 4) \div (x - 3)$
$= 3x^3 + x^2 + 3x + 3$ remainder 5

Q3 **a)**

$$\begin{array}{r} x^2 - 15x + 21 \ \ \text{r} -10 \\ x+1\overline{)\,x^3 - 14x^2 + 6x + 11} \\ -\ (x^3 + x^2) \qquad\qquad \\ \hline -15x^2 + 6x \qquad \\ -\ (-15x^2 - 15x) \qquad \\ \hline 21x + 11 \\ -\ (21x + 21) \\ \hline -10 \end{array}$$

Quotient: $x^2 - 15x + 21$, remainder: -10

b)

$$\begin{array}{r} x^2 + 7x - 6 \ \ \text{r}\ 5 \\ x+3\overline{)\,x^3 + 10x^2 + 15x - 13} \\ -\ (x^3 + 3x^2) \qquad\qquad \\ \hline 7x^2 + 15x \qquad \\ -\ (7x^2 + 21x) \qquad \\ \hline -6x - 13 \\ -\ (-6x - 18) \\ \hline 5 \end{array}$$

Quotient: $x^2 + 7x - 6$, remainder: 5

c)

$$\begin{array}{r} 2x^2 + 9x + 10 \ \ \text{r}\ 3 \\ x-2\overline{)\,2x^3 + 5x^2 - 8x - 17} \\ -\ (2x^3 - 4x^2) \qquad\qquad \\ \hline 9x^2 - 8x \qquad \\ -\ (9x^2 - 18x) \qquad \\ \hline 10x - 17 \\ -\ (10x - 20) \\ \hline 3 \end{array}$$

Quotient: $2x^2 + 9x + 10$, remainder: 3

d)

$$\begin{array}{r} 3x^2 - 15x - 3 \text{ r } 24 \\ x+5\,\overline{)\,3x^3 + 0x^2 - 78x + 9} \\ -\ \underline{(3x^3 + 15x^2)} \\ -15x^2 - 78x \\ -\ \underline{(-15x^2 - 75x)} \\ -3x + 9 \\ -\ \underline{(-3x - 15)} \\ 24 \end{array}$$

Quotient: $3x^2 - 15x - 3$, remainder: 24

e)

$$\begin{array}{r} x^3 + x^2 + x + 1 \\ x-1\,\overline{)\,x^4 + 0x^3 + 0x^2 + 0x - 1} \\ -\ \underline{(x^4 - x^3)} \\ x^3 + 0x^2 \\ -\ \underline{(x^3 - x^2)} \\ x^2 + 0x \\ -\ \underline{(x^2 - x)} \\ x - 1 \\ -\ \underline{(x - 1)} \\ 0 \end{array}$$

Quotient: $x^3 + x^2 + x + 1$, remainder: 0

f)

$$\begin{array}{r} 4x^2 + 3x + 5 \text{ r } 25 \\ 2x-3\,\overline{)\,8x^3 - 6x^2 + x - 10} \\ -\ \underline{(8x^3 - 12x^2)} \\ 6x^2 + x \\ -\ \underline{(6x^2 - 9x)} \\ 10x + 10 \\ -\ \underline{(10x - 15)} \\ 25 \end{array}$$

Quotient: $4x^2 + 3x + 5$, remainder: 25

Q4

$$\begin{array}{r} 5x^2 + x - 3 \text{ r } 24 \\ 2x+1\,\overline{)\,10x^3 + 7x^2 - 5x + 21} \\ -\ \underline{(10x^3 + 5x^2)} \\ 2x^2 - 5x \\ -\ \underline{(2x^2 + x)} \\ -6x + 21 \\ -\ \underline{(-6x - 3)} \\ 24 \end{array}$$

Quotient: $5x^2 + x - 3$, remainder: 24

Q5

$$\begin{array}{r} 8x^3 + 12x^2 + 18x + 27 \text{ r } 81 \\ 2x-3\,\overline{)\,16x^4 + 0x^3 + 0x^2 + 0x + 0} \\ -\ \underline{(16x^4 - 24x^3)} \\ 24x^3 + 0x^2 \\ -\ \underline{(24x^3 - 36x^2)} \\ 36x^2 + 0x \\ -\ \underline{(36x^2 - 54x)} \\ 54x + 0 \\ -\ \underline{(54x - 81)} \\ 81 \end{array}$$

Quotient: $8x^3 + 12x^2 + 18x + 27$, remainder: 81

Q6

$$\begin{array}{r} 3x^3 + 13x^2 + 4x \\ x-2\,\overline{)\,3x^4 + 7x^3 - 22x^2 - 8x} \\ -\ \underline{(3x^4 - 6x^3)} \\ 13x^3 - 22x^2 \\ -\ \underline{(13x^3 - 26x^2)} \\ 4x^2 - 8x \\ -\ \underline{(4x^2 - 8x)} \\ 0 \end{array}$$

So $3x^4 + 7x^3 - 22x^2 - 8x = (3x^3 + 13x^2 + 4x)(x - 2)$
$= x(3x^2 + 13x + 4)(x - 2)$
$= x(3x + 1)(x + 4)(x - 2)$

and the solutions are: $x = 0$, $x = -\frac{1}{3}$, $x = -4$ and $x = 2$

You could have taken out a factor of x at the start, although it wouldn't affect the method too much.

For questions 4-6, you could also have used the formula method — you should get the same quotient and remainder.

2. Mappings and Functions

Exercise 2.1 — Mappings and functions

Q1

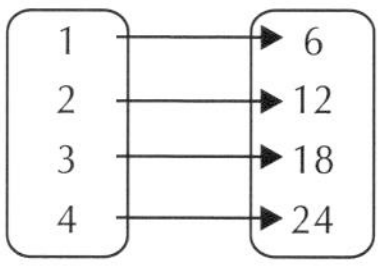

Q2

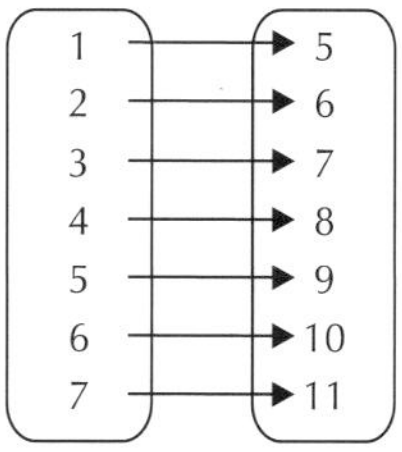

Q3 **a)**

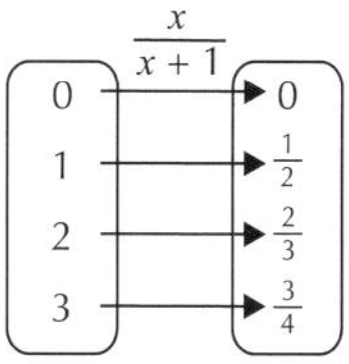

b)

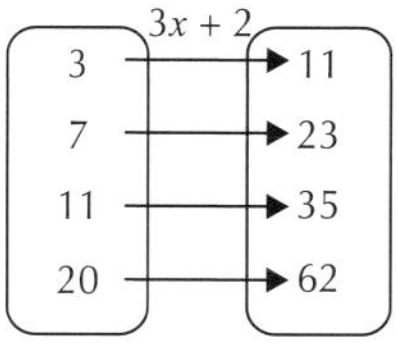

Q4 $g(0) = \frac{1}{2(0)+1} = 1$, $g(2) = \frac{1}{2(2)+1} = \frac{1}{5}$

Q5 $f(1) = \frac{1}{2 + \log_{10} 1} = \frac{1}{2+0} = \frac{1}{2}$

$f(100) = \frac{1}{2 + \log_{10} 100} = \frac{1}{2+2} = \frac{1}{4}$

Q6 **a)** The minimum value of $h(x)$ in the domain is $\sin 0° = 0$, and the maximum is $\sin 90° = 1$.
So the range is $0 \le h(x) \le 1$

b) The minimum value of $j(x)$ in this domain is $\cos 180° = -1$, and the maximum is $\cos 0° = 1$.
So the range is $-1 \le j(x) \le 1$

Q7 **a)** 3^x is defined for all $x \in \mathbb{R}$, so the domain of $f(x)$ is $\mathbb{R}$.
The value of 3^x gets closer to 0 as x becomes more negative, so its range is $3^x > 0$.
So the range of f is $f(x) > -1$

b) $\ln x$ is not defined for $x \le 0$, so the largest possible domain of $g(x)$ is $x > 0$.
The range of $\ln x$ is $\mathbb{R}$, but squaring the result means that $g(x)$ will always be positive.
So the range is $g(x) \ge 0$.

Q8 **a)** Yes, it is a function.

b) No, because the map is not defined for elements 4 and 5 of the domain.

c) No, because a value in the domain can map to more than one value in the range.

Exercise 2.2 — Graphs of functions

Q1 **a)** Yes, it is a function.

b) No, because a value of x can map to more than one value of $f(x)$.

Q2 **a)**

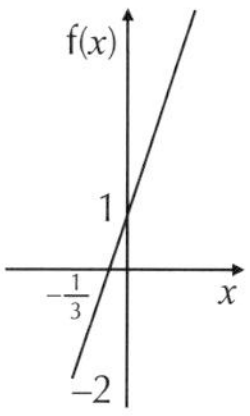

Range: $f(x) \ge -2$

b) 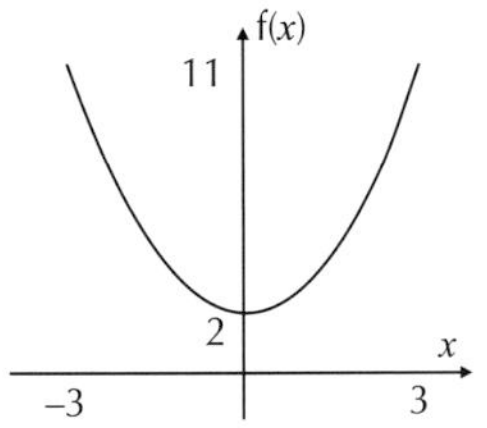

Range: $2 \le f(x) \le 11$

c) 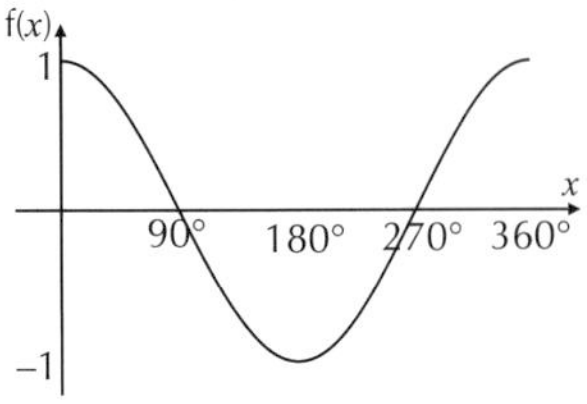

Range: $-1 \le f(x) \le 1$

d) 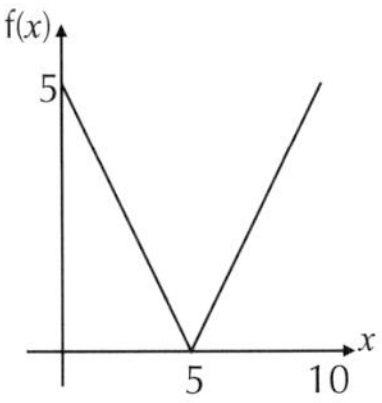

Range: $0 \le f(x) \le 5$

Q3 a) The range shown is $\{f(x) : 3 \le f(x) \le 11\}$, so the domain is: $\frac{3-1}{2} \le x \le \frac{11-1}{2}$, which is $1 \le x \le 5$.
In set notation, this is $\{x : 1 \le x \le 5\}$.

b) The domain shown is $\{x : -1 \le x \le 4\}$.
The range is between f(2) and f(−1):
$((2)^2 - 4(2) + 5) \le f(x) \le ((-1)^2 - 4(-1) + 5)$,
which is $1 \le f(x) \le 10$, i.e. $\{f(x) : 1 \le f(x) \le 10\}$.

Q4 When $x = 0$, $f(x) = 2$.
For large x, $x + 2 \approx x + 1$ so as $x \to \infty$,
$\frac{x+2}{x+1} \to 1$. So $f(x) = 1$ is an asymptote.
So the range is $1 < f(x) \le 2$.

Q5 The function $f(x) = \frac{1}{x-2}$ is undefined when $x = 2$, so $a = 2$.

Q6 The function $f(x) = +\sqrt{9 - x^2}$ is only defined when $x^2 \le 9$, i.e. when $-3 \le x \le 3$. So $a = -3$ and $b = 3$

Q7 The graph of $h(x) = +\sqrt{x+1}$ is shown below.

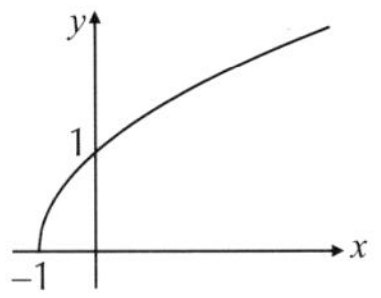

h(x) is undefined when x is less than −1.

Restricting the domain to $\{x : x \ge -1\}$ would make h(x) a function.

Q8 The graph of $k : x \to \tan x$ is shown below.

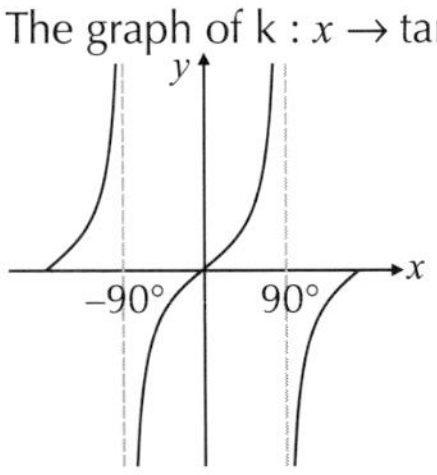

$\tan x$ is undefined at e.g. −90° and 90° (it is also undefined periodically either side).

Restricting the domain to e.g. $-90° < x < 90°$ would make this a function.

Q9 The function $m(x) = \frac{1}{x^2 - 4}$ is undefined when $x = 2$ and when $x = -2$.
So the largest continuous domain that makes it a function would be either $x > 2$ or $x < -2$.

Q10 a) It is not a function because it is not defined for all values of x in the domain (it's not defined at $x = 0$ or $x = 4$).

b) $x \in \mathbb{R}, x \ne 0, x \ne 4$

Exercise 2.3 — Types of function

It helps to sketch the graph of each function to identify its type.

Q1 a) One-to-one:

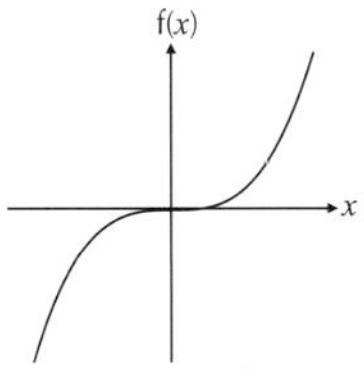

b) Many-to-one:

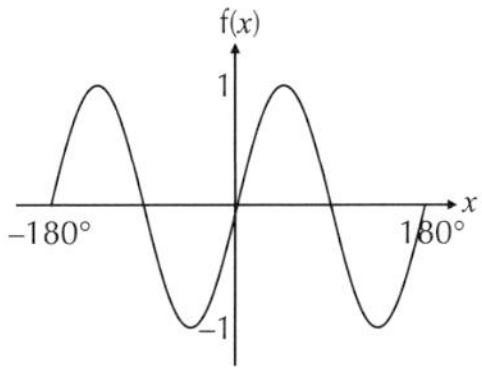

c) One-to-one:

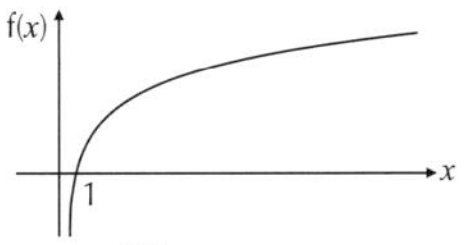

d) Many-to-one:

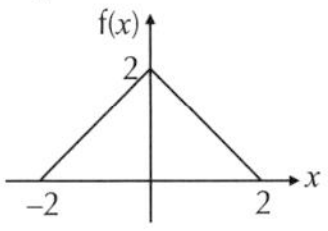

e) Many-to-one:

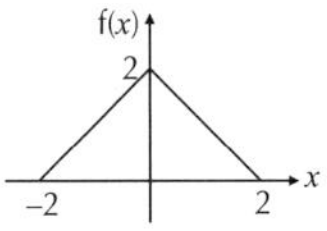

3. Composite Functions

Exercise 3.1 — Composite functions

Q1 a) Do $g(3) = 2(3) + 1 = 7$
then $f(7) = 7^2 = 49$, so $fg(3) = 49$.

b) $gf(3) = g(3^2) = g(9) = 2(9) + 1 = 19$

c) $f^2(5) = f(5^2) = f(25) = 25^2 = 625$

d) $g^2(2) = g(2(2) + 1) = g(5) = 2(5) + 1 = 11$

Q2 $fg(90°) = f(2(90°)) = f(180°) = \sin 180° = 0$
$gf(90°) = g(\sin 90°) = g(1) = 2(1) = 2$

Q3 a) $gf(1) = g\left(\frac{3}{1+2}\right) = g(1) = 2(1) = 2.$
$fg(1) = f(2(1)) = f(2) = \frac{3}{2+2} = \frac{3}{4}.$
$f^2(4) = f\left(\frac{3}{4+2}\right) = f\left(\frac{1}{2}\right) = \frac{3}{\frac{1}{2}+2} = \frac{6}{1+4} = \frac{6}{5}$

b) $g(-1) = -2$, and $f(-2)$ has a denominator of 0 (which is undefined).

Q4 a) $fg(x) = f(2x) = \cos 2x$

b) $gf(x) = g(\cos x) = 2 \cos x$

Q5 a) $fg(x) = f(2^x) = 2(2^x) - 1$ $(= 2^{x+1} - 1)$

b) $gf(x) = g(2x - 1) = 2^{(2x-1)}$

c) $f^2(x) = f(2x - 1) = 2(2x - 1) - 1 = 4x - 3$

Q6 $fg(x) = f(x+4) = \frac{2}{x+4-1} = \frac{2}{x+3}$

$gf(x) = g\left(\frac{2}{x-1}\right) = \frac{2}{x-1} + 4$

$= \frac{2}{x-1} + \frac{4x-4}{x-1} = \frac{4x-2}{x-1} = \frac{2(2x-1)}{x-1}$

Q7 $f^2(x) = f\left(\frac{x}{1-x}\right) = \frac{\frac{x}{1-x}}{1-\frac{x}{1-x}} = \frac{x}{(1-x)-x} = \frac{x}{1-2x}$

$gfg(x) = gf(x^2) = g\left(\frac{x^2}{1-x^2}\right) = \left(\frac{x^2}{1-x^2}\right)^2 = \frac{x^4}{(1-x^2)^2}$

Q8 **a)** $fg(x) = f(2x-3) = (2x-3)^2$, range: $fg(x) \geq 0$.

b) $gf(x) = g(x^2) = 2x^2 - 3$, range: $gf(x) \geq -3$.

Q9 **a)** $gf(x) = g\left(\frac{1}{x}\right) = \ln\left(\frac{1}{x}+1\right)$

Domain: $\{x : x > 0\}$

(domain of g = range of f = $\{x : x > 0\}$)

Range: $\{gf(x) : gf(x) > 0\}$.

b) $fg(x) = f(\ln(x+1)) = \frac{1}{\ln(x+1)}$

Domain: $\{x : x > 0\}$

(domain of f = range of g = $\{x : x > 0\}$)

Range: $\{fg(x) : fg(x) > 0\}$

Q10 $fgh(x) = fg(x^2+1) = f(5(x^2+1)-1)$

$= f(5x^2+4) = 3(5x^2+4) + 2 = 15x^2 + 14$

Exercise 3.2 — Solving composite function equations

Q1 **a)** $fg(x) = f(3x-4) = 2(3x-4) + 1 = 6x - 7$

$6x - 7 = 23 \Rightarrow x = 5$

b) $gf(x) = g\left(\frac{1}{x}\right) = \frac{2}{x} + 5$

$\frac{2}{x} + 5 = 6 \Rightarrow x = 2$

c) $gf(x) = g(x^2) = \frac{x^2}{x^2-3}$

$\frac{x^2}{x^2-3} = 4 \Rightarrow x^2 = 4x^2 - 12 \Rightarrow x^2 = 4$

$x = 2$ or $x = -2$

d) $fg(x) = f(3x-2) = (3x-2)^2 + 1$

$(3x-2)^2 + 1 = 50 \Rightarrow x = \frac{\pm\sqrt{49}+2}{3}$

$x = 3$ or $x = -\frac{5}{3}$

e) $fg(x) = f(\sqrt{x}) = 2(\sqrt{x}) + 1$

$2(\sqrt{x}) + 1 = 17 \Rightarrow \sqrt{x} = 8 \Rightarrow x = 64$

f) $fg(x) = f(3-x) = \log_{10}(3-x)$

$\log_{10}(3-x) = 0 \Rightarrow 3 - x = 1 \Rightarrow x = 2$

g) $fg(x) = f(x^2+2x) = 2^{(x^2+2x)}$

$2^{(x^2+2x)} = 8 \Rightarrow x^2 + 2x = 3 \Rightarrow (x-1)(x+3) = 0$

$x = 1$ or $x = -3$

h) $fg(x) = f(2x-1) = \frac{2x-1}{2x-1+1} = \frac{2x-1}{2x} = 1 - \frac{1}{2x}$

$gf(x) = g\left(\frac{x}{x+1}\right) = 2\left(\frac{x}{x+1}\right) - 1 = \frac{x-1}{x+1}$

$1 - \frac{1}{2x} = \frac{x-1}{x+1} \Rightarrow 2x(x+1) - (x+1) = 2x(x-1)$

$2x^2 + 2x - x - 1 - 2x^2 + 2x = 0 \Rightarrow 3x = 1 \Rightarrow x = \frac{1}{3}$

Q2 **a)** $fg(x) = f(b-3x) = (b-3x)^2 + b$

Range: $(b-3x)^2 \geq 0$, so $fg(x) \geq b$

$gf(x) = g(x^2+b) = b - 3(x^2+b) = -3x^2 - 2b$

Range: $-3x^2 \leq 0$, so $gf(x) \leq -2b$

b) $gf(2) = -3(2)^2 - 2b = -8 \Rightarrow 12 + 2b = 8 \Rightarrow b = -2$

$fg(2) = (-2 - 3(2))^2 - 2 = (-8)^2 - 2 = 64 - 2 = 62$

4. Inverse Functions

Exercise 4.1 — Inverse functions and their graphs

Q1 **a)** Yes, as the graph shows a one-to-one function.

b) No, as it is a many-to-one map, and many-to-one functions do not have inverse functions.

Q2 **a)** No, as $\sin x$ is a many-to-one function over the domain $x \in \mathbb{R}$.

b) No, as it is a many-to-one function over the domain $x \in \mathbb{R}$.

c) Yes, as it is a one-to-one function for the domain $\{x : x \geq 4\}$.

Q3 **a)** $f(x) = 3x + 4$ with domain $x \in \mathbb{R}$ has a range $f(x) \in \mathbb{R}$.

Replace f(x) with y: $y = 3x + 4$

Rearrange: $x = \frac{y-4}{3}$

Replace with $f^{-1}(x)$ and x: $f^{-1}(x) = \frac{x-4}{3}$.

The domain of $f^{-1}(x)$ is $x \in \mathbb{R}$ and the range is $f^{-1}(x) \in \mathbb{R}$.

b) $f(x) = 5(x-2)$ with domain $x \in \mathbb{R}$ has a range $f(x) \in \mathbb{R}$.

Replace f(x) with y: $y = 5(x-2)$

Rearrange: $x = \frac{y}{5} + 2$

Replace with $f^{-1}(x)$ and x: $f^{-1}(x) = \frac{x}{5} + 2$.

The domain of $f^{-1}(x)$ is $x \in \mathbb{R}$ and the range is $f^{-1}(x) \in \mathbb{R}$.

c) $f(x) = \frac{1}{x+2}$ with domain $x > -2$ has a range $f(x) > 0$.

Replace f(x) with y: $y = \frac{1}{x+2}$

Rearrange: $x = \frac{1}{y} - 2$

Replace with $f^{-1}(x)$ and x: $f^{-1}(x) = \frac{1}{x} - 2$.

The domain of $f^{-1}(x)$ is the range of f(x): $x > 0$.

The range of $f^{-1}(x)$ is the domain of f(x): $f^{-1}(x) > -2$.

d) $f(x) = x^2 + 3$ with domain $\{x : x > 0\}$ has a range $\{f(x) : f(x) > 3\}$.

Replace f(x) with y: $y = x^2 + 3$

Rearrange: $x = \sqrt{y-3}$

Replace with $f^{-1}(x)$ and x: $f^{-1}(x) = \sqrt{x-3}$.

The domain of $f^{-1}(x)$ is the range of f(x): $\{x : x > 3\}$.

The range of $f^{-1}(x)$ is the domain of f(x): $\{f^{-1}(x) : f^{-1}(x) > 0\}$.

Q4 **a)** $f(x) = \frac{3x}{x+1}$ with domain $x > -1$ has a range $f(x) < 3$ — you can work this out by sketching the graph. If you consider what happens as $x \to \infty$ you'll see that f(x) approaches 3.

Replace f(x) with y: $y = \frac{3x}{x+1}$

Rearrange: $y(x+1) = 3x \Rightarrow yx + y = 3x$

$\Rightarrow (3-y)x = y \Rightarrow x = \frac{y}{3-y}$

Replace with $f^{-1}(x)$ and x: $f^{-1}(x) = \frac{x}{3-x}$.

The domain of $f^{-1}(x)$ is the range of f(x): $x < 3$.

The range of $f^{-1}(x)$ is the domain of f(x): $f^{-1}(x) > -1$.

b) $f^{-1}(2) = \frac{2}{3-2} = 2$

c) $f^{-1}\left(\frac{1}{2}\right) = \frac{\frac{1}{2}}{3-\frac{1}{2}} = \frac{1}{6-1} = \frac{1}{5}$

Q5 a) $f(x) = \frac{x-4}{x+3}$ with domain $\{x : x > -3\}$ has a range $\{f(x) : f(x) < 1\}$ — again, sketching the graph will help. If you consider what happens as $x \to \infty$ you'll see that f(x) approaches 1.

Replace f(x) with y: $y = \frac{x-4}{x+3}$

Rearrange: $y(x+3) = x - 4 \Rightarrow yx + 3y = x - 4$

$x(1-y) = 3y + 4 \Rightarrow x = \frac{3y+4}{1-y}$

Replace with $f^{-1}(x)$ and x: $f^{-1}(x) = \frac{3x+4}{1-x}$.

The domain of $f^{-1}(x)$ is the range of f(x): $\{x : x < 1\}$.
The range of $f^{-1}(x)$ is the domain of f(x): $\{f(x) : f^{-1}(x) > -3$.

b) $f^{-1}(0) = \frac{3(0)+4}{1-0} = 4$

c) $f^{-1}\left(-\frac{2}{5}\right) = \frac{3\left(-\frac{2}{5}\right)+4}{1+\frac{2}{5}} = \frac{-6+20}{5+2} = \frac{14}{7} = 2$

Q6 a) f(x) has domain $x > 3$, and range $f(x) \in \mathbb{R}$.
So $f^{-1}(x)$ has domain $x \in \mathbb{R}$ and range $f^{-1}(x) > 3$.

b) f(x) has domain of $1 \le x \le 7$, which will give a range of $(4(1) - 2) \le f(x) \le (4(7) - 2)$, $2 \le f(x) \le 26$.
So $f^{-1}(x)$ has domain $2 \le x \le 26$ and range $1 \le f^{-1}(x) \le 7$.

c) f(x) has domain $\{x : x < 2\}$ and range $\{f(x) : f(x) < 1\}$.
So $f^{-1}(x)$ has domain $\{x : x < 1\}$ and range $\{f^{-1}(x) : f^{-1}(x) < 2\}$.

d) f(x) has domain $\{x : x \ge 2\}$ and range $\{f(x) : f(x) \ge 3\}$.
So $f^{-1}(x)$ has domain $\{x : x \ge 3\}$ and range $\{f^{-1}(x) : f^{-1}(x) \ge 2\}$.

e) f(x) has domain $0° \le x < 90°$ and range $f(x) \ge 0$.
So $f^{-1}(x)$ has domain $x \ge 0$ and range $0° \le f^{-1}(x) < 90°$.

f) f(x) has domain $\{x : 3 \le x \le 4\}$, which will give a range of $\{f(x) : \ln 9 \le f(x) \le \ln 16\}$.
So $f^{-1}(x)$ has domain $\{x : \ln 9 \le x \le \ln 16\}$ and range $\{f^{-1}(x) : 3 \le f^{-1}(x) \le 4\}$.

Q7 a) $y = e^{x+1} \Rightarrow \ln y = x + 1 \Rightarrow (\ln y) - 1 = x$
So $f^{-1}(x) = (\ln x) - 1$
f(x) has domain $x \in \mathbb{R}$ and range $f(x) > 0$, so $f^{-1}(x)$ has domain $x > 0$ and range $f^{-1}(x) \in \mathbb{R}$.

b) $y = x^3 \Rightarrow \sqrt[3]{y} = x$
So $f^{-1}(x) = \sqrt[3]{x}$
f(x) has both domain $x < 0$ and range $f(x) < 0$, so $f^{-1}(x)$ also has domain $x < 0$ and range $f^{-1}(x) < 0$.

c) $y = 2 - \log_2(x) \Rightarrow \log_2(x) = 2 - y \Rightarrow x = 2^{2-y}$
So $f^{-1}(x) = 2^{2-x}$
f(x) has domain $x \ge 1$, which means that $\log_2(x) \ge 0$.
So the range of f is $f(x) \le 2$.
So $f^{-1}(x)$ has domain $x \le 2$ and range $f^{-1}(x) \ge 1$.

d) $y = \frac{1}{x-2} \Rightarrow \frac{1}{y} = x - 2 \Rightarrow \frac{1}{y} + 2 = x$
So $f^{-1}(x) = \frac{1}{x} + 2$
f(x) has domain $\{x : x \ne 2\}$ and range $\{f(x) : f(x) \ne 0\}$, so $f^{-1}(x)$ has domain $\{x : x \ne 0\}$ and range $\{f^{-1}(x) : f^{-1}(x) \ne 2\}$.
You can see the range of f(x) more easily from the graph of y = f(x) — there is a horizontal asymptote at y = 0.

Q8

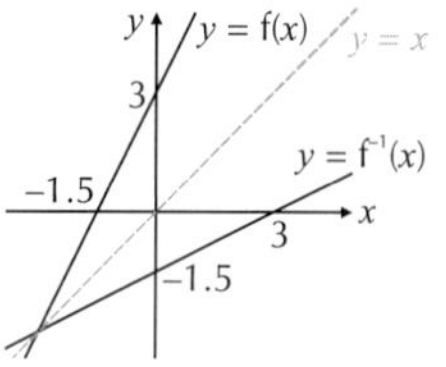

Q9 a)

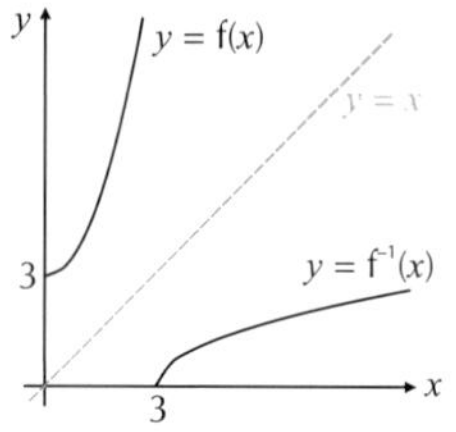

b) f(x) has a domain of $x > 0$, giving a range of $f(x) > 3$.
So $f^{-1}(x)$ has a domain $x > 3$ and range $f^{-1}(x) > 0$.

Q10 a)

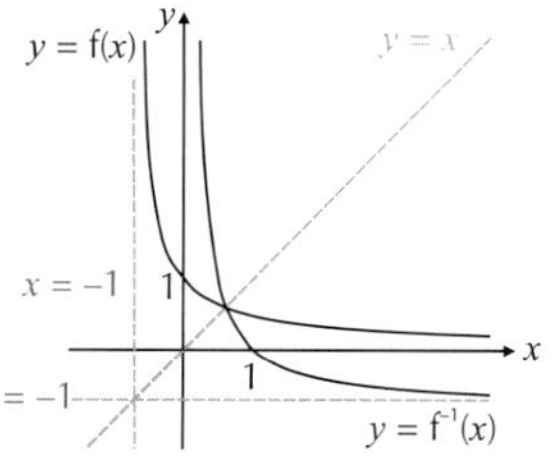

b) There is one point where the graphs intersect.

Q11 a) $f(x) = \frac{1}{x-3}$ with domain $x > 3$ has a range $f(x) > 0$.

Replace f(x) with y: $y = \frac{1}{x-3}$

Rearrange: $x = \frac{1}{y} + 3$

Replace with $f^{-1}(x)$ and x: $f^{-1}(x) = \frac{1}{x} + 3$.

The domain of $f^{-1}(x)$ is the range of f(x): $x > 0$.
The range of $f^{-1}(x)$ is the domain of f(x): $f^{-1}(x) > 3$.

b)

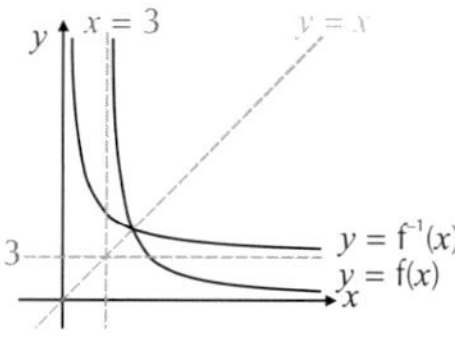

c) There is one solution as the graphs intersect once.

d) $\frac{1}{x-3} = \frac{1}{x} + 3 \Rightarrow x = (x-3) + 3x(x-3)$
$\Rightarrow x^2 - 3x - 1 = 0$
So using the quadratic formula gives $x = \frac{3+\sqrt{13}}{2}$
You can ignore the negative solutions to the quadratic equation because you're only considering the domain x > 3.

5. Modulus

Exercise 5.1 — The modulus function

Q1 a)

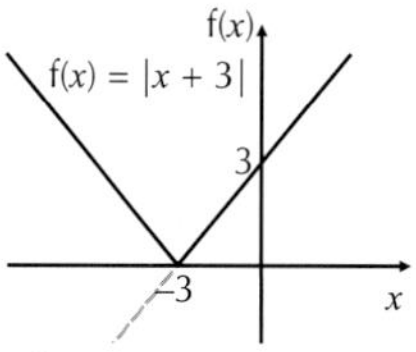

Range: $f(x) \ge 0$

b)

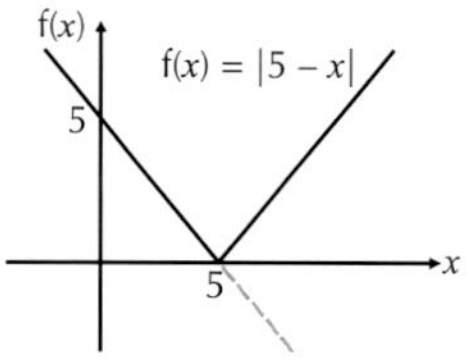

Range: $f(x) \ge 0$

c) 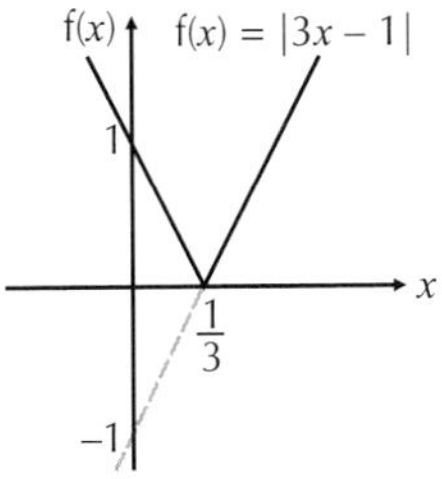

Range: $f(x) \geq 0$

d) 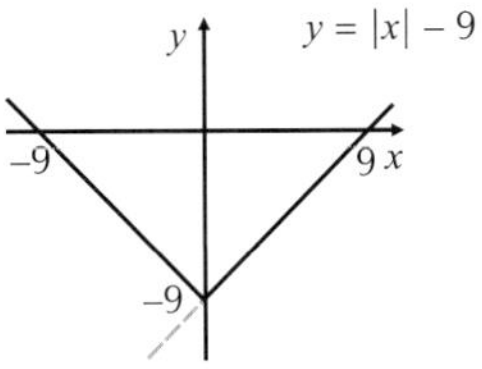

Range: $f(x) \geq -9$

e) 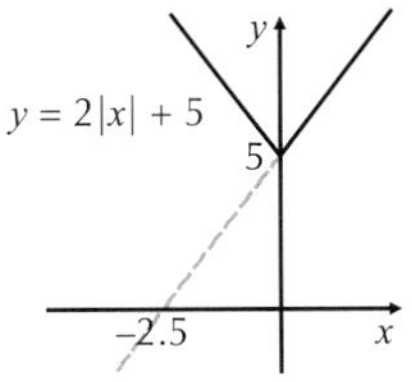

Range: $f(x) \geq 5$

Q2 a) 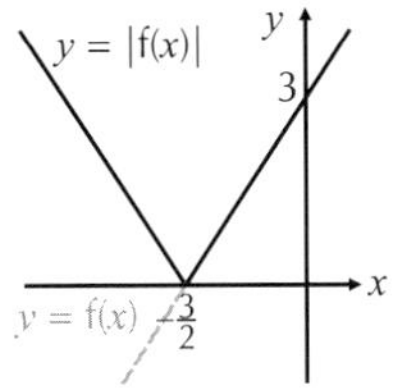

b) 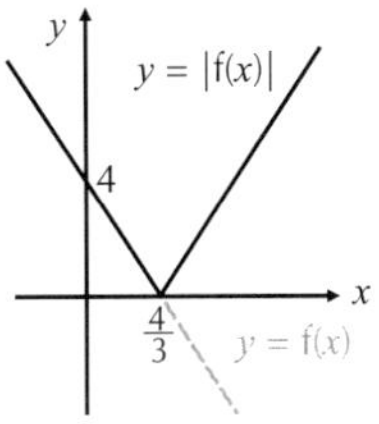

c) 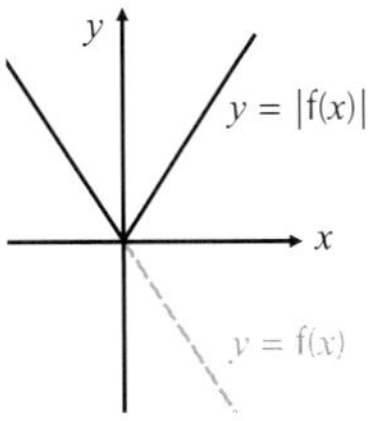

d) 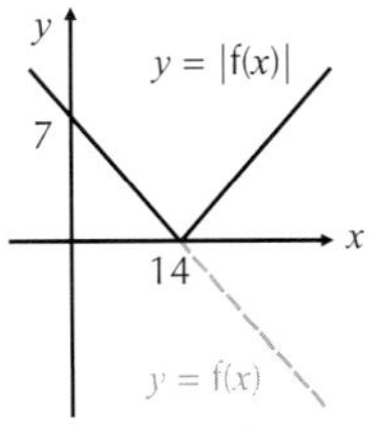

e) 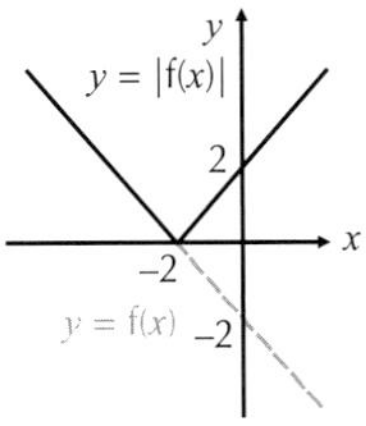

Q3 a) 2 **b)** 1 **c)** 4 **d)** 3

Q4 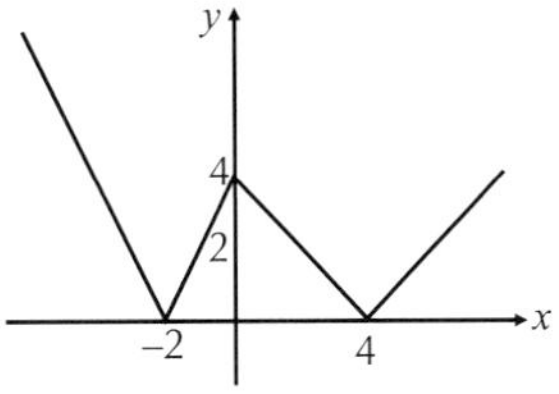

Q5 a) 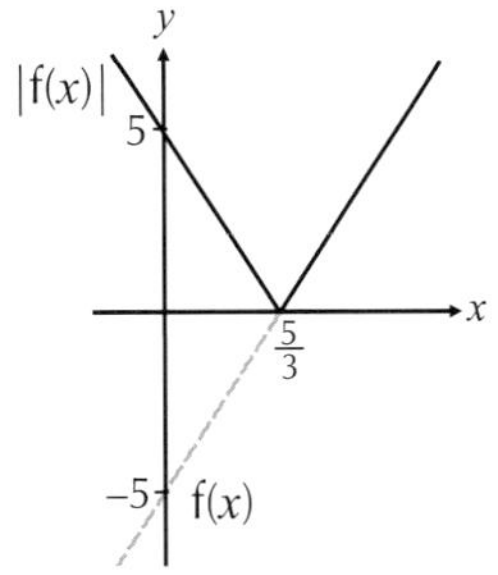

b) 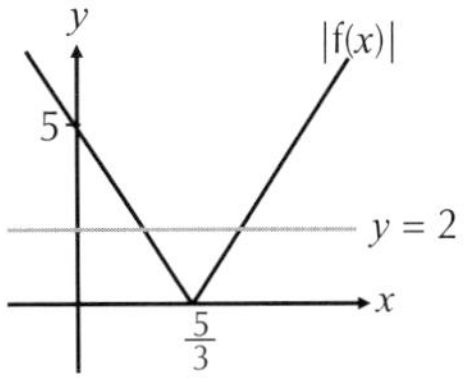

The line $y = 2$ intersects with the line $|3x - 5|$ in two places so there are 2 solutions to $|3x - 5| = 2$.

Q6 a) 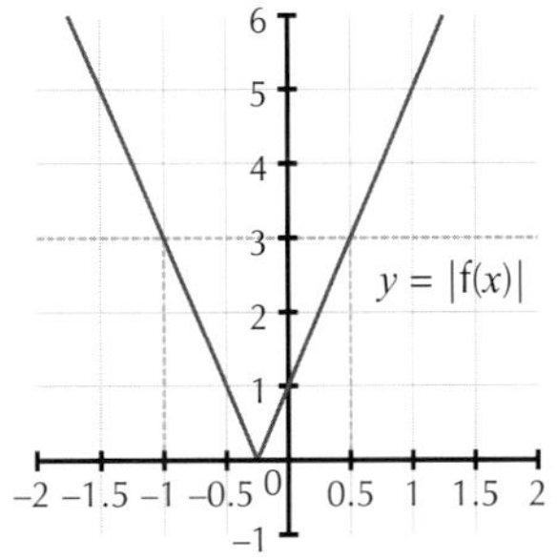

b) There are two solutions to $|4x + 1| = 3$.
Reading off the graph, these are:
$x = 0.5$ and $x = -1$.

Exercise 5.2 — Solving modulus equations and inequalities

Q1 a) 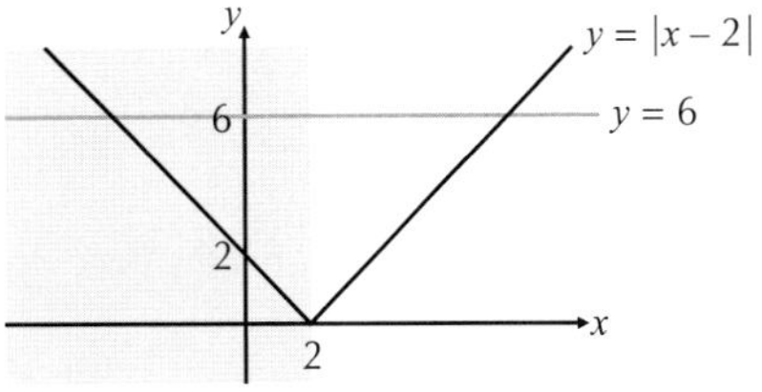

The graph shows that there are two solutions to
$|x - 2| = 6$. $x - 2 \geq 0$ for $x \geq 2$
$x - 2 < 0$ for $x < 2$ (shaded)
So there are two equations to solve:
① $x - 2 = 6 \Rightarrow x = 8$
(this is valid as it's in the range $x \geq 2$)
② (shaded) $-(x - 2) = 6 \Rightarrow x = -4$
(this is also valid as it's in the range $x < 2$)
So the two solutions are $x = 8$ and $x = -4$.

b)

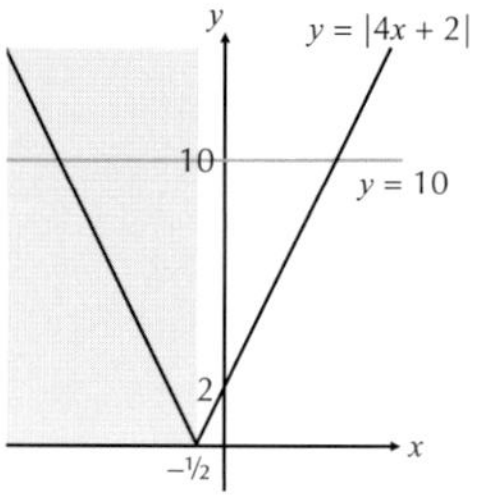

The graph shows that there are two solutions to $|4x + 2| = 10$. $\quad 4x + 2 \geq 0$ for $x \geq -\frac{1}{2}$

$4x + 2 < 0$ for $x < -\frac{1}{2}$ (shaded)

So there are two equations to solve:

① $4x + 2 = 10 \Rightarrow x = 2$

(this is valid as it's in the range $x \geq -\frac{1}{2}$)

② (shaded) $-(4x + 2) = 10 \Rightarrow x = -3$

(this is also valid as it's in the range $x < -\frac{1}{2}$)

So the two solutions are $x = 2$ and $x = -3$.

c) Rearranging the equation gives $|3x - 4| = 1$.

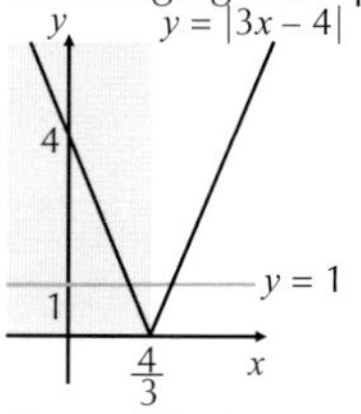

The graph shows that there are two solutions to $|3x - 4| = 1$. $\quad 3x - 4 \geq 0$ for $x \geq \frac{4}{3}$

$3x - 4 < 0$ for $x < \frac{4}{3}$ (shaded)

So there are two equations to solve:

① $3x - 4 = 1 \Rightarrow x = \frac{5}{3}$

(this is valid as it's in the range $x \geq \frac{4}{3}$)

② (shaded) $-(3x - 4) = 1 \Rightarrow x = 1$

(this is also valid as it's in the range $x < \frac{4}{3}$)

So the two solutions are $x = \frac{5}{3}$ and $x = 1$.

Q2 $|x| = 5 \Rightarrow x = 5$ or $x = -5$

If $x = 5$, $|3x + 2| = |15 + 2| = 17$

If $x = -5$, $|3x + 2| = |-15 + 2| = |-13| = 13$

Q3 a)

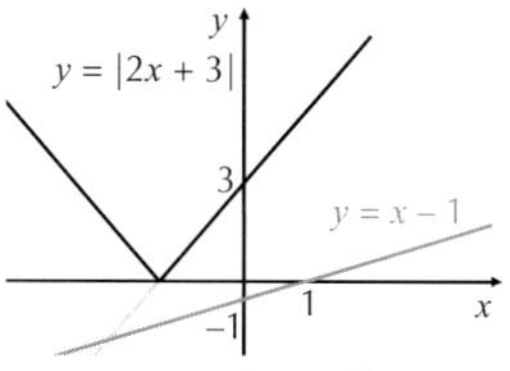

b) You can see from the graph that the lines do not intersect, so there are no solutions to $|f(x)| = g(x)$.

Q4 a)

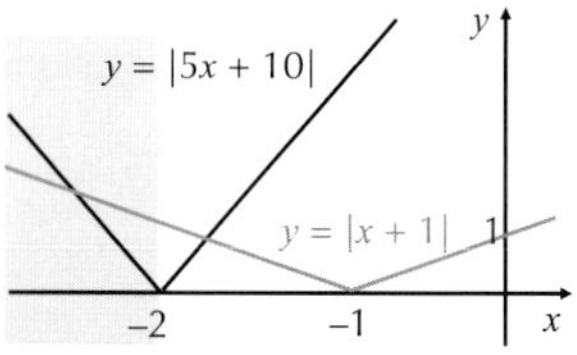

b) There are two solutions to $|f(x)| = |g(x)|$, one where $-2 < x < -1$ and one where $x < -2$ (shaded).

So there are two equations to solve:

① $5x + 10 = -(x + 1) \Rightarrow 6x = -11 \Rightarrow x = -\frac{11}{6}$

(this is valid as it's in the range $-2 < x < -1$)

② (shaded) $-(5x + 10) = -(x + 1)$

$\Rightarrow -4x = 9 \Rightarrow x = -\frac{9}{4}$

(this is also valid as it's in the range $x < -2$)

So the two solutions to $|f(x)| = |g(x)|$ are $x = -\frac{11}{6}$ and $x = -\frac{9}{4}$.

Q5 a)

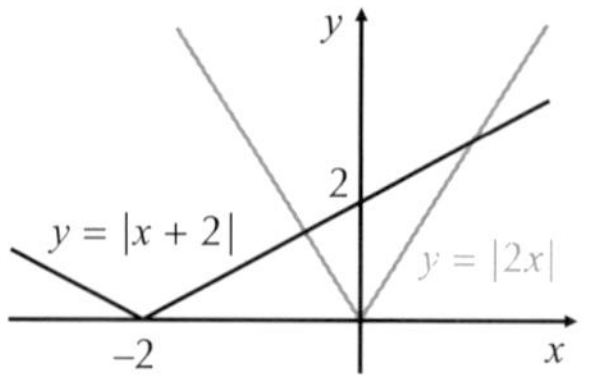

The graph shows that there are two solutions, one where both $x + 2$ and $2x$ are positive, and one where $x + 2$ is positive but $2x$ is negative.

So there are two equations to solve:

① $x + 2 = 2x \Rightarrow x = 2$

② $x + 2 = -2x \Rightarrow x = -\frac{2}{3}$

So the two solutions are $x = 2$ and $x = -\frac{2}{3}$.

Or using the algebraic method, solve:

$(x + 2)^2 = (2x)^2 \Rightarrow x^2 + 4x + 4 = 4x^2$

$\Rightarrow 3x^2 - 4x - 4 = 0 \Rightarrow (3x + 2)(x - 2) = 0$

So $x = 2$ and $x = -\frac{2}{3}$.

b)

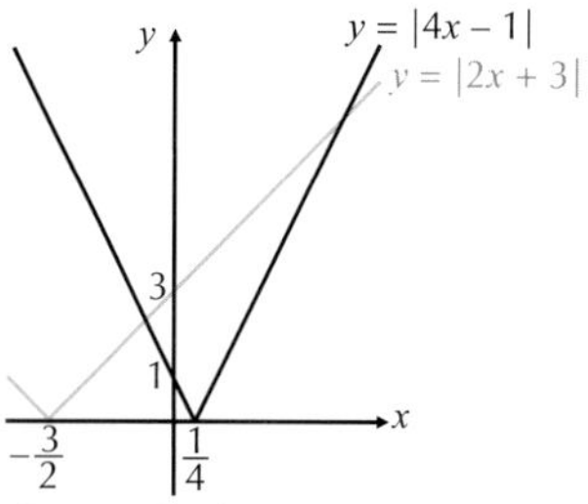

The graph shows that there are two solutions, one where both $2x + 3$ and $4x - 1$ are positive, and one where $2x + 3$ is positive but $4x - 1$ is negative.

So there are two equations to solve:

① $4x - 1 = 2x + 3 \Rightarrow x = 2$

② $-(4x - 1) = 2x + 3 \Rightarrow x = -\frac{1}{3}$

So the two solutions are $x = 2$ and $x = -\frac{1}{3}$.

Or using the algebraic method, solve:

$(2x + 3)^2 = (4x - 1)^2 \Rightarrow 4x^2 + 12x + 9 = 16x^2 - 8x + 1$

$\Rightarrow 12x^2 - 20x - 8 = 0 \Rightarrow 3x^2 - 5x - 2 = 0$

$(3x + 1)(x - 2) = 0$

So $x = 2$ and $x = -\frac{1}{3}$.

c) Solving algebraically:

$|3x - 6| = |10 - 5x| \Rightarrow (3x - 6)^2 = (10 - 5x)^2$

$9x^2 - 36x + 36 = 100 - 100x + 25x^2$

$16x^2 - 64x + 64 = 0$

$\Rightarrow x^2 - 4x + 4 = 0 \Rightarrow (x - 2)^2 = 0$

So there is only one solution at $x = 2$.

If you wanted to solve this graphically, you would see that the graphs look like this:

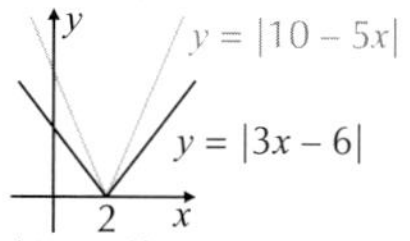

Q6 $|4x + 1| = 3 \Rightarrow 4x + 1 = 3$ or $-(4x + 1) = 3$

$\Rightarrow x = \frac{1}{2}$ or $x = -1$

If $x = \frac{1}{2}$, $2|x - 1| + 3 = 2\left|-\frac{1}{2}\right| + 3 = 1 + 3 = 4$

If $x = -1$, $2|x - 1| + 3 = 2|-2| + 3 = 4 + 3 = 7$

Q7 a) $|x| < 8 \Rightarrow -8 < x < 8$

b) $|x| \geq 5 \Rightarrow x \leq -5$ and $x \geq 5$

c) $|2x| > 12 \Rightarrow 2x < -12$ and $2x > 12$
$\Rightarrow x < -6$ and $x > 6$

d) $|4x + 2| \leq 6 \Rightarrow -6 \leq 4x + 2 \leq 6$
$\Rightarrow -8 \leq 4x \leq 4 \Rightarrow -2 \leq x \leq 1$

e) $3 \geq |3x - 3| \Rightarrow -3 \leq 3x - 3 \leq 3$
$\Rightarrow 0 \leq 3x \leq 6 \Rightarrow 0 \leq x \leq 2$

f) $6 - 2|x + 4| < 0 \Rightarrow 6 < 2|x + 4| \Rightarrow 3 < |x + 4|$
$x + 4 < -3$ and $x + 4 > 3 \Rightarrow x < -7$ and $x > -1$

g)

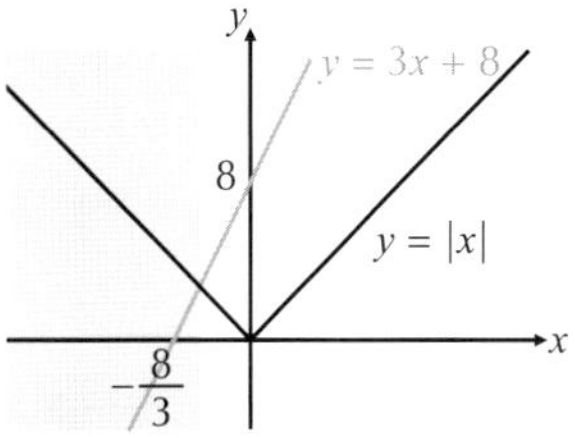

The shaded region shows the values of x that satisfy the inequality.
The graph shows that there is one solution to $3x + 8 = |x|$, where $x < 0$.
So the equation to solve is:
$3x + 8 = -x \Rightarrow 4x = -8 \Rightarrow x = -2$ (valid since $-2 < 0$)
So the region that satisfies the inequality is $x < -2$.

h)

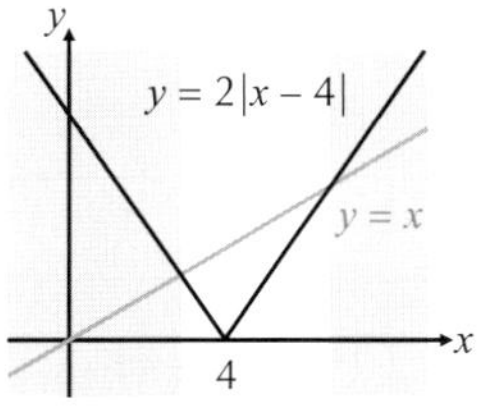

The shaded regions shows the values of x that satisfy the inequality.
The graph shows that there are two solutions to $2|x - 4| = x$, one where $x > 4$ and one where $x < 4$.
So there are two equations to solve:
① $2(x - 4) = x \Rightarrow x = 8$ (valid since $8 > 4$)
② $-2(x - 4) = x \Rightarrow x = \frac{8}{3}$ (valid since $\frac{8}{3} < 4$)
So the regions that satisfy the inequality are $x \leq \frac{8}{3}$ and $x \geq 8$.

i)

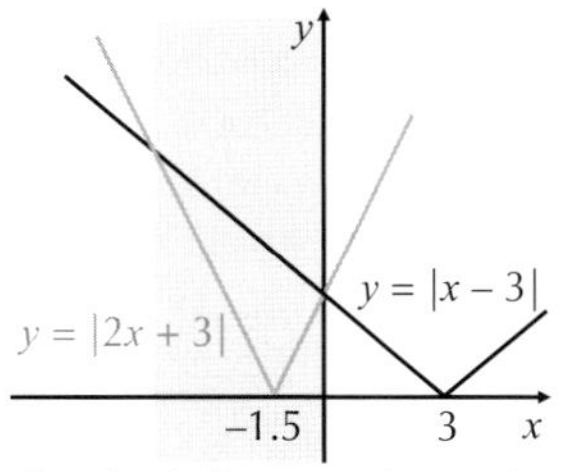

The shaded region shows the values of x that satisfy the inequality. The graph shows that there are two solutions to $|x - 3| = |2x + 3|$, one where both $(x - 3)$ and $(2x + 3)$ are negative (i.e. $x < -1.5$) and one where $(x - 3)$ is negative and $(2x + 3)$ is positive (i.e. $-1.5 < x < 3$).
So there are two equations to solve:
① $-(x - 3) = -(2x + 3) \Rightarrow -x + 3 = -2x - 3$
$\Rightarrow x = -6$ (valid since $-6 < -1.5$)
② $-(x - 3) = 2x + 3 \Rightarrow -x + 3 = 2x + 3$
$\Rightarrow x = 0$ (valid since $-1.5 < 0 < 3$)
So the region that satisfies the inequality is $-6 \leq x \leq 0$.

Q8

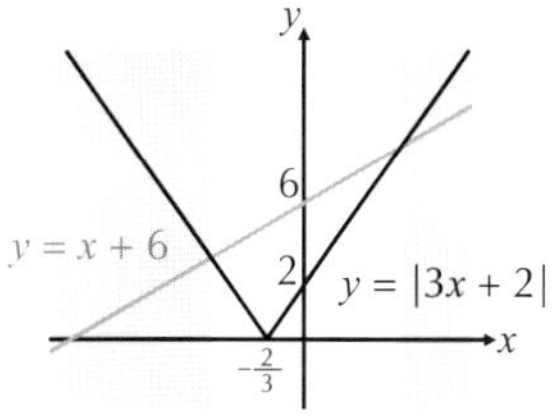

The shaded regions shows the values of x that satisfy the inequality.
The graph shows that there are two solutions to $x + 6 = |3x + 2|$, one where $(3x + 2) > 0$ $\left(x > -\frac{2}{3}\right)$ and one where $(3x + 2) < 0$ $\left(x < -\frac{2}{3}\right)$.
So there are two equations to solve:
① $x + 6 = 3x + 2 \Rightarrow 4 = 2x$
$\Rightarrow x = 2$ (valid since $2 > -\frac{2}{3}$)
② $x + 6 = -3x - 2 \Rightarrow 4x = -8$
$\Rightarrow x = -2$ (valid since $-2 < -\frac{2}{3}$)
So the regions that satisfy the inequality are $x \geq 2$ and $x \leq -2$.
In set notation, this is $\{x : x \geq 2\} \cup \{x : x \leq -2\}$

Q9 $|1 + 2x| \leq 3 \Rightarrow -3 \leq 1 + 2x \leq 3$
$\Rightarrow -4 \leq 2x \leq 2 \Rightarrow -2 \leq x \leq 1$
Draw the graph of $y = |5x + 4|$ for these values:

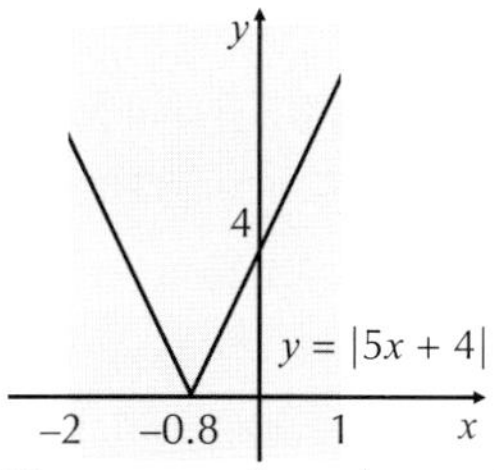

You can see from the graph that the minimum possible value of $|5x + 4|$ is 0, and the maximum is when $x = 1$, i.e. $|5x + 4| = |5 + 4| = |9| = 9$.
So the possible values are $0 \leq |5x + 4| \leq 9$

6. Transformations of Graphs

Exercise 6.1 — Transformations of graphs

Q1

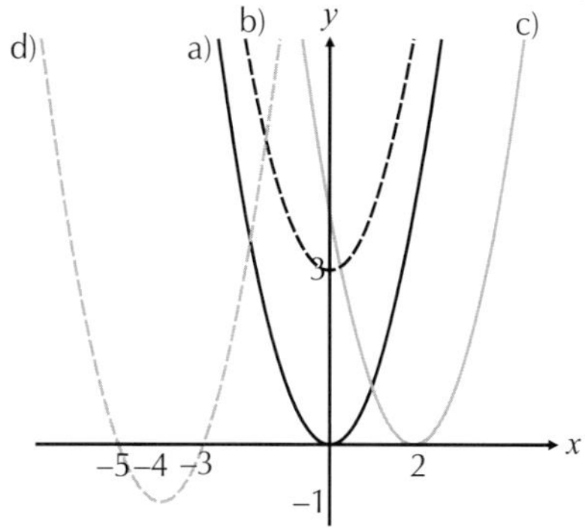

a) Turning point at (0, 0)

b) Turning point at (0, 3), translation vector $\begin{pmatrix} 0 \\ 3 \end{pmatrix}$

c) Turning point at (2, 0), translation vector $\begin{pmatrix} 2 \\ 0 \end{pmatrix}$

d) Turning point at (–4, –1), translation vector $\begin{pmatrix} -4 \\ -1 \end{pmatrix}$

Q2 **a)**

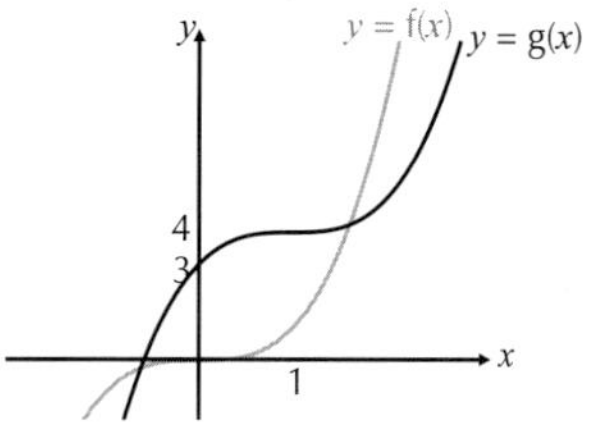

b) $\begin{pmatrix} 1 \\ 4 \end{pmatrix}$

c) $y = (x - 1)^3 + 4$ $(= x^3 - 3x^2 + 3x + 3)$

Q3

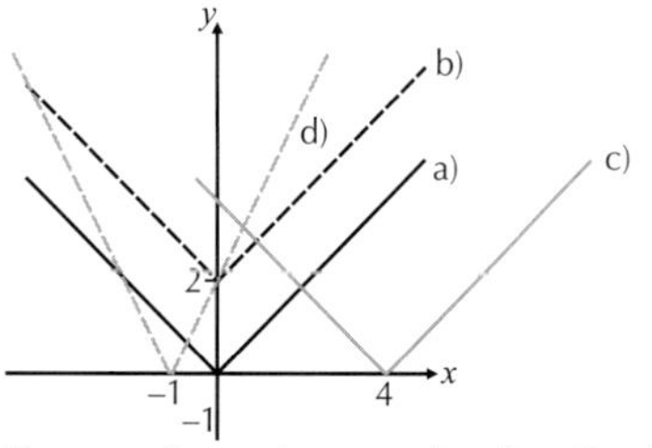

The transformations can be described as follows:

b) A translation of 2 up / by the column vector $\begin{pmatrix} 0 \\ 2 \end{pmatrix}$.

c) A translation of 4 right / by the column vector $\begin{pmatrix} 4 \\ 0 \end{pmatrix}$.

d) A translation of 1 left / by the column vector $\begin{pmatrix} -1 \\ 0 \end{pmatrix}$ and a stretch vertically by a scale factor of 2.

Q4

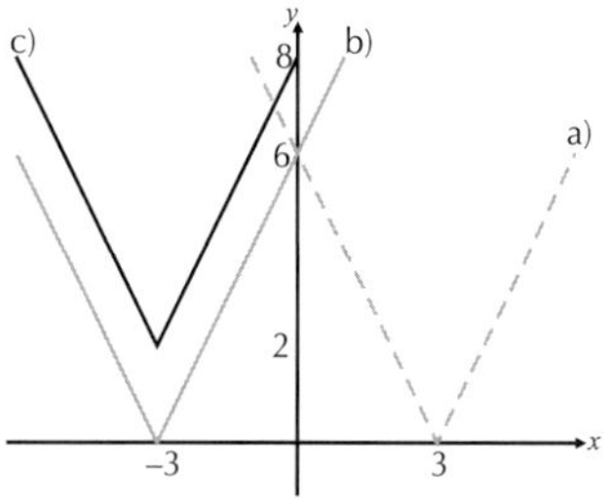

Q5

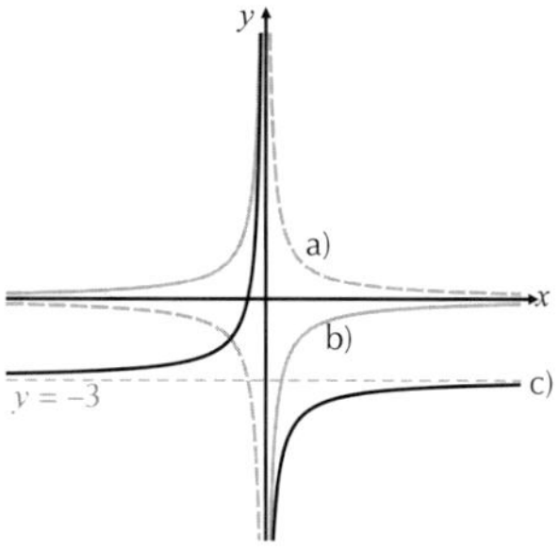

Q6

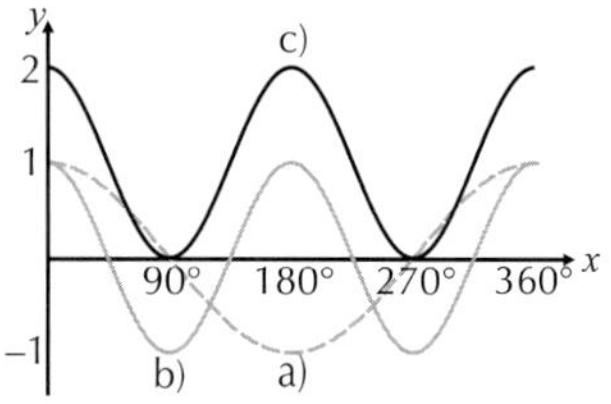

d) The minimum points on c) are (90°, 0) and (270°, 0).

Q7 The maximum value of sin x is 1, and the minimum is –1:

Transformed Function	New equation	Max value	Min value
$f(x) + 2$	$\sin x + 2$	3	1
$f(x - 90°)$	$\sin(x - 90°)$	1	–1
$f(3x)$	$\sin 3x$	1	–1
$4f(x)$	$4 \sin x$	4	–4

Q8 The point of inflection of x^3 is at (0, 0), so:

Transformed Function	New equation	Coordinates of point of inflection
$f(x) + 1$	$x^3 + 1$	(0, 1)
$f(x - 2)$	$(x - 2)^3$	(2, 0)
$-f(x) - 3$	$-x^3 - 3$	(0, –3)
$f(-x) + 4$	$-x^3 + 4$	(0, 4)

Q9 **a)**

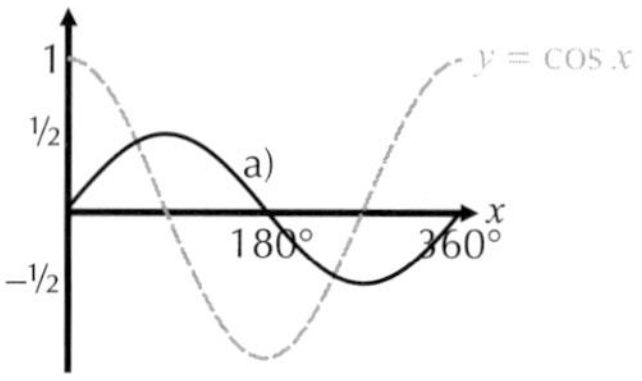

b) $y = \frac{1}{2}\cos(x - 90°)$ $(= \frac{1}{2}\sin x)$

Q10

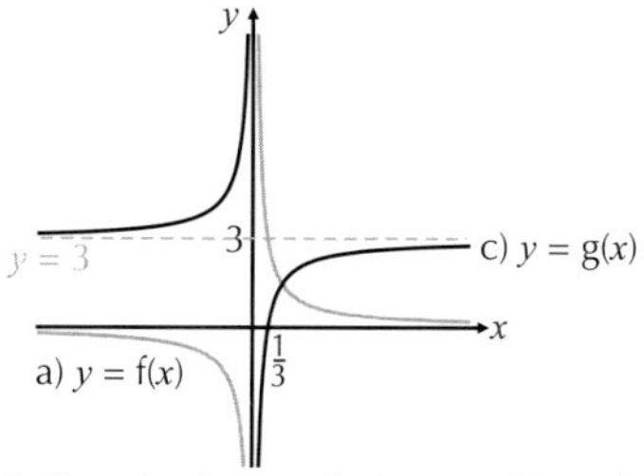

b) Reflect in the y-axis (or x-axis) and translate by $\begin{pmatrix} 0 \\ 3 \end{pmatrix}$ (3 upwards).

Q11

Original graph	New graph	Sequence of transformations
$y = x^3$	$y = (x - 4)^3 + 5$	Translate by $\begin{pmatrix} 4 \\ 5 \end{pmatrix}$ i.e. 4 right and 5 up.
$y = 4^x$	$y = 4^{3x} - 1$	Stretch horizontally by a factor of $\frac{1}{3}$ and translate by $\begin{pmatrix} 0 \\ -1 \end{pmatrix}$ i.e. 1 down.
$y = \lvert x + 1 \rvert$	$y = 1 - \lvert 2x + 1 \rvert$	Stretch horizontally by a factor of $\frac{1}{2}$, reflect in the x-axis and translate by $\begin{pmatrix} 0 \\ 1 \end{pmatrix}$ i.e. 1 up.
$y = \sin x$	$y = -3 \sin 2x + 1$	Stretch horizontally by a factor of $\frac{1}{2}$, stretch vertically by a factor of 3, reflect in the x-axis and translate by $\begin{pmatrix} 0 \\ 1 \end{pmatrix}$ i.e. 1 up.

Q12 **a)** $y = 2x^2 - 4x + 6 = 2[x^2 - 2x + 3] = 2[(x - 1)^2 + 2]$

b) Translate by $\begin{pmatrix} 1 \\ 2 \end{pmatrix}$ i.e. 1 right, then 2 up, then stretch vertically by a factor of 2.

c)

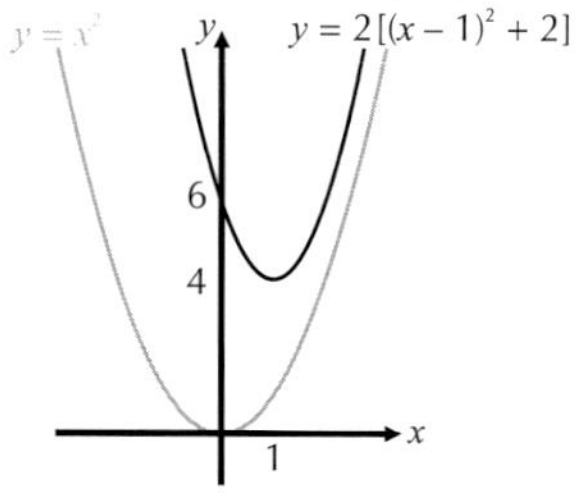

d) The minimum point is at (1, 4).
You can work this out by doing the transformations on the minimum point of the graph $y = x^2$ (which is (0, 0)).

Q13 a) Stretch horizontally, scale factor $\frac{1}{3}$, then stretch vertically, scale factor 4.

b) Stretch horizontally, scale factor $\frac{1}{2}$, then reflect in the x-axis, then translate by $\begin{pmatrix}0\\4\end{pmatrix}$ i.e. 4 up.

c) Translate by $\begin{pmatrix}60°\\0\end{pmatrix}$ i.e. 60° right, then stretch vertically, scale factor 2.

Q14 a)

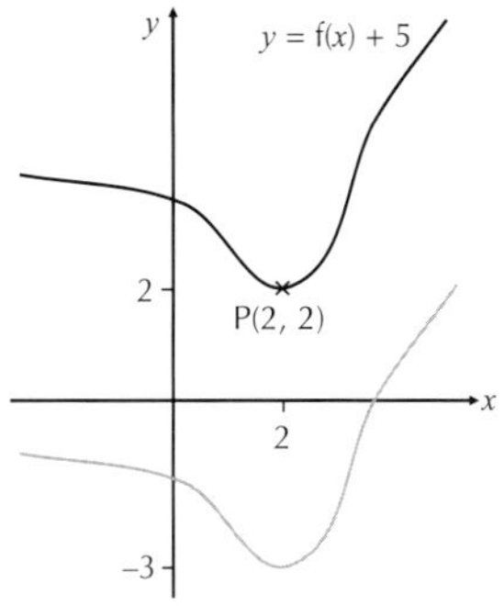

b)

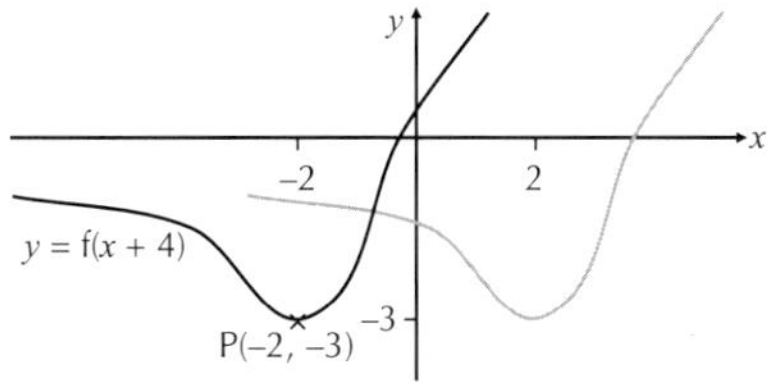

c)

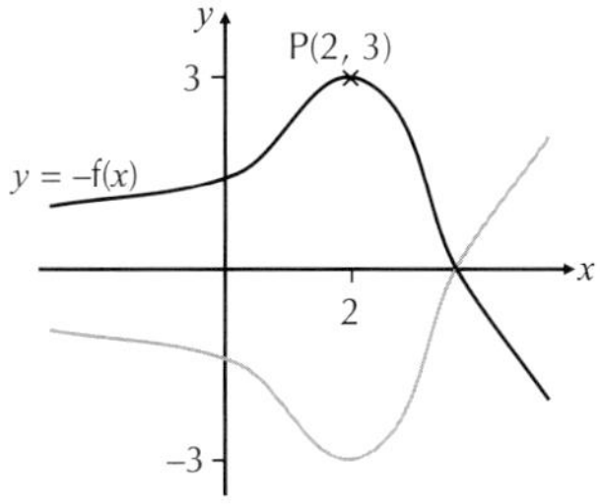

Q15 a)

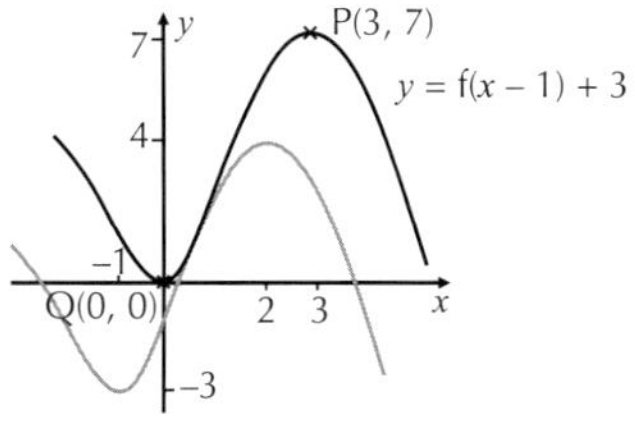

b)

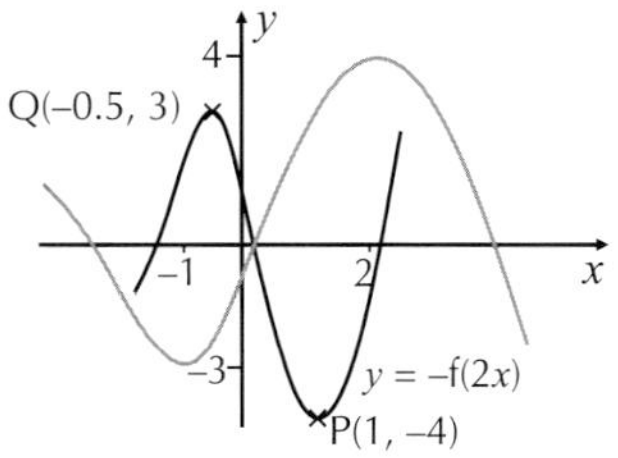

c)

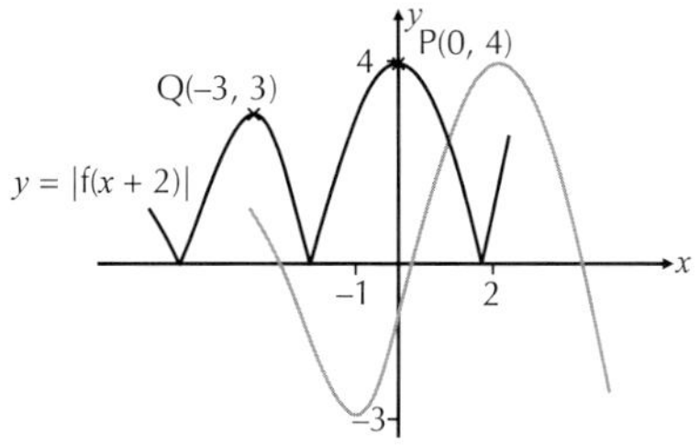

7. Partial Fractions

Exercise 7.1 — Expressing in partial fractions

Q1 $\frac{3x+3}{(x-1)(x-4)} \equiv \frac{A}{(x-1)} + \frac{B}{(x-4)}$

$\Rightarrow \frac{3x+3}{(x-1)(x-4)} \equiv \frac{A(x-4)+B(x-1)}{(x-1)(x-4)}$

$\Rightarrow \quad 3x+3 \equiv A(x-4)+B(x-1)$

Substitution: $x = 4 \Rightarrow 15 = 3B \Rightarrow B = 5$

$x = 1 \Rightarrow 6 = -3A \Rightarrow A = -2$

This gives: $\frac{3x+3}{(x-1)(x-4)} \equiv -\frac{2}{(x-1)} + \frac{5}{(x-4)}$

Q2 $\frac{5x-1}{x(2x+1)} \equiv \frac{A}{x} + \frac{B}{(2x+1)}$

$\Rightarrow 5x - 1 \equiv A(2x+1) + Bx$

Equating coefficients: x terms: $5 = 2A + B$

constants: $-1 = A$

$A = -1$, putting this into the first equation gives:
$5 = -2 + B \Rightarrow B = 7$

This gives: $\frac{5x-1}{x(2x+1)} \equiv -\frac{1}{x} + \frac{7}{(2x+1)}$

Q3 $\frac{3x-2}{x^2+x-12} \equiv \frac{3x-2}{(x+4)(x-3)} \equiv \frac{A}{(x+4)} + \frac{B}{(x-3)}$

$\Rightarrow \frac{3x-2}{(x+4)(x-3)} \equiv \frac{A(x-3)+B(x+4)}{(x+4)(x-3)}$

$\Rightarrow \quad 3x-2 \equiv A(x-3)+B(x+4)$

Equating coefficients: x terms: $3 = A + B$

constants: $-2 = -3A + 4B$

Solving simultaneously gives: $A = 2$, $B = 1$

This gives: $\frac{3x-2}{x^2+x-12} \equiv \frac{2}{(x+4)} + \frac{1}{(x-3)}$

Q4 $\frac{2}{x^2-16} \equiv \frac{2}{(x+4)(x-4)} \equiv \frac{A}{(x+4)} + \frac{B}{(x-4)}$

$\Rightarrow 2 \equiv A(x-4) + B(x+4)$

Substitution: $x = 4 \Rightarrow 2 = 8B \Rightarrow B = \frac{1}{4}$

$x = -4 \Rightarrow 2 = -8A \Rightarrow A = -\frac{1}{4}$

This gives: $\frac{2}{x^2-16} \equiv -\frac{1}{4(x+4)} + \frac{1}{4(x-4)}$

Don't worry if you get fractions for your coefficients — just put the numerator on the top of your partial fraction and the denominator on the bottom.

Q5 $x^2 - x - 6 = (x-3)(x+2)$

$\frac{5}{(x-3)(x+2)} \equiv \frac{A}{(x-3)} + \frac{B}{(x+2)}$

$\Rightarrow 5 \equiv A(x+2) + B(x-3)$

Equating coefficients: x terms: $0 = A + B$

constants: $5 = 2A - 3B$

Solving simultaneously gives: $A = 1$, $B = -1$

This gives: $\frac{5}{x^2-x-6} \equiv \frac{1}{(x-3)} - \frac{1}{(x+2)}$

Q6 $\frac{11x}{2x^2+5x-12} \equiv \frac{11x}{(2x-3)(x+4)} \equiv \frac{A}{(2x-3)} + \frac{B}{(x+4)}$

$\Rightarrow 11x \equiv A(x+4) + B(2x-3)$

Equating coefficients: x terms: $11 = A + 2B$

constants: $0 = 4A - 3B$

Solving simultaneously gives: $A = 3$, $B = 4$

This gives: $\frac{11x}{2x^2+5x-12} \equiv \frac{3}{(2x-3)} + \frac{4}{(x+4)}$

Q7 **a)** $x^3 - 9x = x(x^2 - 9) = x(x + 3)(x - 3)$

b) $\frac{12x+18}{x(x+3)(x-3)} \equiv \frac{A}{x} + \frac{B}{(x+3)} + \frac{C}{(x-3)}$

$\Rightarrow 12x + 18 \equiv A(x + 3)(x - 3) + Bx(x - 3) + Cx(x + 3)$

Substitution: $x = 0 \Rightarrow 18 = -9A \Rightarrow A = -2$

$x = -3 \Rightarrow -18 = 18B \Rightarrow B = -1$

$x = 3 \Rightarrow 54 = 18C \Rightarrow C = 3$

This gives: $\frac{12x+18}{x^3-9x} \equiv -\frac{2}{x} - \frac{1}{(x+3)} + \frac{3}{(x-3)}$

Q8 $\frac{3x+9}{x^3-36x} \equiv \frac{3x+9}{x(x^2-36)} \equiv \frac{3x+9}{x(x+6)(x-6)}$

$\frac{3x+9}{x(x+6)(x-6)} \equiv \frac{A}{x} + \frac{B}{(x+6)} + \frac{C}{(x-6)}$

$\Rightarrow 3x + 9 \equiv A(x + 6)(x - 6) + Bx(x - 6) + Cx(x + 6)$

Substitution: $x = 0 \Rightarrow 9 = -36A \Rightarrow A = -\frac{9}{36} = -\frac{1}{4}$

$x = -6 \Rightarrow -9 = 72B \Rightarrow B = -\frac{9}{72} = -\frac{1}{8}$

$x = 6 \Rightarrow 27 = 72C \Rightarrow C = \frac{27}{72} = \frac{3}{8}$

This gives: $\frac{3x+9}{x^3-36x} \equiv -\frac{1}{4x} - \frac{1}{8(x+6)} + \frac{3}{8(x-6)}$

Q9 **a)** $f(x) = x^3 - 7x - 6$

$f(-1) = -1 + 7 - 6 = 0 \Rightarrow (x + 1)$ is a factor

Once you've found one factor using the Factor Theorem you can use this method:

$(x + 1)$ is a factor of $x^3 - 7x - 6$, so:

$x^3 - 7x - 6 = (x + 1)(\qquad)$

There's an x^3 on the LHS, so there's an x^2 term in the bracket:

$x^3 - 7x - 6 = (x + 1)(x^2 \qquad)$

There's -6 on the LHS, so you need -6 in the bracket to multiply with the 1 in $(x + 1)$ and give -6:

$x^3 - 7x - 6 = (x + 1)(x^2 \qquad - 6)$

Now you've got $-7x$ on the LHS and $-6x$ (from multiplying x by -6) on the RHS. So you need $-x$ in the middle of the bracket to get another $-x$ on the RHS (from multiplying 1 by $-x$):

$x^3 - 7x - 6 = (x + 1)(x^2 - x - 6)$

When you multiply out the RHS you get the LHS (the terms in x^2 cancel), so this is the right quadratic.

Finally, factorise the quadratic:

$x^3 - 7x - 6 = (x + 1)(x^2 - x - 6) = (x + 1)(x - 3)(x + 2)$

This method looks a bit involved, but it's just laid out like this to show you what's going on — if you did it yourself you'd just write:

'$x^3 - 7x - 6 = (x + 1)(x^2 - x - 6) = (x + 1)(x - 3)(x + 2)$'

So $x^3 - 7x - 6 = (x + 1)(x - 3)(x + 2)$

Or you can keep using trial and error to find the other factors, but this could take a while:

$f(3) = 27 - 21 - 6 = 0 \Rightarrow (x - 3)$ is a factor

$f(-2) = -8 + 14 - 6 = 0 \Rightarrow (x + 2)$ is a factor

So $x^3 - 7x - 6 = (x + 1)(x - 3)(x + 2)$

b) $\frac{6x+2}{x^3-7x-6} \equiv \frac{6x+2}{(x+1)(x-3)(x+2)}$

$\frac{6x+2}{(x+1)(x-3)(x+2)} \equiv \frac{A}{(x+1)} + \frac{B}{(x-3)} + \frac{C}{(x+2)}$

$\Rightarrow 6x + 2$

$\equiv A(x - 3)(x + 2) + B(x + 1)(x + 2) + C(x + 1)(x - 3)$

Substitution: $x = -1 \Rightarrow -4 = -4A \Rightarrow A = 1$

$x = 3 \Rightarrow 20 = 20B \Rightarrow B = 1$

$x = -2 \Rightarrow -10 = 5C \Rightarrow C = -2$

This gives: $\frac{6x+2}{x^3-7x-6} \equiv \frac{1}{(x+1)} + \frac{1}{(x-3)} - \frac{2}{(x+2)}$

Q10 $\frac{6x+4}{(x+4)(x-1)(x+1)} \equiv \frac{A}{(x+4)} + \frac{B}{(x-1)} + \frac{C}{(x+1)}$

$\Rightarrow 6x + 4 \equiv A(x - 1)(x + 1) + B(x + 4)(x + 1) + C(x + 4)(x - 1)$

Substitution: $x = -4 \Rightarrow -20 = 15A \Rightarrow A = -\frac{4}{3}$

$x = 1 \Rightarrow 10 = 10B \Rightarrow B = 1$

$x = -1 \Rightarrow -2 = -6C \Rightarrow C = \frac{1}{3}$

This gives:

$\frac{6x+4}{(x+4)(x-1)(x+1)} \equiv -\frac{4}{3(x+4)} + \frac{1}{(x-1)} + \frac{1}{3(x+1)}$

Q11 $\frac{15x-27}{x^3-6x^2+3x+10} \equiv \frac{15x-27}{(x+1)(x-2)(x-5)}$

You get this by using the factor theorem, as in Q9.

$\frac{15x-27}{(x+1)(x-2)(x-5)} \equiv \frac{A}{(x+1)} + \frac{B}{(x-2)} + \frac{C}{(x-5)}$

$\Rightarrow 15x - 27$

$\equiv A(x - 2)(x - 5) + B(x + 1)(x - 5) + C(x + 1)(x - 2)$

Substitution: $x = -1 \Rightarrow -42 = 18A \Rightarrow A = -\frac{7}{3}$

$x = 2 \Rightarrow 3 = -9B \Rightarrow B = -\frac{1}{3}$

$x = 5 \Rightarrow 48 = 18C \Rightarrow C = \frac{8}{3}$

This gives:

$\frac{15x-27}{x^3-6x^2+3x+10} \equiv -\frac{7}{3(x+1)} - \frac{1}{3(x-2)} + \frac{8}{3(x-5)}$

Exercise 7.2 — Repeated factors

Q1 $\frac{3x}{(x+5)^2} \equiv \frac{A}{(x+5)} + \frac{B}{(x+5)^2}$

$\Rightarrow 3x \equiv A(x + 5) + B$

Equating coefficients: x terms: $3 = A$

constants: $0 = 5A + B \Rightarrow B = -15$

This gives: $\frac{3x}{(x+5)^2} \equiv \frac{3}{(x+5)} - \frac{15}{(x+5)^2}$

Q2 $\frac{5x+2}{x^2(x+1)} \equiv \frac{A}{x} + \frac{B}{x^2} + \frac{C}{(x+1)}$

$\Rightarrow 5x + 2 \equiv Ax(x + 1) + B(x + 1) + Cx^2$

Equating coefficients:

constants: $2 = B$

x terms: $5 = A + B \Rightarrow A = 3$

x^2 terms: $0 = A + C \Rightarrow C = -3$

This gives: $\frac{5x+2}{x^2(x+1)} \equiv \frac{3}{x} + \frac{2}{x^2} - \frac{3}{(x+1)}$

Q3 **a)** $\frac{2x-7}{(x-3)^2} \equiv \frac{A}{(x-3)} + \frac{B}{(x-3)^2}$

$\Rightarrow 2x - 7 \equiv A(x - 3) + B$

Substitution: $x = 3 \Rightarrow -1 = B$

Equating coefficients of the x terms: $2 = A$

This gives: $\frac{2x-7}{(x-3)^2} \equiv \frac{2}{(x-3)} - \frac{1}{(x-3)^2}$

b) $\frac{6x+7}{(2x+3)^2} \equiv \frac{A}{(2x+3)} + \frac{B}{(2x+3)^2}$

$\Rightarrow 6x + 7 \equiv A(2x + 3) + B$

Substitution: $x = -\frac{3}{2} \Rightarrow -2 = B$

Equating coefficients of x: $A = 3$

This gives: $\frac{3x+7}{(2x+3)^2} \equiv \frac{3}{(2x+3)} - \frac{2}{(2x+3)^2}$

c) $\frac{7x}{(x+4)^2(x-3)} \equiv \frac{A}{(x+4)} + \frac{B}{(x+4)^2} + \frac{C}{(x-3)}$

$\Rightarrow 7x \equiv A(x + 4)(x - 3) + B(x - 3) + C(x + 4)^2$

Substitution: $x = -4 \Rightarrow -28 = -7B \Rightarrow B = 4$

$x = 3 \Rightarrow 21 = 49C \Rightarrow C = \frac{3}{7}$

Equating coefficients of the x^2 terms:

$0 = A + C \Rightarrow A = -\frac{3}{7}$

This gives:

$\frac{7x}{(x+4)^2(x-3)} \equiv -\frac{3}{7(x+4)} + \frac{4}{(x+4)^2} + \frac{3}{7(x-3)}$

d) $\frac{11x-10}{x(x-5)^2} \equiv \frac{A}{x} + \frac{B}{(x-5)} + \frac{C}{(x-5)^2}$

$\Rightarrow 11x - 10 \equiv A(x-5)^2 + Bx(x-5) + Cx$

Substitution: $x = 5 \Rightarrow 45 = 5C \Rightarrow C = 9$

$x = 0 \Rightarrow -10 = 25A \Rightarrow A = -\frac{2}{5}$

Equating the coefficients of x^2:

$0 = A + B \Rightarrow B = \frac{2}{5}$

This gives: $\frac{11x-10}{x(x-5)^2} \equiv -\frac{2}{5x} + \frac{2}{5(x-5)} + \frac{9}{(x-5)^2}$

Q4 $x^3 - 10x^2 + 25x = x(x^2 - 10x + 25) = x(x-5)(x-5)$

So $\frac{5x+10}{x^3-10x^2+25x} \equiv \frac{5x+10}{x(x-5)^2} \equiv \frac{A}{x} + \frac{B}{(x-5)} + \frac{C}{(x-5)^2}$

$\Rightarrow 5x + 10 \equiv A(x-5)^2 + Bx(x-5) + Cx$

Substitution: $x = 0 \Rightarrow 10 = 25A \Rightarrow A = \frac{2}{5}$

$x = 5 \Rightarrow 35 = 5C \Rightarrow C = 7$

Equating coefficients of x^2 terms:

$0 = A + B \Rightarrow B = -\frac{2}{5}$

This gives: $\frac{5x+10}{x^3-10x^2+25x} \equiv \frac{2}{5x} - \frac{2}{5(x-5)} + \frac{7}{(x-5)^2}$

Q5 $(x-2)(x^2-4) = (x-2)(x+2)(x-2) = (x+2)(x-2)^2$

$\frac{3x+2}{(x-2)(x^2-4)} \equiv \frac{3x+2}{(x+2)(x-2)^2} \equiv \frac{A}{(x+2)} + \frac{B}{(x-2)} + \frac{C}{(x-2)^2}$

$\Rightarrow 3x + 2 \equiv A(x-2)^2 + B(x+2)(x-2) + C(x+2)$

Substitution: $x = 2 \Rightarrow 8 = 4C \Rightarrow C = 2$

$x = -2 \Rightarrow -4 = 16A \Rightarrow A = -\frac{1}{4}$

Equating coefficients of x^2 terms:

$0 = A + B \Rightarrow B = \frac{1}{4}$

This gives: $\frac{3x+2}{(x-2)(x^2-4)} \equiv -\frac{1}{4(x+2)} + \frac{1}{4(x-2)} + \frac{2}{(x-2)^2}$

Q6 $\frac{x+17}{(x+1)(x+c)^2} \equiv \frac{1}{(x+1)} - \frac{1}{(x+c)} + \frac{5}{(x+c)^2}$

$\Rightarrow x + 17 \equiv (x+c)^2 - (x+c)(x+1) + 5(x+1)$

$x + 17 \equiv x^2 + 2cx + c^2 - x^2 - cx - x - c + 5x + 5$

$x + 17 \equiv (2c - c - 1 + 5)x + (c^2 - c + 5)$

$x + 17 \equiv (c + 4)x + (c^2 - c + 5)$

Equating coefficients of x:

$1 = c + 4 \Rightarrow c = -3$

You can check this by equating constant terms:
$(-3)^2 - (-3) + 5 = 9 + 3 + 5 = 17$

Chapter 3: Trigonometry

1. Arcs and Sectors

Exercise 1.1 — Radians

Q1 **a)** π **b)** $\frac{3\pi}{4}$ **c)** $\frac{3\pi}{2}$

d) $\frac{7\pi}{18}$ **e)** $\frac{5\pi}{6}$ **f)** $\frac{5\pi}{12}$

Q2 **a)** 45° **b)** 90° **c)** 60°

d) 450° **e)** 135° **f)** 420°

Exercise 1.2 — Arc length and sector area

Q1 $s = r\theta = 6 \times 2 = 12$ cm

$A = \frac{1}{2}r^2\theta = \frac{1}{2} \times 6^2 \times 2 = 36$ cm^2

Q2 Get the angle in radians:

$46° = \frac{46 \times \pi}{180} = 0.802...$ radians

$s = r\theta = 8 \times 0.802... = 6.4$ cm (1 d.p.)

$A = \frac{1}{2}r^2\theta = \frac{1}{2} \times 8^2 \times 0.802... = 25.7$ cm^2 (1 d.p.)

Q3 $A = \frac{1}{2}r^2\theta \Rightarrow 6\pi = \frac{1}{2} \times 4^2 \times \theta$

$\Rightarrow 6\pi = 8\theta \Rightarrow \theta = \frac{6\pi}{8} = \frac{3\pi}{4}$

Q4 **a)** $s = r\theta = 5 \times 1.2 = 6$ cm

$A = \frac{1}{2}r^2\theta = \frac{1}{2} \times 5^2 \times 1.2 = 15$ cm^2

b) $s = r\theta = 4 \times 0.6 = 2.4$ cm

$A = \frac{1}{2}r^2\theta = \frac{1}{2} \times 4^2 \times 0.6 = 4.8$ cm^2

c) Get the angle in radians:

$80° = \frac{80 \times \pi}{180} = \frac{4\pi}{9}$ radians

$s = r\theta = 9 \times \frac{4\pi}{9} = 4\pi$ cm $= 12.6$ cm (3 s.f.)

$A = \frac{1}{2}r^2\theta = \frac{1}{2} \times 9^2 \times \frac{4\pi}{9} = 18\pi$ cm^2 $= 56.5$ cm^2 (3 s.f.)

d) $s = r\theta = 4 \times \frac{5\pi}{12} = \frac{5\pi}{3}$ cm $= 5.24$ cm (3 s.f.)

$A = \frac{1}{2}r^2\theta = \frac{1}{2} \times 4^2 \times \frac{5\pi}{12} = \frac{10\pi}{3}$ cm^2 $= 10.5$ cm^2 (3 s.f.)

Q5 Find the radius, r:

$A = \frac{1}{2}r^2\theta \Rightarrow 16.2 = \frac{1}{2} \times r^2 \times 0.9$

$16.2 = 0.45r^2 \Rightarrow 36 = r^2 \Rightarrow r = 6$ cm

$s = r\theta = 6 \times 0.9 = 5.4$ cm

Q6 Get the angle in radians:

$20° = \frac{20 \times \pi}{180} = \frac{\pi}{9}$ radians

$s = r\theta = 3 \times \frac{\pi}{9} = \frac{\pi}{3}$ cm

$A = \frac{1}{2}r^2\theta = \frac{1}{2} \times 3^2 \times \frac{\pi}{9} = \frac{\pi}{2}$ cm^2

Q7 Find the radius, r:

$s = r\theta \Rightarrow r = \frac{s}{\theta} = \frac{7}{1.4} = 5$ cm

$A = \frac{1}{2}r^2\theta = \frac{1}{2} \times 5^2 \times 1.4 = 17.5$ cm^2

Q8 $s = r\theta \Rightarrow r\theta = 16\pi$ ①

$A = \frac{1}{2}r^2\theta \Rightarrow \frac{1}{2}r^2\theta = 80\pi \Rightarrow r^2\theta = 160\pi$ ②

② ÷ ①: $\frac{r^2\theta}{r\theta} = \frac{160\pi}{16\pi} \Rightarrow r = 10$ cm

①: $10\theta = 16\pi \Rightarrow \theta = \frac{8\pi}{5} = 5.03$ radians (3 s.f.)

Q9 Area A $= \frac{1}{2}r^2\theta = \frac{1}{2}(2)^2\theta = 2\theta$

Area B $= \frac{1}{2}(2)^2(\pi - \theta) - \frac{1}{2}(1)^2(\pi - \theta)$

$= 2(\pi - \theta) - \frac{1}{2}(\pi - \theta) = \frac{3}{2}(\pi - \theta)$

A = B, so $2\theta = \frac{3}{2}(\pi - \theta)$

$4\theta = 3\pi - 3\theta$

$7\theta = 3\pi \Rightarrow \theta = \frac{3\pi}{7}$

The angle of the missing sector would be $\pi - \theta$, since they lie on a straight line.

2. Small Angle Approximations

Exercise 2.1 — The small angle approximations

Q1 **a)** $\sin 0.23 \approx 0.23$

From a calculator, $\sin 0.23 = 0.228$ (3 d.p.)

b) $\cos 0.01 \approx 1 - \frac{1}{2}(0.01)^2 = 1 - 0.00005 = 0.99995$

From a calculator, $\cos 0.01 = 0.999950$ (6 d.p.)

c) $\tan 0.18 \approx 0.18$

From a calculator, $\tan 0.18 = 0.182$ (3 d.p.)

Q2 $f(\theta) = \sin\theta + \cos\theta \approx \theta + 1 - \frac{1}{2}\theta^2$

a) $f(0.3) \approx 0.3 + 1 - \frac{1}{2}(0.3)^2 = 1.255$

$f(0.3) = 1.2509$ (4 d.p.)

b) $f(0.5) \approx 0.5 + 1 - \frac{1}{2}(0.5)^2 = 1.375$
$f(0.5) = 1.3570$ (4 d.p.)

c) $f(0.25) \approx 0.25 + 1 - \frac{1}{2}(0.25)^2 = 1.21875$
$f(0.25) = 1.216316$ (6 d.p.)

d) $f(0.01) \approx 0.01 + 1 - \frac{1}{2}(0.01)^2 = 1.00995$
$f(0.01) = 1.009950$ (6 d.p.)

Q3 a) $\sin\theta\cos\theta \approx \theta\left(1 - \frac{1}{2}\theta^2\right) = \theta - \frac{1}{2}\theta^3$

b) $\theta\tan 5\theta\sin\theta \approx \theta(5\theta)(\theta) = 5\theta^3$

c) $\dfrac{\sin 4\theta\cos 3\theta}{2\theta} \approx \dfrac{4\theta\left(1 - \frac{1}{2}(3\theta)^2\right)}{2\theta} = 2\left(1 - \frac{9}{2}\theta^2\right) = 2 - 9\theta^2$

d) $3\tan\theta + \cos 2\theta \approx 3\theta + 1 - \frac{1}{2}(2\theta)^2 = 1 + 3\theta - 2\theta^2$

e) $\sin\frac{1}{2}\theta - \cos\theta \approx \frac{1}{2}\theta - 1 + \frac{1}{2}\theta^2 = \frac{1}{2}(\theta^2 + \theta - 2)$

f) $\dfrac{\cos\theta - \cos 2\theta}{1 - (\cos 3\theta + 3\sin\theta\tan\theta)} \approx \dfrac{\left(1 - \frac{1}{2}\theta^2\right) - \left(1 - \frac{1}{2}(2\theta)^2\right)}{1 - \left(1 - \frac{1}{2}(3\theta)^2\right) - 3(\theta)(\theta)}$

$= \dfrac{-\frac{1}{2}\theta^2 + 2\theta^2}{\frac{9}{2}\theta^2 - 3\theta^2} = \dfrac{\frac{3}{2}\theta^2}{\frac{3}{2}\theta^2} = 1$

Q4 a) $\mathbf{d} = 6\sin\theta\,\mathbf{i} + 6(1 - \cos\theta)\mathbf{j}$

$|\mathbf{d}| = \sqrt{(6\sin\theta)^2 + (6 - 6\cos\theta)^2}$
$= \sqrt{36\sin^2\theta + 36 - 72\cos\theta + 36\cos^2\theta}$
$= \sqrt{36(\sin^2\theta + \cos^2\theta) + 36 - 72\cos\theta}$
$= \sqrt{36 + 36 - 72\cos\theta}$
$= \sqrt{72(1 - \cos\theta)} = 6\sqrt{2(1 - \cos\theta)}$ as required

b) Arc length $s = 6\theta$
$|\mathbf{d}| = 6\sqrt{2(1 - \cos\theta)} \approx 6\sqrt{2\left(1 - \left(1 - \frac{1}{2}\theta^2\right)\right)}$
$= 6\sqrt{2\left(\frac{1}{2}\theta^2\right)} = 6\sqrt{\theta^2} = 6\theta = s$ as required

3. Inverse Trig Functions

Exercise 3.1 — Arcsin, arccos and arctan

Q1 a) If $x = \arccos 1$ then $1 = \cos x$ so $x = 0$.

b) If $x = \arcsin\frac{\sqrt{3}}{2}$ then $\frac{\sqrt{3}}{2} = \sin x$ so $x = \frac{\pi}{3}$.

c) If $x = \arctan\sqrt{3}$ then $\sqrt{3} = \tan x$ so $x = \frac{\pi}{3}$.

Q2 a)

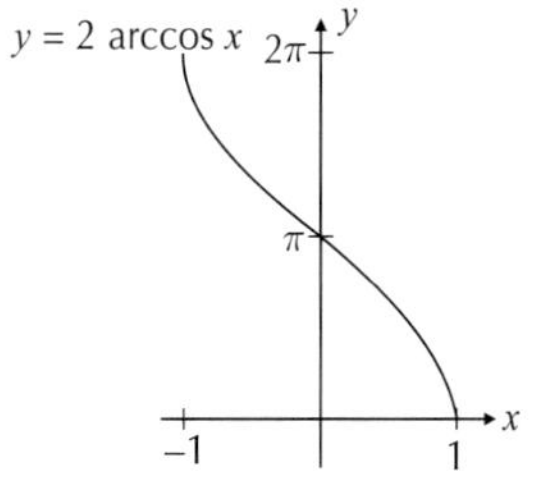

The graph is the same as y = arccos x but stretched vertically by a factor of 2, so the y-coordinates of the endpoints and y-intercept are doubled.

b)

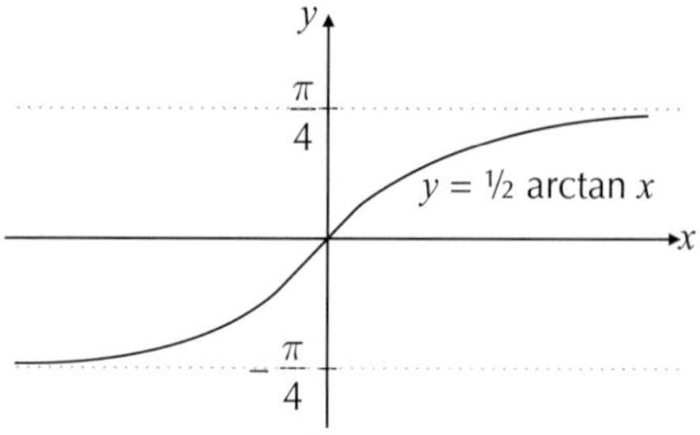

The range is $-\frac{\pi}{4} < \frac{1}{2}\arctan x < \frac{\pi}{4}$.

The graph is the same as y = arctan x but stretched vertically by a factor of $\frac{1}{2}$, so the y-coordinates of the asymptotes are halved.

Q3

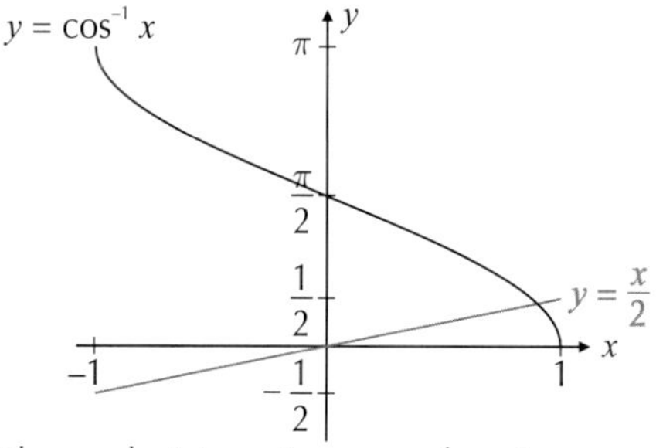

The graphs intersect once, so there is one real root of the equation $\cos^{-1} x = \frac{x}{2}$.

Q4 a) $\sin^{-1}(-1) = -\frac{\pi}{2}$.

This is one of the endpoints of the arcsin x graph.

b) To find $\cos^{-1}\left(-\frac{\sqrt{3}}{2}\right)$, first find the angle a for which $\cos a = \frac{\sqrt{3}}{2}$:

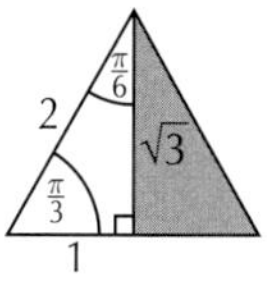

So $\cos\frac{\pi}{6} = \frac{\sqrt{3}}{2}$:

Now use the CAST diagram to find the negative solutions that lie in the domain $0 \le x \le \pi$:

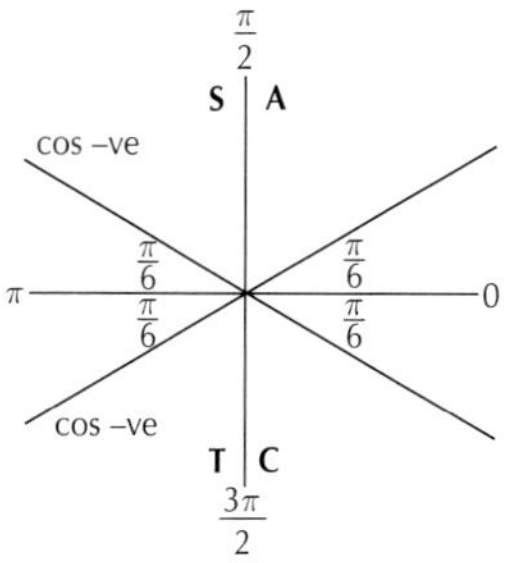

The only negative solution in that domain is $\pi - \frac{\pi}{6} = \frac{5\pi}{6}$.

So $\cos^{-1}\left(-\frac{\sqrt{3}}{2}\right) = \frac{5\pi}{6}$.

Q5 a) $\arcsin\frac{1}{2} = \frac{\pi}{6}$, so $\tan\left(\arcsin\frac{1}{2}\right) = \tan\frac{\pi}{6} = \frac{1}{\sqrt{3}}$.

b) This is just the cos function followed by its inverse function so the answer is $\frac{2\pi}{3}$.

c) $\arcsin\frac{1}{2} = \frac{\pi}{6}$, so $\cos\left(\arcsin\frac{1}{2}\right) = \cos\frac{\pi}{6} = \frac{\sqrt{3}}{2}$.

Q6 To find the inverse of the function, first write as $y = 1 + \sin 2x$, then rearrange to make x the subject:

$\sin 2x = y - 1 \Rightarrow 2x = \sin^{-1}(y - 1) \Rightarrow x = \frac{1}{2}\sin^{-1}(y - 1)$

Now replace x with $f^{-1}(x)$ and y with x:

$f^{-1}(x) = \frac{1}{2}\sin^{-1}(x - 1) = \frac{1}{2}\arcsin(x - 1)$

4. Cosec, Sec and Cot

Exercise 4.1 — Graphs of cosec, sec and cot

Q1 a)

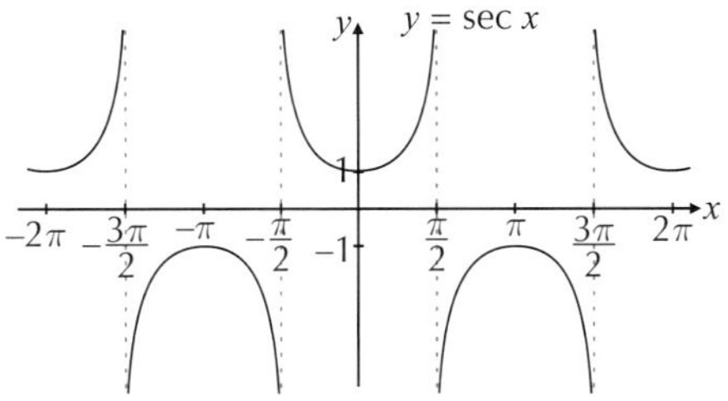

b) The minimum points are at $(-2\pi, 1)$, $(0, 1)$ and $(2\pi, 1)$.

c) The maximum points are at $(-\pi, -1)$ and $(\pi, -1)$.

d) The range is $y \geq 1$ or $y \leq -1$.
You could also say that y is undefined for $-1 < y < 1$.

Q2 a)

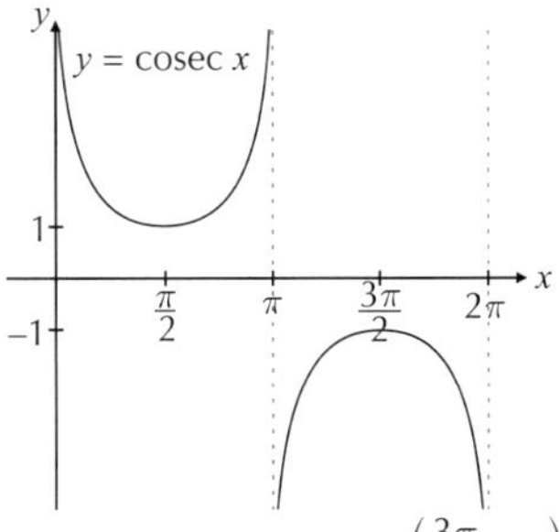

b) There is a maximum at $\left(\frac{3\pi}{2}, -1\right)$ and a minimum at $\left(\frac{\pi}{2}, 1\right)$.

c) The domain is $x \in \mathbb{R}$, $x \neq n\pi$ (where n is an integer).
The range is $y \geq 1$ or $y \leq -1$.
The domain is all real numbers except those for which cosec x is undefined (i.e. at the asymptotes).

Q3 A horizontal translation right by $\frac{\pi}{2}$ (or 90°) or a horizontal translation left by $\frac{3\pi}{2}$ (or 270°).

Q4 a) If $f(x) = \cot x$, then $y = \cot \frac{x}{4} = f\left(\frac{x}{4}\right)$.
This is a horizontal stretch scale factor 4.

b) The period of $y = \cot x$ is 180°, so the period of $y = \cot \frac{x}{4}$ is $180° \times 4 = 720°$.

c)

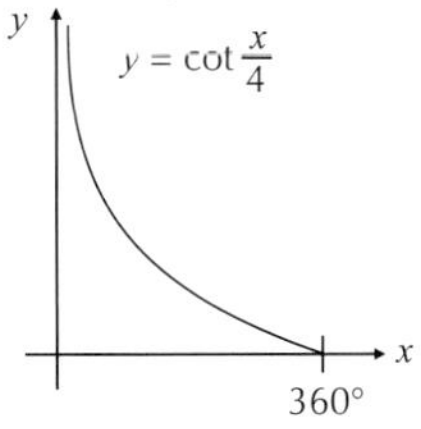

Q5 a) $y = 2 + \sec x$ is the graph of $y = \sec x$ translated vertically up by 2:

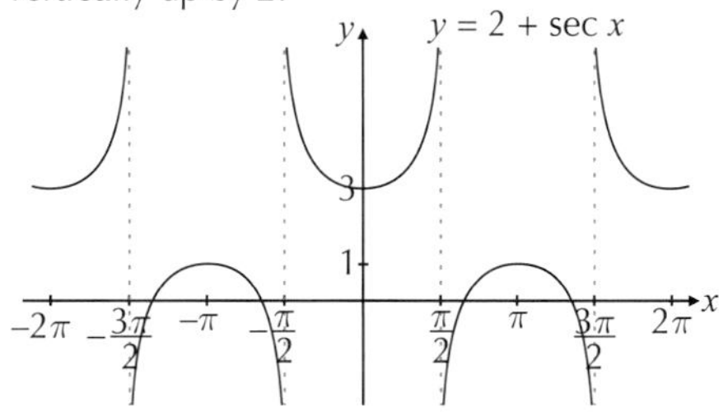

b) The minimum points are at $(-2\pi, 3)$, $(0, 3)$ and $(2\pi, 3)$.
The maximum points are at $(-\pi, 1)$ and $(\pi, 1)$.
The maximum and minimum points have the same x-coordinates as on the graph of y = sec x, but the y-coordinates have all been increased by 2.

c) The domain is $x \in \mathbb{R}$, $x \neq \left(n\pi + \frac{\pi}{2}\right)$ (where n is an integer).
The range is $y \geq 3$ or $y \leq 1$.

Q6 a) $y = 2\,\text{cosec}\, 2x$ is the graph of $y = \text{cosec}\, x$ stretched horizontally by a factor of $\frac{1}{2}$ and stretched vertically by a factor of 2.

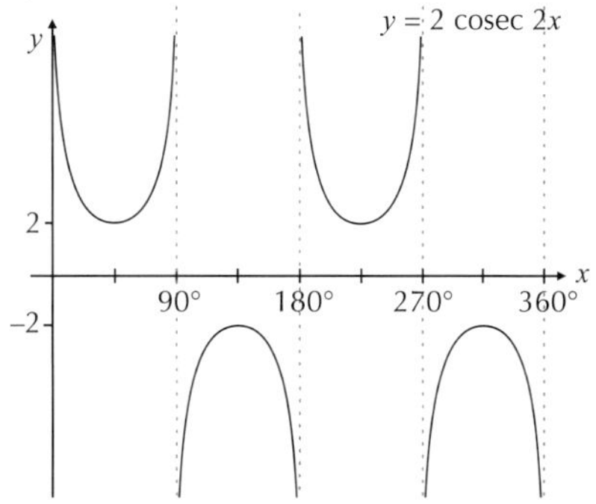

b) The minimum points are at (45°, 2) and (225°, 2).

c) The maximum points are at (135°, –2) and (315°, –2).

d) $y = 2\,\text{cosec}\, 2x$ is undefined when $x = 0°, 90°, 180°, 270°$ and $360°$.

Q7 a) If $f(x) = \text{cosec}\, x$, then $y = 2 + 3\,\text{cosec}\, x = 3f(x) + 2$, which is a vertical stretch scale factor 3, followed by a vertical translation of 2 up. Vertical transformations do not affect the position of the asymptotes, so they are in the same position as for the graph of $y = \text{cosec}\, x$, i.e. at $n\pi$ or $180n°$, where n is an integer.

b) The period of the graph will be the same as for the graph of $y = \text{cosec}\, x$, i.e. 360°.
Vertical transformations will not affect how often the graph repeats itself.

c)

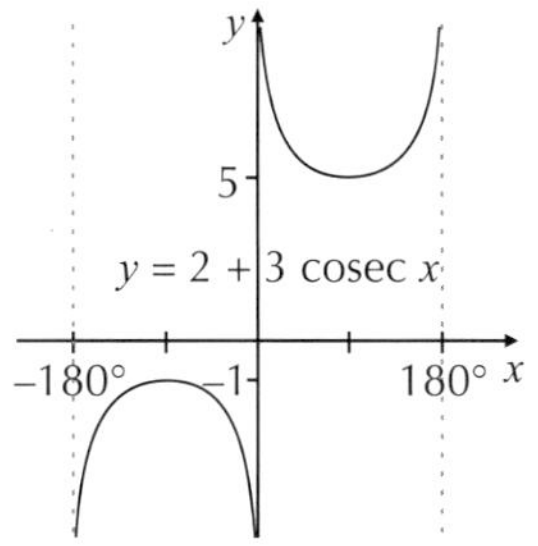

d) The range is $y \geq 5$ or $y \leq -1$.

Exercise 4.2 — Evaluating cosec, sec and cot

Q1 **a)** $\text{cosec } 80° = \frac{1}{\sin 80°} = 1.02$

b) $\sec 75° = \frac{1}{\cos 75°} = 3.86$

c) $\cot 30° = \frac{1}{\tan 30°} = 1.73$

d) $\sec (-70)° = \frac{1}{\cos(-70°)} = 2.92$

e) $3 - \cot 250° = 3 - \frac{1}{\tan 250°} = 2.64$

f) $2 \text{ cosec } 25° = \frac{2}{\sin 25°} = 4.73$

Q2 **a)** $\sec 3 = \frac{1}{\cos 3} = -1.01$

b) $\cot 0.6 = \frac{1}{\tan 0.6} = 1.46$

c) $\text{cosec } 1.8 = \frac{1}{\sin 1.8} = 1.03$

d) $\sec(-1) = \frac{1}{\cos(-1)} = 1.85$

e) $\text{cosec } \frac{\pi}{8} = \frac{1}{\sin \frac{\pi}{8}} = 2.61$

f) $8 + \cot \frac{\pi}{8} = 8 + \frac{1}{\tan \frac{\pi}{8}} = 10.4$

g) $\frac{1}{1 + \sec \frac{\pi}{10}} = \frac{1}{1 + \frac{1}{\cos \frac{\pi}{10}}} = 0.487$

h) $\frac{1}{6 + \cot \frac{\pi}{5}} = \frac{1}{6 + \frac{1}{\tan \frac{\pi}{5}}} = 0.136$

Q3 **a)** $\sec 60° = \frac{1}{\cos 60°} = \frac{1}{\left(\frac{1}{2}\right)} = 2$

b) $\text{cosec } 30° = \frac{1}{\sin 30°} = \frac{1}{\left(\frac{1}{2}\right)} = 2$

c) $\cot 45° = \frac{1}{\tan 45°} = \frac{1}{1} = 1$

d) $\text{cosec } \frac{\pi}{3} = \frac{1}{\sin \frac{\pi}{3}} = \frac{1}{\left(\frac{\sqrt{3}}{2}\right)} = \frac{2}{\sqrt{3}} = \frac{2\sqrt{3}}{3}$

e) $\sec(-180°) = \frac{1}{\cos(-180°)} = \frac{1}{\cos 180°} = -1$

The graph of $y = \cos x$ is symmetrical about the y-axis, so $\cos(-x) = \cos x$.

f) $\text{cosec } 135° = \text{cosec } (180° - 45°) = \frac{1}{\sin(180° - 45°)}$

The CAST diagram below shows that sin 135° is the same size as sin 45°, and also lies in a positive quadrant for sin:

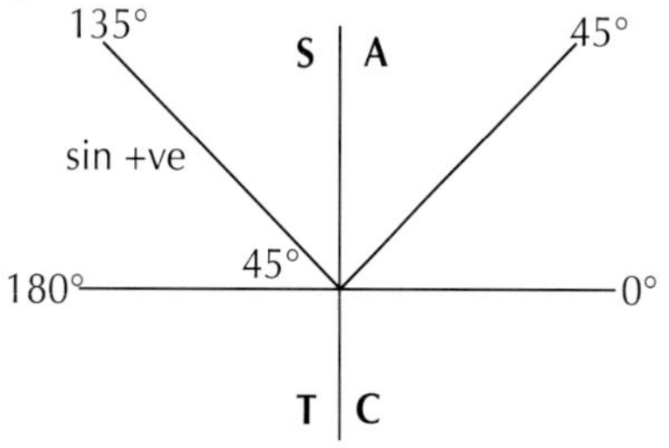

So $\text{cosec } 135° = \frac{1}{\sin 45°} = \frac{1}{\left(\frac{1}{\sqrt{2}}\right)} = \sqrt{2}$

g) $\cot 330° = \cot (360° - 30°) = \frac{1}{\tan(360° - 30°)}$

The CAST diagram below shows that tan 330° is the same size as tan 30°, but lies in a negative quadrant for tan:

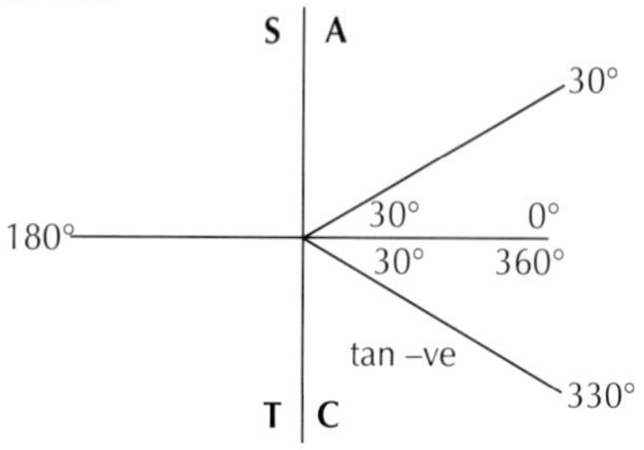

So $\cot 330° = \frac{1}{-\tan 30°} = \frac{1}{\left(-\frac{1}{\sqrt{3}}\right)} = -\sqrt{3}$

h) $\sec \frac{5\pi}{4} = \sec\left(\pi + \frac{\pi}{4}\right) = \frac{1}{\cos\left(\pi + \frac{\pi}{4}\right)}$

The CAST diagram below shows that $\cos\left(\pi + \frac{\pi}{4}\right)$ is the same size as $\cos \frac{\pi}{4}$, but lies in a negative quadrant for cos:

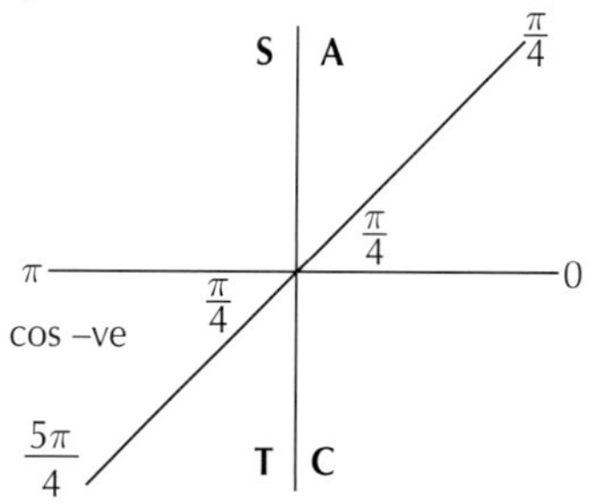

So $\sec \frac{5\pi}{4} = \frac{1}{-\cos \frac{\pi}{4}} = \frac{1}{\left(-\frac{1}{\sqrt{2}}\right)} = -\sqrt{2}$

i) $\text{cosec } \frac{5\pi}{3} = \text{cosec}\left(2\pi - \frac{\pi}{3}\right) = \frac{1}{\sin\left(2\pi - \frac{\pi}{3}\right)}$

The CAST diagram shows that $\sin\left(2\pi - \frac{\pi}{3}\right) = -\sin \frac{\pi}{3}$, so:

$\frac{1}{\sin\left(2\pi - \frac{\pi}{3}\right)} = \frac{1}{-\sin \frac{\pi}{3}} = \frac{1}{-\left(\frac{\sqrt{3}}{2}\right)} = -\frac{2}{\sqrt{3}} = -\frac{2\sqrt{3}}{3}$

j) $\text{cosec } \frac{2\pi}{3} = \text{cosec}\left(\pi - \frac{\pi}{3}\right) = \frac{1}{\sin\left(\pi - \frac{\pi}{3}\right)}$

The CAST diagram shows that $\sin\left(\pi - \frac{\pi}{3}\right) = \sin \frac{\pi}{3}$, so:

$\frac{1}{\sin\left(\pi - \frac{\pi}{3}\right)} = \frac{1}{\sin \frac{\pi}{3}} = \frac{1}{\left(\frac{\sqrt{3}}{2}\right)} = \frac{2}{\sqrt{3}} = \frac{2\sqrt{3}}{3}$

k) $3 - \cot \frac{3\pi}{4} = 3 - \cot\left(\pi - \frac{\pi}{4}\right) = 3 - \frac{1}{\tan\left(\pi - \frac{\pi}{4}\right)}$

$= 3 - \frac{1}{-\tan \frac{\pi}{4}} = 3 - \left(\frac{1}{-1}\right) = 4$

l) $\frac{\sqrt{3}}{\cot \frac{\pi}{6}} = \frac{\sqrt{3}}{\left(\frac{1}{\tan \frac{\pi}{6}}\right)} = \sqrt{3}\left(\tan \frac{\pi}{6}\right) = \sqrt{3} \times \frac{1}{\sqrt{3}} = 1$

Q4 a) $\frac{1}{1+\sec 60°} = \frac{1}{1+\left(\frac{1}{\cos 60°}\right)} = \frac{1}{1+\frac{1}{\left(\frac{1}{2}\right)}} = \frac{1}{3}$

b) $\cot 315° = \cot(360° - 45°) = \frac{1}{\tan(360°-45°)}$
$= \frac{1}{-\tan 45°} = -1$, so:
$\frac{2}{6+\cot 315°} = \frac{2}{6+(-1)} = \frac{2}{5}$

c) $\frac{1}{\sqrt{3}-\sec 30°} = \frac{1}{\sqrt{3}-\left(\frac{1}{\cos 30°}\right)}$

$= \frac{\cos 30°}{\sqrt{3}(\cos 30°)-1} = \frac{\left(\frac{\sqrt{3}}{2}\right)}{\sqrt{3}\left(\frac{\sqrt{3}}{2}\right)-1} = \frac{\sqrt{3}}{3-2} = \sqrt{3}$

d) $1 + \cot 420° = 1 + \cot(360° + 60°) = 1 + \cot 60°$
$= 1 + \frac{1}{\tan 60°} = 1 + \frac{1}{\sqrt{3}} = \frac{3+\sqrt{3}}{3}$

e) $\cot 150° = \cot(180° - 30°) = \frac{1}{\tan(180°-30°)}$
$= \frac{1}{-\tan 30°} = \frac{1}{-\left(\frac{1}{\sqrt{3}}\right)} = -\sqrt{3}$, so:

$\frac{2}{7+\sqrt{3}\cot 150°} = \frac{2}{7+\sqrt{3}(-\sqrt{3})} = \frac{2}{7-3} = \frac{1}{2}$

Exercise 4.3 — Simplifying expressions and solving equations

Q1 a) $\frac{1}{\cos x} = \sec x$, so $\sec x + \sec x = 2\sec x$

b) $(\text{cosec}^2 x)(\sin^2 x) = \frac{1}{\sin^2 x}(\sin^2 x) = 1$

c) $\frac{1}{\tan x} = \cot x$, so $2\cot x + \cot x = 3\cot x$

d) $\frac{\sec x}{\text{cosec}\, x} = \frac{\left(\frac{1}{\cos x}\right)}{\left(\frac{1}{\sin x}\right)} = \frac{\sin x}{\cos x} = \tan x$

e) $(\cos x)(\text{cosec}\, x) = \frac{\cos x}{\sin x} = \frac{1}{\left(\frac{\sin x}{\cos x}\right)} = \frac{1}{\tan x} = \cot x$

f) $\frac{\text{cosec}^2 x}{\cot x} = \frac{\left(\frac{1}{\sin^2 x}\right)}{\left(\frac{1}{\tan x}\right)} = \frac{\tan x}{\sin^2 x} = \frac{\left(\frac{\sin x}{\cos x}\right)}{\sin^2 x}$

$= \frac{\sin x}{\cos x \sin^2 x} = \frac{1}{\cos x \sin x} = \sec x\, \text{cosec}\, x$

Q2 a) $\sin x \cot x = \sin x\left(\frac{1}{\tan x}\right) = \sin x\left(\frac{\cos x}{\sin x}\right) = \cos x$

b) $\sec x - \cos x = \frac{1}{\cos x} - \cos x = \frac{1-\cos^2 x}{\cos x}$
Use the identity $\sin^2 x + \cos^2 x \equiv 1$:
$= \frac{\sin^2 x}{\cos x} = \left(\frac{\sin x}{\cos x}\right)\sin x = \tan x \sin x$

c) $\tan x\, \text{cosec}\, x = \left(\frac{\sin x}{\cos x}\right)\left(\frac{1}{\sin x}\right) = \frac{1}{\cos x} = \sec x$

d) $\frac{(\tan^2 x)(\text{cosec}\, x)}{\sin x} = \frac{\left(\frac{\sin^2 x}{\cos^2 x}\right)\left(\frac{1}{\sin x}\right)}{\sin x}$

$= \frac{\left(\frac{\sin x}{\cos^2 x}\right)}{\sin x} = \frac{1}{\cos^2 x} = \sec^2 x$

Q3 a) $\sec x = 1.9 \Rightarrow \cos x = \frac{1}{1.9} = 0.52631...$
$x = \cos^{-1}(0.52631...) = 58.2°$ (1 d.p.)

There is another positive solution in the interval $0 \le x \le 360°$ at $(360° - 58.2°)$, so $x = 301.8°$ (1 d.p.)

Remember, you can use the graphs or the CAST diagram to find other solutions in the interval.

b) $\cot x = 2.4 \Rightarrow \tan x = 0.41666...$
$\tan^{-1}(0.41666...) = 22.6°$ (1 d.p.)

There is another positive solution in the interval $0 \le x \le 360°$ at $(180° + 22.6°)$, so $x = 202.6°$ (1 d.p.)

c) $\text{cosec}\, x = -2 \Rightarrow \sin x = -0.5$
$\sin^{-1}(-0.5) = -30°$ which is not in the interval $0° \le x \le 360°$.

There are two negative solutions in the interval $0 \le x \le 360°$ at $(180° + 30°)$ and $(360° - 30°)$, so $x = 210°$ and $330°$.

d) $\sec x = -1.3 \Rightarrow \cos x = -0.76923...$
$\cos^{-1}(-0.76923...) = 140.3°$ (1 d.p.)

There are two negative solutions in the interval $0 \le x \le 360°$ at $140.3°$ and $(360° - 140.3°)$, so $x = 140.3°$ and $219.7°$ (1 d.p.)

e) $\cot x = -2.4 \Rightarrow \tan x = -0.41666...$
$\tan^{-1}(-0.41666...) = -22.6°$ (1 d.p.)

There are two negative solutions in the interval $0 \le x \le 360°$ at $(-22.6° + 180°)$ and $(-22.6° + 360°)$, so $x = 157.4°$ and $337.4°$ (1 d.p.)

f) $4\sec 2x = -7 \Rightarrow \cos 2x = -0.57142...$
$\cos^{-1}(-0.57142...) = 124.84990...$

You need to find all solutions for x in the interval $0 \le x \le 360°$ so $0 \le 2x \le (2 \times 360°)$ so you'll need to look for solutions for 2x in the interval $0 \le 2x \le 720°$.

There are four negative solutions for $2x$ in the interval $0 \le x \le 720°$ at $124.849...$, $(360° - 124.849...°)$, $(360° + 124.849...°)$ and $(720° - 124.849...°)$ and each of these needs to be divided by 2 to give x.
So $x = 62.4°$, $117.6°$, $242.4°$ and $297.6°$ (1 d.p.)

Q4 a) $\sec x = 2 \Rightarrow \cos x = 0.5 \Rightarrow x = \cos^{-1}(0.5) = \frac{\pi}{3}$
There is another positive solution in the interval $0 \le x \le 2\pi$ at $\left(2\pi - \frac{\pi}{3}\right)$, so $x = \frac{5\pi}{3}$.

b) $\text{cosec}\, x = -2 \Rightarrow \sin x = -0.5 \Rightarrow \sin^{-1}(-0.5) = -\frac{\pi}{6}$
There are two negative solutions in the interval $0 \le x \le 2\pi$ at $\left(\pi + \frac{\pi}{6}\right)$ and $\left(2\pi - \frac{\pi}{6}\right)$, so $x = \frac{7\pi}{6}$ and $\frac{11\pi}{6}$.

c) $\cot 2x = 1 \Rightarrow \tan 2x = 1 \Rightarrow 2x = \tan^{-1}(1) = \frac{\pi}{4}$
Don't forget to double the interval for the next bit — you're looking for solutions for 2x instead of x.

There are 3 other positive solutions for $2x$ in the interval $0 \le 2x \le 4\pi$, at $\left(\pi + \frac{\pi}{4}\right)$, $\left(2\pi + \frac{\pi}{4}\right)$ and $\left(3\pi + \frac{\pi}{4}\right)$, so $x = \frac{\pi}{8}, \frac{5\pi}{8}, \frac{9\pi}{8}$ and $\frac{13\pi}{8}$.

d) $\sec 5x = -1 \Rightarrow \cos 5x = -1 \Rightarrow 5x = \cos^{-1}(-1) = \pi$
In this case you're looking for solutions for 5x — the interval you'll need to look in is $0 \le 5x \le 10\pi$ since $0 \le x \le 2\pi$. Use the fact that the cos graph repeats itself every 2π. $(\pi, -1)$ is a minimum point on the graph, so this will be repeated every 2π.

There are 4 other solutions for $5x$ in the interval $0 \le 5x \le 10\pi$, at 3π, 5π, 7π, and 9π, so $x = \frac{\pi}{5}, \frac{3\pi}{5}, \pi, \frac{7\pi}{5}$ and $\frac{9\pi}{5}$.

Q5 $\cot 2x - 4 = -5 \Rightarrow \tan 2x = -1 \Rightarrow 2x = \tan^{-1}(-1) = -\frac{\pi}{4}$
There are 4 negative solutions for $2x$ in the interval $0 \le 2x \le 4\pi$, at $\left(\pi - \frac{\pi}{4}\right)$, $\left(2\pi - \frac{\pi}{4}\right)$, $\left(3\pi - \frac{\pi}{4}\right)$ and $\left(4\pi - \frac{\pi}{4}\right)$, so $x = \frac{3\pi}{8}, \frac{7\pi}{8}, \frac{11\pi}{8}$ and $\frac{15\pi}{8}$.

Q6 $2 \operatorname{cosec} 2x = 3 \Rightarrow \sin 2x = \frac{2}{3}$
$2x = \sin^{-1}\left(\frac{2}{3}\right) = 41.81031...°$
There are three other positive solutions for $2x$ in the interval $0 \le 2x \le 720°$ at $(180° - 41.81...°)$, $(360° + 41.81...°)$ and $(540° - 41.81...°)$, so $x = 20.9°, 69.1°, 200.9°$ and $249.1°$ (1 d.p.)

Q7 $-2 \sec x = 4 \Rightarrow \cos x = -0.5 \Rightarrow x = \cos^{-1}(-0.5) = \frac{2\pi}{3}$
There are 2 negative solutions in the interval $0 \le x \le 2\pi$ at $\frac{2\pi}{3}$ and $\left(2\pi - \frac{2\pi}{3}\right)$, so $x = \frac{2\pi}{3}$ and $\frac{4\pi}{3}$.

Q8 $\sqrt{3} \operatorname{cosec} 3x = 2 \Rightarrow \sin 3x = \frac{\sqrt{3}}{2}$
$\Rightarrow 3x = \sin^{-1}\left(\frac{\sqrt{3}}{2}\right) = \frac{\pi}{3}$
There are 5 other positive solutions in the interval $0 \le 3x \le 6\pi$, at $\left(\pi - \frac{\pi}{3}\right)$, $\left(2\pi + \frac{\pi}{3}\right)$, $\left(3\pi - \frac{\pi}{3}\right)$, $\left(4\pi + \frac{\pi}{3}\right)$ and $\left(5\pi - \frac{\pi}{3}\right)$, so $x = \frac{\pi}{9}, \frac{2\pi}{9}, \frac{7\pi}{9}, \frac{8\pi}{9}, \frac{13\pi}{9}$ and $\frac{14\pi}{9}$.

Q9 **a)** $\sec^2 x - 2\sqrt{2} \sec x + 2$ factorises to $(\sec x - \sqrt{2})^2$.
$(\sec x - \sqrt{2})^2 = 0 \Rightarrow \sec x = \sqrt{2} \Rightarrow \cos x = \frac{1}{\sqrt{2}}$
$\Rightarrow x = 45°$

b) $2 \cot^2 x + 3 \cot x - 2$ factorises to give $(2 \cot x - 1)(\cot x + 2)$
So $(2 \cot x - 1)(\cot x + 2) = 0$
$\Rightarrow \cot x = \frac{1}{2}$ or $\cot x = -2$
$\Rightarrow \tan x = 2$ or $\tan x = -\frac{1}{2}$
The solution for $\tan x = 2$ is $x = 63.4°$.
$\tan^{-1} -\frac{1}{2} = -26.6°$, so the other solution is $x = -26.6° + 180° = 153.4°$.
So $x = 63.4°$ and $153.4°$ (1 d.p.)

Q10 $(\operatorname{cosec} x - 3)(2 \tan x + 1) = 0$ means that either $\operatorname{cosec} x = 3$ (and so $\sin x = \frac{1}{3}$) or $\tan x = -\frac{1}{2}$.
The solutions for $\sin x = \frac{1}{3}$ are $x = 19.5°$ or $x = 180° - 19.5° = 160.5°$.
The solutions for $\tan x = -\frac{1}{2}$ are
$x = -26.6° + 180° = 153.4°$ or
$x = -26.6° + 360° = 333.4°$.
So $x = 19.5°, 160.5°, 153.4°, 333.4°$ (1 d.p.) are all solutions.

5. Identities Involving Cosec, Sec and Cot

Exercise 5.1 — Using the identities

Q1 $\operatorname{cosec}^2 x + 2 \cot^2 x = \operatorname{cosec}^2 x + 2(\operatorname{cosec}^2 x - 1)$
$= 3 \operatorname{cosec}^2 x - 2$

Q2 $\tan^2 x - \frac{1}{\cos^2 x} = \tan^2 x - \sec^2 x$
$= \tan^2 x - (1 + \tan^2 x) = -1$

Q3 $x + \frac{1}{x} = \sec\theta + \tan\theta + \frac{1}{\sec\theta + \tan\theta}$
$= \frac{(\sec\theta + \tan\theta)^2 + 1}{\sec\theta + \tan\theta}$
$= \frac{\sec^2\theta + 2\sec\theta\tan\theta + \tan^2\theta + 1}{\sec\theta + \tan\theta}$
But since $\sec^2\theta = \tan^2\theta + 1$:
$x + \frac{1}{x} = \frac{\sec^2\theta + 2\sec\theta\tan\theta + \sec^2\theta}{\sec\theta + \tan\theta}$
$= \frac{2\sec\theta(\sec\theta + \tan\theta)}{\sec\theta + \tan\theta} = 2\sec\theta$ as required.

Q4 **a)** $\tan^2 x = 2 \sec x + 2 \Rightarrow \sec^2 x - 1 = 2 \sec x + 2$
$\Rightarrow \sec^2 x - 2 \sec x - 3 = 0$

b) Solve $\sec^2 x - 2 \sec x - 3 = 0$
This factorises to give: $(\sec x - 3)(\sec x + 1) = 0$
So $\sec x = 3 \Rightarrow \cos x = \frac{1}{3}$
and $\sec x = -1 \Rightarrow \cos x = -1$
Solving these over the interval $0° \le x \le 360°$ gives:
$x = 70.5°$, $180°$ and $289.5°$ (1 d.p.)
Just use the graph of cos x or the CAST diagram as usual to find all the solutions in the interval.

Q5 **a)** $2 \operatorname{cosec}^2 x = 5 - 5 \cot x \Rightarrow 2(1 + \cot^2 x) = 5 - 5 \cot x$
$\Rightarrow 2 \cot^2 x + 5 \cot x - 3 = 0$

b) Solve $2 \cot^2 x + 5 \cot x - 3 = 0$
This factorises to give: $(2 \cot x - 1)(\cot x + 3) = 0$
So $\cot x = \frac{1}{2} \Rightarrow \tan x = 2$
and $\cot x = -3 \Rightarrow \tan x = -\frac{1}{3}$
Solving these over the interval $-\pi \le x \le \pi$ gives:
$x = -2.03, -0.32, 1.11, 2.82$ (2 d.p.)

Q6 **a)** $2 \cot^2 A + 5 \operatorname{cosec} A = 10$
$\Rightarrow 2(\operatorname{cosec}^2 A - 1) + 5 \operatorname{cosec} A = 10$
$\Rightarrow 2 \operatorname{cosec}^2 A - 2 + 5 \operatorname{cosec} A = 10$
$\Rightarrow 2 \operatorname{cosec}^2 A + 5 \operatorname{cosec} A - 12 = 0$

b) Solve $2 \operatorname{cosec}^2 A + 5 \operatorname{cosec} A - 12 = 0$
This factorises to give:
$(\operatorname{cosec} A + 4)(2 \operatorname{cosec} A - 3) = 0$
So $\operatorname{cosec} A = -4 \Rightarrow \sin A = -\frac{1}{4}$
Or $\operatorname{cosec} x = \frac{3}{2} \Rightarrow \sin A = \frac{2}{3}$
The solutions (to 1 d.p.) are
$A = 194.5°, 345.5°, 41.8°, 138.2°$.

Q7 $\sec^2 x + \tan x = 1 \Rightarrow 1 + \tan^2 x + \tan x = 1$
$\Rightarrow \tan^2 x + \tan x = 0 \Rightarrow \tan x(\tan x + 1) = 0$
So $\tan x = 0 \Rightarrow x = 0, \pi, 2\pi$.
And $\tan x = -1 \Rightarrow x = \frac{3\pi}{4}, \frac{7\pi}{4}$.

Q8 **a)** $\operatorname{cosec}^2\theta + 2\cot^2\theta = 2$
$\Rightarrow \operatorname{cosec}^2\theta + 2(\operatorname{cosec}^2\theta - 1) = 2$
$\Rightarrow 3\operatorname{cosec}^2\theta - 2 = 2 \Rightarrow 3\operatorname{cosec}^2\theta = 4$
$\Rightarrow \operatorname{cosec}^2\theta = \frac{4}{3} \Rightarrow \operatorname{cosec}\theta = \pm\frac{2}{\sqrt{3}}$
$\Rightarrow \sin\theta = \pm\frac{\sqrt{3}}{2}$

b) Solving $\sin\theta = \pm\frac{\sqrt{3}}{2}$ over the interval $0° \le x \le 180°$ gives: $\theta = 60°, 120°$.

Q9 $\sec^2 x = 3 + \tan x \Rightarrow (1 + \tan^2 x) = 3 + \tan x$
$\Rightarrow \tan^2 x - \tan x - 2 = 0 \Rightarrow (\tan x - 2)(\tan x + 1) = 0$
So $\tan x = 2$ or $\tan x = -1$.
Solving over the interval $0° \le x \le 360°$ gives:
$x = 63.4°, 135°, 243.4°$ and $315°$ (1 d.p.)

Q10 $\cot^2 x + \text{cosec}^2 x = 7 \Rightarrow \cot^2 x + (1 + \cot^2 x) = 7$

$\Rightarrow 2\cot^2 x + 1 = 7 \Rightarrow \cot^2 x = 3 \Rightarrow \cot x = \pm\sqrt{3}$

$\Rightarrow \tan x = \pm\frac{1}{\sqrt{3}}$

Solving over the interval $0 \le x \le 2\pi$ gives:

$x = \frac{\pi}{6}$ and $\frac{7\pi}{6}$ when $\tan x = +\frac{1}{\sqrt{3}}$

and $x = \frac{5\pi}{6}$ and $\frac{11\pi}{6}$ when $\tan x = -\frac{1}{\sqrt{3}}$.

Q11 $\tan^2 x + 5\sec x + 7 = 0 \Rightarrow (\sec^2 x - 1) + 5\sec x + 7 = 0$

$\Rightarrow \sec^2 x + 5\sec x + 6 = 0 \Rightarrow (\sec x + 2)(\sec x + 3) = 0$

So $\sec x = -2 \Rightarrow \cos x = -\frac{1}{2}$

and $\sec x = -3 \Rightarrow \cos x = -\frac{1}{3}$

Solving over the interval $0 \le x \le 2\pi$ gives:

$x = 1.91, 2.09, 4.19$ and 4.37 (2 d.p.)

Q12 Drawing a right angled triangle will help to solve this question:

$\tan\theta = \frac{60}{11}$

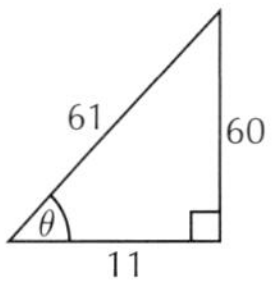

Notice that $180° \le \theta \le 270°$ — this puts us in the 3rd quadrant of the CAST diagram so sin will be –ve, cos will be –ve and tan will be +ve.

a) From the triangle, $\sin\theta = -\frac{\text{opp}}{\text{hyp}} = -\frac{60}{61}$.

b) $\cos\theta = -\frac{\text{adj}}{\text{hyp}} = -\frac{11}{61} \Rightarrow \sec\theta = \frac{1}{\left(-\frac{11}{61}\right)} = -\frac{61}{11}$.

c) $\text{cosec}\,\theta = \frac{1}{\sin\theta} = \frac{1}{\left(-\frac{60}{61}\right)} = -\frac{61}{60}$.

Q13 Drawing a right angled triangle will help to solve this question:

$\text{cosec}\,\theta = -\frac{17}{15}$

$\sin\theta = -\frac{15}{17}$

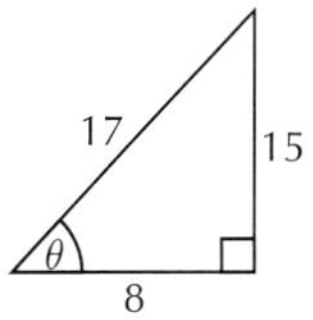

Notice that $180° \le \theta \le 270°$ — this puts us in the 3rd quadrant of the CAST diagram so sin will be –ve, cos will be –ve and tan will be +ve.

a) $\cos\theta = -\frac{\text{adj}}{\text{hyp}} = -\frac{8}{17}$.

b) $\sec\theta = \frac{1}{\cos\theta} = -\frac{17}{8}$.

c) $\tan\theta = \frac{\text{opp}}{\text{adj}} = \frac{15}{8}$ so $\cot\theta = \frac{1}{\tan\theta} = \frac{8}{15}$.

Q14 $\cos x = \frac{1}{6} \Rightarrow \sec x = 6 \Rightarrow \sec^2 x = 36$

So $1 + \tan^2 x = 36 \Rightarrow \tan^2 x = 35 \Rightarrow \tan x = \pm\sqrt{35}$

Exercise 5.2 — Proving other identities

Q1 **a)** $\sec^2\theta - \text{cosec}^2\theta \equiv (1 + \tan^2\theta) - (1 + \cot^2\theta)$

$\equiv \tan^2\theta - \cot^2\theta$

b) $\tan^2\theta - \cot^2\theta$ is the difference of two squares, and so can be written as $(\tan\theta + \cot\theta)(\tan\theta - \cot\theta)$.

So is $\sec^2\theta - \text{cosec}^2\theta$, so it can be written $(\sec\theta + \text{cosec}\,\theta)(\sec\theta - \text{cosec}\,\theta)$.

So using the result from part a),

$(\sec\theta + \text{cosec}\,\theta)(\sec\theta - \text{cosec}\,\theta)$

$\equiv (\tan\theta + \cot\theta)(\tan\theta - \cot\theta)$.

Q2 First expand the bracket:

$(\tan x + \cot x)^2 \equiv \tan^2 x + \cot^2 x + 2\tan x\cot x$

$\equiv \tan^2 x + \cot^2 x + \frac{2\tan x}{\tan x}$

$\equiv \tan^2 x + \cot^2 x + 2$

Split up that '+2' into two lots of '+1' so it starts to resemble the identities...

$\equiv (1 + \tan^2 x) + (1 + \cot^2 x)$

$\equiv \sec^2 x + \text{cosec}^2 x$

Q3 $\cot^2 x + \sin^2 x \equiv (\text{cosec}^2 x - 1) + (1 - \cos^2 x)$

$\equiv \text{cosec}^2 x - \cos^2 x$

This is the difference of two squares...

$\equiv (\text{cosec}\,x + \cos x)(\text{cosec}\,x - \cos x)$

Q4 $\frac{(\sec x - \tan x)(\tan x + \sec x)}{\text{cosec}\,x - \cot x} \equiv \frac{\sec^2 x - \tan^2 x}{\text{cosec}\,x - \cot x}$

$\equiv \frac{(1 + \tan^2 x) - \tan^2 x}{\text{cosec}\,x - \cot x} \equiv \frac{1}{\text{cosec}\,x - \cot x}$

Multiply top and bottom by (cosec x + cot x)...

$\frac{1}{\text{cosec}\,x - \cot x} \equiv \frac{\text{cosec}\,x + \cot x}{(\text{cosec}\,x - \cot x)(\text{cosec}\,x + \cot x)}$

$\equiv \frac{\text{cosec}\,x + \cot x}{\text{cosec}^2 x - \cot^2 x} \equiv \frac{\text{cosec}\,x + \cot x}{(1 + \cot^2 x) - \cot^2 x}$

$\equiv \cot x + \text{cosec}\,x$

Q5 $\frac{\cot x}{1 + \text{cosec}\,x} + \frac{1 + \text{cosec}\,x}{\cot x} \equiv \frac{\cot^2 x + (1 + \text{cosec}\,x)^2}{\cot x(1 + \text{cosec}\,x)}$

$\equiv \frac{(\text{cosec}^2 x - 1) + (1 + 2\,\text{cosec}\,x + \text{cosec}^2 x)}{\cot x(1 + \text{cosec}\,x)}$

$\equiv \frac{2\,\text{cosec}\,x(1 + \text{cosec}\,x)}{\cot x(1 + \text{cosec}\,x)} \equiv \frac{2\,\text{cosec}\,x}{\cot x} \equiv \frac{2\tan x}{\sin x}$

$\equiv \frac{2\sin x}{\sin x\cos x} \equiv \frac{2}{\cos x} \equiv 2\sec x$

Q6 $\frac{\text{cosec}\,x + 1}{\text{cosec}\,x - 1} \equiv \frac{(\text{cosec}\,x + 1)(\text{cosec}\,x + 1)}{(\text{cosec}\,x - 1)(\text{cosec}\,x + 1)}$

$\equiv \frac{\text{cosec}^2 x + 2\,\text{cosec}\,x + 1}{\text{cosec}^2 x - 1} \equiv \frac{\text{cosec}^2 x + 2\,\text{cosec}\,x + 1}{(1 + \cot^2 x) - 1}$

$\equiv \frac{\text{cosec}^2 x + 2\,\text{cosec}\,x + 1}{\cot^2 x} \equiv \frac{\text{cosec}^2 x}{\cot^2 x} + \frac{2\,\text{cosec}\,x}{\cot^2 x} + \frac{1}{\cot^2 x}$

$\equiv \frac{\tan^2 x}{\sin^2 x} + \frac{2\tan^2 x}{\sin x} + \tan^2 x$

$\equiv \frac{\sin^2 x}{\cos^2 x\sin^2 x} + \frac{2\sin^2 x}{\cos^2 x\sin x} + \tan^2 x$

$\equiv \frac{1}{\cos^2 x} + \frac{2\sin x}{\cos x\cos x} + \tan^2 x \equiv \frac{1}{\cos^2 x} + \frac{2\tan x}{\cos x} + \tan^2 x$

$\equiv \sec^2 x + 2\tan x\sec x + (\sec^2 x - 1)$

$\equiv 2\sec^2 x + 2\tan x\sec x - 1$

6. The Addition Formulas

Exercise 6.1 — Finding exact values

Q1 **a)** $\cos 72°\cos 12° + \sin 72°\sin 12°$

$= \cos(72° - 12°) = \cos 60° = \frac{1}{2}$

b) $\cos 13°\cos 17° - \sin 13°\sin 17°$

$= \cos(13° + 17°) = \cos 30° = \frac{\sqrt{3}}{2}$

c) $\frac{\tan 12° + \tan 18°}{1 - \tan 12°\tan 18°} = \tan(12° + 18°)$

$= \tan 30° = \frac{1}{\sqrt{3}}$

d) $\frac{\tan 500° - \tan 140°}{1 + \tan 500°\tan 140°} = \tan(500° - 140°)$

$= \tan 360° = 0$

e) $\sin 35° \cos 10° + \cos 35° \sin 10° = \sin(35° + 10°)$
$= \sin 45° = \frac{1}{\sqrt{2}}$

f) $\sin 69° \cos 9° - \cos 69° \sin 9° = \sin(69° - 9°)$
$= \sin 60° = \frac{\sqrt{3}}{2}$

Q2 a) $\sin\frac{2\pi}{3}\cos\frac{\pi}{2} - \cos\frac{2\pi}{3}\sin\frac{\pi}{2} = \sin\left(\frac{2\pi}{3} - \frac{\pi}{2}\right)$
$= \sin\frac{\pi}{6} = \frac{1}{2}$

b) $\cos 4\pi \cos 3\pi + \sin 4\pi \sin 3\pi = \cos(4\pi - 3\pi)$
$= \cos\pi = -1$

c) $\dfrac{\tan\frac{5\pi}{12} + \tan\frac{5\pi}{4}}{1 - \tan\frac{5\pi}{12}\tan\frac{5\pi}{4}} = \tan\left(\frac{5\pi}{12} + \frac{5\pi}{4}\right)$
$= \tan\frac{5\pi}{3} = \tan\left(2\pi - \frac{\pi}{3}\right) = -\tan\frac{\pi}{3} = -\sqrt{3}$

Q3 a) $\sin(5x - 2x) = \sin 3x$

b) $\cos(4x + 6x) = \cos 10x$

c) $\tan(7x + 3x) = \tan 10x$

d) $5\sin(2x + 3x) = 5\sin 5x$

e) $8\cos(7x - 5x) = 8\cos 2x$

Q4 Before answering a)-d), calculate $\cos x$ and $\sin y$:

$\sin x = \frac{3}{4} \Rightarrow \sin^2 x = \frac{9}{16} \Rightarrow \cos^2 x = 1 - \frac{9}{16} = \frac{7}{16}$

$\Rightarrow \cos x = \frac{\sqrt{7}}{4}$

x is acute, meaning cos x must be positive, so take the positive square root.

$\cos y = \frac{3}{\sqrt{10}} \Rightarrow \cos^2 y = \frac{9}{10} \Rightarrow \sin^2 y = 1 - \frac{9}{10} = \frac{1}{10}$

$\Rightarrow \sin y = \frac{1}{\sqrt{10}}$

Again, y is acute, so sin y must be positive, so you can take the positive square root. If you don't like using this method, you can use the triangle method to work out sin y and cos x.

a) $\sin(x + y) = \sin x \cos y + \cos x \sin y$
$= \left(\frac{3}{4} \times \frac{3}{\sqrt{10}}\right) + \left(\frac{\sqrt{7}}{4} \times \frac{1}{\sqrt{10}}\right)$
$= \frac{9 + \sqrt{7}}{4\sqrt{10}} = \frac{9\sqrt{10} + \sqrt{70}}{40}$

b) $\cos(x - y) = \cos x \cos y + \sin x \sin y$
$= \left(\frac{\sqrt{7}}{4} \times \frac{3}{\sqrt{10}}\right) + \left(\frac{3}{4} \times \frac{1}{\sqrt{10}}\right)$
$= \frac{3\sqrt{7} + 3}{4\sqrt{10}} = \frac{3\sqrt{70} + 3\sqrt{10}}{40}$

c) $\operatorname{cosec}(x + y) = \frac{1}{\sin(x + y)} = \frac{40}{9\sqrt{10} + \sqrt{70}}$
$= \frac{18\sqrt{10} - 2\sqrt{70}}{37}$

d) $\sec(x - y) = \frac{1}{\cos(x - y)} = \frac{40}{3\sqrt{70} + 3\sqrt{10}}$
$= \frac{2\sqrt{70} - 2\sqrt{10}}{9}$

Q5 $\cos\frac{\pi}{12} = \cos\left(\frac{\pi}{4} - \frac{\pi}{6}\right) = \cos\frac{\pi}{4}\cos\frac{\pi}{6} + \sin\frac{\pi}{4}\sin\frac{\pi}{6}$
$= \left(\frac{1}{\sqrt{2}} \times \frac{\sqrt{3}}{2}\right) + \left(\frac{1}{\sqrt{2}} \times \frac{1}{2}\right) = \frac{\sqrt{3} + 1}{2\sqrt{2}}$

Now rationalise the denominator...

$= \frac{(\sqrt{3} + 1) \times \sqrt{2}}{(2\sqrt{2}) \times \sqrt{2}} = \frac{\sqrt{6} + \sqrt{2}}{4}$

Q6 $\sin 75° = \sin(30° + 45°)$
$= \sin 30° \cos 45° + \cos 30° \sin 45°$
$= \left(\frac{1}{2} \times \frac{1}{\sqrt{2}}\right) + \left(\frac{\sqrt{3}}{2} \times \frac{1}{\sqrt{2}}\right)$
$= \frac{1 + \sqrt{3}}{2\sqrt{2}} = \frac{(1 + \sqrt{3}) \times \sqrt{2}}{(2\sqrt{2}) \times \sqrt{2}} = \frac{\sqrt{6} + \sqrt{2}}{4}$

Q7 $\tan 75° = \tan(45° + 30°) = \frac{\tan 45° + \tan 30°}{1 - \tan 45° \tan 30°}$

$= \dfrac{1 + \frac{1}{\sqrt{3}}}{1 - 1 \times \frac{1}{\sqrt{3}}} = \dfrac{\left(\frac{\sqrt{3} + 1}{\sqrt{3}}\right)}{\left(\frac{\sqrt{3} - 1}{\sqrt{3}}\right)} = \frac{\sqrt{3} + 1}{\sqrt{3} - 1}.$

Exercise 6.2 — Simplifying, solving equations and proving identities

Q1 $\tan(A - B) \equiv \frac{\sin(A - B)}{\cos(A - B)}$
$\equiv \frac{\sin A \cos B - \cos A \sin B}{\cos A \cos B + \sin A \sin B}$

Divide through by cos A cos B...

$\equiv \dfrac{\left(\frac{\sin A \cos B}{\cos A \cos B}\right) - \left(\frac{\cos A \sin B}{\cos A \cos B}\right)}{\left(\frac{\cos A \cos B}{\cos A \cos B}\right) + \left(\frac{\sin A \sin B}{\cos A \cos B}\right)}$

$\equiv \dfrac{\left(\frac{\sin A}{\cos A}\right) - \left(\frac{\sin B}{\cos B}\right)}{1 + \left(\frac{\sin A}{\cos A}\right)\left(\frac{\sin B}{\cos B}\right)}$

Now use tan = sin / cos...

$\equiv \frac{\tan A - \tan B}{1 + \tan A \tan B}$

Q2 a) $\frac{\cos(A - B) - \cos(A + B)}{\cos A \sin B}$

$\equiv \frac{(\cos A \cos B + \sin A \sin B) - (\cos A \cos B - \sin A \sin B)}{\cos A \sin B}$

$\equiv \frac{2\sin A \sin B}{\cos A \sin B} \equiv \frac{2\sin A}{\cos A} \equiv 2\tan A$

b) $\frac{1}{2}[\cos(A - B) - \cos(A + B)]$
$\equiv \frac{1}{2}[(\cos A \cos B + \sin A \sin B)$
$- (\cos A \cos B - \sin A \sin B)]$

$\equiv \frac{1}{2}(2\sin A \sin B) \equiv \sin A \sin B$

c) $\sin(x + 90°) \equiv \sin x \cos 90° + \cos x \sin 90°$
$\equiv \sin x\,(0) + \cos x\,(1) \equiv \cos x$

Q3 $4\sin x \cos\frac{\pi}{3} - 4\cos x \sin\frac{\pi}{3} = \cos x$
$\Rightarrow 2\sin x - 2\sqrt{3}\cos x = \cos x$
$\Rightarrow 2\sin x = (1 + 2\sqrt{3})\cos x$
$\Rightarrow \frac{\sin x}{\cos x} = \frac{1 + 2\sqrt{3}}{2} = \tan x$
$\Rightarrow x = -1.99$ and 1.15 (2 d.p.)

Q4 a) $\tan\left(-\frac{\pi}{12}\right) = \tan\left(\frac{\pi}{6} - \frac{\pi}{4}\right) \equiv \dfrac{\tan\frac{\pi}{6} - \tan\frac{\pi}{4}}{1 + \tan\frac{\pi}{6}\tan\frac{\pi}{4}}$

$\equiv \dfrac{\frac{1}{\sqrt{3}} - 1}{1 + \frac{1}{\sqrt{3}}} \equiv \frac{1 - \sqrt{3}}{\sqrt{3} + 1}$

Now rationalise the denominator...

$\equiv \frac{1 - \sqrt{3}}{\sqrt{3} + 1} \times \frac{\sqrt{3} - 1}{\sqrt{3} - 1} \equiv \frac{2\sqrt{3} - 4}{2} \equiv \sqrt{3} - 2$

b) $\cos x = \cos x \cos \frac{\pi}{6} - \sin x \sin \frac{\pi}{6}$

$\Rightarrow \cos x = \frac{\sqrt{3}}{2} \cos x - \frac{1}{2} \sin x$

$\Rightarrow (2-\sqrt{3}) \cos x = -\sin x$

$\Rightarrow \frac{\sin x}{\cos x} = \tan x = \sqrt{3} - 2$

From a), $\tan\left(-\frac{\pi}{12}\right) = \sqrt{3} - 2$,
so one solution for x is $-\frac{\pi}{12}$.
To get an answer in the correct interval, add π, since $\tan x$ repeats itself every π radians.
So $x = \frac{11\pi}{12}$.

Q5 $2 \sin(x + 30°) \equiv 2 \sin x \cos 30° + 2 \cos x \sin 30°$
$\equiv 2 \sin x \left(\frac{\sqrt{3}}{2}\right) + 2 \cos x \left(\frac{1}{2}\right)$
$\equiv \sqrt{3} \sin x + \cos x$

Q6 $\tan\left(\frac{\pi}{3} - x\right) \equiv \frac{\tan\frac{\pi}{3} - \tan x}{1 + \tan\frac{\pi}{3}\tan x} \equiv \frac{\sqrt{3} - \tan x}{1 + \sqrt{3}\tan x}$

Q7 $\tan(A + B) = \frac{\tan A + \tan B}{1 - \tan A \tan B} = \frac{1}{4}$

$\Rightarrow \frac{\frac{3}{8} + \tan B}{1 - \frac{3}{8}\tan B} = \frac{1}{4} \Rightarrow \frac{3}{8} + \tan B = \frac{1}{4}\left(1 - \frac{3}{8}\tan B\right)$

$\Rightarrow \frac{3}{8} + \tan B = \frac{1}{4} - \frac{3}{32} \tan B$

$\Rightarrow \tan B + \frac{3}{32} \tan B = \frac{1}{4} - \frac{3}{8} = -\frac{1}{8}$

$\Rightarrow \frac{35}{32} \tan B = -\frac{1}{8} \Rightarrow \tan B = -\frac{1}{8} \times \frac{32}{35} = -\frac{4}{35}$

Q8 Starting with the addition formulas for sin:
$\sin(x + y) \equiv \sin x \cos y + \cos x \sin y$
$\sin(x - y) \equiv \sin x \cos y - \cos x \sin y$

So $\sin(x + y) + \sin(x - y)$
$\equiv \sin x \cos y + \cos x \sin y + \sin x \cos y - \cos x \sin y$
$\equiv 2 \sin x \cos y$

Substitute $A = x + y$ and $B = x - y$

Then $A + B = 2x$ and $A - B = 2y$,

so $x = \left(\frac{A+B}{2}\right)$ and $y = \left(\frac{A-B}{2}\right)$.

Putting this back into the identity gives:

$\sin A + \sin B \equiv 2 \sin\left(\frac{A+B}{2}\right) \cos\left(\frac{A-B}{2}\right)$

Q9 a) $\sin x \cos y + \cos x \sin y = 4 \cos x \cos y + 4 \sin x \sin y$

Dividing through by $\cos x \cos y$ gives:

$\frac{\sin x}{\cos x} + \frac{\sin y}{\cos y} = 4 + \frac{4 \sin x \sin y}{\cos x \cos y}$

$\Rightarrow \tan x + \tan y = 4 + 4 \tan x \tan y$

$\Rightarrow \tan x - 4 \tan x \tan y = 4 - \tan y$

$\Rightarrow \tan x (1 - 4 \tan y) = 4 - \tan y$

$\Rightarrow \tan x = \frac{4 - \tan y}{1 - 4\tan y}$

b) $\tan x = \frac{4 - \tan\frac{\pi}{4}}{1 - 4\tan\frac{\pi}{4}}$

Comparing the equation you have to solve to the one in part a) you can see that $y = \frac{\pi}{4}$.

$\Rightarrow \tan x = \frac{4-1}{1-4} = -1 \Rightarrow x = \frac{3\pi}{4}$ and $\frac{7\pi}{4}$

Q10 a) Use the sin addition formula on $\sin(\theta + 45°)$:

$\sqrt{2}(\sin\theta \cos 45° + \cos\theta \sin 45°) = 3 \cos\theta$

$\sqrt{2}\left(\frac{1}{\sqrt{2}}\sin\theta + \frac{1}{\sqrt{2}}\cos\theta\right) = 3 \cos\theta$

$\sin\theta + \cos\theta = 3\cos\theta \Rightarrow \sin\theta = 2\cos\theta$

$\frac{\sin\theta}{\cos\theta} = 2 \Rightarrow \tan\theta = 2$

So $\theta = 63.43°$ (2 d.p.)
and $63.43° + 180° = 243.43°$ (2 d.p.).

b) Use the cos addition formula:

$2 \cos\left(\theta - \frac{2\pi}{3}\right) - 5 \sin\theta = 0$

$2\left(\cos\theta\cos\frac{2\pi}{3} + \sin\theta\sin\frac{2\pi}{3}\right) - 5 \sin\theta = 0$

$2\left(-\frac{1}{2}\cos\theta + \frac{\sqrt{3}}{2}\sin\theta\right) - 5 \sin\theta = 0$

$-\cos\theta + \sqrt{3}\sin\theta - 5\sin\theta = 0$

$-\cos\theta + (\sqrt{3} - 5)\sin\theta = 0$

$\cos\theta = (\sqrt{3} - 5)\sin\theta$

$\frac{1}{(\sqrt{3}-5)} = \frac{\sin\theta}{\cos\theta} = \tan\theta$

So $\theta = -0.296... + \pi = 2.84$ (2 d.p.)
and $-0.296... + 2\pi = 5.99$ (2 d.p.).

c) Use the addition formulas:

$\sin(\theta - 30°) - \cos(\theta + 60°) = 0$

$(\sin\theta\cos 30° - \cos\theta\sin 30°)$
$- (\cos\theta\cos 60° - \sin\theta\sin 60°) = 0$

$\left(\frac{\sqrt{3}}{2}\sin\theta - \frac{1}{2}\cos\theta\right) - \left(\frac{1}{2}\cos\theta - \frac{\sqrt{3}}{2}\sin\theta\right) = 0$

$\frac{\sqrt{3}}{2}\sin\theta - \frac{1}{2}\cos\theta - \frac{1}{2}\cos\theta + \frac{\sqrt{3}}{2}\sin\theta = 0$

$\sqrt{3}\sin\theta - \cos\theta = 0$

$\sqrt{3}\sin\theta = \cos\theta$

$\frac{\sin\theta}{\cos\theta} = \frac{1}{\sqrt{3}} \Rightarrow \tan\theta = \frac{1}{\sqrt{3}}$

$\theta = 30°$ and $30° + 180° = 210°$

Q11 Use the sin addition formula for $\sin\left(x + \frac{\pi}{6}\right)$:

$\sin\left(x + \frac{\pi}{6}\right) = \sin x \cos\frac{\pi}{6} + \cos x \sin\frac{\pi}{6}$

$= \frac{\sqrt{3}}{2}\sin x + \frac{1}{2}\cos x$

Now use the small angle approximations for sin and cos:

$\approx \frac{\sqrt{3}}{2}x + \frac{1}{2}\left(1 - \frac{1}{2}x^2\right)$

$= \frac{\sqrt{3}}{2}x + \frac{1}{2} - \frac{1}{4}x^2 = \frac{1}{2} + \frac{\sqrt{3}}{2}x - \frac{1}{4}x^2$

7. The Double Angle Formulas

Exercise 7.1 — Using the double angle formulas

Q1 a) $\sin 2A \equiv 2 \sin A \cos A$

$\Rightarrow 4 \sin A \cos A \equiv 2 \sin 2A$

$\Rightarrow 4 \sin\frac{\pi}{12}\cos\frac{\pi}{12} = 2 \sin\frac{\pi}{6} = 2 \times \frac{1}{2} = 1$

b) $\cos 2A \equiv 2\cos^2 A - 1$

$\Rightarrow \cos\frac{2\pi}{3} = 2\cos^2\frac{\pi}{3} - 1 = 2\left(\frac{1}{2}\right)^2 - 1 = -\frac{1}{2}$

c) $\sin 2A \equiv 2 \sin A \cos A$

$\Rightarrow \frac{\sin 2A}{2} \equiv \sin A \cos A$

$\Rightarrow \frac{\sin 120°}{2} = \sin 60° \cos 60° = \frac{\sqrt{3}}{2} \times \frac{1}{2} = \frac{\sqrt{3}}{4}$

d) $\tan 2A \equiv \frac{2\tan A}{1-\tan^2 A}$

$\Rightarrow \frac{\tan A}{2-2\tan^2 A} \equiv \frac{\tan 2A}{4}$

$\Rightarrow \frac{\tan 15°}{2-2\tan^2 15°} = \frac{\tan 30°}{4} = \frac{1}{4\sqrt{3}} = \frac{\sqrt{3}}{12}$

e) $\cos 2A \equiv 1 - 2\sin^2 A$

$\Rightarrow 2\sin^2 A - 1 \equiv -\cos 2A$

$\Rightarrow 2\sin^2 15° - 1 = -\cos 30° = -\frac{\sqrt{3}}{2}$

Q2 a) $\cos 2A \equiv 1 - 2\sin^2 A$

$\Rightarrow \cos 2x = 1 - 2\sin^2 x = 1 - 2\left(\frac{1}{6}\right)^2 = \frac{17}{18}$

b) First find $\cos x$:

$\cos^2 x = 1 - \sin^2 x = 1 - \left(\frac{1}{6}\right)^2 = \frac{35}{36}$

$\Rightarrow \cos x = \frac{\sqrt{35}}{6}$

x is acute so take the positive root for cos x. Again, if you find it easier you can use the triangle method here.

$\sin 2A \equiv 2\sin A\cos A$

$\Rightarrow \sin 2x = 2\left(\frac{1}{6} \times \frac{\sqrt{35}}{6}\right) = \frac{\sqrt{35}}{18}$

c) $\tan 2x = \frac{\sin 2x}{\cos 2x} = \frac{\sqrt{35}}{17}$

Q3 a) $\cos 2x = 1 - 2\sin^2 x = 1 - 2\left(-\frac{1}{4}\right)^2 = \frac{7}{8}$

b) First find $\cos x$:

$\cos^2 x = 1 - \sin^2 x = 1 - \left(-\frac{1}{4}\right)^2 = \frac{15}{16}$

$\Rightarrow \cos x = -\frac{\sqrt{15}}{4}$

x is in the 3rd quadrant of the CAST diagram where cos x is negative, so take the negative root for cos x.

$\sin 2A \equiv 2\sin A\cos A$

$\Rightarrow \sin 2x = 2\left(-\frac{1}{4} \times -\frac{\sqrt{15}}{4}\right) = \frac{\sqrt{15}}{8}$

c) $\tan 2x = \frac{\sin 2x}{\cos 2x} = \frac{\sqrt{15}}{7}$

Q4 a) Using the sin double angle formula:

$\frac{\sin 3\theta\cos 3\theta}{3} \equiv \frac{\sin 6\theta}{6}$

b) Using the cos double angle formula:

$\sin^2\left(\frac{2y}{3}\right) - \cos^2\left(\frac{2y}{3}\right) \equiv -\cos\left(\frac{4y}{3}\right)$

c) Using the tan double angle formula:

$\frac{1-\tan^2\left(\frac{x}{2}\right)}{2\tan\left(\frac{x}{2}\right)} \equiv \frac{1}{\tan x} \equiv \cot x$

Exercise 7.2 — Solving equations and proving identities

Q1 a) Using the cos double angle formula involving sin:

$4(1 - 2\sin^2 x) - 14\sin x = 0$

$\Rightarrow 4 - 8\sin^2 x - 14\sin x = 0$

$\Rightarrow 8\sin^2 x + 14\sin x - 4 = 0$

$\Rightarrow 4\sin^2 x + 7\sin x - 2 = 0$

$\Rightarrow (4\sin x - 1)(\sin x + 2) = 0$

So $\sin x = \frac{1}{4}$ or $\sin x = -2$ (not valid)

Solving $\sin x = \frac{1}{4}$ in the interval $0 \le x \le 360°$:

$x = 14.5°$ (1 d.p.) and $(180° - 14.5°) = 165.5°$ (1 d.p.)

b) Using the cos double angle formula involving cos:

$5(2\cos^2 x - 1) + 9\cos x + 7 = 0$

$\Rightarrow 10\cos^2 x + 9\cos x + 2 = 0$

$\Rightarrow (2\cos x + 1)(5\cos x + 2) = 0$

So $\cos x = -\frac{1}{2}$ or $\cos x = -\frac{2}{5}$

$\Rightarrow x = 113.6°, 120°, 240°, 246.4°$ (1 d.p.)

c) Using the tan double angle formula:

$\frac{4(1-\tan^2 x)}{2\tan x} + \frac{1}{\tan x} = 5$

$\Rightarrow 2(1 - \tan^2 x) + 1 = 5\tan x$

$\Rightarrow 3 - 2\tan^2 x = 5\tan x$

$\Rightarrow 0 = 2\tan^2 x + 5\tan x - 3$

$\Rightarrow 0 = (2\tan x - 1)(\tan x + 3)$

So $\tan x = \frac{1}{2}$ or $\tan x = -3$

$\Rightarrow x = 26.6°, 108.4°, 206.6°, 288.4°$ (1 d.p.)

d) $\tan x - 5(2\sin x\cos x) = 0$

$\Rightarrow \frac{\sin x}{\cos x} = 10\sin x\cos x$

$\Rightarrow \sin x = 10\sin x\cos^2 x$

$\Rightarrow \sin x - 10\sin x\cos^2 x = 0$

$\Rightarrow \sin x(1 - 10\cos^2 x) = 0$

So $\sin x = 0$ or $\cos x = \pm\frac{1}{\sqrt{10}}$

Don't forget to find cos⁻¹ of both the positive and negative root...

$x = 0°, 71.6°, 108.4°, 180°, 251.6°, 288.4°, 360°$.

Q2 a) $4(2\cos^2 x - 1) - 10\cos x + 1 = 0$

$\Rightarrow 8\cos^2 x - 10\cos x - 3 = 0$

$\Rightarrow (4\cos x + 1)(2\cos x - 3) = 0$

So $\cos x = -\frac{1}{4}$ or $\cos x = \frac{3}{2}$ (not valid)

$\Rightarrow x = 1.82$ and 4.46 (3 s.f.)

b) $\cos 2x - 3 = 6\sin^2 x - 3$

$\Rightarrow (1 - 2\sin^2 x) - 3 - 6\sin^2 x + 3 = 0$

$\Rightarrow 1 - 8\sin^2 x = 0$

So $\sin x = \pm\frac{1}{2\sqrt{2}}$

$\Rightarrow x = 0.361, 2.78, 3.50, 5.92$ (3 s.f.)

Q3 a) $2\cos^2 x - 1 + 7\cos x = -4$

$\Rightarrow 2\cos^2 x + 7\cos x + 3 = 0$

$\Rightarrow (2\cos x + 1)(\cos x + 3) = 0$

So $\cos x = -\frac{1}{2}$ or $\cos x = -3$ (not valid)

$\Rightarrow x = \frac{2\pi}{3}$ and $\frac{4\pi}{3}$

b) $2\sin\frac{x}{2}\cos\frac{x}{2} + \cos\frac{x}{2} = 0$

$\Rightarrow \cos\frac{x}{2}\left(2\sin\frac{x}{2} + 1\right) = 0$

So $\cos\frac{x}{2} = 0$ or $\sin\frac{x}{2} = -\frac{1}{2}$

There are no solutions for $\sin\frac{x}{2} = -\frac{1}{2}$ in the interval $0 \le x \le \pi$ (they both lie in the 3rd and 4th quadrants of the CAST diagram, $\pi \le x \le 2\pi$) so...

$\Rightarrow \frac{x}{2} = \frac{\pi}{2} \Rightarrow x = \pi$

Q4 a) $\sin 2x\sec^2 x \equiv (2\sin x\cos x)\left(\frac{1}{\cos^2 x}\right)$

$\equiv \frac{2\sin x}{\cos x} \equiv 2\tan x$

b) $\frac{2}{1+\cos 2x} \equiv \frac{2}{1+2\cos^2 x-1}$

$\equiv \frac{2}{2\cos^2 x} \equiv \frac{1}{\cos^2 x} \equiv \sec^2 x$

c) $\cot x - 2\cot 2x \equiv \frac{1}{\tan x} - \frac{2}{\tan 2x}$

$\equiv \frac{1}{\tan x} - \frac{2(1-\tan^2 x)}{2\tan x}$

$\equiv \frac{1-(1-\tan^2 x)}{\tan x} \equiv \frac{\tan^2 x}{\tan x} \equiv \tan x$

d) $\tan 2x + \cot 2x \equiv \frac{\sin 2x}{\cos 2x} + \frac{\cos 2x}{\sin 2x}$

$\equiv \frac{\sin^2 2x + \cos^2 2x}{\sin 2x \cos 2x}$

Use $\sin^2 x + \cos^2 x \equiv 1$, and $\sin 4x \equiv 2\sin 2x\cos 2x$...

$\equiv \frac{1}{\frac{1}{2}\sin 4x} \equiv \frac{2}{\sin 4x} \equiv 2\,\text{cosec}\,4x$

Q5 a) $\frac{1+\cos 2x}{\sin 2x} \equiv \frac{1+(2\cos^2 x - 1)}{2\sin x\cos x}$

$\equiv \frac{2\cos^2 x}{2\sin x\cos x} \equiv \frac{\cos x}{\sin x} \equiv \cot x$

b) Use $4\theta = 2x$, so $x = 2\theta$ and so $\frac{1+\cos 4\theta}{\sin 4\theta} \equiv \cot 2\theta$

Solve $\cot 2\theta = 7$

$\Rightarrow \tan 2\theta = \frac{1}{7}$

$\Rightarrow 2\theta = 8.130°, 188.130°, 368.130°, 548.130°$

$\Rightarrow \theta = 4.1°, 94.1°, 184.1°, 274.1°$ (1 d.p.)

Q6 a) $\text{cosec}\,x - \cot\frac{x}{2} \equiv \frac{1}{\sin x} - \frac{\cos\frac{x}{2}}{\sin\frac{x}{2}}$

$\equiv \frac{1}{2\sin\frac{x}{2}\cos\frac{x}{2}} - \frac{\cos\frac{x}{2}}{\sin\frac{x}{2}} \equiv \frac{1-2\cos^2\frac{x}{2}}{2\sin\frac{x}{2}\cos\frac{x}{2}}$

Here we've let $2A = x$ so $x = \frac{A}{2}$.

$\equiv \frac{-\left(2\cos^2\frac{x}{2}-1\right)}{2\sin\frac{x}{2}\cos\frac{x}{2}} \equiv \frac{-\cos x}{\sin x}$

$-\frac{1}{\left(\frac{\sin x}{\cos x}\right)} \equiv -\frac{1}{\tan x} \equiv -\cot x.$

b) Rearranging, $\text{cosec}\,y - \cot\frac{y}{2} = -2 = -\cot y$

$\Rightarrow \cot y = 2 \Rightarrow \tan y = \frac{1}{2}$.

There are 2 solutions in the interval $-\pi \le y \le \pi$, at $y = 0.464$ (3 s.f.) and $y = 0.464 - \pi = -2.68$ (3 s.f.)

Q7 First find $\cos\theta$. You know $\sin\theta = \frac{5}{13}$, so using the triangle method, $\cos\theta = \frac{12}{13}$.

a) (i) $\cos^2\left(\frac{\theta}{2}\right) = \frac{1}{2}(1+\cos\theta) = \frac{1}{2}\left(1+\frac{12}{13}\right) = \frac{25}{26}$

So $\cos\left(\frac{\theta}{2}\right) = \sqrt{\frac{25}{26}} = \frac{5}{\sqrt{26}}$.

θ is acute, so $\frac{\theta}{2}$ is also acute, and $\cos\left(\frac{\theta}{2}\right)$ is +ve so we can ignore the negative root.

(ii) $\sin^2\left(\frac{\theta}{2}\right) = \frac{1}{2}(1-\cos\theta) = \frac{1}{2}\left(1-\frac{12}{13}\right) = \frac{1}{26}$

So $\sin\left(\frac{\theta}{2}\right) = \sqrt{\frac{1}{26}} = \frac{1}{\sqrt{26}}$.

Again, θ is acute so $\sin\left(\frac{\theta}{2}\right)$ must be positive.

b) $\tan\left(\frac{\theta}{2}\right) = \frac{\sin\left(\frac{\theta}{2}\right)}{\cos\left(\frac{\theta}{2}\right)} = \frac{\left(\frac{1}{\sqrt{26}}\right)}{\left(\frac{5}{\sqrt{26}}\right)} = \frac{1}{5}$

8. The R Addition Formulas

Exercise 8.1 — Expressions of the form $a\cos\theta + b\sin\theta$

Q1 $3\sin x - 2\cos x \equiv R\sin(x-\alpha)$

$\Rightarrow 3\sin x - 2\cos x \equiv R\sin x\cos\alpha - R\cos x\sin\alpha$

$\Rightarrow$ ① $R\cos\alpha = 3$ and ② $R\sin\alpha = 2$

② ÷ ① gives $\tan\alpha = \frac{2}{3} \Rightarrow \alpha = 33.7°$ (1 d.p.)

①² + ②² gives:

$R^2\cos^2\alpha + R^2\sin^2\alpha = 3^2 + 2^2 = 13$

$\Rightarrow R^2(\cos^2\alpha + \sin^2\alpha) = 13 \Rightarrow R^2 = 13 \Rightarrow R = \sqrt{13}$

So $3\sin x - 2\cos x \equiv \sqrt{13}\sin(x - 33.7°)$.

Q2 $6\cos x - 5\sin x \equiv R\cos(x+\alpha)$

$\Rightarrow 6\cos x - 5\sin x \equiv R\cos x\cos\alpha - R\sin x\sin\alpha$

$\Rightarrow$ ① $R\cos\alpha = 6$ and ② $R\sin\alpha = 5$

② ÷ ① gives $\tan\alpha = \frac{5}{6} \Rightarrow \alpha = 39.8°$ (1 d.p.)

①² + ②² gives:

$R^2\cos^2\alpha + R^2\sin^2\alpha = 6^2 + 5^2 = 61 \Rightarrow R = \sqrt{61}$

So $6\cos x - 5\sin x \equiv \sqrt{61}\cos(x + 39.8°)$.

Q3 $\sin x + \sqrt{7}\cos x \equiv R\sin(x+\alpha)$

$\Rightarrow \sin x + \sqrt{7}\cos x \equiv R\sin x\cos\alpha + R\cos x\sin\alpha$

$\Rightarrow$ ① $R\cos\alpha = 1$ and ② $R\sin\alpha = \sqrt{7}$

② ÷ ① gives $\tan\alpha = \sqrt{7} \Rightarrow \alpha = 1.21$ (3 s.f.)

①² + ②² gives:

$R^2\cos^2\alpha + R^2\sin^2\alpha = 1^2 + (\sqrt{7})^2 = 8$

$\Rightarrow R = \sqrt{8} = \sqrt{4\times 2} = \sqrt{4}\times\sqrt{2} = 2\sqrt{2}$

So $\sin x + \sqrt{7}\cos x \equiv 2\sqrt{2}\sin(x + 1.21)$.

Q4 $\sqrt{2}\sin x - \cos x \equiv R\sin(x-\alpha)$

$\Rightarrow \sqrt{2}\sin x - \cos x \equiv R\sin x\cos\alpha - R\cos x\sin\alpha$

$\Rightarrow$ ① $R\cos\alpha = \sqrt{2}$ and ② $R\sin\alpha = 1$

② ÷ ① gives $\tan\alpha = \frac{1}{\sqrt{2}}$

①² + ②² gives: $R^2\cos^2\alpha + R^2\sin^2\alpha = (\sqrt{2})^2 + 1^2 = 3$

$\Rightarrow R^2(\cos^2\alpha + \sin^2\alpha) = 3$

$\Rightarrow R^2 = 3 \Rightarrow R = \sqrt{3}$

So $\sqrt{2}\sin x - \cos x \equiv \sqrt{3}\sin(x-\alpha)$, where $\tan\alpha = \frac{1}{\sqrt{2}}$.

Q5 $3\cos 2x + 5\sin 2x \equiv R\cos(2x-\alpha)$

$\Rightarrow 3\cos 2x + 5\sin 2x \equiv R\cos 2x\cos\alpha + R\sin 2x\sin\alpha$

$\Rightarrow$ ① $R\cos\alpha = 3$ and ② $R\sin\alpha = 5$

② ÷ ① gives $\tan\alpha = \frac{5}{3}$

①² + ②² gives:

$R^2\cos^2\alpha + R^2\sin^2\alpha = 3^2 + 5^2 = 34 \Rightarrow R = \sqrt{34}$

So $3\cos 2x + 5\sin 2x \equiv \sqrt{34}\cos(2x-\alpha)$, where $\tan\alpha = \frac{5}{3}$.

Q6 a) $\sqrt{3}\sin x + \cos x \equiv R\sin(x+\alpha)$

$\Rightarrow \sqrt{3}\sin x + \cos x \equiv R\sin x\cos\alpha + R\cos x\sin\alpha$

$\Rightarrow$ ① $R\cos\alpha = \sqrt{3}$ and ② $R\sin\alpha = 1$

② ÷ ① gives $\tan\alpha = \frac{1}{\sqrt{3}} \Rightarrow \alpha = \frac{\pi}{6}$

①² + ②² gives:

$R^2\cos^2\alpha + R^2\sin^2\alpha = (\sqrt{3})^2 + 1^2 = 4$

$\Rightarrow R = \sqrt{4} = 2$

So $\sqrt{3}\sin x + \cos x \equiv 2\sin\left(x + \frac{\pi}{6}\right)$.

b) The graph of $y = 2\sin\left(x + \frac{\pi}{6}\right)$ is the graph of $y = \sin x$ transformed in the following way:
a horizontal translation left by $\frac{\pi}{6}$, then a vertical stretch by a factor of 2.

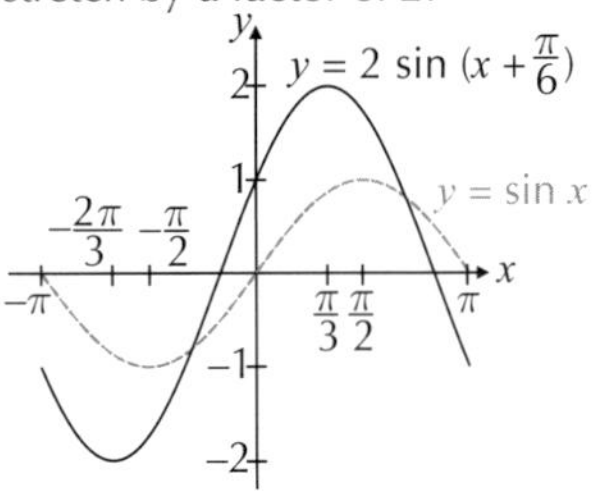

c) The graph of $y = \sin x$ has a minimum at $\left(-\frac{\pi}{2}, -1\right)$, a maximum at $\left(\frac{\pi}{2}, 1\right)$, and cuts the x-axis at $(-\pi, 0)$, $(0, 0)$ and $(\pi, 0)$.
To describe the graph of $y = 2\sin\left(x + \frac{\pi}{6}\right)$, each of these points needs to have $\frac{\pi}{6}$ subtracted from the x-coordinates, and the y-coordinates multiplied by 2.
So the graph of $y = 2\sin\left(x + \frac{\pi}{6}\right)$ has a minimum at $\left(-\frac{2\pi}{3}, -2\right)$, a maximum at $\left(\frac{\pi}{3}, 2\right)$, and cuts the x-axis at $\left(-\frac{\pi}{6}, 0\right)$ and $\left(\frac{5\pi}{6}, 0\right)$.
To find the y-intercept, put $x = 0$ into the equation:
$y = 2\sin\left(0 + \frac{\pi}{6}\right) = 2 \times \frac{1}{2} = 1$
So the y-intercept is at $(0, 1)$.

Exercise 8.2 — Applying the R addition formulas

Q1 a) $5\cos\theta - 12\sin\theta \equiv R\cos(\theta + \alpha)$
$\equiv R\cos\theta\cos\alpha - R\sin\theta\sin\alpha$
$\Rightarrow R\cos\alpha = 5,\ R\sin\alpha = 12$
$\tan\alpha = \frac{12}{5} \Rightarrow \alpha = 67.4°$ (1 d.p.)
$R = \sqrt{5^2 + 12^2} = \sqrt{169} = 13$
$\Rightarrow 5\cos\theta - 12\sin\theta \equiv 13\cos(\theta + 67.4°)$

b) $13\cos(\theta + 67.4°) = 4$, in the interval $67.4° \le (\theta + 67.4°) \le 427.4°$
$\cos(\theta + 67.4°) = \frac{4}{13}$
$\Rightarrow \theta + 67.4° = \cos^{-1}\frac{4}{13}$
$= 72.1°$ and $(360° - 72.1°) = 287.9°$
$\Rightarrow \theta = (72.1° - 67.4°)$ and $(287.9° - 67.4°)$
$= 4.7°$ and $220.5°$ (1 d.p.)

c) The maximum and minimum values of $\cos\theta$ are at ± 1. So the maximum and minimum values of $13\cos(\theta + 67.4°)$ are at ± 13.

Q2 a) $2\sin 2\theta + 3\cos 2\theta \equiv R\sin(2\theta + \alpha)$
$\equiv R\sin 2\theta\cos\alpha + R\cos 2\theta\sin\alpha$
$\Rightarrow R\cos\alpha = 2,\ R\sin\alpha = 3$
$\tan\alpha = \frac{3}{2} \Rightarrow \alpha = 0.983$ (3 s.f.)
$R = \sqrt{2^2 + 3^2} = \sqrt{13}$
$\Rightarrow 2\sin 2\theta + 3\cos 2\theta \equiv \sqrt{13}\sin(2\theta + 0.983)$

b) $\sqrt{13}\sin(2\theta + 0.983) = 1$ in the interval $0.983 \le (2\theta + 0.983) \le 13.549$.
$\sin(2\theta + 0.983) = \frac{1}{\sqrt{13}}$
$\Rightarrow 2\theta + 0.983 = 0.281$ (not in correct interval), $(\pi - 0.281)$, $(2\pi + 0.281)$, $(3\pi - 0.281)$, $(4\pi + 0.281)$.
There will be 4 solutions for θ between 0 and 2π because you're dealing with $\sin 2\theta$.
$\Rightarrow 2\theta + 0.983 = 2.861, 6.564, 9.144, 12.847$
$\Rightarrow \theta = 0.939, 2.79, 4.08, 5.93$ (3 s.f.)

Q3 a) $3\sin\theta - 2\sqrt{5}\cos\theta \equiv R\sin(\theta - \alpha)$
$\equiv R\sin\theta\cos\alpha - R\cos\theta\sin\alpha$
$\Rightarrow R\cos\alpha = 3,\ R\sin\alpha = 2\sqrt{5}$
$\tan\alpha = \frac{2\sqrt{5}}{3} \Rightarrow \alpha = 56.1$ (1 d.p.)
$R = \sqrt{3^2 + (2\sqrt{5})^2} = \sqrt{29}$
$\Rightarrow 3\sin\theta - 2\sqrt{5}\cos\theta \equiv \sqrt{29}\sin(\theta - 56.1°)$

b) $\sqrt{29}\sin(\theta - 56.1°) = 5$ in the interval $-56.1° \le (\theta - 56.1°) \le 303.9°$.
$\sin(\theta - 56.1°) = \frac{5}{\sqrt{29}}$
$\Rightarrow \theta - 56.1° = 68.2°$ and $(180° - 68.2°) = 111.8°$
$\Rightarrow \theta = 124.3°$ and $167.9°$ (1 d.p.)

c) $f(x) = \sqrt{29}\sin(x - 56.1°)$. The maximum of $\sin x$ is 1, so the maximum value of $f(x)$ is $f(x) = \sqrt{29}$.
When $f(x) = \sqrt{29}$, $\sin(x - 56.1°) = 1$
$\Rightarrow x - 56.1° = 90° \Rightarrow x = 146.1°$ (1 d.p.)

Q4 a) $3\sin x + \cos x \equiv R\sin(x + \alpha)$
$\equiv R\sin x\cos\alpha + R\cos x\sin\alpha$
$\Rightarrow R\cos\alpha = 3,\ R\sin\alpha = 1$
$\tan\alpha = \frac{1}{3} \Rightarrow \alpha = 18.4$ (1 d.p.)
$R = \sqrt{3^2 + 1^2} = \sqrt{10}$
$\Rightarrow 3\sin x + \cos x \equiv \sqrt{10}\sin(x + 18.4°)$

b) $\sqrt{10}\sin(x + 18.4°) = 2$ in the interval $18.4° \le (x + 18.4°) \le 378.4°$
$\sin(x + 18.4°) = \frac{2}{\sqrt{10}}$
$\Rightarrow x + 18.4° = 39.2°$ and $(180° - 39.2°) = 140.8°$
$\Rightarrow x = 20.8°$ and $122.4°$ (1 d.p.)

c) The maximum and minimum values of $f(x)$ are $\pm\sqrt{10}$.

Q5 a) $4\sin x + \cos x \equiv R\sin(x + \alpha)$
$\equiv R\sin x\cos\alpha + R\cos x\sin\alpha$
$\Rightarrow R\cos\alpha = 4,\ R\sin\alpha = 1$
$\tan\alpha = \frac{1}{4} \Rightarrow \alpha = 0.245$ (3 s.f.)
$R = \sqrt{4^2 + 1^2} = \sqrt{17}$
$\Rightarrow 4\sin x + \cos x \equiv \sqrt{17}\sin(x + 0.245)$

b) Maximum value of $4\sin x + \cos x = \sqrt{17}$, so the greatest value of $(4\sin x + \cos x)^4 = (\sqrt{17})^4 = 289$.

c) $\sqrt{17}\sin(x + 0.245) = 1$ in the interval $0.245 \le (x + 0.245) \le 3.387$.
$\sin(x + 0.245) = \frac{1}{\sqrt{17}}$
$\Rightarrow x + 0.245 = 0.245$ and $(\pi - 0.245) = 2.897$
$\Rightarrow x = 0$ and 2.65 (3 s.f.)

Q6 a) $8 \cos x + 15 \sin x \equiv R \cos (x - \alpha)$
$\equiv R \cos x \cos \alpha + R \sin x \sin \alpha$
$\Rightarrow R \cos \alpha = 8,\ R \sin \alpha = 15$
$\tan \alpha = \frac{15}{8} \Rightarrow \alpha = 1.08$ (3 s.f.)
$R = \sqrt{8^2 + 15^2} = \sqrt{289} = 17$
$\Rightarrow 8 \cos x + 15 \sin x \equiv 17 \cos (x - 1.08)$.

b) So solve for $17 \cos (x - 1.08) = 5$ in the interval $-1.08 \le (x - 1.08) \le 5.20$.
$\cos (x - 1.08) = \frac{5}{17}$
$\Rightarrow x - 1.08 = 1.27$ and $(2\pi - 1.27) = 5.01$
$\Rightarrow x = 2.35$ and 6.09 (3 s.f.)

c) $g(x) = (8 \cos x + 15 \sin x)^2 = 17^2 \cos^2 (x - 1.08)$
$= 289 \cos^2 (x - 1.08)$
The function $\cos^2 x$ has a minimum value of 0 (since all negative values of $\cos x$ become positive when you square it) so the minimum of g(x) is 0.
This minimum occurs when $\cos^2(x - 1.08) = 0$ so $\cos(x - 1.08) = 0$ so $x - 1.08 = \frac{\pi}{2} \Rightarrow x = 2.65$ (3 s.f.).

Q7 a) $2 \cos x + \sin x \equiv R \cos (x - \alpha)$
$\equiv R \cos x \cos \alpha + R \sin x \sin \alpha$
$\Rightarrow R \cos \alpha = 2,\ R \sin \alpha = 1$
$\tan \alpha = \frac{1}{2} \Rightarrow \alpha = 26.6°$ (3 s.f.)
$R = \sqrt{2^2 + 1^2} = \sqrt{5}$
$\Rightarrow 2 \cos x + \sin x \equiv \sqrt{5} \cos (x - 26.6°)$.

b) The range of g(x) is between the maximum and minimum values, which are at $\pm\sqrt{5}$.
So $-\sqrt{5} \le g(x) \le \sqrt{5}$.

Q8 $3 \sin \theta - \frac{3}{2} \cos \theta \equiv R \sin (\theta - \alpha)$
$\equiv R \sin \theta \cos \alpha - R \cos \theta \sin \alpha$
$\Rightarrow R \cos \alpha = 3,\ R \sin \alpha = \frac{3}{2}$
$\tan \alpha = \frac{1}{2} \Rightarrow \alpha = 0.464$ (3 s.f.)
$R = \sqrt{3^2 + \left(\frac{3}{2}\right)^2} = \sqrt{\frac{45}{4}} = \frac{3\sqrt{5}}{2}$
$\Rightarrow 3 \sin \theta - \frac{3}{2} \cos \theta \equiv \frac{3\sqrt{5}}{2} \sin (\theta - 0.464)$
So solve $\frac{3\sqrt{5}}{2} \sin (\theta - 0.464) = 3$ in the interval $-0.464 \le (\theta - 0.464) \le 5.819$.
$\sin (\theta - 0.464) = \frac{2}{\sqrt{5}}$
$\Rightarrow \theta - 0.464 = 1.107$ and $(\pi - 1.107) = 2.034$
$\Rightarrow \theta = 1.57$ and 2.50 (3 s.f.)

Q9 $4 \sin 2\theta + 3 \cos 2\theta \equiv R \sin (2\theta + \alpha)$
$\equiv R \sin 2\theta \cos \alpha + R \cos 2\theta \sin \alpha$
$\Rightarrow R \cos \alpha = 4,\ R \sin \alpha = 3$
$\tan \alpha = \frac{3}{4} \Rightarrow \alpha = 0.644$ (3 s.f.)
$R = \sqrt{3^2 + 4^2} = 5$
$\Rightarrow 4 \sin 2\theta + 3 \cos 2\theta \equiv 5 \sin (2\theta + 0.644)$
So solve $5 \sin (2\theta + 0.644) = 2$ in the interval $0.644 \le (2\theta + 0.644) \le 6.927$.
$\sin (2\theta + 0.644) = \frac{2}{5}$
$\Rightarrow 2\theta + 0.644 = 0.412$ (not in the interval),
$(\pi - 0.412) = 2.730$, and
$(2\pi + 0.412) = 6.695$
$\Rightarrow \theta = 1.04$ and 3.03 (3 s.f.)
You could also solve this by writing the function in the form $R \cos (2\theta - \alpha)$ ($R = 5$, $\alpha = 0.927$) — you should get the same solutions for θ either way.

9. Modelling with Trig Functions

Exercise 9.1 — Trigonometry in modelling

Q1 For each sector of the circle, $r = 20$ so
Area $= \frac{1}{2}r^2\theta = (200\theta)$ m², and
Perimeter $= 2r + r\theta = (40 + 20\theta)$ m
Convert each angle to radians then find the area and perimeter:
$120° = \frac{120 \times \pi}{180} = \frac{2\pi}{3}$
Area $= 200 \times \frac{2\pi}{3} = 419$ m² (3 s.f.)
Perimeter $= 40 + 20 \times \frac{2\pi}{3} = 81.9$ m (3 s.f.)
$144° = \frac{144 \times \pi}{180} = \frac{4\pi}{5}$
Area $= 200 \times \frac{4\pi}{5} = 503$ m² (3 s.f.)
Perimeter $= 40 + 20 \times \frac{4\pi}{5} = 90.3$ m (3 s.f.)
$96° = \frac{96 \times \pi}{180} = \frac{8\pi}{15}$
Area $= 200 \times \frac{8\pi}{15} = 335$ m² (3 s.f.)
Perimeter $= 40 + 20 \times \frac{8\pi}{15} = 73.5$ m (3 s.f.)

Q2 a) $\cos \theta$ has a maximum of 1 and a minimum of –1.
You want $A + B \cos \theta$ to have a maximum of 17 and a minimum of 7.
So form two equations:
① $A + B(1) = 17$ (at maximum)
② $A + B(-1) = 7$ (at minimum)
① + ②: $2A = 24 \Rightarrow A = 12$
Sub back into ①: $12 + B = 17 \Rightarrow B = 5$
So $f(\theta) = 12 + 5 \cos \theta$
You could also think of this as a transformation of the graph of $\cos \theta$ — a vertical stretch with scale factor 5 followed by a translation up by 12.

b) The longest day ($t = 0$) will have 17 hours of daylight, which is the maximum of the function. This will occur when $\cos (Ct + D) = 1$, i.e. when $Ct + D = 0$.
Similarly, the shortest day ($t = 6$) will occur when $\cos (Ct + D) = -1$ i.e. when $Ct + D = \pi$.
So $C(0) + D = 0$ (longest day) $\Rightarrow D = 0$
$C(6) + D = \pi$ (shortest day)
$\Rightarrow 6C + 0 = \pi \Rightarrow C = \frac{\pi}{6}$
So $g(t) = 12 + 5 \cos \left(\frac{\pi}{6}t\right)$
Because cos is periodic, you might have found $D = 2\pi$. Also, because cos is symmetrical, $C = -\frac{\pi}{6}$ will give you the same answer — either one is fine.

Q3 $h(t) = 14 + 5 \sin t + 5 \cos t \equiv 14 + R \cos (t - \alpha)$
$\equiv 14 + R \cos t \cos \alpha + R \sin t \sin \alpha$
$\Rightarrow R \cos \alpha = 5,\ R \sin \alpha = 5$
$\tan \alpha = 1 \Rightarrow \alpha = \frac{\pi}{4}$
$R = \sqrt{5^2 + 5^2} = \sqrt{50} = 5\sqrt{2}$
$\Rightarrow h(t) \equiv 14 + 5\sqrt{2} \cos \left(t - \frac{\pi}{4}\right)$
The maximum and minimum of cos are 1 and –1, so the maximum of $h(t) = 14 + 5\sqrt{2} = 21.1$ m (1 d.p.) and the minimum of $h(t) = 14 - 5\sqrt{2} = 6.9$ m (1 d.p.)

Q4 a) $H = 10 + \frac{7}{2}\sin t - \frac{7\sqrt{3}}{2}\cos t \equiv 10 + R\sin(t - \alpha)$

$\equiv 10 + R\sin t\cos\alpha - R\cos t\sin\alpha$

$\Rightarrow R\cos\alpha = \frac{7}{2},\ R\sin\alpha = \frac{7\sqrt{3}}{2}$

$\tan\alpha = \sqrt{3} \Rightarrow \alpha = \frac{\pi}{3}$

$R = \sqrt{\left(\frac{7}{2}\right)^2 + \left(\frac{7\sqrt{3}}{2}\right)^2} = \sqrt{\frac{49}{4} + \frac{147}{4}}$

$= \sqrt{\frac{196}{4}} = \sqrt{49} = 7$

$\Rightarrow H \equiv 10 + 7\sin\left(t - \frac{\pi}{3}\right)$

b) Replace $\left(t - \frac{\pi}{3}\right)$ with x to make the working clearer:

$h = 10 + 7\sin x - \cos 2x$

Write $\cos 2x$ in terms of $\sin x$, i.e. $1 - 2\sin^2 x$

$h = 10 + 7\sin x - (1 - 2\sin^2 x)$

$= 10 + 7\sin x - 1 + 2\sin^2 x$

$= 9 + 7\sin x + 2\sin^2 x$

i.e. $A = 9$ and $B = 2$.

c) You need to find $h = 13$, i.e.

$9 + 7\sin x + 2\sin^2 x = 13$

$2\sin^2 x + 7\sin x - 4 = 0$

This is a quadratic in $\sin x$, so factorise:

$(2\sin x - 1)(\sin x + 4) = 0$

$\Rightarrow \sin x = \frac{1}{2}$ or $\sin x = -4$ (not valid since $-1 < \sin x < 1$)

So $\sin x = \frac{1}{2}$ i.e. $\sin\left(t - \frac{\pi}{3}\right) = \frac{1}{2}$

You need to find the smallest t that satisfies this equation.

$\sin\left(t - \frac{\pi}{3}\right) = \frac{1}{2} \Rightarrow t - \frac{\pi}{3} = -\frac{7\pi}{6}, \frac{\pi}{6}, \frac{5\pi}{6}$, etc.

$\Rightarrow t = -\frac{5\pi}{6}, \frac{\pi}{2}, \frac{7\pi}{6}$, etc.

Time cannot be negative, so the smallest solution is $t = \frac{\pi}{2}$ i.e. $t = 1.57$ s (3 s.f.)

Chapter 4: Coordinate Geometry in the (x, y) Plane

1. Parametric Equations of Curves

Exercise 1.1 — Finding coordinates from parametric equations

Q1 a) $x = 3t = 3 \times 5 = 15$

$y = t^2 = 5^2 = 25$

So coordinates are (15, 25).

b) $18 = 3t \Rightarrow t = 6$

c) $36 = t^2 \Rightarrow t = \pm 6 \Rightarrow x = \pm 18$

Q2 a) $x = 2t - 1 = (2 \times 7) - 1 = 13$

$y = 4 - t^2 = 4 - 7^2 = -45$

So coordinates are (13, –45).

b) $2t - 1 = 15 \Rightarrow t = 8$

c) $4 - t^2 = -5 \Rightarrow t = \pm 3 \Rightarrow x = -7$ or 5

Q3 a) $x = 2 + \sin\theta = 2 + \sin\frac{\pi}{4} = \frac{4 + \sqrt{2}}{2}$

$y = -3 + \cos\theta = -3 + \cos\frac{\pi}{4} = \frac{\sqrt{2} - 6}{2}$

So coordinates are $\left(\frac{4 + \sqrt{2}}{2}, \frac{\sqrt{2} - 6}{2}\right)$.

b) $2 + \sin\theta = \frac{4 + \sqrt{3}}{2} = 2 + \frac{\sqrt{3}}{2}$

$\Rightarrow \sin\theta = \frac{\sqrt{3}}{2} \Rightarrow \theta = \frac{\pi}{3}$

c) $-3 + \cos\theta = -\frac{7}{2} \Rightarrow \cos\theta = -\frac{1}{2} \Rightarrow \theta = \frac{2\pi}{3}$

The angle in a) was given in radians so make sure you use radians for b) and c). They're both cos values you should know an angle for off by heart. The cos value in part c) is negative so you can use the CAST diagram (have a look at your Year 1 notes) to figure out which angle you need (cos is negative in the 2nd and 3rd quadrants).

Q4

t	–5	–4	–3	–2	–1	1	2	3	4	5
x	–25	–20	–15	–10	–5	5	10	15	20	25
y	$-\frac{2}{5}$	$-\frac{1}{2}$	$-\frac{2}{3}$	–1	–2	2	1	$\frac{2}{3}$	$\frac{1}{2}$	$\frac{2}{5}$

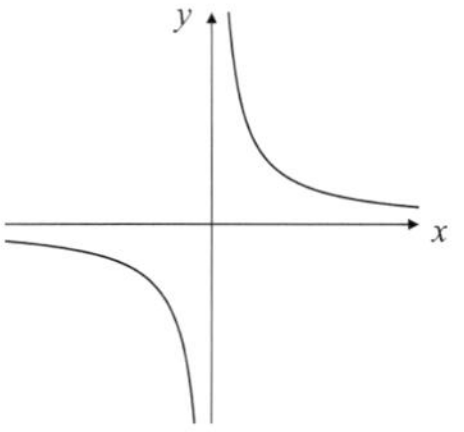

When t = 0, y is undefined, so there must be an asymptote.

Q5

θ	0	$\frac{\pi}{4}$	$\frac{\pi}{3}$	$\frac{\pi}{2}$	$\frac{2\pi}{3}$	$\frac{3\pi}{4}$	π	$\frac{4\pi}{3}$	$\frac{3\pi}{2}$	$\frac{5\pi}{3}$	2π
x	1	1.71	1.87	2	1.87	1.71	1	0.13	0	0.13	1
y	3	2.71	2.5	2	1.5	1.29	1	1.5	2	2.5	3

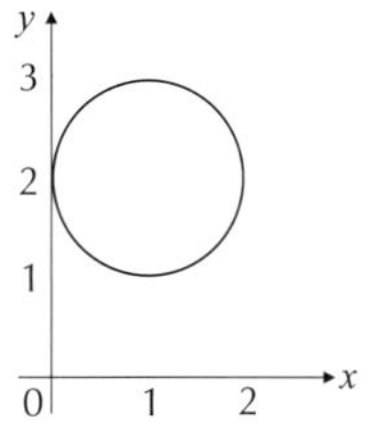

Q6 a) When $\theta = 0$: $x = 3\sin 0 = 3(0) = 0$

$y = 7 + 9\cos 0 = 7 + 9(1) = 16$

The comet is at (0, 16), so the distance to the Sun at (0, 0) is 16 AU.

b) When $\theta = \frac{\pi}{2}$: $x = 3\sin\frac{\pi}{2} = 3(1) = 3$

$y = 7 + 9\cos\frac{\pi}{2} = 7 + 9(0) = 7$

The comet is at (3, 7), so the distance to the Sun at (0, 0) is $\sqrt{3^2 + 7^2} = \sqrt{58} = 7.62$ AU (to 3 s.f.).

Q7 a) When $t = 2$, $x = 2^2 + 4(2) = 4 + 8 = 12$

The plane travels 12 m horizontally in the first 2 s.

b) Sonia is standing at the point $x = 21$.

$x = 21 \Rightarrow t^2 + 4t = 21 \Rightarrow t^2 + 4t - 21 = 0$

$\Rightarrow (t + 7)(t - 3) = 0 \Rightarrow t = -7$ or $t = 3$

The parametric equations are only valid for $0 \le t \le 5$, so the plane is at $x = 21$ when $t = 3$.

When $t = 3$, $y = 25 - 3^2 = 25 - 9 = 16$

The plane passes over Sonia's head at a height of 16 m.

Exercise 1.2 — Finding intersections

Q1 At A: $y = 0 \Rightarrow -2 + t = 0 \Rightarrow t = 2$
$x = 3 + t = 3 + 2 = 5$
At B: $x = 0 \Rightarrow 3 + t = 0 \Rightarrow t = -3$
$y = -2 + (-3) = -5$
So the coordinates are A(5, 0) and B(0, –5).

Q2 **a)** The curve meets the x-axis when $y = 0$, so:
$3t^3 - 24 = 0 \Rightarrow 3(t^3 - 8) = 0 \Rightarrow t^3 = 8 \Rightarrow t = 2$
b) The curve meets the y-axis when $x = 0$, so:
$2t^2 - 50 = 0 \Rightarrow 2(t^2 - 25) = 0 \Rightarrow t^2 = 25 \Rightarrow t = \pm 5$
If you'd have been asked to give the coordinates you would then just put the t values into the parametric equations.

Q3 The curve meets the y-axis when $x = 0$, so:
$64 - t^3 = 0 \Rightarrow t^3 = 64 \Rightarrow t = 4$
$y = \frac{1}{t} = \frac{1}{4}$, so P is $\left(0, \frac{1}{4}\right)$.

Q4 $y = x - 3 \Rightarrow 4t = (2t + 1) - 3 \Rightarrow 2t = -2 \Rightarrow t = -1$
To find the coordinates when t = –1, just put this value back into the parametric equations...
$x = (2 \times -1) + 1 = -1, \; y = 4 \times -1 = -4$
So P is (–1, –4).

Q5 $y = x^2 + 32 \Rightarrow 6t^2 = (2t)^2 + 32 \Rightarrow 6t^2 = 4t^2 + 32$
$\Rightarrow 2t^2 = 32 \Rightarrow t^2 = 16 \Rightarrow t = \pm 4$
When $t = 4$: $x = 2 \times 4 = 8, \; y = 6 \times 4^2 = 96$
So one point of intersection is (8, 96).
When $t = -4$: $x = 2 \times -4 = -8, \; y = 6 \times (-4)^2 = 96$
So the other point of intersection is (–8, 96).

Q6 $x^2 + y^2 = 32 \Rightarrow (t^2)^2 + (2t)^2 = 32 \Rightarrow t^4 + 4t^2 - 32 = 0$
$\Rightarrow (t^2 - 4)(t^2 + 8) = 0 \Rightarrow t^2 = 4 \Rightarrow t = \pm 2$
(there are no real solutions to $t^2 = -8$).
When $t = 2$: $x = 2^2 = 4, \; y = 2 \times 2 = 4$
So one point of intersection is (4, 4).
When $t = -2$: $x = (-2)^2 = 4, \; y = 2 \times -2 = -4$
So the other point of intersection is (4, –4).

Q7 **a)** At the point (0, 4):
$x = 0 \Rightarrow a(t - 2) = 0 \Rightarrow t = 2$ (as $a \neq 0$)
$y = 4 \Rightarrow 2at^2 + 3 = 4 \Rightarrow 2a(2^2) + 3 = 4$
$\Rightarrow 8a + 3 = 4 \Rightarrow a = \frac{1}{8}$.
b) The curve would meet the x-axis when $y = 0$.
$y = 2at^2 + 3 = \frac{2t^2}{8} + 3 = \frac{t^2}{4} + 3$
So at the x-axis, $\frac{t^2}{4} + 3 = 0 \Rightarrow t^2 = -12$.
This has no real solutions,
so the curve does not meet the x-axis.

Q8 **a)** The curve crosses the x-axis when $y = 0$.
$t^2 - 9 = 0 \Rightarrow t = \pm 3$
When $t = 3$, $x = \frac{2}{3}$, so one point is $\left(\frac{2}{3}, 0\right)$.
When $t = -3$, $x = -\frac{2}{3}$, so the other point is $\left(-\frac{2}{3}, 0\right)$.
b) The curve would meet the y-axis when $x = 0$, i.e. when $\frac{2}{t} = 0$. This has no solutions,
so the curve does not meet the y-axis.
c) $y = \frac{10}{x} - 3 \Rightarrow t^2 - 9 = \frac{10t}{2} - 3 \Rightarrow t^2 - 5t - 6 = 0$
$\Rightarrow (t + 1)(t - 6) = 0 \Rightarrow t = -1$ and $t = 6$
When $t = -1$: $x = -2, \; y = (-1)^2 - 9 = -8$
So one point of intersection is (–2, –8).
When $t = 6$: $x = \frac{2}{6} = \frac{1}{3}, \; y = 6^2 - 9 = 27$
So the other point of intersection is $\left(\frac{1}{3}, 27\right)$.

Q9 **a)** The curve meets the x-axis when $y = 0$.
$5 \cos t = 0 \Rightarrow \cos t = 0 \Rightarrow t = \frac{\pi}{2}$ and $\frac{3\pi}{2}$.
When $t = \frac{\pi}{2}$, $x = 3 \sin\left(\frac{\pi}{2}\right) = 3$, so (3, 0).
When $t = \frac{3\pi}{2}$, $x = 3 \sin\left(\frac{3\pi}{2}\right) = -3$, so (–3, 0).
So the curve crosses the x-axis twice in the domain $0 \leq t \leq 2\pi$, at (–3, 0) and (3, 0).
The curve meets the y-axis when $x = 0$.
$3 \sin t = 0 \Rightarrow \sin t = 0 \Rightarrow t = 0, \pi$ and 2π.
When $t = 0$, $y = 5 \cos 0 = 5$, so (0, 5).
When $t = \pi$, $y = 5 \cos \pi = -5$, so (0, –5).
When $t = 2\pi$, $y = 5 \cos 2\pi = 5$, so (0, 5).
So the curve crosses the y-axis twice in the domain $0 \leq t \leq 2\pi$, at (0, –5) and (0, 5).
b) $y = \left(\frac{5\sqrt{3}}{9}\right)x \Rightarrow 5 \cos t = \left(\frac{5\sqrt{3}}{3}\right)\sin t$
$\Rightarrow \frac{15}{5\sqrt{3}} = \frac{\sin t}{\cos t} = \tan t \Rightarrow \tan t = \frac{3}{\sqrt{3}} = \sqrt{3}$
$\Rightarrow t = \frac{\pi}{3}$ and $\left(\pi + \frac{\pi}{3}\right) \Rightarrow t = \frac{\pi}{3}$ and $\frac{4\pi}{3}$
When $t = \frac{\pi}{3}$: $x = 3 \sin \frac{\pi}{3} = \frac{3\sqrt{3}}{2}$
$y = 5 \cos \frac{\pi}{3} = \frac{5}{2}$
So one point of intersection is $\left(\frac{3\sqrt{3}}{2}, \frac{5}{2}\right)$
When $t = \frac{4\pi}{3}$: $x = 3 \sin \frac{4\pi}{3} = -\frac{3\sqrt{3}}{2}$
$y = 5 \cos \frac{4\pi}{3} = -\frac{5}{2}$
So the other point of intersection is $\left(-\frac{3\sqrt{3}}{2}, -\frac{5}{2}\right)$

Q10 The ships will collide if they have the same x- and y-coordinates at the same time. To find the value of t when they both have the same x-coordinate, set the two equations for x equal to each other:
$x_1 = x_2 \Rightarrow 24 - t = t + 10 \Rightarrow 14 = 2t \Rightarrow t = 7$
Now see if the y-coordinates are the same when $t = 7$:
$t = 7 \Rightarrow y_1 = 10 + 3(7) = 31,$
$y_2 = 12 + 2(7) - 0.1(7)^2 = 21.1$
$t = 7$ minutes is the only time when $x_1 = x_2$, but $y_1 \neq y_2$ at this time, so they will not collide.
You could have done this the other way around, by setting $y_1 = y_2$ to find the values of t when the y-coordinates are equal. In this case, that's a much more difficult way to solve the question — it gives you a fairly awkward quadratic to solve, and you end up with two values of t to test instead of one.
This question shows how parametric equations can be useful in modelling — if you drew the two Cartesian graphs, you'd find they cross in two places, so you might think the ships would collide. It's only when you think about how the ships' positions depend on time that you can work out what actually happens.

2. Parametric and Cartesian Equations

Exercise 2.1 — Converting parametric equations to Cartesian equations

Q1 **a)** $x = t + 3 \Rightarrow t = x - 3$, so $y = t^2 = (x - 3)^2 = x^2 - 6x + 9$
b) $x = 3t \Rightarrow t = \frac{x}{3}$ so $y = \frac{6}{t} = \frac{18}{x}$
c) $x = 2t^3 \Rightarrow t = \left(\frac{x}{2}\right)^{\frac{1}{3}}$, so $y = t^2 = \left(\frac{x}{2}\right)^{\frac{2}{3}}$
d) $x = t + 7 \Rightarrow t = x - 7$,
so $y = 12 - 2t = 12 - 2(x - 7) = 26 - 2x$

e) $x = t + 4 \Rightarrow t = x - 4$,
so $y = t^2 - 9 = (x - 4)^2 - 9 = x^2 - 8x + 7$

f) $x = \sin\theta,\ y = \cos\theta$
Use trig identities here rather than rearranging...
$\sin^2\theta + \cos^2\theta \equiv 1 \Rightarrow x^2 + y^2 = 1$

g) $x = 1 + \sin\theta \Rightarrow \sin\theta = x - 1$
$y = 2 + \cos\theta \Rightarrow \cos\theta = y - 2$
$\sin^2\theta + \cos^2\theta \equiv 1 \Rightarrow (x - 1)^2 + (y - 2)^2 = 1$
You can leave this equation in the form it's in — it's the equation of a circle radius 1, centre (1, 2).

h) $x = \sin\theta,\ y = \cos 2\theta$
The 2θ should make you think of the double angle formulae...
$\cos 2\theta \equiv 1 - 2\sin^2\theta \Rightarrow y = 1 - 2x^2$

i) $x = \cos\theta,\ y = \cos 2\theta$
$\cos 2\theta \equiv 2\cos^2\theta - 1 \Rightarrow y = 2x^2 - 1$

j) $x = \cos\theta - 5 \Rightarrow \cos\theta = x + 5$
$y = \cos 2\theta = 2\cos^2\theta - 1 \Rightarrow y = 2(x + 5)^2 - 1$
$= 2x^2 + 20x + 49$

Q2 $x = \tan\theta,\ y = \sec\theta$
Using the identity $\sec^2\theta \equiv 1 + \tan^2\theta$ gives: $y^2 = 1 + x^2$

Q3 $x = 2\cot\theta \Rightarrow \cot\theta = \frac{x}{2}$
$y = 3\,\mathrm{cosec}\,\theta \Rightarrow \mathrm{cosec}\,\theta = \frac{y}{3}$
Using the identity $\mathrm{cosec}^2\theta \equiv 1 + \cot^2\theta$ gives:
$\frac{y^2}{9} = 1 + \frac{x^2}{4} \Rightarrow y^2 = 9 + \frac{9x^2}{4}$

Q4 a) From the parametric equations the centre of the circle is (5, –3), and the radius is 1.

b) $x = 5 + \sin\theta \Rightarrow \sin\theta = x - 5$
$y = -3 + \cos\theta \Rightarrow \cos\theta = y + 3$
Using the identity $\sin^2\theta + \cos^2\theta \equiv 1$ gives:
$(x - 5)^2 + (y + 3)^2 = 1$

Q5 a) $x = \frac{1 + 2t}{t} \Rightarrow xt = 1 + 2t \Rightarrow xt - 2t = 1$
$\Rightarrow t(x - 2) = 1 \Rightarrow t = \frac{1}{(x-2)}$

b) $y = \frac{3 + t}{t^2} = \frac{3 + \frac{1}{(x-2)}}{\frac{1}{(x-2)^2}} = 3(x - 2)^2 + (x - 2)$
$= 3(x^2 - 4x + 4) + x - 2$
$= 3x^2 - 11x + 10$
$= (3x - 5)(x - 2)$

c)

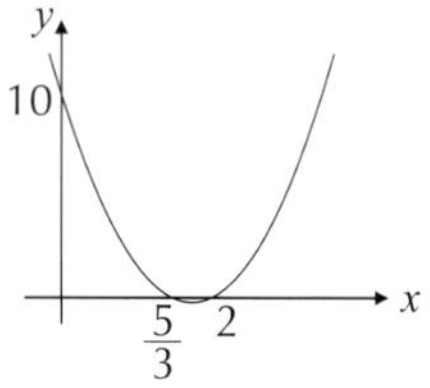

Q6 $x = \frac{2 - 3t}{1 + t} \Rightarrow x(1 + t) = 2 - 3t \Rightarrow x + xt = 2 - 3t$
$\Rightarrow xt + 3t = 2 - x \Rightarrow t(x + 3) = 2 - x \Rightarrow t = \frac{2 - x}{x + 3}$

$$y = \frac{5 - t}{4t + 1} = \frac{5 - \left(\frac{2-x}{x+3}\right)}{4\left(\frac{2-x}{x+3}\right) + 1} = \frac{5(x+3) - (2-x)}{4(2-x) + (x+3)}$$

$\Rightarrow y = \frac{6x + 13}{11 - 3x}$
You could also write this as $6x - 11y + 3xy + 13 = 0$

Q7 $x = 5\sin^2\theta \Rightarrow \sin^2\theta = \frac{x}{5},\ y = \cos\theta$
Using the identity $\sin^2\theta + \cos^2\theta \equiv 1$ gives:
$\frac{x}{5} + y^2 = 1 \Rightarrow y^2 = 1 - \frac{x}{5}$

Q8 a) $x = a\sin\theta \Rightarrow \sin\theta = \frac{x}{a}$
$y = b\cos\theta \Rightarrow \cos\theta = \frac{y}{b}$
Using the identity $\sin^2\theta + \cos^2\theta \equiv 1$ gives:
$\left(\frac{x}{a}\right)^2 + \left(\frac{y}{b}\right)^2 = 1$

b) To sketch the graph, find the x- and y-intercepts.
When $x = 0$, $\left(\frac{y}{b}\right)^2 = 1 \Rightarrow y = \pm b$.
When $y = 0$, $\left(\frac{x}{a}\right)^2 = 1 \Rightarrow x = \pm a$.
So the curve looks like this:

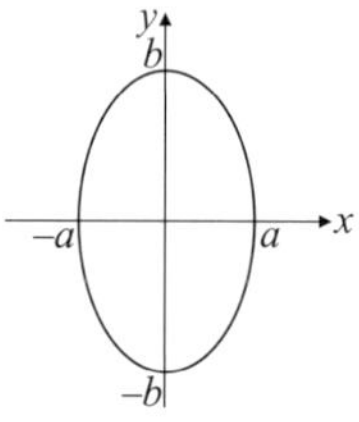

c) An ellipse.
If a and b were equal it would be a circle.

Q9 a) $y = 2t - 1 \Rightarrow t = \frac{y + 1}{2}$
$x = 3t^2 = 3\left(\frac{y+1}{2}\right)^2 = \frac{3}{4}(y + 1)^2$

b) Substitute $y = 4x - 3$ into the equation:
$x = \frac{3}{4}(4x - 3 + 1)^2 \Rightarrow 4x = 3(4x - 2)^2$
$4x = 3(16x^2 - 16x + 4) \Rightarrow 4x = 48x^2 - 48x + 12$
$12x^2 - 13x + 3 = 0 \Rightarrow (3x - 1)(4x - 3) = 0$
$x = \frac{1}{3}$ and $x = \frac{3}{4}$
$y = 4\left(\frac{1}{3}\right) - 3 = -\frac{5}{3}$ and $y = 4\left(\frac{3}{4}\right) - 3 = 0$
So the curve and the line intersect at $\left(\frac{1}{3}, -\frac{5}{3}\right)$ and $\left(\frac{3}{4}, 0\right)$.

Q10 $x = 7t + 2\ \ t = \frac{x - 2}{7}$
$y = \frac{5}{t} = \frac{5}{\left(\frac{x-2}{7}\right)} = \frac{35}{x - 2}$
This is the graph of $y = \frac{1}{x}$ stretched vertically by a factor of 35 and translated right by 2.
The y-intercept is at $x = 0$, so $y = \frac{35}{-2} = -17.5$.

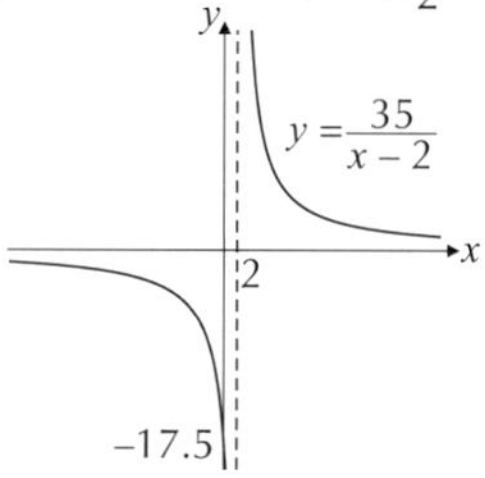

Chapter 5: Sequences and Series

1. Sequences

Exercise 1.1 — n^{th} term

Q1 $a_{20} = 3(20) - 5 = 55$

Q2 4th term $= 4(4 + 2) = 24$

Q3 1st term $= (1 - 1)(1 + 1) = 0$
2nd term $= (2 - 1)(2 + 1) = 3$
Using the same method, 3rd, 4th and 5th terms = 8, 15, 24

Q4 $29 = 4k - 3 \Rightarrow k = 8$

Q5 $a_k = 13 - 6k$ and $a_{k+1} = 13 - 6(k+1) = 7 - 6k$
The sequence is decreasing if $a_{k+1} < a_k$ for all values of k.
$7 - 6k < 13 - 6k \Rightarrow 7 < 13$, which is true
so the sequence is decreasing.

Q6 Form equations for the 2nd and 5th terms:
$15 = a(2^2) + b$
$99 = a(5^2) + b$
Solve the equations simultaneously to get $a = 4$, $b = -1$

Q7 Form equations for the first 3 terms:
$9 = (1^2)e + f + g$
$20 = (2^2)e + 2f + g$
$37 = (3^2)e + 3f + g$
Solve the equations simultaneously
to get $e = 3$, $f = 2$, $g = 4$.
To solve simultaneous equations with 3 unknowns, you use a similar method to when there are 2 unknowns — it just takes a few more steps.

Q8 $49 = (n - 1)^2$, $n = 8$

Q9 This first 8 terms of the sequence are: 13, 11, 9, 7, 5, 3, 1, –1,... The sequence continues to decrease.
So 7 terms are positive.
A different way to solve this one would be to use an inequality — set $15 - 2n > 0$ and solve for n (taking the integer value of n).

Exercise 1.2 — Recurrence relations

Q1 $u_1 = 10$
$u_2 = 3u_1 = 3(10) = 30$
$u_3 = 3u_2 = 3(30) = 90$
$u_4 = 3u_3 = 3(90) = 270$
$u_5 = 3u_4 = 3(270) = 810$

Q2 $u_1 = 2$
$u_2 = u_1^2 = 2^2 = 4$
$u_3 = u_2^2 = 4^2 = 16$
$u_4 = u_3^2 = 16^2 = 256$

For Q3, 4, 5, you can use any letter in place of u.

Q3 Each term is the previous term doubled and the first term is 3.
$u_{n+1} = 2u_n$, $u_1 = 3$

Q4 **a)** Each term is 4 more than the previous term. The first term is 12.
$u_{n+1} = u_n + 4$, $u_1 = 12$

b) Work out the number of 'jumps' of 4 needed to get from 28 to 100:
$100 - 28 = 72$, $72 \div 4 = 18$
Add on the first 5 terms given in the question:
$18 + 5 = 23$ terms

You could have written an nth term expression and used it to find the position of 100 in the sequence (it's the last term). The nth term expression would be 4n + 8.

Q5 $u_{n+1} = 11 - u_n$ or $u_{n+1} = 28 \div u_n$, with $u_1 = 7$.
This one is tricky. It's the sort you suddenly go "aah" with. There are other possible answers here, but these are the two simplest.

Q6 $u_1 = 4$, $u_2 = 3(4) - 1 = 11$, $u_3 = 3(11) - 1 = 32$
$u_4 = 3(32) - 1 = 95$, so $k = 4$

Q7 $x_1 = 9$, $x_2 = (9 + 1) \div 2 = 5$
Keep substituting the result into the formula until...
$x_6 = (\frac{3}{2} + 1) \div 2 = \frac{5}{4}$, so $r = 6$

Q8 $u_1 = 7$, $u_2 = 7 + 1 = 8$, $u_3 = 8 + 2 = 10$
$u_4 = 10 + 3 = 13$, $u_5 = 13 + 4 = 17$

Q9 Form an equation for getting u_2 from u_1, and an equation for getting u_3 from u_2:
$7 = 6a + b$, $8.5 = 7a + b$
Solve the equations simultaneously to get $a = 1.5$, $b = -2$

Q10 First 5 terms:
$u_1 = 8$, $u_2 = \frac{1}{2}(8) = 4$, $u_3 = \frac{1}{2}(4) = 2$
$u_4 = \frac{1}{2}(2) = 1$, $u_5 = \frac{1}{2}(1) = \frac{1}{2}$
The terms are all powers of 2:
$8 = 2^3$, $4 = 2^2$, $2 = 2^1$, $1 = 2^0$, $\frac{1}{2} = 2^{-1}$
so $u_n = 2^{(4-n)}$ or $u_n = 16 \div 2^n$

2. Arithmetic Sequences

Exercise 2.1 — Finding the nth term

Q1 $a = 7$, $d = 5$
n^{th} term $= a + (n - 1)d = 7 + (n - 1)5 = 5n + 2$
10th term $= 5(10) + 2 = 52$

Q2 **a)** $a = 6$, $d = 3$,
so n^{th} term $= 6 + (n - 1)3 = 3n + 3$

b) $a = 4$, $d = 5$, n^{th} term $= 5n - 1$

c) $a = 12$, $d = -4$, n^{th} term $= -4n + 16$

d) $a = 1.5$, $d = 2$, n^{th} term $= 2n - 0.5$

Q3 Form equations for 4th and 10th terms:
n^{th} term $= a + (n - 1)d$
$19 = a + (4 - 1)d$ $\quad$ $19 = a + 3d$
$43 = a + (10 - 1)d$ $\quad$ $43 = a + 9d$
Solving the simultaneous equations: $a = 7$, $d = 4$.

Q4 Form equations for 7th and 11th terms:
n^{th} term $= a + (n - 1)d$
$8 = a + (7 - 1)d$ $\quad$ $8 = a + 6d$
$10 = a + (11 - 1)d$ $\quad$ $10 = a + 10d$
Solving the simultaneous equations: $a = 5$, $d = 0.5$
So $u_3 = 5 + 2(0.5) = 6$

Q5 Form equations for 3rd and 7th terms:
n^{th} term $= a + (n - 1)d$
$15 = a + (3 - 1)d$ $\quad$ $15 = a + 2d$
$27 = a + (7 - 1)d$ $\quad$ $27 = a + 6d$
Solving the simultaneous equations: $a = 9$, $d = 3$.
Now write an equation for the kth term:
$66 = 9 + (k - 1)3 = 6 + 3k$
And solve to find $k = 20$

Q6 The difference between all the terms is the same in an arithmetic sequence. Use this fact to set up an equation:
$\ln(x + 8) - \ln x = \ln(x + 48) - \ln(x + 8)$
$\Rightarrow \ln\left(\frac{x+8}{x}\right) = \ln\left(\frac{x+48}{x+8}\right) \Rightarrow \frac{x+8}{x} = \frac{x+48}{x+8}$
$\Rightarrow (x + 8)^2 = (x + 48)x \Rightarrow x^2 + 16x + 64 = x^2 + 48x$
$\Rightarrow 64 = 32x \Rightarrow x = 2$
So the common difference is $\ln(2 + 8) - \ln 2 = \ln 5$.
Add $\ln 5$ to the third term to get the next term:
$\ln(x + 48) + \ln 5 = \ln(2 + 48) + \ln 5 = \ln 250$.

3. Arithmetic Series

Exercise 3.1 — Sequences and Series

Q1 $a = 8$, $d = 3$

n^{th} term $= a + (n-1)d = 8 + (n-1)3 = 3n + 5$

10th term $= 3(10) + 5 = 35$

$$S_n = \frac{n}{2}[2a + (n-1)d]$$

$$S_{10} = \frac{10}{2}[2(8) + 9(3)] = 215$$

Alternatively, you could have used the $S_n = \frac{1}{2}n(a + l)$ *formula here. You'd worked out the last term earlier in the question.*

Q2 Form equations for 2nd and 5th terms:

n^{th} term $= a + (n-1)d$

$16 = a + d$

$10 = a + 4d$

Solving the simultaneous equations: $a = 18$, $d = -2$.

$$S_n = \frac{n}{2}[2a + (n-1)d]$$

$$S_8 = \frac{8}{2}[2(18) + 7(-2)] = 88$$

Q3 $a = 12$, $d = 6$

n^{th} term $= a + (n-1)d = 12 + (n-1)6 = 6n + 6$

$u_{100} = 6(100) + 6 = 606$

$$S_n = \frac{1}{2}n(a + l)$$

$$S_{100} = \frac{1}{2} \times 100(12 + 606) = 30\,900$$

Q4 $a = 5(1) - 2 = 3$

$l = 5(12) - 2 = 58$

$$S_n = \frac{1}{2}n(a + l)$$

$$S_{12} = \frac{1}{2} \times 12(3 + 58) = 366$$

Q5 $a = 20 - 2(1) = 18$, $l = 20 - 2(9) = 2$

$$S_n = \frac{1}{2}n(a + l)$$

$$S_9 = \frac{1}{2} \times 9(18 + 2) = 90$$

Q6 $a = 3$, $d = 2$

$$S_n = \frac{n}{2}[2a + (n-1)d]$$

$$960 = \frac{n}{2}[2(3) + 2(n-1)]$$

$$960 = n^2 + 2n$$

$$n^2 + 2n - 960 = 0$$

You're expecting a whole number for n, so you should be able to factorise the quadratic — you need two numbers that are 2 apart and multiply to give 960.

$(n + 32)(n - 30) = 0$

Ignore the negative solution since n needs to be positive, so $n = 30$.

Q7 $a = 5(1) + 2 = 7$, $l = 5k + 2$

$$S_n = \frac{1}{2}n(a + l)$$

$$553 = \frac{1}{2}k(7 + 5k + 2)$$

$1106 = 5k^2 + 9k$

$5k^2 + 9k - 1106 = 0$

Factorising gives: $(5k + 79)(k - 14) = 0$

Now we can ignore the negative solution, so $k = 14$.

This one looked very tricky to factorise, but you can cheat a little here — you know you're trying to get to k = 14, so one of the brackets is going to be (k – 14)...

Q8 The first thing to do is use the fact that it's an arithmetic progression to write down some equations — remember, there's a common difference, d, between each term.

$x + 11 + d = 4x + 4 \Rightarrow -3x + d = -7$

$x + 11 + 2d = 9x + 5 \Rightarrow -8x + 2d = -6$

You've now got a pair of simultaneous equations in d and x. Solving these gives $x = -4$, $d = -19$

So the first term is $a = -4 + 11 = 7$.

Now you can put $a = 7$, $d = -19$ and $n = 11$ into the formula for S_n:

$$S_n = \frac{n}{2}[2a + (n-1)d]$$

$$S_{11} = \frac{11}{2}[2(7) + 10(-19)]$$

$$S_{11} = \frac{11}{2} \times -176 = -968$$

Exercise 3.2 — Sum of the first *n* natural numbers

Q1 a) $S_n = \frac{1}{2}n(n+1)$

$$S_{10} = \frac{1}{2} \times 10 \times 11 = 55$$

b) $S_{2000} = \frac{1}{2} \times 2000 \times 2001 = 2\,001\,000$

Q2 $S_{32} = \frac{1}{2} \times 32 \times 33 = 528$

Q3 $\sum_{n=11}^{20} n = \sum_{n=1}^{20} n - \sum_{n=1}^{10} n = S_{20} - S_{10}$

$$S_{10} = \frac{1}{2} \times 10 \times 11 = 55$$

$$S_{20} = \frac{1}{2} \times 20 \times 21 = 210$$

$$\sum_{n=11}^{20} n = 210 - 55 = 155$$

Q4 $66 = \frac{1}{2}n(n+1)$

$132 = n^2 + n$

$0 = n^2 + n - 132$

$(n + 12)(n - 11) = 0$

$n = -12$ or 11 — so ignoring the negative answer, the sum of the first 11 terms is 66, so $n = 11$.

Q5 $S_n = \frac{1}{2}n(n+1)$

$120 = \frac{1}{2}k(k+1)$

$240 = k^2 + k$

$0 = k^2 + k - 240$

$(k + 16)(k - 15) = 0$

Ignoring the negative solution gives $k = 15$.

Q6 Subtract the sum of the first 15 natural numbers from the sum of the first 35:

$$S_{35} = \frac{1}{2} \times 35 \times 36 = 630$$

$$S_{15} = \frac{1}{2} \times 15 \times 16 = 120$$

So the sum of the series is $630 - 120 = 510$.

Q7 $S_n = \frac{1}{2}n(n+1)$

$\frac{1}{2}k(k+1) > 1\,000\,000$

$k^2 + k > 2\,000\,000$

$k^2 + k - 2\,000\,000 > 0$

Put the quadratic equal to zero and solve using the quadratic formula to get $k = 1413.7...$ or $-1414.7...$
It's a u-shaped quadratic, so the quadratic is positive when $k > 1413.7$ (ignoring the negative solution).
So the first natural number for which the sum exceeds 1 000 000 is 1414.

4. Geometric Sequences and Series

Exercise 4.1 — Geometric sequences

Q1 Common ratio $r = \frac{\text{second term}}{\text{first term}} = \frac{3}{2} = 1.5$.
Then:
4th term = 3rd term × 1.5 = 4.5 × 1.5 = 6.75
5th term = 4th term × 1.5 = 6.75 × 1.5 = 10.125,
6th term = 5th term × 1.5 = 10.125 × 1.5 = 15.1875,
7th term = 6th term × 1.5 = 15.1875 × 1.5 = 22.78125.
This method isn't as slow as it looks because you can use a scientific calculator to get the terms of the series quickly: press '2 =' then '× 1.5 =' to get the second term. Pressing '=' repeatedly will give you the following terms. Even so, the method below is quicker, so unless you're asked to find the term after one you've already got, you're better off doing this:
Or: First term $a = 2$, common ratio $r = 1.5$
n^{th} term $= ar^{n-1} = 2 \times (1.5)^{n-1}$
7th term $= 2 \times (1.5)^6 = 2 \times 11.390625 = 22.78125$

Q2 Common ratio $r = \frac{7^{th}\text{ term}}{6^{th}\text{ term}} = \frac{6561}{2187} = 3$.
The 6th term is $2187 = ar^5 = a \times 3^5 \Rightarrow a = \frac{2187}{3^5} = 9$.
So the 1st term is 9.

Q3 Common ratio: $r = \frac{\text{second term}}{\text{first term}} = \frac{12}{24} = 0.5$.
First term: $a = 24$.
n^{th} term: $u_n = ar^{n-1} = 24 \times (0.5)^{n-1}$
9th term: $u_9 = ar^8 = 24 \times (0.5)^8 = 0.09375$

Q4 First term: $a = 1.125$
n^{th} term: $u_n = ar^{n-1} = 1.125r^{n-1}$
14th term: $9216 = u_{14} = 1.125r^{13}$
$9216 = 1.125r^{13} \Rightarrow r^{13} = \frac{9216}{1.125} = 8192$
$\Rightarrow r = \sqrt[13]{8192} = 2$

Q5 Common ratio: $r = \frac{1.1}{1} = 1.1$, first term: $a = 1$.
n^{th} term: $u_n = ar^{n-1} = 1 \times 1.1^{n-1} = 1.1^{n-1}$
So to find the number of terms in the sequence that are less than 4, solve: $u_n = 1.1^{n-1} < 4$
$\Rightarrow \log 1.1^{n-1} < \log 4$
Use the log law, $\log x^n = n(\log x)$:
$\Rightarrow (n-1)\log 1.1 < \log 4$
log 1.1 > 0 so dividing through by log 1.1 doesn't change the direction of the inequality:
$\Rightarrow n - 1 < \frac{\log 4}{\log 1.1}$
$\Rightarrow n - 1 < 14.54$
$\Rightarrow n < 15.54$
so u_n is less than 4 when n is less than 15.54 (2 d.p.).
Therefore u_{15} is the last term that's less than 4, so there are 15 terms that are less than 4.
You could solve this as an equation instead, finding the value of n such that $u_n = 4$ and rounding down.

Q6 $a = 5$, $r = 0.6$, n^{th} term: $u_n = ar^{n-1} = 5 \times (0.6)^{n-1}$
10th term: $u_{10} = 5 \times (0.6)^9 = 0.050388$ (6 d.p.)
15th term: $u_{15} = 5 \times (0.6)^{14} = 0.003918$ (6 d.p.)
Difference: 0.003918 – 0.050388 = –0.04647 (5 d.p.)
You could also have:
Difference: 0.050388 – 0.003918 = 0.04647 (5 d.p.)

Q7 $a = 25\,000$, $r = 0.8$
n^{th} term: $u_n = ar^{n-1} = 25\,000 \times (0.8)^{n-1}$
to find the first term in the sequence less than 1000,
solve: $u_n = 25\,000 \times (0.8)^{n-1} < 1000$
$25\,000 \times (0.8)^{n-1} < 1000$
$\Rightarrow (0.8)^{n-1} < \frac{1000}{25\,000} = 0.04$
$\Rightarrow \log(0.8)^{n-1} < \log 0.04$
$\Rightarrow (n-1)\log(0.8) < \log 0.04$
$\Rightarrow n - 1 > \frac{\log 0.04}{\log 0.8}$
$\Rightarrow n - 1 > 14.425...$
$\Rightarrow n > 15.425...$
so u_n is less than 1000 when n is greater than 15.425..., therefore u_{16} is the first term that's less than 1000.
0.8 < 1 so log 0.8 < 0 and dividing through by log 0.8 changes the direction of the inequality because log 0.8 is negative. Again, you could solve this as an equation by finding n such that $u_n = 1000$ and rounding up.

Q8 Divide consecutive terms to find the common ratio r:
e.g. $r = \frac{\text{second term}}{\text{first term}} = \frac{-5}{5} = -1$

Q9 **a)** Common ratio $r = \frac{\text{second term}}{\text{first term}} = \frac{\frac{3}{16}}{\frac{1}{4}} = \frac{3}{4}$
b) First term: $a = \frac{1}{4}$
n^{th} term: $u_n = ar^{n-1} = \frac{1}{4} \times \left(\frac{3}{4}\right)^{n-1}$
8th term: $u_8 = \frac{1}{4} \times \left(\frac{3}{4}\right)^7 = \frac{1}{4} \times \frac{2187}{16384} = \frac{2187}{65536}$

Q10 $r = 0.8$, n^{th} term: $u_n = ar^{n-1} = a(0.8)^{n-1}$
7th term: $196.608 = u_7 = a(0.8)^6$
$196.608 = a(0.8)^6 \Rightarrow a = \frac{196.608}{0.8^6} = 750$

Q11 **a)** Common ratio $r = \frac{\text{second term}}{\text{first term}} = \frac{-2.4}{3} = -0.8$
b) Continuing the sequence gives –1.536, 1.2288, –0.98304. So there are 5 terms in the series before a term has modulus less than 1.
You could also answer this part of the question by writing a new series where each term is the modulus of the old series, then using logs to find the first term less than 1. But in this case it's much easier to just find the next few terms.

Exercise 4.2 — Geometric series

Q1 The sum of the first n terms is $S_n = \frac{a(1-r^n)}{(1-r)}$, $a = 8$ and $r = 1.2$, so the sum of the first 15 terms is:
$S_{15} = \frac{a(1-r^{15})}{(1-r)} = \frac{8(1-(1.2)^{15})}{(1-1.2)} = 576.28$ to 2 d.p.

Q2 For a geometric series with first term a and common ratio r: $\sum_{k=0}^{n-1} ar^k = \frac{a(1-r^n)}{1-r}$,
so: $\sum_{k=0}^{9} ar^k = \sum_{k=0}^{9} 25(0.7)^k = \frac{25(1-(0.7)^{10})}{1-0.7} = 80.98$ (2 d.p.)

Q3 $a = 3$ and $r = 2$, the sum of the first n terms is:

$$\frac{3(1-2^n)}{(1-2)} = -3(1-2^n)$$

$196\,605 = S_n = -3(1 - 2^n) \Rightarrow -65\,535 = 1 - 2^n$

$\Rightarrow 65\,536 = 2^n \Rightarrow \log 65\,536 = \log 2^n$

$\Rightarrow \log 65\,536 = n\log 2 \Rightarrow n = \frac{\log 65\,536}{\log 2} = 16$

Q4 The first term is $a = 4$,
the common ratio is $r = \frac{\text{second term}}{\text{first term}} = \frac{5}{4} = 1.25$
The sum of the first x terms is

$$S_x = \frac{a(1-r^x)}{(1-r)} = \frac{4(1-(1.25)^x)}{(1-1.25)} = 16(1-(1.25)^x)$$

So: $103.2 = -16(1 - (1.25)^x) \Rightarrow -6.45 = 1 - (1.25)^x$

$\Rightarrow 7.45 = 1.25^x \Rightarrow \log 7.45 = x\log 1.25$

$\Rightarrow x = \frac{\log 7.45}{\log 1.25} = 9.00$ to 2 d.p.

Q5 **a)** 3rd term $= ar^2 = 6$, 8th term $= ar^7 = 192$.
Dividing the two equations gives:

$$\frac{ar^7}{ar^2} = r^5 = \frac{192}{6} = 32 \Rightarrow r = \sqrt[5]{32} = 2$$

b) 3rd term $= ar^2 = 6$ and $r = 2$, so $a = \frac{6}{r^2} = \frac{6}{2^2} = 1.5$

c) The sum of the first 15 terms is:

$$S_{15} = \frac{a(1-r^{15})}{(1-r)} = \frac{1.5(1-2^{15})}{(1-2)} = 49\,150.5$$

Q6 **a)** Common ratio: $\frac{\text{2nd term}}{\text{1st term}} = \frac{\text{3rd term}}{\text{2nd term}}$, so:

$\frac{m}{m+10} = \frac{2m-21}{m} \Rightarrow m^2 = (m + 10)(2m - 21)$

$\Rightarrow m^2 = 2m^2 - 21m + 20m - 210$

$\Rightarrow 0 = m^2 - m - 210$

b) Factorising $m^2 - m - 210 = 0$ gives:
$(m - 15)(m + 14) = 0$, so $m = 15$ or $m = -14$,
since $m > 0$, $m = 15$.

c) $m = 15$ gives the first three terms 25, 15, 9. Common ratio $= \frac{\text{second term}}{\text{first term}} = \frac{15}{25} = 0.6$

d) $a = 25$ and $r = 0.6$, so sum of first 10 terms is:

$$S_{10} = \frac{a(1-r^{10})}{(1-r)} = \frac{25(1-0.6^{10})}{(1-0.6)} = 62.12 \text{ to 2 d.p.}$$

Q7 **a)** $1 + x + x^2 = 3 \Rightarrow x^2 + x - 2 = 0$

$\Rightarrow (x - 1)(x + 2) = 0$

$\Rightarrow x = 1$ or $x = -2$.

Since the terms are all different $x \neq 1$ (as 1 is the first term and $x = 1 \Rightarrow x^2 = 1^2 = 1$), hence $x = -2$.

b) $a = 1$ and $r = \frac{\text{second term}}{\text{first term}} = \frac{-2}{1} = -2$,
so the sum of the first 7 terms is:

$$S_7 = \frac{a(1-r^7)}{(1-r)} = \frac{1(1-(-2)^7)}{(1-(-2))} = 43$$

Q8 $a = 7.2$ and $r = 0.38$, so:

$$\sum_{k=0}^{9} ar^k = \frac{a(1-r^{10})}{(1-r)} = \frac{7.2(1-0.38^{10})}{(1-0.38)} = 11.61 \text{ to 2 d.p.}$$

Q9 $$1.2 = S_8 = \frac{a(1-r^8)}{(1-r)} = \frac{a\left(1-\left(-\frac{1}{3}\right)^8\right)}{\left(1-\left(-\frac{1}{3}\right)\right)} = a(0.749...)$$

$\Rightarrow a = 1.60$ to 2 d.p.

Q10 The geometric sequence $a, -2a, 4a, ...$ has first term a and common ratio $r = -2$, so $\sum_{k=0}^{12} a(-2)^k$ is the sum of the first 13 terms of the sequence. Therefore:

$$\sum_{k=0}^{12} a(-2)^k = -5735.1$$

$$\sum_{k=0}^{12} a(-2)^k = \frac{a(1-r^{13})}{(1-r)} = \frac{a(1-(-2)^{13})}{(1-(-2))} = 2731a$$

$\Rightarrow -5735.1 = 2731a \Rightarrow a = -2.1$

Exercise 4.3 — Convergent geometric series

Q1 **a)** $r = \frac{1.1}{1} = 1.1$, $|r| = |1.1| = 1.1 > 1$,
so the sequence does not converge.

b) $r = \frac{0.8^2}{0.8} = 0.8$, $|r| = |0.8| = 0.8 < 1$,
so the sequence converges.

c) $r = \frac{\frac{1}{4}}{1} = \frac{1}{4}$, $|r| = \left|\frac{1}{4}\right| = \frac{1}{4} < 1$,
so the sequence converges.

d) $r = \frac{\frac{9}{2}}{3} = \frac{3}{2}$, $|r| = \left|\frac{3}{2}\right| = \frac{3}{2} > 1$,
so the sequence does not converge.

e) $r = \frac{-\frac{1}{2}}{1} = -\frac{1}{2}$, $|r| = \left|-\frac{1}{2}\right| = \frac{1}{2} < 1$,
so the sequence converges.

f) $r = \frac{5}{5} = 1$, $|r| = |1| = 1$ (and 1 is not less than 1),
so the sequence does not converge.

Q2 Find the common ratio: $r = \frac{2^{\text{nd}} \text{ term}}{1^{\text{st}} \text{ term}} = \frac{8.1}{9} = 0.9$.
The first term is $a = 9$.
The sum to infinity is:

$$S_\infty = \frac{a}{1-r} = \frac{9}{1-0.9} = \frac{9}{0.1} = 90$$

Q3 $S_\infty = \frac{a}{1-r} = 2a \Rightarrow \frac{1}{1-r} = 2 \Rightarrow 1 = 2 - 2r$

$\Rightarrow 2r = 1 \Rightarrow r = 0.5$

Q4 **a)** Sum to infinity is $13.5 = S_\infty = \frac{a}{1-r}$

Sum of the first three terms is $13 = S_3 = \frac{a(1-r^3)}{(1-r)}$

Divide S_3 by S_∞: $\frac{13}{13.5} = \frac{a(1-r^3)}{1-r} \div \frac{a}{1-r}$

$= \frac{a(1-r^3)}{1-r} \times \frac{1-r}{a} = 1 - r^3$

You can cancel the $1 - r$ because r can't be 1 (as the series converges), so $1 - r \neq 0$.

So $1 - r^3 = \frac{13}{13.5} = \frac{26}{27} \Rightarrow r^3 = \frac{1}{27} \Rightarrow r = \frac{1}{3}$.

b) $13.5 = S_\infty = \frac{a}{1-r} \Rightarrow a = 13.5(1 - r)$

$= 13.5 \times \frac{2}{3} = 9$.

Q5 $ar = 3, 12 = S_\infty = \frac{a}{1-r} \Rightarrow 12 - 12r = a$

The first equation gives $a = \frac{3}{r}$,
plugging this into the second equation gives:

$12 - 12r = \frac{3}{r} \Rightarrow 12r - 12r^2 = 3$

$\Rightarrow 12r^2 - 12r + 3 = 0$

$\Rightarrow 4r^2 - 4r + 1 = 0$

$4r^2 - 4r + 1 = 0$ factorises to $(2r - 1)(2r - 1) = 0$

Hence $2r - 1 = 0 \Rightarrow r = 0.5$

Then $a = \frac{3}{r} = \frac{3}{0.5} = 6$.

Q6 **a)** $a = 6$, $10 = S_\infty = \frac{a}{1-r} = \frac{6}{1-r}$
$\Rightarrow 1 - r = \frac{6}{10} = 0.6 \Rightarrow r = 0.4$

b) 5th term: $u_5 = ar^4 = 6 \times 0.4^4 = 0.1536$.

Q7 **a)** Second term $= ar = -48$, 5th term $= ar^4 = 0.75$.
Dividing gives: $r^3 = \frac{ar^4}{ar} = \frac{0.75}{-48} = -0.015625$
$\Rightarrow r = -0.25$

b) $ar = -48 \Rightarrow a = \frac{-48}{r} = \frac{-48}{-0.25} = 192$

c) $|r| < 1$ so you can find the sum to infinity:
$S_\infty = \frac{a}{1-r} = \frac{192}{1-(-0.25)} = \frac{192}{1.25} = 153.6.$

Q8 The sum of terms after the 10th is $S_\infty - S_{10}$.
So the question tells you that $S_\infty - S_{10} < \frac{1}{100}S_\infty$
$\Rightarrow \frac{99}{100}S_\infty - S_{10} < 0 \Rightarrow \frac{99}{100}S_\infty < S_{10}$
Then: $0.99(S_\infty) = \frac{0.99a}{1-r} < \frac{a(1-r^{10})}{(1-r)} = S_{10}$
You can cancel and keep the inequality sign because the series is convergent so $|r| < 1 \Rightarrow 1 - r > 0$, and you know $a > 0$ from the question:
$\Rightarrow 0.99 < 1 - r^{10} \quad \Rightarrow r^{10} < 0.01$
$\Rightarrow |r| < \sqrt[10]{0.01} \quad \Rightarrow |r| < 0.631$ (to 3 s.f.)

Q9 Using the formulas for S_∞ and S_4
$\frac{a}{1-r} = \frac{9}{8} \times \frac{a(1-r^4)}{1-r}$
Cancelling $(1 - r)$ and a gives
$\frac{8}{9} = 1 - r^4 \Rightarrow r^4 = \frac{1}{9} \Rightarrow r = \sqrt[4]{\frac{1}{9}}$
So, given that r is positive and real, $r = \frac{1}{\sqrt{3}} = \frac{\sqrt{3}}{3}$.

5. Modelling Problems

Exercise 5.1 — Real-life problems

Q1 $a = 60$, $d = 3$
n^{th} term $= a + (n-1)d = 60 + (n-1)3 = 3n + 57$
12th term $= 3(12) + 57 = 93$
So she'll earn £93 in her 12th week.
With wordy problems, don't forget to check what the units should be and include them.

Q2 The heights form a geometric sequence (starting from the smallest doll), with $a = 3$, $r = 1.25$.
The nth term is $ar^{n-1} = 3 \times 1.25^{n-1}$,
The first term is the height of the first doll, so the height of the 8th doll is the 8th term.
The 8th term is: $3 \times 1.25^7 = 14.3$ cm (to 1 d.p.)

Q3 $a = 40$, $d = 5$
n^{th} term $= a + (n-1)d = 40 + (n-1)5 = 5n + 35$
$80 = 5n + 35 \Rightarrow n = (80 - 35) \div 5 = 9$
So he'll sell 80 sandwiches on the 9th day.

Q4 $a = 300\,000$, $d = -30\,000$
n^{th} term $= a + (n-1)d$
$= 300\,000 - 30\,000(n-1)$
$= 330\,000 - 30\,000n$
$330\,000 - 30\,000n < 50\,000$
$n > 280\,000 \div 30\,000 = 9.33...$
You want the smallest integer value of n that satisfies the inequality, so round up. Sales will have fallen below £50 000 after 10 months.
$n = 9$ doesn't satisfy the inequality.

Q5 The value decreases by 15% each year so multiply by $1 - 0.15 = 0.85$ to get from one term to the next, so the common ratio $r = 0.85$.
So the nth term $u_n = ar^{n-1} = a(0.85)^{n-1}$
The price when new is the first term in the series and the value after 10 years is the 11th term.
11th term: $2362 = u_{11} = a(0.85)^{10}$
$2362 = a(0.85)^{10} \Rightarrow a = \frac{2362}{0.85^{10}} = 11\,997.496...$
$= 11\,997.50$ to 2 d.p.
When new, the car cost £11 997.50

Q6 **a)** The cost increases by 3% each year so multiply by $1 + 0.03 = 1.03$ to get from one term to the next, so
2006 cost $= 1.03 \times$ (2011 cost)
$= 1.03 \times$ £120 = £123.60

b) The costs each year form a geometric sequence with common ratio $r = 1.03$ and first term $a = 120$, so the total cost between 2011 and 2016 (including 2011 and 2016) is the sum of the first 6 terms:
$S_6 = \frac{a(1-r^6)}{(1-r)} = \frac{120(1-1.03^6)}{(1-1.03)} = 776.21$ to 2 d.p.
Nigel paid £776.21 (the cost to the nearest penny).

Q7 $a = 6000$, $d = 2000$ and $n = 12$ (12 months in the year)
$S_n = \frac{n}{2}[2a + (n-1)d]$
$S_{12} = \frac{12}{2}[2(6000) + 11(2000)] = 204\,000$
204 000 copies will be sold.

Q8 4 weeks is 28 days. The claim is that the leeks' height increases by 15% every 2 days.
$28 \div 2 = 14$, so there are 14 lots of 2 days in 28 days
In 28 days, the height should increase 14 times by 15%.
So if the claim is correct, the height of the leeks after 28 days will be the 15th term of a geometric progression with first term 5 and common ratio 1.15.
The first term is the initial height, the second term is the height after one 15% increase and so on, so the 15th term is the height after fourteen 15% increases.
The common ratio is 1.15 because multiplying something by 1.15 is the same as increasing it by 15%.
So $a = 5$ and $r = 1.15$.
The nth term is: $u_n = ar^{n-1} = 5(1.15)^{n-1}$
So the 15th term is: $5 \times 1.15^{14} = 35.3785...$ cm.
So if the claim were true, the leeks would be 35.4 cm tall (to 1 d.p.). Since the leeks only reach a height of 25 cm, the claim on the compost is not justified.

Q9 **a)** The gnome value goes up by 2% each year, so the price after 1 year is 102% of £80 000:
$80\,000 \times 1.02 = 81\,600$
The value after 1 year is £ 81 600.

b) To get from one term to the next you multiply by 1.02 (to increase by 2% each time), so the common ratio $r = 1.02$.

c) Price at start $= a = 80\,000$, $r = 1.02$
n^{th} term $= u_n = ar^{n-1} = 80\,000 \times (1.02)^{n-1}$
The value at the start is the 1st term, the value after 1 year is the 2nd term and so on, so the value after 10 years is the 11th term:
11th term $= u_{11} = 80\,000 \times (1.02)^{10}$
$= 97\,519.55$ (2 d.p.)
The value after 10 years is £97 519.55 (to the nearest penny)

d) The value after k years is the $(k + 1)^{\text{th}}$ term:
$(k + 1)^{\text{th}}$ term $= u_{k+1} = ar^k = 80\,000 \times (1.02)^k$
After k years the value is more than 120 000, so:
$u_{k+1} = 80\,000 \times (1.02)^k > 120\,000$
$\Rightarrow \quad (1.02)^k > \frac{120\,000}{80\,000} = 1.5$
$\Rightarrow \log(1.02)^k > \log 1.5$
$\Rightarrow k\log(1.02) > \log 1.5$
$\Rightarrow \quad k > \frac{\log 1.5}{\log 1.02} = 20.475...$
So u_{k+1} (the value after k years) is more than 120 000 when $k > 20.475...$, therefore the value exceeds 120 000 after 21 years.

Q10 Frazer's series is the natural numbers up to 31.
$S_n = \frac{1}{2}n(n+1)$

$S_{31} = \frac{1}{2} \times 31 \times 32 = 496$

So Frazer draws 496 dots.

Q11 The thickness of the paper doubles every time you fold it in half.
So the paper thickness forms a geometric progression:
After 1 fold, thickness = 0.01×2 cm.
After 2 folds, thickness = 0.01×2^2 cm
After n folds, thickness = 0.01×2^n cm
Distance to the moon:
384 000 km $= 3.84 \times 10^5$ km
$= (3.84 \times 10^5 \times 1000)$ m
$= (3.84 \times 10^5 \times 1000 \times 100)$ cm
$= 3.84 \times 10^{10}$ cm
Therefore the paper reaches the moon when:
$0.01 \times 2^n = 3.84 \times 10^{10}$
$2^n = 3.84 \times 10^{12}$.
Taking logs: $n \log 2 = \log(3.84 \times 10^{12})$
$\Rightarrow n = \log(3.84 \times 10^{12}) \div \log 2 = 41.80...$
So after approximately 42 folds the paper would reach the moon.

Q12 a) The distances she runs form a geometric sequence with $a = 12$, $r = 1.03$.
The nth term is $ar^{n-1} = 12 \times 1.03^{n-1}$,
so the 10th term is: $12 \times 1.03^9 = 15.7$ miles (to 1 d.p.).

b) The total distance she runs in 20 days is the sum of the first 20 terms of the sequence.
Using: $S_n = \frac{a(1-r^n)}{(1-r)}$, where $a = 12$, $r = 1.03$, $n = 20$,

$S_{20} = \frac{12(1-1.03^{20})}{(1-1.03)} = 322.44...$

In 20 days she runs a total of 322 miles (to the nearest mile).

Q13 Laura's series is the natural numbers. You need to find how many are needed to exceed 1000 (£10 in pence).
$S_n = \frac{1}{2}n(n+1)$
$\frac{1}{2}(n^2 + n) > 1000$
$n^2 + n > 2000 \Rightarrow n^2 + n - 2000 > 0$
Putting the quadratic equal to zero and solving using the quadratic formula gives $n = 44.2$ (ignoring the negative solution), which by looking at the shape of the quadratic graph gives $n > 44.2$ as the solution to the inequality.
So on the 45th day she'll have over £10.

Q14 After 0 years, $u_1 = a$ and after 1 year, $u_2 = ar$.
So after 10 years, $u_{11} = ar^{10}$. She wants her investment to double so $u_{11} = ar^{10} = 2a \Rightarrow r^{10} = 2$
$\Rightarrow |r| = \sqrt[10]{2} = 1.071773$.
So the interest rate needed is 7.17 % (3 s.f.).

Chapter 6: The Binomial Expansion

1. The Binomial Expansion

Exercise 1.1 — Expansions where n is a positive integer

Q1 $(1+x)^3 = 1 + 3x + \frac{3(3-1)}{1\times 2}x^2 + \frac{3(3-1)(3-2)}{1\times 2\times 3}x^3$
$= 1 + 3x + 3x^2 + x^3$

Q2 $(1+x)^7 = 1 + 7x + \frac{7(7-1)}{1\times 2}x^2 + \frac{7(7-1)(7-2)}{1\times 2\times 3}x^3 + ...$
$= 1 + 7x + 21x^2 + 35x^3 + ...$
This isn't the full expansion, so keep the dots at the end to show it carries on.

Q3 $(1-x)^4 = 1 + 4(-x) + \frac{4(4-1)}{1\times 2}(-x)^2 + \frac{4(4-1)(4-2)}{1\times 2\times 3}(-x)^3$
$+ \frac{4(4-1)(4-2)(4-3)}{1\times 2\times 3\times 4}(-x)^4$
$= 1 - 4x + 6x^2 - 4x^3 + x^4$

Q4 $(1+3x)^6 = 1 + 6(3x) + \frac{6(6-1)}{1\times 2}(3x)^2 + ...$
$= 1 + 6(3x) + 15(9x^2) + ...$
$= 1 + 18x + 135x^2 + ...$

Q5 $(1+2x)^8 = 1 + 8(2x) + \frac{8(8-1)}{1\times 2}(2x)^2$
$+ \frac{8(8-1)(8-2)}{1\times 2\times 3}(2x)^3 + ...$
$= 1 + 8(2x) + 28(4x^2) + 56(8x^3) + ...$
$= 1 + 16x + 112x^2 + 448x^3 + ...$

Q6 $(1-5x)^5 = 1 + 5(-5x) + \frac{5(5-1)}{1\times 2}(-5x)^2 + ...$
$= 1 + 5(-5x) + 10(25x^2) + ...$
$= 1 - 25x + 250x^2 - ...$

Q7 $(1-4x)^3 = 1 + 3(-4x) + \frac{3(3-1)}{1\times 2}(-4x)^2$
$+ \frac{3(3-1)(3-2)}{1\times 2\times 3}(-4x)^3$
$= 1 + 3(-4x) + 3(16x^2) + (-64x^3)$
$= 1 - 12x + 48x^2 - 64x^3$

Q8 $(1+6x)^6 = 1 + 6(6x) + \frac{6(6-1)}{1\times 2}(6x)^2$
$+ \frac{6(6-1)(6-2)}{1\times 2\times 3}(6x)^3 + ...$
$= 1 + 6(6x) + 15(36x^2) + 20(216x^3)$
$= 1 + 36x + 540x^2 + 4320x^3 + ...$

Exercise 1.2 — Expansions where n is negative or a fraction

Q1 $(1+x)^{-4} = 1 + (-4)x + \frac{-4(-4-1)}{1\times 2}x^2$
$+ \frac{-4(-4-1)(-4-2)}{1\times 2\times 3}x^3 + ...$
$= 1 + (-4)x + \frac{-4\times -5}{2}x^2 + \frac{-4\times -5\times -6}{6}x^3 + ...$
$= 1 - 4x + 10x^2 - 20x^3 + ...$

Q2 **a)** $(1-6x)^{-3} = 1 + (-3)(-6x) + \frac{-3(-3-1)}{1\times2}(-6x)^2$

$+ \frac{-3(-3-1)(-3-2)}{1\times2\times3}(-6x)^3 + \ldots$

$= 1 + 18x + \frac{-3\times-4}{2}(36x^2)$

$+ \frac{-3\times-4\times-5}{6}(-216x^3) + \ldots$

$= 1 + 18x + 6(36x^2) + (-10)(-216x^3) + \ldots$

$= 1 + 18x + 216x^2 + 2160x^3 + \ldots$

b) The expansion is valid for $\left|\frac{-6x}{1}\right| < 1$, so $|x| < \frac{1}{6}$.

Q3 **a)** $(1+4x)^{\frac{1}{3}} = 1 + \frac{1}{3}(4x) + \frac{\frac{1}{3}\left(\frac{1}{3}-1\right)}{1\times2}(4x)^2 + \ldots$

$= 1 + \frac{1}{3}(4x) + \frac{\frac{1}{3}\times-\frac{2}{3}}{2}(16x^2) + \ldots$

$= 1 + \frac{1}{3}(4x) + \left(-\frac{1}{9}\right)(16x^2) + \ldots$

$= 1 + \frac{4x}{3} - \frac{16x^2}{9} + \ldots$

b) $(1+4x)^{-\frac{1}{2}} = 1 + \left(-\frac{1}{2}\right)(4x) + \frac{-\frac{1}{2}\left(-\frac{1}{2}-1\right)}{1\times2}(4x)^2 + \ldots$

$= 1 - \frac{1}{2}(4x) + \frac{-\frac{1}{2}\times-\frac{3}{2}}{2}(16x^2) + \ldots$

$= 1 - \frac{1}{2}(4x) + \left(\frac{3}{8}\right)(16x^2) + \ldots$

$= 1 - 2x + 6x^2 - \ldots$

c) Both a) and b) are valid for $|x| < \frac{1}{4}$.

Q4 **a)** $\frac{1}{(1-4x)^2} = (1-4x)^{-2}$

$= 1 + (-2)(-4x) + \frac{-2(-2-1)}{1\times2}(-4x)^2$

$+ \frac{-2(-2-1)(-2-2)}{1\times2\times3}(-4x)^3 + \ldots$

$= 1 + (-2)(-4x) + \frac{-2\times-3}{2}(16x^2)$

$+ \frac{-2\times-3\times-4}{6}(-64x^3) + \ldots$

$= 1 + 8x + 3(16x^2) + (-4)(-64x^3) + \ldots$

$= 1 + 8x + 48x^2 + 256x^3 + \ldots$

b) $\sqrt{1+6x} = (1+6x)^{\frac{1}{2}}$

$= 1 + \frac{1}{2}(6x) + \frac{\frac{1}{2}\left(\frac{1}{2}-1\right)}{1\times2}(6x)^2$

$+ \frac{\frac{1}{2}\left(\frac{1}{2}-1\right)\left(\frac{1}{2}-2\right)}{1\times2\times3}(6x)^3 + \ldots$

$= 1 + \frac{1}{2}(6x) + \frac{\frac{1}{2}\times-\frac{1}{2}}{2}(36x^2)$

$+ \frac{\frac{1}{2}\times-\frac{1}{2}\times-\frac{3}{2}}{6}(216x^3) + \ldots$

$= 1 + \frac{1}{2}(6x) + \left(-\frac{1}{8}\right)(36x^2) + \frac{1}{16}(216x^3) + \ldots$

$= 1 + 3x - \frac{9x^2}{2} + \frac{27x^3}{2} - \ldots$

c) $\frac{1}{\sqrt{1-3x}} = (1-3x)^{-\frac{1}{2}}$

$= 1 + \left(-\frac{1}{2}\right)(-3x) + \frac{-\frac{1}{2}\left(-\frac{1}{2}-1\right)}{1\times2}(-3x)^2$

$+ \frac{-\frac{1}{2}\left(-\frac{1}{2}-1\right)\left(-\frac{1}{2}-2\right)}{1\times2\times3}(-3x)^3 + \ldots$

$= 1 + \left(-\frac{1}{2}\right)(-3x) + \frac{-\frac{1}{2}\times-\frac{3}{2}}{2}(9x^2)$

$+ \frac{-\frac{1}{2}\times-\frac{3}{2}\times-\frac{5}{2}}{6}(-27x^3) + \ldots$

$= 1 + \frac{3x}{2} + \frac{3}{8}(9x^2) + \left(-\frac{5}{16}\right)(-27x^3) + \ldots$

$= 1 + \frac{3x}{2} + \frac{27x^2}{8} + \frac{135x^3}{16} + \ldots$

d) $\sqrt[3]{1+\frac{x}{2}} = \left(1+\frac{1}{2}x\right)^{\frac{1}{3}}$

$= 1 + \frac{1}{3}\left(\frac{x}{2}\right) + \frac{\frac{1}{3}\left(\frac{1}{3}-1\right)}{1\times2}\left(\frac{x}{2}\right)^2$

$+ \frac{\frac{1}{3}\left(\frac{1}{3}-1\right)\left(\frac{1}{3}-2\right)}{1\times2\times3}\left(\frac{x}{2}\right)^3\ldots$

$= 1 + \frac{1}{3}\left(\frac{x}{2}\right) + \frac{\frac{1}{3}\times-\frac{2}{3}}{2}\left(\frac{x}{2}\right)^2$

$+ \frac{\frac{1}{3}\times-\frac{2}{3}\times-\frac{5}{3}}{6}\left(\frac{x}{2}\right)^3\ldots$

$= 1 + \frac{1}{3}\left(\frac{x}{2}\right) + \left(-\frac{1}{9}\right)\left(\frac{x}{2}\right)^2 + \frac{5}{81}\left(\frac{x}{2}\right)^3\ldots$

$= 1 + \frac{x}{6} - \frac{x^2}{36} + \frac{5x^3}{648}\ldots$

Q5 **a)** $\frac{1}{(1+7x)^4} = (1+7x)^{-4}$

The x^3 term is $\frac{-4(-4-1)(-4-2)}{1\times2\times3}(7x)^3$

$= \frac{-4\times-5\times-6}{6}(343x^3)$

$= -20(343x^3) = -6860x^3$

So the coefficient of the x^3 term is -6860.

b) The expansion is valid for $|x| < \frac{1}{7}$.

Q6 **a)** $\sqrt[4]{1-4x} = (1-4x)^{\frac{1}{4}}$

The x^5 term is:

$\frac{\frac{1}{4}\left(\frac{1}{4}-1\right)\left(\frac{1}{4}-2\right)\left(\frac{1}{4}-3\right)\left(\frac{1}{4}-4\right)}{1\times2\times3\times4\times5}(-4x)^5$

$= \frac{\frac{1}{4}\times-\frac{3}{4}\times-\frac{7}{4}\times-\frac{11}{4}\times-\frac{15}{4}}{120}(-4)^5x^5$

$= \frac{1\times-3\times-7\times-11\times-15}{120}\times\left(\frac{-4}{4}\right)^5\times x^5$

$= \frac{3465}{120}\times-1\times x^5 = -\frac{231x^5}{8}$

So the coefficient of the x^5 term is $-\frac{231}{8}$.

b) The expansion is valid for $|x| < \frac{1}{4}$.

Q7 **a)** $(1-5x)^{\frac{1}{6}} = 1 + \frac{1}{6}(-5x) + \frac{\frac{1}{6}\left(\frac{1}{6}-1\right)}{1\times2}(-5x)^2 + \ldots$

$= 1 + \frac{1}{6}(-5x) + \frac{\frac{1}{6}\times-\frac{5}{6}}{2}(25x^2) + \ldots$

$= 1 - \frac{5x}{6} + \left(-\frac{5}{72}\right)(25x^2) + \ldots$

$= 1 - \frac{5x}{6} - \frac{125x^2}{72} - \ldots$

b) $(1+4x)^4 = 1 + 16x + 96x^2 + \ldots$

So $(1+4x)^4(1-5x)^{\frac{1}{6}}$

$= (1 + 16x + 96x^2 + \ldots)\left(1 - \frac{5x}{6} - \frac{125x^2}{72} + \ldots\right)$

$= 1(1 + 16x + 96x^2) - \frac{5x}{6}(1 + 16x) - \frac{125x^2}{72}(1) + \ldots$

$= 1 + 16x + 96x^2 - \frac{5x}{6} - \frac{40x^2}{3} - \frac{125x^2}{72} + \ldots$

$= 1 + \frac{91x}{6} + \frac{5827x^2}{72} + \ldots$

c) The expansion of $(1-5x)^{\frac{1}{6}}$ is valid for $|x| < \frac{1}{5}$.
The expansion of $(1+4x)^4$ is valid for all values of x, since n is a positive integer.
So overall, the expansion of $(1+4x)^4(1-5x)^{\frac{1}{6}}$ is valid for the narrower of these ranges, i.e. $|x| < \frac{1}{5}$.

Q8 a) $\frac{(1+3x)^4}{(1+x)^3} = (1 + 3x)^4(1 + x)^{-3}$

$(1 + 3x)^4 = 1 + 12x + 54x^2 + ...$

$(1 + x)^{-3} = 1 - 3x + 6x^2 - ...$

So $(1 + 3x)^4(1 + x)^{-3}$

$= 1 - 3x + 6x^2 + [12x(1 - 3x)] + 54x^2 + ...$

$= 1 - 3x + 6x^2 + 12x - 36x^2 + 54x^2 + ...$

$= 1 + 9x + 24x^2 + ...$

b) The expansion of $(1 + 3x)^4$ is valid for all values of x, since n is a positive integer.

The expansion of $(1 + x)^{-3}$ is valid for $|x| < 1$.

So overall, the expansion of $\frac{(1+3x)^4}{(1+x)^3}$ is valid for the narrower of these ranges, i.e. $|x| < 1$.

Q9 a) (i) $(1 + ax)^4 = 1 + 4ax + \frac{4\times3}{1\times2}(ax)^2 + \frac{4\times3\times2}{1\times2\times3}(ax)^3 + \frac{4\times3\times2\times1}{1\times2\times3\times4}(ax)^4$

$= 1 + 4ax + 6a^2x^2 + 4a^3x^3 + a^4x^4$

(ii) $(1 - bx)^{-3} = 1 + (-3)(-bx) + \frac{-3\times-4}{1\times2}(-bx)^2 + ...$

$= 1 + 3bx + 6b^2x^2 + ...$

b) $\frac{(1+ax)^4}{(1-bx)^3} = (1 + ax)^4(1 - bx)^{-3}$

$= (1 + 4ax + 6a^2x^2 + ...)(1 + 3bx + 6b^2x^2 + ...)$

$= 1 + 3bx + 6b^2x^2 + 4ax + 12abx^2 + 6a^2x^2 + ...$

$= 1 + (4a + 3b)x + (6a^2 + 12ab + 6b^2)x^2 + ...$

You're only asked to expand up to the x^2 term, so you don't need to write down any x^3 or higher power terms here.

c) Equating coefficients — x^2 terms:

$6a^2 + 12ab + 6b^2 = 24 \Rightarrow 6(a + b)^2 = 24$

$\Rightarrow (a + b)^2 = 4$

$\Rightarrow a + b = \pm2$

Equating coefficients — x terms:

$4a + 3b = 1 \Rightarrow a + 3(a + b) = 1$

So $a + b = 2 \Rightarrow a + 3 \times 2 = 1 \Rightarrow a = -5$

$\Rightarrow -5 + b = 2 \Rightarrow b = 7$

$a + b = -2 \Rightarrow a + 3 \times -2 = 1 \Rightarrow a = 7$

$\Rightarrow 7 + b = -2 \Rightarrow b = -9$

So the two pairs of values are:

$a = -5$, $b = 7$ and $a = 7$, $b = -9$

You could also have rearranged to get either a or b in terms of the other, and then substituted that into the other equation. You would get the same answer, but the working is a bit longer.

Exercise 1.3 — Expanding $(p + qx)^n$

Q1 a) $(2 + 4x)^3 = 2^3(1 + 2x)^3 = 8(1 + 2x)^3$

$(1 + 2x)^3 = 1 + 3(2x) + \frac{3\times2}{1\times2}(2x)^2 + \frac{3\times2\times1}{1\times2\times3}(2x)^3$

$= 1 + 6x + 3(4x^2) + 8x^3 = 1 + 6x + 12x^2 + 8x^3$

So $(2 + 4x)^3 = 8(1 + 6x + 12x^2 + 8x^3)$

$= 8 + 48x + 96x^2 + 64x^3$

b) $(3 + 4x)^5 = 3^5\left(1 + \frac{4}{3}x\right)^5 = 243\left(1 + \frac{4}{3}x\right)^5$

$\left(1 + \frac{4}{3}x\right)^5 = 1 + 5\left(\frac{4}{3}x\right) + \frac{5\times4}{1\times2}\left(\frac{4}{3}x\right)^2 + \frac{5\times4\times3}{1\times2\times3}\left(\frac{4}{3}x\right)^3 + ...$

$= 1 + \frac{20x}{3} + \frac{160x^2}{9} + \frac{640x^3}{27} + ...$

$(3 + 4x)^5 = 243(1 + \frac{20x}{3} + \frac{160x^2}{9} + \frac{640x^3}{27} + ...)$

$= 243 + 1620x + 4320x^2 + 5760x^3 + ...$

c) $(4 + x)^{\frac{1}{2}} = 4^{\frac{1}{2}}\left(1 + \frac{x}{4}\right)^{\frac{1}{2}} = 2\left(1 + \frac{x}{4}\right)^{\frac{1}{2}}$

$\left(1 + \frac{x}{4}\right)^{\frac{1}{2}} = 1 + \frac{1}{2}\left(\frac{x}{4}\right) + \frac{\frac{1}{2}\times-\frac{1}{2}}{1\times2}\left(\frac{x}{4}\right)^2 + \frac{\frac{1}{2}\times-\frac{1}{2}\times-\frac{3}{2}}{1\times2\times3}\left(\frac{x}{4}\right)^3 + ...$

$= 1 + \frac{x}{8} - \frac{x^2}{128} + \frac{x^3}{1024} - ...$

$(4 + x)^{\frac{1}{2}} = 2\left(1 + \frac{x}{4}\right)^{\frac{1}{2}}$

$= 2(1 + \frac{x}{8} - \frac{x^2}{128} + \frac{x^3}{1024} - ...)$

$= 2 + \frac{x}{4} - \frac{x^2}{64} + \frac{x^3}{512} - ...$

d) $(8 + 2x)^{-\frac{1}{3}} = 8^{-\frac{1}{3}}\left(1 + \frac{2x}{8}\right)^{-\frac{1}{3}} = \frac{1}{2}\left(1 + \frac{x}{4}\right)^{-\frac{1}{3}}$

$\left(1 + \frac{x}{4}\right)^{-\frac{1}{3}} = 1 + \left(-\frac{1}{3}\right)\left(\frac{x}{4}\right) + \frac{-\frac{1}{3}\times-\frac{4}{3}}{1\times2}\left(\frac{x}{4}\right)^2 + \frac{-\frac{1}{3}\times-\frac{4}{3}\times-\frac{7}{3}}{1\times2\times3}\left(\frac{x}{4}\right)^3 + ...$

$= 1 - \frac{x}{12} + \frac{x^2}{72} - \frac{7x^3}{2592} + ...$

$(8 + 2x)^{-\frac{1}{3}} = \frac{1}{2}\left(1 + \frac{x}{4}\right)^{-\frac{1}{3}}$

$= \frac{1}{2}(1 - \frac{x}{12} + \frac{x^2}{72} - \frac{7x^3}{2592} + ...)$

$= \frac{1}{2} - \frac{x}{24} + \frac{x^2}{144} - \frac{7x^3}{5184} + ...$

Q2 $(a + 5x)^5 = a^5\left(1 + \frac{5}{a}x\right)^5$

So the x^2 term is:

$a^5\left(\frac{5\times4}{1\times2}\right)\left(\frac{5}{a}x\right)^2 = 10a^5\left(\frac{25}{a^2}x^2\right) = \left(\frac{250a^5}{a^2}\right)x^2 = 250a^3x^2$

So $250a^3 = 2000 \Rightarrow a^3 = 8 \Rightarrow a = 2$

Q3 a) $(2 - 5x)^7 = 2^7\left(1 - \frac{5}{2}x\right)^7 = 128\left(1 - \frac{5}{2}x\right)^7$

$\left(1 - \frac{5}{2}x\right)^7 = 1 + 7\left(-\frac{5}{2}x\right) + \frac{7\times6}{1\times2}\left(-\frac{5}{2}x\right)^2 + ...$

$= 1 - \frac{35x}{2} + \frac{525x^2}{4} - ...$

$(2 - 5x)^7 = 128(1 - \frac{35x}{2} + \frac{525x^2}{4} - ...)$

$= 128 - 2240x + 16\,800x^2 - ...$

b) $(1 + 6x)^3 = 1 + 18x + 108x^2 + ...$

So $(1 + 6x)^3(2 - 5x)^7 =$

$128 - 2240x + 16\,800x^2 + 18x(128 - 2240x) + 108x^2(128) + ...$

$= 128 + 64x - 9696x^2 + ...$

Q4 a) $\left(1 + \frac{6}{5}x\right)^{-\frac{1}{2}} = 1 + \left(-\frac{1}{2}\right)\left(\frac{6}{5}x\right) + \frac{\left(-\frac{1}{2}\right)\times\left(-\frac{3}{2}\right)}{1\times2}\left(\frac{6}{5}x\right)^2 + \frac{\left(-\frac{1}{2}\right)\times\left(-\frac{3}{2}\right)\times\left(-\frac{5}{2}\right)}{1\times2\times3}\left(\frac{6}{5}x\right)^3 + ...$

$= 1 - \frac{3x}{5} + \frac{27x^2}{50} - \frac{27x^3}{50} + ...$

The expansion is valid for $\left|\frac{6x}{5}\right| < 1$, so $|x| < \frac{5}{6}$.

b) $\sqrt{\frac{20}{5+6x}} = \frac{\sqrt{20}}{\sqrt{5+6x}} = 20^{\frac{1}{2}}(5 + 6x)^{-\frac{1}{2}}$

$= (20^{\frac{1}{2}})(5^{-\frac{1}{2}})\left(1 + \frac{6}{5}x\right)^{-\frac{1}{2}} = \left(\frac{20^{\frac{1}{2}}}{5^{\frac{1}{2}}}\right)\left(1 + \frac{6}{5}x\right)^{-\frac{1}{2}}$

$= \left(\frac{20}{5}\right)^{\frac{1}{2}}\left(1 + \frac{6}{5}x\right)^{-\frac{1}{2}} = 4^{\frac{1}{2}}\left(1 + \frac{6}{5}x\right)^{-\frac{1}{2}}$

$= 2\left(1 + \frac{6}{5}x\right)^{-\frac{1}{2}}$

So, using the expansion in part a):

$\sqrt{\frac{20}{5+6x}} = 2(1 - \frac{3x}{5} + \frac{27x^2}{50} - \frac{27x^3}{50} + ...)$

$= 2 - \frac{6x}{5} + \frac{27x^2}{25} - \frac{27x^3}{25} + ...$

Q5 a) $\frac{1}{\sqrt{5-2x}} = (5-2x)^{-\frac{1}{2}} = 5^{-\frac{1}{2}}\left(1-\frac{2}{5}x\right)^{-\frac{1}{2}}$

$\left(1-\frac{2}{5}x\right)^{-\frac{1}{2}} = 1 + \left(-\frac{1}{2}\right)\left(-\frac{2}{5}x\right) + \frac{\left(-\frac{1}{2}\right)\times\left(-\frac{3}{2}\right)}{1\times 2}\left(-\frac{2}{5}x\right)^2 + \ldots$

$= 1 + \frac{x}{5} + \left(\frac{3}{8}\right)\left(\frac{4}{25}x^2\right) + \ldots$

$= 1 + \frac{x}{5} + \frac{3x^2}{50} + \ldots$

So $\frac{1}{\sqrt{5-2x}} = 5^{-\frac{1}{2}}\left(1-\frac{2}{5}x\right)^{-\frac{1}{2}}$

$= \frac{1}{\sqrt{5}}(1 + \frac{x}{5} + \frac{3x^2}{50} \ldots)$

$= \frac{1}{\sqrt{5}} + \frac{x}{5\sqrt{5}} + \frac{3x^2}{50\sqrt{5}} + \ldots$

b) $\frac{3+x}{\sqrt{5-2x}} \approx (3+x)\left(\frac{1}{\sqrt{5}} + \frac{x}{5\sqrt{5}} + \frac{3x^2}{50\sqrt{5}}\right)$

$\approx 3\left(\frac{1}{\sqrt{5}} + \frac{x}{5\sqrt{5}} + \frac{3x^2}{50\sqrt{5}}\right) + x\left(\frac{1}{\sqrt{5}} + \frac{x}{5\sqrt{5}}\right)$

$= \frac{3}{\sqrt{5}} + \frac{3x}{5\sqrt{5}} + \frac{9x^2}{50\sqrt{5}} + \frac{x}{\sqrt{5}} + \frac{x^2}{5\sqrt{5}}$

$= \frac{3}{\sqrt{5}} + \frac{8x}{5\sqrt{5}} + \frac{19x^2}{50\sqrt{5}}$

Q6 a) $(9+4x)^{-\frac{1}{2}} = 9^{-\frac{1}{2}}\left(1+\frac{4}{9}x\right)^{-\frac{1}{2}}$

$\left(1+\frac{4}{9}x\right)^{-\frac{1}{2}} = 1 + \left(-\frac{1}{2}\right)\left(\frac{4}{9}x\right) + \frac{\left(-\frac{1}{2}\right)\times\left(-\frac{3}{2}\right)}{1\times 2}\left(\frac{4}{9}x\right)^2 + \ldots$

$= 1 - \frac{2x}{9} + \frac{2x^2}{27} - \ldots$

So $(9+4x)^{-\frac{1}{2}} = \frac{1}{\sqrt{9}}\left(1 - \frac{2x}{9} + \frac{2x^2}{27} - \ldots\right)$

$= \frac{1}{3} - \frac{2x}{27} + \frac{2x^2}{81} - \ldots$

b) $(1 + 6x)^4 = 1 + 24x + 216x^2 + \ldots$

So $\frac{(1+6x)^4}{\sqrt{9+4x}} = (1+6x)^4(9+4x)^{-\frac{1}{2}}$

$\approx (1 + 24x + 216x^2)\left(\frac{1}{3} - \frac{2x}{27} + \frac{2x^2}{81}\right)$

$\approx \frac{1}{3} - \frac{2x}{27} + \frac{2x^2}{81} + 24x\left(\frac{1}{3} - \frac{2x}{27}\right) + 216x^2\left(\frac{1}{3}\right)$

$= \frac{1}{3} - \frac{2x}{27} + \frac{2x^2}{81} + 8x - \frac{16x^2}{9} + 72x^2$

$= \frac{1}{3} + \frac{214x}{27} + \frac{5690x^2}{81}$

2. Using the Binomial Expansion as an Approximation

Exercise 2.1 — Approximating with binomial expansions

Q1 a) $(1 + 6x)^{-1} = 1 + (-1)(6x) + \frac{-1\times -2}{1\times 2}(6x)^2 + \ldots$

$= 1 - 6x + 36x^2 - \ldots$

b) The expansion is valid if $|x| < \frac{1}{6}$.

c) $\frac{100}{106} = \frac{1}{1.06} = \frac{1}{1+0.06} = (1 + 0.06)^{-1}$

This is the same as $(1 + 6x)^{-1}$ with $x = 0.01$.

$0.01 < \frac{1}{6}$ so the approximation is valid.

$(1 + 6(0.01))^{-1} \approx 1 - 6(0.01) + 36(0.01^2)$

$= 1 - 0.06 + 0.0036 = 0.9436$

d) $\left|\frac{\left(\frac{100}{106}\right) - 0.9436}{\left(\frac{100}{106}\right)}\right| \times 100 = 0.02\%$ (1 s.f.)

In the answers that follow, the expansions will just be stated. Look back at the previous sections of this chapter to check how to set out the working if you need to.

Q2 a) $(1+3x)^{\frac{1}{4}} = 1 + \frac{3x}{4} - \frac{27x^2}{32} + \frac{189x^3}{128} - \ldots$

b) The expansion is valid if $|x| < \frac{1}{3}$.

c) $\sqrt[4]{1.9} = 1.9^{\frac{1}{4}} = (1+0.9)^{\frac{1}{4}}$

This is the same as $(1 + 3x)^{\frac{1}{4}}$ with $x = 0.3$.

$0.3 < \frac{1}{3}$ so the approximation is valid.

$(1+3(0.3))^{\frac{1}{4}} \approx 1 + \frac{3(0.3)}{4} - \frac{27(0.3^2)}{32} + \frac{189(0.3^3)}{128}$

$= 1.1889$ (4 d.p.)

d) $\left|\frac{(\sqrt[4]{1.9}) - 1.1889}{(\sqrt[4]{1.9})}\right| \times 100 = 1.26\%$ (3 s.f.)

Q3 a) $(1-2x)^{-\frac{1}{2}} = 1 + x + \frac{3x^2}{2} + \frac{5x^3}{2} + \ldots$

b) The expansion is valid if $|x| < \frac{1}{2}$.

c) Using $x = \frac{1}{10}$ gives:

$\left(1 - \frac{2}{10}\right)^{-\frac{1}{2}} \approx 1 + \frac{1}{10} + \frac{3}{200} + \frac{5}{2000}$

$\left(\frac{4}{5}\right)^{-\frac{1}{2}} \approx 1 + 0.1 + 0.015 + 0.0025$

$\left(\frac{5}{4}\right)^{\frac{1}{2}} \approx 1.1175$

$\left(\frac{5}{4}\right)^{\frac{1}{2}} = \sqrt{\frac{5}{4}} = \frac{\sqrt{5}}{\sqrt{4}} = \frac{\sqrt{5}}{2}$

So $\sqrt{5} \approx 2 \times 1.1175 = 2.235$

d) $\left|\frac{\sqrt{5} - 2.235}{\sqrt{5}}\right| \times 100 = 0.048\%$ (2 s.f.)

Q4 a) $(2 - 5x)^6 = 2^6\left(1 - \frac{5}{2}x\right)^6$

$= 64(1 - 15x + \frac{375x^2}{4} - \ldots)$

$= 64 - 960x + 6000x^2 - \ldots$

b) $1.95^6 = (2 - 0.05)^6$

This is the same as $(2 - 5x)^6$ with $x = 0.01$.

$(2 - 5(0.01))^6 \approx 64 - 960(0.01) + 6000(0.01)^2$

$= 64 - 9.6 + 0.6 = 55$

c) $\left|\frac{1.95^6 - 55}{1.95^6}\right| \times 100 = 0.036\%$ (2 s.f.)

Q5 a) $\sqrt{3-4x} = (3-4x)^{\frac{1}{2}} = \sqrt{3}\left(1 - \frac{4}{3}x\right)^{\frac{1}{2}}$

$= \sqrt{3}\left(1 - \frac{2x}{3} - \frac{2x^2}{9} - \ldots\right)$

$= \sqrt{3} - \frac{2\sqrt{3}x}{3} - \frac{2\sqrt{3}x^2}{9} - \ldots$

Remember to factorise the original expression to get it in the form $(1 + ax)^n$.

b) The expansion is valid if $|x| < \frac{3}{4}$.

c) $\sqrt{3 - 4\left(\frac{3}{40}\right)} \approx \sqrt{3} - \frac{2\sqrt{3}\times 3}{3\times 40} - \frac{2\sqrt{3}\times 9}{9\times 1600}$

$\sqrt{3 - \frac{3}{10}} \approx \sqrt{3} - \frac{\sqrt{3}}{20} - \frac{\sqrt{3}}{800}$

$\sqrt{\frac{27}{10}} \approx \frac{759\sqrt{3}}{800}$

$\frac{3\sqrt{3}}{\sqrt{10}} \approx \frac{759\sqrt{3}}{800} \Rightarrow \frac{3}{\sqrt{10}} \approx \frac{759}{800}$

d) $\left|\dfrac{\left(\frac{3}{\sqrt{10}}\right)-\left(\frac{759}{800}\right)}{\left(\frac{3}{\sqrt{10}}\right)}\right| \times 100 = 0.007\%$ (1 s.f.)

3. Binomial Expansion and Partial Fractions

Exercise 3.1 — Finding binomial expansions using partial fractions

Q1 a) $5 - 12x \equiv A(4 + 3x) + B(1 + 6x)$

Using the substitution method:

When $x = -\frac{4}{3}$, $5 + 16 = -7B \Rightarrow B = -3$.

When $x = -\frac{1}{6}$, $5 + 2 = \frac{7A}{2} \Rightarrow A = 2$.

You could also have used the 'equating coefficients' method.

b) (i) $(1 + 6x)^{-1} = 1 - 6x + 36x^2 - \ldots$

(ii) $(4 + 3x)^{-1} = 4^{-1}\left(1 + \frac{3}{4}x\right)^{-1} = \frac{1}{4}\left(1 - \frac{3x}{4} + \frac{9x^2}{16} - \ldots\right)$

$= \frac{1}{4} - \frac{3x}{16} + \frac{9x^2}{64} - \ldots$

c) From a): $\dfrac{5-12x}{(1+6x)(4+3x)} \equiv \dfrac{2}{(1+6x)} - \dfrac{3}{(4+3x)}$

$\equiv 2(1 + 6x)^{-1} - 3(4 + 3x)^{-1}$

$2(1 + 6x)^{-1} - 3(4 + 3x)^{-1}$

$\approx 2(1 - 6x + 36x^2) - 3\left(\frac{1}{4} - \frac{3x}{16} + \frac{9x^2}{64}\right)$

$= 2 - 12x + 72x^2 - \frac{3}{4} + \frac{9x}{16} - \frac{27x^2}{64}$

$= \frac{5}{4} - \frac{183x}{16} + \frac{4581x^2}{64}$

d) $(1 + 6x)^{-1}$ is valid if $|x| < \frac{1}{6}$.

$(4 + 3x)^{-1}$ is valid if $|x| < \frac{4}{3}$.

So the full expansion is valid if $|x| < \frac{1}{6}$.

Q2 a) $\dfrac{6}{(1-x)(1+x)(1+2x)} \equiv \dfrac{A}{(1-x)} + \dfrac{B}{(1+x)} + \dfrac{C}{(1+2x)}$

$6 \equiv A(1 + x)(1 + 2x) + B(1 - x)(1 + 2x) + C(1 - x)(1 + x)$

Using the substitution method:

When $x = 1$, $6 = 6A \Rightarrow A = 1$

When $x = -1$, $6 = -2B \Rightarrow B = -3$

When $x = -\frac{1}{2}$, $6 = \frac{3}{4}C \Rightarrow C = 8$

Putting these values back into the expression gives:

$\dfrac{6}{(1-x)(1+x)(1+2x)} \equiv \dfrac{1}{(1-x)} - \dfrac{3}{(1+x)} + \dfrac{8}{(1+2x)}$

b) f(x) can also be expressed as:

$(1 - x)^{-1} - 3(1 + x)^{-1} + 8(1 + 2x)^{-1}$

Expanding these three parts separately:

$(1 - x)^{-1} = 1 + x + x^2 + \ldots$

$(1 + x)^{-1} = 1 - x + x^2 + \ldots.$

$(1 + 2x)^{-1} = 1 - 2x + 4x^2 + \ldots$

So $f(x) \approx 1 + x + x^2 - 3(1 - x + x^2) + 8(1 - 2x + 4x^2)$

$= 1 + x + x^2 - 3 + 3x - 3x^2 + 8 - 16x + 32x^2$

$= 6 - 12x + 30x^2$

c) $f(0.01) = \dfrac{6}{(1-0.01)(1+0.01)(1+2(0.01))} = 5.8829\ldots$

From the expansion:

$f(0.01) \approx 6 - 12(0.01) + 30(0.01)^2$

$= 6 - 0.12 + 0.003 = 5.883$

So the % error is:

$\left|\dfrac{5.8829\ldots - 5.883}{5.8829\ldots}\right| \times 100 = 0.0010\%$ (2 s.f.)

Q3 a) $2x^3 + 5x^2 - 3x = x(2x - 1)(x + 3)$

b) $\dfrac{5x-6}{2x^3+5x^2-3x} \equiv \dfrac{5x-6}{x(2x-1)(x+3)}$

$\equiv \dfrac{A}{x} + \dfrac{B}{(2x-1)} + \dfrac{C}{(x+3)}$

$5x - 6 \equiv A(2x - 1)(x + 3) + Bx(x + 3) + Cx(2x - 1)$

Using the substitution method:

When $x = 0$, $-6 = -3A \Rightarrow A = 2$

When $x = \frac{1}{2}$, $\frac{5}{2} - 6 = \frac{7}{4}B \Rightarrow B = -2$

When $x = -3$, $-15 - 6 = 21C \Rightarrow C = -1$

So $\dfrac{5x-6}{2x^3+5x^2-3x} \equiv \dfrac{2}{x} - \dfrac{2}{(2x-1)} - \dfrac{1}{(x+3)}$

c) $(2x - 1)^{-1} = -(1 - 2x)^{-1} = -(1 + 2x + 4x^2 + \ldots)$

$(x + 3)^{-1} = \frac{1}{3}\left(1 + \frac{1}{3}x\right)^{-1} = \frac{1}{3}\left(1 - \frac{x}{3} + \frac{x^2}{9} + \ldots\right)$

$= \frac{1}{3} - \frac{x}{9} + \frac{x^2}{27} + \ldots$

$\dfrac{5x-6}{2x^3+5x^2-3x} \equiv \dfrac{2}{x} - 2(2x - 1)^{-1} - (x + 3)^{-1}$

$\approx \frac{2}{x} - 2[-(1 + 2x + 4x^2)] - \left(\frac{1}{3} - \frac{x}{9} + \frac{x^2}{27}\right)$

$= \frac{2}{x} + 2 + 4x + 8x^2 - \frac{1}{3} + \frac{x}{9} - \frac{x^2}{27}$

$= \frac{2}{x} + \frac{5}{3} + \frac{37x}{9} + \frac{215x^2}{27}$

d) $(2x - 1)^{-1}$ is valid for $|x| < \frac{1}{2}$.

$(x + 3)^{-1}$ is valid for $|x| < 3$.

$\frac{2}{x}$ is valid for $x \neq 0$.

So the full expansion is valid for $|x| < \frac{1}{2}$, $x \neq 0$.

Make sure you don't get caught out here — the brackets are in the form $(qx + p)^n$, not $(p + qx)^n$.

Q4 a) $\dfrac{55x+7}{(2x-5)(3x+1)^2} \equiv \dfrac{A}{(2x-5)} + \dfrac{B}{(3x+1)} + \dfrac{C}{(3x+1)^2}$

$55x + 7 \equiv A(3x + 1)^2 + B(2x - 5)(3x + 1) + C(2x - 5)$

Using the substitution method:

When $x = \frac{5}{2}$, $\frac{275}{2} + 7 = A\left(\frac{17}{2}\right)^2$

$\Rightarrow \frac{289}{2} = \frac{289}{4}A \Rightarrow A = 2$

When $x = -\frac{1}{3}$, $-\frac{55}{3} + 7 = C\left(-\frac{2}{3} - 5\right)$

$-\frac{34}{3} = -\frac{17}{3}C \Rightarrow C = 2$

Equating coefficients of x^2:

$0 = 9A + 6B \Rightarrow 6B = -18 \Rightarrow B = -3$

So $\dfrac{55x+7}{(2x-5)(3x+1)^2} \equiv \dfrac{2}{(2x-5)} - \dfrac{3}{(3x+1)} + \dfrac{2}{(3x+1)^2}$

b) $(2x - 5)^{-1} = -\frac{1}{5}\left(1 - \frac{2}{5}x\right)^{-1}$

$= -\frac{1}{5}\left(1 + \frac{2x}{5} + \frac{4x^2}{25} + \ldots\right)$

$= -\frac{1}{5} - \frac{2x}{25} - \frac{4x^2}{125} - \ldots$

$(3x + 1)^{-1} = 1 - 3x + 9x^2 + \ldots$

$(3x + 1)^{-2} = 1 - 6x + 27x^2 + \ldots$

$f(x) = 2(2x - 5)^{-1} - 3(3x + 1)^{-1} + 2(3x + 1)^{-2}$

$\approx 2\left(-\frac{1}{5} - \frac{2x}{25} - \frac{4x^2}{125}\right) - 3(1 - 3x + 9x^2)$

$+ 2(1 - 6x + 27x^2)$

$= -\frac{2}{5} - 3 + 2 - \frac{4x}{25} + 9x - 12x - \frac{8x^2}{125} - 27x^2 + 54x^2$

$= -\frac{7}{5} - \frac{79x}{25} - \frac{3367x^2}{125}$

Chapter 7: Differentiation

1. Points of Inflection

Exercise 1.1 — Points of inflection

Q1 a) $y = \frac{1}{6}x^3 - \frac{5}{2}x^2 + \frac{1}{4}x + \frac{1}{9}$

$\Rightarrow \frac{dy}{dx} = \frac{1}{2}x^2 - 5x + \frac{1}{4} \Rightarrow \frac{d^2y}{dx^2} = x - 5$

The graph is concave when $\frac{d^2y}{dx^2} < 0$

$\Rightarrow x - 5 < 0 \Rightarrow x < 5$

b) $y = 4x^2 - x^4 \Rightarrow \frac{dy}{dx} = 8x - 4x^3 \Rightarrow \frac{d^2y}{dx^2} = 8 - 12x^2$

The graph is concave when $\frac{d^2y}{dx^2} < 0 \Rightarrow 8 - 12x^2 < 0$

$\Rightarrow 8 < 12x^2 \Rightarrow \frac{2}{3} < x^2 \Rightarrow x < -\sqrt{\frac{2}{3}}$ or $x > \sqrt{\frac{2}{3}}$

Q2 $y = \frac{3}{2}x^4 - x^2 - 3x$

$\Rightarrow \frac{dy}{dx} = 6x^3 - 2x - 3 \Rightarrow \frac{d^2y}{dx^2} = 18x^2 - 2$

At the points of inflection, $\frac{d^2y}{dx^2} = 0$

$\Rightarrow 18x^2 - 2 = 0 \Rightarrow x^2 = \frac{1}{9} \Rightarrow x = \pm\frac{1}{3}$

Q3 $f(x) = \frac{1}{16}x^4 + \frac{3}{4}x^3 - \frac{21}{8}x^2 - 6x + 20$

$\Rightarrow f'(x) = \frac{1}{4}x^3 + \frac{9}{4}x^2 - \frac{21}{4}x - 6$

$\Rightarrow f''(x) = \frac{3}{4}x^2 + \frac{9}{2}x - \frac{21}{4}$

So f(x) is convex when $f''(x) > 0 \Rightarrow \frac{3}{4}x^2 + \frac{9}{2}x - \frac{21}{4} > 0$

$\Rightarrow 3x^2 + 18x - 21 > 0 \Rightarrow x^2 + 6x - 7 > 0$

$\Rightarrow (x + 7)(x - 1) > 0 \Rightarrow x < -7$ or $x > 1$

So f(x) is convex for $x < -7$ and $x > 1$, and concave for $-7 < x < 1$.

Sketch the graph of f''(x) if you're not sure which way the inequalities should go.

Q4 $y = x^2 - \frac{1}{x} \Rightarrow \frac{dy}{dx} = 2x + \frac{1}{x^2} \Rightarrow \frac{d^2y}{dx^2} = 2 - \frac{2}{x^3}$

At a point of inflection, $\frac{d^2y}{dx^2} = 0 \Rightarrow 2 - \frac{2}{x^3} = 0$

$\Rightarrow 2 = \frac{2}{x^3} \Rightarrow x^3 = 1 \Rightarrow x = 1$

If $x < 1$, $x^3 < 1$, so $\frac{2}{x^3} > 2$, so $\frac{d^2y}{dx^2}$ is negative.

If $x > 1$, $x^3 > 1$, so $\frac{2}{x^3} < 2$, so $\frac{d^2y}{dx^2}$ is positive.

So at $x = 1$, $\frac{d^2y}{dx^2} = 0$ and the sign of $\frac{d^2y}{dx^2}$ changes, so this is a point of inflection.

When $x = 1$, $y = 1^2 - \frac{1}{1} = 0$, so (1, 0) is a point of inflection of $y = x^2 - \frac{1}{x}$.

Q5 a) $f(x) = x^3 + 2x^2 + 3x + 3 \Rightarrow f'(x) = 3x^2 + 4x + 3$

$\Rightarrow f''(x) = 6x + 4$

At a point of inflection, $f''(x) = 0$

$\Rightarrow 6x + 4 = 0 \Rightarrow x = -\frac{2}{3}$

When $x > -\frac{2}{3}$, $6x + 4 > 0$

and when $x < -\frac{2}{3}$, $6x + 4 < 0$

So the graph of $y = f(x)$ has one point of inflection, at $x = -\frac{2}{3}$.

b) $f'(x) = 3x^2 + 4x + 3$, so at $x = -\frac{2}{3}$,

$f'(x) = 3\left(-\frac{2}{3}\right)^2 + 4\left(-\frac{2}{3}\right) + 3 = \frac{4}{3} - \frac{8}{3} + 3 = \frac{5}{3}$

$f'(x) \neq 0$ at the point of inflection, so it is not a stationary point.

Q6 $y = \frac{1}{10}x^5 - \frac{1}{3}x^3 + \frac{1}{2}x + 4$

$\Rightarrow \frac{dy}{dx} = \frac{1}{2}x^4 - x^2 + \frac{1}{2} \Rightarrow \frac{d^2y}{dx^2} = 2x^3 - 2x$

At a stationary point, $\frac{dy}{dx} = 0$

$\Rightarrow \frac{1}{2}x^4 - x^2 + \frac{1}{2} = 0 \Rightarrow x^4 - 2x^2 + 1 = 0$

$\Rightarrow (x^2 - 1)^2 = 0 \Rightarrow [(x + 1)(x - 1)]^2 = 0 \Rightarrow x = \pm 1$

When $x = 1$, $y = \frac{1}{10} - \frac{1}{3} + \frac{1}{2} + 4 = \frac{64}{15}$

and $\frac{d^2y}{dx^2} = 2 - 2 = 0$.

When $x = -1$, $y = -\frac{1}{10} + \frac{1}{3} - \frac{1}{2} + 4 = \frac{56}{15}$

and $\frac{d^2y}{dx^2} = -2 + 2 = 0$.

$\frac{d^2y}{dx^2} = 0$ at both stationary points, so check what happens to $\frac{d^2y}{dx^2}$ around $x = 1$ and $x = -1$.

$\frac{d^2y}{dx^2} = 2x^3 - 2x = 2x(x^2 - 1) = 2x(x + 1)(x - 1)$

When $x < -1$, $2x$ is negative, $x + 1$ is negative and $x - 1$ is negative, so $\frac{d^2y}{dx^2}$ is negative for $x < -1$.

When $-1 < x < 0$, $2x$ is negative, $x + 1$ is positive and $x - 1$ is negative, so $\frac{d^2y}{dx^2}$ is positive for $-1 < x < 0$.

So $\frac{d^2y}{dx^2}$ changes sign at $x = -1$.

When $0 < x < 1$, $2x$ is positive, $x + 1$ is positive and $x - 1$ is negative, so $\frac{d^2y}{dx^2}$ is negative for $0 < x < 1$.

When $1 < x$, $2x$ is positive, $x + 1$ is positive and $x - 1$ is positive, so $\frac{d^2y}{dx^2}$ is positive for $1 < x$.

So $\frac{d^2y}{dx^2}$ changes sign at $x = 1$.

The graph of $y = \frac{1}{10}x^5 - \frac{1}{3}x^3 + \frac{1}{2}x + 4$ has stationary points at $\left(-1, \frac{56}{15}\right)$ and $\left(1, \frac{64}{15}\right)$, and both stationary points are points of inflection.

Again, it might help to sketch the graph of $\frac{d^2y}{dx^2}$.

You could also use the fact that $\frac{dy}{dx} = \frac{1}{2}(x^2 - 1)^2$ to show that the gradient is always ≥ 0, which means that any stationary points must be points of inflection.

2. Chain Rule

Exercise 2.1 — The chain rule

Q1 a) $y = (x + 7)^2$, so let $y = u^2$ where $u = x + 7$

$\Rightarrow \frac{dy}{du} = 2u = 2(x + 7)$, $\frac{du}{dx} = 1$

$\frac{dy}{dx} = \frac{dy}{du} \times \frac{du}{dx} = 2(x + 7) \times 1 = 2(x + 7)$

b) $y = (2x - 1)^5$, so let $y = u^5$ where $u = 2x - 1$

$\Rightarrow \frac{dy}{du} = 5u^4 = 5(2x - 1)^4$, $\frac{du}{dx} = 2$

$\frac{dy}{dx} = \frac{dy}{du} \times \frac{du}{dx} = 5(2x - 1)^4 \times 2 = 10(2x - 1)^4$

c) $y = 3(4 - x)^8$, so let $y = 3u^8$ where $u = 4 - x$

$\Rightarrow \frac{dy}{du} = 24u^7 = 24(4 - x)^7$, $\frac{du}{dx} = -1$

$\frac{dy}{dx} = \frac{dy}{du} \times \frac{du}{dx} = 24(4 - x)^7 \times (-1) = -24(4 - x)^7$

d) $y = (3 - 2x)^7$, so let $y = u^7$ where $u = 3 - 2x$

$\Rightarrow \frac{dy}{du} = 7u^6 = 7(3 - 2x)^6$, $\frac{du}{dx} = -2$

$\frac{dy}{dx} = \frac{dy}{du} \times \frac{du}{dx} = 7(3 - 2x)^6 \times (-2) = -14(3 - 2x)^6$

e) $y = (x^2 + 3)^5$, so let $y = u^5$ where $u = x^2 + 3$

$\Rightarrow \frac{dy}{du} = 5u^4 = 5(x^2 + 3)^4$, $\frac{du}{dx} = 2x$

$\frac{dy}{dx} = \frac{dy}{du} \times \frac{du}{dx} = 5(x^2 + 3)^4 \times 2x = 10x(x^2 + 3)^4$

f) $y = (5x^2 + 3)^2$, so let $y = u^2$ where $u = 5x^2 + 3$

$\Rightarrow \frac{dy}{du} = 2u = 2(5x^2 + 3)$, $\frac{du}{dx} = 10x$

$\frac{dy}{dx} = \frac{dy}{du} \times \frac{du}{dx} = 2(5x^2 + 3) \times 10x = 20x(5x^2 + 3)$

Q2 a) $f(x) = (4x^3 - 9)^8$, so let $y = u^8$ where $u = 4x^3 - 9$

$\Rightarrow \frac{dy}{du} = 8u^7 = 8(4x^3 - 9)^7$, $\frac{du}{dx} = 12x^2$

$f'(x) = \frac{dy}{du} \times \frac{du}{dx} = 8(4x^3 - 9)^7 \times 12x^2$
$= 96x^2(4x^3 - 9)^7$

b) $f(x) = (6 - 7x^2)^4$, so let $y = u^4$ where $u = 6 - 7x^2$

$\Rightarrow \frac{dy}{du} = 4u^3 = 4(6 - 7x^2)^3$, $\frac{du}{dx} = -14x$

$f'(x) = \frac{dy}{du} \times \frac{du}{dx} = 4(6 - 7x^2)^3 \times (-14x)$
$= -56x(6 - 7x^2)^3$

c) $f(x) = (x^2 + 5x + 7)^6$, so let $y = u^6$
where $u = x^2 + 5x + 7$

$\Rightarrow \frac{dy}{du} = 6u^5 = 6(x^2 + 5x + 7)^5$, $\frac{du}{dx} = 2x + 5$

$f'(x) = \frac{dy}{du} \times \frac{du}{dx} = 6(x^2 + 5x + 7)^5 \times (2x + 5)$
$= (x^2 + 5x + 7)^5 (12x + 30)$

d) $f(x) = (x + 4)^{-3}$, so let $y = u^{-3}$ where $u = x + 4$

$\Rightarrow \frac{dy}{du} = -3u^{-4} = -3(x + 4)^{-4}$, $\frac{du}{dx} = 1$

$f'(x) = \frac{dy}{du} \times \frac{du}{dx} = -3(x + 4)^{-4} \times 1 = -3(x + 4)^{-4}$

e) $f(x) = (5 - 3x)^{-2}$, so let $y = u^{-2}$ where $u = 5 - 3x$

$\Rightarrow \frac{dy}{du} = -2u^{-3} = -2(5 - 3x)^{-3}$, $\frac{du}{dx} = -3$

$f'(x) = \frac{dy}{du} \times \frac{du}{dx} = -2(5 - 3x)^{-3} \times (-3) = 6(5 - 3x)^{-3}$

f) $f(x) = \frac{1}{(5-3x)^4} = (5 - 3x)^{-4}$,
so let $y = u^{-4}$ where $u = 5 - 3x$

$\Rightarrow \frac{dy}{du} = -4u^{-5} = -4(5 - 3x)^{-5}$, $\frac{du}{dx} = -3$

$f'(x) = \frac{dy}{du} \times \frac{du}{dx} = -4(5 - 3x)^{-5} \times (-3) = \frac{12}{(5-3x)^5}$

g) $f(x) = (3x^2 + 4)^{\frac{3}{2}}$, so let $y = u^{\frac{3}{2}}$ where $u = 3x^2 + 4$

$\Rightarrow \frac{dy}{du} = \frac{3}{2}u^{\frac{1}{2}} = \frac{3}{2}(3x^2 + 4)^{\frac{1}{2}}$, $\frac{du}{dx} = 6x$

$f'(x) = \frac{dy}{du} \times \frac{du}{dx} = \frac{3}{2}(3x^2 + 4)^{\frac{1}{2}} \times 6x = 9x(3x^2 + 4)^{\frac{1}{2}}$

h) $f(x) = \frac{1}{\sqrt{5-3x}} = (5 - 3x)^{-\frac{1}{2}}$,

so let $y = u^{-\frac{1}{2}}$ where $u = 5 - 3x$

$\Rightarrow \frac{dy}{du} = -\frac{1}{2}u^{-\frac{3}{2}} = -\frac{1}{2}(5 - 3x)^{-\frac{3}{2}}$, $\frac{du}{dx} = -3$

$f'(x) = \frac{dy}{du} \times \frac{du}{dx} = -\frac{1}{2}(5 - 3x)^{-\frac{3}{2}} \times (-3) = \frac{3}{2(\sqrt{5-3x})^3}$

Q3 a) $y = (5x - 3x^2)^{-\frac{1}{2}}$, so let $y = u^{-\frac{1}{2}}$ where $u = 5x - 3x^2$

$\Rightarrow \frac{dy}{du} = -\frac{1}{2}u^{-\frac{3}{2}} = -\frac{1}{2}(5x - 3x^2)^{-\frac{3}{2}}$, $\frac{du}{dx} = 5 - 6x$

$\frac{dy}{dx} = \frac{dy}{du} \times \frac{du}{dx} = -\frac{1}{2}(5x - 3x^2)^{-\frac{3}{2}} \times (5 - 6x)$
$= -\frac{5-6x}{2(\sqrt{5x-3x^2})^3}$

When $x = 1$, $\frac{dy}{dx} = -\frac{5-(6\times1)}{2(\sqrt{(5\times1)-(3\times1^2)})^3} = \frac{1}{4\sqrt{2}}$

b) $y = \frac{12}{\sqrt[3]{x+6}} = 12(x + 6)^{-\frac{1}{3}}$,

so let $y = 12u^{-\frac{1}{3}}$ where $u = x + 6$

$\Rightarrow \frac{dy}{du} = -4u^{-\frac{4}{3}} = -4(x + 6)^{-\frac{4}{3}}$, $\frac{du}{dx} = 1$

$\frac{dy}{dx} = \frac{dy}{du} \times \frac{du}{dx} = -4(x + 6)^{-\frac{4}{3}} \times 1 = -\frac{4}{\sqrt[3]{(x+6)^4}}$

When $x = 1$, $\frac{dy}{dx} = -\frac{4}{\sqrt[3]{(x+6)^4}} = -\frac{4}{7(\sqrt[3]{7})}$

Q4 a) $\left(\sqrt{x} + \frac{1}{\sqrt{x}}\right)^2 = \sqrt{x}\sqrt{x} + 2\sqrt{x}\frac{1}{\sqrt{x}} + \frac{1}{\sqrt{x}}\frac{1}{\sqrt{x}} = x + \frac{1}{x} + 2$

$\frac{d}{dx}(x + \frac{1}{x} + 2) = 1 - \frac{1}{x^2}$

Remember $\frac{1}{x} = x^{-1}$

b) $y = \left(\sqrt{x} + \frac{1}{\sqrt{x}}\right)^2$, so let $y = u^2$ where $u = \sqrt{x} + \frac{1}{\sqrt{x}}$

$\Rightarrow \frac{dy}{du} = 2u = 2\left(\sqrt{x} + \frac{1}{\sqrt{x}}\right)$,

$\frac{du}{dx} = \frac{1}{2}x^{-\frac{1}{2}} - \frac{1}{2}x^{-\frac{3}{2}} = \frac{1}{2\sqrt{x}} - \frac{1}{2(\sqrt{x})^3}$

$\frac{dy}{dx} = \frac{dy}{du} \times \frac{du}{dx} = 2\left(\sqrt{x} + \frac{1}{\sqrt{x}}\right) \times \left(\frac{1}{2\sqrt{x}} - \frac{1}{2(\sqrt{x})^3}\right)$

$= 2\left(\frac{1}{2} + \frac{1}{2x} - \frac{1}{2x} - \frac{1}{2x^2}\right) = 1 - \frac{1}{x^2}$

Using powers notation makes this question easier to handle.

Q5 $y = (x - 3)^5$, so let $y = u^5$ where $u = x - 3$

$\Rightarrow \frac{dy}{du} = 5u^4 = 5(x - 3)^4$, $\frac{du}{dx} = 1$

$\frac{dy}{dx} = \frac{dy}{du} \times \frac{du}{dx} = 5(x - 3)^4 \times 1 = 5(x - 3)^4$

At the point (1, –32), gradient $= 5(1 - 3)^4 = 80$

The equation of a straight line is $y = mx + c$

$\Rightarrow -32 = (80 \times 1) + c \Rightarrow c = -112$

So the equation of the tangent is $y = 80x - 112$.

Q6 $y = \frac{1}{4}(x - 7)^4$, so let $y = \frac{1}{4}u^4$ where $u = x - 7$

$\Rightarrow \frac{dy}{du} = u^3 = (x - 7)^3$, $\frac{du}{dx} = 1$

$\frac{dy}{dx} = \frac{dy}{du} \times \frac{du}{dx} = (x - 7)^3 \times 1 = (x - 7)^3$

When $x = 6$, $y = \frac{1}{4}(6 - 7)^4 = \frac{1}{4}$ and $\frac{dy}{dx} = (6 - 7)^3 = -1$

The gradient of the normal is $-1 \div -1 = 1$.

The equation of a straight line is $y = mx + c$

$\Rightarrow \frac{1}{4} = (1 \times 6) + c \Rightarrow c = -\frac{23}{4}$

So the equation of the normal is $y = x - \frac{23}{4}$

Q7 $y = (7x^2 - 3)^{-4}$, so let $y = u^{-4}$ where $u = 7x^2 - 3$

$\Rightarrow \frac{dy}{du} = -4u^{-5} = -4(7x^2 - 3)^{-5}$, $\frac{du}{dx} = 14x$

$\frac{dy}{dx} = \frac{dy}{du} \times \frac{du}{dx} = -4(7x^2 - 3)^{-5} \times 14x = -56x(7x^2 - 3)^{-5}$

When $x = 1$, $\frac{dy}{dx} = -56(1)(7(1)^2 - 3)^{-5} = -56(4^{-5}) = -\frac{7}{128}$

Q8 $y = \frac{7}{\sqrt[3]{3-2x}}$, so let $y = 7u^{-\frac{1}{3}}$ where $u = 3 - 2x$

$\Rightarrow \frac{dy}{du} = -\frac{7}{3}u^{-\frac{4}{3}} = -\frac{7}{3(\sqrt[3]{3-2x})^4}$, $\frac{du}{dx} = -2$

$f'(x) = \frac{dy}{du} \times \frac{du}{dx} = -\frac{7}{3(\sqrt[3]{3-2x})^4} \times -2 = \frac{14}{3(\sqrt[3]{3-2x})^4}$

f'(x) could also be written as $\frac{14}{3}(3-2x)^{-\frac{4}{3}}$.

Q9 $y = \sqrt{5x-1}$, so let $y = u^{\frac{1}{2}}$ where $u = 5x - 1$

$\Rightarrow \frac{dy}{du} = \frac{1}{2}u^{-\frac{1}{2}} = \frac{1}{2}\frac{1}{\sqrt{5x-1}}$, $\frac{du}{dx} = 5$

$\frac{dy}{dx} = \frac{dy}{du} \times \frac{du}{dx} = \frac{1}{2}\frac{1}{\sqrt{5x-1}} \times 5 = \frac{5}{2\sqrt{5x-1}}$

when $x = 2$, $\frac{dy}{dx} = \frac{5}{2\sqrt{10-1}} = \frac{5}{6}$

and $y = \sqrt{10-1} = 3$

$y = mx + c \Rightarrow 3 = (\frac{5}{6} \times 2) + c \Rightarrow c = \frac{4}{3}$

So the equation of the tangent is $y = \frac{5}{6}x + \frac{4}{3}$.

In the form $ax + by + c = 0$, $5x - 6y + 8 = 0$.

Q10 $y = \sqrt[3]{3x-7}$, so let $y = u^{\frac{1}{3}}$ where $u = 3x - 7$

$\Rightarrow \frac{dy}{du} = \frac{1}{3}u^{-\frac{2}{3}} = \frac{1}{3(\sqrt[3]{3x-7})^2}$, $\frac{du}{dx} = 3$

$\frac{dy}{dx} = \frac{dy}{du} \times \frac{du}{dx} = \frac{1}{3(\sqrt[3]{3x-7})^2} \times 3 = \frac{1}{(\sqrt[3]{3x-7})^2}$

When $x = 5$, $\frac{dy}{dx} = \frac{1}{(\sqrt[3]{(3\times5)-7})^2} = \frac{1}{4}$, $y = \sqrt[3]{15-7} = 2$

Gradient of normal $= \frac{-1}{\frac{1}{4}} = -4$

$y = mx + c \Rightarrow 2 = (-4 \times 5) + c \Rightarrow c = 22$

So the equation of the normal is $y = 22 - 4x$.

Q11 $y = (x^4 + x^3 + x^2)^2$, so let $y = u^2$ where $u = x^4 + x^3 + x^2$

$\Rightarrow \frac{dy}{du} = 2u = 2(x^4 + x^3 + x^2)$, $\frac{du}{dx} = 4x^3 + 3x^2 + 2x$

$\frac{dy}{dx} = \frac{dy}{du} \times \frac{du}{dx} = 2(x^4 + x^3 + x^2) \times (4x^3 + 3x^2 + 2x)$

At $x = -1$, $\frac{dy}{dx} = 2(1 - 1 + 1) \times (-4 + 3 + -2) = -6$

$y = ((-1)^4 + (-1)^3 + (-1)^2)^2 = 1$

$y = mx + c \Rightarrow 1 = (-6 \times -1) + c \Rightarrow c = -5$

So the equation of the tangent is $y = -6x - 5$.

Q12 $y = (2x - 3)^7$, so let $y = u^7$ where $u = 2x - 3$

$\Rightarrow \frac{dy}{du} = 7u^6 = 7(2x - 3)^6$, $\frac{du}{dx} = 2$

$\frac{dy}{dx} = \frac{dy}{du} \times \frac{du}{dx} = 7(2x - 3)^6 \times 2 = 14(2x - 3)^6$

Use the chain rule again to find $\frac{d^2y}{dx^2}$:

$\frac{dy}{dx} = 14(2x - 3)^6$, so let $\frac{dy}{dx} = 14u^6$ where $u = 2x - 3$

$\Rightarrow \frac{d}{du}\left(\frac{dy}{dx}\right) = 84u^5 = 84(2x - 3)^5$, $\frac{du}{dx} = 2$

$\frac{d^2y}{dx^2} = \frac{d}{dx}\left(\frac{dy}{dx}\right) = \frac{d}{du}\left(\frac{dy}{dx}\right) \times \frac{du}{dx} = 84(2x - 3)^5 \times 2$
$= 168(2x - 3)^5$

At a point of inflection, $\frac{d^2y}{dx^2} = 0 \Rightarrow 2x - 3 = 0$
$\Rightarrow x = \frac{3}{2}$

When $x < \frac{3}{2}$, $(2x - 3) < 0 \Rightarrow \frac{d^2y}{dx^2} < 0$

When $x > \frac{3}{2}$, $(2x - 3) > 0 \Rightarrow \frac{d^2y}{dx^2} > 0$

So the sign of $\frac{d^2y}{dx^2}$ changes at $x = \frac{3}{2}$,
so this is a point of inflection.

$x = \frac{3}{2} \Rightarrow y = (2 \times \frac{3}{2} - 3)^7 = 0$

So the coordinates of the point of inflection are $\left(\frac{3}{2}, 0\right)$.

Q13 $y = (\frac{x}{4} - 2)^3$, so let $y = u^3$ where $u = \frac{x}{4} - 2$

$\Rightarrow \frac{dy}{du} = 3u^2 = 3(\frac{x}{4} - 2)^2$, $\frac{du}{dx} = \frac{1}{4}$

$\frac{dy}{dx} = \frac{dy}{du} \times \frac{du}{dx} = 3(\frac{x}{4} - 2)^2 \times \frac{1}{4} = \frac{3}{4}(\frac{x}{4} - 2)^2$

Use the chain rule again to find $\frac{d^2y}{dx^2}$:

$\frac{dy}{dx} = \frac{3}{4}(\frac{x}{4} - 2)^2$, so let $\frac{dy}{dx} = \frac{3}{4}u^2$ where $u = \frac{x}{4} - 2$

$\Rightarrow \frac{d}{du}\left(\frac{dy}{dx}\right) = \frac{3}{2}u = \frac{3}{2}(\frac{x}{4} - 2) = \frac{3}{8}x - 3$, $\frac{du}{dx} = \frac{1}{4}$

$\frac{d^2y}{dx^2} = \frac{d}{dx}\left(\frac{dy}{dx}\right) = \frac{d}{du}\left(\frac{dy}{dx}\right) \times \frac{du}{dx} = \left(\frac{3}{8}x - 3\right) \times \frac{1}{4}$
$= \frac{3}{32}x - \frac{3}{4}$

The curve is convex when $\frac{d^2y}{dx^2} > 0 \Rightarrow \frac{3}{32}x - \frac{3}{4} > 0$

$\Rightarrow \frac{3}{32}x > \frac{3}{4} \Rightarrow x > 8$

The curve is concave when $\frac{d^2y}{dx^2} < 0 \Rightarrow \frac{3}{32}x - \frac{3}{4} < 0$

$\Rightarrow \frac{3}{32}x < \frac{3}{4} \Rightarrow x < 8$

So the curve is convex for $x > 8$ and concave for $x < 8$.

Exercise 2.2 — Finding $\frac{dy}{dx}$ when $x = f(y)$

Q1 a) $\frac{dx}{dy} = 6y + 5 \Rightarrow \frac{dy}{dx} = \frac{1}{6y+5}$

At $(5, -1)$, $y = -1$ so $\frac{dy}{dx} = \frac{1}{-1} = -1$

b) $\frac{dx}{dy} = 3y^2 - 2 \Rightarrow \frac{dy}{dx} = \frac{1}{3y^2-2}$

At $(-4, -2)$, $y = -2$ so $\frac{dy}{dx} = \frac{1}{10} = 0.1$

c) $x = (2y + 1)(y - 2) = 2y^2 - 3y - 2$

$\frac{dx}{dy} = 4y - 3 \Rightarrow \frac{dy}{dx} = \frac{1}{4y-3}$

At $(3, -1)$ $y = -1$ so $\frac{dy}{dx} = -\frac{1}{7}$.

d) $x = \frac{4+y^2}{y} = 4y^{-1} + y$

$\frac{dx}{dy} = -4y^{-2} + 1 \Rightarrow \frac{dy}{dx} = \frac{1}{1-\frac{4}{y^2}} = \frac{y^2}{y^2-4}$

At $(5, 4)$, $y = 4$ so $\frac{dy}{dx} = \frac{4}{3}$.

Q2 $x = (2y^3 - 5)^3$, so let $x = u^3$ where $u = 2y^3 - 5$

$\Rightarrow \frac{dx}{du} = 3u^2 = 3(2y^3 - 5)^2$, $\frac{du}{dy} = 6y^2$

$\frac{dx}{dy} = \frac{dx}{du} \times \frac{du}{dy} = 3(2y^3 - 5)^2 \times 6y^2 = 18y^2(2y^3 - 5)^2$

$\frac{dy}{dx} = \frac{1}{18y^2(2y^3-5)^2}$

Q3 a) $x = \sqrt{4+y} \Rightarrow x = u^{\frac{1}{2}}$, $u = 4 + y$

$\Rightarrow \frac{dx}{du} = \frac{1}{2}u^{-\frac{1}{2}} = \frac{1}{2}\frac{1}{\sqrt{4+y}}$, $\frac{du}{dy} = 1$

$\frac{dx}{dy} = \frac{dx}{du} \times \frac{du}{dy} = \frac{1}{2}\frac{1}{\sqrt{4+y}} \times 1 = \frac{1}{2\sqrt{4+y}} = \frac{1}{2x}$

$\Rightarrow \frac{dy}{dx} = 2x$

Use $x = \sqrt{4+y}$ *from the question to get dx/dy in terms of x.*

b) $x = \sqrt{4+y} \Rightarrow x^2 = 4 + y$
$\Rightarrow y = x^2 - 4 \Rightarrow \frac{dy}{dx} = 2x$

3. Differentiation of e^x, $\ln x$ and a^x

Exercise 3.1 — Differentiating e^x

Q1 a) $y = e^{f(x)}$, where $f(x) = 3x$, so $f'(x) = 3$

$\frac{dy}{dx} = f'(x)e^{f(x)} = 3 \times e^{3x} = 3e^{3x}$

b) $\frac{dy}{dx} = f'(x)e^{f(x)} = 2 \times e^{2x-5} = 2e^{2x-5}$

c) $\frac{dy}{dx} = f'(x)e^{f(x)} = 1 \times e^{x+7} = e^{x+7}$

d) $\frac{dy}{dx} = f'(x)e^{f(x)} = 3 \times e^{3x+9} = 3e^{3x+9}$

e) $\frac{dy}{dx} = f'(x)e^{f(x)} = (-2) \times e^{7-2x} = -2e^{7-2x}$

f) $\frac{dy}{dx} = f'(x)e^{f(x)} = 3x^2 \times e^{x^3} = 3x^2e^{x^3}$

Q2 a) $f'(x) = g'(x)e^{g(x)} = (3x^2 + 3)e^{x^3+3x}$

b) $f'(x) = g'(x)e^{g(x)} = (3x^2 - 3)e^{x^3-3x-5}$

c) $f(x) = e^{2x^2+x}$, so:

$f'(x) = g'(x)e^{g(x)} = (4x + 1) \times e^{2x^2+x} = (4x + 1)e^{x(2x+1)}$

Q3 a) e^x differentiates to e^x and e^{-x} differentiates to $-e^{-x}$ so

$f'(x) = \frac{1}{2} \times (e^x - (-e^{-x})) = \frac{1}{2}(e^x + e^{-x})$

b) $f(x) = e^{x^2+7x+12}$

$f'(x) = g'(x)e^{g(x)} = (2x + 7)e^{x^2+7x+12}$

c) $\frac{d}{dx}(e^{x^4+3x^2}) = (4x^3 + 6x)e^{x^4+3x^2}$ and $\frac{d}{dx}(2e^{2x}) = 4e^{2x}$

So $f'(x) = (4x^3 + 6x)e^{x^4+3x^2} + 4e^{2x}$

Q4 $\frac{dy}{dx} = f'(x)e^{f(x)} = 2 \times e^{2x} = 2e^{2x}$

At $x = 0$, $\frac{dy}{dx} = 2 \times e^{2 \times 0} = 2 \times 1 = 2$

$y = mx + c \Rightarrow 1 = (2 \times 0) + c \Rightarrow c = 1$

So the equation of the tangent is $y = 2x + 1$.

There's no real need to use the y = mx + c formula here as we already know the place it crosses the y-axis is (O, 1) (it's given in the question), but it's good to be safe.

Q5 $\frac{dy}{dx} = x - f'(x)e^{f(x)} = x - 2e^{2x-6}$

$\frac{d^2y}{dx^2} = 1 - 2 \times f'(x)e^{f(x)} = 1 - 4e^{2x-6}$

At the point of inflection, $\frac{d^2y}{dx^2} = 0$

$\Rightarrow 1 - 4e^{2x-6} = 0 \Rightarrow 1 = 4e^{2x-6} \Rightarrow \ln(1) = \ln(4e^{2x-6})$

$\Rightarrow 0 = \ln 4 + \ln(e^{2x-6}) = \ln 4 + 2x - 6 \Rightarrow 2x = 6 - \ln 4$

$\Rightarrow x = 3 - \frac{1}{2}\ln 4 = 3 - \ln 2$

Remember from laws of logs that $\frac{1}{2}\ln 4 = \ln 4^{\frac{1}{2}} = \ln 2$. There are other ways you could have used the log laws in this question — you should get the same answer.

Q6 $\frac{dy}{dx} = f'(x)e^{f(x)} = 4xe^{2x^2}$

At $x = 1$, $\frac{dy}{dx} = 4 \times 1 \times e^2 = 4e^2$ and $y = e^2$

$y = mx + c \Rightarrow e^2 = (4e^2 \times 1) + c \Rightarrow c = -3e^2$

So the equation of the tangent is $y = 4e^2x - 3e^2$.

Q7 $\frac{dy}{dx} = f'(x)e^{f(x)} - 1 = 2e^{2x-4} - 1$, $\frac{d^2y}{dx^2} = 2 \times f'(x)e^{f(x)} = 4e^{2x-4}$

$4e^{2x-4} > 0$ for all values of x, so $\frac{d^2y}{dx^2}$ is always positive, which means the graph of $y = e^{2x-4} - x$ is always convex.

Q8 $\frac{dy}{dx} = f'(x)e^{f(x)} = (3 \times e^{3x}) = 3e^{3x}$

When it crosses the y-axis, $x = 0$, so $y = e^{3 \times 0} + 3 = 4$

$\frac{dy}{dx} = (3 \times e^{3 \times 0}) = (3 \times 1) = 3$

$\Rightarrow$ Gradient of the normal $= -\frac{1}{3}$

$y = mx + c \Rightarrow 4 = (-\frac{1}{3} \times 0) + c \Rightarrow c = 4$

So the equation of the normal is $y = -\frac{1}{3}x + 4$.

Q9 Let $f(x) = 2x$ and $g(x) = 3 - 4x$, then

$\frac{dy}{dx} = 2 \times f'(x)e^{f(x)} - \frac{1}{2} \times g'(x)e^{g(x)} = 2 \times 2e^{2x} - \frac{1}{2} \times -4e^{3-4x}$
$= 4e^{2x} + 2e^{3-4x}$

$\frac{d^2y}{dx^2} = 4 \times f'(x)e^{f(x)} + 2 \times g'(x)e^{g(x)} = 4 \times 2e^{2x} + 2 \times -4e^{3-4x}$
$= 8e^{2x} - 8e^{3-4x}$

At a point of inflection, $\frac{d^2y}{dx^2} = 0 \Rightarrow 8e^{2x} - 8e^{3-4x} = 0$
$\Rightarrow e^{2x} = e^{3-4x}$

$\Rightarrow 2x = 3 - 4x \Rightarrow 6x = 3 \Rightarrow x = \frac{1}{2}$

$\frac{d^2y}{dx^2} > 0$ if $x > \frac{1}{2}$ and $\frac{d^2y}{dx^2} < 0$ if $x < \frac{1}{2}$

$\frac{d^2y}{dx^2}$ changes sign from negative to positive at $x = \frac{1}{2}$.

So $x = \frac{1}{2}$ is a point of inflection.

At $x = \frac{1}{2}$, $y = 2e^{2x} - \frac{1}{2}e^{3-4x} = 2e^1 - \frac{1}{2}e^1 = \frac{3}{2}e$.

The graph has one point of inflection, at $\left(\frac{1}{2}, \frac{3}{2}e\right)$.

Q10 If y has a stationary point, the gradient $\frac{dy}{dx}$ will be 0.

$\frac{dy}{dx} = f'(x)e^{f(x)} = (3x^2 - 3)e^{x^3-3x-5}$

So if $\frac{dy}{dx} = 0$, either $3x^2 - 3 = 0$ or $e^{x^3-3x-5} = 0$.

If $3x^2 - 3 = 0 \Rightarrow 3(x^2 - 1) = 0 \Rightarrow x^2 = 1 \Rightarrow x = \pm 1$

and if $e^{x^3-3x-5} = 0$, there are no solutions.

So the gradient is 0 when $x = \pm 1$

$\Rightarrow$ the curve has stationary points at $x = \pm 1$.

Q11 Stationary points occur when the gradient is 0.

$\frac{dy}{dx} = f'(x)e^{f(x)} - 6 = 3e^{3x} - 6$

$3e^{3x} - 6 = 0 \Rightarrow e^{3x} = 2 \Rightarrow 3x = \ln 2 \Rightarrow x = \frac{1}{3}\ln 2$

Take ln of both sides to get rid of the exponential.

To find the nature of the stationary point, calculate $\frac{d^2y}{dx^2}$

$\frac{d^2y}{dx^2} = 3f'(x)e^{f(x)} = 9e^{3x}$

When $x = \frac{1}{3}\ln 2$, $\frac{d^2y}{dx^2} = 9e^{\ln 2} = 9 \times 2 = 18$

$\frac{d^2y}{dx^2}$ is positive, so it's a minimum point.

You could also use the fact that $9e^{3x}$ is always positive to show that it's a minimum without needing to substitute.

Exercise 3.2 — Differentiating $\ln x$

Q1 a) $y = \ln(3x) = \ln 3 + \ln x \Rightarrow \frac{dy}{dx} = \frac{1}{x}$

You could use that the derivative of ln (f(x)) is $\frac{f'(x)}{f(x)}$ — it'd give the same answer.

b) $\frac{dy}{dx} = \frac{f'(x)}{f(x)} = \frac{1}{1+x}$

The coefficient of x is 1 so it's a 1 on top of the fraction.

c) $\frac{dy}{dx} = \frac{f'(x)}{f(x)} = \frac{5}{1+5x}$

d) $\frac{dy}{dx} = 4 \times \frac{f'(x)}{f(x)} = 4 \times \frac{4}{4x-2} = \frac{16}{4x-2} = \frac{8}{2x-1}$
Don't forget to simplify your answers if you can.

Q2 a) $\frac{dy}{dx} = \frac{f'(x)}{f(x)} = \frac{2x}{1+x^2}$

b) $y = \ln(2+x)^2 = 2\ln(2+x)$
$\frac{dy}{dx} = 2\frac{f'(x)}{f(x)} = 2 \times \frac{1}{2+x} = \frac{2}{2+x}$

c) $3\ln x^3 = 9\ln x \Rightarrow \frac{dy}{dx} = \frac{9}{x}$

d) $\frac{dy}{dx} = \frac{f'(x)}{f(x)} = \frac{3x^2+2x}{x^3+x^2} = \frac{3x+2}{x^2+x}$

Q3 a) $f(x) = \ln\frac{1}{x} = \ln x^{-1} = -\ln x \Rightarrow f'(x) = -\frac{1}{x}$

b) $f(x) = \ln\sqrt{x} = \ln x^{\frac{1}{2}} = \frac{1}{2}\ln x \Rightarrow f'(x) = \frac{1}{2x}$

Q4 $\ln((2x+1)^2\sqrt{x-4}) = \ln(2x+1)^2 + \ln\sqrt{x-4}$
$= 2\ln(2x+1) + \frac{1}{2}\ln(x-4)$

First part: $f'(x) = 2 \times \frac{2}{2x+1} = \frac{4}{2x+1}$

Second part: $f'(x) = \frac{1}{2} \times \frac{1}{x-4} = \frac{1}{2(x-4)}$

Putting it all together:
$f'(x) = \frac{4}{2x+1} + \frac{1}{2(x-4)} = \frac{8(x-4)+2x+1}{2(2x+1)(x-4)}$
$= \frac{10x-31}{2(2x+1)(x-4)}$

Q5 $g(x) = x - \sqrt{x-4} \Rightarrow g'(x) = 1 - \left(\frac{dy}{du} \times \frac{du}{dx}\right)$
$= 1 - \frac{1}{2\sqrt{x-4}}$

$f'(x) = \frac{g'(x)}{g(x)} = \frac{1 - \frac{1}{2\sqrt{x-4}}}{x - \sqrt{x-4}} = \frac{2\sqrt{x-4}-1}{2(x\sqrt{x-4}-x+4)}$

Q6 $\ln\left(\frac{(3x+1)^2}{\sqrt{2x+1}}\right) = \ln(3x+1)^2 - \ln\sqrt{2x+1}$
$= 2\ln(3x+1) - \frac{1}{2}\ln(2x+1)$

First part: $f'(x) = 2 \times \frac{3}{3x+1} = \frac{6}{3x+1}$

Second part: $f'(x) = \frac{1}{2} \times \frac{2}{2x+1} = \frac{1}{2x+1}$

Putting it all together:
$f'(x) = \frac{6}{3x+1} - \frac{1}{2x+1} = \frac{6(2x+1)-3x-1}{(3x+1)(2x+1)}$
$= \frac{9x+5}{(3x+1)(2x+1)}$

You could have left your answer as 2 fractions.

Q7 a) $y = \ln(3x)^2 = 2\ln(3x) = 2\ln 3 + 2\ln x \Rightarrow \frac{dy}{dx} = \frac{2}{x}$
When $x = -2$, $\frac{dy}{dx} = -1$ and $y = \ln 36$
$y = mx + c \Rightarrow \ln 36 = [(-1) \times (-2)] + c$
$\Rightarrow c = \ln 36 - 2$
So the equation of the tangent is $y = -x + \ln 36 - 2$

b) When $x = 2$, $\frac{dy}{dx} = 1$ and $y = \ln 36$
$y = mx + c \Rightarrow \ln 36 = [1 \times 2] + c \Rightarrow c = \ln 36 - 2$
So the equation of the tangent is $y = x + \ln 36 - 2$

Q8 a) $y = \ln(x+6)^2 = 2\ln(x+6) \Rightarrow \frac{dy}{dx} = 2\frac{f'(x)}{f(x)} = \frac{2}{x+6}$
When $x = -3$, $\frac{dy}{dx} = \frac{2}{3}$ and $y = \ln 9$
So the gradient of the normal $= -\frac{1}{\left(\frac{2}{3}\right)} = -\frac{3}{2}$
$y = mx + c \Rightarrow \ln 9 = (-\frac{3}{2} \times -3) + c \Rightarrow c = \ln 9 - \frac{9}{2}$
so the equation of the normal is $y = -\frac{3}{2}x + \ln 9 - \frac{9}{2}$.

b) When $x = 0$, $\frac{dy}{dx} = \frac{2}{6} = \frac{1}{3}$ and $y = \ln 36$
So the gradient of the normal $= -\frac{1}{\left(\frac{1}{3}\right)} = -3$
$y = mx + c \Rightarrow \ln 36 = (-3 \times 0) + c \Rightarrow c = \ln 36$
So the equation of the normal is $y = -3x + \ln 36$

Q9 Stationary points occur when the gradient is 0.
$\frac{dy}{dx} = \frac{f'(x)}{f(x)} = \frac{3x^2-6x+3}{x^3-3x^2+3x}$
so $3x^2 - 6x + 3 = 0 \Rightarrow 3(x-1)(x-1) = 0$
You can ignore the denominator here, as it's only the top part that affects when it's equal to 0.
So the gradient is 0 when $x = 1$.
When $x = 1$, $y = 0$ so the stationary point is at (1, 0).

Exercise 3.3 — Differentiating a^x

Q1 a) $\frac{dy}{dx} = 5^x \ln 5$

b) Let $u = 2x$ and $y = 3^u$, then
$\frac{dy}{du} = 3^u \ln 3 = 3^{2x}\ln 3$ and $\frac{du}{dx} = 2$
So $\frac{dy}{dx} = \frac{dy}{du} \times \frac{du}{dx} = 3^{2x}\ln 3 \times 2 = (2\ln 3)3^{2x}$

c) Let $u = -x$ and $y = 10^u$, then
$\frac{dy}{du} = 10^u \ln 10 = 10^{-x}\ln 10$ and $\frac{du}{dx} = -1$
So $\frac{dy}{dx} = 10^{-x}\ln 10 \times (-1) = -(10^{-x}\ln 10)$

d) Let $u = qx$ and $y = p^u$, then
$\frac{dy}{du} = p^u \ln p = p^{qx}\ln p$ and $\frac{du}{dx} = q$
So $\frac{dy}{dx} = p^{qx}\ln p \times q = (q\ln p)p^{qx}$

Q2 a) Let $u = 4x$, then $y = 2^u$ and
$\frac{dy}{dx} = \frac{d}{du}(2^u) \times \frac{d}{dx}(4x) = 4(2^{4x}\ln 2)$

b) When $x = 2$, $\frac{dy}{dx} = 4(2^8\ln 2) = 1024\ln 2$
$y = 2^8 = 256$
Putting this into $y = mx + c$ gives:
$256 = 2048\ln 2 + c \Rightarrow c = 256 - 2048\ln 2$.
So the equation of the tangent is:
$y = (1024\ln 2)x + 256 - 2048\ln 2$
or $y = (1024\ln 2)x + 256(1 - 8\ln 2)$

Q3 a) When $x = 1$, $2^p = 32$, so $p = 5$.
You should just know this result, but if not you can take logs ($p \ln 2 = \ln 32$, so $p = \ln 32 \div \ln 2 = 5$).

b) $\frac{dy}{dx} = 5(2^{5x}\ln 2)$
When $x = 1$, $\frac{dy}{dx} = 5(32\ln 2) = 160\ln 2$

Q4 a) Let $u = x^3$, then $y = p^u$, and so
$\frac{dy}{dx} = \frac{d}{du}(p^u) \times \frac{d}{dx}(x^3) = 3x^2(p^{x^3}\ln p)$

b) When $x = 2$, $y = p^8 = 6561$.
So $p = \sqrt[8]{6561} = 3$.
You can also work this out by taking logs.
$8\ln p = \ln 6561$, so $p = \exp\left(\frac{\ln 6561}{8}\right) = 3$.

c) When $x = 1$, $y = 3^1 = 3$.
The gradient at $x = 1$ is $3(3 \ln 3) = 9 \ln 3$.
Putting this into $y = mx + c$ gives:
$3 = 9 \ln 3 + c \Rightarrow c = 3 - 9 \ln 3$
So the equation of the tangent at (1, 3) is
$y = (9 \ln 3)x + (3 - 9 \ln 3)$

Q5 When $x = 25$, $y = 4^5 = 1024$, so $a = 1024$.
Let $u = x^{\frac{1}{2}}$ and $y = 4^u$, so
$\frac{dy}{dx} = \frac{d}{du}(4^u)\frac{d}{dx}(x^{\frac{1}{2}}) = \frac{1}{2}x^{-\frac{1}{2}}(4^{\sqrt{x}} \ln 4)$
The gradient of the tangent when $x = 25$ is
$\frac{1}{10}(1024 \ln 4) = 102.4 \ln 4$
Putting this into $y = mx + c$ gives:
$1024 = 2560 \ln 4 + c \Rightarrow c = 1024 - 2560 \ln 4$.
So the equation of the tangent is
$y = (102.4 \ln 4)x + (1024 - 2560 \ln 4)$
$\Rightarrow y = 142x - 2520$ to 3 s.f.

Q6 a) $\frac{dy}{dx} = -3(2^{-3x} \ln 2)$

b) When $x = 2$, $y = b = 2^{-6} = \frac{1}{64}$ and $\frac{dy}{dx} = -\frac{3}{64} \ln 2$
At $\left(2, \frac{1}{64}\right)$, the gradient of the tangent is $-\frac{3}{64} \ln 2$.

c) Putting this into $y = mx + c$ gives:
$\frac{1}{64} = -\frac{6}{64} \ln 2 + c \Rightarrow c = \frac{1}{64} + \frac{6}{64} \ln 2$.
So the equation of the tangent is
$y = \frac{1 + 6\ln 2}{64} - \left(\frac{3 \ln 2}{64}\right)x$
$\Rightarrow 64y = 1 + 6 \ln 2 - (3 \ln 2)x$.

4. Differentiating Trig Functions

Exercise 4.1 — Differentiating sin, cos and tan

Q1 a) $y = \sin(3x)$, so let $y = \sin u$ where $u = 3x$
$\Rightarrow \frac{dy}{du} = \cos u = \cos(3x)$, $\frac{du}{dx} = 3$
$\frac{dy}{dx} = \frac{dy}{du} \times \frac{du}{dx} = 3 \cos(3x)$

b) $y = \cos(-2x)$, so let $y = \cos u$ where $u = -2x$
$\Rightarrow \frac{dy}{du} = -\sin(u) = -\sin(-2x)$, $\frac{du}{dx} = -2$
$\frac{dy}{dx} = \frac{dy}{du} \times \frac{du}{dx} = (-\sin(-2x)) \times (-2) = 2 \sin(-2x)$

c) $\frac{dy}{dx} = \frac{dy}{du} \times \frac{du}{dx} = \frac{1}{2} \times -\sin \frac{x}{2} = -\frac{1}{2} \sin \frac{x}{2}$

d) $\frac{dy}{dx} = \frac{dy}{du} \times \frac{du}{dx} = \cos(x + \frac{\pi}{4}) \times 1 = \cos(x + \frac{\pi}{4})$

e) $\frac{dy}{dx} = \frac{dy}{du} \times \frac{du}{dx} = 6 \times \sec^2 \frac{x}{2} \times \frac{1}{2} = 3 \sec^2 \frac{x}{2}$

f) $\frac{dy}{dx} = \frac{dy}{du} \times \frac{du}{dx} = 3 \times \sec^2(5x) \times 5 = 15 \sec^2(5x)$

Q2 $f'(x) = \frac{dy}{du} \times \frac{du}{dx} = 3 \times \sec^2(2x - 1) \times 2 = 6 \sec^2(2x - 1)$

Q3 First part: $\frac{dy}{dx} = 3 \sec^2 x$
Second part: $\frac{dy}{dx} = \frac{dy}{du} \times \frac{du}{dx} = 3 \sec^2(3x)$
Putting it all together: $f'(x) = 3(\sec^2 x + \sec^2(3x))$

Q4 $f'(x) = \frac{dy}{du} \times \frac{du}{dx} = 2x \cos(x^2 + \frac{\pi}{3})$

Q5 $f(x) = \sin^2 x$, so let $y = u^2$ where $u = \sin x$
$f'(x) = \frac{dy}{du} \times \frac{du}{dx} = (2 \sin x) \times \cos x = 2 \sin x \cos x$

Q6 $f(x) = 2 \sin^3 x$, so let $y = 2u^3$ where $u = \sin x$
$f'(x) = \frac{dy}{du} \times \frac{du}{dx} = 6 \sin^2 x \cos x$

Q7 a) $f'(x) = 3 \cos x - 2 \sin x$

b) $f'(x) = 0 \Rightarrow 3 \cos x - 2 \sin x = 0$
$\Rightarrow 3 \cos x = 2 \sin x$
$\Rightarrow \frac{3}{2} = \tan x \Rightarrow x = \tan^{-1} \frac{3}{2} = 0.983$ (3 s.f.)
Remember that $\tan x = \frac{\sin x}{\cos x}$.

Q8 $y = \frac{1}{\cos x} = (\cos x)^{-1}$, so let $y = u^{-1}$ where $u = \cos x$
$\Rightarrow \frac{dy}{du} = -u^{-2} = -\frac{1}{\cos^2 x}$, $\frac{du}{dx} = -\sin x$
$\frac{dy}{dx} = \frac{dy}{du} \times \frac{du}{dx} = -\frac{1}{\cos^2 x} \times -\sin x = \frac{\sin x}{\cos^2 x} = \sec x \tan x$
Remember that $\frac{1}{\cos x} = \sec x$.

Q9 Let $y = f(x)$, where $f(x) = \cos x$.
Then $\frac{dy}{dx} = \lim_{h \to 0}\left[\frac{f(x+h) - f(x)}{(x+h) - x}\right] = \lim_{h \to 0}\left[\frac{\cos(x+h) - \cos x}{(x+h) - x}\right]$
Using the cos addition formula:
$\frac{dy}{dx} = \lim_{h \to 0}\left[\frac{\cos x \cos h - \sin x \sin h - \cos x}{(x+h) - x}\right]$
$= \lim_{h \to 0}\left[\frac{\cos x(\cos h - 1) - \sin x \sin h}{h}\right]$
$= \lim_{h \to 0}\left[\frac{\cos x(\cos h - 1)}{h} - \frac{\sin x \sin h}{h}\right]$
Using small angle approximations,
$\cos h \approx 1 - \frac{1}{2}h^2$ and $\sin h \approx h$, so:
$\frac{dy}{dx} = \lim_{h \to 0}\left[\frac{\cos x \times \left(-\frac{1}{2}h^2\right)}{h} - \frac{\sin x \times h}{h}\right]$
$= \lim_{h \to 0}\left[-\frac{h \cos x}{2} - \sin x\right] = -\sin x$

Q10 a) $y = \cos^2 x$, so let $y = u^2$ where $u = \cos x$.
$\frac{dy}{dx} = \frac{dy}{du} \times \frac{du}{dx} = (2 \cos x) \times (-\sin x)$
$= -2 \sin x \cos x$

b) Using the double angle formula:
$\cos(2x) \equiv 2 \cos^2 x - 1 \Rightarrow \cos^2 x = \frac{1}{2}(\cos(2x) + 1)$
$\frac{dy}{dx} = \frac{1}{2} \times 2 \times (-\sin(2x)) = -\sin(2x)$
As the original function was in terms of cos x rather than cos (2x), it would be better to rearrange this.
From the double angle formula for sin:
$-\sin(2x) \equiv -2 \sin x \cos x$

Q11 First part:
$y = 6 \cos^2 x = 6(\cos x)^2$, so let $y = 6u^2$ where $u = \cos x$
$\Rightarrow \frac{dy}{du} = 12u = 12 \cos x$, $\frac{du}{dx} = -\sin x$
$\frac{dy}{dx} = \frac{dy}{du} \times \frac{du}{dx} = -12 \sin x \cos x$

Second part:
$y = 2 \sin(2x)$, so let $y = 2 \sin u$ where $u = 2x$
$\Rightarrow \frac{dy}{du} = 2 \cos u = 2 \cos(2x)$, $\frac{du}{dx} = 2$
$\frac{dy}{dx} = \frac{dy}{du} \times \frac{du}{dx} = 4 \cos(2x)$
Putting it all together:
$\frac{dy}{dx} = -12 \sin x \cos x - 4 \cos(2x)$
Double angle formula: $2 \sin x \cos x \equiv \sin(2x)$
$\Rightarrow -12 \sin x \cos x - 4 \cos(2x) = -6 \sin(2x) - 4 \cos(2x)$

Q12 $\frac{dy}{dx} = \cos x$. When $x = \frac{\pi}{4}$, $\frac{dy}{dx} = \frac{1}{\sqrt{2}}$

Q13 $\frac{dy}{dx} = -2\sin(2x)$

When $x = \frac{\pi}{4}$, $y = 0$ and $\frac{dy}{dx} = -2$

So the gradient of the normal is $\frac{-1}{-2} = \frac{1}{2}$.

$y = mx + c \Rightarrow 0 = \left(\frac{1}{2} \times \frac{\pi}{4}\right) + c \Rightarrow c = -\frac{\pi}{8}$

So the equation of the normal is

$y = \frac{1}{2}x - \frac{\pi}{8}$ (or $4x - 8y - \pi = 0$)

Q14 a) $\frac{dx}{dy} = 2\cos(2y)$, $\frac{dy}{dx} = \frac{1}{2\cos(2y)} = \frac{1}{2}\sec(2y)$

At the point $\left(\frac{\sqrt{3}}{2}, \frac{\pi}{6}\right)$, $\frac{dy}{dx} = \frac{1}{2\cos\frac{\pi}{3}} = 1$

$y = mx + c \Rightarrow \frac{\pi}{6} = \frac{\sqrt{3}}{2} + c \Rightarrow c = \frac{\pi}{6} - \frac{\sqrt{3}}{2}$

So the equation of the tangent is $y = x + \frac{\pi}{6} - \frac{\sqrt{3}}{2}$.

b) From part a), $\frac{dy}{dx} = 1$, so the normal gradient is -1.

$y = mx + c \Rightarrow \frac{\pi}{6} = -\frac{\sqrt{3}}{2} + c \Rightarrow c = \frac{\pi}{6} + \frac{\sqrt{3}}{2}$

So the equation of the normal is $y = -x + \frac{\pi}{6} + \frac{\sqrt{3}}{2}$.

Q15 a) $y = 2\sin(2x)\cos x \Rightarrow y = 4\sin x\cos^2 x$

(from double angle formula $\sin(2x) \equiv 2\sin x\cos x$)

$\Rightarrow y = 4\sin x(1 - \sin^2 x)$ *(from $\sin^2 x + \cos^2 x \equiv 1$)*

$\Rightarrow y = 4\sin x - 4\sin^3 x$

b) First part: $\frac{dy}{dx} = 4\cos x$

Second part: $y = 4\sin^3 x = 4(\sin x)^3$,

so let $y = 4u^3$ where $u = \sin x$

$\frac{dy}{dx} = \frac{dy}{du} \times \frac{du}{dx} = 12\sin^2 x\cos x$

Putting it all together: $\frac{dy}{dx} = 4\cos x - 12\sin^2 x\cos x$

Exercise 4.2 — Differentiating by using the chain rule twice

Q1 a) $y = \sin(\cos(2x))$, so let $y = \sin u$ where $u = \cos(2x)$

$\frac{dy}{du} = \cos u = \cos(\cos(2x))$

Using the chain rule again on $\frac{du}{dx}$ gives:

$u = \cos(2x) \Rightarrow \frac{du}{dx} = -2\sin(2x)$

Putting it all together:

$\frac{dy}{dx} = \frac{dy}{du} \times \frac{du}{dx} = -2\sin(2x)\cos(\cos(2x))$

b) $y = 2\ln f(x) \Rightarrow \frac{dy}{dx} = 2\frac{f'(x)}{f(x)}$

$f(x) = \cos(3x) \Rightarrow f'(x) = -3\sin(3x)$

$\frac{dy}{dx} = 2\frac{-3\sin(3x)}{\cos(3x)} = -6\tan(3x)$

c) $y = \ln(\tan^2 x) = \ln(f(x)) \Rightarrow \frac{dy}{dx} = \frac{f'(x)}{f(x)}$

$f(x) = (\tan x)^2$, so let $f(x) = v = u^2$ where $u = \tan x$

$\frac{dv}{du} = 2u = 2\tan x$, $\frac{du}{dx} = \sec^2 x$

$f'(x) = \frac{dv}{dx} = \frac{dv}{du} \times \frac{du}{dx} = 2\tan x\sec^2 x$

Putting it all together:

$\frac{dy}{dx} = \frac{2\tan x\sec^2 x}{\tan^2 x} = 2\sec x\operatorname{cosec} x$

You could've written $\ln(\tan^2 x)$ as $2\ln(\tan x)$ using the laws of logs and then differentiated — you'd end up with the same answer.

d) $y = e^{f(x)} \Rightarrow \frac{dy}{dx} = f'(x)e^{f(x)}$

$f(x) = \tan(2x)$, so let $f(x) = v = \tan u$ where $u = 2x$

$\frac{dv}{du} = \sec^2 u = \sec^2(2x)$, $\frac{du}{dx} = 2$

$\Rightarrow f'(x) = \frac{dv}{dx} = \frac{dv}{du} \times \frac{du}{dx} = 2\sec^2(2x)$

$\Rightarrow \frac{dy}{dx} = 2\sec^2(2x)e^{\tan(2x)}$

Q2 a) $\frac{dy}{dx} = \sin^4 x^2 = (\sin x^2)^4$, so let $y = u^4$ where $u = \sin x^2$

$\frac{dy}{du} = 4u^3 = 4\sin^3 x^2$

For $\frac{du}{dx}$, set up another chain rule:

$u = \sin x^2$ so let $u = \sin v$, $v = x^2$

$\frac{du}{dv} = \cos v = \cos x^2$, $\frac{dv}{dx} = 2x$

$\frac{du}{dx} = \frac{du}{dv} \times \frac{dv}{dx} = 2x\cos x^2$

Putting it all together:

$\frac{dy}{dx} = \frac{dy}{du} \times \frac{du}{dx} = 8x\sin^3 x^2\cos x^2$

b) $\frac{dy}{dx} = f'(x)e^{f(x)}$

$f(x) = \sin^2 x = (\sin x)^2$, so let $y = u^2$ where $u = \sin x$

$f'(x) = \frac{dy}{du} \times \frac{du}{dx} = (2\sin x) \times (\cos x)$
$= 2\sin x\cos x$

$\Rightarrow \frac{dy}{dx} = 2e^{\sin^2 x}\sin x\cos x$

c) First part:

$y = \tan^2(3x) = (\tan(3x))^2$,

so let $y = u^2$ where $u = \tan(3x)$

$\frac{dy}{du} = 2u = 2\tan(3x)$

For $\frac{du}{dx}$ set up the chain rule again:

$u = \tan(3x)$, so let $u = \tan v$ where $v = 3x$

$\frac{du}{dv} = \sec^2 v = \sec^2(3x)$, $\frac{dv}{dx} = 3$

$\Rightarrow \frac{du}{dx} = \frac{du}{dv} \times \frac{dv}{dx} = 3\sec^2(3x)$

$\Rightarrow \frac{dy}{dx} = \frac{dy}{du} \times \frac{du}{dx} = 6\tan(3x)\sec^2(3x)$

Second part:

$\frac{dy}{dx} = \cos x$

Putting it all together:

$\frac{dy}{dx} = 6\tan(3x)\sec^2(3x) + \cos x$

With practice, you should be able to do some of the simpler chain rule calculations in your head, e.g. $\frac{d}{dx}\tan^2(3x) = 6\tan(3x)\sec^2(3x)$.

d) First part: $y = e^{f(x)}$, $f(x) = 2\cos(2x) \Rightarrow \frac{dy}{dx} = f'(x)e^{f(x)}$

$f(x) = 2\cos(2x) \Rightarrow f'(x) = -4\sin(2x)$

So $\frac{dy}{dx} = -4\sin(2x)e^{2\cos(2x)}$

Second part: $y = \cos^2(2x) = (\cos(2x))^2$,

so let $y = u^2$ where $u = \cos(2x)$

$\frac{dy}{du} = 2u = 2\cos(2x)$

For $\frac{du}{dx}$, set up the chain rule again:

$u = \cos(2x) \Rightarrow \frac{du}{dx}\ -2\sin(2x)$

$\Rightarrow \frac{dy}{dx} = \frac{dy}{du} \times \frac{du}{dx} = -4\sin(2x)\cos(2x)$

Putting it all together:

$\frac{dy}{dx} = -4\sin(2x)\,e^{2\cos(2x)} - 4\sin(2x)\cos(2x)$

5. Product Rule

Exercise 5.1 — Differentiating functions multiplied together

Q1 **a)** $y = x(x + 2) = x^2 + 2x$

$\frac{dy}{dx} = 2x + 2$

b) $u = x,\ v = x + 2 \Rightarrow \frac{du}{dx} = 1,\ \frac{dv}{dx} = 1$

$\frac{dy}{dx} = u\frac{dv}{dx} + v\frac{du}{dx} = x + (x + 2) = 2x + 2$

Q2 **a)** $u = x^2,\ v = (x + 6)^3 \Rightarrow \frac{du}{dx} = 2x,\ \frac{dv}{dx} = 3(x + 6)^2$

$\frac{dy}{dx} = u\frac{dv}{dx} + v\frac{du}{dx} = [x^2 \times 3(x + 6)^2] + [(x + 6)^3 \times 2x]$

$= 3x^2(x + 6)^2 + 2x(x + 6)^3 = x(x + 6)^2[3x + 2(x + 6)]$

$= x(x + 6)^2(5x + 12)$

Here the chain rule was used to find $\frac{dv}{dx}$ — write out all the steps if you're struggling.

b) $u = x^3,\ v = (5x + 2)^4 \Rightarrow \frac{du}{dx} = 3x^2,\ \frac{dv}{dx} = 20(5x + 2)^3$

$\frac{dy}{dx} = u\frac{dv}{dx} + v\frac{du}{dx} = [x^3 \times 20(5x + 2)^3] + [(5x + 2)^4 \times 3x^2]$

$= 20x^3(5x + 2)^3 + 3x^2(5x + 2)^4$

$= x^2(5x + 2)^3[20x + 3(5x + 2)]$

$= x^2(5x + 2)^3(35x + 6)$

c) $u = x^3,\ v = e^x \Rightarrow \frac{du}{dx} = 3x^2,\ \frac{dv}{dx} = e^x$

$\frac{dy}{dx} = u\frac{dv}{dx} + v\frac{du}{dx} = x^3e^x + e^x3x^2 = x^2e^x(x + 3)$

d) $u = x,\ v = e^{4x} \Rightarrow \frac{du}{dx} = 1,\ \frac{dv}{dx} = 4e^{4x}$

$\frac{dy}{dx} = u\frac{dv}{dx} + v\frac{du}{dx} = 4xe^{4x} + e^{4x} = e^{4x}(4x + 1)$

e) $u = x,\ v = e^{x^2} \Rightarrow \frac{du}{dx} = 1,\ \frac{dv}{dx} = 2xe^{x^2}$

$\frac{dy}{dx} = u\frac{dv}{dx} + v\frac{du}{dx} = x \times 2xe^{x^2} + e^{x^2} = e^{x^2}(2x^2 + 1)$

f) $u = e^{2x},\ v = \sin x \Rightarrow \frac{du}{dx} = 2e^{2x},\ \frac{dv}{dx} = \cos x$

$\frac{dy}{dx} = u\frac{dv}{dx} + v\frac{du}{dx} = e^{2x} \times \cos x + \sin x \times 2e^{2x}$

$= e^{2x}(\cos x + 2\sin x)$

Q3 **a)** $u = x^3,\ v = (x + 3)^{\frac{1}{2}}$

$\Rightarrow \frac{du}{dx} = 3x^2,\ \frac{dv}{dx} = \frac{1}{2}(x + 3)^{-\frac{1}{2}}$

$f'(x) = u\frac{dv}{dx} + v\frac{du}{dx} = \left[x^3 \times \frac{1}{2}(x + 3)^{-\frac{1}{2}}\right] + \left[(x + 3)^{\frac{1}{2}} \times 3x^2\right]$

$= \frac{x^3}{2(x + 3)^{\frac{1}{2}}} + 3x^2(x + 3)^{\frac{1}{2}}$

b) $u = x^2,\ v = (x - 7)^{-\frac{1}{2}} \Rightarrow \frac{du}{dx} = 2x,\ \frac{dv}{dx} = -\frac{1}{2}(x - 7)^{-\frac{3}{2}}$

$f'(x) = u\frac{dv}{dx} + v\frac{du}{dx}$

$= \left[x^2 \times \left(-\frac{1}{2}\right) \times (x - 7)^{-\frac{3}{2}}\right] + \left[(x - 7)^{-\frac{1}{2}} \times 2x\right]$

$= -\frac{x^2}{2(\sqrt{x - 7})^3} + \frac{2x}{\sqrt{x - 7}}$

c) $u = x^4,\ v = \ln x \Rightarrow \frac{du}{dx} = 4x^3,\ \frac{dv}{dx} = \frac{1}{x}$

$f'(x) = u\frac{dv}{dx} + v\frac{du}{dx} = \left[x^4 \times \frac{1}{x}\right] + [\ln x \times 4x^3]$

$= x^3 + 4x^3\ln x = x^3(1 + 4\ln x)$

d) $u = 4x,\ v = \ln x^2 = 2\ln x \Rightarrow \frac{du}{dx} = 4,\ \frac{dv}{dx} = \frac{2}{x}$

$f'(x) = u\frac{dv}{dx} + v\frac{du}{dx} = \left[4x \times \frac{2}{x}\right] + [\ln x^2 \times 4] = 8 + 4\ln x^2$

e) $u = 2x^3,\ v = \cos x \Rightarrow \frac{du}{dx} = 6x^2,\ \frac{dv}{dx} = -\sin x$

$f'(x) = u\frac{dv}{dx} + v\frac{du}{dx} = (-2x^3 \times \sin x) + (\cos x \times 6x^2)$

$= 2x^2(3\cos x - x\sin x)$

f) $u = x^2,\ v = \cos(2x) \Rightarrow \frac{du}{dx} = 2x,\ \frac{dv}{dx} = -2\sin(2x)$

$f'(x) = u\frac{dv}{dx} + v\frac{du}{dx}$

$= [x^2 \times -2\sin(2x)] + [\cos(2x) \times 2x]$

$= 2x\cos(2x) - 2x^2\sin(2x)$

$= 2x(\cos(2x) - x\sin(2x))$

Q4 **a)** $u = (x + 1)^2,\ v = x^2 - 1 \Rightarrow \frac{du}{dx} = 2(x + 1),\ \frac{dv}{dx} = 2x$

$\frac{dy}{dx} = u\frac{dv}{dx} + v\frac{du}{dx}$

$= [(x + 1)^2 \times 2x] + [(x^2 - 1) \times 2(x + 1)]$

$= 2x(x + 1)^2 + 2(x^2 - 1)(x + 1)$

$= 2x^3 + 4x^2 + 2x + 2x^3 + 2x^2 - 2x - 2$

$= 4x^3 + 6x^2 - 2$

b) $u = (x + 1)^3,\ v = x - 1 \Rightarrow \frac{du}{dx} = 3(x + 1)^2,\ \frac{dv}{dx} = 1$

$\frac{dy}{dx} = u\frac{dv}{dx} + v\frac{du}{dx}$

$= [(x + 1)^3 \times 1] + [(x - 1) \times 3(x + 1)^2]$

$= (x + 1)^3 + 3(x - 1)(x + 1)^2$

$= (x^3 + 3x^2 + 3x + 1) + (3x^3 + 6x^2 + 3x - 3x^2 - 6x - 3)$

$= 4x^3 + 6x^2 - 2$

Use the binomial formula to expand $(x + 1)^3$ — you'll get the coefficients 1, 3, 3, 1.

c) $y = (x + 1)^2(x^2 - 1) = (x + 1)^2(x + 1)(x - 1)$

$= (x + 1)^3(x - 1)$

$x^2 - 1$ is the difference of two squares.

Q5 When $y = xe^x$ is concave, $\frac{d^2y}{dx^2}$ is negative.

$u = x,\ v = e^x \Rightarrow \frac{du}{dx} = 1,\ \frac{dv}{dx} = e^x$

$\frac{dy}{dx} = u\frac{dv}{dx} + v\frac{du}{dx} = xe^x + e^x$

$\frac{d^2y}{dx^2} = \frac{d}{dx}(xe^x) + \frac{d}{dx}(e^x) = [xe^x + e^x] + [e^x] = xe^x + 2e^x$

$\frac{d^2y}{dx^2} < 0 \Rightarrow xe^x + 2e^x < 0$

$\Rightarrow e^x(x + 2) < 0$

Since $e^x > 0$ for all x, this means that $(x + 2) < 0$

$\Rightarrow x < -2$

So the curve $y = xe^x$ is concave when $x < -2$.

Q6 $u = \sqrt{x+2},\ v = \sqrt{x+7}$

$\Rightarrow \frac{du}{dx} = \frac{1}{2\sqrt{x+2}},\ \frac{dv}{dx} = \frac{1}{2\sqrt{x+7}}$

$\frac{dy}{dx} = u\frac{dv}{dx} + v\frac{du}{dx} = \frac{\sqrt{x+2}}{2\sqrt{x+7}} + \frac{\sqrt{x+7}}{2\sqrt{x+2}}$

At the point (2, 6), $\frac{dy}{dx} = \frac{\sqrt{4}}{2\sqrt{9}} + \frac{\sqrt{9}}{2\sqrt{4}} = \frac{13}{12}$

$y = mx + c \Rightarrow 6 = \left(2 \times \frac{13}{12}\right) + c \Rightarrow c = \frac{23}{6}$

So the equation of the tangent is $y = \frac{13}{12}x + \frac{23}{6}$.

To write this in the form ax + by + c = 0 where a, b and c are integers, multiply by 12 and rearrange.

$y = \frac{13}{12}x + \frac{23}{6} \Rightarrow 12y = 13x + 46$

$\Rightarrow 13x - 12y + 46 = 0$

Q7 a) $u = (x-1)^{\frac{1}{2}},\ v = (x+4)^{-\frac{1}{2}}$

$\Rightarrow \frac{du}{dx} = \frac{1}{2\sqrt{x-1}},\ \frac{dv}{dx} = -\frac{1}{2(\sqrt{x+4})^3}$

$\frac{dy}{dx} = u\frac{dv}{dx} + v\frac{du}{dx}$

$= \frac{1}{2\sqrt{x-1}\sqrt{x+4}} - \frac{\sqrt{x-1}}{2(\sqrt{x+4})^3}$

When $x = 5$, $\frac{dy}{dx} = \frac{1}{2\sqrt{4}\sqrt{9}} - \frac{\sqrt{4}}{2(\sqrt{9})^3} = \frac{5}{108}$

and $y = \frac{\sqrt{4}}{\sqrt{9}} = \frac{2}{3}$

$y = mx + c \Rightarrow \frac{2}{3} = \left(5 \times \frac{5}{108}\right) + c \Rightarrow c = \frac{47}{108}$

So the equation of the tangent is $y = \frac{5}{108}x + \frac{47}{108}$

To write this in the form ax + by + c = 0 where a, b and c are integers, multiply by 108 and rearrange.

$y = \frac{5}{108}x + \frac{47}{108} \Rightarrow 108y = 5x + 47$

$\Rightarrow 5x - 108y + 47 = 0$

b) Gradient of the normal $= -\frac{1}{\left(\frac{5}{108}\right)} = -\frac{108}{5}$

$y = mx + c \Rightarrow \frac{2}{3} = \left(5 \times \left(-\frac{108}{5}\right)\right) + c \Rightarrow c = \frac{326}{3}$

So the equation of the normal is $y = -\frac{108}{5}x + \frac{326}{3}$.

To write this in the form ax + by + c = 0 where a, b and c are integers, multiply by 15 and rearrange.

$y = -\frac{108}{5}x + \frac{326}{3} \Rightarrow 15y = -324x + 1630$

$\Rightarrow 324x + 15y - 1630 = 0$

Q8 First use the chain rule:

$\frac{dy}{dx} = f'(x)e^{f(x)}$ where $f(x) = x^2\sqrt{x+3}$.

Then use the product rule to find $f'(x)$:

$u = x^2,\ v = \sqrt{x+3} \Rightarrow \frac{du}{dx} = 2x,\ \frac{dv}{dx} = \frac{1}{2\sqrt{x+3}}$

$\frac{dy}{dx} = u\frac{dv}{dx} + v\frac{du}{dx} = \left[x^2 \times \frac{1}{2\sqrt{x+3}}\right] + \left[\sqrt{x+3} \times 2x\right]$

$= \frac{x^2 + 4x(x+3)}{2\sqrt{x+3}} = \frac{5x^2 + 12x}{2\sqrt{x+3}}$

Now putting it all together:

$\frac{dy}{dx} = \frac{5x^2 + 12x}{2\sqrt{x+3}} e^{x^2\sqrt{x+3}}$

Q9 Stationary points occur when the gradient is 0. Differentiate with the product rule:

$u = x,\ v = e^{x-x^2} \Rightarrow \frac{du}{dx} = 1,\ \frac{dv}{dx} = (1-2x)e^{x-x^2}$

$\frac{dy}{dx} = u\frac{dv}{dx} + v\frac{du}{dx} = \left[x(1-2x)e^{x-x^2}\right] + \left[e^{x-x^2} \times 1\right]$

$= e^{x-x^2}(x - 2x^2 + 1)$

e^{x-x^2} cannot be 0, so the stationary points occur when $-2x^2 + x + 1 = 0$

$\Rightarrow (2x+1)(-x+1) = 0 \Rightarrow x = 1$ or $x = -\frac{1}{2}$

When $x = 1$, $y = 1 \times e^0 = 1$.

When $x = -\frac{1}{2}$, $y = -\frac{1}{2} \times e^{-\frac{3}{4}} = -\frac{e^{-\frac{3}{4}}}{2}$.

So the stationary points are (1, 1) and $\left(-\frac{1}{2}, -\frac{e^{-\frac{3}{4}}}{2}\right)$.

Q10 a) Stationary points occur when the gradient is 0.

$u = (x-2)^2,\ v = (x+4)^3$

$\Rightarrow \frac{du}{dx} = 2(x-2),\ \frac{dv}{dx} = 3(x+4)^2$

$\frac{dy}{dx} = u\frac{dv}{dx} + v\frac{du}{dx}$

$= [(x-2)^2 \times 3(x+4)^2] + [(x+4)^3 \times 2(x-2)]$

$= 3(x-2)^2(x+4)^2 + 2(x-2)(x+4)^3$

$= (x-2)(x+4)^2[3x - 6 + 2x + 8]$

$= (x-2)(x+4)^2(5x+2)$

So the stationary points occur when:

$x - 2 = 0 \Rightarrow x = 2$

and $x + 4 = 0 \Rightarrow x = -4$

and $5x + 2 = 0 \Rightarrow x = -\frac{2}{5} = -0.4$

When $x = 2$, $y = 0 \times 6^3 = 0$

When $x = -4$, $y = (-6)^2 \times 0 = 0$

When $x = -0.4$, $y = (-2.4)^2(3.6)^3 = 268.74$ (2 d.p.)

So the stationary points are (2, 0), (–4, 0) and (–0.4, 268.74).

b) To write $\frac{dy}{dx}$ in the form $(Ax^2 + Bx + C)(x + D)^n$, multiply out the brackets $(x - 2)$ and $(5x + 2)$, which will give you the quadratic bracket:

$\frac{dy}{dx} = (x-2)(x+4)^2(5x+2)$

$= [5x^2 - 10x + 2x - 4](x+4)^2$

$= (5x^2 - 8x - 4)(x+4)^2$

Now you can differentiate this using the product rule:

$u = (5x^2 - 8x - 4),\ v = (x+4)^2$

$\Rightarrow \frac{du}{dx} = (10x - 8),\ \frac{dv}{dx} = 2(x+4)$

$\frac{d^2y}{dx^2} = u\frac{dv}{dx} + v\frac{du}{dx}$

$= [(5x^2 - 8x - 4) \times 2(x+4)] + [(x+4)^2 \times (10x - 8)]$

$= (x+4)(10x^2 - 16x - 8) + (x+4)(10x^2 + 32x - 32)$

$= (x+4)(20x^2 + 16x - 40)$

$= 4(x+4)(5x^2 + 4x - 10)$

When $x = 2$, $\frac{d^2y}{dx^2} = 4(2+4)(5(2)^2 + 4(2) - 10)$

$= 4(6)(20 + 8 - 10)$

$= 432\ (> 0)$

So (2, 0) is a minimum point.

When $x = -4$, $\frac{d^2y}{dx^2} = 4(-4+4)(5(-4)^2 + 4(-4) - 10)$

$= 4(0)(80 - 16 - 10) = 0$

Check $\frac{d^2y}{dx^2}$ either side of $x = -4$:

When $x > -4$, $(x + 4) > 0$ and $(5x^2 + 4x - 10) > 0$

$\Rightarrow \frac{d^2y}{dx^2} > 0$

When $x < -4$, $(x + 4) < 0$ and $(5x^2 + 4x - 10) > 0$

$\Rightarrow \frac{d^2y}{dx^2} < 0$

So (–4, 0) is a point of inflection.

When $x = -0.4$,

$\frac{d^2y}{dx^2} = 4(-0.4+4)(5(-0.4)^2 + 4(-0.4) - 10)$

$= 4(3.6)(0.8 - 1.6 - 10)$

$= -155.52\ (< 0)$

So (–0.4, 268.74) is a maximum point.

6. Quotient Rule

Exercise 6.1 — Differentiating a function divided by a function

Q1 a) $u = x + 5$, $v = x - 3 \Rightarrow \frac{du}{dx} = 1$, $\frac{dv}{dx} = 1$

$$\frac{dy}{dx} = \frac{v\frac{du}{dx} - u\frac{dv}{dx}}{v^2} = \frac{((x-3)\times 1) - ((x+5)\times 1)}{(x-3)^2}$$
$$= -\frac{8}{(x-3)^2}$$

b) $u = (x - 7)^4$, $v = (5 - x)^3$

$\Rightarrow \frac{du}{dx} = 4(x-7)^3$, $\frac{dv}{dx} = -3(5-x)^2$

$$\frac{dy}{dx} = \frac{v\frac{du}{dx} - u\frac{dv}{dx}}{v^2}$$
$$= \frac{[(5-x)^3 \times 4(x-7)^3] - [(x-7)^4 \times (-3)(5-x)^2]}{(5-x)^6}$$
$$= \frac{(5-x)^2(x-7)^3[4(5-x) + 3(x-7)]}{(5-x)^6} = \frac{(x-7)^3(-x-1)}{(5-x)^4}$$

c) $u = e^x$, $v = x^2 \Rightarrow \frac{du}{dx} = e^x$, $\frac{dv}{dx} = 2x$

$$\frac{dy}{dx} = \frac{v\frac{du}{dx} - u\frac{dv}{dx}}{v^2} = \frac{x^2e^x - e^x 2x}{x^4} = \frac{xe^x(x-2)}{x^4} = \frac{e^x(x-2)}{x^3}$$

d) $u = 3x$, $v = (x - 1)^2 \Rightarrow \frac{du}{dx} = 3$, $\frac{dv}{dx} = 2(x - 1)$

$$\frac{dy}{dx} = \frac{v\frac{du}{dx} - u\frac{dv}{dx}}{v^2} = \frac{[(x-1)^2 \times 3] - [3x \times 2(x-1)]}{(x-1)^4}$$
$$= \frac{(x-1)[3(x-1) - 6x]}{(x-1)^4} = \frac{-3x-3}{(x-1)^3}$$

Q2 a) $u = x^3$, $v = (x + 3)^3 \Rightarrow \frac{du}{dx} = 3x^2$, $\frac{dv}{dx} = 3(x + 3)^2$

$$f'(x) = \frac{v\frac{du}{dx} - u\frac{dv}{dx}}{v^2} = \frac{[(x+3)^3 \times 3x^2] - [x^3 \times 3(x+3)^2]}{(x+3)^6}$$
$$= \frac{3x^2(x+3) - 3x^3}{(x+3)^4} = \frac{9x^2}{(x+3)^4}$$

b) $u = x^2$, $v = \sqrt{x-7} \Rightarrow \frac{du}{dx} = 2x$, $\frac{dv}{dx} = \frac{1}{2\sqrt{x-7}}$

$$f'(x) = \frac{v\frac{du}{dx} - u\frac{dv}{dx}}{v^2} = \frac{[\sqrt{x-7} \times 2x] - \left[x^2 \frac{1}{2\sqrt{x-7}}\right]}{x-7}$$
$$= \frac{4x(x-7) - x^2}{2(\sqrt{x-7})^3} = \frac{3x^2 - 28x}{2(\sqrt{x-7})^3}$$

c) $u = e^{2x}$, $v = e^{2x} + e^{-2x} \Rightarrow \frac{du}{dx} = 2e^{2x}$, $\frac{dv}{dx} = 2e^{2x} - 2e^{-2x}$

$$f'(x) = \frac{v\frac{du}{dx} - u\frac{dv}{dx}}{v^2}$$
$$= \frac{[(e^{2x} + e^{-2x})2e^{2x}] - [e^{2x}(2e^{2x} - 2e^{-2x})]}{(e^{2x} + e^{-2x})^2}$$
$$= \frac{2e^{4x} + 2 - 2e^{4x} + 2}{e^{4x} + e^{-4x} + 2} = \frac{4}{e^{4x} + e^{-4x} + 2}$$

d) $u = x$, $v = \sin x \Rightarrow \frac{du}{dx} = 1$, $\frac{dv}{dx} = \cos x$

$$f'(x) = \frac{v\frac{du}{dx} - u\frac{dv}{dx}}{v^2} = \frac{\sin x - x\cos x}{\sin^2 x}$$

e) $u = \sin x$, $v = x \Rightarrow \frac{du}{dx} = \cos x$, $\frac{dv}{dx} = 1$

$$f'(x) = \frac{v\frac{du}{dx} - u\frac{dv}{dx}}{v^2} = \frac{x\cos x - \sin x}{x^2}$$

Q3 $u = x^2$, $v = \tan x \Rightarrow \frac{du}{dx} = 2x$, $\frac{dv}{dx} = \sec^2 x$

$$f'(x) = \frac{v\frac{du}{dx} - u\frac{dv}{dx}}{v^2} = \frac{[\tan x \times 2x] - [x^2 \times \sec^2 x]}{\tan^2 x}$$
$$= \frac{2x\tan x - x^2\sec^2 x}{\tan^2 x} = \frac{2x}{\tan x} - x^2\frac{\frac{1}{\cos^2 x}}{\frac{\sin^2 x}{\cos^2 x}}$$
$$= 2x\cot x - x^2\operatorname{cosec}^2 x$$

Q4 $u = 5x - 4$, $v = 2x^2 \Rightarrow \frac{du}{dx} = 5$, $\frac{dv}{dx} = 4x$

$$\frac{dy}{dx} = \frac{v\frac{du}{dx} - u\frac{dv}{dx}}{v^2} = \frac{[2x^2 \times 5] - [(5x-4) \times 4x]}{(2x^2)^2}$$
$$= \frac{10x^2 - (20x^2 - 16x)}{4x^4} = \frac{16x - 10x^2}{4x^4} = \frac{8-5x}{2x^3}$$

There is a stationary point when $\frac{dy}{dx} = 0$:

$\frac{dy}{dx} = 0 \Rightarrow 8 - 5x = 0 \Rightarrow 5x = 8 \Rightarrow x = \frac{8}{5}$

When $x = \frac{8}{5}$, $y = \frac{5\left(\frac{8}{5}\right) - 4}{2\left(\frac{8}{5}\right)^2} = \frac{8-4}{2\left(\frac{64}{25}\right)} = \frac{25}{32}$

So the coordinates of the stationary point are $\left(\frac{8}{5}, \frac{25}{32}\right)$.

To determine the nature of the stationary point, differentiate again to find $\frac{d^2y}{dx^2}$:

$\frac{dy}{dx} = \frac{8-5x}{2x^3}$

$u = 8 - 5x$, $v = 2x^3 \Rightarrow \frac{du}{dx} = -5$, $\frac{dv}{dx} = 6x^2$

$$\frac{d^2y}{dx^2} = \frac{v\frac{du}{dx} - u\frac{dv}{dx}}{v^2} = \frac{[2x^3 \times (-5)] - [(8-5x) \times 6x^2]}{(2x^3)^2}$$
$$= \frac{-10x^3 - (48x^2 - 30x^3)}{4x^6} = \frac{20x^3 - 48x^2}{4x^6} = \frac{5x - 12}{x^4}$$

When $x = \frac{8}{5}$, $\frac{d^2y}{dx^2} = \frac{5\left(\frac{8}{5}\right) - 12}{\left(\frac{8}{5}\right)^4} = \frac{8-12}{\left(\frac{8}{5}\right)^4} = \frac{-4}{\left(\frac{8}{5}\right)^4}$

Since the denominator is positive, $\frac{d^2y}{dx^2}$ is negative, and so the stationary point must be a maximum.

Q5 $u = x$, $v = e^x \Rightarrow \frac{du}{dx} = 1$, $\frac{dv}{dx} = e^x$

$$\frac{dy}{dx} = \frac{v\frac{du}{dx} - u\frac{dv}{dx}}{v^2} = \frac{[e^x \times 1] - [x \times e^x]}{(e^x)^2} = \frac{e^x - xe^x}{(e^x)^2} = \frac{1-x}{e^x}$$

Use the quotient rule again to find $\frac{d^2y}{dx^2}$:

$u = 1 - x$, $v = e^x \Rightarrow \frac{du}{dx} = -1$, $\frac{dv}{dx} = e^x$

$$\frac{d^2y}{dx^2} = \frac{v\frac{du}{dx} - u\frac{dv}{dx}}{v^2} = \frac{[e^x \times -1] - [(1-x) \times e^x]}{(e^x)^2}$$
$$= \frac{-e^x - e^x + xe^x}{(e^x)^2} = \frac{-1-1+x}{e^x} = \frac{x-2}{e^x}$$

$\frac{d^2y}{dx^2} = 0$ when $x - 2 = 0 \Rightarrow x = 2$

Check $\frac{d^2y}{dx^2}$ either side of $x = 2$:

When $x > 2$, $(x - 2) > 0 \Rightarrow \frac{d^2y}{dx^2} > 0$

When $x < 2$, $(x - 2) < 0 \Rightarrow \frac{d^2y}{dx^2} < 0$

$\frac{d^2y}{dx^2}$ changes sign at $x = 2$, so it's a point of inflection.

When $x = 2$, $y = \frac{2}{e^2}$, so the only point of inflection of the graph $y = \frac{x}{e^x}$ occurs at the point $\left(2, \frac{2}{e^2}\right)$.

Q6 a) $u = x,\ v = \cos(2x) \Rightarrow \frac{du}{dx} = 1,\ \frac{dv}{dx} = -2\sin(2x)$

$$\frac{dy}{dx} = \frac{v\frac{du}{dx} - u\frac{dv}{dx}}{v^2} = \frac{\cos(2x) - [x \times -2\sin(2x)]}{\cos^2(2x)} = \frac{\cos(2x) + 2x\sin(2x)}{\cos^2(2x)}$$

b) $\frac{dy}{dx} = 0$ if $\cos(2x) + 2x\sin(2x) = 0$
$\Rightarrow -\cos(2x) = 2x\sin(2x)$
$\Rightarrow 2x = -\frac{\cos(2x)}{\sin(2x)} = -\cot(2x) \Rightarrow x = -\frac{1}{2}\cot(2x)$

Remember cos x/sin x = 1/tan x = cot x.

Q7 a) $u = 1,\ v = 1 + 4\cos x \Rightarrow \frac{du}{dx} = 0,\ \frac{dv}{dx} = -4\sin x$

$$\frac{dy}{dx} = \frac{v\frac{du}{dx} - u\frac{dv}{dx}}{v^2} = \frac{[(1 + 4\cos x) \times 0] - [1 \times (-4\sin x)]}{(1 + 4\cos x)^2} = \frac{4\sin x}{(1 + 4\cos x)^2}$$

When $x = \frac{\pi}{2}$, $\frac{dy}{dx} = \frac{4}{(1)^2} = 4$ and $y = \frac{1}{1} = 1$

$y = mx + c \Rightarrow 1 = 4\frac{\pi}{2} + c \Rightarrow c = 1 - 2\pi$

So the equation of the tangent is $y = 4x + 1 - 2\pi$.

b) From part a), the gradient of the normal must be $-\frac{1}{4}$. Equation of a straight line:

$y = mx + c \Rightarrow 1 = -\frac{1}{4}\frac{\pi}{2} + c \Rightarrow c = 1 + \frac{\pi}{8}$

So the equation of the normal is $y = -\frac{1}{4}x + 1 + \frac{\pi}{8}$

You could also differentiate by writing y as $(1 + 4\cos x)^{-1}$, and then using the chain rule. In general, you don't have to use the quotient rule when the numerator is just a number.

Q8 $u = 2x,\ v = \cos x \Rightarrow \frac{du}{dx} = 2,\ \frac{dv}{dx} = -\sin x$

$$\frac{dy}{dx} = \frac{v\frac{du}{dx} - u\frac{dv}{dx}}{v^2} = \frac{[\cos x \times 2] - [2x \times (-\sin x)]}{\cos^2 x} = \frac{2\cos x + 2x\sin x}{\cos^2 x}$$

When $x = \frac{\pi}{3}$, $\frac{dy}{dx} = \frac{1 + \frac{\pi\sqrt{3}}{3}}{\left(\frac{1}{2}\right)^2} = 4 + \frac{4\pi\sqrt{3}}{3}$

Q9 $u = x - \sin x,\ v = 1 + \cos x,$
$\Rightarrow \frac{du}{dx} = 1 - \cos x,\ \frac{dv}{dx} = -\sin x$

$$\frac{dy}{dx} = \frac{v\frac{du}{dx} - u\frac{dv}{dx}}{v^2} = \frac{[(1 + \cos x)(1 - \cos x)] - [(x - \sin x)(-\sin x)]}{(1 + \cos x)^2}$$
$$= \frac{1 - \cos^2 x - \sin^2 x + x\sin x}{(1 + \cos x)^2} = \frac{1 - 1 + x\sin x}{(1 + \cos x)^2} = \frac{x\sin x}{(1 + \cos x)^2}$$

Use the identity $\sin^2 x + \cos^2 x \equiv 1$ to simplify the expression.

Q10 Stationary points occur when the gradient is 0.

$u = \cos x,\ v = 4 - 3\cos x, \Rightarrow \frac{du}{dx} = -\sin x,\ \frac{dv}{dx} = 3\sin x$

$$\frac{dy}{dx} = \frac{v\frac{du}{dx} - u\frac{dv}{dx}}{v^2} = \frac{[(4 - 3\cos x)(-\sin x)] - [\cos x(3\sin x)]}{(4 - 3\cos x)^2} = \frac{-4\sin x}{(4 - 3\cos x)^2}$$

$\frac{dy}{dx} = 0 \Rightarrow -4\sin x = 0 \Rightarrow x = \sin^{-1} 0 = 0,\ \pi$ and 2π.

When $x = 0$, $y = \frac{1}{4 - 3} = 1$

When $x = \pi$, $y = \frac{-1}{4 - (-3)} = -\frac{1}{7}$

When $x = 2\pi$, $y = \frac{1}{4 - 3} = 1$

So the stationary points are $(0, 1)$, $\left(\pi, -\frac{1}{7}\right)$ and $(2\pi, 1)$

Q11 First use the chain rule: $y = e^{f(x)} \Rightarrow \frac{dy}{dx} = f'(x)e^{f(x)}$

Then use the quotient rule to find $f'(x)$:

$u = 1 + x,\ v = 1 - x \Rightarrow \frac{du}{dx} = 1,\ \frac{dv}{dx} = -1$

$$\frac{dy}{dx} = \frac{v\frac{du}{dx} - u\frac{dv}{dx}}{v^2} = \frac{[(1 - x)(1)] - [(1 + x)(-1)]}{(1 - x)^2} = \frac{2}{(1 - x)^2}$$

So $\frac{dy}{dx} = f'(x)e^{f(x)} = \frac{2e^{\frac{1+x}{1-x}}}{(1 - x)^2}$

Q12 $y = \frac{2 + 3x^2}{3x - 1}$ is increasing when $\frac{dy}{dx} > 0$.

$u = 2 + 3x^2,\ v = 3x - 1 \Rightarrow \frac{du}{dx} = 6x,\ \frac{dv}{dx} = 3$

$$\frac{dy}{dx} = \frac{v\frac{du}{dx} - u\frac{dv}{dx}}{v^2} = \frac{[(3x - 1) \times 6x] - [(2 + 3x^2) \times 3]}{(3x - 1)^2}$$
$$= \frac{(18x^2 - 6x) - (6 + 9x^2)}{(3x - 1)^2} = \frac{9x^2 - 6x - 6}{(3x - 1)^2}$$

Since the denominator is squared, it is always positive.

So $\frac{dy}{dx} > 0$ when $9x^2 - 6x - 6 > 0 \Rightarrow 3x^2 - 2x - 2 > 0$

Use the quadratic formula to solve $\frac{dy}{dx} = 0$:

$$x = \frac{2 \pm \sqrt{4 - (4 \times 3 \times -2)}}{6} = \frac{2 \pm \sqrt{28}}{6} = \frac{1 \pm \sqrt{7}}{3}$$

Since the coefficient of x^2 in $9x^2 - 6x - 6$ is positive, the graph of $\frac{dy}{dx}$ is u-shaped. So the values of x where $\frac{dy}{dx} > 0$ and where $\frac{2 + 3x^2}{3x - 1}$ is increasing are

$x < \frac{1 - \sqrt{7}}{3}$ and $x > \frac{1 + \sqrt{7}}{3}$.

The function y is undefined when $3x - 1 = 0$, i.e. $x = \frac{1}{3}$ (the graph has an asymptote), but this isn't in the range where the function is increasing, so it doesn't affect the answer.

7. More Differentiation

Exercise 7.1 — Differentiating cosec, sec and cot

Q1 a) $\frac{dy}{dx} = -2\mathrm{cosec}(2x)\cot(2x)$

b) $\frac{dy}{dx} = \frac{dy}{du} \times \frac{du}{dx} = (2\mathrm{cosec}\, x)(-\mathrm{cosec}\, x\cot x) = -2\mathrm{cosec}^2 x\cot x$

c) $\frac{dy}{dx} = \frac{dy}{du} \times \frac{du}{dx} = -7\mathrm{cosec}^2(7x)$

d) $\frac{dy}{dx} = \frac{dy}{du} \times \frac{du}{dx} = (7\cot^6 x)(-\mathrm{cosec}^2 x) = -7\cot^6 x\,\mathrm{cosec}^2 x$

e) $\frac{dy}{dx} = u\frac{dv}{dx} + v\frac{du}{dx} = [x^4 \times (-\mathrm{cosec}^2 x)] + [\cot x \times 4x^3]$
$= 4x^3\cot x - x^4\mathrm{cosec}^2 x$
$= x^3(4\cot x - x\mathrm{cosec}^2 x)$

f) $\frac{dy}{dx} = \frac{dy}{du} \times \frac{du}{dx} = 2(x + \sec x)(1 + \sec x\tan x)$

g) $\frac{dy}{dx} = \frac{dy}{du} \times \frac{du}{dx} = [-\mathrm{cosec}(x^2 + 5)\cot(x^2 + 5)] \times 2x$
$= -2x\,\mathrm{cosec}(x^2 + 5)\cot(x^2 + 5)$

h) $\frac{dy}{dx} = u\frac{dv}{dx} + v\frac{du}{dx} = [e^{3x} \times \sec x \tan x] + [\sec x \times 3e^{3x}]$
$= e^{3x} \sec x (\tan x + 3)$

i) $\frac{dy}{dx} = \frac{dy}{du} \times \frac{du}{dx} = 3(2x + \cot x)^2(2 - \text{cosec}^2 x)$

Q2 a) $f'(x) = \frac{v\frac{du}{dx} - u\frac{dv}{dx}}{v^2} = \frac{(x+3)\sec x \tan x - \sec x}{(x+3)^2}$

b) $f'(x) = \frac{dy}{du} \times \frac{du}{dx} = \left(\sec\frac{1}{x}\tan\frac{1}{x}\right)\left(-\frac{1}{x^2}\right)$
$= -\frac{\sec\frac{1}{x}\tan\frac{1}{x}}{x^2}$

c) $f'(x) = \frac{dy}{du} \times \frac{du}{dx} = (\sec\sqrt{x}\tan\sqrt{x})\left(\frac{1}{2} \times \frac{1}{\sqrt{x}}\right)$
$= \frac{\sec\sqrt{x}\tan\sqrt{x}}{2\sqrt{x}} = \frac{\tan\sqrt{x}}{2\sqrt{x}\cos\sqrt{x}}$

Q3 $f'(x) = \frac{dy}{du} \times \frac{du}{dx}$
$= 2(\sec x + \text{cosec}\, x)(\sec x \tan x - \text{cosec}\, x \cot x)$

Q4 $f(x) = \frac{1}{x\cot x} = \frac{\tan x}{x}$
$f'(x) = \frac{v\frac{du}{dx} - u\frac{dv}{dx}}{v^2} = \frac{x\sec^2 x - \tan x}{x^2}$

Q5 $f'(x) = u\frac{dv}{dx} + v\frac{du}{dx} = [e^x \times (-\text{cosec}\, x \cot x)] + [\text{cosec}\, x \times e^x]$
$= e^x \text{cosec}\, x(1 - \cot x)$

Q6 $f'(x) = u\frac{dv}{dx} + v\frac{du}{dx} = [e^{3x} \times (-4\,\text{cosec}^2\, 4x)] + [\cot 4x \times 3e^{3x}]$
$= e^{3x}(3\cot 4x - 4\,\text{cosec}^2\, 4x)$

The chain rule was used here to differentiate cot 4x — use y = cot u and u = 4x.

Q7 $f'(x) = u\frac{dv}{dx} + v\frac{du}{dx}$
$= [e^{-2x} \times (-4\,\text{cosec}\, 4x \cot 4x)] + [\text{cosec}\, 4x \times (-2)e^{-2x}]$
$= -2e^{-2x}\text{cosec}\, 4x\, [2\cot 4x + 1]$

Q8 $f'(x) = u\frac{dv}{dx} + v\frac{du}{dx} = [\ln x \times (-\text{cosec}\, x \cot x)] + \left[\text{cosec}\, x \times \frac{1}{x}\right]$
$= \text{cosec}\, x\left(\frac{1}{x} - \ln x \cot x\right)$

Q9 $f'(x) = \frac{dy}{du} \times \frac{du}{dx} = \left(\frac{1}{2} \times \frac{1}{\sqrt{\sec x}}\right) \times \sec x \tan x$
$= \frac{\sec x \tan x}{2\sqrt{\sec x}} = \frac{1}{2}\tan x\sqrt{\sec x}$

Q10 $f'(x) = g'(x)e^{g(x)} = e^{\sec x}\sec x \tan x$

Q11 a) $f'(x) = \frac{g'(x)}{g(x)} = \frac{-\text{cosec}\, x \cot x}{\text{cosec}\, x} = -\cot x$

b) $\ln(\text{cosec}\, x) = \ln\left(\frac{1}{\sin x}\right) = \ln 1 - \ln(\sin x)$
$= -\ln(\sin x)$ (as $\ln 1 = 0$)

Here you could also rearrange by saying ln (1/sin x) = ln (sin x)⁻¹ = −ln (sin x).

$f'(x) = -\frac{\cos x}{\sin x} = -\frac{1}{\tan x} = -\cot x$

Q12 $f'(x) = \frac{g'(x)}{g(x)} = \frac{1 + \sec x \tan x}{x + \sec x}$

Q13 $\frac{dy}{dx} = \frac{dy}{du} \times \frac{du}{dx} = \sec\sqrt{x^2+5}\,\tan\sqrt{x^2+5} \times \frac{2x}{2\sqrt{x^2+5}}$
$= \frac{x\sec\sqrt{x^2+5}\,\tan\sqrt{x^2+5}}{\sqrt{x^2+5}}$

8. Connected Rates of Change

Exercise 8.1 — Connected rates of change

Q1 $\frac{dx}{dt} = -0.1$ (it's negative as the cube is shrinking),
and $V = x^3$, so $\frac{dV}{dx} = 3x^2$
$\frac{dV}{dt} = \frac{dV}{dx} \times \frac{dx}{dt} = -0.3x^2$ cm³ min⁻¹

Q2 $V = 30x^3$ so $\frac{dV}{dx} = 90x^2$, and $\frac{dx}{d\theta} = 0.15$.
$\frac{dV}{d\theta} = \frac{dV}{dx} \times \frac{dx}{d\theta} = 13.5x^2$ cm³ °C⁻¹
When $x = 3$, $\frac{dV}{d\theta} = 121.5$ cm³ °C⁻¹

Q3 $A = 4\pi r^2$ so $\frac{dA}{dr} = 8\pi r$, and $\frac{dr}{dt} = -1.6$.
$\frac{dA}{dt} = \frac{dA}{dr} \times \frac{dr}{dt} = -12.8\pi r$ cm² h⁻¹
When $r = 5.5$ cm, $\frac{dA}{dt} = -221.17$ cm² h⁻¹ (2 d.p.)

Q4 $\frac{dr}{d\theta} = 2 \times 10^{-2}$ mm °C⁻¹ $= 2 \times 10^{-5}$ m °C⁻¹,
and $V = \frac{4}{3}\pi r^3$ m³, so $\frac{dV}{dr} = 4\pi r^2$ m³ m⁻¹.
$\frac{dV}{d\theta} = \frac{dV}{dr} \times \frac{dr}{d\theta} = 8 \times 10^{-5}\pi r^2$ m³ °C⁻¹
You could have converted to mm instead, and given your answer in mm³ °C⁻¹.

Q5 The surface area of the tank,
$A = 2(\pi r^2) + 3r(2\pi r) = 8\pi r^2$, so $\frac{dA}{dr} = 16\pi r$,
and $\frac{dH}{dA} = -2$ (it's negative as heat is lost).
$\frac{dH}{dr} = \frac{dH}{dA} \times \frac{dA}{dr} = -32\pi r$ J cm⁻¹
When $r = 12.3$, $\frac{dH}{dr} = -1236.53$ J cm⁻¹ (2 d.p.)

Q6 $\frac{dH}{dt} = -0.5$ mm h⁻¹ $= -0.05$ cm h⁻¹ (length is decreasing).
$V = \pi r^2 H$, so $\frac{dV}{dH} = \pi r^2$.
$\frac{dV}{dt} = \frac{dV}{dH} \times \frac{dH}{dt} = -0.05\pi r^2$ cm³ h⁻¹

Q7 a) Using Pythagoras, the height of the triangle is
$\sqrt{x^2 - \left(\frac{x}{2}\right)^2} = \frac{\sqrt{3}x}{2}$, so the area of the end is
$A = \frac{1}{2}\left(\frac{\sqrt{3}x}{2}\right)x = \frac{\sqrt{3}x^2}{4}$.

b) $V = A \times h = 5\sqrt{3}\,x^2$, so $\frac{dV}{dx} = 10\sqrt{3}\,x$.
$\frac{dx}{dt} = 0.6$.
$\frac{dV}{dt} = \frac{dV}{dx} \times \frac{dx}{dt} = 6\sqrt{3}\,x$ mm³ per day

c) $x = 0.5 \Rightarrow \frac{dV}{dt} = 3\sqrt{3} = 5.20$ mm³ per day (2 d.p.)

Q8 a) n is directly proportional to D, so you can write this as $n = kD$, where k is a constant.
Use the condition that when $D = 2$, $n = 208$:
$208 = 2k \Rightarrow k = 104$
$n = 104D \Rightarrow \frac{dn}{dD} = 104$
$D = 1 + 2^{\lambda t}$, so $\frac{dD}{dt} = \lambda 2^{\lambda t}\ln 2$.
Use the rule for differentiating aˣ from earlier in the chapter.
So $\frac{dn}{dt} = \frac{dn}{dD} \times \frac{dD}{dt} = 104\lambda 2^{\lambda t}\ln 2$ per day.

b) If $t = 1$ day and $\lambda = 5$,

$\frac{dn}{dt} = 104 \times 5 \times 2^5 \times \ln 2 = 16640 \ln 2$
$= 1.15 \times 10^4$ per day (3 s.f.)

Q9 a) Volume of water $V = \pi r^2 h$, so $\frac{dV}{dh} = \pi r^2$.

$\frac{dV}{dt} = -0.3$.

It's negative because the volume of water remaining is decreasing with time.

$\frac{dh}{dt} = \frac{dh}{dV} \times \frac{dV}{dt}$

$= \frac{1}{\frac{dV}{dh}} \times \frac{dV}{dt} = \frac{1}{\pi r^2} \times -0.3 = -\frac{3}{10\pi r^2}$ cm s^{-1}

b) $r = 6$, so $\frac{dh}{dt} = -\frac{1}{120\pi}$ cm s^{-1}.

So the water level falls at a constant rate of $\frac{1}{120\pi}$ cm s^{-1} and $\frac{1}{2\pi} = 0.159$ cm min^{-1} (3 s.f.)

You don't actually need to use h in the calculation.

Q10 a) $\frac{dV}{d\theta} = k$, $V = \frac{2}{3}\pi r^3$, so $\frac{dV}{dr} = 2\pi r^2$.

$\frac{dr}{d\theta} = \frac{dr}{dV} \times \frac{dV}{d\theta}$

$= \frac{1}{\frac{dV}{dr}} \times \frac{dV}{d\theta} = \frac{1}{2\pi r^2} \times k = \frac{k}{2\pi r^2}$ cm °C^{-1}

b) $V = 4$, so $r = \sqrt[3]{\frac{6}{\pi}} = 1.240...$, and $k = 1.5$.

So $\frac{dr}{d\theta} = \frac{1.5}{2\pi(1.240...)^2} = 0.155$ cm °C^{-1} (3 s.f.)

9. Differentiation with Parametric Equations

Exercise 9.1 — Differentiating parametric equations

Q1 a) $\frac{dx}{dt} = 2t$, $\frac{dy}{dt} = 3t^2 - 1$.

Using the chain rule, $\frac{dy}{dx} = \frac{dy}{dt} \div \frac{dx}{dt}$

so $\frac{dy}{dx} = \frac{3t^2 - 1}{2t}$

b) $\frac{dx}{dt} = 3t^2 + 1$, $\frac{dy}{dt} = 4t$, so $\frac{dy}{dx} = \frac{4t}{3t^2 + 1}$

c) $\frac{dx}{dt} = 4t^3$, $\frac{dy}{dt} = 3t^2 - 2t$,

so $\frac{dy}{dx} = \frac{3t^2 - 2t}{4t^3} = \frac{3t - 2}{4t^2}$

d) $\frac{dx}{dt} = -\sin t$, $\frac{dy}{dt} = 4 - 2t$, so $\frac{dy}{dx} = \frac{2t - 4}{\sin t}$

Q2 a) $\frac{dx}{dt} = 2t$, $\frac{dy}{dt} = 2e^{2t}$, so $\frac{dy}{dx} = \frac{e^{2t}}{t}$

b) When $t = 1$, $\frac{dy}{dx} = e^2$.

Q3 a) $\frac{dy}{dt} = 12t^2 - 4t$, $\frac{dx}{dt} = 3e^{3t}$, so $\frac{dy}{dx} = \frac{12t^2 - 4t}{3e^{3t}}$

b) When $t = 0$, $\frac{dy}{dx} = 0$.

Q4 a) $\frac{dx}{dt} = 3t^2$, $\frac{dy}{dt} = 2t\cos t - t^2 \sin t$,

so $\frac{dy}{dx} = \frac{2t\cos t - t^2 \sin t}{3t^2} = \frac{2\cos t - t\sin t}{3t}$

b) When $t = \pi$, $\frac{dy}{dx} = -\frac{2}{3\pi}$

Q5 a) $\frac{dx}{dt} = 2t\sin t + t^2 \cos t$,

$\frac{dy}{dt} = t^3 \cos t + 3t^2 \sin t - \sin t$,

so $\frac{dy}{dx} = \frac{t^3 \cos t + (3t^2 - 1)\sin t}{2t\sin t + t^2 \cos t}$

b) When $t = \pi$, $\frac{dy}{dx} = \pi$.

Q6 a) $\frac{dx}{dt} = \frac{1}{t}$, $\frac{dy}{dt} = 6t - 3t^2$, so $\frac{dy}{dx} = 6t^2 - 3t^3$

b) When $t = -1$, $\frac{dy}{dx} = 9$.

c) At the stationary points $6t^2 - 3t^3 = 3t^2(2 - t) = 0$, so the stationary points occur at $t = 0$ and $t = 2$.
At $t = 0$, x is not defined.
At $t = 2$, the coordinates are $(\ln 2, 4)$.

Exercise 9.2 — Finding tangents and normals

Q1 $\frac{dx}{dt} = 2t$, $\frac{dy}{dt} = 3t^2 - 6$.

Using the chain rule, $\frac{dy}{dx} = \frac{dy}{dt} \div \frac{dx}{dt}$, so $\frac{dy}{dx} = \frac{3t^2 - 6}{2t}$.

When $t = 3$:

$\frac{dy}{dx} = \frac{21}{6} = \frac{7}{2}$, $x = 3^2 = 9$ and $y = 3^3 - 6(3) = 9$.

Putting this into $y = mx + c$ gives:

$9 = \frac{7}{2}(9) + c \Rightarrow c = -\frac{45}{2}$

So the equation of the tangent is:

$y = \frac{7}{2}x - \frac{45}{2} \Rightarrow 7x - 2y - 45 = 0$.

Q2 $\frac{dy}{dx} = \frac{3t^2 - 2t + 5}{3t^2 - 4t}$

When $t = -1$:

$\frac{dy}{dx} = \frac{10}{7}$, $x = -3$ and $y = -7$.

Putting this into $y = mx + c$ gives:

$-7 = \frac{10}{7}(-3) + c \Rightarrow c = -\frac{19}{7}$

So the equation of the tangent is $10x - 7y - 19 = 0$.

Q3 $\frac{dy}{dx} = \frac{3\cos t - t\sin t}{2\cos 2t}$

When $t = \pi$:

$\frac{dy}{dx} = -\frac{3}{2}$, $x = 0$ and $y = -\pi$.

The gradient of the normal is $\frac{2}{3}$.

Putting this into $y = mx + c$ gives:

$-\pi = \frac{2}{3}(0) + c \Rightarrow c = -\pi$

So the equation of the normal is $y = \frac{2}{3}x - \pi$.

Q4 $\frac{dy}{dx} = \frac{3t^2 - 2t}{1 + \ln t}$

When $t = 1$:

$\frac{dy}{dx} = \frac{3 - 2}{1 + 0} = 1$, $x = 0$ and $y = 3$.

Putting this into $y = mx + c$ gives:

$3 = 0 + c \Rightarrow c = 3$

So the equation of the tangent is: $y = x + 3$

Q5 $\frac{dy}{dx} = \frac{2\theta + \cos\theta - \theta\sin\theta}{\sin 2\theta + 2\theta\cos 2\theta}$

When $\theta = \frac{\pi}{2}$:

$\frac{dy}{dx} = -\frac{1}{2}$, $x = 0$ and $y = \frac{\pi^2}{4}$.

The gradient of the normal is 2.

Putting this into $y = mx + c$ gives:

$\frac{\pi^2}{4} = 2(0) + c \Rightarrow c = \frac{\pi^2}{4}$

So the equation of the normal is $y = 2x + \frac{\pi^2}{4}$.

Q6 a) $\frac{dy}{dx} = \frac{3 - 3t^2}{2t - 1}$

When $t = 2$, $\frac{dy}{dx} = -3$, $x = 2$ and $y = -2$.

Putting this into $y = mx + c$ gives:

$-2 = -3(2) + c \Rightarrow c = 4$

So the equation of the tangent is $y = 4 - 3x$.

b) The gradient of the normal at $t = 2$ is $\frac{1}{3}$.

Putting this into $y = mx + c$ gives:

$-2 = \frac{1}{3}(2) + c \Rightarrow c = -\frac{8}{3}$

So the equation of the normal is $3y = x - 8$.

This crosses the x-axis at $y = 0$, so $0 = x - 8 \Rightarrow x = 8$ and so the coordinates are (8, 0).

Q7 a) $\frac{dy}{dx} = \frac{\theta\cos\theta + \sin\theta}{2\cos 2\theta - 2\sin\theta}$

b) When $\theta = \frac{\pi}{2}$: $\frac{dy}{dx} = -\frac{1}{4}$, $x = 0$ and $y = \frac{\pi}{2}$.

Putting this into $y = mx + c$ gives:

$\frac{\pi}{2} = -\frac{1}{4}(0) + c \Rightarrow c = \frac{\pi}{2}$

So the equation of the tangent is $y = \frac{\pi}{2} - \frac{1}{4}x$.

The gradient of the normal is 4. The normal also goes through the point $\left(0, \frac{\pi}{2}\right)$, so again $c = \frac{\pi}{2}$.

So the equation of the normal is $y = 4x + \frac{\pi}{2}$.

Q8 a) The path cuts the y-axis when $x = 0$, so $s^3 \ln s = 0$.

Since ln 0 is undefined, $s = 0$ cannot be a solution, so the only solution is $s = 1$ (i.e. when $\ln s = 0$).

b) $\frac{dy}{dx} = \frac{3s^2 - s - 2s\ln s}{s^2 + 3s^2\ln s} = \frac{3s - 1 - 2\ln s}{s + 3s\ln s}$.

From a), when $x = 0$, $s = 1$.

When $s = 1$: $\frac{dy}{dx} = 2$, $x = 0$ and $y = 1$.

Putting this into $y = mx + c$ gives:

$1 = 2(0) + c \Rightarrow c = 1$.

So the equation of the tangent is $y = 2x + 1$.

Q9 a) Using the quotient rule, $\frac{dy}{d\theta} = \frac{-\theta^3\sin\theta - 3\theta^2\cos\theta}{\theta^6}$

$= \frac{-\theta\sin\theta - 3\cos\theta}{\theta^4}$

$\frac{dx}{d\theta} = \theta^2\cos\theta + 2\theta\sin\theta$

Hence $\frac{dy}{dx} = \frac{-\theta\sin\theta - 3\cos\theta}{\theta^6\cos\theta + 2\theta^5\sin\theta}$

This looks a bit complicated but leave it as it is — you'll find the cos and sin terms usually disappear when you substitute.

When $\theta = \pi$, $\frac{dy}{dx} = \frac{-\pi(0) - 3(-1)}{\pi^6(-1) + 2\pi^5(0)} = -\frac{3}{\pi^6}$.

b) The gradient of the normal when $\theta = \pi$ is $\frac{\pi^6}{3}$, and $y = -\frac{1}{\pi^3}$, $x = 0$.

Putting this into $y = mx + c$ gives:

$-\frac{1}{\pi^3} = \frac{\pi^6}{3}(0) + c \Rightarrow c = -\frac{1}{\pi^3}$

So the equation of the normal is $y = \frac{\pi^6}{3}x - \frac{1}{\pi^3}$.

10. Implicit Differentiation

Exercise 10.1 — Implicit differentiation

Q1 a) $\frac{d}{dx}(y) + \frac{d}{dx}(y^3) = \frac{d}{dx}(x^2) + \frac{d}{dx}(4)$

$\frac{dy}{dx} + 3y^2\frac{dy}{dx} = 2x + 0$

$(1 + 3y^2)\frac{dy}{dx} = 2x \Rightarrow \frac{dy}{dx} = \frac{2x}{1 + 3y^2}$

b) $2x + 2y\frac{dy}{dx} = 2 + 2\frac{dy}{dx}$

$(2y - 2)\frac{dy}{dx} = 2 - 2x \Rightarrow \frac{dy}{dx} = \frac{2 - 2x}{2y - 2} = \frac{1 - x}{y - 1}$

c) $9x^2 - 4\frac{dy}{dx} = 2y\frac{dy}{dx} + 1$

$9x^2 - 1 = (2y + 4)\frac{dy}{dx} \Rightarrow \frac{dy}{dx} = \frac{9x^2 - 1}{2y + 4}$

d) $5 - 2y\frac{dy}{dx} = 5x^4 - 6\frac{dy}{dx}$

$5 - 5x^4 = (2y - 6)\frac{dy}{dx} \Rightarrow \frac{dy}{dx} = \frac{5 - 5x^4}{2y - 6}$

e) $-\sin x + \cos y\frac{dy}{dx} = 2x + 3y^2\frac{dy}{dx}$

$(\cos y - 3y^2)\frac{dy}{dx} = 2x + \sin x \Rightarrow \frac{dy}{dx} = \frac{2x + \sin x}{\cos y - 3y^2}$

f) $3x^2y^2 + 2x^3y\frac{dy}{dx} - \sin x = 4y + 4x\frac{dy}{dx}$

$(2x^3y - 4x)\frac{dy}{dx} = 4y - 3x^2y^2 + \sin x$

$\Rightarrow \frac{dy}{dx} = \frac{4y - 3x^2y^2 + \sin x}{2x^3y - 4x}$

g) $e^x + e^y\frac{dy}{dx} = 3x^2 - \frac{dy}{dx}$

$(e^y + 1)\frac{dy}{dx} = 3x^2 - e^x \Rightarrow \frac{dy}{dx} = \frac{3x^2 - e^x}{e^y + 1}$

h) $3y^2 + 6xy\frac{dy}{dx} + 4xy + 2x^2\frac{dy}{dx} = 3x^2 + 4$

$(6xy + 2x^2)\frac{dy}{dx} = 3x^2 + 4 - 3y^2 - 4xy$

$\Rightarrow \frac{dy}{dx} = \frac{3x^2 + 4 - 3y^2 - 4xy}{6xy + 2x^2}$

Q2 a) $3x^2 + 2y + 2x\frac{dy}{dx} = 4y^3\frac{dy}{dx}$

$(4y^3 - 2x)\frac{dy}{dx} = 3x^2 + 2y \Rightarrow \frac{dy}{dx} = \frac{3x^2 + 2y}{4y^3 - 2x}$

b) $2xy + x^2\frac{dy}{dx} + 2y\frac{dy}{dx} = 3x^2$

$(x^2 + 2y)\frac{dy}{dx} = 3x^2 - 2xy \Rightarrow \frac{dy}{dx} = \frac{3x^2 - 2xy}{x^2 + 2y}$

c) $y^3 + 3xy^2\frac{dy}{dx} + \frac{dy}{dx} = \cos x$

$(3xy^2 + 1)\frac{dy}{dx} = \cos x - y^3 \Rightarrow \frac{dy}{dx} = \frac{\cos x - y^3}{3xy^2 + 1}$

d) $-y\sin x + \cos x\frac{dy}{dx} + \sin y + x\cos y\frac{dy}{dx} = y + x\frac{dy}{dx}$

$(\cos x + x\cos y - x)\frac{dy}{dx} = y + y\sin x - \sin y$

$\Rightarrow \frac{dy}{dx} = \frac{y + y\sin x - \sin y}{\cos x + x\cos y - x}$

e) $e^x + e^y\frac{dy}{dx} = y + x\frac{dy}{dx}$

$(e^y - x)\frac{dy}{dx} = y - e^x \Rightarrow \frac{dy}{dx} = \frac{y - e^x}{e^y - x}$

f) $\frac{1}{x} + 2x = 3y^2\frac{dy}{dx} + \frac{dy}{dx}$

$(3y^2 + 1)\frac{dy}{dx} = \frac{1}{x} + 2x \Rightarrow \frac{dy}{dx} = \frac{\frac{1}{x} + 2x}{3y^2 + 1} = \frac{1 + 2x^2}{3xy^2 + x}$

g) $2e^{2x} + 3e^{3y}\frac{dy}{dx} = 6xy^2 + 6x^2y\frac{dy}{dx}$

$(3e^{3y} - 6x^2y)\frac{dy}{dx} = 6xy^2 - 2e^{2x} \Rightarrow \frac{dy}{dx} = \frac{6xy^2 - 2e^{2x}}{3e^{3y} - 6x^2y}$

h) $\ln x + 1 + \frac{y}{x} + \ln x \frac{dy}{dx} = 5x^4 + 3y^2\frac{dy}{dx}$

$(\ln x - 3y^2)\frac{dy}{dx} = 5x^4 - \ln x - 1 - \frac{y}{x}$

$\Rightarrow \frac{dy}{dx} = \frac{5x^4 - \ln x - 1 - \frac{y}{x}}{\ln x - 3y^2} = \frac{5x^5 - x\ln x - x - y}{x\ln x - 3xy^2}$

Q3 a) At (0, 1): LHS: $e^0 + 2\ln 1 = 1$
RHS: $1^3 = 1$
So (0, 1) is a point on the curve.

b) $e^x + \frac{2}{y}\frac{dy}{dx} = 3y^2\frac{dy}{dx} \Rightarrow \frac{dy}{dx} = \frac{e^x}{3y^2 - \frac{2}{y}} = \frac{ye^x}{3y^3 - 2}$.

At (0, 1) the gradient is $\frac{1e^0}{3(1^3) - 2} = 1$

Q4 a) $3x^2 + 2y\frac{dy}{dx} - 2y - 2x\frac{dy}{dx} = 0 \Rightarrow \frac{dy}{dx} = \frac{2y - 3x^2}{2y - 2x}$

b) Putting $x = -2$ into the equation gives:
$-8 + y^2 + 4y = 0$
Complete the square to solve...
$(y + 2)^2 - 4 - 8 = 0 \Rightarrow y + 2 = \pm\sqrt{12} = \pm 2\sqrt{3}$
so $y = -2 \pm 2\sqrt{3}$

c) At $(-2, -2 + 2\sqrt{3})$:

$\frac{dy}{dx} = \frac{(-4 + 4\sqrt{3}) - 3(-2)^2}{(-4 + 4\sqrt{3}) - 2(-2)} = \frac{-16 + 4\sqrt{3}}{4\sqrt{3}}$

$= \frac{\sqrt{3} - 4}{\sqrt{3}}$

Rationalise the denominator: $= \frac{3 - 4\sqrt{3}}{3}$

$= 1 - \frac{4}{3}\sqrt{3}$

Q5 a) Putting $x = 1$ into the equation gives:
$1 - y = 2y^2 \Rightarrow 2y^2 + y - 1 = 0$
$\Rightarrow (2y - 1)(y + 1) = 0$
So $y = -1$ (given as the other point) and $y = \frac{1}{2}$.
So a $= \frac{1}{2}$.

b) $3x^2 - y - x\frac{dy}{dx} = 4y\frac{dy}{dx}$

$\frac{dy}{dx} = \frac{3x^2 - y}{4y + x}$

At (1, –1), $\frac{dy}{dx} = \frac{3(1^2) - (-1)}{4(-1) + 1} = -\frac{4}{3}$

At $\left(1, \frac{1}{2}\right)$, $\frac{dy}{dx} = \frac{3(1^2) - \left(\frac{1}{2}\right)}{4\left(\frac{1}{2}\right) + 1} = \frac{5}{6}$

Q6 a) Putting $x = 1$ into the equation gives:
$y + y^2 - y - 4 = 0 \Rightarrow y^2 - 4 = 0$
so it cuts the curve at $y = 2$ and $y = -2$.

b) $2xy + x^2\frac{dy}{dx} + y^2 + 2xy\frac{dy}{dx} = y + x\frac{dy}{dx} + 0$

$(x^2 + 2xy - x)\frac{dy}{dx} = y - 2xy - y^2$

$\frac{dy}{dx} = \frac{y - 2xy - y^2}{x^2 + 2xy - x}$

At (1, 2), $\frac{dy}{dx} = \frac{2 - 2(1)(2) - 2^2}{1^2 + 2(1)(2) - 1} = -\frac{3}{2}$

At (1, –2), $\frac{dy}{dx} = \frac{(-2) - 2(1)(-2) - (-2)^2}{1^2 + 2(1)(-2) - 1} = \frac{1}{2}$

Exercise 10.2 — Applications of implicit differentiation

Q1 a) Differentiating:

$2x + 2 + 3\frac{dy}{dx} - 2y\frac{dy}{dx} = 0 \Rightarrow \frac{dy}{dx} = \frac{2x + 2}{2y - 3}$.

At the stationary points, $\frac{dy}{dx} = 0$, so:
$2x + 2 = 0 \Rightarrow x = -1$
When $x = -1$, $y^2 - 3y + 1 = 0 \Rightarrow y = \frac{3 \pm \sqrt{5}}{2}$
= 2.62 or 0.38 (2 d.p.)
So there are 2 stationary points with coordinates (–1, 2.62) and (–1, 0.38).

b) Putting $x = 0$ into the equation gives:
$3y - y^2 = 0 \Rightarrow y(3 - y) = 0$
$\Rightarrow y = 0$ and $y = 3$

At (0, 0), $\frac{dy}{dx} = \frac{2(0) + 2}{2(0) - 3} = -\frac{2}{3}$

The y-intercept is 0 (as it goes through (0, 0)).
So the equation of the tangent is $y = -\frac{2}{3}x$
or $3y = -2x$.

At (0, 3), $\frac{dy}{dx} = \frac{2(0) + 2}{2(3) - 3} = \frac{2}{3}$

The y-intercept is 3 (as it goes through (0, 3)).
Putting this into $y = mx + c$ gives
the equation of the tangent as $y = \frac{2}{3}x + 3$.

Q2 a) Differentiating:

$3x^2 + 2x + \frac{dy}{dx} = 2y\frac{dy}{dx}$

$\Rightarrow \frac{dy}{dx} = \frac{3x^2 + 2x}{2y - 1}$.

At the stationary points, $\frac{dy}{dx} = 0$, so:
$3x^2 + 2x = 0 \Rightarrow x(3x + 2) = 0$
$\Rightarrow x = 0$ and $x = -\frac{2}{3}$
When $x = 0$, $y = y^2 \Rightarrow y(y - 1) = 0$
$\Rightarrow y = 0$ and $y = 1$.
When $x = -\frac{2}{3}$, $-\frac{8}{27} + \frac{4}{9} + y = y^2$
$\Rightarrow \frac{4}{27} + y = y^2 \Rightarrow 27y^2 - 27y - 4 = 0$
This has solutions $y = 1.13$ and -0.13 (2 d.p.)
So there are 4 stationary points with coordinates (0, 0), (0, 1), $\left(-\frac{2}{3}, 1.13\right)$ and $\left(-\frac{2}{3}, -0.13\right)$.

b) Putting $x = 2$ into the equation gives:
$8 + 4 + y = y^2 \Rightarrow y^2 - y - 12 = 0$
$\Rightarrow (y - 4)(y + 3) = 0$
$\Rightarrow y = 4$ and $y = -3$

At (2, 4), $\frac{dy}{dx} = \frac{3(2^2) + 2(2)}{2(4) - 1} = \frac{16}{7}$

Putting this into $y = mx + c$ gives:
$4 = \frac{16}{7}(2) + c \Rightarrow c = -\frac{4}{7}$
So the equation of the tangent is $y = \frac{16}{7}x - \frac{4}{7}$
or $16x - 7y - 4 = 0$.

At (2, –3), $\frac{dy}{dx} = \frac{3(2^2) + 2(2)}{2(-3) - 1} = -\frac{16}{7}$

Putting this into $y = mx + c$ gives:
$-3 = -\frac{16}{7}(2) + c \Rightarrow c = \frac{11}{7}$
So the equation of the tangent is $y = \frac{11}{7} - \frac{16}{7}x$
or $16x + 7y - 11 = 0$.

Q3 **a)** Putting $y = 1$ into the equation gives:
$x^2 + 1 = x + 7 \Rightarrow x^2 - x - 6 = 0$
$\Rightarrow (x - 3)(x + 2) = 0 \Rightarrow x = 3$ and $x = -2$.
Differentiating:
$$2xy + x^2\frac{dy}{dx} + 3y^2\frac{dy}{dx} = 1 \Rightarrow \frac{dy}{dx} = \frac{1 - 2xy}{x^2 + 3y^2}$$
At (–2, 1), $\frac{dy}{dx} = \frac{1 - 2(-2)(1)}{(-2)^2 + 3(1^2)} = \frac{5}{7}$
so the gradient of the normal is $-\frac{7}{5}$.
Putting this into $y = mx + c$ gives:
$1 = -\frac{7}{5}(-2) + c \Rightarrow c = -\frac{9}{5}$
and so the equation of the normal is $y = -\frac{7}{5}x - \frac{9}{5}$
or $7x + 5y + 9 = 0$.
At (3, 1), $\frac{dy}{dx} = \frac{1 - 2(3)(1)}{(3)^2 + 3(1^2)} = -\frac{5}{12}$
so the gradient of the normal is $\frac{12}{5}$.
Putting this into $y = mx + c$ gives:
$1 = \frac{12}{5}(3) + c \Rightarrow c = -\frac{31}{5}$
so the equation of the normal is $y = \frac{12}{5}x - \frac{31}{5}$
or $12x - 5y - 31 = 0$.

b) The normals intersect when:
$-7x - 9 = 12x - 31 \Rightarrow 22 = 19x \Rightarrow x = \frac{22}{19}$
And so $5y = \frac{264}{19} - 31 = -\frac{325}{19} \Rightarrow y = -\frac{65}{19}$
So they intersect at $\left(\frac{22}{19}, -\frac{65}{19}\right)$.

Q4 **a)** Putting $x = 0$ into the equation gives:
$1 + y^2 = 5 - 3y \Rightarrow y^2 + 3y - 4 = 0$
$\Rightarrow (y + 4)(y - 1) = 0 \Rightarrow y = -4$ and $y = 1$.
So a = –4 and b = 1 (a < b).
Differentiating:
$$e^x + 2y\frac{dy}{dx} - y - x\frac{dy}{dx} = -3\frac{dy}{dx} \Rightarrow \frac{dy}{dx} = \frac{y - e^x}{2y - x + 3}$$
At (0, 1), $\frac{dy}{dx} = \frac{1 - e^0}{2(1) - 0 + 3} = 0$
so this is a stationary point.

b) At (0, –4), $\frac{dy}{dx} = \frac{(-4) - e^0}{2(-4) - 0 + 3} = 1$.
So the gradient of the tangent is 1 and the gradient of the normal is –1.
$-4 = 0 + c$ (so $c = -4$) for both, since $x = 0$,
so the equation of the tangent is $y = x - 4$
and the equation of the normal is $y = -x - 4$.

Q5 $y = \arccos x \Rightarrow \cos y = x$
Differentiating:
$$-\sin y \frac{dy}{dx} = 1 \Rightarrow \frac{dy}{dx} = -\frac{1}{\sin y} = -\frac{1}{\sqrt{\sin^2 y}} = -\frac{1}{\sqrt{1 - \cos^2 y}}$$
Using $\cos y = x$: $\frac{dy}{dx} = -\frac{1}{\sqrt{1 - x^2}}$

Q6 $y = \arctan x \Rightarrow \tan y = x$
Differentiating:
$$\sec^2 y \frac{dy}{dx} = 1 \Rightarrow \frac{dy}{dx} = \frac{1}{\sec^2 y}$$
Using the identity $\sec^2 y \equiv 1 + \tan^2 y$:
$$\frac{dy}{dx} = \frac{1}{1 + \tan^2 y}$$
Using $\tan y = x$: $\frac{dy}{dx} = \frac{1}{1 + x^2}$ as required.

Q7 **a)** When $x = 1$:
$0 + y^2 = y + 6 \Rightarrow y^2 - y - 6 = 0$
$\Rightarrow (y - 3)(y + 2) = 0 \Rightarrow y = 3$ and $y = -2$.
So the curve passes through (1, 3) and (1, –2).

b) Differentiating:
$$\frac{1}{x} + 2y\frac{dy}{dx} = 2xy + x^2\frac{dy}{dx}$$
$$\frac{dy}{dx} = \frac{2xy - \frac{1}{x}}{2y - x^2} = \frac{2x^2y - 1}{2xy - x^3}$$
At (1, 3), $\frac{dy}{dx} = \frac{2(1^2)(3) - 1}{2(1)(3) - (1)^3} = 1$
so the gradient of the normal is –1.
Putting this into $y = mx + c$ gives:
$3 = -1(1) + c \Rightarrow c = 4$
so the equation of the normal is $y = 4 - x$.
At (1, –2), $\frac{dy}{dx} = \frac{2(1^2)(-2) - 1}{2(1)(-2) - (1)^3} = 1$,
so the gradient of the normal is also –1.
$-2 = -1(1) + c \Rightarrow c = -1$
so the equation of the normal is $y = -x - 1$.
Because the gradients are the same,
these lines are parallel and can never intersect.

Q8 When $y = 0$: $1 + x^2 = 0 + 4x \Rightarrow x^2 - 4x + 1 = 0$
Complete the square to solve...
$(x - 2)^2 - 4 + 1 = 0 \Rightarrow x - 2 = \pm\sqrt{3} \Rightarrow x = 2 \pm \sqrt{3}$
So $a = 2 + \sqrt{3}$ and $b = 2 - \sqrt{3}$.
Differentiating:
$$e^y\frac{dy}{dx} + 2x = 3y^2\frac{dy}{dx} + 4 \Rightarrow \frac{dy}{dx} = \frac{2x - 4}{3y^2 - e^y}$$
At $(2 + \sqrt{3}, 0)$, $\frac{dy}{dx} = \frac{4 + 2\sqrt{3} - 4}{3(0) - e^0} = -2\sqrt{3}$
$0 = -2\sqrt{3}(2 + \sqrt{3}) + c \Rightarrow c = 4\sqrt{3} + 6$
So the tangent at this point is $y = 4\sqrt{3} + 6 - 2\sqrt{3}x$.
At $(2 - \sqrt{3}, 0)$, $\frac{dy}{dx} = \frac{4 - 2\sqrt{3} - 4}{3(0) - e^0} = 2\sqrt{3}$
$0 = 2\sqrt{3}(2 - \sqrt{3}) + c \Rightarrow c = 6 - 4\sqrt{3}$
So the tangent at this point is $y = 2\sqrt{3}x + 6 - 4\sqrt{3}$.

Q9 Differentiating:
$$\ln x\frac{dy}{dx} + \frac{y}{x} + 2x = 2y\frac{dy}{dx} - \frac{dy}{dx}$$
$$\frac{dy}{dx} = \frac{\frac{y}{x} + 2x}{2y - \ln x - 1} = \frac{y + 2x^2}{2xy - x\ln x - x}$$
$y + 2x^2 = 0 \Rightarrow \frac{dy}{dx} = \frac{y + 2x^2}{2xy - x\ln x - x} = 0$
So if a point on the curve satisfies $y + 2x^2 = 0$,
then it's a stationary point.

Q10 Use the chain rule to find $f'(x)$:
$y = \arccos u$, $u = x^2$
$$\frac{dy}{dx} = \frac{dy}{du} \times \frac{du}{dx} = -\frac{1}{\sqrt{1 - u^2}} \times 2x = -\frac{2x}{\sqrt{1 - x^4}}$$
When $x = \frac{1}{\sqrt{2}}$, $y = \arccos x^2 = \arccos \frac{1}{2} = \frac{\pi}{3}$
This is the only solution for y in the given interval.
$$\frac{dy}{dx} = -\frac{2\left(\frac{1}{\sqrt{2}}\right)}{\sqrt{1 - \left(\frac{1}{\sqrt{2}}\right)^4}} = -\frac{\frac{2}{\sqrt{2}}}{\sqrt{1 - \frac{1}{(\sqrt{2})^4}}} = -\frac{\sqrt{2}}{\sqrt{1 - \frac{1}{4}}}$$
$$= -\frac{\sqrt{2}}{\sqrt{\frac{3}{4}}} = -\frac{\sqrt{2}}{\frac{\sqrt{3}}{2}} = -\frac{2\sqrt{2}}{\sqrt{3}}$$

Rationalising the denominator: $= -\frac{2\sqrt{2} \times \sqrt{3}}{\sqrt{3} \times \sqrt{3}} = -\frac{2\sqrt{6}}{3}$

So the tangent has a gradient of $-\frac{2\sqrt{6}}{3}$ and passes through the point $\left(\frac{1}{\sqrt{2}}, \frac{\pi}{3}\right)$.

$y = mx + c \Rightarrow \frac{\pi}{3} = -\frac{2\sqrt{6}}{3} \times \frac{1}{\sqrt{2}} + c$

$c = \frac{\pi}{3} + \frac{2\sqrt{3}}{3} = \frac{\pi + 2\sqrt{3}}{3}$

So the equation of the tangent is $y = -\frac{2\sqrt{6}}{3}x + \frac{\pi + 2\sqrt{3}}{3}$

Q11 a) Differentiating:

$2e^{2y}\frac{dy}{dx} + e^x = 2x\frac{dy}{dx} + 2y \Rightarrow \frac{dy}{dx} = \frac{2y - e^x}{2e^{2y} - 2x}$

When $y = 0$: $1 + e^x - e^4 = 1 \Rightarrow e^x = e^4 \Rightarrow x = 4$

$\frac{dy}{dx} = \frac{2(0) - e^4}{2e^0 - 2(4)} = \frac{e^4}{6}$ (gradient of the tangent)

$0 = \frac{e^4}{6}(4) + c \Rightarrow c = -\frac{2e^4}{3}$.

So the equation of the tangent is $y = \frac{e^4}{6}(x - 4)$.

b) The gradient of the normal is $-6e^{-4}$.

$0 = -6e^{-4}(4) + c \Rightarrow c = 24e^{-4}$

So the normal is $y = 6e^{-4}(4 - x)$.

c) The lines intersect when

$\frac{e^4}{6}(x - 4) = 6e^{-4}(4 - x) \Rightarrow e^8(x - 4) = 36(4 - x)$

$\Rightarrow e^8x + 36x = 4e^8 + 144 \Rightarrow x(e^8 + 36) = 4e^8 + 144$

$\Rightarrow x = \frac{4e^8 + 144}{e^8 + 36}$

Q12 When $x = 2$:

$2y^2 + 4y - 24 = 4 + 2 \Rightarrow 2y^2 + 4y - 30 = 0$

$\Rightarrow y^2 + 2y - 15 = 0 \Rightarrow (y + 5)(y - 3) = 0$

$\Rightarrow y = -5$ and $y = 3$.

Differentiating:

$y^2 + 2yx\frac{dy}{dx} + 2y + 2x\frac{dy}{dx} - 9x^2 = 2x$

$\frac{dy}{dx} = \frac{2x - y^2 - 2y + 9x^2}{2yx + 2x}$

At $(2, -5)$, $\frac{dy}{dx} = \frac{4 - 25 + 10 + 36}{-20 + 4} = -\frac{25}{16}$

This is the gradient of the tangent at $(2, -5)$, so:

$-5 = -\frac{25}{16}(2) + c \Rightarrow c = -\frac{15}{8}$

So the equation of the tangent at $(2, -5)$ is:

$y = -\frac{25}{16}x - \frac{15}{8}$ or $16y = -25x - 30$

At $(2, 3)$, $\frac{dy}{dx} = \frac{4 - 9 - 6 + 36}{12 + 4} = \frac{25}{16}$

This is the gradient of the tangent at $(2, 3)$, so:

$3 = \frac{25}{16}(2) + c \Rightarrow c = -\frac{1}{8}$

So the equation of the tangent at $(2, 3)$ is:

$y = \frac{25}{16}x - \frac{1}{8}$ or $16y = 25x - 2$

The two tangents intersect when:

$-25x - 30 = 25x - 2 \Rightarrow 50x = -28 \Rightarrow x = -\frac{14}{25}$

And $16y = 25\left(-\frac{14}{25}\right) - 2 \Rightarrow 16y = -16 \Rightarrow y = -1$.

So they intersect at $\left(-\frac{14}{25}, -1\right)$.

Q13 a) When $x = \frac{\pi}{2}$:

$\cos y \cos \frac{\pi}{2} + \cos y \sin \frac{\pi}{2} = \frac{1}{2}$,

$0 + \cos y = \frac{1}{2} \Rightarrow y = \frac{\pi}{3}$

This is the only solution for y in the given interval.

When $x = \pi$:

$\cos y \cos \pi + \cos y \sin \pi = \frac{1}{2}$,

$-\cos y + 0 = \frac{1}{2} \Rightarrow y = \frac{2\pi}{3}$

You can use the CAST diagram or the graph of cos x to find this solution.

b) Differentiating:

$-\cos y \sin x - \sin y \cos x\frac{dy}{dx} + \cos y \cos x - \sin y \sin x\frac{dy}{dx} = 0$

$\frac{dy}{dx} = \frac{\cos y \cos x - \cos y \sin x}{\sin y \cos x + \sin y \sin x}$

At $\left(\frac{\pi}{2}, \frac{\pi}{3}\right)$, $\frac{dy}{dx} = \frac{\cos\frac{\pi}{3}\cos\frac{\pi}{2} - \cos\frac{\pi}{3}\sin\frac{\pi}{2}}{\sin\frac{\pi}{3}\cos\frac{\pi}{2} + \sin\frac{\pi}{3}\sin\frac{\pi}{2}}$

$= \frac{0 - \frac{1}{2}}{0 + \frac{\sqrt{3}}{2}} = -\frac{1}{\sqrt{3}}$

This is the gradient of the tangent, so:

$\frac{\pi}{3} = -\frac{1}{\sqrt{3}}\left(\frac{\pi}{2}\right) + c \Rightarrow c = \frac{(2 + \sqrt{3})\pi}{6}$

So the equation of the tangent at $\left(\frac{\pi}{2}, \frac{\pi}{3}\right)$ is

$y = -\frac{1}{\sqrt{3}}x + \frac{(2 + \sqrt{3})\pi}{6}$.

At $\left(\pi, \frac{2\pi}{3}\right)$, $\frac{dy}{dx} = \frac{\cos\frac{2\pi}{3}\cos\pi - \cos\frac{2\pi}{3}\sin\pi}{\sin\frac{2\pi}{3}\cos\pi + \sin\frac{2\pi}{3}\sin\pi}$

$= \frac{\frac{1}{2} - 0}{-\frac{\sqrt{3}}{2} + 0} = -\frac{1}{\sqrt{3}}$

This is the gradient of the tangent, so:

$\frac{2\pi}{3} = -\frac{1}{\sqrt{3}}(\pi) + c \Rightarrow c = \frac{(2 + \sqrt{3})\pi}{3}$

So the equation of the tangent at $\left(\pi, \frac{2\pi}{3}\right)$ is

$y = -\frac{1}{\sqrt{3}}x + \frac{(2 + \sqrt{3})\pi}{3}$.

Q14 Differentiating: $\left(\frac{2}{3}y\right)\frac{dy}{dx} = 18x^2 - \left(2y + 2x\frac{dy}{dx}\right)$

$\left(\frac{2}{3}y + 2x\right)\frac{dy}{dx} = 18x^2 - 2y$

$\frac{dy}{dx} = \frac{18x^2 - 2y}{\frac{2}{3}y + 2x} = \frac{9x^2 - y}{\frac{1}{3}y + x} = \frac{27x^2 - 3y}{y + 3x}$

When $\frac{dy}{dx} = 0$, $27x^2 - 3y = 0 \Rightarrow 3y = 27x^2 \Rightarrow y = 9x^2$

Substitute $y = 9x^2$ back into the original equation:

$\frac{1}{3}(9x^2)^2 = 6x^3 - 2x(9x^2) \Rightarrow 27x^4 = 6x^3 - 18x^3$

$\Rightarrow 27x^4 + 12x^3 = 0$

$\Rightarrow 3x^3(9x + 4) = 0$

So the stationary points occur when $x = 0$ and $x = -\frac{4}{9}$.

When $x = 0$, $y = 9(0)^2 = 0$

When $x = -\frac{4}{9}$, $y = 9\left(-\frac{4}{9}\right)^2 = \frac{16}{9}$

So the stationary points are at $(0, 0)$ and $\left(-\frac{4}{9}, \frac{16}{9}\right)$.

Chapter 8: Integration

1. Integration of $(ax+b)^n$

Exercise 1.1 — Integrating $(ax+b)^n$, $n \neq -1$

Q1 a) $\int (x+10)^{10}\,dx = \frac{1}{1\times 11}(x+10)^{11} + C$

$= \frac{1}{11}(x+10)^{11} + C$

b) $\int (5x)^7\,dx = \frac{1}{5\times 8}(5x)^8 + C = \frac{1}{40}(5x)^8 + C$

$= \frac{1}{40}5^8x^8 + C = \frac{5^8}{40}x^8 + C = \frac{78125x^8}{8} + C$

c) $\int (3-5x)^{-2}\,dx = \frac{1}{-5\times -1}(3-5x)^{-1} + C$

$= \frac{1}{5}(3-5x)^{-1} + C = \frac{1}{5(3-5x)} + C$

d) $\int (3x-4)^{-\frac{4}{3}}\,dx = \frac{1}{3\times\left(-\frac{1}{3}\right)}(3x-4)^{-\frac{1}{3}} + C$

$= \frac{1}{-1}(3x-4)^{-\frac{1}{3}} + C = -(3x-4)^{-\frac{1}{3}} + C = \frac{-1}{\sqrt[3]{3x-4}} + C$

Q2 a) Begin by taking the constant of 8 outside of the integration and then integrate $\int (2x-4)^4\,dx$ as usual.

$\int 8(2x-4)^4\,dx = 8\int (2x-4)^4\,dx$

$= 8\times\left(\frac{1}{2\times 5}(2x-4)^5 + c\right)$

$= \frac{8}{10}(2x-4)^5 + C$

$= \frac{4}{5}(2x-4)^5 + C = \frac{4(2x-4)^5}{5} + C$

b) *Use your answer to part a). The integral you found will be the same but without the constant of integration — it'll have limits instead.*

$\int_{\frac{3}{2}}^{\frac{5}{2}} 8(2x-4)^4\,dx = \frac{4}{5}[(2x-4)^5]_{\frac{3}{2}}^{\frac{5}{2}}$

$= \frac{4}{5}\left(\left[\left(2\left(\frac{5}{2}\right)-4\right)^5\right] - \left[\left(2\left(\frac{3}{2}\right)-4\right)^5\right]\right)$

$= \frac{4}{5}([(1)^5] - [(-1)^5])$

$= \frac{4}{5}([1]-[-1]) = \frac{8}{5}$

Q3 $\int_0^1 (6x+1)^{-3}\,dx = \left[\frac{1}{6\times -2}(6x+1)^{-2}\right]_0^1$

$= -\frac{1}{12}[(6x+1)^{-2}]_0^1$

$= -\frac{1}{12}([(7)^{-2}] - [(1)^{-2}]) = -\frac{1}{12}\left(\frac{1}{49} - 1\right) = \frac{4}{49}$

Q4 You've been given that $f'(x) = (8-7x)^4$ and you need to find $f(x)$, so integrate with respect to x.

$f(x) = \int f'(x)\,dx = \int (8-7x)^4\,dx$

$= \frac{1}{-7\times 5}(8-7x)^5 + C = -\frac{1}{35}(8-7x)^5 + C$

Substitute in the values of x and y at the point given to find the value of C.

$\frac{3}{35} = -\frac{1}{35}(8-(7\times 1))^5 + C$

$\Rightarrow \frac{3}{35} = -\frac{1}{35}(1)^5 + C \Rightarrow C = \frac{4}{35}$

So $f(x) = -\frac{1}{35}(8-7x)^5 + \frac{4}{35}$

2. Integration of e^x and $\frac{1}{x}$

Exercise 2.1 — Integrating e^x and e^{ax+b}

Q1 a) $\int 2e^x\,dx = 2\int e^x\,dx = 2e^x + C$

b) $\int 4x + 7e^x\,dx = \int 4x\,dx + \int 7e^x\,dx$

$= \int 4x\,dx + 7\int e^x\,dx = 2x^2 + 7e^x + C$

c) $\int e^{10x}\,dx = \frac{1}{10}e^{10x} + C$

d) $\int e^{-3x} + x\,dx = \int e^{-3x}\,dx + \int x\,dx$

$= -\frac{1}{3}e^{-3x} + \frac{1}{2}x^2 + C$

e) $\int e^{\frac{7}{2}x}\,dx = \frac{1}{\left(\frac{7}{2}\right)}e^{\frac{7}{2}x} + C = \frac{2}{7}e^{\frac{7}{2}x} + C$

f) $\int e^{4x-2}\,dx = \frac{1}{4}e^{4x-2} + C$

g) $\int \frac{1}{2}e^{2-\frac{3}{2}x}\,dx = \frac{1}{2}\int e^{2-\frac{3}{2}x}\,dx$

$= \frac{1}{2}\left(\frac{1}{\left(-\frac{3}{2}\right)}e^{2-\frac{3}{2}x} + c\right) = \left(\frac{1}{2}\times -\frac{2}{3}e^{2-\frac{3}{2}x}\right) + C$

$= -\frac{1}{3}e^{2-\frac{3}{2}x} + C$

h) $\int e^{4\left(\frac{x}{3}+1\right)}\,dx = \int e^{\frac{4}{3}x+4}\,dx$

$= \frac{1}{\left(\frac{4}{3}\right)}e^{\frac{4}{3}x+4} + C = \frac{3}{4}e^{4\left(\frac{x}{3}+1\right)} + C$

Q2 You've been given the derivative of the curve, so integrate it to get the equation of the curve.

$y = \int \frac{dy}{dx}\,dx = \int 10e^{-5x-1}\,dx = 10\int e^{-5x-1}\,dx$

$= 10\left(\frac{1}{-5}e^{-5x-1} + c\right) = -2e^{-5x-1} + C$

To find C, use the fact that the curve goes through the origin (0, 0). The equation of the curve is $y = -2e^{-5x-1} + C$ and substituting in $x = 0$ and $y = 0$ gives:

$0 = -2e^{-(5\times 0)-1} + C = -2e^{-1} + C = -\frac{2}{e} + C$ so $C = \frac{2}{e}$.

So the curve has equation $y = -2e^{-5x-1} + \frac{2}{e}$.

Q3 $\int e^{8y+5}\,dy = \frac{1}{8}e^{8y+5} + C$

Q4 a) $\int_2^3 e^{2x}\,dx = \frac{1}{2}[e^{2x}]_2^3 = \frac{1}{2}([e^6] - [e^4]) = \frac{1}{2}(e^6 - e^4)$

b) $\int_{-1}^0 12e^{12x+12}\,dx = \left[12\times\frac{1}{12}e^{12x+12}\right]_{-1}^0$

$= [e^{12x+12}]_{-1}^0 = e^{12} - e^0 = e^{12} - 1$

c) $\int_{-\frac{\pi}{2}}^{\frac{\pi}{2}} e^{\pi-2x}\,dx = \left[\frac{1}{-2}e^{\pi-2x}\right]_{-\frac{\pi}{2}}^{\frac{\pi}{2}}$

$= -\frac{1}{2}[e^{\pi-2x}]_{-\frac{\pi}{2}}^{\frac{\pi}{2}}$

$= -\frac{1}{2}([e^{\pi-\pi}] - [e^{\pi+\pi}])$

$= -\frac{1}{2}(e^0 - e^{2\pi}) = \frac{1}{2}(e^{2\pi} - 1)$

d) $\int_3^6 \sqrt[6]{e^x} + \frac{1}{\sqrt[3]{e^x}}\,dx = \int_3^6 (e^x)^{\frac{1}{6}} + (e^x)^{-\frac{1}{3}}\,dx$

$= \int_3^6 e^{\frac{x}{6}} + e^{-\frac{x}{3}}\,dx = \left[\frac{1}{\left(\frac{1}{6}\right)}e^{\frac{x}{6}} + \frac{1}{\left(-\frac{1}{3}\right)}e^{-\frac{x}{3}}\right]_3^6$

$= [6e^{\frac{x}{6}} - 3e^{-\frac{x}{3}}]_3^6$

$= [6e^1 - 3e^{-2}] - [6e^{\frac{1}{2}} - 3e^{-1}]$

$= 6e - \frac{3}{e^2} - 6\sqrt{e} + \frac{3}{e}$

Exercise 2.2 — Integrating $\frac{1}{x}$ and $\frac{1}{ax+b}$

Q1 **a)** $\int \frac{19}{x}\,dx = 19\int \frac{1}{x}\,dx = 19\ln|x| + C$

b) $\int \frac{1}{7x}\,dx = \frac{1}{7}\int \frac{1}{x}\,dx = \frac{1}{7}\ln|x| + C$

An equivalent answer to b) would be $\frac{1}{7}\ln|7x| + C$ if you used the general formula for integrating $\frac{1}{ax+b}$ instead.

c) There is no constant term to take out here, so use the general formula:

$\int \frac{1}{7x+2}\,dx = \frac{1}{7}\ln|7x+2| + C$

d) $\int \frac{4}{1-3x}\,dx = 4\left(\frac{1}{-3}\ln|1-3x| + c\right)$

$= -\frac{4}{3}\ln|1-3x| + C$

Q2 $\int \frac{1}{8x} - \frac{20}{x}\,dx = \frac{1}{8}\int \frac{1}{x}\,dx - 20\int \frac{1}{x}\,dx$

$= \frac{1}{8}\ln|x| - 20\ln|x| + C = -\frac{159}{8}\ln|x| + C$

You could also notice that $\frac{1}{8x} - \frac{20}{x} = \frac{1-160}{8x} = \frac{-159}{8x}$ and integrate $\frac{-159}{8x}$ using the method for $\frac{1}{x}$.

Q3 **a)** $\int \frac{6}{x} - \frac{3}{x}\,dx = \int \frac{3}{x}\,dx = 3\ln|x| + C$

$= \ln|x|^3 + C = \ln|x^3| + C$

b) $\int_4^5 \frac{6}{x} - \frac{3}{x}\,dx = [\ln|x^3|]_4^5$

$= [\ln 5^3] - [\ln 4^3] = 3\ln 5 - 3\ln 4$

$= 3(\ln 5 - \ln 4) = 3\ln\left(\frac{5}{4}\right)$

You could also have written your answer as $\ln\frac{125}{64}$.

Q4 $\int_b^a 15(5+3x)^{-1}\,dx = \int_b^a \frac{15}{(5+3x)}\,dx$

$= \left[15 \times \frac{1}{3}\ln|5+3x|\right]_b^a = 5[\ln|5+3x|]_b^a$

$= 5(\ln|5+3a| - \ln|5+3b|)$

$= 5\ln\left(\frac{|5+3a|}{|5+3b|}\right) = 5\ln\left|\frac{5+3a}{5+3b}\right| = \ln\left|\frac{5+3a}{5+3b}\right|^5$

Q5 Integrate the derivative to find f(x).

$f(x) = \int f'(x)\,dx = \int \frac{4}{10-9x}\,dx$

$= 4 \times \frac{1}{-9}\ln|10-9x| + C = -\frac{4}{9}\ln|10-9x| + C$

The curve passes through the point (1, 2), so substitute these values to find C.

$2 = -\frac{4}{9}\ln|10 - (9 \times 1)| + C$

$2 = -\frac{4}{9}\ln|1| + C \Rightarrow 2 = 0 + C$

So C = 2 and the equation of f(x) is $f(x) = -\frac{4}{9}\ln|10-9x| + 2$.

Q6 **a)** The area required is the shaded area below:

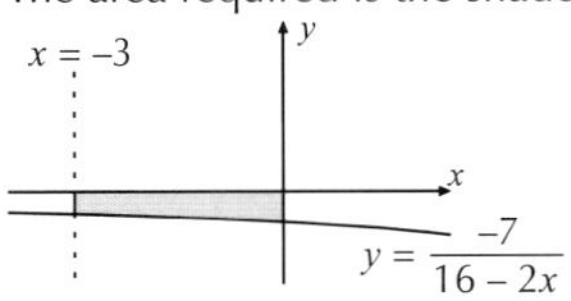

This is found by integrating the curve with respect to x between the limits $x = -3$ and $x = 0$.
So the area is expressed by the integral $\int_{-3}^{0} \frac{-7}{16-2x}\,dx$

b) $\int_{-3}^{0} \frac{-7}{16-2x} = \left[-7 \times \frac{1}{-2}\ln|16-2x|\right]_{-3}^{0}$

$= \frac{7}{2}[\ln|16-2x|]_{-3}^{0} = \frac{7}{2}(\ln 16 - \ln 22)$

$= \frac{7}{2}\ln\frac{16}{22} = \frac{7}{2}\ln\frac{8}{11} = \ln\left[\left(\frac{8}{11}\right)^{\frac{7}{2}}\right]$

Q7 Work out the integral and put in the limits to find A.

$\int_1^A \frac{4}{6x-5}\,dx = 10$

$\left[4 \times \frac{1}{6}\ln|6x-5|\right]_1^A = 10$

$\frac{2}{3}[\ln|6x-5|]_1^A = 10$

$\frac{2}{3}(\ln|6A-5| - \ln|1|) = 10$

$\frac{2}{3}(\ln|6A-5| - 0) = 10$

$\frac{2}{3}\ln|6A-5| = 10$

$\ln|6A-5| = \frac{10}{\left(\frac{2}{3}\right)} = 15$

Take the exponential of both sides to get rid of the ln.

$|6A-5| = e^{15}$

$A = \frac{e^{15}+5}{6}$

As $A \geq 1$, then $6A - 5$ must be greater than $6 - 5 = 1$, so the modulus can be removed as it'll always be positive.

3. Integration of Trigonometric Functions

Exercise 3.1 — Integration of sin x and cos x

Q1 **a)** $\int \frac{1}{7}\cos x\,dx = \frac{1}{7}\int \cos x\,dx = \frac{1}{7}\sin x + C$

b) $\int -3\sin x\,dx = -3\int \sin x\,dx$

$= -3(-\cos x) + C = 3\cos x + C$

c) $\int -3\cos x - 3\sin x\,dx = -3\int \cos x + \sin x\,dx$

$= -3(\sin x - \cos x + c)$

$= -3\sin x + 3\cos x + C$

d) $\int \sin 5x\,dx = -\frac{1}{5}\cos 5x + C$

e) $\int \cos\left(\frac{x}{7}\right)dx = \frac{1}{\left(\frac{1}{7}\right)}\sin\left(\frac{x}{7}\right) + C = 7\sin\left(\frac{x}{7}\right) + C$

f) $\int 2\sin(-3x)\,dx = 2\int \sin(-3x)\,dx$

$= 2\left(-\left(-\frac{1}{3}\right)\cos(-3x) + c\right)$

$= \frac{2}{3}\cos(-3x) + C$

An alternative solution would be $\frac{2}{3}\cos(3x) + C$ since $\cos(x) = \cos(-x)$.

g) $\int 5\cos\left(3x + \frac{\pi}{5}\right)dx = 5\left(\frac{1}{3}\sin\left(3x + \frac{\pi}{5}\right) + c\right)$

$= \frac{5}{3}\sin\left(3x + \frac{\pi}{5}\right) + C$

h) $\int -4\sin\left(4x - \frac{\pi}{3}\right)dx = -4\left(-\frac{1}{4}\cos\left(4x - \frac{\pi}{3}\right) + c\right)$

$= \cos\left(4x - \frac{\pi}{3}\right) + C$

i) $\int \cos(4x+3) + \sin(3-4x)\,dx$

$= \frac{1}{4}\sin(4x+3) + \frac{1}{-4}(-\cos(3-4x)) + C$

$= \frac{1}{4}\sin(4x+3) + \frac{1}{4}\cos(3-4x) + C$

Q2 $\int \frac{1}{2}\cos 3\theta - \sin\theta\,d\theta = \frac{1}{2}\left(\frac{1}{3}\sin 3\theta\right) - (-\cos\theta) + C$

$= \frac{1}{6}\sin 3\theta + \cos\theta + C$

Q3 a) $\int_0^{\frac{\pi}{2}} \sin x\,dx = [-\cos x]_0^{\frac{\pi}{2}} = -\cos\frac{\pi}{2} + \cos 0 = 0 + 1 = 1$

b) $\int_{\frac{\pi}{6}}^{\frac{\pi}{3}} \sin 3x\,dx = -\frac{1}{3}[\cos 3x]_{\frac{\pi}{6}}^{\frac{\pi}{3}}$

$= -\frac{1}{3}\left(\left[\cos\left(3\times\frac{\pi}{3}\right)\right] - \left[\cos\left(3\times\frac{\pi}{6}\right)\right]\right)$

$= -\frac{1}{3}\left(\cos\pi - \cos\frac{\pi}{2}\right) = -\frac{1}{3}(-1-0) = \frac{1}{3}$

c) $\int_{-1}^{2} 3\sin(\pi x + \pi)\,dx = -\frac{3}{\pi}[\cos(\pi x + \pi)]_{-1}^{2}$

$= -\frac{3}{\pi}(\cos 3\pi - \cos 0)$

$= -\frac{3}{\pi}(-1-1) = \frac{6}{\pi}$

Q4 a) Integrate the function with respect to x within the limits 1 and 2:

$\int_1^2 2\pi\cos\left(\frac{\pi x}{2}\right)dx = \frac{2\pi}{\left(\frac{\pi}{2}\right)}\left[\sin\left(\frac{\pi x}{2}\right)\right]_1^2 = 4\left[\sin\left(\frac{\pi x}{2}\right)\right]_1^2$

$= 4\left(\sin(\pi) - \sin\left(\frac{\pi}{2}\right)\right) = 4(0-1) = -4$

b) Since the function doesn't cross the x-axis for $1 < x < 2$, the described area has to be either entirely above the x-axis or entirely below it. So, because the integral gives a negative value, this means that all of the area is below the x-axis.

Q5 $\int_{\frac{\pi}{3}}^{\frac{\pi}{2}} \sin(-x) + \cos(-x)\,dx = \int_{\frac{\pi}{3}}^{\frac{\pi}{2}} -\sin x + \cos x\,dx$

$= [-(-\cos x) + \sin x]_{\frac{\pi}{3}}^{\frac{\pi}{2}}$

$= [\cos x + \sin x]_{\frac{\pi}{3}}^{\frac{\pi}{2}}$

$= \left[\cos\left(\frac{\pi}{2}\right) + \sin\left(\frac{\pi}{2}\right)\right] - \left[\cos\left(\frac{\pi}{3}\right) + \sin\left(\frac{\pi}{3}\right)\right]$

$= [0+1] - \left[\frac{1}{2} + \frac{\sqrt{3}}{2}\right]$

$= 1 - \frac{1}{2} - \frac{\sqrt{3}}{2} = \frac{1}{2} - \frac{\sqrt{3}}{2} = \frac{1-\sqrt{3}}{2}$

$\sin(-x) = -\sin x$ and $\cos(-x) = \cos x$ are used in the first step of this solution.

Q6 Integrate the function between -2π and π:

$\int_{-2\pi}^{\pi} 5\cos\frac{x}{6}\,dx = \left[\frac{5}{\left(\frac{1}{6}\right)}\sin\left(\frac{x}{6}\right)\right]_{-2\pi}^{\pi} = 30\left[\sin\left(\frac{x}{6}\right)\right]_{-2\pi}^{\pi}$

$= 30\left(\sin\left(\frac{\pi}{6}\right) - \sin\left(-\frac{\pi}{3}\right)\right)$

$= 30\left(\frac{1}{2} - \left(-\frac{\sqrt{3}}{2}\right)\right)$

$= 30\left(\frac{1+\sqrt{3}}{2}\right) = 15(1+\sqrt{3})$

Exercise 3.2 — Integration of $\sec^2 x$

Q1 a) $\int 2\sec^2 x + 1\,dx = 2\tan x + x + C$

b) $\int \sec^2 9x\,dx = \frac{1}{9}\tan 9x + C$

c) $\int 20\sec^2 3y\,dy = 20\times\frac{1}{3}\tan 3y + C$

$= \frac{20}{3}\tan 3y + C$

d) $\int \sec^2\frac{x}{7}\,dx = \frac{1}{\left(\frac{1}{7}\right)}\tan\left(\frac{x}{7}\right) + C$

$= 7\tan\left(\frac{x}{7}\right) + C$

e) $\int_0^{\frac{\pi}{3}} -\frac{1}{\cos^2\theta}\,d\theta = \int_0^{\frac{\pi}{3}} -\sec^2\theta\,d\theta = [-\tan\theta]_0^{\frac{\pi}{3}}$

$= -\sqrt{3} + 0 = -\sqrt{3}$

f) $\int_0^{\frac{\pi}{4}} 3\sec^2(-3x)\,dx = \left[\frac{3}{-3}\tan(-3x)\right]_0^{\frac{\pi}{4}}$

$= [-\tan(-3x)]_0^{\frac{\pi}{4}}$

$= \left[-\tan\left(-3\times\frac{\pi}{4}\right)\right] - [-\tan(0)]$

$= -1 + 0 = -1$

Q2 Integrate the function between the limits:

$\int_{\frac{2}{3}\pi}^{\pi} \sec^2 x\,dx = [\tan x]_{\frac{2}{3}\pi}^{\pi} = \tan\pi - \tan\frac{2\pi}{3}$

$= 0 - (-\sqrt{3}) = \sqrt{3}$

Q3 The constants α and β do not affect the integration. You only need to worry about the coefficients of x.

$\int \sec^2(x+\alpha) + \sec^2(3x+\beta)\,dx$

$= \tan(x+\alpha) + \frac{1}{3}\tan(3x+\beta) + C$

Q4 $\int_{\frac{\pi}{12}}^{\frac{\pi}{6}} 5A\sec^2\left(\frac{\pi}{3} - 2\theta\right)d\theta = \left[-\frac{5A}{2}\tan\left(\frac{\pi}{3} - 2\theta\right)\right]_{\frac{\pi}{12}}^{\frac{\pi}{6}}$

$= -\frac{5A}{2}\left[\tan\left(\frac{\pi}{3} - 2\theta\right)\right]_{\frac{\pi}{12}}^{\frac{\pi}{6}}$

$= -\frac{5A}{2}\left(\left[\tan\left(\frac{\pi}{3} - \frac{\pi}{3}\right)\right] - \left[\tan\left(\frac{\pi}{3} - \frac{\pi}{6}\right)\right]\right)$

$= -\frac{5A}{2}\left(\tan(0) - \tan\left(\frac{\pi}{6}\right)\right)$

$= -\frac{5A}{2}\left(0 - \frac{\sqrt{3}}{3}\right)$

$= \frac{5\sqrt{3}A}{6}$

Exercise 3.3 — Integration of other trigonometric functions

Q1 a) $\int \operatorname{cosec}^2 11x\,dx = -\frac{1}{11}\cot 11x + C$

b) $\int 5\sec 10\theta\tan 10\theta\,d\theta = 5\times\frac{1}{10}\sec 10\theta + C$

$= \frac{1}{2}\sec 10\theta + C$

c) $\int -\operatorname{cosec}(x+17)\cot(x+17)\,dx$

$= -(-\operatorname{cosec}(x+17)) + C$

$= \operatorname{cosec}(x+17) + C$

d) $\int -3\operatorname{cosec}3x\cot 3x\,dx = -3\left(-\frac{1}{3}\operatorname{cosec}3x\right) + C$
$= \operatorname{cosec}3x + C$

e) $\int 13\sec\left(\frac{\pi}{4} - x\right)\tan\left(\frac{\pi}{4} - x\right)dx$
$= 13\left(\frac{1}{-1}\sec\left(\frac{\pi}{4} - x\right)\right) + C$
$= -13\sec\left(\frac{\pi}{4} - x\right) + C$

Q2 $\int 10\operatorname{cosec}^2\left(\alpha - \frac{x}{2}\right) - 60\sec(\alpha - 6x)\tan(\alpha - 6x)\,dx$
$= -\frac{10}{\left(-\frac{1}{2}\right)}\cot\left(\alpha - \frac{x}{2}\right) - \frac{60}{-6}\sec(\alpha - 6x) + C$
$= 20\cot\left(\alpha - \frac{x}{2}\right) + 10\sec(\alpha - 6x) + C$

Q3 $\int_{\frac{\pi}{12}}^{\frac{\pi}{8}} 6\sec 2x\tan 2x + 6\operatorname{cosec}2x\cot 2x\,dx$
$= \left[\frac{6}{2}\sec 2x - \frac{6}{2}\operatorname{cosec}2x\right]_{\frac{\pi}{12}}^{\frac{\pi}{8}}$
$= 3[\sec 2x - \operatorname{cosec}2x]_{\frac{\pi}{12}}^{\frac{\pi}{8}}$
$= 3\left(\left[\sec\frac{\pi}{4} - \operatorname{cosec}\frac{\pi}{4}\right] - \left[\sec\frac{\pi}{6} - \operatorname{cosec}\frac{\pi}{6}\right]\right)$
$= 3\left(\sqrt{2} - \sqrt{2} - \frac{2}{\sqrt{3}} + 2\right)$
$= 3\left(2 - \frac{2}{\sqrt{3}}\right) = 6 - 2\sqrt{3}$

Q4 $\int_{\frac{\pi}{12}}^{\frac{\pi}{6}} \operatorname{cosec}^2(3x)\,dx = \left[-\frac{1}{3}\cot(3x)\right]_{\frac{\pi}{12}}^{\frac{\pi}{6}}$
$= -\frac{1}{3}\left(\cot\left(\frac{\pi}{2}\right) - \cot\left(\frac{\pi}{4}\right)\right) = -\frac{1}{3}(0 - 1) = \frac{1}{3}$

If you know the tan values for the common angles, you can work out the cot values using cot x = 1/tan x.

4. Integration of $\frac{f'(x)}{f(x)}$

Exercise 4.1 — Integrating $\frac{f'(x)}{f(x)}$

Q1 a) Differentiating the denominator:
$\frac{d}{dx}(x^4 - 1) = 4x^3$ = numerator
$\int \frac{4x^3}{x^4 - 1}dx = \ln|x^4 - 1| + C$

b) $\frac{d}{dx}(x^2 - x) = 2x - 1$ = numerator
$\int \frac{2x - 1}{x^2 - x}dx = \ln|x^2 - x| + C$

c) $\frac{d}{dx}(3x^5 + 6) = 15x^4$
$\int \frac{x^4}{3x^5 + 6}dx = \frac{1}{15}\int \frac{15x^4}{3x^5 + 6}dx$
$= \frac{1}{15}\ln|3x^5 + 6| + C$

d) $\frac{d}{dx}(x^4 + 2x^3 - x) = 4x^3 + 6x^2 - 1$
$\int \frac{12x^3 + 18x^2 - 3}{x^4 + 2x^3 - x}dx = \int \frac{3(4x^3 + 6x^2 - 1)}{x^4 + 2x^3 - x}dx$
$= 3\int \frac{4x^3 + 6x^2 - 1}{x^4 + 2x^3 - x}dx = 3\ln|x^4 + 2x^3 - x| + C$

Q2 a) $\frac{d}{dx}(e^x + 6) = e^x$
$\int \frac{e^x}{e^x + 6}dx = \ln|e^x + 6| + C$

b) $\frac{d}{dx}(e^{2x} + 6e^x) = 2e^{2x} + 6e^x = 2(e^{2x} + 3e^x)$
$\int \frac{2(e^{2x} + 3e^x)}{e^{2x} + 6e^x}dx = \ln|e^{2x} + 6e^x| + C$

c) $\frac{d}{dx}(e^x + 3) = e^x$
$\int \frac{e^x}{3(e^x + 3)}dx = \frac{1}{3}\int \frac{e^x}{(e^x + 3)}dx = \frac{1}{3}\ln|e^x + 3| + C$

Q3 a) $\frac{d}{dx}(1 + \sin 2x) = 2\cos 2x$
$\int \frac{2\cos 2x}{1 + \sin 2x}dx = \ln|1 + \sin 2x| + C$

b) $\frac{d}{dx}(\cos 3x - 1) = -3\sin 3x$
$\int \frac{\sin 3x}{\cos 3x - 1}dx = -\frac{1}{3}\int \frac{-3\sin 3x}{\cos 3x - 1}dx$
$= -\frac{1}{3}\ln|\cos 3x - 1| + C$

c) $\frac{d}{dx}(\operatorname{cosec}x - x^2 + 4) = -\operatorname{cosec}x\cot x - 2x$
$\int \frac{3\operatorname{cosec}x\cot x + 6x}{\operatorname{cosec}x - x^2 + 4}dx$
$= \int \frac{-3(-\operatorname{cosec}x\cot x - 2x)}{\operatorname{cosec}x - x^2 + 4}dx$
$= -3\int \frac{-\operatorname{cosec}x\cot x - 2x}{\operatorname{cosec}x - x^2 + 4}dx$
$= -3\ln|\operatorname{cosec}x - x^2 + 4| + C$

d) $\frac{d}{dx}(\tan x) = \sec^2 x$
$\int \frac{\sec^2 x}{\tan x}dx = \ln|\tan x| + C$

e) $\frac{d}{dx}(\sec x + 5) = \sec x\tan x$
$\int \frac{\sec x\tan x}{\sec x + 5}dx = \ln|\sec x + 5| + C$

Q4 $\frac{d}{dx}(\sin(2x + 7)) = 2\cos(2x + 7)$
$\int \frac{4\cos(2x + 7)}{\sin(2x + 7)}dx = 2\int \frac{2\cos(2x + 7)}{\sin(2x + 7)}dx$
$= 2(\ln|\sin(2x + 7)| + C)$
$= 2(\ln|\sin(2x + 7)| + \ln k)$
$= 2\ln|k\sin(2x + 7)|$

Q5 a) Using the hint, multiply the inside of the integral by a fraction which is the same on the top and bottom (it's equal to 1, so it'll make no difference).
$\int \sec x\,dx = \int \sec x\left(\frac{\sec x + \tan x}{\sec x + \tan x}\right)dx$
$= \int \frac{\sec^2 x + \sec x\tan x}{\sec x + \tan x}dx$
Now differentiating the denominator of this integral gives:
$\frac{d}{dx}(\sec x + \tan x) = \sec x\tan x + \sec^2 x$
So the numerator is the derivative of the denominator, so use the result:
$\int \sec x\,dx = \int \frac{\sec^2 x + \sec x\tan x}{\sec x + \tan x}dx$
$= \ln|\sec x + \tan x| + C$

b) Use the same method as part a), this time using $\frac{\operatorname{cosec} x + \cot x}{\operatorname{cosec} x + \cot x}$:

$$\int \operatorname{cosec} x \, dx = \int \operatorname{cosec} x\left(\frac{\operatorname{cosec} x + \cot x}{\operatorname{cosec} x + \cot x}\right) dx$$

$$= \int \frac{\operatorname{cosec}^2 x + \operatorname{cosec} x \cot x}{\operatorname{cosec} x + \cot x} dx$$

Differentiating the denominator of this integral:

$$\frac{d}{dx}(\operatorname{cosec} x + \cot x) = -\operatorname{cosec} x \cot x - \operatorname{cosec}^2 x$$
$$= -(\operatorname{cosec} x \cot x + \operatorname{cosec}^2 x)$$

So the numerator is $-1 \times$ the derivative of the denominator, so use the result:

$$\int \operatorname{cosec} x \, dx = \int \frac{\operatorname{cosec}^2 x + \operatorname{cosec} x \cot x}{\operatorname{cosec} x + \cot x} dx$$

$$= -\int \frac{-\operatorname{cosec}^2 x - \operatorname{cosec} x \cot x}{\operatorname{cosec} x + \cot x} dx$$

$$= -\ln|\operatorname{cosec} x + \cot x| + C$$

Q6 a) $\int 2 \tan x \, dx = 2\int \frac{\sin x}{\cos x} dx = -2\int \frac{-\sin x}{\cos x}$
$= -2 \ln|\cos x| + C$

b) $\int \tan 2x \, dx = \int \frac{\sin 2x}{\cos 2x} dx = -\frac{1}{2}\int \frac{-2\sin 2x}{\cos 2x} dx$
$= -\frac{1}{2}\ln|\cos 2x| + C$

c) $\int 4 \operatorname{cosec} x \, dx = 4\int \operatorname{cosec} x \, dx$
$= -4 \ln|\operatorname{cosec} x + \cot x| + C$

d) $\int \cot 3x \, dx = \frac{1}{3} \ln|\sin 3x| + C$

e) $\int \frac{1}{2}\sec 2x \, dx = \frac{1}{2}\left(\frac{1}{2}\ln|\sec 2x + \tan 2x| + C\right)$
$= \frac{1}{4}\ln|\sec 2x + \tan 2x| + C$

f) $\int 3 \operatorname{cosec} 6x \, dx = 3\left(-\frac{1}{6}\ln|\operatorname{cosec} 6x + \cot 6x| + C\right)$
$= -\frac{1}{2}\ln|\operatorname{cosec} 6x + \cot 6x| + C$

Q7 *This one looks really complicated, but if you split it into parts and use some standard results it's actually pretty simple.*

$$\int \frac{\sec^2 x}{2\tan x} - 4 \sec 2x \tan 2x + \frac{\operatorname{cosec} 2x \cot 2x - 1}{\operatorname{cosec} 2x + 2x} dx$$

$$= \int \frac{\sec^2 x}{2\tan x} dx - \int 4 \sec 2x \tan 2x \, dx + \int \frac{\operatorname{cosec} 2x \cot 2x - 1}{\operatorname{cosec} 2x + 2x} dx$$

$$= \frac{1}{2}\int \frac{\sec^2 x}{\tan x} dx - 2\int 2 \sec 2x \tan 2x \, dx - \frac{1}{2}\int \frac{-2\operatorname{cosec} 2x \cot 2x + 2}{\operatorname{cosec} 2x + 2x} dx$$

$$= \frac{1}{2}\ln|\tan x| - 2 \sec 2x - \frac{1}{2}\ln|\operatorname{cosec} 2x + 2x| + C$$

The first and third integrals were put in the form $\int \frac{f'(x)}{f(x)} dx$*, and the second one you can tackle by reversing the result* $\frac{d}{dx}(\sec 2x) = 2 \sec 2x \tan 2x$.

5. Integrating $\frac{du}{dx} f'(u)$

Exercise 5.1 — Integrating using the reverse of the chain rule

Q1 Let $u = x^2$ so $\frac{du}{dx} = 2x$ and $f'(u) = e^u$ so $f(u) = e^u$.
Using the formula: $\int 2xe^{x^2} dx = e^{x^2} + C$

Q2 Let $u = 2x^3$ so $\frac{du}{dx} = 6x^2$ and $f'(u) = e^u$ so $f(u) = e^u$.
Using the formula: $\int 6x^2 e^{2x^3} dx = e^{2x^3} + C$

Q3 Let $u = \sqrt{x}$ so $\frac{du}{dx} = \frac{1}{2\sqrt{x}}$, and $f'(u) = e^u$ so $f(u) = e^u$.
Using the formula: $\int \frac{1}{2\sqrt{x}} e^{\sqrt{x}} dx = e^{\sqrt{x}} + C$

Q4 Let $u = x^4$ so $\frac{du}{dx} = 4x^3$, and $f'(u) = e^u$ so $f(u) = e^u$.
Use the formula: $\int 4x^3 e^{x^4} dx = e^{x^4} + C$
So divide by 4 to get the original integral:
$\int x^3 e^{x^4} dx = \frac{1}{4}\int 4x^3 e^{x^4} dx = \frac{1}{4}e^{x^4} + C$

Q5 Let $u = x^2 - \frac{1}{2}x$ so $\frac{du}{dx} = 2x - \frac{1}{2} = \frac{1}{2}(4x - 1)$, and $f'(u) = e^u$ so $f(u) = e^u$.
Use the formula: $\int \left(2x - \frac{1}{2}\right)e^{\left(x^2 - \frac{1}{2}x\right)} dx = e^{\left(x^2 - \frac{1}{2}x\right)} + C$
Multiply by 2 to get the original integral:

$$\int (4x - 1)e^{\left(x^2 - \frac{1}{2}x\right)} dx = \int 2\left(2x - \frac{1}{2}\right)e^{\left(x^2 - \frac{1}{2}x\right)} dx$$
$$= 2\int \left(2x - \frac{1}{2}\right)e^{\left(x^2 - \frac{1}{2}x\right)} dx$$
$$= 2e^{\left(x^2 - \frac{1}{2}x\right)} + C$$

Q6 Let $u = x^2 + 1$ so $\frac{du}{dx} = 2x$, and $f'(u) = \sin u$ so $f(u) = -\cos u$.
Use the formula: $\int 2x \sin(x^2 + 1) \, dx = -\cos(x^2 + 1) + C$

Q7 Let $u = x^4$ so $\frac{du}{dx} = 4x^3$, and $f'(u) = \cos u$ so $f(u) = \sin u$.
Use the formula: $\int 4x^3 \cos(x^4) \, dx = \sin(x^4) + C$
Now divide by 4 to get the original integral:
$\int x^3 \cos(x^4) \, dx = \frac{1}{4}\int 4x^3 \cos(x^4) \, dx = \frac{1}{4}\sin(x^4) + C$

Q8 Let $u = x^2$ so $\frac{du}{dx} = 2x$, and $f'(u) = \sec^2 u$ so $f(u) = \tan u$.
Use the formula: $\int 2x \sec^2(x^2) \, dx = \tan(x^2) + C$
Now divide by 2 to get the original integral:
$\int x \sec^2(x^2) \, dx = \frac{1}{2}\int 2x \sec^2(x^2) \, dx = \frac{1}{2}\tan(x^2) + C$

Q9 *It's less obvious which function to choose as u in this one — keep looking out for a function and its derivative. Here we have cos x and sin x. Remember to make u the one which is within another function, i.e. cos x.*
Let $u = \cos x$ so $\frac{du}{dx} = -\sin x$, and $f'(u) = e^u$ so $f(u) = e^u$.
Use the formula: $\int -\sin x \, e^{\cos x} dx = e^{\cos x} + C$
Multiply by -1 to get the original integral:
$\int \sin x \, e^{\cos x} dx = -\int -\sin x \, e^{\cos x} dx = -e^{\cos x} + C$

Q10 Let $u = \sin 2x$ so $\frac{du}{dx} = 2\cos 2x$, and $f'(u) = e^u$ so $f(u) = e^u$.
Use the formula: $\int 2 \cos 2x \, e^{\sin 2x} dx = e^{\sin 2x} + C$
Divide by 2 to get the original integral:
$\int \cos 2x \, e^{\sin 2x} dx = \frac{1}{2}\int 2\cos 2x \, e^{\sin 2x} dx = \frac{1}{2}e^{\sin 2x} + C$

Q11 Let $u = \tan x$ then $\frac{du}{dx} = \sec^2 x$, and $f'(u) = e^u$ so $f(u) = e^u$.
Use the formula: $\int \sec^2 x \, e^{\tan x} dx = e^{\tan x} + C$

Q12 Let $u = \sec x$ then $\frac{du}{dx} = \sec x \tan x$, and $f'(u) = e^u$ so $f(u) = e^u$.
Use the formula: $\int \sec x \tan x \, e^{\sec x} dx = e^{\sec x} + C$

Exercise 5.2 — Integrating $f'(x) \times [f(x)]^n$

Q1 a) Let $f(x) = x^2 + 5$ so $f'(x) = 2x$. $n = 2$ so $n + 1 = 3$.
Using the formula: $\int 3 \times 2x(x^2 + 5)^2\,dx = (x^2 + 5)^3 + C$
So $\int 6x(x^2 + 5)^2\,dx = (x^2 + 5)^3 + C$

b) Let $f(x) = x^2 + 7x$ so $f'(x) = 2x + 7$. $n = 4$ so $n + 1 = 5$.
Using the formula:
$\int 5(2x + 7)(x^2 + 7x)^4\,dx = (x^2 + 7x)^5 + C$
Divide by 5 to get the original integral:
$\int (2x + 7)(x^2 + 7x)^4\,dx = \frac{1}{5}(x^2 + 7x)^5 + C$

c) Let $f(x) = x^4 + 4x^2$ so $f'(x) = 4x^3 + 8x$.
$n = 3$ so $n + 1 = 4$. Using the formula:
$\int 4(4x^3 + 8x)(x^4 + 4x^2)^3\,dx = (x^4 + 4x^2)^4 + C$
Divide by 16 to get the original integral:
$$\int (x^3 + 2x)(x^4 + 4x^2)^3\,dx$$
$$= \frac{1}{16}\int 4(4x^3 + 8x)(x^4 + 4x^2)^3\,dx$$
$$= \frac{1}{16}(x^4 + 4x^2)^4 + C$$

d) Let $f(x) = x^2 - 1$, so $f'(x) = 2x$. $n = -3$ so $n + 1 = -2$.
Using the formula:
$\int -2(2x)(x^2 - 1)^{-3}\,dx = (x^2 - 1)^{-2} + C$
Divide by -2 to get the original integral:
$$\int \frac{2x}{(x^2-1)^3}\,dx = \int (2x)(x^2-1)^{-3}\,dx$$
$$= -\frac{1}{2}(x^2-1)^{-2} + C = -\frac{1}{2(x^2-1)^2} + C$$

e) Let $f(x) = e^{3x} - 5$, so $f'(x) = 3e^{3x}$. $n = -2$ so $n + 1 = -1$.
Using the formula:
$\int -1(3e^{3x})(e^{3x} - 5)^{-2}\,dx = (e^{3x} - 5)^{-1} + C$
Multiply by -2 to get the original integral:
$$\int \frac{6e^{3x}}{(e^{3x}-5)^2}\,dx = \int 6e^{3x}(e^{3x}-5)^{-2}\,dx$$
$$= -2\int -3e^{3x}(e^{3x}-5)^{-2}\,dx$$
$$= -2(e^{3x}-5)^{-1} + C$$

f) $\sin x \cos^5 x = \sin x(\cos x)^5$
Let $f(x) = \cos x$ so $f'(x) = -\sin x$. $n = 5$ so $n + 1 = 6$.
Using the formula:
$\int 6(-\sin x)(\cos x)^5\,dx = (\cos x)^6 + C = \cos^6 x + C$
Divide by -6 to get the original integral:
$$\int \sin x \cos^5 x\,dx = \frac{1}{-6}\int 6(-\sin x)(\cos x)^5\,dx$$
$$= -\frac{1}{6}\cos^6 x + C$$

g) *It's a bit more difficult to tell which is the derivative and which is the function here — both functions are to a power. Remember that the derivative of tan x is sec² x.*
$2\sec^2 x \tan^3 x = 2\sec^2 x(\tan x)^3$
Let $f(x) = \tan x$ so $f'(x) = \sec^2 x$. $n = 3$ so $n + 1 = 4$.
Using the formula:
$\int 4\sec^2 x(\tan x)^3\,dx = (\tan x)^4 + C = \tan^4 x + C$
Divide by 2 to get the original integral:
$$\int 2\sec^2 x \tan^3 x\,dx = \frac{1}{2}\int 4\sec^2 x(\tan x)^3\,dx$$
$$= \frac{1}{2}\tan^4 x + C$$

h) Let $f(x) = e^x + 4$ so $f'(x) = e^x$. $n = 2$ so $n + 1 = 3$.
Using the formula: $\int 3e^x(e^x + 4)^2\,dx = (e^x + 4)^3 + C$

i) Let $f(x) = e^{4x} - 3x^2$ so $f'(x) = 4e^{4x} - 6x$.
$n = 7$ so $n + 1 = 8$.
Using the formula:
$\int 8(4e^{4x} - 6x)(e^{4x} - 3x^2)^7\,dx = (e^{4x} - 3x^2)^8 + C$
So $\int 16(2e^{4x} - 3x)(e^{4x} - 3x^2)^7\,dx = (e^{4x} - 3x^2)^8 + C$
Multiply by 2 to get the original integral:
$$\int 32(2e^{4x} - 3x)(e^{4x} - 3x^2)^7\,dx$$
$$= 2\int 16(2e^{4x} - 3x)(e^{4x} - 3x^2)^7\,dx$$
$$= 2(e^{4x} - 3x^2)^8 + C$$

j) Let $f(x) = 2 + \sin x$, so $f'(x) = \cos x$.
$n = -4$, so $n + 1 = -3$.
Using the formula:
$\int -3(\cos x)(2 + \sin x)^{-4}\,dx = (2 + \sin x)^{-3} + C$
Divide by -3 to get the original integral:
$$\int \frac{\cos x}{(2+\sin x)^4}\,dx = \int \cos x(2+\sin x)^{-4}\,dx$$
$$= -\frac{1}{3}(2+\sin x)^{-3} + C = -\frac{1}{3(2+\sin x)^3} + C$$

k) *Using the hint, you know that the derivative of cosec x is −cosec x cot x.*
Let $f(x) = \text{cosec}\, x$ so $f'(x) = -\text{cosec}\, x \cot x$.
$n = 4$ so $n + 1 = 5$.
Using the formula:
$\int 5(-\text{cosec}\, x \cot x)(\text{cosec}\, x)^4\,dx = (\text{cosec}\, x)^5 + C$
Multiply by -1:
$\int 5\,\text{cosec}\, x \cot x\, \text{cosec}^4 x\,dx = -\text{cosec}^5 x + C$

l) *Using the hint, cot x differentiates to −cosec² x.*
Let $f(x) = \cot x$ so $f'(x) = -\text{cosec}^2 x$. $n = 3$ so $n + 1 = 4$.
Using the formula:
$\int 4(-\text{cosec}^2 x)\cot^3 x\,dx = \cot^4 x + C$
Divide by -2 to get the original integral:
$$\int 2\,\text{cosec}^2 x \cot^3 x\,dx = \frac{1}{-2}\int -4\,\text{cosec}^2 x \cot^3 x\,dx$$
$$= -\frac{1}{2}\cot^4 x + C$$

Q2 a) *sec x differentiates to sec x tan x, so try to write the function as a product of sec x tan x and sec x to a power.*
$6\tan x \sec^6 x = 6\tan x \sec x \sec^5 x$
Let $f(x) = \sec x$ so $f'(x) = \sec x \tan x$.
$n = 5$ so $n + 1 = 6$.
Using the formula:
$\int 6(\tan x \sec x)(\sec^5 x)\,dx = \sec^6 x + C$
So $\int 6\tan x \sec^6 x\,dx = \sec^6 x + C$

b) *cosec x differentiates to −cot x cosec x so do the same as you did in part a).*
$\cot x\, \text{cosec}^3 x = \cot x\, \text{cosec}\, x\, \text{cosec}^2 x$
Let $f(x) = \text{cosec}\, x$ so $f'(x) = -\cot x\, \text{cosec}\, x$.
$n = 2$ so $n + 1 = 3$.
Using the formula:
$\int 3(-\cot x\, \text{cosec}\, x)(\text{cosec}^2 x)\,dx = \text{cosec}^3 x + C$
So $\int -3\cot x\, \text{cosec}^3 x\,dx = \text{cosec}^3 x + C$
Divide by -3 to get the original integral.
$$\int \cot x\, \text{cosec}^3 x\,dx = \frac{1}{-3}\int -3\cot x\, \text{cosec}^3 x\,dx$$
$$= -\frac{1}{3}\text{cosec}^3 x + C$$

Q3 a) *This one looks really complicated, but if you differentiate the bracket ($e^{\sin x} - 5$) using the chain rule, you'll get the function at the front.*

Let $f(x) = e^{\sin x} - 5$ so $f'(x) = \cos x\, e^{\sin x}$.

$n = 3$ so $n + 1 = 4$.

Using the formula:

$\int 4(\cos x\, e^{\sin x})(e^{\sin x} - 5)^3\, dx = (e^{\sin x} - 5)^4 + C$

b) Let $f(x) = e^{\cos x} + 4x$ so $f'(x) = -\sin x\, e^{\cos x} + 4$.

$n = 6$ so $n + 1 = 7$.

Using the formula:

$\int 7(-\sin x\, e^{\cos x} + 4)(e^{\cos x} + 4x)^6\, dx = (e^{\cos x} + 4x)^7 + C$

So divide by –7 to get the original integral:

$\int (\sin x\, e^{\cos x} - 4)(e^{\cos x} + 4x)^6\, dx$

$= -\frac{1}{7}\int 7(-\sin x\, e^{\cos x} + 4)(e^{\cos x} + 4x)^6\, dx$

$= -\frac{1}{7}(e^{\cos x} + 4x)^7 + C$

Q4 a) Start by writing the function as $\sec^2 x \tan^{-4} x$.

Let $f(x) = \tan x$ so $f'(x) = \sec^2 x$. $n = -4$ so $n + 1 = -3$.

Using the formula: $\int -3 \sec^2 x \tan^{-4} x\, dx = \tan^{-3} x + C$

Divide by –3 to get the original integral:

$\int \frac{\sec^2 x}{\tan^4 x}\, dx = -\frac{1}{3}\tan^{-3} x + C = -\frac{1}{3\tan^3 x} + C$

b) Start by writing the function as $\cot x \operatorname{cosec} x (\operatorname{cosec} x)^{\frac{1}{2}}$.

Let $f(x) = \operatorname{cosec} x$ so $f'(x) = -\cot x \operatorname{cosec} x$.

$n = \frac{1}{2}$ so $n + 1 = \frac{3}{2}$.

Using the formula:

$\int \frac{3}{2}(-\cot x \operatorname{cosec} x)(\operatorname{cosec} x)^{\frac{1}{2}}\, dx = (\operatorname{cosec} x)^{\frac{3}{2}} + C$

Divide by $-\frac{3}{2}$ to get the original integral:

$\int \cot x \operatorname{cosec} x \sqrt{\operatorname{cosec} x}\, dx = -\frac{2}{3}(\operatorname{cosec} x)^{\frac{3}{2}} + C$

$= -\frac{2}{3}(\sqrt{\operatorname{cosec} x})^3 + C$

6. Using Trigonometric Identities in Integration

Exercise 6.1 — Integrating using trig identities

Q1 a) Using the cos double angle formula:

$\cos^2 x = \frac{1}{2}(\cos 2x + 1)$

So the integral is:

$\int \cos^2 x\, dx = \int \frac{1}{2}(\cos 2x + 1)\, dx$

$= \frac{1}{2}\left(\frac{1}{2}\sin 2x + x\right) + C$

$= \frac{1}{4}\sin 2x + \frac{1}{2}x + C$

b) $6 \sin x \cos x = 3(2 \sin x \cos x) = 3 \sin 2x$

So the integral is:

$\int 6 \sin x \cos x\, dx = \int 3 \sin 2x\, dx = -\frac{3}{2}\cos 2x + C$

c) $\sin^2 6x = \frac{1}{2}(1 - \cos(2 \times 6x)) = \frac{1}{2}(1 - \cos 12x)$

So the integral is:

$\int \sin^2 6x\, dx = \int \frac{1}{2}(1 - \cos 12x)\, dx$

$= \frac{1}{2}\left(x - \frac{1}{12}\sin 12x\right) + C$

$= \frac{1}{2}x - \frac{1}{24}\sin 12x + C$

d) Using the tan double angle formula:

$\frac{2\tan 2x}{1 - \tan^2 2x} = \tan 4x$

So the integral is:

$\int \frac{2\tan 2x}{1 - \tan^2 2x}\, dx = \int \tan 4x\, dx$

$= -\frac{1}{4}\ln|\cos 4x| + C$

$\left(\text{or} = \frac{1}{4}\ln|\sec 4x| + C\right)$

e) $2 \sin 4x \cos 4x = \sin 8x$

So the integral is:

$\int 2 \sin 4x \cos 4x\, dx = \int \sin 8x\, dx = -\frac{1}{8}\cos 8x + C$

f) $2\cos^2 4x = 2\left(\frac{1}{2}(\cos 8x + 1)\right) = \cos 8x + 1$

So the integral is:

$\int 2\cos^2 4x\, dx = \int \cos 8x + 1\, dx = \frac{1}{8}\sin 8x + x + C$

g) $\cos x \sin x = \frac{1}{2}(2\cos x \sin x) = \frac{1}{2}\sin 2x$

So the integral is:

$\int \cos x \sin x\, dx = \int \frac{1}{2}\sin 2x\, dx$

$= \frac{1}{2}\left(-\frac{1}{2}\cos 2x\right) + C$

$= -\frac{1}{4}\cos 2x + C$

h) $\sin 3x \cos 3x = \frac{1}{2}(2\sin 3x \cos 3x) = \frac{1}{2}\sin 6x$

So the integral is:

$\int \sin 3x \cos 3x\, dx = \int \frac{1}{2}\sin 6x\, dx$

$= \frac{1}{2}\left(-\frac{1}{6}\cos 6x\right) + C$

$= -\frac{1}{12}\cos 6x + C$

i) $\frac{6\tan 3x}{1 - \tan^2 3x} = 3\left(\frac{2\tan 3x}{1 - \tan^2 3x}\right) = 3\tan 6x$

So the integral is:

$\int \frac{6\tan 3x}{1 - \tan^2 3x}\, dx = \int 3\tan 6x\, dx$

$= 3\left(-\frac{1}{6}\ln|\cos 6x|\right) + C$

$= -\frac{1}{2}\ln|\cos 6x| + C$

$\left(\text{or} = \frac{1}{2}\ln|\sec 6x| + C\right)$

j) $5 \sin 2x \cos 2x = \frac{5}{2}(2 \sin 2x \cos 2x) = \frac{5}{2}\sin 4x$

So the integral is:

$\int 5\sin 2x \cos 2x\, dx = \int \frac{5}{2}\sin 4x\, dx$

$= \frac{5}{2}\left(-\frac{1}{4}\cos 4x\right) + C$

$= -\frac{5}{8}\cos 4x + C$

k) $(\sin x + \cos x)^2 = \sin^2 x + 2 \sin x \cos x + \cos^2 x$

$= \sin^2 x + \cos^2 x + 2 \sin x \cos x$

$= 1 + 2 \sin x \cos x$

$= 1 + \sin 2x$

$\sin^2 x + \cos^2 x \equiv 1$ has been used to simplify here.

So the integral is:

$\int (\sin x + \cos x)^2\, dx = \int 1 + \sin 2x\, dx$

$= x - \frac{1}{2}\cos 2x + C$

l) $4 \sin x \cos x \cos 2x = 2(2 \sin x \cos x) \cos 2x$
$= 2 \sin 2x \cos 2x$
$= \sin 4x$

So the integral is:

$$\int 4 \sin x \cos x \cos 2x \, dx = \int \sin 4x \, dx = -\frac{1}{4}\cos 4x + C$$

m) $(\cos x + \sin x)(\cos x - \sin x)$
$= \cos^2 x - \cos x \sin x + \sin x \cos x - \sin^2 x$
$= \cos^2 x - \sin^2 x$
$= \cos 2x$

So the integral is:

$$\int (\cos x + \sin x)(\cos x - \sin x) \, dx = \int \cos 2x \, dx$$
$$= \frac{1}{2}\sin 2x + C$$

n) $\sin^2 x \cot x = \sin^2 x \frac{1}{\tan x} = \sin^2 x \frac{\cos x}{\sin x}$
$= \sin x \cos x = \frac{1}{2}\sin 2x$

So the integral is:

$$\int \sin^2 x \cot x \, dx = \int \frac{1}{2}\sin 2x \, dx$$
$$= \frac{1}{2}\left(-\frac{1}{2}\cos 2x\right) + C$$
$$= -\frac{1}{4}\cos 2x + C$$

Q2 a) $\sin^2 x = \frac{1}{2}(1 - \cos 2x)$

So the integral is:

$$\int_0^{\frac{\pi}{4}} \sin^2 x \, dx = \int_0^{\frac{\pi}{4}} \frac{1}{2}(1 - \cos 2x) \, dx$$
$$= \frac{1}{2}\left[\left(x - \frac{1}{2}\sin 2x\right)\right]_0^{\frac{\pi}{4}}$$
$$= \frac{1}{2}\left(\left(\frac{\pi}{4} - \frac{1}{2}\sin\frac{\pi}{2}\right) - \left(-\frac{1}{2}\sin 0\right)\right)$$
$$= \frac{1}{2}\left(\left(\frac{\pi}{4} - \left(\frac{1}{2} \times 1\right)\right) - \left(-\frac{1}{2} \times 0\right)\right)$$
$$= \frac{1}{2}\left(\frac{\pi}{4} - \frac{1}{2}\right) = \frac{\pi}{8} - \frac{1}{4}$$

b) $\cos^2 2x = \frac{1}{2}(\cos 4x + 1)$

So the integral is:

$$\int_0^{\pi} \frac{1}{2}(\cos 4x + 1) \, dx = \left[\frac{1}{2}\left(\frac{1}{4}\sin 4x + x\right)\right]_0^{\pi}$$
$$= \left[\left(\frac{1}{8}\sin 4x + \frac{x}{2}\right)\right]_0^{\pi}$$
$$= \left(\frac{1}{8}\sin 4\pi + \frac{\pi}{2}\right) - \left(\frac{1}{8}\sin 0 + \frac{0}{2}\right)$$
$$= \left(\frac{1}{8} \times 0 + \frac{\pi}{2}\right) - (0 + 0) = \frac{\pi}{2}$$

c) $\sin\frac{x}{2}\cos\frac{x}{2} = \frac{1}{2}\left(2\sin\frac{x}{2}\cos\frac{x}{2}\right) = \frac{1}{2}\sin x$

So the integral is:

$$\int_0^{\pi} \sin\frac{x}{2}\cos\frac{x}{2} \, dx = \int_0^{\pi} \frac{1}{2}\sin x \, dx = -\frac{1}{2}[\cos x]_0^{\pi}$$
$$= -\frac{1}{2}(\cos \pi - \cos 0) = -\frac{1}{2}(-1 - 1) = 1$$

d) $\sin^2 2x = \frac{1}{2}(1 - \cos 4x)$

So the integral is:

$$\int_{\frac{\pi}{4}}^{\frac{\pi}{2}} \sin^2 2x \, dx = \int_{\frac{\pi}{4}}^{\frac{\pi}{2}} \frac{1}{2}(1 - \cos 4x) \, dx$$
$$= \frac{1}{2}\left[x - \frac{1}{4}\sin 4x\right]_{\frac{\pi}{4}}^{\frac{\pi}{2}}$$
$$= \frac{1}{2}\left(\left[\frac{\pi}{2} - \frac{1}{4}\sin 2\pi\right] - \left[\frac{\pi}{4} - \frac{1}{4}\sin \pi\right]\right)$$
$$= \frac{1}{2}\left(\left[\frac{\pi}{2} - 0\right] - \left[\frac{\pi}{4} - 0\right]\right) = \frac{\pi}{8}$$

e) $\cos 2x \sin 2x = \frac{1}{2}(2 \sin 2x \cos 2x) = \frac{1}{2}\sin 4x$

So the integral is:

$$\int_0^{\frac{\pi}{4}} \cos 2x \sin 2x \, dx = \int_0^{\frac{\pi}{4}} \frac{1}{2}\sin 4x \, dx$$
$$= \frac{1}{2}\left[\left(-\frac{1}{4}\cos 4x\right)\right]_0^{\frac{\pi}{4}}$$
$$= -\frac{1}{8}[\cos 4x]_0^{\frac{\pi}{4}}$$
$$= -\frac{1}{8}(\cos \pi - \cos 0)$$
$$= -\frac{1}{8}(-1 - 1) = \frac{1}{4}$$

f) $\sin^2 x - \cos^2 x = -(\cos^2 x - \sin^2 x) = -\cos 2x$

So the integral is:

$$\int_{\frac{\pi}{4}}^{\frac{\pi}{2}} \sin^2 x - \cos^2 x \, dx = \int_{\frac{\pi}{4}}^{\frac{\pi}{2}} -\cos 2x \, dx = -\frac{1}{2}[\sin 2x]_{\frac{\pi}{4}}^{\frac{\pi}{2}}$$
$$= -\frac{1}{2}\left(\sin \pi - \sin\frac{\pi}{2}\right) = -\frac{1}{2}(0 - 1) = \frac{1}{2}$$

Q3 a) $\cot^2 x - 4 = (\text{cosec}^2 x - 1) - 4 = \text{cosec}^2 x - 5$

So the integral is:

$$\int \cot^2 x - 4 \, dx = \int \text{cosec}^2 x - 5 \, dx$$
$$= -\cot x - 5x + C$$

b) $\tan^2 x = \sec^2 x - 1$

So the integral is:

$$\int \tan^2 x \, dx = \int \sec^2 x - 1 \, dx$$
$$= \tan x - x + C$$

c) $3 \cot^2 x = 3(\text{cosec}^2 x - 1) = 3 \text{cosec}^2 x - 3$

So the integral is:

$$\int 3 \cot^2 x \, dx = \int 3 \text{cosec}^2 x - 3 \, dx$$
$$= -3 \cot x - 3x + C$$

d) $\tan^2 4x = \sec^2 4x - 1$

So the integral is:

$$\int \tan^2 4x \, dx = \int \sec^2 4x - 1 \, dx$$
$$= \frac{1}{4}\tan 4x - x + C$$

Q4 $\tan^2 x + \cos^2 x - \sin^2 x = (\sec^2 x - 1) + \cos 2x$

So the integral is:

$$\int_0^{\frac{\pi}{4}} \tan^2 x + \cos^2 x - \sin^2 x \, dx$$
$$= \int_0^{\frac{\pi}{4}} \sec^2 x - 1 + \cos 2x \, dx$$
$$= \left[\tan x - x + \frac{1}{2}\sin 2x\right]_0^{\frac{\pi}{4}}$$
$$= \left[\tan\frac{\pi}{4} - \frac{\pi}{4} + \frac{1}{2}\sin\frac{2\pi}{4}\right] - \left[\tan 0 - 0 + \frac{1}{2}\sin 0\right]$$
$$= \left[1 - \frac{\pi}{4} + \frac{1}{2}\right] - [0 - 0 + 0]$$
$$= \frac{3}{2} - \frac{\pi}{4}$$

Q5 $(\sec x + \tan x)^2 = \sec^2 x + 2 \tan x \sec x + \tan^2 x$
$= \sec^2 x + 2 \tan x \sec x + (\sec^2 x - 1)$
$= 2 \sec^2 x + 2 \tan x \sec x - 1$

Remember that the derivative of sec x is sec x tan x.

So the integral is:

$$\int (\sec x + \tan x)^2 \, dx = \int 2 \sec^2 x + 2 \tan x \sec x - 1 \, dx$$
$$= 2 \tan x + 2 \sec x - x + C$$

Q6 $(\cot x + \operatorname{cosec} x)^2 = \cot^2 x + 2 \cot x \operatorname{cosec} x + \operatorname{cosec}^2 x$
$= (\operatorname{cosec}^2 x - 1) + 2 \cot x \operatorname{cosec} x + \operatorname{cosec}^2 x$
$= 2 \operatorname{cosec}^2 x + 2 \cot x \operatorname{cosec} x - 1$

Just keep using the identities that you know until you get to something that you know how to integrate.

So the integral is:

$$\int (\cot x + \operatorname{cosec} x)^2 \, dx$$
$$= \int 2\operatorname{cosec}^2 x + 2\cot x \operatorname{cosec} x - 1 \, dx$$
$$= -2\cot x - 2\operatorname{cosec} x - x + C$$

Q7 $4 + \cot^2 3x = 4 + (\operatorname{cosec}^2 3x - 1) = 3 + \operatorname{cosec}^2 3x$

So the integral is:

$$\int 4 + \cot^2 3x \, dx = \int 3 + \operatorname{cosec}^2 3x \, dx$$
$$= 3x - \frac{1}{3}\cot 3x + C$$

Q8 $\cos^2 4x + \cot^2 4x = \frac{1}{2}(\cos 8x + 1) + (\operatorname{cosec}^2 4x - 1)$
$= \frac{1}{2}\cos 8x + \operatorname{cosec}^2 4x - \frac{1}{2}$

So the integral is:

$$\int \cos^2 4x + \cot^2 4x \, dx = \int \frac{1}{2}\cos 8x + \operatorname{cosec}^2 4x - \frac{1}{2} \, dx$$
$$= \frac{1}{2}\left(\frac{1}{8}\sin 8x\right) - \frac{1}{4}\cot 4x - \frac{1}{2}x + C$$
$$= \frac{1}{16}\sin 8x - \frac{1}{4}\cot 4x - \frac{1}{2}x + C$$

Q9 **a)** $\tan^3 x + \tan^5 x = \tan^3 x(1 + \tan^2 x)$
$= \tan^3 x \sec^2 x = \sec^2 x \tan^3 x$

This is a product containing $\tan x$ to a power, and its derivative $\sec^2 x$.

Using the formula with $f(x) = \tan x$, $f'(x) = \sec^2 x$, $n = 3$ and $n + 1 = 4$ gives:

$$\int 4\sec^2 x \tan^3 x \, dx = \tan^4 x + C$$

So the integral is:

$$\int \tan^3 x + \tan^5 x \, dx = \int \sec^2 x \tan^3 x \, dx$$
$$= \frac{1}{4}\int 4\sec^2 x \tan^3 x \, dx$$
$$= \frac{1}{4}\tan^4 x + C$$

b) $\cot^5 x + \cot^3 x = \cot^3 x(\cot^2 x + 1) = \cot^3 x \operatorname{cosec}^2 x$

Again, this is a product of a function to a power and its derivative so use the formula with $f(x) = \cot x$, $f'(x) = -\operatorname{cosec}^2 x$, $n = 3$ and $n + 1 = 4$.

$$\int -4\operatorname{cosec}^2 x \cot^3 x \, dx = \cot^4 x + C$$

So the integral is:

$$\int \cot^5 x + \cot^3 x \, dx = \int \operatorname{cosec}^2 x \cot^3 x \, dx$$
$$= -\frac{1}{4}\int -4\operatorname{cosec}^2 x \cot^3 x \, dx$$
$$= -\frac{1}{4}\cot^4 x + C$$

c) $\sin^3 x = \sin x \sin^2 x = \sin x(1 - \cos^2 x)$
$= \sin x - \sin x \cos^2 x$

The second term of this function is a product of a function to a power and its derivative. Using the result with $f(x) = \cos x$, $f'(x) = -\sin x$, $n = 2$ and $n + 1 = 3$ gives: $\int -3\sin x \cos^2 x \, dx = \cos^3 x + c$

So the integral is:

$$\int \sin^3 x \, dx = \int \sin x - \sin x \cos^2 x \, dx$$
$$= \int \sin x \, dx + \int -\sin x \cos^2 x \, dx$$
$$= -\cos x + \frac{1}{3}\cos^3 x + C$$

Q10 You want to find A and B, where $\frac{A+B}{2} = 4x$, and $\frac{A-B}{2} = x$.

Solve simultaneously:

$\frac{A+B}{2} + \frac{A-B}{2} = 4x + x \Rightarrow A = 5x$. So $B = 3x$. Then

$$2 \sin 4x \cos x \equiv 2 \sin\left(\frac{5x + 3x}{2}\right) \cos\left(\frac{5x - 3x}{2}\right)$$
$$\equiv \sin 5x + \sin 3x$$

So $\int 2\sin 4x \cos x \, dx = \int \sin 5x + \sin 3x \, dx$

$$= -\frac{1}{5}\cos 5x - \frac{1}{3}\cos 3x + C$$

7. Finding Area using Integration

Exercise 7.1 — Finding enclosed areas

Q1 **a)** Start by finding the points where the curve and the line intersect. Solve $3x^2 + 4 = 16$:
$\Rightarrow 3x^2 = 12 \Rightarrow x^2 = 4 \Rightarrow x = -2$ or 2.
So they intersect at $x = -2$ and $x = 2$.

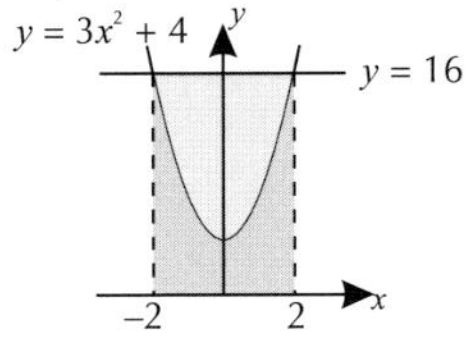

So the area is found by subtracting the integral of $3x^2 + 4$ between -2 and 2 from the integral of 16 between -2 and 2.

The area under the line is just a rectangle which is 16 by 4 so the area is $16 \times 4 = 64$.

$$\int_{-2}^{2} (3x^2 + 4) \, dx = [x^3 + 4x]_{-2}^{2}$$
$$= (2^3 + 4(2)) - ((-2)^3 + 4(-2))$$
$$= (8 + 8) - (-8 - 8)$$
$$= 16 + 16 = 32$$

So the area is $64 - 32 = 32$.

b) Solving to find the point of intersection of $y = 4$ and $y = \frac{1}{x^2}$: $4 = \frac{1}{x^2} \Rightarrow 4x^2 = 1 \Rightarrow x^2 = \frac{1}{4} \Rightarrow x = \pm\frac{1}{2}$
so the intersection in the diagram is $x = \frac{1}{2}$:

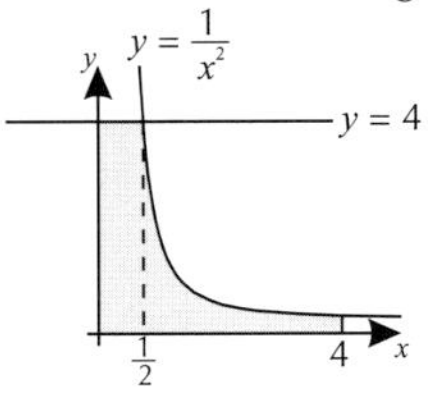

So required area is the area under the line $y = 4$ from $x = 0$ and $x = \frac{1}{2}$ added to the area under the curve $y = \frac{1}{x^2}$ from $x = \frac{1}{2}$ and $x = 4$.

The first area is just a rectangle which is $\frac{1}{2}$ by 4 so the area is $\frac{1}{2} \times 4 = 2$.

Area under curve $= \int_{\frac{1}{2}}^{4} \frac{1}{x^2} \, dx = \int_{\frac{1}{2}}^{4} x^{-2} \, dx$

$$= \left[\frac{x^{-1}}{-1}\right]_{\frac{1}{2}}^{4} = \left[-\frac{1}{x}\right]_{\frac{1}{2}}^{4}$$
$$= \left(-\frac{1}{4}\right) - \left(-\frac{1}{\left(\frac{1}{2}\right)}\right)$$
$$= -\frac{1}{4} + 2 = \frac{7}{4}$$

So the area is $2 + \frac{7}{4} = \frac{15}{4}$.

c) Solve $-1 - (x-3)^2 = -5 \Rightarrow 4 = (x-3)^2$
$\Rightarrow x - 3 = \pm 2 \Rightarrow x = 1$ and $x = 5$.
Be careful — this area is below the x-axis so the integrals will give negative areas.
So the area you want is the area above the line between 1 and 5 minus the area between the curve and the x-axis between 1 and 5.

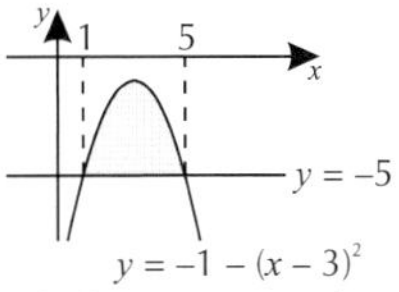

The first area is just a rectangle which is 5 by 4 so the area is $5 \times 4 = 20$. So the area between the line and the x-axis between 0 and 5 is 20.
Area between curve and x-axis

$$= \int_1^5 (-1-(x-3)^2)\,dx$$
$$= \int_1^5 (-1-(x^2-6x+9))\,dx$$
$$= \int_1^5 (6x - x^2 - 10)\,dx$$
$$= \left[\frac{6x^2}{2} - \frac{x^3}{3} - 10x\right]_1^5 = \left[3x^2 - \frac{x^3}{3} - 10x\right]_1^5$$
$$= \left((3\times5^2) - \frac{5^3}{3} - (10\times5)\right) - \left((3\times1^2) - \frac{1^3}{3} - (10\times1)\right)$$
$$= \left(75 - \frac{125}{3} - 50\right) - \left(3 - \frac{1}{3} - 10\right) = -\frac{28}{3}$$

So the area between the curve and the x-axis between 1 and 5 is $\frac{28}{3}$.
So the area is $20 - \frac{28}{3} = \frac{32}{3}$.

d) Solve $x^3 = 4x \Rightarrow x^3 - 4x = 0 \Rightarrow x(x^2 - 4) = 0$
$\Rightarrow x(x+2)(x-2) = 0 \Rightarrow x = 0$ and $x = \pm 2$.

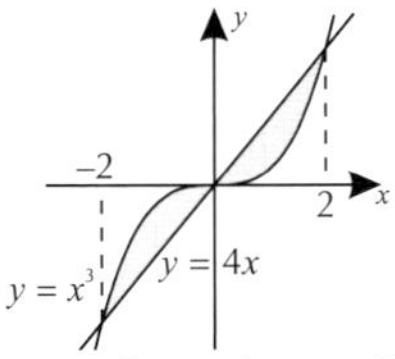

The area above the x-axis is the same as the area below the x-axis (from the symmetry of the graph).
For the area above the x-axis, find the area under the line from $x = 0$ to $x = 2$, minus the area under the curve from $x = 0$ to $x = 2$.

$\int_0^2 4x\,dx = [2x^2]_0^2 = (2 \times 2^2) - (2 \times 0^2) = 8$

$\int_0^2 x^3\,dx = \left[\frac{x^4}{4}\right]_0^2 = \left(\frac{2^4}{4}\right) - \left(\frac{0^2}{4}\right) = 4$

So the area above the axis is $8 - 4 = 4$.
So the area below the x-axis is also 4.
The total area is $4 + 4 = 8$.
For part d) you could have used the formula for finding the area of a triangle rather than integrating.

Q2 a) Solve $x^2 + 4 = x + 4 \Rightarrow x^2 - x = 0 \Rightarrow x(x-1) = 0$
$\Rightarrow x = 0$ and $x = 1$.
Draw a diagram:

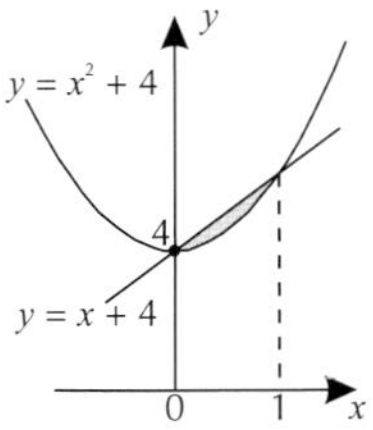

So the area you want is the area under the line minus the area under the curve, between $x = 0$ and $x = 1$.

$$\int_0^1 (x+4)\,dx = \left[\frac{x^2}{2} + 4x\right]_0^1$$
$$= \left(\frac{1^2}{2} + (4\times1)\right) - \left(\frac{0^2}{2} + (4\times0)\right)$$
$$= \frac{1}{2} + 4 = \frac{9}{2}$$
$$\int_0^1 (x^2+4)\,dx = \left[\frac{x^3}{3} + 4x\right]_0^1$$
$$= \left(\frac{1^3}{3} + (4\times1)\right) - \left(\frac{0^3}{3} + (4\times0)\right)$$
$$= \frac{1}{3} + 4 = \frac{13}{3}$$

So the area is $\frac{9}{2} - \frac{13}{3} = \frac{1}{6}$.

b) Solve $x^2 + 2x - 3 = 4x \Rightarrow x^2 - 2x - 3 = 0$
$\Rightarrow (x-3)(x+1) = 0 \Rightarrow x = -1$ and $x = 3$.
It'll also help to find where the curve meets the x-axis by solving $x^2 + 2x - 3 = 0 \Rightarrow (x+3)(x-1)$
$\Rightarrow x = 1$ and $x = -3$.
Draw a diagram:

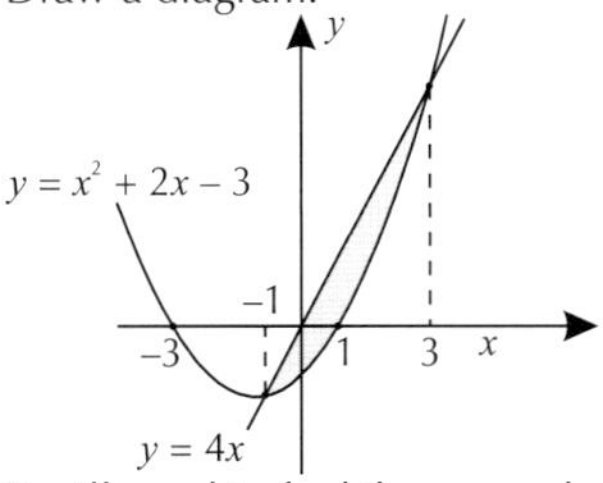

You'll need to find the areas above and below the x-axis separately.
For the area above the x-axis, integrate $4x$ from $x = 0$ to $x = 3$ and subtract the integral of $x^2 + 2x - 3$ from $x = 1$ to $x = 3$.

$\int_0^3 4x\,dx = [2x^2]_0^3 = (2(3)^2) - (2(0)^2) = 18$

$$\int_1^3 (x^2 + 2x - 3)\,dx = \left[\frac{x^3}{3} + x^2 - 3x\right]_1^3$$
$$= \left(\frac{(3)^3}{3} + (3)^2 - 3(3)\right) - \left(\frac{(1)^3}{3} + (1)^2 - 3(1)\right)$$
$$= \left(\frac{27}{3} + 9 - 9\right) - \left(\frac{1}{3} + 1 - 3\right) = \frac{32}{3}$$

So the area above the x-axis is $18 - \frac{32}{3} = \frac{22}{3}$
To find the area below the x-axis you need to find the positive area between the curve and the axis between –1 and 1 minus the positive area between the line and the axis between –1 and 0.

$$\int_{-1}^1 (x^2 + 2x - 3)\,dx = \left[\frac{x^3}{3} + x^2 - 3x\right]_{-1}^1$$
$$= \left(\frac{(1)^3}{3} + (1)^2 - 3(1)\right) - \left(\frac{(-1)^3}{3} + (-1)^2 - 3(-1)\right)$$
$$= \left(\frac{1}{3} + 1 - 3\right) - \left(\frac{-1}{3} + 1 + 3\right) = -\frac{16}{3}$$

So the area between the curve and the axis between –1 and 1 is $\frac{16}{3}$.

$\int_{-1}^0 4x\,dx = [2x^2]_{-1}^0 = (2(0)^2) - (2(-1)^2) = -2$

So the area between the line and the axis between –1 and 1 is 2.
So the area under the x-axis is $\frac{16}{3} - 2 = \frac{10}{3}$
So the total area enclosed by the curve and the line is $\frac{22}{3} + \frac{10}{3} = \frac{32}{3}$

Q3 Solve $\cos x = 0.5 \Rightarrow x = \cos^{-1} 0.5 = \frac{\pi}{3}$ and $\frac{5\pi}{3}$.

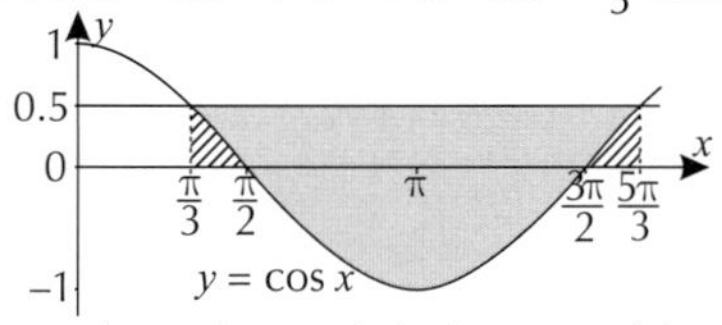

The area above the x-axis is the area of the rectangle between $\frac{\pi}{3}$ and $\frac{5\pi}{3}$, minus the striped areas.

Rectangle = $(\frac{5\pi}{3} - \frac{\pi}{3}) \times 0.5 = \frac{2\pi}{3}$

One striped area = $\int_{\frac{\pi}{3}}^{\frac{\pi}{2}} \cos x \, dx = [\sin x]_{\frac{\pi}{3}}^{\frac{\pi}{2}}$

$= \sin\left(\frac{\pi}{2}\right) - \sin\left(\frac{\pi}{3}\right) = 1 - \frac{\sqrt{3}}{2}$.

The two striped areas are the same (because of the symmetry of the cosine curve). So the area above the x-axis is: $\frac{2\pi}{3} - 2\left(1 - \frac{\sqrt{3}}{2}\right) = \frac{2\pi}{3} - 2 + \sqrt{3}$

Integrate to find the area below the x-axis:

$\int_{\frac{\pi}{2}}^{\frac{3\pi}{2}} \cos x \, dx = [\sin x]_{\frac{\pi}{2}}^{\frac{3\pi}{2}} = \sin\left(\frac{3\pi}{2}\right) - \sin\left(\frac{\pi}{2}\right) = -1 - 1 = -2$

So the area below the x-axis is 2.

The shaded area is $\frac{2\pi}{3} - 2 + \sqrt{3} + 2 = \frac{2\pi}{3} + \sqrt{3}$

Q4 The area of each grey section is the integral of the upper function minus the integral of the lower function. So the total grey area is:

$$\left(\int_0^\pi 5\sin x \, dx - \int_0^\pi 4\sin x \, dx\right) + \left(\int_0^\pi 3\sin x \, dx - \int_0^\pi 2\sin x \, dx\right) + \left(\int_0^\pi \sin x \, dx\right)$$

$$= 5\int_0^\pi \sin x \, dx - 4\int_0^\pi \sin x \, dx + 3\int_0^\pi \sin x \, dx - 2\int_0^\pi \sin x \, dx + \int_0^\pi \sin x \, dx$$

$$= (5 - 4 + 3 - 2 + 1)\int_0^\pi \sin x \, dx$$

$$= 3\int_0^\pi \sin x \, dx = 3[-\cos x]_0^\pi$$

$= 3[(-\cos\pi) - (-\cos 0)] = 3[(-(-1) - (-1)] = 3(1 + 1) = 6$

Exercise 7.2 — Parametric integration

Q1 **a)** $x = 3t^{-1} \Rightarrow \frac{dx}{dt} = -3t^{-2} = -\frac{3}{t^2}$

$y = 4t^2$

$\int y \, dx = \int y \frac{dx}{dt} \, dt = \int (4t^2) \times \left(-\frac{3}{t^2}\right) dt = \int -12 \, dt$

You could take the constant outside the integral — i.e. this could be written as $-12\int 1 \, dt$.

b) $x = \tan 5\theta \Rightarrow \frac{dx}{d\theta} = 5\sec^2 5\theta$

$y = \sec^2 5\theta$

$\int y \, dx = \int y \frac{dx}{d\theta} \, d\theta = \int (\sec^2 5\theta)(5\sec^2 5\theta) \, d\theta$

$= \int 5\sec^4 5\theta \, d\theta$

Q2 **a)** $x = (4t - 5)^2 \Rightarrow \frac{dx}{dt} = 2(4t - 5) \times 4 = 32t - 40$

$y = t^2 - 3t$

$\int y \, dx = \int y \frac{dx}{dt} \, dt = \int (t^2 - 3t) \times (32t - 40) \, dt$

$= \int (32t^3 - 136t^2 + 120t) \, dt$

$= 8t^4 - \frac{136}{3}t^3 + 60t^2 + C$

b) $x = t^2 + 3 \Rightarrow \frac{dx}{dt} = 2t$

$y = 4t - 1$

$\int y \, dx = \int y \frac{dx}{dt} \, dt = \int (4t - 1) \times (2t) \, dt$

$= \int (8t^2 - 2t) \, dt$

$= \frac{8}{3}t^3 - t^2 + C$

Q3 $x = 3t^2 \Rightarrow \frac{dx}{dt} = 6t$

$y = \frac{5}{t}$

So $y\frac{dx}{dt} = \frac{5}{t} \times 6t = 30$

When $x = 75$, $3t^2 = 75 \Rightarrow t = 5$ (since $t > 0$).

When $x = 3$, $3t^2 = 3 \Rightarrow t = 1$ (since $t > 0$).

$\int_3^{75} y \, dx = \int_1^5 y \frac{dx}{dt} \, dt = \int_1^5 30 \, dt = 30\int_1^5 1 \, dt$

$= 30[t]_1^5$

$= 30(5 - 1)$

$= 120$

Q4 **a)** When $x = 8$:

$4t(t + 1) = 8 \Rightarrow t(t + 1) = 2 \Rightarrow t^2 + t - 2 = 0$

$(t - 1)(t + 2) = 0 \Rightarrow t = 1$ $(t > 0)$.

When $x = 120$:

$4t(t + 1) = 120 \Rightarrow t(t + 1) = 30 \Rightarrow t^2 + t - 30 = 0$

$(t - 5)(t + 6) = 0 \Rightarrow t = 5$ $(t > 0)$.

b) $x = 4t^2 + 4t \Rightarrow \frac{dx}{dt} = 8t + 4$

$y = 3t^3$

So $y\frac{dx}{dt} = 24t^4 + 12t^3$

$\int_8^{120} y \, dx = \int_1^5 y \frac{dx}{dt} \, dt$

$= \int_1^5 24t^4 + 12t^3 \, dt = \left[\frac{24t^5}{5} + 3t^4\right]_1^5$

$= (15\,000 + 1875) - \left(\frac{24}{5} + 3\right)$

$= \frac{84\,336}{5} = 16\,867.2$

8. Integration by Substitution

Exercise 8.1 — Integration by substitution

Q1 **a)** $u = x + 3 \Rightarrow \frac{du}{dx} = 1 \Rightarrow dx = du$

So $\int 12(x + 3)^5 \, dx = \int 12u^5 \, du$

$= 2u^6 + C$

$= 2(x + 3)^6 + C$

You could also have solved this by using the rule given on p180, as it's of the form $(ax + b)^n$.

b) $u = 11 - x \Rightarrow \frac{du}{dx} = -1 \Rightarrow dx = -du$

So $\int (11 - x)^4 \, dx = -\int u^4 \, du$

$= -\frac{1}{5}u^5 + C$

$= -\frac{1}{5}(11 - x)^5 + C$

c) $u = x^2 + 4 \Rightarrow \frac{du}{dx} = 2x \Rightarrow dx = \frac{1}{2x} \, du$

So $\int 24x(x^2 + 4)^3 \, dx = \int 24x \times u^3 \times \frac{1}{2x} \, du$

$= \int 12u^3 \, du$

$= 3u^4 + C$

$= 3(x^2 + 4)^4 + C$

d) $u = \sin x \Rightarrow \frac{du}{dx} = \cos x \Rightarrow dx = \frac{1}{\cos x}\,du$

So $\int \sin^5 x \cos x\,dx = \int u^5 \cos x \times \frac{1}{\cos x}\,du$

$= \int u^5\,du$

$= \frac{1}{6}u^6 + C$

$= \frac{1}{6}\sin^6 x + C$

e) $u = x - 1 \Rightarrow \frac{du}{dx} = 1 \Rightarrow dx = du$

and $u = x - 1 \Rightarrow x = u + 1$

So $\int x(x-1)^5\,dx = \int (u+1)u^5\,du = \int u^6 + u^5\,du$

$= \frac{1}{7}u^7 + \frac{1}{6}u^6 + C$

$= \frac{1}{7}(x-1)^7 + \frac{1}{6}(x-1)^6 + C$

Q2 a) Let $u = x + 2 \Rightarrow \frac{du}{dx} = 1 \Rightarrow dx = du$

So $\int 21(x+2)^6\,dx = \int 21u^6\,du$

$= 3u^7 + C$

$= 3(x+2)^7 + C$

b) Let $u = 5x + 4 \Rightarrow \frac{du}{dx} = 5 \Rightarrow dx = \frac{1}{5}du$

So $\int (5x+4)^3\,dx = \int \frac{1}{5}u^3\,du$

$= \frac{1}{20}u^4 + C$

$= \frac{1}{20}(5x+4)^4 + C$

c) Let $u = 2x + 3 \Rightarrow \frac{du}{dx} = 2 \Rightarrow dx = \frac{1}{2}\,du$

and $u = 2x + 3 \Rightarrow x = \frac{u-3}{2}$

So $\int x(2x+3)^3\,dx = \int \frac{u-3}{2} \times u^3 \times \frac{1}{2}\,du$

$= \frac{1}{4}\int u^4 - 3u^3\,du$

$= \frac{1}{4}\left(\frac{1}{5}u^5 - \frac{3}{4}u^4\right) + C$

$= \frac{1}{20}u^5 - \frac{3}{16}u^4 + C$

$= \frac{1}{20}(2x+3)^5 - \frac{3}{16}(2x+3)^4 + C$

d) Let $u = x^2 - 5 \Rightarrow \frac{du}{dx} = 2x \Rightarrow dx = \frac{1}{2x}\,du$

So $\int 24x(x^2-5)^7\,dx = \int 24x \times u^7 \times \frac{1}{2x}\,du$

$= \int 12u^7\,du$

$= \frac{3}{2}u^8 + C$

$= \frac{3}{2}(x^2-5)^8 + C$

Q3 a) $u = \sqrt{x+1} \Rightarrow \frac{du}{dx} = \frac{1}{2\sqrt{x+1}} = \frac{1}{2u} \Rightarrow dx = 2u\,du$

$u = \sqrt{x+1} \Rightarrow x = u^2 - 1$

So $\int 6x\sqrt{x+1}\,dx = \int 6(u^2-1) \times u \times 2u\,du$

$= \int 12u^4 - 12u^2\,du$

$= \frac{12}{5}u^5 - 4u^3 + C$

$= \frac{12}{5}(\sqrt{x+1})^5 - 4(\sqrt{x+1})^3 + C$

b) $u = \sqrt{4-x} \Rightarrow \frac{du}{dx} = -\frac{1}{2\sqrt{4-x}} = -\frac{1}{2u}$

$\Rightarrow dx = -2u\,du$

$u = \sqrt{4-x} \Rightarrow x = 4 - u^2$

So $\int \frac{x}{\sqrt{4-x}}\,dx = \int \frac{4-u^2}{u} \times -2u\,du$

$= \int -2(4-u^2)\,du = \int 2u^2 - 8\,du$

$= \frac{2}{3}u^3 - 8u + C$

$= \frac{2}{3}(\sqrt{4-x})^3 - 8(\sqrt{4-x}) + C$

c) $u = \ln x \Rightarrow \frac{du}{dx} = \frac{1}{x} \Rightarrow dx = x\,du$

So $\int \frac{15(\ln x)^4}{x}\,dx = \int \frac{15u^4}{x} \times x\,du = \int 15u^4\,du$

$= 3u^5 + C$

$= 3(\ln x)^5 + C$

Q4 a) Let $u = \sqrt{2x-1} \Rightarrow \frac{du}{dx} = \frac{1}{\sqrt{2x-1}} = \frac{1}{u} \Rightarrow dx = u\,du$

and $u = \sqrt{2x-1} \Rightarrow x = \frac{u^2+1}{2}$

So $\int \frac{4x}{\sqrt{(2x-1)}}\,dx = \int 4\left(\frac{u^2+1}{2}\right) \times \frac{1}{u} \times u\,du$

$= \int 2u^2 + 2\,du$

$= \frac{2}{3}u^3 + 2u + C$

$= \frac{2}{3}(\sqrt{2x-1})^3 + 2(\sqrt{2x-1}) + C$

b) Let $u = 4 - \sqrt{x} \Rightarrow \frac{du}{dx} = -\frac{1}{2\sqrt{x}} \Rightarrow -2\sqrt{x}\,du = dx$

and $u = 4 - \sqrt{x} \Rightarrow x = (4-u)^2$

$\Rightarrow dx$ can be written $-2(4-u)\,du = (2u-8)du$

So $\int \frac{1}{4-\sqrt{x}}\,dx = \int \frac{1}{u}(2u-8)\,du$

$= \int 2 - \frac{8}{u}\,du$

$= 2u - 8\ln|u| + C$

$= 2(4-\sqrt{x}) - 8\ln|4-\sqrt{x}| + C$

$= -2\sqrt{x} - 8\ln|4-\sqrt{x}| + C$

You can leave out any constant terms that appear after you integrate (like the 8 you get out of the term $2(4-\sqrt{x})$ here) — they just get absorbed into the constant of integration, C.

c) Let $u = 1 + e^x \Rightarrow \frac{du}{dx} = e^x \Rightarrow dx = \frac{1}{e^x}\,du$

and $u = 1 + e^x \Rightarrow e^x = u - 1$

So $\int \frac{e^{2x}}{1+e^x}\,dx = \int \frac{e^{2x}}{u} \times \frac{1}{e^x}\,du$

$= \int \frac{e^x}{u}\,du$

$= \int \frac{u-1}{u}\,du$

$= \int 1 - \frac{1}{u}\,du$

$= u - \ln|u| + c$

$= 1 + e^x - \ln|1+e^x| + c$

$= e^x - \ln(1+e^x) + C$

You can remove the modulus because e^x is always positive, so $1 + e^x > 1$.

In Q4, you might have used different substitutions, but as long as your substitution is valid and you end up with the same final answer then it's okay.

Q5 Let $u = f(x) \Rightarrow \frac{du}{dx} = f'(x) \Rightarrow dx = \frac{1}{f'(x)}\,du$

So $\int (n+1)f'(x)[f(x)]^n\,dx = \int (n+1)f'(x)u^n \times \frac{1}{f'(x)}\,du$

$= \int (n+1)u^n\,du$

$= u^{n+1} + C$

$= [f(x)]^{n+1} + C$, as required

Exercise 8.2 — Definite integrals

Q1 a) $u = 3x - 2 \Rightarrow \frac{du}{dx} = 3 \Rightarrow dx = \frac{1}{3}\,du$

$x = \frac{2}{3} \Rightarrow u = 2 - 2 = 0$

$x = 1 \Rightarrow u = 3 - 2 = 1$

So $\int_{\frac{2}{3}}^{1} (3x-2)^4\,dx = \int_0^1 \frac{1}{3}u^4\,du$

$= \left[\frac{1}{15}u^5\right]_0^1 = \frac{1}{15}$

b) $u = x + 3 \Rightarrow \frac{du}{dx} = 1 \Rightarrow dx = 1\,du$

$u = x + 3 \Rightarrow x = u - 3$

$x = -2 \Rightarrow u = -2 + 3 = 1$

$x = 1 \Rightarrow u = 1 + 3 = 4$

So $\int_{-2}^{1} 2x(x+3)^4\,dx = \int_1^4 2(u-3) \times u^4 \times 1\,du$

$= \int_1^4 (2u - 6)u^4\,du$

$= \int_1^4 (2u^5 - 6u^4)\,du$

$= \left[\frac{u^6}{3} - \frac{6u^5}{5}\right]_1^4$

$= \left(\frac{4^6}{3} - \frac{6(4)^5}{5}\right) - \left(\frac{1^6}{3} - \frac{6(1)^5}{5}\right)$

$= \frac{687}{5}$ (= 137.4)

c) $u = \sin x \Rightarrow \frac{du}{dx} = \cos x \Rightarrow dx = \frac{1}{\cos x}\,du$

$x = 0 \Rightarrow u = \sin(0) = 0$

$x = \frac{\pi}{6} \Rightarrow u = \sin\frac{\pi}{6} = \frac{1}{2}$

So $\int_0^{\frac{\pi}{6}} 8\sin^3 x \cos x\,dx = \int_0^{\frac{1}{2}} 8u^3 \cos x \times \frac{1}{\cos x}\,du$

$= \int_0^{\frac{1}{2}} 8u^3\,du$

$= [2u^4]_0^{\frac{1}{2}}$

$= 2\left(\frac{1}{2}\right)^4 - 0 = \frac{1}{8}$

d) $u = \sqrt{x+1} \Rightarrow \frac{du}{dx} = \frac{1}{2\sqrt{x+1}} = \frac{1}{2u} \Rightarrow dx = 2u\,du$

$u = \sqrt{x+1} \Rightarrow x = u^2 - 1$

$x = 0 \Rightarrow u = \sqrt{x+1} = \sqrt{1} = 1$

$x = 3 \Rightarrow u = \sqrt{4} = 2$

So $\int_0^3 x\sqrt{x+1}\,dx = \int_1^2 (u^2 - 1) \times u \times 2u\,du$

$= \int_1^2 2u^4 - 2u^2\,du$

$= \left[\frac{2}{5}u^5 - \frac{2}{3}u^3\right]_1^2$

$= \left[\left(\frac{64}{5} - \frac{16}{3}\right) - \left(\frac{2}{5} - \frac{2}{3}\right)\right] = \frac{116}{15}$

In Q2, 3 and 5 below, there are different substitutions you could use — again, if you get the final answer right and you use a valid substitution, then your method is fine.

Q2 a) Let $u = x^2 - 3 \Rightarrow \frac{du}{dx} = 2x \Rightarrow dx = \frac{1}{2x}\,du$

$x = 2 \Rightarrow u = 4 - 3 = 1$

$x = \sqrt{5} \Rightarrow u = 5 - 3 = 2$

So $\int_2^{\sqrt{5}} x(x^2-3)^4\,dx = \int_1^2 x \times u^4 \times \frac{1}{2x}\,du$

$= \int_1^2 \frac{1}{2}u^4\,du$

$= \left[\frac{1}{10}u^5\right]_1^2$

$= \frac{32}{10} - \frac{1}{10} = \frac{31}{10}$ (= 3.1)

b) Let $u = 3x - 4 \Rightarrow \frac{du}{dx} = 3 \Rightarrow dx = \frac{1}{3}\,du$

and $u = 3x - 4 \Rightarrow x = \frac{u+4}{3}$

$x = 1 \Rightarrow u = 3 - 4 = -1$

$x = 2 \Rightarrow u = 6 - 4 = 2$

So $\int_1^2 x(3x-4)^3\,dx = \int_{-1}^2 \frac{u+4}{3} \times u^3 \times \frac{1}{3}\,du$

$= \frac{1}{9}\int_{-1}^2 u^4 + 4u^3\,du$

$= \frac{1}{9}\left[\frac{u^5}{5} + u^4\right]_{-1}^2$

$= \frac{1}{9}\left[\left(\frac{32}{5} + 16\right) - \left(\frac{-1}{5} + 1\right)\right]$

$= \frac{1}{9}\left[\frac{108}{5}\right] = \frac{12}{5}$ (= 2.4)

c) Let $u = \sqrt{x-1} \Rightarrow \frac{du}{dx} = \frac{1}{2\sqrt{x-1}} = \frac{1}{2u} \Rightarrow dx = 2u\,du$

and $u = \sqrt{x-1} \Rightarrow x = u^2 + 1$

Using $u = \sqrt{x-1}$, $x = 2 \Rightarrow u = \sqrt{1} = 1$

and $x = 10 \Rightarrow u = \sqrt{9} = 3$

So $\int_2^{10} \frac{x}{\sqrt{x-1}}\,dx = \int_1^3 \frac{u^2+1}{u} \times 2u\,du$

$= \int_1^3 2u^2 + 2\,du$

$= \left[\frac{2}{3}u^3 + 2u\right]_1^3$

$= \left[(18+6) - \left(\frac{2}{3} + 2\right)\right] = \frac{64}{3}$

Q3 Let $u = 3 - \sqrt{x} \Rightarrow \frac{du}{dx} = -\frac{1}{2\sqrt{x}} \Rightarrow dx = -2\sqrt{x}\,du$

$u = 3 - \sqrt{x} \Rightarrow x = (3-u)^2 \Rightarrow dx = (2u - 6)\,du$

So $x = 1 \Rightarrow u = 3 - 1 = 2$

and $x = 4 \Rightarrow u = 3 - 2 = 1$

So $\int_1^4 \frac{1}{3-\sqrt{x}}\,dx = \int_2^1 \frac{1}{u} \times (2u - 6)\,du$

$= \int_2^1 2 - \frac{6}{u}\,du$

$= [2u - 6\ln|u|]_2^1$

$= (2 - 6\ln 1) - (4 - 6\ln 2)$

$= 2 - 0 - 4 + 6\ln 2 = -2 + 6\ln 2$

You could have put 2 as the upper limit and 1 as the lower limit in the integral with respect to u, and put a minus sign in front. Both methods give the right answer, but whichever you use, be careful not to lose any minus signs.

Q4 $u = 1 + e^x \Rightarrow \frac{du}{dx} = e^x \Rightarrow dx = \frac{1}{e^x}\,du$

$x = 0 \Rightarrow u = 1 + e^0 = 2$

$x = 1 \Rightarrow u = 1 + e$

So $\int_0^1 2e^x(1+e^x)^3\,dx = \int_2^{1+e} 2e^x u^3 \times \frac{1}{e^x}\,du$

$= \int_2^{1+e} 2u^3\,du = \left[\frac{u^4}{2}\right]_2^{1+e}$

$= \frac{(1+e)^4}{2} - 8 = 87.6$ (1 d.p.)

Q5 Let $u = \sqrt{3x+1} \Rightarrow \frac{du}{dx} = \frac{3}{2\sqrt{3x+1}} = \frac{3}{2u}$
$\Rightarrow dx = \frac{2}{3}u\, du$
and $u = \sqrt{3x+1} \Rightarrow x = \frac{u^2-1}{3}$
$x = 1 \Rightarrow u = \sqrt{4} = 2$
$x = 5 \Rightarrow u = \sqrt{16} = 4$
So $\int_1^5 \frac{x}{\sqrt{3x+1}}\, dx = \int_2^4 \frac{u^2-1}{3u} \times \frac{2}{3}u\, du$
$= \frac{2}{9}\int_2^4 u^2 - 1\, du$
$= \frac{2}{9}\left[\frac{1}{3}u^3 - u\right]_2^4 = \frac{2}{9}\left[\left(\frac{64}{3} - 4\right) - \left(\frac{8}{3} - 2\right)\right]$
$= \frac{2}{9}\left[\frac{50}{3}\right] = \frac{100}{27}$

Exercise 8.3 — Trig identities

Q1 $x = \tan\theta \Rightarrow \frac{dx}{d\theta} = \sec^2\theta \Rightarrow dx = \sec^2\theta\, d\theta$
$x = 0 \Rightarrow \tan\theta = 0 \Rightarrow \theta = 0$
$x = 1 \Rightarrow \tan\theta = 1 \Rightarrow \theta = \frac{\pi}{4}$
So, using the identity $\sec^2\theta \equiv 1 + \tan^2\theta$:
$\int_0^1 \frac{1}{1+x^2}\, dx = \int_0^{\frac{\pi}{4}} \frac{1}{1+\tan^2\theta} \times \sec^2\theta\, d\theta$
$= \int_0^{\frac{\pi}{4}} \frac{\sec^2\theta}{\sec^2\theta}\, d\theta = \int_0^{\frac{\pi}{4}} 1\, d\theta = [\theta]_0^{\frac{\pi}{4}} = \frac{\pi}{4}$

Q2 $u = \sin x \Rightarrow \frac{du}{dx} = \cos x \Rightarrow dx = \frac{1}{\cos x}\, du$
$x = 0 \Rightarrow u = \sin 0 = 0$
$x = \frac{\pi}{6} \Rightarrow u = \sin\frac{\pi}{6} = \frac{1}{2}$
So, using the identity $\sin 2x \equiv 2\sin x\cos x$:
$\int_0^{\frac{\pi}{6}} 3\sin x \sin 2x\, dx \equiv \int_0^{\frac{\pi}{6}} 6\sin^2 x\cos x\, dx$
$= \int_0^{\frac{1}{2}} 6u^2\cos x \times \frac{1}{\cos x}\, du$
$= \int_0^{\frac{1}{2}} 6u^2\, du = [2u^3]_0^{\frac{1}{2}} = \frac{1}{4}$

Q3 $x = 2\sin\theta \Rightarrow \frac{dx}{d\theta} = 2\cos\theta \Rightarrow dx = 2\cos\theta\, d\theta$
$x = 1 \Rightarrow \sin\theta = \frac{1}{2} \Rightarrow \theta = \frac{\pi}{6}$
$x = \sqrt{3} \Rightarrow \sin\theta = \frac{\sqrt{3}}{2} \Rightarrow \theta = \frac{\pi}{3}$
So, using the identity $\sin^2\theta + \cos^2\theta \equiv 1$:
$\int_1^{\sqrt{3}} \frac{1}{(4-x^2)^{\frac{3}{2}}}\, dx = \int_{\frac{\pi}{6}}^{\frac{\pi}{3}} \frac{1}{(4-4\sin^2\theta)^{\frac{3}{2}}} \times 2\cos\theta\, d\theta$
$= \int_{\frac{\pi}{6}}^{\frac{\pi}{3}} \frac{2\cos\theta}{(4-4+4\cos^2\theta)^{\frac{3}{2}}}\, d\theta$
$= \int_{\frac{\pi}{6}}^{\frac{\pi}{3}} \frac{2\cos\theta}{8\cos^3\theta}\, d\theta$
$= \int_{\frac{\pi}{6}}^{\frac{\pi}{3}} \frac{1}{4}\sec^2\theta\, d\theta = \frac{1}{4}[\tan\theta]_{\frac{\pi}{6}}^{\frac{\pi}{3}}$
$= \frac{1}{4}\left(\sqrt{3} - \frac{1}{\sqrt{3}}\right) = \frac{\sqrt{3}}{6}$

Q4 $x = \cos\theta \Rightarrow \frac{dx}{d\theta} = -\sin\theta \Rightarrow dx = -\sin\theta\, d\theta$
$x = \frac{1}{2} \Rightarrow \cos\theta = \frac{1}{2} \Rightarrow \theta = \frac{\pi}{3}$
$x = 1 \Rightarrow \cos\theta = 1 \Rightarrow \theta = 0$
So, using the identity $\sin^2\theta + \cos^2\theta \equiv 1$:
$\int_{\frac{1}{2}}^1 \frac{1}{x^2\sqrt{1-x^2}}\, dx = \int_{\frac{\pi}{3}}^0 \frac{1}{\cos^2\theta\sqrt{1-\cos^2\theta}} \times -\sin\theta\, d\theta$
$= \int_{\frac{\pi}{3}}^0 -\frac{\sin\theta}{\cos^2\theta\sin\theta}\, d\theta = \int_0^{\frac{\pi}{3}} \frac{\sin\theta}{\cos^2\theta\sin\theta}\, d\theta$
$= \int_0^{\frac{\pi}{3}} \frac{1}{\cos^2\theta}\, d\theta = \int_0^{\frac{\pi}{3}} \sec^2\theta\, d\theta$
$= [\tan\theta]_0^{\frac{\pi}{3}} = \sqrt{3} - 0 = \sqrt{3}$

Q5 $u = \sec^2 x \Rightarrow \frac{du}{dx} = 2\sec^2 x\tan x$
$\Rightarrow dx = \frac{1}{2\sec^2 x\tan x}\, du = \frac{1}{2u\tan x}\, du$
And using the identity
$\sec^2 x \equiv 1 + \tan^2 x \Rightarrow \tan^2 x \equiv \sec^2 x - 1 = u - 1$
$\int 2\tan^3 x\, dx = \int 2\tan x(u-1) \times \frac{1}{2u\tan x}\, du$
$= \int \frac{u-1}{u}\, du = \int 1 - \frac{1}{u}\, du$
$= u - \ln|u| + C$
$= \sec^2 x - \ln(\sec^2 x) + C$

9. Integration by Parts

Exercise 9.1 — Integration by parts

Q1 a) Let $u = x$ and $\frac{dv}{dx} = e^x$.
Then $\frac{du}{dx} = 1$ and $v = e^x$.
So $\int xe^x\, dx = xe^x - \int e^x\, dx = xe^x - e^x + C$

b) Let $u = x$ and $\frac{dv}{dx} = e^{-x}$.
Then $\frac{du}{dx} = 1$ and $v = -e^{-x}$.
So $\int xe^{-x}\, dx = -xe^{-x} - \int -e^{-x}\, dx = -xe^{-x} - e^{-x} + C$

c) Let $u = x$ and $\frac{dv}{dx} = e^{-\frac{x}{3}}$.
Then $\frac{du}{dx} = 1$ and $v = -3e^{-\frac{x}{3}}$.
So $\int xe^{-\frac{x}{3}}\, dx = -3xe^{-\frac{x}{3}} - \int -3e^{-\frac{x}{3}}\, dx$
$= -3xe^{-\frac{x}{3}} - 9e^{-\frac{x}{3}} + C$

d) Let $u = x$ and $\frac{dv}{dx} = e^x + 1$.
Then $\frac{du}{dx} = 1$ and $v = e^x + x$.
So $\int x(e^x + 1)\, dx = x(e^x + x) - \int e^x + x\, dx$
$= xe^x + x^2 - e^x - \frac{1}{2}x^2 + C$
$= xe^x - e^x + \frac{1}{2}x^2 + C$

You could also have solved this one by expanding the brackets and integrating $xe^x + x$ separately.

You might have spotted a pattern here — all the parts of this question had $u = x$ and $\frac{dv}{dx}$ as a function involving e.
Your answers might look a bit different if you factorised them.

Q2 a) Let $u = x$ and $\frac{dv}{dx} = \sin x$.
Then $\frac{du}{dx} = 1$ and $v = -\cos x$.
So $\int_0^\pi x\sin x\, dx = [-x\cos x]_0^\pi - \int_0^\pi -\cos x\, dx$
$= [-x\cos x]_0^\pi + [\sin x]_0^\pi$
$= (\pi - 0) + (0 - 0)$
$= \pi$

b) Let $u = 2x$ and $\frac{dv}{dx} = \cos x$.
Then $\frac{du}{dx} = 2$ and $v = \sin x$.
So $\int 2x\cos x\, dx = 2x\sin x - \int 2\sin x\, dx$
$= 2x\sin x + 2\cos x + C$

c) Let $u = 3x$ and $\frac{dv}{dx} = \cos\frac{1}{2}x$.
Then $\frac{du}{dx} = 3$ and $v = 2\sin\frac{1}{2}x$.
So $\int 3x\cos\frac{1}{2}x\, dx = 6x\sin\frac{1}{2}x - \int 6\sin\frac{1}{2}x\, dx$
$= 6x\sin\frac{1}{2}x + 12\cos\frac{1}{2}x + C$

d) Let $u = 2x$ and $\frac{dv}{dx} = 1 - \sin x$.
Then $\frac{du}{dx} = 2$ and $v = x + \cos x$.
So $\int_{-\frac{\pi}{2}}^{\frac{\pi}{2}} 2x(1 - \sin x)\,dx$
$= \left[2x(x+\cos x)\right]_{-\frac{\pi}{2}}^{\frac{\pi}{2}} - \int_{-\frac{\pi}{2}}^{\frac{\pi}{2}} 2(x + \cos x)\,dx$
$= \left[2x(x+\cos x)\right]_{-\frac{\pi}{2}}^{\frac{\pi}{2}} - \left[x^2 + 2\sin x\right]_{-\frac{\pi}{2}}^{\frac{\pi}{2}}$
$= \left[\pi\left(\frac{\pi}{2}+0\right) - (-\pi)\left(-\frac{\pi}{2}+0\right)\right] - \left[\left(\frac{\pi^2}{4}+2\right) - \left(\frac{\pi^2}{4}-2\right)\right]$
$= -4$

Q3 a) Let $u = \ln x$ and $\frac{dv}{dx} = 2$.
Then $\frac{du}{dx} = \frac{1}{x}$ and $v = 2x$.
So $\int 2 \ln x\,dx = 2x \ln x - \int 2\,dx$
$= 2x \ln x - 2x + C$

b) Let $u = \ln x$ and $\frac{dv}{dx} = x^4$.
Then $\frac{du}{dx} = \frac{1}{x}$ and $v = \frac{1}{5}x^5$.
So $\int x^4 \ln x\,dx = \frac{1}{5}x^5 \ln x - \int \frac{1}{5}x^4\,dx$
$= \frac{1}{5}x^5 \ln x - \frac{1}{25}x^5 + C$

c) Let $u = \ln 4x$ and $\frac{dv}{dx} = 1$.
Then $\frac{du}{dx} = \frac{1}{x}$ and $v = x$.
So $\int \ln 4x\,dx = x \ln 4x - \int 1\,dx$
$= x \ln 4x - x + C$

d) Let $u = \ln x^3$ and $\frac{dv}{dx} = 1$.
Then $\frac{du}{dx} = \frac{3}{x}$ and $v = x$.
So $\int \ln x^3\,dx = x \ln x^3 - \int 3\,dx$
$= x \ln x^3 - 3x + C$

For parts a), c) and d), if the question hadn't told you to use integration by parts, you could have just used the result for integrating ln x shown on p.216 (you'd need to rewrite the logs in parts c) and d) as ln 4 + ln x and 3 ln x).

Q4 a) Let $u = 20x$ and $\frac{dv}{dx} = (x + 1)^3$.
Then $\frac{du}{dx} = 20$ and $v = \frac{1}{4}(x + 1)^4$.
So $\int_{-1}^{1} 20x(x + 1)^3\,dx = [5x(x+1)^4]_{-1}^{1} - \int_{-1}^{1} 5(x + 1)^4\,dx$
$= [5x(x+1)^4]_{-1}^{1} - [(x+1)^5]_{-1}^{1}$
$= [5(2^4) - 0] - [(2^5) - 0]$
$= 80 - 32 = 48$

b) Let $u = 30x$ and $\frac{dv}{dx} = (2x+1)^{\frac{1}{2}}$.
Then $\frac{du}{dx} = 30$ and $v = \frac{1}{3}(2x+1)^{\frac{3}{2}}$.
So $\int_0^{1.5} 30x\sqrt{2x+1}\,dx$
$= \left[10x(2x+1)^{\frac{3}{2}}\right]_0^{1.5} - \int_0^{1.5} 10(2x+1)^{\frac{3}{2}}\,dx$
$= \left[10x(2x+1)^{\frac{3}{2}}\right]_0^{1.5} - \left[2(2x+1)^{\frac{5}{2}}\right]_0^{1.5}$
$= \left[15(4)^{\frac{3}{2}} - 0\right] - \left[2(4)^{\frac{5}{2}} - 2(1)^{\frac{5}{2}}\right]$
$= 120 - 62 = 58$

Q5 a) Let $u = x$ and $\frac{dv}{dx} = 12e^{2x}$.
Then $\frac{du}{dx} = 1$ and $v = 6e^{2x}$.
So $\int_0^1 12xe^{2x}\,dx = [6xe^{2x}]_0^1 - \int_0^1 6e^{2x}\,dx$
$= 6e^2 - [3e^{2x}]_0^1$
$= 6e^2 - (3e^2 - 3e^0)$
$= 3e^2 + 3$

b) Let $u = x$ and $\frac{dv}{dx} = 18 \sin 3x$.
Then $\frac{du}{dx} = 1$ and $v = -6 \cos 3x$.
So $\int_0^{\frac{\pi}{3}} 18x \sin 3x\,dx = [-6x\cos 3x]_0^{\frac{\pi}{3}} - \int_0^{\frac{\pi}{3}} -6 \cos 3x\,dx$
$= -2\pi \cos \pi + [2\sin 3x]_0^{\frac{\pi}{3}}$
$= 2\pi + [2 \sin \pi - 2 \sin 0]$
$= 2\pi$

c) Let $u = \ln x$ and $\frac{dv}{dx} = \frac{1}{x^2}$.
Then $\frac{du}{dx} = \frac{1}{x}$ and $v = -\frac{1}{x}$.
So $\int_1^2 \frac{1}{x^2} \ln x\,dx = \left[-\frac{1}{x}\ln x\right]_1^2 - \int_1^2 -\frac{1}{x^2}\,dx$
$= -\frac{1}{2} \ln 2 + \ln 1 - \left[\frac{1}{x}\right]_1^2$
$= -\frac{1}{2} \ln 2 - \frac{1}{2} + 1$
$= \frac{1}{2} - \frac{1}{2} \ln 2$

Q6 Let $u = x$ and $\frac{dv}{dx} = e^{-2x}$.
Then $\frac{du}{dx} = 1$ and $v = -\frac{1}{2}e^{-2x}$.
So $\int \frac{x}{e^{2x}}\,dx = -\frac{x}{2}e^{-2x} - \int -\frac{1}{2}e^{-2x}\,dx$
$= -\frac{x}{2e^{2x}} - \frac{1}{4e^{2x}} + C$

Q7 Let $u = x + 1$ and $\frac{dv}{dx} = (x+2)^{\frac{1}{2}}$.
Then $\frac{du}{dx} = 1$ and $v = \frac{2}{3}(x+2)^{\frac{3}{2}}$.
So $\int (x+1)\sqrt{x+2}\,dx$
$= \frac{2}{3}(x+1)(x+2)^{\frac{3}{2}} - \int \frac{2}{3}(x+2)^{\frac{3}{2}}\,dx$
$= \frac{2}{3}(x+1)(x+2)^{\frac{3}{2}} - \frac{4}{15}(x+2)^{\frac{5}{2}} + C$

Q8 Let $u = \ln(x + 1)$ and $\frac{dv}{dx} = 1$.
Then $\frac{du}{dx} = \frac{1}{x+1}$ and $v = x$.
So $\int \ln(x + 1)\,dx = x \ln(x + 1) - \int \frac{x}{x+1}\,dx$
$= x \ln(x + 1) - \int \frac{x+1-1}{x+1}\,dx$
$= x \ln(x + 1) - \int 1 - \frac{1}{x+1}\,dx$
$= x \ln|x + 1| - x + \ln|x + 1| + C$
$= (x + 1) \ln|x + 1| - x + C$

Exercise 9.2 — Repeated use of integration by parts

Q1 **a)** Let $u = x^2$ and $\frac{dv}{dx} = e^x$.

Then $\frac{du}{dx} = 2x$ and $v = e^x$.

So $\int x^2e^x\,dx = x^2e^x - \int 2xe^x\,dx$

Integrate by parts again to find $\int 2xe^x\,dx$:

Let $u = 2x$ and $\frac{dv}{dx} = e^x$.

Then $\frac{du}{dx} = 2$ and $v = e^x$.

So $\int 2xe^x\,dx = 2xe^x - \int 2e^x\,dx$
$= 2xe^x - 2e^x + c$

So $\int x^2e^x\,dx = x^2e^x - \int 2xe^x\,dx$
$= x^2e^x - (2xe^x - 2e^x + c)$
$= x^2e^x - 2xe^x + 2e^x + C$

b) Let $u = x^2$ and $\frac{dv}{dx} = \cos x$.

Then $\frac{du}{dx} = 2x$ and $v = \sin x$.

So $\int x^2\cos x\,dx = x^2\sin x - \int 2x\sin x\,dx$

Integrate by parts again to find $\int 2x\sin x\,dx$:

Let $u = 2x$ and $\frac{dv}{dx} = \sin x$.

Then $\frac{du}{dx} = 2$ and $v = -\cos x$.

So $\int 2x\sin x\,dx = -2x\cos x - \int -2\cos x\,dx$
$= -2x\cos x + 2\sin x + c$

So $\int x^2\cos x\,dx = x^2\sin x - \int 2x\sin x\,dx$
$= x^2\sin x - (-2x\cos x + 2\sin x + c)$
$= x^2\sin x + 2x\cos x - 2\sin x + C$

c) Let $u = x^2$ and $\frac{dv}{dx} = 4\sin 2x$.

Then $\frac{du}{dx} = 2x$ and $v = -2\cos 2x$.

So $\int 4x^2\sin 2x\,dx = -2x^2\cos 2x + \int 4x\cos 2x\,dx$

Integrate by parts again to find $\int 4x\cos 2x\,dx$:

Let $u = x$ and $\frac{dv}{dx} = 4\cos 2x$.

Then $\frac{du}{dx} = 1$ and $v = 2\sin 2x$.

So $\int 4x\cos 2x\,dx = 2x\sin 2x - \int 2\sin 2x\,dx$
$= 2x\sin 2x + \cos 2x + C$

So $\int 4x^2\sin 2x\,dx = -2x^2\cos 2x + \int 4x\cos 2x\,dx$
$= -2x^2\cos 2x + (2x\sin 2x + \cos 2x + C)$
$= -2x^2\cos 2x + 2x\sin 2x + \cos 2x + C$

d) Let $u = 40x^2$ and $\frac{dv}{dx} = (2x - 1)^4$.

Then $\frac{du}{dx} = 80x$ and $v = \frac{1}{10}(2x - 1)^5$.

So $\int 40x^2(2x - 1)^4\,dx = 4x^2(2x - 1)^5 - \int 8x(2x - 1)^5\,dx$

Integrate by parts again to find $\int 8x(2x - 1)^5\,dx$:

Let $u = 8x$ and $\frac{dv}{dx} = (2x - 1)^5$.

Then $\frac{du}{dx} = 8$ and $v = \frac{1}{12}(2x - 1)^6$

So $\int 8x(2x - 1)^5\,dx = \frac{2x}{3}(2x - 1)^6 - \int \frac{2}{3}(2x - 1)^6\,dx$
$= \frac{2x}{3}(2x - 1)^6 - \frac{1}{21}(2x - 1)^7 + C$

So $\int 40x^2(2x - 1)^4\,dx$
$= 4x^2(2x - 1)^5 - \frac{2x}{3}(2x - 1)^6 + \frac{1}{21}(2x - 1)^7 + C$

Q2 Let $u = x^2$ and $\frac{dv}{dx} = (x + 1)^4$

Then $\frac{du}{dx} = 2x$ and $v = \frac{1}{5}(x + 1)^5$.

So $\int_{-1}^{0} x^2(x + 1)^4\,dx = \left[\frac{x^2}{5}(x+1)^5\right]_{-1}^{0} - \int_{-1}^{0} \frac{2}{5}x(x + 1)^5\,dx$
$= 0 - \int_{-1}^{0} \frac{2}{5}x(x + 1)^5\,dx$

Integrate by parts again to find $\int_{-1}^{0} \frac{2}{5}x(x + 1)^5\,dx$:

Let $u = x$ and $\frac{dv}{dx} = \frac{2}{5}(x + 1)^5$.

Then $\frac{du}{dx} = 1$ and $v = \frac{1}{15}(x + 1)^6$.

So $\int_{-1}^{0} \frac{2}{5}x(x + 1)^5\,dx = \left[\frac{x}{15}(x+1)^6\right]_{-1}^{0} - \int_{-1}^{0} \frac{1}{15}(x + 1)^6\,dx$
$= 0 - \left[\frac{1}{105}(x+1)^7\right]_{-1}^{0} = -\frac{1}{105}$

So $\int_{-1}^{0} x^2(x + 1)^4\,dx = 0 - \int_{-1}^{0} \frac{2}{5}x(x + 1)^5\,dx$
$= -\left(-\frac{1}{105}\right) = \frac{1}{105}$

Q3 Let $u = x^2$ and $\frac{dv}{dx} = e^{-2x}$.

Then $\frac{du}{dx} = 2x$ and $v = -\frac{1}{2}e^{-2x}$.

So area $= \int_0^1 x^2e^{-2x}\,dx = \left[-\frac{x^2}{2}e^{-2x}\right]_0^1 + \int_0^1 xe^{-2x}\,dx$

Integrate by parts again to find $\int_0^1 xe^{-2x}\,dx$:

Let $u = x$ and $\frac{dv}{dx} = e^{-2x}$.

Then $\frac{du}{dx} = 1$ and $v = -\frac{1}{2}e^{-2x}$.

So $\int_0^1 xe^{-2x}\,dx = \left[-\frac{x}{2}e^{-2x}\right]_0^1 + \int_0^1 \frac{1}{2}e^{-2x}\,dx$
$= \left[-\frac{x}{2}e^{-2x}\right]_0^1 + \left[-\frac{1}{4}e^{-2x}\right]_0^1$

So area $= \left[-\frac{x^2}{2}e^{-2x}\right]_0^1 + \left[-\frac{x}{2}e^{-2x}\right]_0^1 + \left[-\frac{1}{4}e^{-2x}\right]_0^1$
$= -\frac{1}{2}e^{-2} - \frac{1}{2}e^{-2} - \frac{1}{4}e^{-2} + \frac{1}{4} = \frac{1}{4} - \frac{5}{4}e^{-2}$

10. Integrating Using Partial Fractions

Exercise 10.1 — Use of partial fractions

Q1 **a)** First write the function as partial fractions. Factorise the denominator and write as an identity:

$$\frac{24(x-1)}{9-4x^2} \equiv \frac{24(x-1)}{(3-2x)(3+2x)} \equiv \frac{A}{(3-2x)} + \frac{B}{(3+2x)}$$

Add the fractions and cancel denominators:

$$\frac{24(x-1)}{(3-2x)(3+2x)} \equiv \frac{A(3+2x)+B(3-2x)}{(3-2x)(3+2x)}$$

$\Rightarrow 24(x - 1) \equiv A(3 + 2x) + B(3 - 2x)$

Substituting $x = -\frac{3}{2}$ gives: $-60 = 6B \Rightarrow B = -10$.

Substituting $x = \frac{3}{2}$ gives: $12 = 6A \Rightarrow A = 2$.

So $\frac{24(x-1)}{9-4x^2} \equiv \frac{2}{(3-2x)} - \frac{10}{(3+2x)}$.

So the integral can be expressed:

$\int \frac{24(x-1)}{9-4x^2}\,dx = \int \frac{2}{(3-2x)} - \frac{10}{(3+2x)}\,dx$
$= \frac{2}{-2}\ln|3 - 2x| - \frac{10}{2}\ln|3 + 2x| + C$
$= -\ln|3 - 2x| - 5\ln|3 + 2x| + C$

b) $\dfrac{21x-82}{(x-2)(x-3)(x-4)} \equiv \dfrac{A}{x-2} + \dfrac{B}{x-3} + \dfrac{C}{x-4}$

$\equiv \dfrac{A(x-3)(x-4)+B(x-2)(x-4)+C(x-2)(x-3)}{(x-2)(x-3)(x-4)}$

$21x - 82$

$\equiv A(x-3)(x-4) + B(x-2)(x-4) + C(x-2)(x-3)$

Substituting $x = 2$ gives: $-40 = 2A \Rightarrow A = -20$

Substituting $x = 3$ gives: $-19 = -B \Rightarrow B = 19$

Substituting $x = 4$ gives: $2 = 2C \Rightarrow C = 1$

So $\dfrac{21x-82}{(x-2)(x-3)(x-4)} \equiv -\dfrac{20}{x-2} + \dfrac{19}{x-3} + \dfrac{1}{x-4}$

$\equiv \dfrac{1}{x-4} + \dfrac{19}{x-3} - \dfrac{20}{x-2}$

So the integral can be expressed as:

$\displaystyle\int \frac{21x-82}{(x-2)(x-3)(x-4)}\,dx = \int \frac{1}{x-4} + \frac{19}{x-3} - \frac{20}{x-2}\,dx$

$= \ln|x-4| + 19\ln|x-3| - 20\ln|x-2| + C$

Q2 First write the function as partial fractions:

$\dfrac{x}{(x-2)(x-3)} \equiv \dfrac{A}{x-2} + \dfrac{B}{x-3}$

$\equiv \dfrac{A(x-3)+B(x-2)}{(x-2)(x-3)}$

$\Rightarrow x \equiv A(x-3) + B(x-2)$

Substituting $x = 3$: $B = 3$

Substituting $x = 2$: $-A = 2 \Rightarrow A = -2$

So $\dfrac{x}{(x-2)(x-3)} \equiv \dfrac{-2}{x-2} + \dfrac{3}{x-3}$

$\equiv \dfrac{3}{x-3} - \dfrac{2}{x-2}$

$\Rightarrow \displaystyle\int_0^1 \frac{x}{(x-2)(x-3)}\,dx = \int_0^1 \frac{3}{x-3} - \frac{2}{x-2}\,dx$

$= [3\ln|x-3| - 2\ln|x-2|]_0^1$

$= [3\ln|1-3| - 2\ln|1-2|] - [3\ln|0-3| - 2\ln|0-2|]$

$= 3\ln 2 - 2\ln 1 - 3\ln 3 + 2\ln 2$

$= 0 + 5\ln 2 - 3\ln 3$

$= \ln 2^5 - \ln 3^3 = \ln\dfrac{32}{27}$

Note that the modulus is important in this question, for example $|1-2| = |-1| = 1$.

Q3 **a)** Factorise the denominator, then write as an identity:

$\dfrac{6}{2x^2-5x+2} = \dfrac{6}{(2x-1)(x-2)}$

$\equiv \dfrac{A}{2x-1} + \dfrac{B}{x-2}$

$\Rightarrow 6 \equiv A(x-2) + B(2x-1)$

Substituting $x = 2$ gives $6 = 3B$ so $B = 2$.

Substituting $x = \frac{1}{2}$ gives $6 = -\frac{3}{2}A$ so $A = -4$.

So $\dfrac{6}{2x^2-5x+2} \equiv -\dfrac{4}{2x-1} + \dfrac{2}{x-2}$

$= \dfrac{2}{x-2} - \dfrac{4}{2x-1}$

b) Using part a)

$\displaystyle\int \frac{6}{2x^2-5x+2}\,dx = \int \frac{2}{x-2} - \frac{4}{2x-1}\,dx$

$= 2\ln|x-2| - \frac{4}{2}\ln|2x-1| + C$

$= 2\ln|x-2| - 2\ln|2x-1| + C$

$x > 2$ so $x - 2 > 0$ and $2x - 1 > 3$ so the modulus signs can be removed. So:

$\displaystyle\int \frac{6}{2x^2-5x+2}\,dx = 2\ln(x-2) - 2\ln(2x-1) + C$

$= 2\ln\left(\dfrac{x-2}{2x-1}\right) + C = \ln\left[\left(\dfrac{x-2}{2x-1}\right)^2\right] + C$

c) Using part b)

$\displaystyle\int_3^5 \frac{6}{2x^2-5x+2}\,dx = \left[\ln\left[\left(\frac{x-2}{2x-1}\right)^2\right]\right]_3^5$

$= \ln\left[\left(\dfrac{5-2}{10-1}\right)^2\right] - \ln\left[\left(\dfrac{3-2}{6-1}\right)^2\right]$

$= \ln\dfrac{1}{9} - \ln\dfrac{1}{25} = \ln\dfrac{25}{9}$

Q4 First express the fraction as an identity:

$\dfrac{f(x)}{g(x)} = \dfrac{3x+5}{x(x+10)} \equiv \dfrac{A}{x} + \dfrac{B}{x+10}$

$\Rightarrow 3x + 5 \equiv A(x+10) + Bx$

Now use the equating coefficients method:

Equating constant terms: $10A = 5 \Rightarrow A = \frac{1}{2}$

Equating x coefficients: $A + B = 3 \Rightarrow \frac{1}{2} + B = 3$

$\Rightarrow B = \frac{5}{2}$

So $\dfrac{3x+5}{x(x+10)} \equiv \dfrac{1}{2x} + \dfrac{5}{2(x+10)}$

Now the integration can be expressed:

$\displaystyle\int_1^2 \frac{f(x)}{g(x)}\,dx = \int_1^2 \frac{3x+5}{x(x+10)}\,dx = \int_1^2 \frac{1}{2x} + \frac{5}{2(x+10)}\,dx$

$= \left[\frac{1}{2}\ln|x| + \frac{5}{2}\ln|x+10|\right]_1^2$

$= \left[\frac{1}{2}\ln|2| + \frac{5}{2}\ln|2+10|\right] - \left[\frac{1}{2}\ln|1| + \frac{5}{2}\ln|1+10|\right]$

$= \left[\frac{1}{2}\ln 2 + \frac{5}{2}\ln 12\right] - \left[0 + \frac{5}{2}\ln 11\right]$

$= \frac{1}{2}\ln 2 + \frac{5}{2}\ln 12 - \frac{5}{2}\ln 11 = \frac{1}{2}\left(\ln 2 + 5\ln\frac{12}{11}\right)$

$= \frac{1}{2}\left(\ln 2 + \ln\left(\frac{12}{11}\right)^5\right) = 0.564$ (3 d.p.)

Q5 Begin by writing the function as partial fractions.

$\dfrac{-(t+3)}{(3t+2)(t+1)} \equiv \dfrac{A}{(3t+2)} + \dfrac{B}{(t+1)}$

$\Rightarrow -(t+3) \equiv A(t+1) + B(3t+2)$

Substituting $t = -1$ gives: $-2 = -B \Rightarrow B = 2$

Equating coefficients of t gives

$A + 3B = -1 \Rightarrow A + 6 = -1 \Rightarrow A = -7$

So $\dfrac{-(t+3)}{(3t+2)(t+1)} \equiv \dfrac{2}{(t+1)} - \dfrac{7}{(3t+2)}$

The integral can be expressed:

$\displaystyle\int_0^{\frac{2}{3}} \frac{-(t+3)}{(3t+2)(t+1)}\,dt = \int_0^{\frac{2}{3}} \frac{2}{(t+1)} - \frac{7}{(3t+2)}\,dt$

$= \left[2\ln|t+1| - \frac{7}{3}\ln|3t+2|\right]_0^{\frac{2}{3}}$

$= \left[2\ln\left|\frac{5}{3}\right| - \frac{7}{3}\ln|4|\right] - \left[2\ln|1| - \frac{7}{3}\ln|2|\right]$

$= \left[2\ln\left(\frac{5}{3}\right) - \frac{7}{3}\ln(4)\right] - \left[0 - \frac{7}{3}\ln(2)\right]$

$= 2\ln\left(\frac{5}{3}\right) - \frac{7}{3}\ln(4) + \frac{7}{3}\ln(2)$

$= 2\ln\left(\frac{5}{3}\right) - \frac{7}{3}(\ln(4) - \ln(2))$

$= 2\ln\left(\frac{5}{3}\right) - \frac{7}{3}\ln\left(\frac{4}{2}\right) = 2\ln\frac{5}{3} - \frac{7}{3}\ln 2$

Q6 $u = \sqrt{x} \Rightarrow \dfrac{du}{dx} = \dfrac{1}{2\sqrt{x}} \Rightarrow dx = 2\sqrt{x}\,du = 2u\,du$

and $u = \sqrt{x} \Rightarrow x = u^2$.

Using $u = \sqrt{x}$, $x = 9 \Rightarrow u = \sqrt{9} = 3$

and $x = 16 \Rightarrow u = \sqrt{16} = 4$

So $\displaystyle\int_9^{16} \frac{4}{\sqrt{x}(9x-4)}\,dx = \int_3^4 \frac{4}{u(9u^2-4)}\,2u\,dx$

$= \displaystyle\int_3^4 \frac{8}{(9u^2-4)}\,dx$

This function still needs some work before it can be integrated. The next step is to write it as partial fractions.

$$\frac{8}{(9u^2-4)} \equiv \frac{8}{(3u+2)(3u-2)} \equiv \frac{A}{(3u+2)} + \frac{B}{(3u-2)}$$

So $8 = A(3u - 2) + B(3u + 2)$

Using substitution:

$$u = \frac{2}{3} \Rightarrow 8 = 4B \Rightarrow 2 = B$$

$$u = -\frac{2}{3} \Rightarrow 8 = -4A \Rightarrow -2 = A$$

So $\frac{8}{(9u^2-4)} \equiv \frac{-2}{(3u+2)} + \frac{2}{(3u-2)} \equiv \frac{2}{(3u-2)} - \frac{2}{(3u+2)}$

So the integral can be expressed:

$$\int_3^4 \frac{8}{(9u^2-4)}\,dx = \int_3^4 \frac{2}{(3u-2)} - \frac{2}{(3u+2)}\,dx$$

$$= \left[\frac{2}{3}\ln|3u-2| - \frac{2}{3}\ln|3u+2|\right]_3^4$$

$$= \left[\frac{2}{3}\ln\left|\frac{3u-2}{3u+2}\right|\right]_3^4$$

$$= \frac{2}{3}\ln\frac{10}{14} - \frac{2}{3}\ln\frac{7}{11} = \frac{2}{3}\ln\frac{55}{49}$$

11. Differential Equations

Exercise 11.1 — Differential equations

Q1 The rate of change of N with respect to t is $\frac{dN}{dt}$.
So $\frac{dN}{dt} \propto N \Rightarrow \frac{dN}{dt} = kN$, for some $k > 0$.

Q2 The rate of change of x with respect to t is $\frac{dx}{dt}$.
So $\frac{dx}{dt} \propto \frac{1}{x^2} \Rightarrow \frac{dx}{dt} = \frac{k}{x^2}$, for some $k > 0$.

Q3 Let the variable t represent time.
Then the rate of change of A with respect to t is $\frac{dA}{dt}$.
So $\frac{dA}{dt} \propto \sqrt{A} \Rightarrow \frac{dA}{dt} = -k\sqrt{A}$, for some $k > 0$.
Don't forget to include a minus sign when the situation involves a rate of decrease.

Q4 Let the variable t represent time.
Then the rate of change of y with respect to t is $\frac{dy}{dt}$.
So $\frac{dy}{dt} \propto (y - \lambda) \Rightarrow \frac{dy}{dt} = -k(y - \lambda)$, for some $k > 0$.

Q5 Let the variable t represent time. V is the volume in the container and it is equal to $V_{in} - V_{out}$.
Then the rate of change of V with respect to t is $\frac{dV}{dt} = \frac{dV_{in}}{dt} - \frac{dV_{out}}{dt}$. $\frac{dV_{in}}{dt}$ is directly proportional to V, so $\frac{dV_{in}}{dt} = kV$ for some constant k, $k > 0$ and $\frac{dV_{out}}{dt} = 20$.
So the overall rate of change of V is $\frac{dV}{dt} = kV - 20$, for some $k > 0$.

Exercise 11.2 — Solving differential equations

Q1 **a)** $\frac{dy}{dx} = 8x^3 \Rightarrow dy = 8x^3\,dx$
$\Rightarrow \int 1\,dy = \int 8x^3\,dx$
$\Rightarrow y = 2x^4 + C$

b) $\frac{dy}{dx} = 5y \Rightarrow \frac{1}{y}\,dy = 5\,dx$
$\Rightarrow \int \frac{1}{y}\,dy = \int 5\,dx$
$\Rightarrow \ln|y| = 5x + \ln k$
$\Rightarrow y = e^{5x + \ln k} = ke^{5x}$

c) $\frac{dy}{dx} = 6x^2y \Rightarrow \frac{1}{y}\,dy = 6x^2dx$
$\Rightarrow \int \frac{1}{y}\,dy = \int 6x^2\,dx$
$\Rightarrow \ln|y| = 2x^3 + \ln k$
$\Rightarrow y = e^{2x^3 + \ln k} = ke^{2x^3}$

d) $\frac{dy}{dx} = \frac{y}{x} \Rightarrow \frac{1}{y}\,dy = \frac{1}{x}\,dx$
$\Rightarrow \int \frac{1}{y}\,dy = \int \frac{1}{x}\,dx$
$\Rightarrow \ln|y| = \ln|x| + \ln k = \ln|kx|$
$\Rightarrow y = kx$

e) $\frac{dy}{dx} = (y + 1)\cos x \Rightarrow \frac{1}{y+1}\,dy = \cos x\,dx$
$\Rightarrow \int \frac{1}{y+1}\,dy = \int \cos x\,dx$
$\Rightarrow \ln|y + 1| = \sin x + \ln k$
$\Rightarrow y + 1 = e^{\sin x + \ln k}$
$\Rightarrow y = ke^{\sin x} - 1$

f) $\frac{dy}{dx} = \frac{3xy - 6y}{(x-4)(2x-5)} = \frac{(3x-6)y}{(x-4)(2x-5)}$
$\Rightarrow \frac{1}{y}\,dy = \frac{(3x-6)}{(x-4)(2x-5)}\,dx$
$\Rightarrow \int \frac{1}{y}\,dy = \int \frac{(3x-6)}{(x-4)(2x-5)}\,dx$

The integration on the right hand side needs to be split into partial fractions before you can integrate.

$\frac{(3x-6)}{(x-4)(2x-5)} \equiv \frac{A}{(x-4)} + \frac{B}{(2x-5)}$
$\Rightarrow (3x - 6) \equiv A(2x - 5) + B(x - 4)$

Substitution: $x = 4$: $6 = 3A \Rightarrow A = 2$
$x = \frac{5}{2}$: $\frac{3}{2} = -\frac{3}{2}B \Rightarrow B = -1$

So $\int \frac{(3x-6)}{(x-4)(2x-5)}\,dx \equiv \int \frac{2}{(x-4)} - \frac{1}{(2x-5)}\,dx$

So $\int \frac{1}{y}\,dy = \int \frac{2}{(x-4)} - \frac{1}{(2x-5)}\,dx$
$\Rightarrow \ln|y| = 2\ln|x - 4| - \frac{1}{2}\ln|2x - 5| + \ln k$
$\Rightarrow \ln|y| = \ln|(x - 4)^2| - \ln|\sqrt{2x-5}| + \ln k$
$\Rightarrow \ln|y| = \ln\left|\frac{k(x-4)^2}{\sqrt{2x-5}}\right|$
$\Rightarrow y = \frac{k(x-4)^2}{\sqrt{2x-5}}$

Q2 **a)** $\frac{dy}{dx} = -\frac{x}{y} \Rightarrow y\,dy = -x\,dx$
$\Rightarrow \int y\,dy = \int -x\,dx$
$\Rightarrow \frac{1}{2}y^2 = -\frac{1}{2}x^2 + c \Rightarrow y^2 = -x^2 + C$

So when $x = 0$ and $y = 2$, $C = 4 \Rightarrow y^2 + x^2 = 4$

b) $\frac{dx}{dt} = \frac{2}{\sqrt{x}} \Rightarrow \sqrt{x}\,dx = 2\,dt$
$\Rightarrow \int x^{\frac{1}{2}}\,dx = \int 2\,dt$
$\Rightarrow \frac{2}{3}x^{\frac{3}{2}} = 2t + C$

So when $t = 5$ and $x = 9$,
$\frac{2}{3}(27) = 10 + C \Rightarrow 18 = 10 + C \Rightarrow C = 8$
$\Rightarrow \frac{2}{3}x^{\frac{3}{2}} = 2t + 8$
$\Rightarrow x^{\frac{3}{2}} = 3t + 12$
$\Rightarrow x^3 = (3t + 12)^2$

You could have left out the last couple of steps here, as the question didn't specify the form of the answer.

c) $\frac{dV}{dt} = 3(V-1) \Rightarrow \frac{1}{V-1}\,dV = 3\,dt$

$\Rightarrow \int \frac{1}{V-1}\,dV = \int 3\,dt$

$\Rightarrow \ln|V-1| = 3t + \ln k$

$\Rightarrow V = ke^{3t} + 1$

So when $t = 0$ and $V = 5$,

$5 = k + 1 \quad \Rightarrow k = 4$

$\Rightarrow V = 4e^{3t} + 1$

d) $\frac{dy}{dx} = \frac{\tan y}{x} \Rightarrow \frac{1}{\tan y}\,dy = \frac{1}{x}\,dx$

$\Rightarrow \int \cot y\,dy = \int \frac{1}{x}\,dx$

$\Rightarrow \ln|\sin y| = \ln|x| + \ln k$

$\Rightarrow \sin y = kx$

So when $x = 2$ and $y = \frac{\pi}{2}$,

$1 = 2k \Rightarrow k = \frac{1}{2} \Rightarrow \sin y = \frac{x}{2}$

e) $\frac{dx}{dt} = 10x(x+1) \Rightarrow \frac{1}{x(x+1)}\,dx = 10\,dt$

Using partial fractions, $\frac{1}{x(x+1)} \equiv \frac{1}{x} - \frac{1}{x+1}$,

so $\int \frac{1}{x} - \frac{1}{x+1}\,dx = \int 10\,dt$

$\Rightarrow \ln|x| - \ln|x+1| = 10t + \ln k$

$\Rightarrow \ln\left|\frac{x}{x+1}\right| = 10t + \ln k$

$\Rightarrow \frac{x}{x+1} = ke^{10t}$

So when $t = 0$ and $x = 1$, $\frac{1}{2} = k$

$\Rightarrow \frac{x}{x+1} = \frac{1}{2}e^{10t}$

Q3 a) $\frac{dV}{dt} = a - bV \Rightarrow \frac{1}{a-bV}\,dV = dt$

$\Rightarrow \int \frac{1}{a-bV}\,dV = \int 1\,dt$

$\Rightarrow -\frac{1}{b}\ln|a-bV| = t + C$

$\Rightarrow \ln|a-bV| = -bt - bC$

b and C are both constants, so let $-bC = \ln k$:

$\Rightarrow \ln|a-bV| = -bt + \ln k$

$\Rightarrow a - bV = ke^{-bt}$

$\Rightarrow bV = a - ke^{-bt}$

$\Rightarrow V = \frac{a}{b} - Ae^{-bt}$ (letting $A = k \div b$)

b) When $t = 0$ and $V = \frac{a}{4b}$,

$\frac{a}{4b} = \frac{a}{b} - A \Rightarrow A = \frac{a}{b} - \frac{a}{4b} = \frac{4a-a}{4b} = \frac{3a}{4b}$

c) As t gets very large, e^{-bt} gets very close to zero, so V approaches $\frac{a}{b}$.

Exercise 11.3 — Applying differential equations to real-life problems

Q1 a) $\frac{dN}{dt} = kN \Rightarrow \frac{1}{N}\,dN = k\,dt$

$\Rightarrow \int \frac{1}{N}\,dN = \int k\,dt$

$\Rightarrow \ln N = kt + \ln A$

$\Rightarrow N = e^{kt + \ln A} = Ae^{kt}$

Note that you don't need to put modulus signs in ln N here, as N can't be negative — you can't have a negative number of germs in your body. The same principle will apply to a lot of real-life differential equations questions.

b) $t = 0, N = 200 \Rightarrow 200 = Ae^0 = A$

$\Rightarrow N = 200e^{kt}$

$t = 8, N = 400 \Rightarrow 400 = 200e^{8k}$

$\Rightarrow \ln 2 = 8k$

$\Rightarrow k = \frac{1}{8}\ln 2$

$\Rightarrow N = 200e^{\frac{t}{8}\ln 2}$

So $t = 24 \Rightarrow N = 200e^{3\ln 2} = 1600$

c) Some possible answers are:

- The number of germs doubles every 8 hours, so over a long time the model becomes unrealistic because the total number of germs will become very large.
- The number of germs is a discrete variable (you can't have half of a germ) but the model is a continuous function.
- The model does not account for the differences between patients or other conditions such as the presence of other germs or chemicals.

Q2 a) $\frac{dV}{dt} \propto V \Rightarrow \frac{dV}{dt} = -kV$, for some $k > 0$

$\Rightarrow \frac{1}{V}\,dV = -k\,dt$

$\Rightarrow \int \frac{1}{V}\,dV = \int -k\,dt$

$\Rightarrow \ln V = -kt + \ln A$

$\Rightarrow V = Ae^{-kt}$

$t = 0, V = V_0 \Rightarrow V_0 = Ae^0 = A$

$\Rightarrow V = V_0e^{-kt}$

b) Using years as the unit of time:

$t = 1, V = \frac{1}{2}V_0 \Rightarrow \frac{1}{2}V_0 = V_0e^{-k}$

$\Rightarrow \frac{1}{2} = e^{-k}$

$\Rightarrow \ln\frac{1}{2} = -k$

$\Rightarrow k = \ln 2 \Rightarrow V = V_0e^{-t\ln 2}$

So $V = 0.05V_0 \Rightarrow 0.05V_0 = V_0e^{-t\ln 2}$

$\Rightarrow 0.05 = e^{-t\ln 2}$

$\Rightarrow \ln 0.05 = -t\ln 2$

$\Rightarrow t = \ln 0.05 \div -\ln 2 = 4.322$ years

$\Rightarrow t = 4$ years, 4 months (or 52 months)

You could have used months as the units of time instead, and started with t = 12. You'd get the same answer.

Q3 a) $\frac{dN}{dt} \propto N \Rightarrow \frac{dN}{dt} = kN$

b) $\frac{dN}{dt} = kN \Rightarrow \int \frac{1}{N}\,dN = \int k\,dt$

$\Rightarrow \ln N = kt + \ln A$

$\Rightarrow N = e^{kt + \ln A} = Ae^{kt}$

$N = 20$ at $t = 0 \Rightarrow 20 = Ae^0 = A \Rightarrow N = 20e^{kt}$

$N = 30$ at $t = 4 \Rightarrow 30 = 20e^{4k}$

$\Rightarrow k = 0.25\ln 1.5$

$\Rightarrow N = 20e^{0.25t\ln 1.5}$

So $N = 1000 \Rightarrow 1000 = 20e^{0.25t\ln 1.5}$

$\Rightarrow \ln 50 = 0.25t\ln 1.5$

$\Rightarrow t = 4\ln 50 \div \ln 1.5 = 38.59$

So the field will be over-run in 39 weeks.

c) $\frac{dN}{dt} \propto \sqrt{N} \Rightarrow \frac{dN}{dt} = k\sqrt{N}$

$\Rightarrow \int \frac{1}{\sqrt{N}}\, dN = \int k\, dt$

$\Rightarrow 2\sqrt{N} = kt + C$

$N = 20$ at $t = 0 \Rightarrow 2\sqrt{20} = 4\sqrt{5} = C$

$\Rightarrow 2\sqrt{N} = kt + 4\sqrt{5}$

$N = 30$ at $t = 4 \Rightarrow 2\sqrt{30} = 4k + 4\sqrt{5}$

$\Rightarrow k = \frac{\sqrt{30} - 2\sqrt{5}}{2}$

$\Rightarrow 2\sqrt{N} = \frac{\sqrt{30} - 2\sqrt{5}}{2}t + 4\sqrt{5}$

So $N = 1000 \Rightarrow 2\sqrt{1000} = \frac{\sqrt{30} - 2\sqrt{5}}{2}t + 4\sqrt{5}$

$\Rightarrow t = \frac{4\sqrt{1000} - 8\sqrt{5}}{\sqrt{30} - 2\sqrt{5}} = 108.05$

So the field will be over-run in 108 weeks.

Be careful with all these square roots knocking about — it's easy to make a mistake.

d) Some possible answers are:

- The model could be adjusted to use a discrete-valued function, since the number of mice is a discrete variable.
- When there is only 1 mouse (i.e. $N = 1$), the population still increases, which is unrealistic. The model could be adjusted to be more accurate for very low values of N.
- As t increases, N gets larger without any limit. Introducing an upper limit on t or N would show at which point the model stops being accurate.

Q4 a) $\frac{dx}{dt} = \frac{1}{x^2(t+1)}$

$V = x^3 \Rightarrow \frac{dV}{dx} = 3x^2$

So $\frac{dV}{dt} = \frac{dV}{dx} \times \frac{dx}{dt} = \frac{3x^2}{x^2(t+1)} = \frac{3}{t+1}$

b) $\frac{dV}{dt} = \frac{3}{t+1} \Rightarrow \int 1\, dV = \int \frac{3}{t+1}\, dt$

$\Rightarrow V = 3 \ln (t + 1) + C$

$V = 15$ at $t = 0 \Rightarrow 15 = 3 \ln (1) + C \Rightarrow C = 15$

$\Rightarrow V = 3 \ln (t + 1) + 15$

So $V = 18 \Rightarrow 18 = 3 \ln (t + 1) + 15$

$\Rightarrow \frac{3}{3} = \ln (t + 1)$

$\Rightarrow t = e^1 - 1 = 1.72$ seconds (3 s.f.)

Q5 a) $\frac{dy}{dt} = k(p - y) \Rightarrow \int \frac{1}{p-y}\, dy = \int k\, dt$

$\Rightarrow -\ln (p - y) = kt + \ln a$

$\Rightarrow \ln (p - y) = -kt - \ln a$

$\Rightarrow p - y = e^{-kt - \ln a} = e^{-kt}e^{-\ln a}$

$e^{-kt}e^{\ln\frac{1}{a}} = \frac{1}{a}e^{-kt} = Ae^{-kt}$

$\Rightarrow y = p - Ae^{-kt}$

b) If $p = 30\,000$ and $y = 10\,000$ at $t = 0$, then

$10\,000 = 30\,000 - Ae^0 = 30\,000 - A$

$\Rightarrow A = 20\,000$

$\Rightarrow y = 30\,000 - 20\,000e^{-kt}$

$t = 5$, $y = 12\,000$

$\Rightarrow 12\,000 = 30\,000 - 20\,000e^{-5k}$

$\Rightarrow e^{-5k} = 18\,000 \div 20\,000 = 0.9$

$\Rightarrow -5k = \ln 0.9$

$\Rightarrow k = -0.2 \ln 0.9$

$\Rightarrow y = 30\,000 - 20\,000e^{0.2t \ln 0.9}$

So $y = 25\,000 \Rightarrow 20\,000e^{0.2t \ln 0.9} = 5000$

$\Rightarrow e^{0.2t \ln 0.9} = 0.25$

$\Rightarrow 0.2t \ln 0.9 = \ln 0.25$

$\Rightarrow t = 5 \ln 0.25 \div \ln 0.9$

$= 65.79 = 66$ days

c)

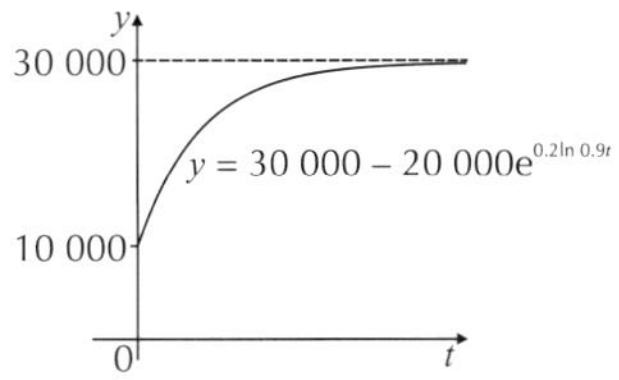

Remember that (0.2 ln 0.9) is negative when sketching the graph.

d) $t = 92 \Rightarrow y = 30\,000 - 20\,000e^{18.4 \ln 0.9}$

$= 30\,000 - 20\,000e^{-1.939}$

$= 30\,000 - 20\,000(0.1439)$

$= 27\,122$ signatures

So no, the target will not be achieved.

Don't forget, you'll often need to relate your answer back to the question when you've finished calculating.

e) Some possible answers are:

- The model suggests that the number of signatures eventually reaches the entire population of 30 000 (if you round to the nearest whole number). It may be more accurate to assume that there are some people who will never sign the petition, no matter how much time elapses, and allow for this in the model.
- The number of signatures is a discrete variable, and the model uses a continuous function. Adjusting the function to only allow integer values of y would fix this.
- The model does not account for people entering or leaving the town. Adjusting the function so that p varies with time may make the model more accurate.

Chapter 9: Numerical Methods

1. Location of Roots

Exercise 1.1 — Locating roots by changes of sign

Q1 $f(2) = 2^3 - 5 \times 2 + 1 = -1$

$f(3) = 3^3 - 5 \times 3 + 1 = 13$

There is a sign change (and the function is continuous in this interval) so there is a root in this interval.

They wouldn't ask you if there's a root if the function wasn't continuous in this interval, but it's worth saying anyway just to keep your answer 'strictly true'.

Q2 $f(0.9) = \sin (1.8) - 0.9 = 0.0738...$

$f(1.0) = \sin (2.0) - 1.0 = -0.0907...$

There is a sign change (and the function is continuous in this interval) so there is a root in this interval.

Q3 $f(1.2) = 1.2^3 + \ln 1.2 - 2 = -0.089...$

$f(1.3) = 1.3^3 + \ln 1.3 - 2 = 0.459...$

There is a sign change (and the function is continuous in this interval) so there is a root in this interval.

Q4 f(1.2) = (2 × 1.2²) – (8 × 1.2) + 7 = 0.28
f(1.3) = (2 × 1.3²) – (8 × 1.3) + 7 = –0.02
There is a sign change (and the function is continuous in this interval) so there is a root in this interval
— i.e. the bird hits water between 1.2 and 1.3 seconds.

Q5 f(1.6) = (3 × 1.6) – 1.6^4 + 3 = 1.24...
f(1.7) = (3 × 1.7) – 1.7^4 + 3 = –0.25...
There is a sign change (and the function is continuous in this interval) so there is a root in this interval.
f(–1) = (3 × (–1)) – $(-1)^4$ + 3 = –1
f(0) = (3 × 0) – 0^4 + 3 = 3
There is a sign change (and the function is continuous in this interval) so there is a root in this interval.

Q6 f(0.01) = $e^{0.01-2} - \sqrt{0.01}$ = 0.0366...
f(0.02) = $e^{0.02-2} - \sqrt{0.02}$ = –0.0033...
There is a sign change (and the function is continuous in this interval) so there is a root in this interval.
f(2.4) = $e^{2.4-2} - \sqrt{2.4}$ = –0.057...
f(2.5) = $e^{2.5-2} - \sqrt{2.5}$ = 0.067...
There is a sign change (and the function is continuous in this interval) so there is a root in this interval.

Q7 The upper and lower bounds are 2.75 and 2.85.
f(2.75) = 2.75^3 – (7 × 2.75) – 2 = –0.45...
f(2.85) = 2.85^3 – (7 × 2.85) – 2 = 1.19...
There is a sign change between the upper and lower bounds (and the function is continuous in this interval), so a solution to 1 d.p. is x = 2.8.

Q8 Upper and lower bounds are 0.65 and 0.75
f(0.65) = $(2 \times 0.65) - \frac{1}{0.65}$ = –0.23...
f(0.75) = $(2 \times 0.75) - \frac{1}{0.75}$ = 0.16...
There is a sign change between the upper and lower bounds (and the function is continuous in this interval), so a solution to 1 d.p. is x = 0.7.

Q9 f(0.245) = $e^{0.245}$ – 0.245^3 – (5 × 0.245) = 0.037...
f(0.255) = $e^{0.255}$ – 0.255^3 – (5 × 0.255) = –0.001...
There is a sign change between the upper and lower bounds (and the function is continuous in this interval), so a solution to 2 d.p. is x = 0.25.

Q10 Rearrange the equation to get f(x) = $4x - 2x^3 - 15 = 0$
f(–2.3) = (4 × –2.3) – (2 × $(-2.3)^3$) – 15 = 0.134
f(–2.2) = (4 × –2.2) – (2 × $(-2.2)^3$) – 15 = –2.504
There is a sign change (and the function is continuous in this interval) so there is a root in this interval.

Q11 Rearrange the equation to get f(x) = ln(x + 3) – 5x = 0
f(0.23) = ln(0.23 + 3) – (5 × 0.23) = 0.022...
f(0.24) = ln(0.24 + 3) – (5 × 0.24) = –0.024...
There is a sign change (and the function is continuous in this interval) so there is a root in this interval.

Q12 Rearrange the equation to get f(x) = $e^{3x}\sin x - 5 = 0$
f(0) = $e^{3 \times 0}$ sin 0 – 5 = – 5
f(1) = $e^{3 \times 1}$ sin 1 – 5 = 11.9...
There is a sign change (and the function is continuous in this interval) so there is a root in this interval.

In Q10-12, it's not actually strictly necessary to rearrange to f(x) = 0. It's enough to show that the left-hand side of the original equation is greater than the right-hand side for one of the values of x (x_1 or x_2), and less than the right-hand side for the other.

Exercise 1.2 — Sketching graphs to find approximate roots

Q1 a)

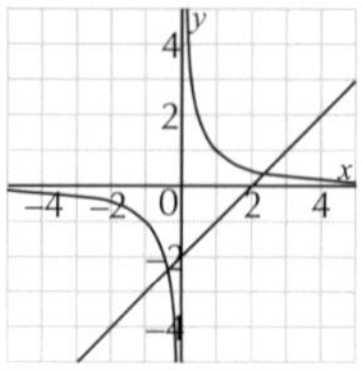

b) The graphs cross twice, so the equation has 2 roots.

c) Rearranging, f(x) = $x - \frac{1}{x} - 2$, and the roots are at f(x) = 0

f(2.4) = $2.4 - \frac{1}{2.4} - 2$ = –0.016...

f(2.5) = $2.5 - \frac{1}{2.5} - 2$ = 0.1

There is a sign change (and the function is continuous in this interval) so there is a root in this interval.

Q2 a)

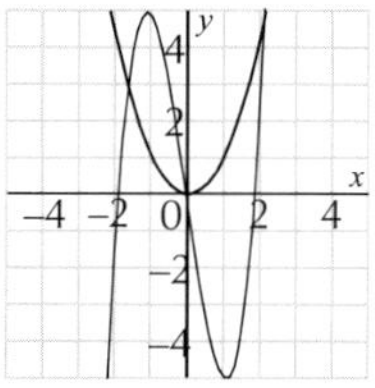

b) The graphs cross 3 times, so the equation has 3 roots.

c) f(–2) = (2 × $(-2)^3$) – $(-2)^2$ – (7 × –2) = –6
f(–1) = (2 × $(-1)^3$) – $(-1)^2$ – (7 × –1) = 4
There is a sign change (and the function is continuous in this interval) so there is a root in this interval.

Q3 a)

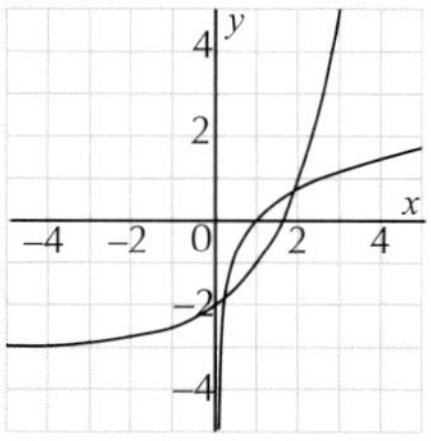

b) The graphs cross twice so the equation has 2 roots.

c) f(1.8) = ln (1.8) – $2^{1.8}$ + 3 = 0.105...
f(2.2) = ln (2.2) – $2^{2.2}$ + 3 = –0.806...
There is a sign change (and the function is continuous in this interval), so there is a root in this interval.
f(2.0) = ln (2.0) – $2^{2.0}$ + 3 = –0.306...
So the root is between 1.8 and 2.0
f(1.9) = ln (1.9) – $2^{1.9}$ + 3 = –0.090...
So the root is between 1.8 and 1.9
f(1.85) = ln (1.85) – $2^{1.85}$ + 3 = 0.010...
There is a sign change (and the function is continuous) between 1.85 and 1.9, so the root is at x = 1.9 (to 1 d.p.).

Q4 a)

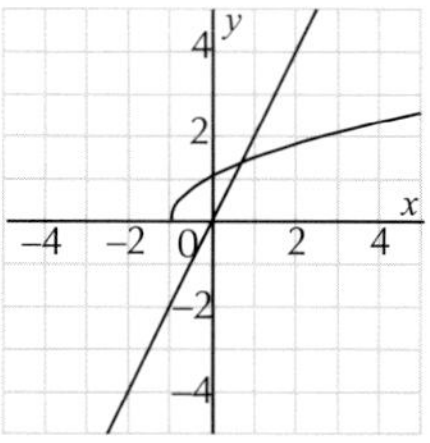

b) The graphs cross once so the equation has 1 root.

c) The roots are at $f(x) = \sqrt{x+1} - 2x = 0$
$f(0.6) = \sqrt{0.6+1} - (2 \times 0.6) = 0.064...$
$f(0.7) = \sqrt{0.7+1} - (2 \times 0.7) = -0.096...$
There is a sign change (and the function is continuous in this interval), so there is a root in this interval.

d) $\sqrt{x+1} = 2x \Rightarrow x + 1 = 4x^2 \Rightarrow 4x^2 - x - 1 = 0$
From quadratic formula root is $x = 0.640$ to 3 s.f.
You can ignore the other solution to this quadratic equation — you want the root between 0.6 and 0.7.

Q5 a)

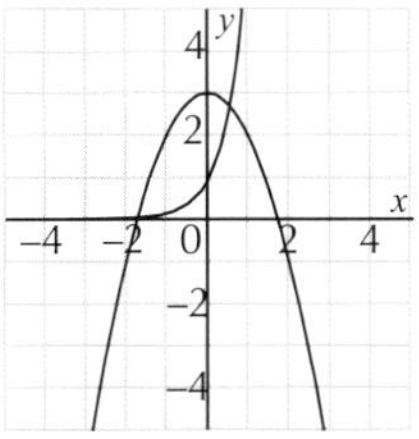

b) If the two functions are set equal to each other they can be rearranged to make $e^{2x} + x^2 = 3$. The graphs cross twice, so the equation has 2 roots.

c) Rearrange to $f(x) = 0$ so $e^{2x} + x^2 - 3 = 0$
$f(-2) = e^{(2 \times -2)} + (-2)^2 - 3 = 1.01...$
$f(-1) = e^{(2 \times -1)} + (-1)^2 - 3 = -1.86...$
There is a sign change (and the function is continuous in this interval) so there is a root in this interval.
$f(-1.5) = e^{(2 \times -1.5)} + (-1.5)^2 - 3 = -0.70...$
So the root is between –1.5 and –2
$f(-1.7) = e^{(2 \times -1.7)} + (-1.7)^2 - 3 = -0.07...$
So the root is between –1.7 and –2
$f(-1.8) = e^{(2 \times -1.8)} + (-1.8)^2 - 3 = 0.26...$
So the root is between –1.7 and –1.8
$f(-1.75) = e^{(2 \times -1.75)} + (-1.75)^2 - 3 = 0.09...$
So the root is between –1.7 and –1.75, so the value to 1 decimal place is $x = -1.7$.

2. Iterative Methods

Exercise 2.1 — Using iteration formulas

Q1 a) $f(1) = 1^3 + 3 \times 1^2 - 7 = -3$
$f(2) = 2^3 + 3 \times 2^2 - 7 = 13$
There is a sign change (and the function is continuous in this interval) so there is a root in this interval.

b) $x_1 = \sqrt{\frac{7 - x_0^3}{3}} = \sqrt{\frac{7 - 1^3}{3}} = 1.414,$
$x_2 = 1.179,\ x_3 = 1.337,\ x_4 = 1.240$

Q2 a) $x_1 = 2 + \ln x_0 = 2 + \ln 3.1 = 3.1314,$
$x_2 = 3.1415,\ x_3 - 3.1447,\ x_4 = 3.1457,$
$x_5 = 3.1460$

b) $\alpha = 3.146$

Q3 a) $f(1.4) = 1.4^4 - (5 \times 1.4) + 3 = -0.1584$
$f(1.5) = 1.5^4 - (5 \times 1.5) + 3 = 0.5625$
There is a sign change (and the function is continuous in this interval) so there is a root in this interval.

b) $x_1 = \sqrt[3]{5 - \frac{3}{x_0}} = \sqrt{5 - \frac{3}{1.4}} = 1.419,$
$x_2 = 1.424,\ x_3 = 1.425,\ x_4 = 1.425,$
$x_5 = 1.425,\ x_6 = 1.425$

c) The last 4 iterations round to the same answer, so to 2 d.p. $x = 1.43$

Q4 a) $f(5) = 5^2 - (5 \times 5) - 2 = -2$
$f(6) = 6^2 - (5 \times 6) - 2 = 4$
There is a sign change (and the function is continuous in this interval) so there is a root in this interval.

b) $x_1 = \frac{2}{x_0} + 5 = \frac{2}{5} + 5 = 5.4,$
$x_2 = 5.370,\ x_3 = 5.372,\ x_4 = 5.372$

Q5 $x_1 = 2 - \ln x_0 = 2 - \ln 1.5 = 1.595,$
$x_2 = 1.533,\ x_3 = 1.573,\ x_4 = 1.547,\ x_5 = 1.563,$
$x_6 = 1.553,\ x_7 = 1.560,\ x_8 = 1.555,\ x_9 = 1.558$ (all 4 s.f.)
The iterative sequence is bouncing up and down but closing in on the correct answer.
The last three iterations round to the same answer to 2 d.p., so to 2 d.p. $x = 1.56$

Q6 a) $f(3) = e^3 - (10 \times 3) = -9.914...$
$f(4) = e^4 - (10 \times 4) = 14.598...$
There is a sign change (and the function is continuous in this interval) so there is a root in this interval.

b) Using starting value $x_0 = 3$:
$x_1 = \ln(10x_0) = \ln(10 \times 3) = 3.401$
$x_2 = 3.527,\ x_3 = 3.563,\ x_4 = 3.573$
You could have started with x_0 as anything between 3 and 4, in which case you'd get different values for $x_1 - x_4$

c) The upper and lower bounds are 3.5765 and 3.5775.
$f(3.5765) = e^{3.5765} - (10 \times 3.5765) = -0.016...$
$f(3.5775) = e^{3.5775} - (10 \times 3.5775) = 0.0089...$
There is a sign change between the upper and lower bounds (and the function is continuous in this interval), so the root to 3 d.p. is $x = 3.577$.

d) Using the new iterative formula with $x_0 = 3$
$x_1 = 2.00855...,\ x_2 = 0.74525..,\ x_3 = 0.21069...,$
$x_4 = 0.12345...,\ x_5 = 0.11313...,\ x_6 = 0.11197...,$
$x_7 = 0.11184...,\ x_8 = 0.11183...$
The formula appears to converge to another root at $x = 0.112$ (3 d.p.)

Q7 a) $x_1 = \frac{x_0^2 - 3x_0}{2} - 5 = \frac{(-1)^2 - (3 \times (-1))}{2} - 5 = -3$
$x_2 = 4,\ x_3 = -3,\ x_4 = 4$
The sequence is alternating between –3 and 4.

b) Using the iterative formula given:
$x_1 = 6.32455...,\ x_2 = 6.45157...,\ x_3 = 6.50060...,$
$x_4 = 6.51943...,\ x_5 = 6.52665...,\ x_6 = 6.52941...,$
$x_7 = 6.53047...,\ x_8 = 6.53087...$
The last 4 iterations all round to 6.53, so the value of the root is $x = 6.53$ to 3 s.f. To verify this, check it's between the upper and lower bounds of 6.53:
$f(6.525) = 6.525^2 - (5 \times 6.525) - 10 = -0.049...$
$f(6.535) = 6.535^2 - (5 \times 6.535) - 10 = 0.031...$
There is a sign change between the upper and lower bounds (and the function is continuous in this interval) so this value is correct to 3 s.f.

Exercise 2.2 — Finding iteration formulas

Q1 **a)** $x^4 + 7x - 3 = 0 \Rightarrow x^4 = 3 - 7x \Rightarrow x = \sqrt[4]{3-7x}$

b) $x^4 + 7x - 3 = 0 \Rightarrow x^4 + 5x + 2x - 3 = 0$

$\Rightarrow 2x = 3 - 5x - x^4 \Rightarrow x = \frac{3-5x-x^4}{2}$

c) $x^4 + 7x - 3 = 0 \Rightarrow x^4 = 3 - 7x$

$\Rightarrow x^2 = \sqrt{3-7x} \Rightarrow x = \frac{\sqrt{3-7x}}{x}$

Q2 **a)** $x^3 - 2x^2 - 5 = 0 \Rightarrow x^3 = 2x^2 + 5 \Rightarrow x = 2 + \frac{5}{x^2}$

b) $x_1 = 2 + \frac{5}{2^2} = 3.25$

$x_2 = 2.473...,\ x_3 = 2.817...,$

$x_4 = 2.629...,\ x_5 = 2.722...,$

So $x_5 = 2.7$ to 1 d.p.

c) $f(2.65) = 2.65^3 - (2 \times 2.65^2) - 5 = -0.43...$

$f(2.75) = 2.75^3 - (2 \times 2.75^2) - 5 = 0.67...$

There is a sign change between the upper and lower bounds (and the function is continuous in this interval) so this value is correct to 1 d.p.

Q3 **a)** $x^2 + 3x - 8 = 0 \Rightarrow x^2 = 8 - 3x \Rightarrow x = \frac{8}{x} - 3$

b) $f(-5) = (-5)^2 + (3 \times -5) - 8 = 2$

$f(-4) = (-4)^2 + (3 \times -4) - 8 = -4$

There is a sign change (and the function is continuous in this interval) so there is a root in this interval.

c) $x_1 = \frac{a}{x_0} + b = \frac{8}{-5} - 3 = -4.6$

$x_2 = -4.739,\ x_3 = -4.688,\ x_4 = -4.706,$

$x_5 = -4.700,\ x_6 = -4.702$

So $x = -4.70$ to 2 d.p.

Q4 **a)** $2^{x-1} = 4\sqrt{x} \Rightarrow 2^{x-1} = 2^2 x^{\frac{1}{2}}$

$\Rightarrow 2^{x-1} \times 2^{-2} = x^{\frac{1}{2}} \Rightarrow 2^{x-3} = x^{\frac{1}{2}}$

$\Rightarrow (2^{x-3})^2 = x \Rightarrow x = 2^{2x-6}$

b) Using the iterative formula given:

$x_1 = 0.0625,\ x_2 = 0.0170,\ x_3 = 0.0160,$

$x_4 = 0.0160$

c) $f(0.01595) = 2^{0.01595-1} - 4\sqrt{0.01595} = 0.00038...$

$f(0.01605) = 2^{0.01605-1} - 4\sqrt{0.01605} = -0.00116...$

There is a sign change between the upper and lower bounds (and the function is continuous in this interval) so $x = 0.0160$ is correct to 4 d.p.

Q5 **a)** $f(0.4) = \ln(2 \times 0.4) + 0.4^3 = -0.159...$

$f(0.5) = \ln(2 \times 0.5) + 0.5^3 = 0.125$

There is a sign change (and the function is continuous in this interval) so there is a root in this interval.

b) $\ln 2x + x^3 = 0 \Rightarrow \ln 2x = -x^3$

$\Rightarrow 2x = e^{-x^3} \Rightarrow x = \frac{e^{-x^3}}{2}$

c) Using iterative formula $x_{n+1} = \frac{e^{-x_n^3}}{2}$ with starting value $x_0 = 0.4$:

You know the root is between 0.4 and 0.5, so it's a good idea to use one of these as your starting value.

$x_1 = \frac{e^{-x_0^3}}{2} = \frac{e^{-0.4^3}}{2} = 0.4690...,$

$x_2 = 0.4509...,\ x_3 = 0.4561...,\ x_4 = 0.4547...,$

$x_5 = 0.4551...,\ x_6 = 0.4550...$

So the value of the root is $x = 0.455$ to 3 d.p.

Q6 **a)** $x^2 - 9x - 20 = 0 \Rightarrow x^2 = 9x + 20 \Rightarrow x = \sqrt{9x+20}$

So an iterative formula is $x_{n+1} = \sqrt{9x_n + 20}$

b) $x_1 = \sqrt{9x_0 + 20} = \sqrt{(9 \times 10) + 20} = 10.488...,$

$x_2 = 10.695...,\ x_3 = 10.782...,\ x_4 = 10.818...,$

$x_5 = 10.833...,\ x_6 = 10.839...$

The last 4 iterations round to 10.8, so the value of the root is $x = 10.8$ to 3 s.f.

c) $x^2 - 9x - 20 = 0 \Rightarrow x^2 - 5x - 4x - 20 = 0$

$\Rightarrow 5x = x^2 - 4x - 20 \Rightarrow x = \frac{x^2 - 4x}{5} - 4$

So an iterative formula is $x_{n+1} = \frac{x_n^2 - 4x_n}{5} - 4$

d) $x_1 = \frac{x_0^2 - 4x_0}{5} - 4 = \frac{1^2 - (4 \times 1)}{5} - 4 = -4.6,$

$x_2 = 3.912,\ x_3 = -4.0688...,\ x_4 = 2.5661...,$

$x_5 = -4.7358...,\ x_6 = 4.2744...,\ x_7 = -3.7653...,$

$x_8 = 1.8479...$

e) The iterations seem to be bouncing up and down without converging to any particular root.

3. Sketching Iterations

Exercise 3.1 — Cobweb and staircase diagrams

Q1 **a)**

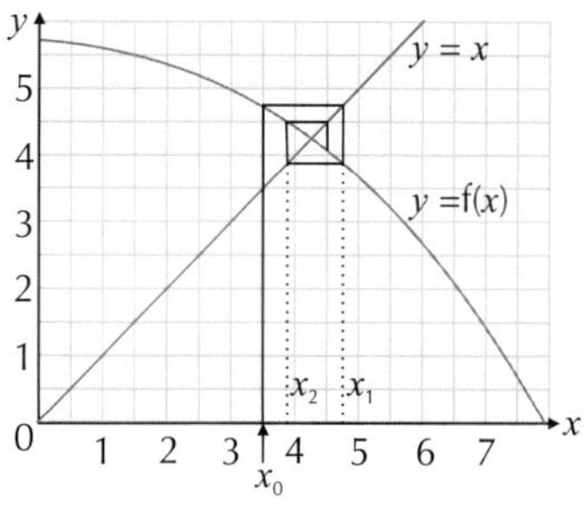

A convergent cobweb diagram.

b)

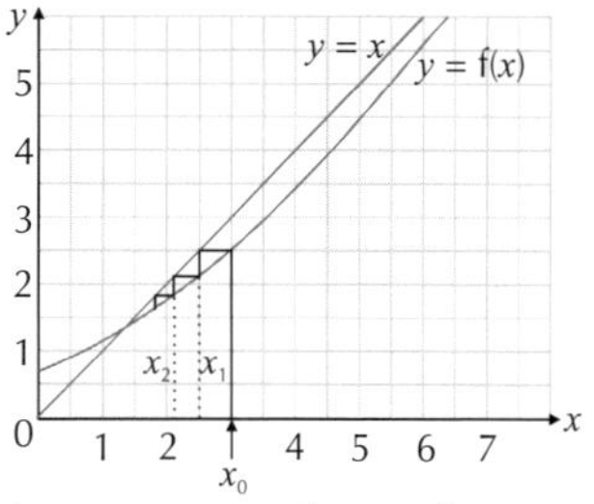

A convergent staircase diagram.

c)

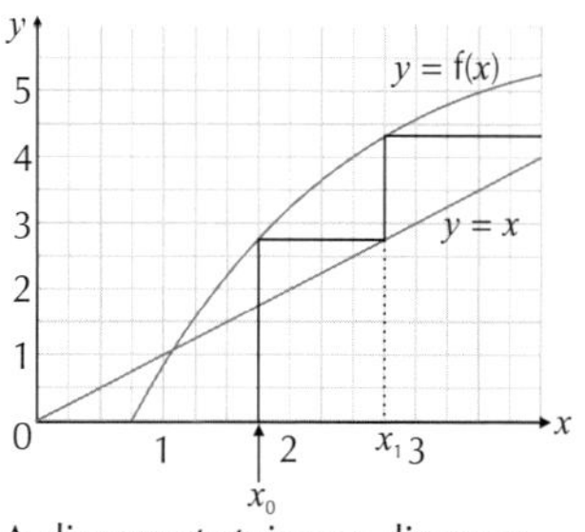

A divergent staircase diagram.

x_2 goes off the scale of the graph so you can't label it.

d)

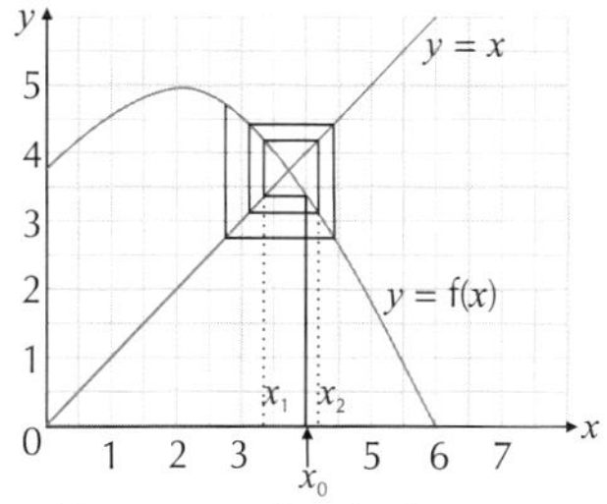

A divergent cobweb diagram.

4. The Newton-Raphson Method

Exercise 4.1 — The Newton-Raphson method

Q1 a) Differentiating f(x) gives f′(x) = 10x.
Putting this into the Newton-Raphson formula gives:
$$x_{n+1} = x_n - \frac{f(x_n)}{f'(x_n)} = x_n - \frac{5x_n^2 - 6}{10x_n}$$

b) Differentiating g(x) gives g′(x) = $3e^{3x} - 8x$.
Putting this into the Newton-Raphson formula gives:
$$x_{n+1} = x_n - \frac{g(x_n)}{g'(x_n)} = x_n - \frac{e^{3x_n} - 4x_n^2 - 1}{3e^{3x_n} - 8x_n}$$

c) Differentiating h(x) gives h′(x) = cos x + $3x^2$.
Putting this into the Newton-Raphson formula gives:
$$x_{n+1} = x_n - \frac{h(x_n)}{h'(x_n)} = x_n - \frac{\sin(x_n) + x_n^3 - 1}{\cos(x_n) + 3x_n^2}$$

Q2 Differentiating f(x) gives f′(x) = $4x^3 - 6x^2$.
Putting this into the Newton-Raphson formula gives:
$$x_{n+1} = x_n - \frac{x_n^4 - 2x_n^3 - 5}{4x_n^3 - 6x_n^2}.$$
Starting with $x_0 = 2.5$,
$$x_1 = 2.5 - \frac{2.5^4 - 2(2.5)^3 - 5}{4(2.5)^3 - 6(2.5)^2} = 2.3875$$
$x_2 = 2.373982...$, $x_3 = 2.373799...$, $x_4 = 2.373799...$
so $x = 2.3738$ (5 s.f.)

Q3 Differentiating f(x) gives f′(x) = $2x - 5$.
Putting this into the Newton-Raphson formula gives:
$$x_{n+1} = x_n - \frac{x_n^2 - 5x_n - 12}{2x_n - 5}$$
Starting with $x_0 = -1$,
$$x_1 = -1 - \frac{(-1)^2 - 5(-1) - 12}{2(-1) - 5} = -1.857142...$$
$x_2 = -1.772833...$, $x_3 = -1.772001...$, $x_4 = -1.772001...$
so $x = -1.7720$ (5 s.f.)
Here, you could find the exact roots using the quadratic formula.

Q4 Rearrange the equation to get $x^2 \ln x - 5 = 0$.
Differentiating using the product rule gives $x(1 + 2\ln x)$.
Putting this into the Newton-Raphson formula gives:
$$x_{n+1} = x_n - \frac{x_n^2 \ln x_n - 5}{x_n(1 + 2\ln x_n)}$$
Starting with $x_0 = 2$, $x_1 = 2 - \frac{2^2 \ln 2 - 5}{2(1 + 2\ln 2)} = 2.466709...$
$x_2 = 2.395369...$, $x_3 = 2.393518...$, $x_4 = 2.393517...$
so $x = 2.3935$ (5 s.f.)

Q5 Differentiating the function gives $6x^2 - 30x$.
Putting this into the Newton-Raphson formula:
$$x_{n+1} = x_n - \frac{2x_n^3 - 15x_n^2 + 109}{6x_n^2 - 30x_n}$$
Starting with $x_0 = 1$:
$$x_1 = 1 - \frac{2(1)^3 - 15(1)^2 + 109}{6(1)^2 - 30(1)} = 1 - \frac{96}{-24} = 1 + 4 = 5$$
$$x_2 = 5 - \frac{2(5)^3 - 15(5)^2 + 109}{6(5)^2 - 30(5)} = 5 - \frac{-16}{0}$$
Since this involves dividing by 0, the method fails, so $x_0 = 1$ cannot be used to find a root.
f′(5) = 0, which means that the tangent line at x_1 is horizontal, and it won't intercept the x-axis.

Exercise 4.2 — Combining the methods

Q1 a) $(x + 2)g(x) = (x + 2)(x^3 - 4x + 1)$
$= x^4 - 4x^2 + x + 2x^3 - 8x + 2$
$= x^4 + 2x^3 - 4x^2 - 7x + 2 = f(x)$
So $x = -2$ is a root of f(x).

b) $g(1) = 1^3 - 4(1) + 1 = 1 - 4 + 1 = -2$
$g(2) = 2^3 - 4(2) + 1 = 8 - 8 + 1 = 1$
There is a sign change (and the function is continuous in this interval) so there is a root in this interval.

c) $x^3 - 4x + 1 = 0 \Rightarrow x^3 = 4x - 1 \Rightarrow x = \sqrt[3]{4x - 1}$

d) $x_1 = \sqrt[3]{4x_0 - 1} = \sqrt[3]{4(2) - 1} = 1.91293...$
$x_2 = 1.88066...$, $x_3 = 1.86842...$, $x_4 = 1.86373...$,
$x_5 = 1.86193...$, so $\alpha = 1.86$ (3 s.f.)

e) Differentiating g(x) gives g′(x) = $3x^2 - 4$.
Putting this into the Newton-Raphson formula gives:
$$x_{n+1} = x_n - \frac{x_n^3 - 4x_n + 1}{3x_n^2 - 4}$$
$$x_1 = -2 - \frac{(-2)^3 - 4(-2) + 1}{3(-2)^2 - 4} = -2.125$$
$x_2 = -2.114975...$, $x_3 = -2.114907...$
so $\beta = -2.115$ (4 s.f.)

f) The Newton-Raphson method fails when $g'(x_n) = 0$ because it involves dividing by zero in the formula. Graphically, it means that the tangent at x_n is horizontal, and so it won't cut the x-axis to find the next iteration.
When $x_n = \frac{2\sqrt{3}}{3}$,
$$g'\left(\frac{2\sqrt{3}}{3}\right) = 3\left(\frac{2\sqrt{3}}{3}\right)^2 - 4 = 3\left(\frac{12}{9}\right) - 4 = 4 - 4 = 0$$

5. The Trapezium Rule

Exercise 5.1 — The trapezium rule

Q1 a) $h = (2 - 0) \div 2 = 1$, so the x-values are 0, 1, 2.
$$\int_0^2 \sqrt{x+2}\, dx \approx \frac{h}{2}[y_0 + 2y_1 + y_2] = \frac{1}{2}[\sqrt{2} + 2\sqrt{3} + \sqrt{4}]$$
$$= \frac{1}{2}[1.4142 + 3.4641 + 2] = 3.44 \text{ (3 s.f.)}$$

b) $h = (3 - 1) \div 4 = 0.5$,
so the x-values are 1, 1.5, 2, 2.5, 3.
$$\int_1^3 2(\ln x)^2\, dx \approx \frac{h}{2}[y_0 + 2(y_1 + y_2 + y_3) + y_4]$$
$$= \frac{1}{4}[2(\ln 1)^2 + 2(2(\ln 1.5)^2 + 2(\ln 2)^2 + 2(\ln 2.5)^2) + 2(\ln 3)^2]$$
$$= \frac{1}{4}[0 + 2(0.3288 + 0.9609 + 1.6792) + 2.4139]$$
$= 2.09$ (3 s.f.)

c) $h = (0.4 - 0) \div 2 = 0.2$,
so the x-values are 0, 0.2, 0.4.

$$\int_0^{0.4} e^{x^2}\,dx \approx \frac{h}{2}[y_0 + 2y_1 + y_2]$$
$$= \frac{0.2}{2}[e^0 + 2e^{0.04} + e^{0.16}]$$
$$= 0.1[1 + 2.0816 + 1.1735]$$
$$= 0.426 \text{ (3 s.f.)}$$

d) $h = \left(\frac{\pi}{4} - -\frac{\pi}{4}\right) \div 4 = \frac{\pi}{8}$,
so the x-values are $-\frac{\pi}{4}, -\frac{\pi}{8}, 0, \frac{\pi}{8}, \frac{\pi}{4}$.

$$\int_{-\frac{\pi}{4}}^{\frac{\pi}{4}} 4x\tan x\,dx \approx \frac{h}{2}[y_0 + 2(y_1 + y_2 + y_3) + y_4]$$
$$= \frac{\pi}{16}\left[-\pi\tan\left(-\frac{\pi}{4}\right) + 2\left(-\frac{\pi}{2}\tan\left(-\frac{\pi}{8}\right) + 0 + \frac{\pi}{2}\tan\left(\frac{\pi}{8}\right)\right) + \pi\tan\left(\frac{\pi}{4}\right)\right]$$
$$= \frac{\pi}{16}[\pi + 2(0.6506 + 0.6506) + \pi]$$
$$= 1.74 \text{ (3 s.f.)}$$

e) $h = (0.3 - 0) \div 6 = 0.05$,
so the x-values are 0, 0.05, 0.1, 0.15, 0.2, 0.25, 0.3.

$$\int_0^{0.3} \sqrt{e^x + 1}\,dx \approx \frac{h}{2}[y_0 + 2(y_1 + y_2 + y_3 + y_4 + y_5) + y_6]$$
$$= \frac{0.05}{2}[\sqrt{e^0 + 1} + 2(\sqrt{e^{0.05} + 1} + \sqrt{e^{0.1} + 1} + \sqrt{e^{0.15} + 1} + \sqrt{e^{0.2} + 1} + \sqrt{e^{0.25} + 1}) + \sqrt{e^{0.3} + 1}]$$
$$= 0.025[\sqrt{2} + 2(\sqrt{2.0513} + \sqrt{2.1052} + \sqrt{2.1618} + \sqrt{2.2214} + \sqrt{2.2840}) + \sqrt{2.3499}]$$
$$= 0.025[1.4142 + 2(1.4322 + 1.4509 + 1.4703 + 1.4904 + 1.5113) + 1.5329]$$
$$= 0.441 \text{ (3 s.f.)}$$

f) $h = \frac{\pi}{6}$, so the x-values are $0, \frac{\pi}{6}, \frac{\pi}{3}, \frac{\pi}{2}, \frac{2\pi}{3}, \frac{5\pi}{6}, \pi$.

$$\int_0^{\pi} \ln(2 + \sin x)\,dx \approx \frac{h}{2}[y_0 + 2(y_1 + y_2 + y_3 + y_4 + y_5) + y_6]$$
$$= \frac{\pi}{12}[\ln(2 + \sin 0) + 2(\ln(2 + \sin\tfrac{\pi}{6}) + \ln(2 + \sin\tfrac{\pi}{3}) + \ln(2 + \sin\tfrac{\pi}{2}) + \ln(2 + \sin\tfrac{2\pi}{3}) + \ln(2 + \sin\tfrac{5\pi}{6})) + \ln(2 + \sin\pi)]$$
$$= \frac{\pi}{12}[0.6931 + 2(0.9163 + 1.0529 + 1.0986 + 1.0529 + 0.9163) + 0.6931]$$
$$= 3.00 \text{ (3 s.f.)}$$

Q2 a) $h = 1$, so the x-values are 0, 1, 2 and the y-values are:
$y_0 = \sqrt{2},\ y_1 = \sqrt{3},\ y_2 = \sqrt{4}$

Use the simplified versions of the trapezium rule.
The lower bound is found using:
$$\int_0^2 \sqrt{x + 2}\,dx \approx h[y_0 + y_1] = 1[\sqrt{2} + \sqrt{3}] = 3.146 \text{ (4 s.f.)}$$
The upper bound is found using:
$$\int_0^2 \sqrt{x + 2}\,dx \approx h[y_1 + y_2] = 1[\sqrt{3} + \sqrt{4}] = 3.732 \text{ (4 s.f.)}$$

b) $h = 0.5$, so the x-values are 1, 1.5, 2, 2.5, 3
Substitute these into the function to get y-values:
$y_0 = 0$, $y_1 = 0.3288...$, $y_2 = 0.9609...$,
$y_3 = 1.6791...$, $y_4 = 2.4138...$

Use the simplified versions of the trapezium rule.
The lower bound is found using:
$$\int_1^3 2(\ln x)^2\,dx \approx h[y_0 + y_1 + y_2 + y_3]$$
$$= 0.5[0 + 0.3288... + 0.9609... + 1.6791...] = 1.484 \text{ (4 s.f.)}$$
The upper bound is found using:
$$\int_1^3 2(\ln x)^2\,dx \approx h[y_1 + y_2 + y_3 + y_4]$$
$$= 0.5[0.3288... + 0.9609... + 1.6791... + 2.4138...] = 2.691 \text{ (4 s.f.)}$$

The corresponding trapezium rule answers from Q1 lie inside the lower and upper bounds.

Q3 $h = \frac{\pi}{2} \div 3 = \frac{\pi}{6}$, so the x-values are $0, \frac{\pi}{6}, \frac{\pi}{3}, \frac{\pi}{2}$.

$$\int_0^{\frac{\pi}{2}} \sin^3\theta\,d\theta \approx \frac{h}{2}[y_0 + 2(y_1 + y_2) + y_3]$$
$$= \frac{\pi}{12}\left[\sin^3 0 + 2\left(\sin^3\frac{\pi}{6} + \sin^3\frac{\pi}{3}\right) + \sin^3\frac{\pi}{2}\right]$$
$$= \frac{\pi}{12}[0 + 2(0.125 + 0.6495) + 1]$$
$$= 0.667 \text{ (3 d.p.)}$$

Q4 $h = (7 - 2) \div 5 = 1$, so the x-values are 2, 3, 4, 5, 6, 7.

$$\int_2^7 \sqrt{\ln x}\,dx \approx \frac{h}{2}[y_0 + 2(y_1 + y_2 + y_3 + y_4) + y_5]$$
$$= \frac{1}{2}[\sqrt{\ln 2} + 2(\sqrt{\ln 3} + \sqrt{\ln 4} + \sqrt{\ln 5} + \sqrt{\ln 6}) + \sqrt{\ln 7}]$$
$$= \frac{1}{2}[0.8326 + 2(1.0481 + 1.1774 + 1.2686 + 1.3386) + 1.3950]$$
$$= 5.947 \text{ m}^2 \text{ (3 d.p.)}$$

Q5 a)

x	0	$\frac{\pi}{8}$	$\frac{\pi}{4}$	$\frac{3\pi}{8}$	$\frac{\pi}{2}$
y	1	1.466	2.028	2.519	2.718

b) (i) $$\int_0^{\frac{\pi}{2}} e^{\sin x}\,dx \approx \frac{\pi}{8}[1 + 2(2.028) + 2.718]$$
$$= 3.05 \text{ (2 d.p.)}$$

(ii) $$\int_0^{\frac{\pi}{2}} e^{\sin x}\,dx$$
$$\approx \frac{\pi}{16}[1 + 2(1.466 + 2.028 + 2.519) + 2.718]$$
$$= 3.09 \text{ (2 d.p.)}$$

c) 3.09 is the better estimate as more intervals have been used in the calculation.

Q6 a) $h = (4 - 2) \div 4 = 0.5$,
so the x-values are 2, 2.5, 3, 3.5, 4.

$$\int_2^4 \frac{3}{\ln x}\,dx \approx \frac{1}{4}\left[\frac{3}{\ln 2} + 2\left(\frac{3}{\ln 2.5} + \frac{3}{\ln 3} + \frac{3}{\ln 3.5}\right) + \frac{3}{\ln 4}\right]$$
$$= 0.25[4.3281 + 2(3.2741 + 2.7307 + 2.3947) + 2.1640]$$
$$= 5.82 \text{ (2 d.p.)}$$

b) It is an over-estimate as the top of each trapezium lies above the curve.

Q7 a) There are 6 intervals of $h = \frac{\pi}{6}$ between $-\frac{\pi}{2}$ and $\frac{\pi}{2}$.

x	$-\frac{\pi}{2}$	$-\frac{\pi}{3}$	$-\frac{\pi}{6}$	0	$\frac{\pi}{6}$	$\frac{\pi}{3}$	$\frac{\pi}{2}$
$y = \cos x$	0	0.5	$\frac{\sqrt{3}}{2}$	1	$\frac{\sqrt{3}}{2}$	0.5	0

$$\int_{-\frac{\pi}{2}}^{\frac{\pi}{2}} \cos x\,dx \approx \frac{h}{2}[y_0 + 2(y_1 + y_2 + y_3 + y_4 + y_5) + y_6]$$
$$= \frac{\left(\frac{\pi}{6}\right)}{2}\left[0 + 2\left(0.5 + \frac{\sqrt{3}}{2} + 1 + \frac{\sqrt{3}}{2} + 0.5\right) + 0\right]$$
$$= \frac{\pi}{12}[0 + 2(2 + \sqrt{3}) + 0]$$
$$= \pi\left(\frac{2(2 + \sqrt{3})}{12}\right)$$
$$= \frac{\pi(2 + \sqrt{3})}{6} \text{ as required}$$

b) It's an under-estimate because between $-\frac{\pi}{2}$ and $\frac{\pi}{2}$ the curve of the graph $y = \cos x$ is concave, and so the top of each trapezium lies below the curve.

Chapter 10: Vectors

1. Vectors in Three Dimensions

Exercise 1.1 — Three-dimensional position vectors

Q1 a) $\overrightarrow{OR} = \begin{pmatrix} 4 \\ -5 \\ 1 \end{pmatrix}$ $\quad \overrightarrow{OS} = \begin{pmatrix} -3 \\ 0 \\ -1 \end{pmatrix}$

b) $\overrightarrow{OR} = 4\mathbf{i} - 5\mathbf{j} + \mathbf{k}$ $\quad \overrightarrow{OS} = -3\mathbf{i} - \mathbf{k}$

There's no **j** *component written for* $\overrightarrow{OS}$ *because its* **j** *component is zero (you don't write* $-3\mathbf{i} + 0\mathbf{j} - \mathbf{k}$*).*

Q2 $\overrightarrow{GH} = \overrightarrow{OH} - \overrightarrow{OG} = \begin{pmatrix} -1 \\ 4 \\ 9 \end{pmatrix} - \begin{pmatrix} 2 \\ -3 \\ 4 \end{pmatrix} = \begin{pmatrix} -3 \\ 7 \\ 5 \end{pmatrix}$

$\overrightarrow{HG} = -\overrightarrow{GH} = -\begin{pmatrix} -3 \\ 7 \\ 5 \end{pmatrix} = \begin{pmatrix} 3 \\ -7 \\ -5 \end{pmatrix}$

Q3 $\overrightarrow{OJ} = 4\mathbf{i} - 3\mathbf{k}$, $\overrightarrow{OK} = -\mathbf{i} + 3\mathbf{j}$, and $\overrightarrow{OL} = 2\mathbf{i} + 2\mathbf{j} + 7\mathbf{k}$

$\overrightarrow{JK} = \overrightarrow{OK} - \overrightarrow{OJ} = -\mathbf{i} + 3\mathbf{j} - (4\mathbf{i} - 3\mathbf{k}) = -5\mathbf{i} + 3\mathbf{j} + 3\mathbf{k}$

$\overrightarrow{KL} = \overrightarrow{OL} - \overrightarrow{OK} = 2\mathbf{i} + 2\mathbf{j} + 7\mathbf{k} - (-\mathbf{i} + 3\mathbf{j}) = 3\mathbf{i} - \mathbf{j} + 7\mathbf{k}$

$\overrightarrow{LJ} = \overrightarrow{OJ} - \overrightarrow{OL} = 4\mathbf{i} - 3\mathbf{k} - (2\mathbf{i} + 2\mathbf{j} + 7\mathbf{k}) = 2\mathbf{i} - 2\mathbf{j} - 10\mathbf{k}$

Q4 a) Point B is twice the distance of A from the origin in each direction, so $\overrightarrow{OB} = 2\overrightarrow{OA}$.

The position vector of B can also be described as:

$\overrightarrow{OB} = \overrightarrow{AB} + \overrightarrow{OA}$

Substituting for $\overrightarrow{OB}$:

$2\overrightarrow{OA} = \overrightarrow{AB} + \overrightarrow{OA}$

$2\overrightarrow{OA} - \overrightarrow{OA} = \overrightarrow{AB}$

So $\overrightarrow{AB} = \overrightarrow{OA}$ — as required.

b) $\overrightarrow{OB} = 2\overrightarrow{OA} = 2(3\mathbf{i} + \mathbf{j} + 2\mathbf{k})$
$= 6\mathbf{i} + 2\mathbf{j} + 4\mathbf{k}$

Q5 The position vector of the midpoint of CD is at

$\overrightarrow{OC} + \frac{1}{2}\overrightarrow{CD} = \begin{pmatrix} -1 \\ 3 \\ -5 \end{pmatrix} + \frac{1}{2}\begin{pmatrix} 4 \\ -4 \\ 6 \end{pmatrix} = \begin{pmatrix} -1 \\ 3 \\ -5 \end{pmatrix} + \begin{pmatrix} 2 \\ -2 \\ 3 \end{pmatrix} = \begin{pmatrix} 1 \\ 1 \\ -2 \end{pmatrix} = \overrightarrow{OM}$

So M is the midpoint of CD.

Q6 A vector is parallel to another if it is multiplied by the same scalar in each direction.

$\mathbf{a} = \left(3 \times \frac{1}{4}\right)\mathbf{i} + \left(\frac{1}{3} \times 1\right)\mathbf{j} + \left(3 \times -\frac{2}{3}\right)\mathbf{k}$

$\mathbf{a} = 3\mathbf{b}$ in the **i** and **k** directions, but $\frac{1}{3}\mathbf{b}$ in the **j** direction, so they are not parallel.

Q7 Model the counters as points A, B and C, with position vectors $\overrightarrow{OA} = \mathbf{i} + 3\mathbf{k}$, $\overrightarrow{OB} = 3\mathbf{i} + \mathbf{j} + 2\mathbf{k}$, and $\overrightarrow{OC} = 7\mathbf{i} + 3\mathbf{j}$.

$\overrightarrow{AB} = \overrightarrow{OB} - \overrightarrow{OA} = 2\mathbf{i} + \mathbf{j} - \mathbf{k}$

$\overrightarrow{BC} = \overrightarrow{OC} - \overrightarrow{OB} = 4\mathbf{i} + 2\mathbf{j} - 2\mathbf{k}$
$= 2 \times (2\mathbf{i} + \mathbf{j} - \mathbf{k}) = 2\overrightarrow{AB}$

$\overrightarrow{BC}$ is a scalar multiple of $\overrightarrow{AB}$, therefore the vectors are parallel. They also share a point (B), so they must be collinear. So the counters must lie in a straight line.

2. Calculating with Vectors

Exercise 2.1 — Calculating with 3D vectors

Q1 a) $\sqrt{1^2 + 4^2 + 8^2} = \sqrt{1 + 16 + 64} = \sqrt{81} = 9$

b) $\sqrt{4^2 + 2^2 + 4^2} = \sqrt{36} = 6$

c) $\sqrt{(-4)^2 + (-5)^2 + 20^2} = \sqrt{441} = 21$

d) $\sqrt{7^2 + 1^2 + (-7)^2} = \sqrt{99} = 3\sqrt{11}$

e) $\sqrt{(-2)^2 + 4^2 + (-6)^2} = \sqrt{56} = \sqrt{4 \times 14} = 2\sqrt{14}$

Q2 a) The resultant is:

$(\mathbf{i} + \mathbf{j} + 2\mathbf{k}) + (\mathbf{i} + 2\mathbf{j} + 4\mathbf{k}) = 2\mathbf{i} + 3\mathbf{j} + 6\mathbf{k}$

Its magnitude is: $\sqrt{2^2 + 3^2 + 6^2} = \sqrt{49} = 7$

b) resultant: $2\mathbf{i} + 14\mathbf{j} + 23\mathbf{k}$

magnitude: $\sqrt{2^2 + 14^2 + 23^2} = \sqrt{729} = 27$

c) resultant: $\begin{pmatrix} 2 \\ 6 \\ 9 \end{pmatrix}$, magnitude: $\sqrt{2^2 + 6^2 + 9^2} = 11$

d) resultant: $\begin{pmatrix} 2 \\ 5 \\ 14 \end{pmatrix}$, magnitude: $\sqrt{2^2 + 5^2 + 14^2} = 15$

e) resultant: $\begin{pmatrix} 10 \\ 2 \\ 14 \end{pmatrix}$, magnitude: $\sqrt{10^2 + 2^2 + 14^2} = 10\sqrt{3}$

Q3 a) $\sqrt{(5-3)^2 + (6-4)^2 + (6-5)^2}$
$= \sqrt{2^2 + 2^2 + 1^2} = \sqrt{4 + 4 + 1} = \sqrt{9} = 3$

b) $\sqrt{(-11-7)^2 + (1-2)^2 + (15-9)^2}$
$= \sqrt{324 + 1 + 36} = \sqrt{361} = 19$

c) $\sqrt{(6-10)^2 + (10-(-2))^2 + (-4-(-1))^2}$
$= \sqrt{16 + 144 + 9} = \sqrt{169} = 13$

d) $\sqrt{(7-0)^2 + (0-(-4))^2 + (14-10)^2}$
$= \sqrt{49 + 16 + 16} = \sqrt{81} = 9$

e) $\sqrt{(2-(-4))^2 + (4-7)^2 + (-12-10)^2}$
$= \sqrt{36 + 9 + 484} = \sqrt{529} = 23$

f) $\sqrt{(30-7)^2 + (9-(-1))^2 + (-6-4)^2}$
$= \sqrt{529 + 100 + 100} = \sqrt{729} = 27$

Q4 $|2\mathbf{m} - \mathbf{n}| = |-6\mathbf{i} - 5\mathbf{j} + 10\mathbf{k}|$
$= \sqrt{(-6)^2 + (-5)^2 + 10^2} = 12.68857... = 12.7$ m (1 d.p.)

Q5 $|\overrightarrow{AO}| = |\overrightarrow{OA}| = \sqrt{1^2 + (-4)^2 + 3^2} = \sqrt{26}$

$|\overrightarrow{BO}| = |\overrightarrow{OB}| = \sqrt{(-1)^2 + (-3)^2 + 5^2} = \sqrt{35}$

It's pretty clear that the magnitude of $\overrightarrow{AO}$ *is going to be the same as* $\overrightarrow{OA}$ *so there's no need to find* $\overrightarrow{AO}$*.*

$\overrightarrow{BA} = \overrightarrow{OA} - \overrightarrow{OB} = (\mathbf{i} - 4\mathbf{j} + 3\mathbf{k}) - (-\mathbf{i} - 3\mathbf{j} + 5\mathbf{k})$
$= 2\mathbf{i} - \mathbf{j} - 2\mathbf{k}$

$|\overrightarrow{BA}| = \sqrt{2^2 + (-1)^2 + (-2)^2} = 3$

Triangle AOB is right-angled because:

$|\overrightarrow{AO}|^2 + |\overrightarrow{BA}|^2 = (\sqrt{26})^2 + 3^2 = 26 + 9 = 35$
$= (\sqrt{35})^2$
$= |\overrightarrow{BO}|^2$

Q6 $|\mathbf{v}| = \sqrt{4^2 + (-4)^2 + (-7)^2} = 9$

so the unit vector is $\frac{\mathbf{v}}{|\mathbf{v}|} = \frac{1}{9}\mathbf{v} = \frac{4}{9}\mathbf{i} - \frac{4}{9}\mathbf{j} - \frac{7}{9}\mathbf{k}$

Q7 $|\overrightarrow{PQ}| = \sqrt{(q-4)^2 + 6^2 + (2q-3)^2}$

So: $\sqrt{(q-4)^2 + 6^2 + (2q-3)^2} = 11$

$$q^2 - 8q + 16 + 36 + 4q^2 - 12q + 9 = 121$$
$$5q^2 - 20q - 60 = 0$$
$$q^2 - 4q - 12 = 0$$
$$(q - 6)(q + 2) = 0$$

Either $q = 6$, then Q is (4, 5, 13),
or $q = -2$, then Q is (–4, 5, –3).

Q8 **a)** Form a triangle between the two vectors (call them **a** and **b**) with angle θ between them. The triangle has side lengths $|\mathbf{a}|$, $|\mathbf{b}|$ and $|\mathbf{b} - \mathbf{a}|$.

$$\mathbf{a} = \begin{pmatrix}1\\3\\2\end{pmatrix},\ |\mathbf{a}| = \sqrt{1^2+3^2+2^2} = \sqrt{14}$$

$$\mathbf{b} = \begin{pmatrix}3\\2\\1\end{pmatrix},\ |\mathbf{b}| = \sqrt{3^2+2^2+1^2} = \sqrt{14}$$

$$\mathbf{b} - \mathbf{a} = \begin{pmatrix}2\\-1\\-1\end{pmatrix},\ |\mathbf{b} - \mathbf{a}| = \sqrt{2^2+(-1)^2+(-1)^2} = \sqrt{6}$$

Using the cosine rule,

$$\cos\theta = \frac{(\sqrt{14})^2+(\sqrt{14})^2-(\sqrt{6})^2}{2\times\sqrt{14}\times\sqrt{14}} = \frac{22}{28} = \frac{11}{14}$$

$$\theta = \cos^{-1}\left(\frac{11}{14}\right) = 38.2132... = 38.2° \text{ (1 d.p.)}$$

b) $|\mathbf{a}| = \sqrt{14}$ as before

$$\mathbf{b} = \begin{pmatrix}0\\1\\0\end{pmatrix},\ |\mathbf{b}| = 1$$

$$\mathbf{b} - \mathbf{a} = \begin{pmatrix}-1\\-2\\-2\end{pmatrix},\ |\mathbf{b} - \mathbf{a}| = \sqrt{(-1)^2+(-2)^2+(-2)^2} = 3$$

Using the cosine rule,

$$\cos\theta = \frac{(\sqrt{14})^2+1^2-3^2}{2\times\sqrt{14}\times 1} = \frac{6}{2\sqrt{14}} = \frac{3}{\sqrt{14}}$$

$$\theta = \cos^{-1}\left(\frac{3}{\sqrt{14}}\right) = 36.6992... = 36.7° \text{ (1 d.p.)}$$

Q9 Speed is the magnitude of velocity, so you want to find $|\mathbf{v}|$. Form a triangle between vector **v** and a vector $\mathbf{p} = \mathbf{i}$, with an angle of 60° between them. The triangle has side lengths $|\mathbf{v}|$, $|\mathbf{p}|$ and $|\mathbf{p} - \mathbf{v}|$.

$|\mathbf{v}| = \sqrt{1^2 + a^2 + 1^2} = \sqrt{2 + a^2}$

$|\mathbf{p}| = 1$

$\mathbf{p} - \mathbf{v} = -a\mathbf{j} - \mathbf{k}$, $|\mathbf{p} - \mathbf{v}| = \sqrt{0^2+(-a)^2+(-1)^2} = \sqrt{a^2+1}$

Using the cosine rule,

$$\cos 60° = \frac{(\sqrt{2+a^2})^2 + 1^2 - (\sqrt{a^2+1})^2}{2\times\sqrt{2+a^2}}$$

$$\frac{1}{2} = \frac{2+a^2+1-a^2-1}{2\sqrt{2+a^2}}$$

$$\frac{1}{2} = \frac{2}{2\sqrt{2+a^2}}$$

$$\sqrt{2+a^2} = 2$$
$$2 + a^2 = 4$$
$$a = \pm\sqrt{2}$$

Therefore the launch speed $|\mathbf{v}| = \sqrt{2+(\pm\sqrt{2})^2} = 2 \text{ ms}^{-1}$

Chapter 11: Probability

1. Conditional Probability

Exercise 1.1 — Conditional probability

Q1 **a)** P(A′) = 1 – P(A) = 1 – 0.3 = 0.7

b) P(A ∪ B) = P(A) + P(B) – P(A ∩ B), so
P(A ∩ B) = P(A) + P(B) – P(A ∪ B)
= 0.3 + 0.5 – 0.7 = 0.1

c) P(A) = P(A ∩ B) + P(A ∩ B′), so
P(A ∩ B′) = P(A) – P(A ∩ B) = 0.3 – 0.1 = 0.2

Q2 The probability that neither X nor Y occurs is P(X′ ∩ Y′)
Using that P(X ∪ Y) + P(X′ ∩ Y′) = 1,
P(X ∪ Y) = 1 – P(X′ ∩ Y′)
Since X and Y are independent, P(X ∩ Y) = P(X)P(Y)
P(X ∪ Y) = P(X) + P(Y) – P(X)P(Y)
= 0.8 + 0.15 – (0.8 × 0.15) = 0.83
So P(X′ ∩ Y′) = 1 – 0.83 = 0.17

Q3 **a)** $P(G|H) = \frac{P(G\cap H)}{P(H)} = \frac{0.24}{0.63} = \frac{24}{63} = \frac{8}{21}$

You could give the answer as a decimal instead, but using a fraction means you can give an exact answer.

b) $P(H|G) = \frac{P(G\cap H)}{P(G)} = \frac{0.24}{0.7} = \frac{24}{70} = \frac{12}{35}$

Q4 **a)** $P(B|A) = \frac{P(A\cap B)}{P(A)} = \frac{0.34}{0.68} = 0.5$

b) $P(A|C) = \frac{P(A\cap C)}{P(C)} = \frac{0.16}{0.44} = \frac{16}{44} = \frac{4}{11}$

c) $P(C'|B) = \frac{P(B\cap C')}{P(B)} = \frac{0.49}{1-0.44} = 0.875$

Q5 **a)** P(J ∩ K) = P(J)P(K|J) = 0.4 × 0.2 = 0.08

$P(K) = \frac{P(J\cap K)}{P(J|K)} = \frac{0.08}{0.64} = 0.125$

P(J ∪ K) = P(J) + P(K) – P(J ∩ K)
= 0.4 + 0.125 – 0.08 = 0.445

b) P(J ∪ K) + P(J′ ∩ K′) = 1, so
P(J′ ∩ K′) = 1 – P(J ∪ K) = 1 – 0.445 = 0.555

Q6 Let F = 'over 6 feet tall' and G = 'can play in goal'.

a) $P(G|F) = \frac{P(F\cap G)}{P(F)}$

$P(F\cap G) = \frac{2}{11}$ and $P(F) = \frac{5}{11}$

So, $P(G|F) = \frac{P(F\cap G)}{P(F)} = \frac{\frac{2}{11}}{\frac{5}{11}} = \frac{2}{5}$

b) $P(F|G) = \frac{P(F\cap G)}{P(G)}$ and $P(G) = \frac{3}{11}$.

So $P(F|G) = \frac{P(F\cap G)}{P(G)} = \frac{\frac{2}{11}}{\frac{3}{11}} = \frac{2}{3}$

Q7 **a)** $P(M|A) = \frac{P(M\cap A)}{P(A)} = \frac{\frac{10}{100}}{\frac{33}{100}} = \frac{10}{33}$

b) $P(A|E\cap M) = \frac{P(A\cap E\cap M)}{P(E\cap M)} = \frac{\frac{2}{100}}{\frac{9}{100}} = \frac{2}{9}$

c) $P(E|M') = \frac{P(E\cap M')}{P(M')} = \frac{\frac{39}{100}}{\frac{60}{100}} = \frac{39}{60} = \frac{13}{20}$

d) $P(A|E') = \frac{P(A \cap E')}{P(E')} = \frac{\frac{18}{100}}{\frac{52}{100}} = \frac{18}{52} = \frac{9}{26}$

e) $P(M|A \cap E) = \frac{P(M \cap A \cap E)}{P(A \cap E)} = \frac{\frac{2}{100}}{\frac{15}{100}} = \frac{2}{15}$

Q8 a) $P(Y) = 1 - P(Y') = 1 - 0.72 = 0.28$

b) $P(X \cap Y) = P(Y)P(X|Y) = 0.28 \times 0.75 = 0.21$

c) $P(X \cap Z) = P(X)P(Z|X) = 0.44 \times 0.25 = 0.11$

d) $P(Y|Z') = \frac{P(Y \cap Z')}{P(Z')} = \frac{0.2}{1 - 0.61} = \frac{20}{39}$

e) $P(X \cap Y \cap Z) = P(Z)P(X \cap Y|Z)$

$= \frac{61}{100} \times \frac{7}{61} = \frac{7}{100} = 0.07$

There are lots of ways to write an expression for $P(X \cap Y \cap Z)$ — e.g. $P(X \cap Y)P(Z|X \cap Y)$ or $P(Y \cap Z)P(X|Y \cap Z)$. You have to choose the way that makes best use of the information in the question.

Exercise 1.2 — Using conditional probability

Q1 $P(A') = 1 - P(A) = 1 - 0.75 = 0.25$
$P(A' \cap B) = P(A')\,P(B|A') = 0.25 \times 0.4 = 0.1$
$P(A \cap B) = P(A) - P(A \cap B') = 0.75 - 0.45 = 0.3$
$P(B) = P(A \cap B) + P(A' \cap B) = 0.3 + 0.1 = 0.4$
$P(A)P(B) = 0.75 \times 0.4 = 0.3 = P(A \cap B)$,
so A and B are independent.

Q2 a)

	Action	Comedy	Total
Under 20	16	10	26
20 and over	6	8	14
Total	22	18	40

or in terms of the probabilities,

	Action	Comedy	Total
Under 20	0.4	0.25	0.65
20 and over	0.15	0.2	0.35
Total	0.55	0.45	1

b) You're told that they saw a comedy, so look at the Comedy column — the total is 18 (or 0.45).
The Under 20 entry in this column is 10 (or 0.25),
so $P(\text{Under } 20|\text{Comedy}) = \frac{10}{18}\left(\text{or } \frac{0.25}{0.45}\right) = \frac{5}{9}$

Q3 a) (i) The circles for R and T do not overlap, so they are mutually exclusive (i.e. $P(R \cap T) = 0$).

(ii) Adding up the probabilities on the Venn diagram:
$P(R) = 0.25$, $P(S) = 0.2$, $P(T) = 0.46$
$P(R)P(S) = 0.25 \times 0.2 = 0.05 = P(R \cap S)$
so R and S are independent.
$P(S)P(T) = 0.2 \times 0.46 = 0.092 \neq P(S \cap T)$
so S and T are not independent.
$P(R)P(T) = 0.25 \times 0.46 = 0.115 \neq P(R \cap T)$
so R and T are not independent.

b) (i) $P(R|S) = \frac{P(R \cap S)}{P(S)} = \frac{0.05}{0.2} = 0.25$

(ii) $P(R|S \cap T') = \frac{P(R \cap S \cap T')}{P(S \cap T')}$
$P(R \cap S \cap T') = P(R \cap S) = 0.05$
$P(S \cap T') = P(S) - P(S \cap T)$
$= 0.2 - 0.11 = 0.09$
So $P(R|S \cap T') = \frac{P(R \cap S \cap T')}{P(S \cap T')} = \frac{0.05}{0.09} = \frac{5}{9}$

Q4 a) $P(X \cap Y) = P(X)P(Y) = 0.84 \times 0.68 = 0.5712$

b) $P(Y' \cap Z') = P(Y')P(Z') = 0.32 \times 0.52 = 0.1664$

c) Since Y and Z are independent, $P(Y|Z) = P(Y) = 0.68$

d) Since Z' and Y' are independent,
$P(Z'|Y') = P(Z') = 1 - 0.48 = 0.52$

e) Since Y and X' are independent, $P(Y|X') = P(Y) = 0.68$

Q5 a) $P(S) = 0.45$, $P(P|S) = 0.6$, $P(P|S') = 0.2$
$P(S \cap P) = P(S)P(P|S)$
$= 0.45 \times 0.6 = 0.27$
$P(S \cap P') = P(S) - P(S \cap P)$
$= 0.45 - 0.27 = 0.18$
$P(S') = 1 - P(S)$
$= 1 - 0.45 = 0.55$
$P(S' \cap P) = P(S')P(P|S')$
$= 0.55 \times 0.2 = 0.11$
$P(S' \cap P') = P(S') - P(S' \cap P)$
$= 0.55 - 0.11 = 0.44$
So the Venn diagram looks like this:

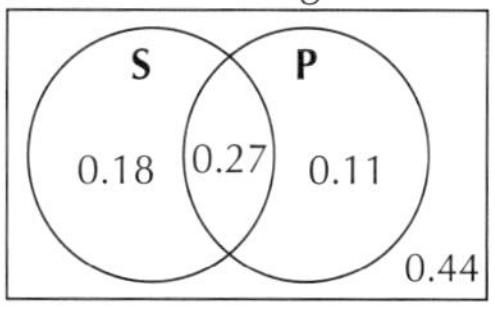

b) $P(P) = 0.27 + 0.11 = 0.38$
$P(S|P) = \frac{P(S \cap P)}{P(P)} = \frac{0.27}{0.38} = \frac{27}{38} = 0.711$ (3 s.f.)

Q6 Let S = 'owns smartphone' and let C = 'has contract costing more than £25 a month'. Then you know the following probabilities: $P(S) = 0.62$, $P(C) = 0.539$ and $P(C'|S) = 0.29$, and you want to find $P(S|C)$.
Using the conditional probability formula:
$P(S|C) = \frac{P(S \cap C)}{P(C)}$.
Use a tree diagram to help you find $P(S \cap C)$. You don't need to label all the branches, just the ones that help you answer the question:

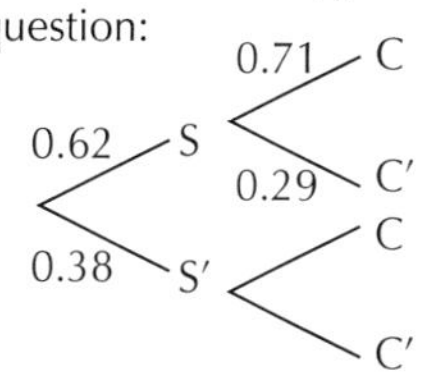

So $P(S \cap C) = 0.62 \times 0.71 = 0.4402$, and so:
$P(S|C) = \frac{P(S \cap C)}{P(C)} = \frac{0.4402}{0.539} = 0.817$ (3 s.f.)

2. Modelling with Probability

Exercise 2.1 — Modelling assumptions and probability

There are many possible answers to these questions — here are some examples:

Q1 – She has assumed that the probability of it landing on each face is equally likely — given that the shape is a cuboid, this is probably not a very good assumption.

– She has assumed that she throws it the same way each time and that her throw gives an equal probability of each face coming up, which might not be the case.

Q2 – She has assumed that the volunteer is equally likely to choose any of the 10 cards, but some people may tend to choose cards in a particular way e.g. from the middle or from the ends more often.

– She has assumed that the card is put back at random, and that the volunteer can't see and track the card before choosing the second one — if they could, they might intentionally try to choose the same one twice, which would affect the probability.

Q3 – He has assumed that his estimate of the probability (using the relative frequency from his data) is valid for children from any other school — there are likely to be lots of reasons why this is not the case. For example, children in urban areas may be more or less likely to walk than children in rural areas.

– The model doesn't account for the age of the child — it might be that older children are more likely to walk while younger children get driven to school.

– He has assumed that the probability holds on any particular day, but factors like weather will affect the probability — children would probably be more likely to walk to school on warm, sunny days than on cold, rainy days.

Chapter 12: The Normal Distribution

1. The Normal Distribution

Exercise 1.1 — The normal distribution

These questions are done using a calculator with an upper bound of 9999 or a lower bound of −9999 where needed.

Q1 **a)** $P(X < 50) = 0.977$ (3 s.f.)
b) $P(X \leq 43) = 0.726$ (3 s.f.)

Q2 **a)** $P(X \geq 28) = 0.0512$ (3 s.f.)
b) $P(X > 25) = 0.342$ (3 s.f.)

Q3 **a)** $P(X > 107) = 0.980$ (3 s.f.)
b) $P(X > 115) = 0.785$ (3 s.f.)

Q4 **a)** $P(X \leq 15) = 0.252$ (3 s.f.)
b) $P(X < 12) = 0.0478$ (3 s.f.)

Q5 **a)** $P(52 < X < 63) = 0.340$ (3 s.f.)
b) $P(57 \leq X < 66) = 0.0801$ (3 s.f.)

Q6 **a)** $P(0.45 \leq X \leq 0.55) = 0.175$ (3 s.f.)
b) $P(0.53 < X < 0.58) = 0.0970$ (3 s.f.)

Q7 **a)** $P(240 < X \leq 280) = 0.818$ (3 s.f.)
b) $P(232 < X < 288) = 0.938$ (3 s.f.)

Exercise 1.2 — The standard normal distribution, Z

These questions are done using a calculator with an upper bound of 9999 or a lower bound of −9999 where needed.

Q1 **a)** $P(Z \leq 1.87) = 0.969$ (3 s.f.)
b) $P(Z < 0.99) = 0.839$ (3 s.f.)
c) $P(Z > 2.48) = 0.00657$ (3 s.f.)
d) $P(Z \geq 0.14) = 0.444$ (3 s.f.)
e) $P(Z > -0.24) = 0.595$ (3 s.f.)
f) $P(Z > -1.21) = 0.887$ (3 s.f.)
g) $P(Z < -0.62) = 0.268$ (3 s.f.)
h) $P(Z \leq -2.06) = 0.0197$ (3 s.f.)
i) $P(Z \geq 1.23) = 0.109$ (3 s.f.)

Q2 **a)** $P(1.34 < Z < 2.18) = 0.0755$ (3 s.f.)
b) $P(0.76 \leq Z < 1.92) = 0.196$ (3 s.f.)
c) $P(-1.45 \leq Z \leq 0.17) = 0.494$ (3 s.f.)
d) $P(-2.14 < Z < 1.65) = 0.934$ (3 s.f.)
e) $P(-1.66 < Z \leq 1.66) = 0.903$ (3 s.f.)
f) $P(-0.34 \leq Z < 0.34) = 0.266$ (3 s.f.)
g) $P(-3.25 \leq Z \leq -2.48) = 0.00599$ (3 s.f.)
h) $P(-1.11 < Z < -0.17) = 0.299$ (3 s.f.)

For some of these questions, you could have used the symmetry of the distribution to work out the answer.

Exercise 1.3 — Converting to the Z distribution

Q1 **a)** Standard deviation $= \sqrt{100} = 10$ and mean $= 106$.
So $P(X \leq 100) = P\left(Z \leq \frac{100-106}{10}\right)$
$= P(Z \leq -0.6) = 0.274$ (3 s.f.)

b) $P(X > 122) = P\left(Z > \frac{122-106}{10}\right)$
$= P(Z > 1.6) = 0.0548$ (3 s.f.)

Q2 **a)** $P(X \geq 9) = P\left(Z \geq \frac{9-11}{2}\right)$
$= P(Z \geq -1) = 0.841$ (3 s.f.)

b) $P(10 \leq X \leq 12) = P\left(\frac{10-11}{2} \leq Z \leq \frac{12-11}{2}\right)$
$= P(-0.5 \leq Z \leq 0.5) = 0.383$ (3 s.f.)

Q3 **a)** $P(240 < X < 280) = P\left(\frac{240-260}{15} < Z < \frac{280-260}{15}\right)$
$= P(-1.3333 < Z < 1.3333) = 0.818$ (3 s.f.)

b) $P(232 < X < 288)$
$= P\left(\frac{232-260}{15} < Z < \frac{288-260}{15}\right)$
$= P(-1.867 < Z < 1.867) = 0.938$ (3 s.f.)

Exercise 1.4 — Finding x-values

These questions are done using a calculator with an upper bound of 9999 or a lower bound of −9999 where needed.

Q1 **a)** $P(X < x) = 0.8944$ for $x = 13$
b) $P(X \leq x) = 0.0304$ for $x = 10.5$
c) Using the fact that the area under the curve is 1:
$P(X < x) = 1 - P(X \geq x) = 1 - 0.2660 = 0.7340$
$P(X < x) = 0.7340$ for $x = 12.5$
d) Using the fact that the area under the curve is 1:
$P(X \leq x) = 1 - P(X > x) = 1 - 0.7917 = 0.2083$
$P(X \leq x) = 0.2083$ for $x = 11.35$

Q2 **a)** $P(X < x) = 0.975$ for $x = 32.84$ (2 d.p.)
b) Using the fact that the area under the curve is 1:
$P(X \leq x) = 1 - P(X > x) = 1 - 0.4 = 0.6$
$P(X \leq x) = 0.6$ for $x = 26.01$ (2 d.p.)

Q3 **a)** $P(53 < X < x) = 0.05$
$\Rightarrow P(X < x) - P(X \leq 53) = 0.05$
$\Rightarrow P(X < x) = 0.05 + P(X \leq 53)$
$= 0.05 + 0.8413 = 0.8913$
$P(X < x) = 0.8913$ for $x = 54.17$ (2 d.p.)

b) $P(x < X < 49) = 0.4312$
$\Rightarrow P(X < 49) - P(X \leq x) = 0.4312$
$\Rightarrow P(X \leq x) = P(X < 49) - 0.4312$
$= 0.5793 - 0.4312 = 0.1481$
$P(X < x) = 0.1481$ for $x = 42.78$ (2 d.p.)

Q4 a) $P(Z < z) = 0.7$ for $z = 0.524$ (3 s.f.)

b) $P(Z < z) = 0.999$ for $z = 3.09$ (3 s.f.)

c) Using the fact that the area under the curve is 1:
$P(Z \leq z) = 1 - P(Z > z) = 1 - 0.005 = 0.995$
$\Phi(z) = 0.995$ for $z = 2.58$ (3 s.f.)

d) Using the fact that the area under the curve is 1:
$P(Z \leq z) = 1 - P(Z > z) = 1 - 0.2 = 0.8$
$\Phi(z) = 0.8$ for $z = 0.842$ (3 s.f.)

Q5 a) $\Phi(z) = 0.004$ for $z = -2.65$ (3 s.f.)

b) Using the fact that the area under the curve is 1:
$P(Z \leq z) = 1 - P(Z > z) = 1 - 0.0951 = 0.9049$
$\Phi(z) = 0.9049$ for $z = 1.31$ (3 s.f.)

c) $P(-z < Z < z) = 0.9426$, so the remaining area is $1 - 0.9426 = 0.0574$.
Using symmetry, $P(Z \leq -z) = 0.0574 \div 2 = 0.0287$
$\Rightarrow P(Z \leq -z) = 0.0287$ for $z = -1.90$ (3 s.f.)
So $z = 1.90$ (3 s.f.)

d) $P(z < Z < -1.25) = 0.0949$
$\Rightarrow P(Z < -1.25) - P(Z \leq z) = 0.0949$
$\Rightarrow P(Z \leq z) = P(Z < -1.25) - 0.0949$
$\Rightarrow P(Z \leq z) = 0.1056 - 0.0949 = 0.0107$
$\Phi(z) = 0.0107$ for $z = -2.30$ (3 s.f.)

Exercise 1.5 — Finding the mean and standard deviation of a normal distribution

Q1 a) $P(X < 23) = 0.9332 \Rightarrow P\left(Z < \frac{23-\mu}{6}\right) = 0.9332$
$\Phi(z) = 0.9332$ for $z = 1.50$ (2 d.p.)
$\Rightarrow \frac{23-\mu}{6} = 1.5 \Rightarrow \mu = 23 - 1.5 \times 6 = 14$

b) $P(X < 57) = 0.9970 \Rightarrow P\left(Z < \frac{57-\mu}{8}\right) = 0.9970$
$\Phi(z) = 0.9970$ for $z = 2.75$ (2 d.p.)
$\Rightarrow \frac{57-\mu}{8} = 2.75 \Rightarrow \mu = 57 - 2.75 \times 8 = 35$

c) $P(X > 528) = 0.1292 \Rightarrow P\left(Z > \frac{528-\mu}{100}\right) = 0.1292$
$\Rightarrow P\left(Z \leq \frac{528-\mu}{100}\right) = 1 - 0.1292 = 0.8708$
$\Phi(z) = 0.8708$ for $z = 1.13$ (2 d.p.)
$\Rightarrow \frac{528-\mu}{100} = 1.13 \Rightarrow \mu = 528 - 1.13 \times 100 = 415$

d) $P(X < 11.06) = 0.0322 \Rightarrow P\left(Z < \frac{11.06-\mu}{0.4}\right) = 0.0322$
$\Phi(z) = 0.0322$ for $z = -1.85$ (2 d.p.)
$\Rightarrow \frac{11.06-\mu}{0.4} = -1.85$
$\Rightarrow \mu = 11.06 - (-1.85) \times 0.4 = 11.8$

e) $P(X > 1.52) = 0.9938 \Rightarrow P\left(Z > \frac{1.52-\mu}{0.02}\right) = 0.9938$
$\Rightarrow P\left(Z < \frac{\mu-1.52}{0.02}\right) = 0.9938$
$\Phi(z) = 0.9938$ for $z = 2.50$ (2 d.p.)
$\Rightarrow \frac{\mu-1.52}{0.02} = 2.5 \Rightarrow \mu = 2.5 \times 0.02 + 1.52 = 1.57$
Here you use the fact that $P(Z > z) = P(Z < -z)$ for the standard normal distribution.

Q2 Start with a sketch showing what you know about X.

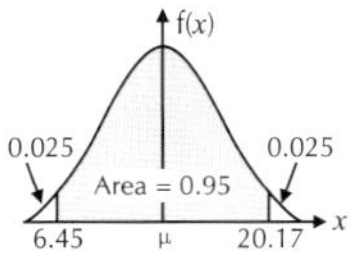

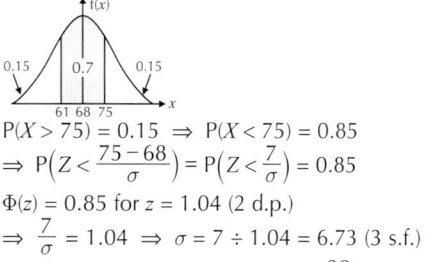

So, $P(X < 20.17) = 0.95 + 0.025 = 0.975$
$\Rightarrow P\left(Z < \frac{20.17-\mu}{3.5}\right) = 0.975$
$\Phi(z) = 0.975$ for $z = 1.96$ (2 d.p.)
$\Rightarrow \frac{20.17-\mu}{3.5} = 1.96 \Rightarrow \mu = 20.17 - 1.96 \times 3.5 = 13.31$
There are different ways you could do this question. For example, using the symmetry of the graph, you know that μ is exactly in the middle of 6.45 and 20.17, so you can simply find the average of these 2 values.

Q3 a) $P(X < 53) = 0.8944$
$\Rightarrow P\left(Z < \frac{53-48}{\sigma}\right) = P\left(Z < \frac{5}{\sigma}\right) = 0.8944$
$\Phi(z) = 0.8944$ for $z = 1.25$ (2 d.p.)
$\Rightarrow \frac{5}{\sigma} = 1.25 \Rightarrow \sigma = 5 \div 1.25 = 4$

b) $P(X < 528) = 0.7734$
$\Rightarrow P\left(Z < \frac{528-510}{\sigma}\right) = P\left(Z < \frac{18}{\sigma}\right) = 0.7734$
$\Phi(z) = 0.7734$ for $z = 0.75$ (2 d.p.)
$\Rightarrow \frac{18}{\sigma} = 0.75 \Rightarrow \sigma = 18 \div 0.75 = 24$

c) $P(X > 24) = 0.0367$
$\Rightarrow P\left(Z > \frac{24-17}{\sigma}\right) = P\left(Z > \frac{7}{\sigma}\right) = 0.0367$
$\Rightarrow P\left(Z \leq \frac{7}{\sigma}\right) = 0.9633$
$\Phi(z) = 0.9633$ for $z = 1.79$ (2 d.p.)
$\Rightarrow \frac{7}{\sigma} = 1.79 \Rightarrow \sigma = 7 \div 1.79 = 3.91$ (3 s.f.)

d) $P(X < 0.95) = 0.3085$
$\Rightarrow P\left(Z < \frac{0.95-0.98}{\sigma}\right) = P\left(Z < -\frac{0.03}{\sigma}\right) = 0.3085$
$\Phi(z) = 0.3085$ for $z = -0.5$.
$\Rightarrow -\frac{0.03}{\sigma} = -0.5 \Rightarrow \sigma = 0.03 \div 0.5 = 0.06$

e) $P(X > 4.85) = 0.8365$
$\Rightarrow P\left(Z > \frac{4.85-5.6}{\sigma}\right) = P\left(Z > -\frac{0.75}{\sigma}\right) = 0.8365$
$\Rightarrow P\left(Z < \frac{0.75}{\sigma}\right) = 0.8365$
$\Phi(z) = 0.8365$ for $z = 0.98$ (2 d.p.)
$\Rightarrow \frac{0.75}{\sigma} = 0.98 \Rightarrow \sigma = 0.75 \div 0.98 = 0.765$ (3 s.f.)

Q4 Start with a sketch showing what you know about X.

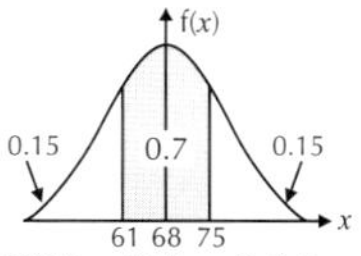

$P(X > 75) = 0.15 \Rightarrow P(X < 75) = 0.85$
$\Rightarrow P\left(Z < \frac{75-68}{\sigma}\right) = P\left(Z < \frac{7}{\sigma}\right) = 0.85$
$\Phi(z) = 0.85$ for $z = 1.04$ (2 d.p.)
$\Rightarrow \frac{7}{\sigma} = 1.04 \Rightarrow \sigma = 7 \div 1.04 = 6.73$ (3 s.f.)

Q5 a) $P(X < 30) = 0.9192 \Rightarrow P\left(Z < \frac{30-\mu}{\sigma}\right) = 0.9192$
$\Phi(z) = 0.9192$ for $z = 1.40$ (2 d.p.)
$\Rightarrow \frac{30-\mu}{\sigma} = 1.4 \Rightarrow 30 - \mu = 1.4\sigma$ ①
$P(X < 36) = 0.9953 \Rightarrow P\left(Z < \frac{36-\mu}{\sigma}\right) = 0.9953$
$\Phi(z) = 0.9953$ for $z = 2.60$ (2 d.p.)
$\Rightarrow \frac{36-\mu}{\sigma} = 2.6 \Rightarrow 36 - \mu = 2.6\sigma$ ②
Subtracting equation ① from equation ② gives:
$36 - 30 - \mu - (-\mu) = 2.6\sigma - 1.4\sigma$
$\Rightarrow 6 = 1.2\sigma \Rightarrow \sigma = 5$
Putting $\sigma = 5$ into equation ① gives:
$\mu = 30 - 1.4 \times 5 = 23$
So $\mu = 23$ and $\sigma = 5$

b) $P(X < 4) = 0.9332 \Rightarrow P\left(Z < \frac{4-\mu}{\sigma}\right) = 0.9332$
$\Phi(z) = 0.9332$ for $z = 1.50$ (2 d.p.)
$\Rightarrow \frac{4-\mu}{\sigma} = 1.5 \Rightarrow 4 - \mu = 1.5\sigma$ ①
$P(X < 4.3) = 0.9987 \Rightarrow P\left(Z < \frac{4.3-\mu}{\sigma}\right) = 0.9987$
$\Phi(z) = 0.9987$ for $z = 3.01$ (2 d.p.)
$\Rightarrow \frac{4.3-\mu}{\sigma} = 3.01 \Rightarrow 4.3 - \mu = 3.01\sigma$ ②
Subtracting equation ① from equation ② gives:
$4.3 - 4 - \mu - (-\mu) = 3.01\sigma - 1.5\sigma$
$\Rightarrow 0.3 = 1.51\sigma \Rightarrow \sigma = 0.198675...$
Putting $\sigma = 0.198675...$ into equation ① gives:
$\mu = 4 - 1.5 \times 0.198675... = 3.70$ (2 d.p.)
So $\mu = 3.7$ and $\sigma = 0.2$

c) $P(X < 20) = 0.7881 \Rightarrow P\left(Z < \frac{20-\mu}{\sigma}\right) = 0.7881$
$\Phi(z) - 0.7881$ for $z = 0.80$ (2 d.p.)
$\Rightarrow \frac{20-\mu}{\sigma} = 0.8 \Rightarrow 20 - \mu = 0.8\sigma$ ①
$P(X < 14) = 0.0548 \Rightarrow P\left(Z < \frac{14-\mu}{\sigma}\right) = 0.0548$
$\Phi(z) = 0.0548$ for $z = -1.60$ (2 d.p.)
$\Rightarrow \frac{14-\mu}{\sigma} = -1.6 \Rightarrow \mu - 14 = 1.6\sigma$ ②
Adding equations ① and ② gives:
$20 - 14 - \mu + \mu = 0.8\sigma + 1.6\sigma$
$\Rightarrow 6 = 2.4\sigma \Rightarrow \sigma = 2.5$
Putting $\sigma = 2.5$ into equation ② gives:
$\mu = 1.6 \times 2.5 + 14 = 18$
So $\mu = 18$ and $\sigma = 2.5$

d) $P(X < 696) = 0.9713 \Rightarrow P\left(Z < \frac{696-\mu}{\sigma}\right) = 0.9713$
$\Phi(z) = 0.9713$ for $z = 1.90$ (2 d.p.)
$\Rightarrow \frac{696-\mu}{\sigma} = 1.9 \Rightarrow 696 - \mu = 1.9\sigma$ ①
$P(X < 592) = 0.2420$
$\Rightarrow P\left(Z < \frac{592-\mu}{\sigma}\right) = 0.2420$
$\Phi(z) = 0.242$ for $z = -0.70$ (2 d.p.)
$\Rightarrow \frac{592-\mu}{\sigma} = -0.7 \Rightarrow \mu - 592 = 0.7\sigma$ ②
Adding equations ① and ② gives:
$696 - 592 - \mu + \mu = 1.9\sigma + 0.7\sigma$
$\Rightarrow 104 = 2.6\sigma \Rightarrow \sigma = 40$
Putting $\sigma = 40$ into equation ② gives:
$\mu = 0.7 \times 40 + 592 = 620$
So $\mu = 620$ and $\sigma = 40$

e) $P(X > 33) = 0.1056 \Rightarrow P\left(Z > \frac{33-\mu}{\sigma}\right) = 0.1056$
$\Rightarrow P\left(Z \leq \frac{33-\mu}{\sigma}\right) = 1 - 0.1056 = 0.8944$
$\Phi(z) = 0.8944$ for $z = 1.25$ (2 d.p.)
$\Rightarrow \frac{33-\mu}{\sigma} = 1.25 \Rightarrow 33 - \mu = 1.25\sigma$ ①
$P(X > 21) = 0.9599 \Rightarrow P\left(Z > \frac{21-\mu}{\sigma}\right) = 0.9599$
$\Rightarrow P\left(Z < \frac{\mu-21}{\sigma}\right) = 0.9599$
$\Phi(z) = 0.9599$ for $z = 1.75$ (2 d.p.)
$\Rightarrow \frac{\mu-21}{\sigma} = 1.75 \Rightarrow \mu - 21 = 1.75\sigma$ ②
Adding equations ① and ② gives:
$33 - 21 - \mu + \mu = 1.25\sigma + 1.75\sigma$
$\Rightarrow 12 = 3\sigma \Rightarrow \sigma = 4$
Putting $\sigma = 4$ into equation ② gives:
$\mu = 1.75 \times 4 + 21 = 28$
So $\mu = 28$ and $\sigma = 4$

f) $P(X > 66) = 0.3632 \Rightarrow P\left(Z > \frac{66-\mu}{\sigma}\right) = 0.3632$
$\Rightarrow P\left(Z \leq \frac{66-\mu}{\sigma}\right) = 0.6368$
$\Phi(z) = 0.6368$ for $z = 0.35$ (2 d.p.)
$\Rightarrow \frac{66-\mu}{\sigma} = 0.35 \Rightarrow 66 - \mu = 0.35\sigma$ ①
$P(X < 48) = 0.3446 \Rightarrow P\left(Z < \frac{48-\mu}{\sigma}\right) = 0.3446$
$\Phi(z) = 0.3446$ for $z = -0.40$ (2 d.p.)
$\Rightarrow \frac{48-\mu}{\sigma} = -0.4 \Rightarrow \mu - 48 = 0.4\sigma$ ②
Adding equations ① and ② gives:
$66 - 48 - \mu + \mu = 0.35\sigma + 0.4\sigma$
$\Rightarrow 18 = 0.75\sigma \Rightarrow \sigma = 24$
Putting $\sigma = 24$ into equation ② gives:
$\mu = 0.4 \times 24 + 48 = 57.6$
So $\mu = 57.6$ and $\sigma = 24$

Exercise 1.6 — The normal distribution in real-life situations

Q1 a) Let $T \sim N(36, 6^2)$ represent the length of time taken to replace red blood cells, in days. Using a calculator: $P(T < 28) = 0.0912$ (3 s.f.)

b) $56 \times 0.0912 = 5.1702$
So approximately 5 blood donors replace their blood cells more quickly than Edward.

c) Let b = the number of days taken by Bella.
Then, $P(T > b) = 0.063$.
$\Rightarrow P(T \leq b) = 1 - 0.063 = 0.9370$
$P(X \leq x) = 0.9370$ for $x = 45.18$ days
Don't forget to say 'days' in your answer — you need to answer the question in the context in which it was asked.

Q2 a) Let $T \sim N(51, 2.1^2)$ represent the personal best time taken to run 400 m in seconds. Using a calculator: $P(T > 49.3) = 0.791$ (3 s.f.)
So, 79.1% are slower than Gary.

b) Let a = the time to beat. Then, $P(T < a) = 0.2$.
$P(T < a) = 0.2$ for $a = 49.2$ s (3 s.f.)

Q3 a) Let V = volume of vinegar in ml. Then $V \sim N(\mu, 5^2)$.
$P(V < 506) = 0.719 \Rightarrow P\left(Z < \frac{506-\mu}{5}\right) = 0.719$
$\Phi(z) = 0.719$ for $z = 0.58$ (2 d.p.)
$\Rightarrow \frac{506-\mu}{5} = 0.58$
$\Rightarrow \mu = 506 - 0.58 \times 5 = 503.1$ ml

b) Using $\mu = 503.1$ gives $P(V < 500) = 0.268$ (3 s.f.)

c) $P(V \geq 500) = 1 - P(V < 500) = 1 - 0.268 = 0.732$
$0.732 \times 1303 = 953.796$
So approximately 954 bottles contain at least 500 ml.

Q4 a) Let $L \sim N(300, 50^2)$ represent the lifetime of a battery in hours. Then $P(L < 200) = 0.0228$ (3 s.f.)

b) $P(L > 380) = 0.0548$ (3 s.f.)

c) Assuming that the lifetimes of the batteries are independent, the probability that both last at least 380 hours $= 0.0548 \times 0.0548 = 0.00300$ (3 s.f.).
If two events are independent, it means you can multiply their probabilities together.

d) $P(160 < L < h) = 0.9746$

$\Rightarrow P\left(\frac{160-300}{50} < Z < \frac{h-300}{50}\right) = 0.9746$

$\Rightarrow P\left(-2.8 < Z < \frac{h-300}{50}\right) = 0.9746$

$\Rightarrow P\left(Z < \frac{h-300}{50}\right) - P(Z \leq -2.8) = 0.9746$

$\Rightarrow P\left(Z < \frac{h-300}{50}\right) = 0.9746 + P(Z \leq -2.8)$

$\Rightarrow P\left(Z < \frac{h-300}{50}\right) = 0.9746 + 0.0026$

$\Rightarrow P\left(Z < \frac{h-300}{50}\right) = 0.9772$

$\Phi(z) = 0.9772$ for $z = 2.00$ (3 s.f.)

So $\frac{h-300}{50} = 2 \Rightarrow h = 2 \times 50 + 300 = 400$

Q5 a) Let H = height in cm. Then $H \sim N(175, \sigma^2)$.

$P(H > 170) = 0.8 \Rightarrow P\left(Z > \frac{170-175}{\sigma}\right) = 0.8$

$\Rightarrow P\left(Z > -\frac{5}{\sigma}\right) = P\left(Z < \frac{5}{\sigma}\right) = 0.8$

$\Phi(z) = 0.8$ for $z = 0.842$ (3 s.f.)

So $\frac{5}{\sigma} = 0.842 \Rightarrow \sigma = 5 \div 0.842 = 5.94$ cm (3 s.f.)

b) Lower bound = 175 – 4 = 171 and upper bound = 175 + 4 = 179.
So $P(171 < H < 179) = 0.499$.

Q6 a) Let $M \sim N(60, 3^2)$ be the mass of an egg in grams.
$P(M > 60 - m) = 0.9525$

$\Rightarrow P\left(Z > \frac{60-m-60}{3}\right) = 0.9525$

$\Rightarrow P\left(Z > \frac{-m}{3}\right) = 0.9525$

$\Rightarrow P\left(Z < \frac{m}{3}\right) = 0.9525$

$\Phi(z) = 0.9525$ for $z = 1.67$

So $\frac{m}{3} = 1.67 \Rightarrow m = 1.67 \times 3 = 5.01$
$= 5$ g to the nearest gram.

b) Let c = the maximum mass of an egg in one of farmer Elizabeth's sponge cakes. Then, $P(M \leq c) = 0.1$.

So $P\left(Z \leq \frac{c-60}{3}\right) = 0.1 \Rightarrow P\left(Z \leq \frac{60-c}{3}\right) = 0.9$

$\Phi(z) = 0.9$ for $z = 1.28 \Rightarrow \frac{60-c}{3} = 1.28$

$\Rightarrow c = 60 - 3 \times 1.28 = 56.16$
$= 56$ g to the nearest gram

So the maximum mass is 56 grams.

Q7 Let R = rainfall in cm. Then $R \sim N(\mu, \sigma^2)$.

$P(R < 4) = 0.102 \Rightarrow P\left(Z < \frac{4-\mu}{\sigma}\right) = 0.102$

$\Rightarrow P\left(Z > \frac{\mu-4}{\sigma}\right) = 0.102 \Rightarrow P\left(Z \leq \frac{\mu-4}{\sigma}\right) = 0.898$

$\Phi(z) = 0.898$ for $z = 1.27$

$\Rightarrow \frac{\mu-4}{\sigma} = 1.27 \Rightarrow \mu - 4 = 1.27\sigma$ ①

$P(R > 7) = 0.648 \Rightarrow P\left(Z > \frac{7-\mu}{\sigma}\right) = 0.648$

$\Rightarrow P\left(Z < \frac{\mu - 7}{\sigma}\right) = 0.648$

$\Phi(z) = 0.648$ for $z = 0.38$

$\Rightarrow \frac{\mu-7}{\sigma} = 0.38 \Rightarrow \mu - 7 = 0.38\sigma$ ②

Subtracting equation ② from equation ① gives:
$\mu - \mu - 4 - (-7) = 1.27\sigma - 0.38\sigma$
$\Rightarrow 3 = 0.89\sigma \Rightarrow \sigma = 3.37$ (3 s.f.)

Putting $\sigma = 3.37$ cm into equation ① gives:
$\mu = 1.27 \times 3.37 + 4 = 8.28$ cm (3 s.f.)
So $\mu = 8.28$ cm and $\sigma = 3.37$ cm (3 s.f.)

2. Normal Approximation to a Binomial Distribution

Exercise 2.1 — Continuity corrections

Q1 a) $P(4.5 < Y < 5.5)$ **b)** $P(11.5 < Y < 15.5)$
c) $P(Y < 10.5)$

Q2 a) $P(49.5 < Y < 50.5)$ **b)** $P(Y < 299.5)$
c) $P(Y > 98.5)$

Q3 a) $P(X = 200)$ is approximated by $P(199.5 < Y < 200.5)$
b) $P(X \geq 650)$ is approximated by $P(Y > 649.5)$.
c) $P(X < 300)$ is approximated by $P(Y < 299.5)$.
d) $P(499 \leq X \leq 501)$ is approximated by $P(498.5 < Y < 501.5)$
e) $P(250 \leq X \leq 750)$ is approximated by $P(249.5 < Y < 750.5)$
f) $P(100 \leq X < 900)$ is approximated by $P(99.5 < Y < 899.5)$

Exercise 2.2 — Normal approximation to a binomial distribution

Q1 a) n is large (600) and $p = 0.51 \approx 0.5$, so a normal approximation would be suitable.
b) n is large (100) but $p = 0.98$ is not close to 0.5. It might still be OK if np and $n(1 - p)$ are > 5. $np = 98$ but $n(1 - p) = 2$, so a normal approximation would not be suitable.
c) n is large (100) but $p = 0.85$ is not close to 0.5. It might still be OK if np and $n(1 - p)$ are > 5. $np = 85$ and $n(1 - p) = 15$, so a normal approximation would be suitable.
d) n is not large (6), so a normal approximation would not be suitable.

Q2 a) $\mu = np = 350 \times 0.45 = 157.5$
$\sigma^2 = np(1 - p) = 350 \times 0.45 \times 0.55 = 86.625$
b) $\mu = np = 250 \times 0.35 = 87.5$
$\sigma^2 = np(1 - p) = 250 \times 0.35 \times 0.65 = 56.875$
c) $\mu = np = 70 \times 0.501 = 35.07$
$\sigma^2 = np(1 - p) = 70 \times 0.501 \times 0.499 = 17.500$ (3 d.p.)

Q3 a) n is very large and even though p is not close to 0.5, $np = 50 > 5$ and $n(1 - p) = 950 > 5$, so the normal distribution is a suitable approximation for this distribution.
b) The normal approximation has
$\mu = np = 1000 \times 0.05 = 50$
$\sigma^2 = np(1 - p) = 1000 \times 0.05 \times 0.95 = 47.5$
So approximate X with $Y \sim N(50, 47.5)$.
You're looking for the probability that $X > 75$.
$P(X > 75) \approx P(Y > 75) = 0.000143$ (3 s.f.)
If you used a continuity correction in this question, you'd work out P(Y > 75.5) = 0.000108 (3 s.f.) instead.

3. Choosing Probability Distributions

Exercise 3.1 — Choosing probability distributions

Q1 a) Most of the data values are in the middle and the number of values tails off towards the ends. It is also fairly symmetrical.
b) $\mu = 46$. There is a point of inflection at approximately $s = 50$. This occurs when $s = \mu + \sigma \Rightarrow 50 = 46 + \sigma \Rightarrow \sigma = 4$
You could have used any of the other conditions involving μ and σ, so any answer between 3 and 4 is acceptable.

Q2 a) There are 10 'trials', each with a probability of success of 0.65. So the number of successful first-serves in one game, G, can be modelled with a binomial distribution — $G \sim B(10, 0.65)$

b) There are 100 'trials', each with a probability of success of 0.65. So the number of successful first-serves in a match, M, can be modelled with a binomial distribution — $M \sim B(100, 0.65)$
Alternatively, n is large, and 0.65 is close to 0.5 (to check, for $n = 100$, $np = 65 > 5$ and $n(1 - p) = 35 > 5$), so M could be approximated by a normal distribution Y, where $\mu = 65$ and $\sigma^2 = 22.75$ — $Y \sim N(65, 22.75)$.
Either a binomial model or a normal approximation is OK in this case, since n is sufficiently large.

Chapter 13: Hypothesis Testing

1. The Product Moment Correlation Coefficient

Exercise 1.1 — PMCC hypothesis testing

Q1 a) This is a one-tailed test, so the hypotheses are $H_0: \rho = 0$ and $H_1: \rho > 0$.
Find the critical value for a one-tailed significance level of $\alpha = 2.5\%$ and a sample size of $n = 8$.
Using the critical values table, the critical value is 0.7067, so the critical region is $r \geq 0.7067$.

b) 0.994 is very close to 1, so it suggests strong positive correlation in the sample, i.e. it suggests that longer leaves tend to also be wider.
Since $0.994 > 0.7067$ the result is significant.
There is evidence at the 2.5% level of significance to reject H_0 and to support the alternative hypothesis that the length and width of leaves are positively correlated.

Q2 This is a one-tailed test, so the hypotheses are $H_0: \rho = 0$ and $H_1: \rho > 0$. Find the critical value for a one-tailed significance level of $\alpha = 5\%$ and a sample size of $n = 13$.
Using the critical values table, the critical value is 0.4762, so you would reject H_0 if $r \geq 0.4762$.
The test statistic is $r = 0.4655$, which is less than the critical value, so the result is not significant.
There is insufficient evidence at the 5% significance level to reject H_0, the hypothesis that maths and science test results are not correlated in the population.

Q3 a) This is a two-tailed test, so the hypotheses are $H_0: \rho = 0$ and $H_1: \rho \neq 0$. Find the critical value for a two-tailed significance level of $\alpha = 1\%$ and a sample size of $n = 9$.
Using the critical values table, the critical value is 0.7977, so the critical region is $r \geq 0.7977$ or $r \leq -0.7977$.

b) $r = -0.8925$ is close to -1, so it suggests strong negative correlation in the sample, i.e. higher weight tends to mean worse kidney function.
You would reject H_0 if $r \geq 0.7977$ or $r \leq -0.7977$.
$r = -0.8925$ is inside the critical region so there is sufficient evidence to reject H_0 and to support the alternative hypothesis that kidney function and weight are correlated.

2. Hypothesis Tests of the Mean of a Population

Exercise 2.1 — Hypothesis tests of a population mean

Q1 a) Let the mean weight in grams of all plums from the tree this year be μ.
Then $H_0: \mu = 42$ and $H_1: \mu > 42$. The significance level is $\alpha = 0.05$ and it is a one-tailed test.
Let $\overline{X}$ be the sample mean of the weight of plums in grams. Then under H_0, $X \sim N(42, 16)$,
so $\overline{X} \sim N\left(42, \frac{16}{25}\right) = N(42, 0.64)$.
So under H_0, $Z = \frac{\overline{X} - 42}{\sqrt{0.64}} \sim N(0, 1)$.
Now $\overline{x} = 43.5$, so $z = \frac{43.5 - 42}{\sqrt{0.64}} = 1.875$
You're interested in the higher end of the distribution, so find the critical value, z, such that $P(Z > z) = 0.05$.
From your calculator, $P(Z \leq z) = 0.95$ for $z = 1.645$.
So the critical region is $Z > 1.645$.
Since $1.875 > 1.645$, the result lies in the critical region and is significant.
There is significant evidence at the 5% level to reject H_0 in favour of the alternative hypothesis that the weight of plums from the tree has increased.

b) At the 1% level, you'll have a different critical region, but everything else is the same. So you're looking for a critical value, z, such that $P(Z > z) = 0.01$.
From your calculator, $P(Z \leq z) = 0.99$ for $z = 2.326$.
So the critical region is $Z > 2.326$.
$1.875 < 2.326$, so the result is not significant. There is insufficient evidence at the 1% level to reject H_0.

Q2 Let the mean of all the weekly winnings after the new person joins the games be £μ.
Then $H_0: \mu = 24$ and $H_1: \mu \neq 24$. The significance level is $\alpha = 0.01$, and it is a two-tailed test.
Let $\overline{X}$ be the sample mean of Bree's winnings.
Then under H_0, $X \sim N(24, 4)$,
so $\overline{X} \sim N\left(24, \frac{4}{12}\right) = N\left(24, \frac{1}{3}\right)$.
So under H_0, $Z = \frac{\overline{X} - 24}{\sqrt{\frac{1}{3}}} \sim N(0, 1)$.
Now $\overline{x} = 22.5$, so $z = \frac{22.5 - 24}{\sqrt{\frac{1}{3}}} = -2.598$ (3 d.p.)
This is a two-tailed test, so the critical region will be split. You want to find critical values, $\pm z$, such that $P(Z > z) = \frac{\alpha}{2} = 0.005$. From your calculator, $P(Z \leq z) = 0.995$ for $z = 2.575$. The other critical value will be -2.575. So the critical region is $Z < -2.575$ or $Z > 2.575$.
Since $z = -2.598 < -2.575$, the result lies in the critical region and is significant. There is significant evidence at the 1% level to reject H_0 in favour of the alternative hypothesis that Bree's mean winnings have changed.

Q3 Let the mean mark of all the students in the school be μ.
Then $H_0: \mu = 70$ and $H_1: \mu < 70$. The significance level is $\alpha = 0.05$, and it is a one-tailed test.
Let $\overline{X}$ be the sample mean of the scores.
Then under H_0, $X \sim N(70, 64)$,
so $\overline{X} \sim N\left(70, \frac{64}{40}\right) = N(70, 1.6)$.

So under H_0, $Z = \frac{\overline{X} - 70}{\sqrt{1.6}} \sim N(0, 1)$.

$\overline{x} = 67.7$, so $z = \frac{67.7 - 70}{\sqrt{1.6}} = -1.818$ (3 d.p.)

You're interested in the lower end of the distribution, so find the critical value, z, such that $P(Z \le z) = 0.05$. From your calculator, $P(Z \le z) = 0.05$ for $z = -1.645$, so the critical region is $Z < -1.645$.

Since $-1.818 < -1.645$, the result lies in the critical region, so there is significant evidence at the 5% level to reject H_0 in favour of the alternative hypothesis that the mean score this year is lower.

Q4 **a)** Let the mean weight in grams of all the pigeons in the city centre this year be μ.
Then H_0: $\mu = 300$ and H_1: $\mu > 300$.

b) This is a one-tailed test.

c) Let $\overline{X}$ be the sample mean of the pigeon weights.
Then under H_0, $X \sim N(300, 45^2)$,
so $\overline{X} \sim N\left(300, \frac{45^2}{25}\right) = N(300, 81)$.

So under H_0, $Z = \frac{\overline{X} - 300}{\sqrt{81}} \sim N(0, 1)$.

$\overline{x} = 314$, so $z = \frac{314 - 300}{\sqrt{81}} = 1.556$ (3 d.p.)

You're interested in the higher end of the distribution, so find the critical value, z, such that $P(Z > z) = 0.01$. From your calculator, $P(Z \le z) = 0.99$ for $z = 2.326$. So the critical region is $Z > 2.326$.
Since $1.556 < 2.326$, the result does not lie in the critical region and is not significant. There is insufficient evidence at the 1% level to reject H_0 in favour of the alternative hypothesis that the mean pigeon weight has gone up.

Q5 Let the mean number of hours that Kyle has spent studying each week this year be μ. Then H_0: $\mu = 32$ and H_1: $\mu > 32$. This is a one-tailed test.

Let $\overline{X}$ be the sample mean number of hours.
Then under H_0, $X \sim N(32, 5^2)$,
so $\overline{X} \sim N\left(32, \frac{5^2}{20}\right) = N(32, 1.25)$.

So under H_0, $Z = \frac{\overline{X} - 32}{\sqrt{1.25}} \sim N(0, 1)$.

$\overline{x} = 34$, so $z = \frac{34 - 32}{\sqrt{1.25}} = 1.789$ (3 d.p.)

First test at the 5% level:
You're interested in the higher end of the distribution, so find the critical value, z, such that $P(Z > z) = 0.05$. From your calculator, $P(Z \le z) = 0.95$ for $z = 1.645$. So the critical region is $Z > 1.645$. Since $1.789 > 1.645$, the result lies in the critical region and is significant. There is significant evidence at the 5% level to reject H_0 in favour of the alternative hypothesis that the mean amount of time spent studying has gone up.

Now test at the 1% level:
Again, you're interested in the higher end of the distribution, so find the critical value, z, such that $P(Z > z) = 0.01$. From your calculator, $P(Z \le z) = 0.99$ for $z = 2.326$. So the critical region is $Z > 2.326$. $1.789 < 2.326$, so the result does not lie in the critical region and is not significant.
There is insufficient evidence at the 1% level to reject H_0 in favour of the alternative hypothesis that the mean amount of time spent studying has gone up.

So the null hypothesis can be rejected at the 5% level, but not at the 1% level.

Chapter 14: Kinematics

1. Projectiles

Exercise 1.1 — The two components of velocity

Q1 **a)** Horizontal component:
$10 \cos 20° = 9.40$ ms^{-1} (3 s.f.)
Vertical component:
$10 \sin 20° = 3.42$ ms^{-1} (3 s.f.)

b) Horizontal component:
$18 \cos 65° = 7.61$ ms^{-1} (3 s.f.)
Vertical component:
$18 \sin 65° = 16.3$ ms^{-1} (3 s.f.)

c) Horizontal component:
$-6.8 \cos 21.6° = -6.32$ ms^{-1} (3 s.f.)
Vertical component:
$6.8 \sin 21.6° = 2.50$ ms^{-1} (3 s.f.)

d) Horizontal component:
$9.7 \cos 19.7° = 9.13$ ms^{-1} (3 s.f.)
Vertical component:
$-9.7 \sin 19.7° = -3.27$ ms^{-1} (3 s.f.)

e) Horizontal component:
$-24 \cos 84° = -2.51$ ms^{-1} (3 s.f.)
Vertical component:
$-24 \sin 84° = -23.9$ ms^{-1} (3 s.f.)

f) Horizontal component:
$16 \cos 123° = -8.71$ ms^{-1} (3 s.f.)
Vertical component:
$16 \sin 123° = 13.4$ ms^{-1} (3 s.f.)

Q2 Horizontal component:
$8 \cos 35° = 6.55$ ms^{-1} (3 s.f.)
Vertical component:
$8 \sin 35° = 4.59$ ms^{-1} (3 s.f.)

Q3 Horizontal component = 0 ms^{-1}
Vertical component = 45 ms^{-1}

Q4

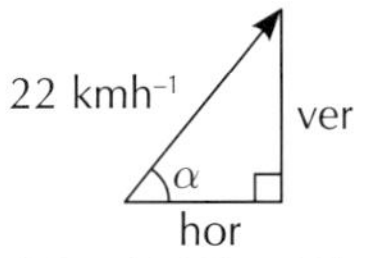

$(22 \times 1000) \div 60^2 = 6.1111... = 6.11$ ms^{-1} (3 s.f.)
Horizontal component = $6.1111... \cos \alpha$ ms^{-1}
Vertical component = $6.1111... \sin \alpha$ ms^{-1}

Q5

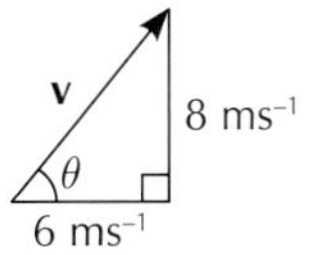

speed = $\sqrt{6^2 + 8^2} = 10$ ms^{-1}
$\theta = \tan^{-1}\left(\frac{8}{6}\right) = 53.1°$ (3 s.f.) above the horizontal

Q6

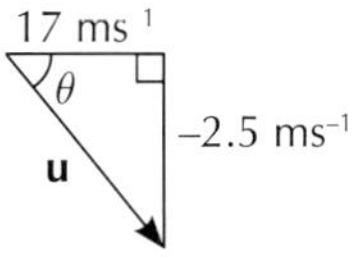

speed = $\sqrt{17^2 + (-2.5)^2} = 17.2$ ms^{-1} (3 s.f.)
$\theta = \tan^{-1}\left(\frac{-2.5}{17}\right) = -8.37°$ (3 s.f.)
i.e. 8.37° below the horizontal

Exercise 1.2 — The constant acceleration equations

Q1 a) Resolving vertically, taking up as positive:
$u = 15 \sin 50°,\ v = 0,\ a = -9.8,\ t = t$
$v = u + at$
$0 = 15 \sin 50° - 9.8t$
$t = 15 \sin 50° \div 9.8 = 1.17$ s (3 s.f.)

b) Resolving vertically, taking up as positive:
$s = s,\ u = 15 \sin 50°,\ v = 0,\ a = -9.8,$
$v^2 = u^2 + 2as$
$0 = (15 \sin 50°)^2 - 19.6s$
$s = (15 \sin 50°)^2 \div 19.6 = 6.74$ m (3 s.f.)

Q2 a) Resolving horizontally, taking right as positive:
$s = s,\ u = 12 \cos 37°,\ a = 0,\ t = 0.5$
$s = ut + \frac{1}{2}at^2$
$s = 12 \cos 37° \times 0.5 = 4.791... = 4.79$ m (3 s.f.)

b) Resolving vertically, taking up as positive:
$s = s,\ u = 12 \sin 37°,\ a = -9.8,\ t = 0.5$
$s = ut + \frac{1}{2}at^2$
$s = (12 \sin 37° \times 0.5) + (\frac{1}{2} \times -9.8 \times 0.5^2) = 2.385...$
$= 2.39$ m (3 s.f.)

c) $\sqrt{(4.791...)^2 + (2.385...)^2} = 5.35$ m (3 s.f.)

Q3 a) Resolving vertically, taking up as positive:
$s = 0,\ u = 8 \sin 59°,\ a = -9.8,\ t = t$
$s = ut + \frac{1}{2}at^2$
$0 = (8 \sin 59°)t - 4.9t^2$
$0 = t(8 \sin 59° - 4.9t)$
$\Rightarrow t = 0$ or $(8 \sin 59° - 4.9t) = 0$
$\Rightarrow t = 8 \sin 59° \div 4.9$
$= 1.399... = 1.40$ s (3 s.f.)

b) Resolving horizontally, taking right as positive:
$s = s,\ u = 8 \cos 59°,\ a = 0,\ t = 1.399...$
$s = ut + \frac{1}{2}at^2$
$s = 8 \cos 59° \times 1.399... = 5.77$ m (3 s.f.)

Q4 a) $\tan\theta = \frac{\text{opp}}{\text{adj}}$, so draw a right-angled triangle with $\sqrt{3}$ as the 'opposite' side and 1 as the 'adjacent' side.
Using Pythagoras' theorem, the hypotenuse will be 2:

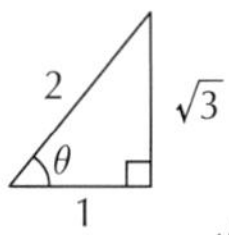

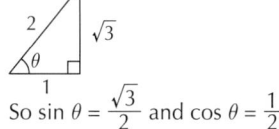

So $\sin\theta = \frac{\sqrt{3}}{2}$ and $\cos\theta = \frac{1}{2}$
Resolving vertically, taking up as positive:
$u = 18 \sin\theta = 9\sqrt{3},\ v = v,\ a = -9.8,\ t = 2$
$v = u + at$
$v = 9\sqrt{3} - 19.6 = -4.011...$
Resolving horizontally, taking right as positive:
$v = u = 18 \cos\theta = 9$
So speed $= \sqrt{(-4.011...)^2 + 9^2} = 9.85$ ms^{-1} (3 s.f.)
You could also have solved this by spotting that θ = 60°, since tan 60° is one of the common values from p.53.

b) $\alpha = \tan^{-1}\left(\frac{-4.011...}{9}\right) = -24.0°$ (3 s.f.)
i.e. 24.0° below the horizontal.

Q5 a) Resolving vertically, taking down as positive:
$s = 40,\ u = 0,\ a = 9.8,\ t = t$
$s = ut + \frac{1}{2}at^2$
$40 = 4.9t^2$
$t^2 = 8.163... \Rightarrow t = 2.857... = 2.86$ s (3 s.f.)

b) Resolving horizontally, taking right as positive:
$s = s,\ u = 80,\ a = 0,\ t = 2.857...$
$s = ut + \frac{1}{2}at^2 = 80 \times 2.857... = 229$ m (3 s.f.)

c) E.g. The acceleration should take into account the increased drag from the water — vertical acceleration can no longer be assumed to be equal to g, and horizontal acceleration can no longer be assumed to be 0.

Q6 Resolving vertically, taking up as positive:
$s = 11 - 4 = 7,\ u = 20 \sin 45°,\ a = -9.8,\ t = t$
$s = ut + \frac{1}{2}at^2$
$7 = (20 \sin 45°)t - 4.9t^2$
$4.9t^2 - (20 \sin 45°)t + 7 = 0$
Using the quadratic formula:
$t = \frac{20\sin 45° \pm \sqrt{(-20\sin 45°)^2 - (4 \times 4.9 \times 7)}}{9.8}$
$= \frac{14.142... \pm \sqrt{62.8}}{9.8}$
$\Rightarrow t = 2.251...$ or $t = 0.634...$
So the object is higher than 11 m above the ground for $2.251... - 0.634... = 1.62$ s (3 s.f.)

Q7 First find the ball's vertical displacement from its starting point when it passes through the first target.
Resolving vertically, taking up as positive:
$s = s_1,\ u = 24 \sin 70°,\ a = -9.8,\ t = 3$
$s = ut + \frac{1}{2}at^2$
$s_1 = (24 \sin 70° \times 3) + (\frac{1}{2} \times -9.8 \times 3^2) = 23.557...$ m
Now find the ball's vertical displacement from its starting point when it passes through the second target.
Resolving vertically, taking up as positive:
$s = s_2,\ u = 24 \sin 70°,\ a = -9.8,\ t = 4$
$s = ut + \frac{1}{2}at^2$
$s_2 = (24 \sin 70° \times 4) + (\frac{1}{2} \times -9.8 \times 4^2) = 11.810...$ m
So $h = 23.557... - 11.810... = 11.7$ m (3 s.f.)

Q8 First find the value of α by considering the particle's motion from its point of projection to its maximum height. Resolving vertically, taking up as positive:
$s = 2.8 - 0.6 = 2.2,\ u = 7.5 \sin\alpha,\ v = 0,\ a = -9.8$
$v^2 = u^2 + 2as$
$0 = (7.5 \sin\alpha)^2 + (2 \times -9.8 \times 2.2)$
$\sin^2\alpha = 43.12 \div 56.25$
$\sin\alpha = 0.875... \Rightarrow \alpha = 61.109...°$
Now resolving horizontally, taking right as positive:
$s = s,\ u = 7.5 \cos(61.109...°) = 3.623...,\ a = 0,\ t = 1.2$
$s = ut + \frac{1}{2}at^2 \Rightarrow s = 3.623... \times 1.2 = 4.35$ m (3 s.f.)

Q9 $\tan\theta = \frac{\text{opp}}{\text{adj}}$, so draw a right-angled triangle with 3 as the 'opposite' side and 4 as the 'adjacent' side.
Using Pythagoras' theorem, the hypotenuse will be 5:

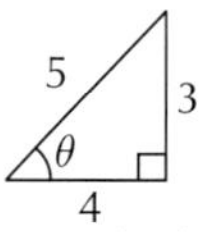

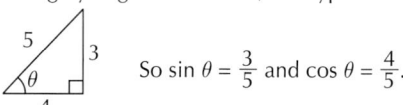

So $\sin\theta = \frac{3}{5}$ and $\cos\theta = \frac{4}{5}$.

Now find an expression for the ball's time of flight.
Resolving horizontally, taking right as positive:
$s = 50,\ u = V\cos\theta = \frac{4}{5}V,\ a = 0,\ t = t$
$s = ut + \frac{1}{2}at^2$
$50 = \frac{4}{5}V \times t \Rightarrow t = \frac{5 \times 50}{4V} = \frac{125}{2V}$

Now resolving vertically, taking up as positive:

$s = 0,\ u = V\sin\theta = \frac{3}{5}V,\ a = -9.8,\ t = \frac{125}{2V}$

$s = ut + \frac{1}{2}at^2$

$0 = \frac{3V \times 125}{5 \times 2V} - 4.9\left(\frac{125}{2V}\right)^2 \Rightarrow 0 = 37.5 - \frac{76\,562.5}{4V^2}$

$\frac{76\,562.5}{4V^2} = 37.5 \Rightarrow \frac{76\,562.5}{4 \times 37.5} = V^2$

$V^2 = \frac{76\,562.5}{150} \Rightarrow V = 22.6\ \text{ms}^{-1}$ (3 s.f.)

Q10 a) Resolving vertically, taking down as positive:

$s = 16,\ u = 3\sin 7°,\ a = 9.8,\ t = t$

$s = ut + \frac{1}{2}at^2$

$16 = (3\sin 7°)t + 4.9t^2$

$4.9t^2 + (3\sin 7°)t - 16 = 0$

Using the quadratic formula:

$t = \frac{-3\sin 7° \pm \sqrt{(3\sin 7°)^2 - (4 \times 4.9 \times -16)}}{9.8}$

$= \frac{-0.3656... \pm \sqrt{313.73...}}{9.8}$

$\Rightarrow t = 1.770...$ or $t = -1.844...$

So the particle's time of flight is $t = 1.770...$ s

Resolving horizontally, taking right as positive:

$s = s,\ u = 3\cos 7°,\ a = 0,\ t = 1.770...$

$s = ut + \frac{1}{2}at^2 = 3\cos 7° \times 1.770... = 5.27$ m (3 s.f.)

b) Resolving vertically, taking down as positive:

$s = 16,\ u = 3\sin 7°,\ v = v_y,\ a = 9.8$

$v^2 = u^2 + 2as$

$v_y^2 = (3\sin 7°)^2 + (2 \times 9.8 \times 16) = 313.733...$

$\Rightarrow v_y = 17.712...\ \text{ms}^{-1}$

Horizontally, $v_x = u_x = 3\cos 7°\ \text{ms}^{-1}$

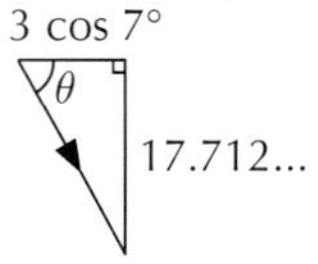

Using Pythagoras' theorem:

speed $= \sqrt{(3\cos 7°)^2 + (17.712...)^2} = 18.0\ \text{ms}^{-1}$ (3 s.f.)

Using trigonometry:

$\theta = \tan^{-1}\left(\frac{17.712...}{3\cos 7°}\right)$

$= 80.5°$ (3 s.f.) below the horizontal

Q11 a) Resolving horizontally, taking right as positive:

$u = U\cos 40°,\ v = V\cos 10°,\ a = 0,\ t = 2$

There is no acceleration horizontally, so $u = v$:

$U\cos 40° = V\cos 10°$

$\Rightarrow V = \frac{U\cos 40°}{\cos 10°}$

Resolving vertically, taking upwards as positive:

$u = U\sin 40°,\ v = V\sin 10°,\ a = -9.8,\ t = 2$

$v = u + at$

$V\sin 10° = U\sin 40° - 19.6$

Substituting $V = \frac{U\cos 40°}{\cos 10°}$ into this equation:

$\frac{U\cos 40°}{\cos 10°} \times \sin 10° = U\sin 40° - 19.6$

$U(\cos 40° \tan 10° - \sin 40°) = -19.6$

$\Rightarrow U = 38.604... = 38.6\ \text{ms}^{-1}$ (3 s.f.)

b) Resolving vertically, taking up as positive:

$s = s,\ u = (38.604...)\sin 40° = 24.814...,$

$a = -9.8,\ t = 2$

$s = ut + \frac{1}{2}at^2$

$s = (24.814... \times 2) + (\frac{1}{2} \times -9.8 \times 2^2)$

$s = 30.0$ m (3 s.f.)

Q12 a) Resolving vertically, taking up as positive:

$s = 2.5 - 0.5 = 2,\ u = 19\sin 28°,\ a = -9.8,\ t = t$

$s = ut + \frac{1}{2}at^2$

$2 = (19\sin 28°)t - 4.9t^2$

$4.9t^2 - (19\sin 28°)t + 2 = 0$

Using the quadratic formula:

$t = \frac{19\sin 28° \pm \sqrt{(-19\sin 28°)^2 - (4 \times 4.9 \times 2)}}{9.8}$

$= \frac{8.9199... \pm \sqrt{40.365...}}{9.8}$

$\Rightarrow t = 0.261...$ or $t = 1.558...$

So the ball is 2.5 m above the ground 0.261... s after being hit (when it is rising) and 1.558... s after being hit (when it is on its descent).

It is caught when it is on its descent, so it is in the air for 1.558... s = 1.56 s (3 s.f.)

b) Resolving horizontally, taking right as positive:

$s = s,\ u = 19\cos 28°,\ a = 0,\ t = 1.558...$

$s = ut + \frac{1}{2}at^2$

$s = 19\cos 28° \times 1.558... = 26.1$ m (3 s.f.)

c) Resolving vertically, taking up as positive:

$s = 2,\ u = 19\sin 28°,\ v = v_y,\ a = -9.8$

$v^2 = u^2 + 2as$

$v_y^2 = (19\sin 28°)^2 + (2 \times -9.8 \times 2) = 40.365...$

$\Rightarrow v_y = 6.353...\ \text{ms}^{-1}$

Horizontally, $v_x = u_x = 19\cos 28°$

Using Pythagoras' theorem:

speed $= \sqrt{(19\cos 28°)^2 + (6.353...)^2}$

$= 17.9\ \text{ms}^{-1}$ (3 s.f.)

Q13 a) Resolving vertically, taking up as positive:

$s = s,\ u = U\sin\theta,\ v = 0,\ a = -g$

$v^2 = u^2 + 2as$

$0 = U^2\sin^2\theta - 2gs \Rightarrow s = \frac{U^2\sin^2\theta}{2g}$ m

b) Resolving vertically, taking up as positive:

$u = U\sin\theta,\ v = 0,\ a = -g,\ t = t$

$v = u + at$

$0 = U\sin\theta - gt \Rightarrow t = \frac{U\sin\theta}{g}$ s

Q14 a) Resolving vertically, taking down as positive:

$s = 0.02,\ u = 0,\ a = 9.8,\ t = t$

$s = ut + \frac{1}{2}at^2$

$0.02 = 4.9t^2 \Rightarrow t = 0.063...$ s

Resolving horizontally, taking right as positive:

$s = s,\ u = 15,\ a = 0,\ t = 0.063...$

$s = ut + \frac{1}{2}at^2$

$s = 15 \times 0.063... = 0.958... = 0.958$ m (3 s.f.)

b) Resolving horizontally, taking right as positive:

$s = 0.958...,\ u = U\cos 5°,\ a = 0,\ t = t$

$s = ut + \frac{1}{2}at^2$

$0.958... = U\cos 5° \times t \Rightarrow t = \frac{0.958...}{U\cos 5°}$

Resolving vertically, taking up as positive:

$s = -0.02,\ u = U\sin 5°,\ a = -9.8,\ t = \frac{0.958...}{U\cos 5°}$

$s = ut + \frac{1}{2}at^2$

$-0.02 = U\sin 5°\frac{0.958...}{U\cos 5°} - 4.9\left(\frac{0.958...}{U\cos 5°}\right)^2$

$4.9\left(\frac{0.958...}{U\cos 5°}\right)^2 = (0.958... \times \tan 5°) + 0.02$

$4.9 \times (0.958...)^2 = (U\cos 5°)^2 \times ((0.958...)\tan 5° + 0.02)$

$U^2 = \frac{4.9 \times (0.958...)^2}{(\cos 5°)^2 \times ((0.958...)\tan 5° + 0.02)} = 43.666...$

$\Rightarrow U = 6.61\ \text{ms}^{-1}$ (3 s.f.)

Q15 The coordinates (3, 1) give you the components of the projectile's displacement — when it passes through this point, its horizontal displacement is 3 m and its vertical displacement is 1 m.

Resolving horizontally, taking right as positive:

$s = 3,\ u = 3\cos\alpha,\ a = 0,\ t = t$

$s = ut + \frac{1}{2}at^2$

$3 = (3\cos\alpha) \times t \Rightarrow t = \frac{1}{\cos\alpha}$

Resolving vertically, taking up as positive:

$s = 1,\ u = 3\sin\alpha,\ a = -g,\ t = \frac{1}{\cos\alpha}$

$s = ut + \frac{1}{2}at^2$

$1 = \left(3\sin\alpha \times \frac{1}{\cos\alpha}\right) + \left(\frac{1}{2} \times -g \times \left(\frac{1}{\cos\alpha}\right)^2\right)$

$\Rightarrow 1 = 3\tan\alpha - \frac{g}{2\cos^2\alpha}$, as required

Q16 a) Resolving horizontally for particle A, taking right as positive:

$s = s,\ u = U,\ a = 0,\ t = 2$

$s = ut + \frac{1}{2}at^2 \Rightarrow s = 2U$

Resolving horizontally for particle B, taking right as positive:

$s = s,\ u = 2U\cos\theta,\ a = 0,\ t = 2$

$s = ut + \frac{1}{2}at^2 \Rightarrow s = 4U\cos\theta$

When the particles collide, their horizontal displacements will be the same:

$2U = 4U\cos\theta$

$\cos\theta = \frac{1}{2} \Rightarrow \theta = 60°$

b) Resolving vertically for particle A, taking down as positive:

$s = s,\ u = 0,\ a = 9.8,\ t = 2$

$s = ut + \frac{1}{2}at^2$

$s = \frac{1}{2} \times 9.8 \times 2^2 = 19.6$ m

So, when the particles collide, A has fallen 19.6 m vertically downwards from a height of 45 m above the ground. Therefore, the particles collide at a height of 25.4 m above the ground.

Resolving vertically for particle B, taking up as positive:

$s = s,\ u = 2U\sin 60° = \sqrt{3}U,\ a = -9.8,\ t = 2$

$s = ut + \frac{1}{2}at^2$

$s = 2\sqrt{3}U + (\frac{1}{2} \times -9.8 \times 2^2) = 2\sqrt{3}U - 19.6$

So, when the particles collide (at a height of 25.4 m above the ground), B has travelled $(2\sqrt{3}U - 19.6)$ m vertically upwards from a height of 15 m above the ground. So:

$15 + (2\sqrt{3}U - 19.6) = 25.4$

$2\sqrt{3}U = 30 \Rightarrow U = \frac{15}{\sqrt{3}} = 8.6602...$

So, the speed of projection of A is 8.66 ms^{-1} (3 s.f.), and the speed of projection of B is 17.3 ms^{-1} (3 s.f.)

Exercise 1.3 — i and j vectors

Q1 a) $\mathbf{u} = (12\mathbf{i} + 16\mathbf{j}),\ \mathbf{v} = \mathbf{v},\ \mathbf{a} = -9.8\mathbf{j},\ t = 2$

$\mathbf{v} = \mathbf{u} + \mathbf{a}t$

$\mathbf{v} = (12\mathbf{i} + 16\mathbf{j}) - 19.6\mathbf{j}$

$\mathbf{v} = (12\mathbf{i} - 3.6\mathbf{j})$ ms^{-1}

b) When the particle reaches its maximum height, the vertical component of its velocity will be zero, and the horizontal component of its velocity will be the same as when it was projected (there is no horizontal acceleration), so its velocity will be $12\mathbf{i}$ ms^{-1}.

c) The particle follows a symmetric curved path, so when it hits the ground, the vertical component of its velocity will have the same magnitude as when it was projected, but in the opposite direction. The horizontal component of velocity remains constant, so the particle's velocity will be $(12\mathbf{i} - 16\mathbf{j})$ ms^{-1}.

You could answer this question by resolving horizontally and vertically and using the suvat equations to find each component if you wanted.

Q2 a) Resolving vertically, taking up as positive:

$s = s,\ u = 10,\ v = 0,\ a = -9.8$

$v^2 = u^2 + 2as$

$0 = 100 - 19.6s$

$s = 5.102...$

So the projectile reaches a maximum height of $5 + 5.102... = 10.1$ m (3 s.f.) above the ground.

b) Resolving vertically, taking up as positive:

$s = -5,\ u = 10,\ v = v_y,\ a = -9.8$

$v^2 = u^2 + 2as$

$v_y^2 = 100 + (2 \times -9.8 \times -5) = 198$

$v_y = -14.071...$ ms^{-1}

v_y is negative because the projectile is travelling downwards when it hits the ground.

Horizontally, $v_x = u_x = 17$ ms^{-1}

Using Pythagoras' theorem:

speed $= \sqrt{17^2 + 198} = 22.1$ ms^{-1} (3 s.f.)

c) Using trigonometry:

$\tan\theta = \left(\frac{-14.071...}{17}\right) \Rightarrow \theta = -39.615...°$

i.e. 39.6° (3 s.f.) below the horizontal.

Q3 a) The stone is thrown from a point 2.5 m above the ground, so when it is 6 m above the ground, it is $6 - 2.5 = 3.5$ m above its point of projection.

Resolving vertically, taking up as positive:

$s = 3.5,\ u = 9,\ a = -9.8,\ t = t$

$s = ut + \frac{1}{2}at^2$

$3.5 = 9t - 4.9t^2 \Rightarrow 4.9t^2 - 9t + 3.5 = 0$

Using the quadratic formula:

$t = \frac{9 \pm \sqrt{(-9)^2 - (4 \times 4.9 \times 3.5)}}{9.8}$

$\Rightarrow t = 0.559...$ or $t = 1.277...$

$1.277... - 0.559... = 0.7186...$

So the stone is at least 6 m above the ground for 0.719 s (3 s.f.)

b) Resolving vertically, taking up as positive:

$s = -2.5,\ u = 9,\ a = -9.8,\ t = t$

$s = ut + \frac{1}{2}at^2$

$-2.5 = 9t - 4.9t^2 \Rightarrow 4.9t^2 - 9t - 2.5 = 0$

Using the quadratic formula:

$t = \frac{9 \pm \sqrt{(-9)^2 - (4 \times 4.9 \times -2.5)}}{9.8}$

$\Rightarrow t = 2.081...$ or $t = -0.245...$

Resolving horizontally, taking right as positive:

$s = s,\ u = 6,\ a = 0,\ t = 2.081...$

$s = ut + \frac{1}{2}at^2$

$s = 6 \times 2.081... = 12.490...$ m

$20 - 12.490... = 7.509...$ m

So it falls short of the target by 7.51 m (3 s.f.)

Q4 a) $\mathbf{s} = (200\mathbf{i} - 40\mathbf{j})$, $\mathbf{u} = (a\mathbf{i} + b\mathbf{j})$, $\mathbf{a} = -9.8\mathbf{j}$, $t = 5$

$\mathbf{s} = \mathbf{u}t + \frac{1}{2}\mathbf{a}t^2$

$(200\mathbf{i} - 40\mathbf{j}) = 5(a\mathbf{i} + b\mathbf{j}) + (\frac{1}{2} \times -9.8\mathbf{j} \times 5^2)$

$5(a\mathbf{i} + b\mathbf{j}) = (200\mathbf{i} + 82.5\mathbf{j})$

$(a\mathbf{i} + b\mathbf{j}) = (40\mathbf{i} + 16.5\mathbf{j}) \Rightarrow a = 40$ and $b = 16.5$

You could also have answered this by considering the horizontal and vertical components separately, and using the suvat equations as usual.

b) The horizontal component of velocity remains constant at 40 ms^{-1}. So you only need to consider the vertical component.

$u = 16.5$, $v = v$, $a = -9.8$, $t = 5$

$v = u + at$

$v = 16.5 + (-9.8 \times 5) = -32.5$

$\Rightarrow \mathbf{v} = (40\mathbf{i} - 32.5\mathbf{j})$ ms^{-1}

Q5 a) $\mathbf{s} = (15\mathbf{i} + 6\mathbf{j})$, $\mathbf{u} = \mathbf{u}$, $\mathbf{a} = -9.8\mathbf{j}$, $t = 3$

$\mathbf{s} = \mathbf{u}t + \frac{1}{2}\mathbf{a}t^2$

$(15\mathbf{i} + 6\mathbf{j}) = 3\mathbf{u} + (\frac{1}{2} \times -9.8\mathbf{j} \times 3^2)$

$3\mathbf{u} = (15\mathbf{i} + 6\mathbf{j}) + 44.1\mathbf{j}$

$\Rightarrow \mathbf{u} = (5\mathbf{i} + 16.7\mathbf{j})$ ms^{-1}

b) $\mathbf{u} = (5\mathbf{i} + 16.7\mathbf{j})$, $\mathbf{v} = \mathbf{v}$, $\mathbf{a} = -9.8\mathbf{j}$, $t = 3$

$\mathbf{v} = \mathbf{u} + \mathbf{a}t$

$\mathbf{v} = (5\mathbf{i} + 16.7\mathbf{j}) + (-9.8\mathbf{j} \times 3)$

$\Rightarrow \mathbf{v} = (5\mathbf{i} - 12.7\mathbf{j})$ ms^{-1}

c) Velocity immediately following impact with wall is $(-2.5\mathbf{i} - 12.7\mathbf{j})$ ms^{-1}.

Resolving vertically, taking down as positive:

$s = 6$, $u = 12.7$, $a = 9.8$, $t = t$

$s = ut + \frac{1}{2}at^2$

$6 = 12.7t + 4.9t^2 \Rightarrow 4.9t^2 + 12.7t - 6 = 0$

Using the quadratic formula:

$$t = \frac{-12.7 \pm \sqrt{12.7^2 - (4 \times 4.9 \times -6)}}{9.8}$$

$\Rightarrow t = 0.4081...$ or $t = -3$

Resolving horizontally, taking left as positive:

$s = s$, $u = 2.5$, $a = 0$, $t = 0.4081...$

$s = ut + \frac{1}{2}at^2$

$s = 2.5 \times 0.4081... = 1.02$ m (3 s.f.)

d) E.g. The model could treat the ball as a three-dimensional shape with a significant diameter, instead of a particle, to better estimate the distance from the wall. The model could include other factors, such as air resistance or the spin of the ball. The model could describe the ball's motion in three dimensions rather than two, as it's unlikely to rebound perpendicular to the wall.

Q6 a) When the toy is moving due north, the velocity in the **i**-direction is zero. So resolving east:

$u = -0.6$, $v = 0$, $a = 0.1$, $t = t$

$v = u + at$

$0 = -0.6 + 0.1t \Rightarrow t = 6$ s

Check that the velocity in the **j**-direction is positive at this time (otherwise the toy could be moving due south). Resolving north:

$u = 3$, $v = v$, $a = 0.5$, $t = 6$

$v = u + at$

$v = 3 + (0.5 \times 6) = 6$

So it <u>is</u> moving due north at $t = 6$ s.

b) Find the velocity when $t = 2.5$:

$\mathbf{u} = (-0.6\mathbf{i} + 3\mathbf{j})$, $\mathbf{v} = \mathbf{v}$, $\mathbf{a} = (0.1\mathbf{i} + 0.5\mathbf{j})$, $t = 2.5$

$\mathbf{v} = \mathbf{u} + \mathbf{a}t$

$\mathbf{v} = (-0.6\mathbf{i} + 3\mathbf{j}) + 2.5(0.1\mathbf{i} + 0.5\mathbf{j})$

$\mathbf{v} = -0.6\mathbf{i} + 3\mathbf{j} + 0.25\mathbf{i} + 1.25\mathbf{j} = -0.35\mathbf{i} + 4.25\mathbf{j}$

Draw a triangle to find the bearing:

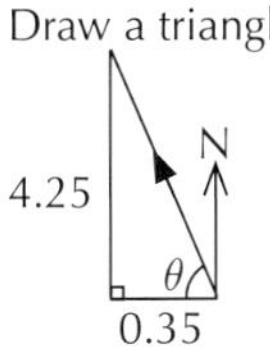

$\theta = \tan^{-1}\left(\frac{4.25}{0.35}\right) = 85.29214...°$

Bearing = $270° + \theta = 355°$ (nearest degree)

c) Find the position when $t = 60$ s:

$\mathbf{s} = \mathbf{s}$, $\mathbf{u} = (-0.6\mathbf{i} + 3\mathbf{j})$, $\mathbf{a} = (0.1\mathbf{i} + 0.5\mathbf{j})$, $t = 60$

$\mathbf{s} = \mathbf{u}t + \frac{1}{2}\mathbf{a}t^2$

$\mathbf{s} = [(-0.6\mathbf{i} + 3\mathbf{j}) \times 60] + [\frac{1}{2} \times (0.1\mathbf{i} + 0.5\mathbf{j}) \times 60^2]$

$\mathbf{s} = 144\mathbf{i} + 1080\mathbf{j}$

Using Pythagoras' theorem:

distance = $\sqrt{144^2 + 1080^2} = 1090$ m (3 s.f.)

d) For example:

- The model could allow for a non-uniform acceleration due to changing resistance force.
- 3D vectors could be used to describe the variation in the toy's depth underwater.

Q7 At $t = 0$, the particle is moving north-west:

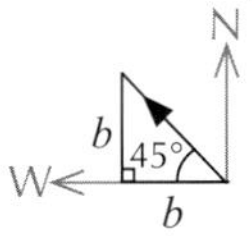

The initial velocity, **u**, can be written as a vector $-b\mathbf{i} + b\mathbf{j}$, with a magnitude of 5 ms^{-1}. Using Pythagoras' theorem:

speed = $\sqrt{(-b)^2 + b^2} = 5$ ms^{-1}

$\Rightarrow \sqrt{2b^2} = 5$

$2b^2 = 5^2$

$b^2 = \frac{5^2}{2} \Rightarrow b = \pm\frac{5}{\sqrt{2}} = \pm\frac{5\sqrt{2}}{2}$

You know you want the positive root to give a north-west direction, so initial velocity $\mathbf{u} = -\frac{5\sqrt{2}}{2}\mathbf{i} + \frac{5\sqrt{2}}{2}\mathbf{j}$

At $t = 10$, the particle is moving south, so final velocity $\mathbf{v} = -\sqrt{2}\mathbf{j}$.

$\mathbf{u} = (-\frac{5\sqrt{2}}{2}\mathbf{i} + \frac{5\sqrt{2}}{2}\mathbf{j})$, $\mathbf{v} = -\sqrt{2}\mathbf{j}$, $\mathbf{a} = (p\mathbf{i} + q\mathbf{j})$, $t = 10$

$\mathbf{v} = \mathbf{u} + \mathbf{a}t$

$-\sqrt{2}\mathbf{j} = (-\frac{5\sqrt{2}}{2}\mathbf{i} + \frac{5\sqrt{2}}{2}\mathbf{j}) + 10(p\mathbf{i} + q\mathbf{j})$

$\frac{5\sqrt{2}}{2}\mathbf{i} - \frac{7\sqrt{2}}{2}\mathbf{j} = 10p\mathbf{i} + 10q\mathbf{j}$

$\Rightarrow 10p = \frac{5\sqrt{2}}{2} \Rightarrow p = \frac{\sqrt{2}}{4}$

$10q = -\frac{7\sqrt{2}}{2} \Rightarrow q = -\frac{7\sqrt{2}}{20}$

2. Non-Uniform Acceleration in 2 Dimensions

Exercise 2.1 — Using vectors

Q1 a) (i) $\mathbf{v} = \dot{\mathbf{r}} = [(3t^2 - 3)\mathbf{i} + 2t\mathbf{j}]$ ms^{-1}

(ii) When $t = 3$, $\mathbf{v} = (3(3)^2 - 3)\mathbf{i} + 2(3)\mathbf{j}$
$= (24\mathbf{i} + 6\mathbf{j})$ ms^{-1}

Using Pythagoras' theorem:
speed $= \sqrt{24^2 + 6^2} = 24.7$ ms^{-1} (3 s.f.)
Using trigonometry:
$\theta = \tan^{-1}\left(\frac{6}{24}\right)$
$= 14.0°$ (3 s.f.) from the positive **i**-direction.

b) (i) $\mathbf{a} = \dot{\mathbf{v}} = (6t\mathbf{i} + 2\mathbf{j})$ ms^{-2}

(ii) When $t = 4$, $\mathbf{a} = 6(4)\mathbf{i} + 2\mathbf{j} = (24\mathbf{i} + 2\mathbf{j})$ ms^{-2}
Using Pythagoras' theorem:
magnitude $= \sqrt{24^2 + 2^2} = 24.1$ ms^{-2} (3 s.f.)

Q2 $\mathbf{a} = \dot{\mathbf{v}} = (t^2 + 4t - 12)\mathbf{i}$ ms^{-2}
At the maximum velocity, $\mathbf{a} = 0$, so:
$t^2 + 4t - 12 = 0 \Rightarrow (t + 6)(t - 2) = 0 \Rightarrow t = -6$ and $t = 2$
$0 \le t \le 4$, so ignore the solution at $t = -6$.
When $t = 2$, $\mathbf{v} = [\frac{1}{3}(2)^3 + 2(2)^2 - 12(2)]\mathbf{i} + 14\mathbf{j}$
$= \frac{-40}{3}\mathbf{i} + 14\mathbf{j}$ ms^{-1}

Q3 a) It's easier to work with the vectors in component form whilst integrating, so use $\mathbf{a} = 3e^{3t}\mathbf{i} + \frac{6}{\sqrt{t}}\mathbf{j}$

$\mathbf{v} = \int \mathbf{a}\,dt = \left(\int 3e^{3t}\,dt\right)\mathbf{i} + \left(\int \frac{6}{\sqrt{t}}\,dt\right)\mathbf{j}$
$= \left(\int 3e^{3t}\,dt\right)\mathbf{i} + \left(\int 6t^{-\frac{1}{2}}\,dt\right)\mathbf{j}$
$= e^{3t}\mathbf{i} + 12t^{\frac{1}{2}}\mathbf{j} + \mathbf{C}$
$= e^{3t}\mathbf{i} + 12\sqrt{t}\,\mathbf{j} + \mathbf{C}$

When $t = 0$, $\mathbf{v} = (4\mathbf{i} + 6\mathbf{j})$, so:
$4\mathbf{i} + 6\mathbf{j} = e^0\mathbf{i} + 12\sqrt{0}\mathbf{j} + \mathbf{C}$
$4\mathbf{i} + 6\mathbf{j} = \mathbf{i} + \mathbf{C}$
$\Rightarrow \mathbf{C} = 3\mathbf{i} + 6\mathbf{j}$

So $\mathbf{v} = (e^{3t} + 3)\mathbf{i} + (6 + 12\sqrt{t})\mathbf{j} = \begin{pmatrix} e^{3t} + 3 \\ 6 + 12\sqrt{t} \end{pmatrix}$ ms^{-1}

b) $\mathbf{r} = \int \mathbf{v}\,dt = \left[\int (e^{3t} + 3)\,dt\right]\mathbf{i} + \left[\int (6 + 12\sqrt{t})\,dt\right]\mathbf{j}$
$= \left[\int (e^{3t} + 3)\,dt\right]\mathbf{i} + \left[\int (6 + 12t^{\frac{1}{2}})\,dt\right]\mathbf{j}$
$= \left(\frac{e^{3t}}{3} + 3t\right)\mathbf{i} + (6t + 8t^{\frac{3}{2}})\mathbf{j} + \mathbf{D}$

When $t = 0$, $\mathbf{r} = (2\mathbf{i} - 9\mathbf{j})$, so:
$2\mathbf{i} - 9\mathbf{j} = (\frac{e^0}{3} + 0)\mathbf{i} + (0 + 0)\mathbf{j} + \mathbf{D}$
$2\mathbf{i} - 9\mathbf{j} = \frac{1}{3}\mathbf{i} + \mathbf{D} \Rightarrow \mathbf{D} = \frac{5}{3}\mathbf{i} - 9\mathbf{j}$
So $\mathbf{r} = (\frac{e^{3t}}{3} + 3t + \frac{5}{3})\mathbf{i} + (6t + 8t^{\frac{3}{2}} - 9)\mathbf{j}$
$= \begin{pmatrix} \frac{1}{3}e^{3t} + 3t + \frac{5}{3} \\ 6t + 8\sqrt{t^3} - 9 \end{pmatrix}$ m

Q4 a) $\mathbf{v} = \dot{\mathbf{r}} = [(\cos t)\mathbf{i} - (\sin t)\mathbf{j}]$ ms^{-1}
The object is moving east when the **j** component is 0, and the **i** component is positive.
$\sin t = 0$ when $t = 0, \pi, 2\pi...$
When $t = \pi$, the **i** component $\cos t = -1$, so the object is moving west, not east.
When $t = 2\pi$, $\cos t = 1$.
So the first non-zero value of t for which the object is moving east is $t = 2\pi$ seconds.

b) At $t = 2\pi$, $\mathbf{r} = (\sin 2\pi)\mathbf{i} + (\cos 2\pi)\mathbf{j} = (0\mathbf{i} + \mathbf{j})$ m.
So it is 1 m north of O.

Q5 $\mathbf{r} = \int \mathbf{v}\,dt = \left[\int (t^2 - 6t)\,dt\right]\mathbf{i} + \left[\int (4t + 5)\,dt\right]\mathbf{j}$
$= (\frac{1}{3}t^3 - 3t^2)\mathbf{i} + (2t^2 + 5t)\mathbf{j} + \mathbf{C}$
When $t = 0$, $\mathbf{r} = 0\mathbf{i} + 0\mathbf{j}$, so:
$0\mathbf{i} + 0\mathbf{j} = (\frac{1}{3}(0)^3 - 3(0)^2)\mathbf{i} + (2(0)^2 + 5(0))\mathbf{j} + \mathbf{C}$
$\Rightarrow \mathbf{C} = 0\mathbf{i} + 0\mathbf{j}$
So $\mathbf{r} = [(\frac{1}{3}t^3 - 3t^2)\mathbf{i} + (2t^2 + 5t)\mathbf{j}]$ m

Equating **j**-components at position vector $(b\mathbf{i} + 12\mathbf{j})$ m:
$2t^2 + 5t = 12$
$2t^2 + 5t - 12 = 0$
$(2t - 3)(t + 4) = 0 \Rightarrow t = 1.5$ and $t = -4$
The time can't be negative, so the particle passes through the point $(b\mathbf{i} + 12\mathbf{j})$ m at time $t = 1.5$ s.
Equating **i**-components and substituting $t = 1.5$:
$\frac{1}{3}t^3 - 3t^2 = b$
$\frac{1}{3}(1.5)^3 - 3(1.5)^2 = b \Rightarrow b = -5.625$

Q6 a) $\mathbf{v}_A = \dot{\mathbf{r}}_A = [(3t^2 - 2t - 4)\mathbf{i} + (3t^2 - 4t + 3)\mathbf{j}]$ ms^{-1}
$(3t^2 - 2t - 4)\mathbf{i} + (3t^2 - 4t + 3)\mathbf{j} = -3\mathbf{i} + 2\mathbf{j}$
Equating **i**-components:
$3t^2 - 2t - 4 = -3$
$3t^2 - 2t - 1 = 0$
$(3t + 1)(t - 1) = 0 \Rightarrow t = -\frac{1}{3}$ and $t = 1$
When $t = 1$, the **j**-component of $\mathbf{v}_A$ is:
$3(1)^2 - 4(1) + 3 = 2$
So the particle's velocity is $(-3\mathbf{i} + 2\mathbf{j})$ ms^{-1} at time $t = 1$ second.

b) When the particle's direction of motion is 45° above **i**, the horizontal and vertical components of the particle's velocity will be equal.
A good way to picture this is that if the particle was moving relative to a pair of coordinate axes, it would be moving along the line $y = x$ (for positive x and y).
$3t^2 - 2t - 4 = 3t^2 - 4t + 3$
$2t = 7 \Rightarrow t = 3.5$ s
This is the value of t for which the horizontal and vertical components of velocity are equal.
You need to check that the particle is moving at 45° above **i**, not 45° below **–i** at this time.
(The components will also be equal if the particle is moving along a path 45° below **–i**.)
From part **a)**,
$\mathbf{v}_A = [(3t^2 - 2t - 4)\mathbf{i} + (3t^2 - 4t + 3)\mathbf{j}]$ ms^{-1}.
When $t = 3.5$,
$\mathbf{v}_A = [(3(3.5)^2 - 2(3.5) - 4)\mathbf{i} + (3(3.5)^2 - 4(3.5) + 3)\mathbf{j}]$
$= (25.75\mathbf{i} + 25.75\mathbf{j})$ ms^{-1}
Both components are positive, so the particle is moving along a path 45° above **i** when $t = 3.5$ s.
You don't actually need to check that both components are positive — if one is, the other one will be too.

c) $\mathbf{v}_B = \int \mathbf{a}_B\,dt = \left(\int 6t\,dt\right)\mathbf{i} + \left(\int 6t\,dt\right)\mathbf{j} = 3t^2\mathbf{i} + 3t^2\mathbf{j} + \mathbf{C}$
When $t = 1$, $\mathbf{v}_B = (4\mathbf{i} - \mathbf{j})$, so:
$4\mathbf{i} - \mathbf{j} = 3(1)^2\mathbf{i} + 3(1)^2\mathbf{j} + \mathbf{C}$
$4\mathbf{i} - \mathbf{j} = 3\mathbf{i} + 3\mathbf{j} + \mathbf{C} \Rightarrow \mathbf{C} = \mathbf{i} - 4\mathbf{j}$
So $\mathbf{v}_B = [(3t^2 + 1)\mathbf{i} + (3t^2 - 4)\mathbf{j}]$ ms^{-1}
$\mathbf{r}_B = \int \mathbf{v}_B\,dt = \left[\int (3t^2 + 1)\,dt\right]\mathbf{i} + \left[\int (3t^2 - 4)\,dt\right]\mathbf{j}$
$= (t^3 + t)\mathbf{i} + (t^3 - 4t)\mathbf{j} + \mathbf{D}$
When $t = 1$, $\mathbf{r}_B = (2\mathbf{i} + 3\mathbf{j})$, so:
$2\mathbf{i} + 3\mathbf{j} = (1^3 + 1)\mathbf{i} + (1^3 - 4(1))\mathbf{j} + \mathbf{D}$
$2\mathbf{i} + 3\mathbf{j} = 2\mathbf{i} - 3\mathbf{j} + \mathbf{D} \Rightarrow \mathbf{D} = 6\mathbf{j}$
So $\mathbf{r}_B = [(t^3 + t)\mathbf{i} + (t^3 - 4t + 6)\mathbf{j}]$ m

d) $\mathbf{r}_B - \mathbf{r}_A = [(t^3 + t)\mathbf{i} + (t^3 - 4t + 6)\mathbf{j}] -$
$[(t^3 - t^2 - 4t + 3)\mathbf{i} + (t^3 - 2t^2 + 3t - 7)\mathbf{j}]$
$= [(t^2 + 5t - 3)\mathbf{i} + (2t^2 - 7t + 13)\mathbf{j}]$ m

e) When $t = 4$,
$\mathbf{r}_B - \mathbf{r}_A = (4^2 + 5(4) - 3)\mathbf{i} + (2(4)^2 - 7(4) + 13)\mathbf{j}$
$= (33\mathbf{i} + 17\mathbf{j})$ m
Using Pythagoras' theorem:
distance $= \sqrt{33^2 + 17^2} = 37.1$ m (3 s.f.)

Q7 a) $\frac{dy}{dt} = \frac{1}{t}$, and $\frac{dx}{dt} = -\frac{3}{t^4}$,
so $\frac{dy}{dx} = \frac{dy}{dt} \div \frac{dx}{dt} = \frac{1}{t} \div \frac{-3}{t^4} = -\frac{t^3}{3}$
When $t = \frac{1}{2}$, $\frac{dy}{dx} = -\frac{1}{3 \times 2^3} = -\frac{1}{24}$

b) $\mathbf{r} = x\mathbf{i} + y\mathbf{j} \Rightarrow \mathbf{v} = \frac{d\mathbf{r}}{dt} = \frac{dx}{dt}\mathbf{i} + \frac{dy}{dt}\mathbf{j} = -\frac{3}{t^4}\mathbf{i} + \frac{1}{t}\mathbf{j}$
$\Rightarrow \mathbf{a} = \frac{d\mathbf{v}}{dt} = \frac{d^2x}{dt^2}\mathbf{i} + \frac{d^2y}{dt^2}\mathbf{j} = \frac{12}{t^5}\mathbf{i} - \frac{1}{t^2}\mathbf{j}$
So when $t = \frac{1}{2}$,
$\mathbf{a} = \frac{12}{\left(\frac{1}{2}\right)^5}\mathbf{i} - \frac{1}{\left(\frac{1}{2}\right)^2}\mathbf{j} = 384\mathbf{i} - 4\mathbf{j}$ (as required)

Q8 Use double angle identities to change the components of **v** into expressions that are easier to integrate:
$\cos 2t \equiv 1 - 2\sin^2 t \Rightarrow 2\sin^2 t \equiv 1 - \cos 2t$
$\sin 2t \equiv 2 \sin t \cos t \Rightarrow 4 \sin 2t \cos 2t \equiv 2 \sin 4t$
$\Rightarrow \mathbf{v} = (1 - \cos 2t)\mathbf{i} + (2 \sin 4t)\mathbf{j}$
$\mathbf{r} = \int \mathbf{v}\, dt = \left[\int (1 - \cos 2t)\, dt\right]\mathbf{i} + \left[\int (2 \sin 4t)\, dt\right]\mathbf{j}$
$= \left(t - \frac{1}{2}\sin 2t\right)\mathbf{i} - \left(\frac{1}{2}\cos 4t\right)\mathbf{j} + \mathbf{C}$
When $t = \frac{\pi}{4}$, $\mathbf{r} = \frac{\pi}{4}\mathbf{i} + \mathbf{j}$, so:
$\frac{\pi}{4}\mathbf{i} + \mathbf{j} = \left(\frac{\pi}{4} - \frac{1}{2}\sin\frac{\pi}{2}\right)\mathbf{i} - \left(\frac{1}{2}\cos \pi\right)\mathbf{j} + \mathbf{C}$
$\frac{\pi}{4}\mathbf{i} + \mathbf{j} = \left(\frac{\pi}{4} - \frac{1}{2}\right)\mathbf{i} + \frac{1}{2}\mathbf{j} + \mathbf{C} \Rightarrow \mathbf{C} = \frac{1}{2}\mathbf{i} + \frac{1}{2}\mathbf{j}$
$\mathbf{r} = \left(t - \frac{1}{2}\sin 2t + \frac{1}{2}\right)\mathbf{i} - \left(\frac{1}{2}\cos 4t + \frac{1}{2}\right)\mathbf{j}$
$= \frac{1}{2}(2t - \sin 2t + 1)\mathbf{i} - \frac{1}{2}(\cos 4t - 1)\mathbf{j}$
So when $t = 0$, $\mathbf{r} = \frac{1}{2}(0 - 0 + 1)\mathbf{i} - \frac{1}{2}(1 - 1)\mathbf{j} = \frac{1}{2}\mathbf{i}$,
so distance from the origin $= \frac{1}{2}$ m.

Q9 a) $\frac{dn}{dt} = -2$ when $n = -4 \Rightarrow k(-4) - 10 = -2 \Rightarrow k = -2$
$\frac{dn}{dt} < 0 \Rightarrow -2n - 10 < 0 \Rightarrow 2n + 10 > 0$
$\Rightarrow n + 5 > 0$ as required

b) In the **i** direction, the acceleration is a constant 0.1 ms^{-2}, so use the constant acceleration equations to find an expression for m in terms of t:
$u = 0,\ v = m,\ a = 0.1,\ t = t$
$v = u + at \Rightarrow m = 0.1t$
In the **j**-direction, form the differential equation to find n in terms of t:
$\frac{dn}{dt} = kn - 10 = -2n - 10 = -2(n + 5)$
$\Rightarrow \int \frac{1}{n+5}\, dn = \int -2\, dt$
$\ln|n + 5| = -2t + C$
When $t = 0$, $n = 0 \Rightarrow \ln|5| = 0 + C \Rightarrow C = \ln 5$
Rearrange to give n in terms of t:
$\ln|n + 5| = -2t + \ln 5$
$\Rightarrow |n + 5| = e^{(-2t + \ln 5)}$
From part a), n + 5 > 0, so you can drop the modulus.
$\Rightarrow n + 5 = e^{-2t}e^{\ln 5}$
$\Rightarrow n = 5e^{-2t} - 5 = 5(e^{-2t} - 1)$
Replace both components in $\mathbf{v} = m\mathbf{i} + n\mathbf{j}$ to give:
$\mathbf{v} = \left[0.1t\mathbf{i} + 5(e^{-2t} - 1)\mathbf{j}\right]$ ms^{-1} as required.

Chapter 15: Dynamics

1. Resolving Forces

Exercise 1.1 — Components of forces

Q1 a) Horizontal component (→) = 6.5 N
Vertical component (↑) = 0 N

b) Horizontal component (→) = 20 cos 40°
= 15.3 N (3 s.f.)
Vertical component (↑) = 20 sin 40° = 12.9 N (3 s.f.)

c) Horizontal component (→) = 17 sin 25°
= 7.18 N (3 s.f.)
Vertical component (↑) = 17 cos 25° = 15.4 N (3 s.f.)

d) Horizontal component (→) = –5 cos 10°
= –4.92 N (3 s.f.)
Vertical component (↑) = 5 sin 10° = 0.868 N (3 s.f.)

e) Horizontal component (→) = 3 cos 30°
= 2.60 N (3 s.f.)
Vertical component (↑) = –3 sin 30° = –1.5 N

f) Horizontal component (→) = –13 sin 60°
= –11.3 N (3 s.f.)
Vertical component (↑) = –13 cos 60° = –6.5 N

Q2 a) 10**j** N
b) –3.2**i** N
c) $(21 \cos 45°)\mathbf{i} + (21 \sin 45°)\mathbf{j}\ \left(= \frac{21}{2}\sqrt{2}\mathbf{i} + \frac{21}{2}\sqrt{2}\mathbf{j}\right)$

Q3 a)

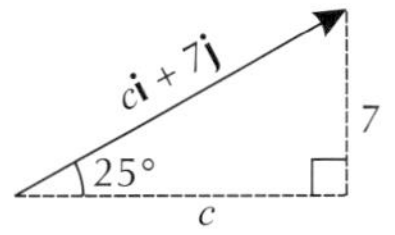

$\tan 25° = \frac{7}{c}$
$c = \frac{7}{\tan 25°} = 15.011... = 15.0$ (to 3 s.f.)

b) magnitude $= \sqrt{(15.011...)^2 + 7^2} = 16.6$ N (to 3 s.f.)

Q4

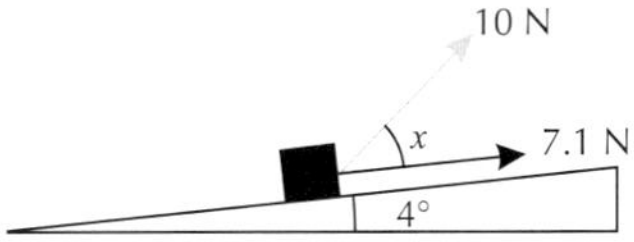

a) $10 \cos x = 7.1$
$x = \cos^{-1}\left(\frac{7.1}{10}\right) = 44.8°$ (to 3 s.f.)

b) $10 \cos (x + 4) = 6.59$ N (to 3 s.f.)

Exercise 1.2 — Resultant forces and equilibrium

Q1 a)

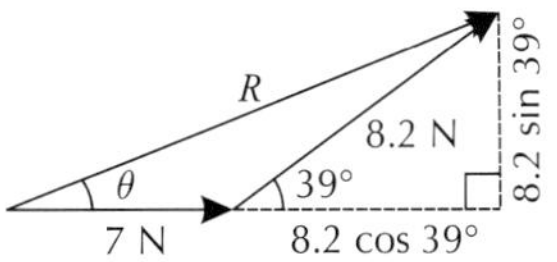

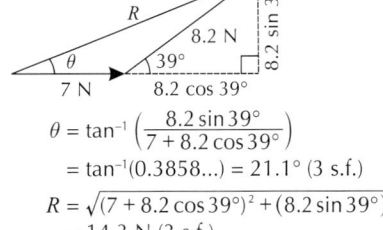

$\theta = \tan^{-1}\left(\frac{8.2 \sin 39°}{7 + 8.2 \cos 39°}\right)$
$= \tan^{-1}(0.3858...) = 21.1°$ (3 s.f.)
$R = \sqrt{(7 + 8.2\cos 39°)^2 + (8.2 \sin 39°)^2}$
$= 14.3$ N (3 s.f.)

b)

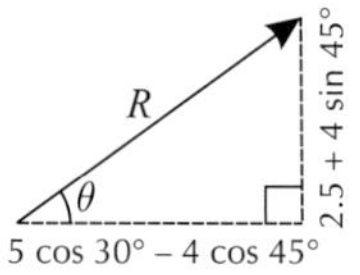

Resolving horizontally:
$R(\rightarrow) = 5\cos 30° - 4\cos 45°$
Resolving vertically:
$R(\uparrow) = 5\sin 30° + 4\sin 45°$
$= 2.5 + 4\sin 45°$

$\theta = \tan^{-1}\left(\frac{2.5 + 4\sin 45°}{5\cos 30° - 4\cos 45°}\right)$
$= \tan^{-1}(3.548...) = 74.3°$ (3 s.f.)

$R = \sqrt{(2.5 + 4\sin 45°)^2 + (5\cos 30° - 4\cos 45°)^2}$
$= 5.54$ N (3 s.f.)

c) Resolving horizontally:
$R(\rightarrow) = 10 - 11\cos 75° - 12\cos 80°$
$= 5.069...$ N
Resolving vertically:
$R(\uparrow) = 11\sin 75° - 12\sin 80°$
$= -1.192...$ N

$R = \sqrt{(5.069...)^2 + (-1.192...)^2}$
$= 5.21$ N (to 3 s.f.)

$\theta = \tan^{-1}\left(\frac{-1.192...}{5.069...}\right) = -13.237...°$
i.e. 13.237...° below the positive horizontal
So direction = 360° – 13.237...° = 347° (3 s.f.)

d) Resolving horizontally:
$R(\rightarrow) = 6.5\cos 69° + 10\cos 47° - 8\sin 61°$
$= 2.152...$ N
Resolving vertically:
$R(\uparrow) = 6.5\sin 69° + 8\cos 61° - 10\sin 47°$
$= 2.633...$ N

$R = \sqrt{(2.152...)^2 + (2.633...)^2}$
$= 3.40$ N (to 3 s.f.)

$\theta = \tan^{-1}\left(\frac{2.633...}{2.152...}\right)$
$= 50.7°$ (to 3 s.f.)

Q2 a)

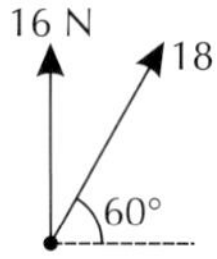

b)

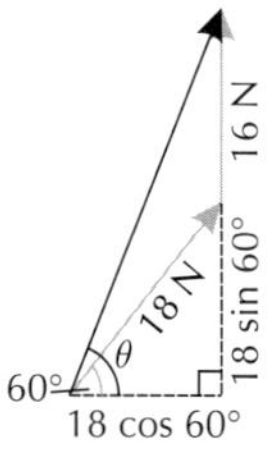

$\theta = \tan^{-1}\left(\frac{16 + 18\sin 60°}{18\cos 60°}\right) = \tan^{-1}(3.509...)$
$= 74.1°$ (3 s.f.) above the positive horizontal

$R = \sqrt{(18\cos 60°)^2 + (16 + 18\sin 60°)^2}$
$= 32.8$ N (3 s.f.)

Q3 Resolving vertically (↑):
$F\cos 31° + F\cos 31° - 14 = 0$
$2F\cos 31° = 14$
$F\cos 31° = 7 \Rightarrow F = 8.17$ N (3 s.f.)

Q4

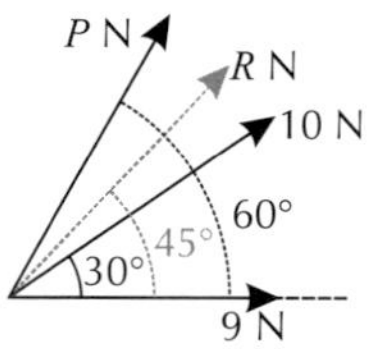

Resolving horizontally (→):
$P\cos 60° + 10\cos 30° + 9 = R\cos 45°$ ①
Resolving vertically (↑):
$P\sin 60° + 10\sin 30° = R\sin 45°$ ②
Because the resultant acts at 45°, its horizontal and vertical components are equal.
i.e. $\cos 45° = \sin 45° \Rightarrow R\cos 45° = R\sin 45°$
So, equating the left hand sides of ① and ②:
$P\cos 60° + 10\cos 30° + 9 = P\sin 60° + 10\sin 30°$
$P\cos 60° - P\sin 60° = 10\sin 30° - 10\cos 30° - 9$
$\Rightarrow P = \frac{10\sin 30° - 10\cos 30° - 9}{\cos 60° - \sin 60°}$
$= \frac{-12.6602...}{-0.36602...} = 34.6$ N (3 s.f.)

Q5

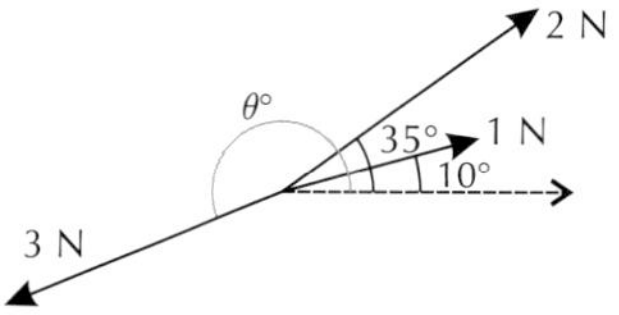

a) Resolving horizontally (→):
$1\cos 10° + 2\cos 35° + 3\cos\theta = 0$
$\theta = \cos^{-1}\left(\frac{-\cos 10° - 2\cos 35°}{3}\right) = \cos^{-1}(-0.874...)$
$= 150.970...° = 151°$ (to 3 s.f.)
θ must lie in the range $180° \leq \theta \leq 360°$
$\cos\theta = \cos(360° - \theta)$, therefore
$\theta = 360° - 150.970...° = 209.029...° = 209°$ (to 3 s.f.)

b) $R = 1\sin 10° + 2\sin 35° + 3\sin(209.029...°)$
$= -0.135$ N (to 3 s.f.)
So, R has magnitude 0.135 N (to 3 s.f.)
(and acts in the negative **j**-direction)

Q6 Resolving horizontally: $2P = 20\sin\theta$ ①
Resolving vertically: $P = 20\cos\theta$ ②
Divide ① by ②: $\tan\theta = \frac{2P}{P} = 2$,
$\theta = \tan^{-1}(2) = 63.43...° = 63.4°$ (to 3 s.f.)
Substituting into ② gives
$P = 20\cos\theta = 20\cos(63.43...°) = 8.94$ N (to 3 s.f)

Exercise 1.3 — Harder equilibrium problems

Q1 a) Resolving horizontally:
$T\cos 50° = 14$ N
$T = \frac{14}{\cos 50°} = 21.78... = 21.8$ N (to 3 s.f.)
Resolving vertically:
$T\sin 50° = mg$
$m = \frac{T\sin 50°}{g} = \frac{21.78...\sin 50°}{9.8} = 1.70$ kg (to 3 s.f.)

b) Resolving horizontally:
$T = 8\sin 25° = 3.380... = 3.38$ N (to 3 s.f.)
Resolving vertically:
$mg = 8\cos 25°$
$m = \frac{8\cos 25°}{g} = 0.740$ kg (to 3 s.f.)

Q2 Resolving horizontally:
$15 \sin x = 13 \sin 32°$
$x = \sin^{-1}\left(\frac{13 \sin 32°}{15}\right) = 27.339...° = 27.3°$ (to 3 s.f.)
Resolving vertically:
$W = 13 \cos 32° + 15 \cos (27.339...°) = 24.3$ N (to 3 s.f.)

Q3 **a)**

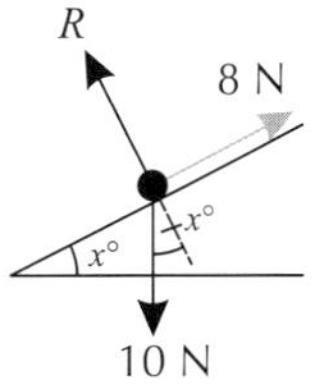

Resolving parallel to the plane:
$10 \sin x = 8$
$x = \sin^{-1}\left(\frac{8}{10}\right) = 53.130... = 53.1°$ (to 3 s.f.)

b) Resolving perpendicular to the plane:
$R = 10 \cos x = 6$ N

Q4

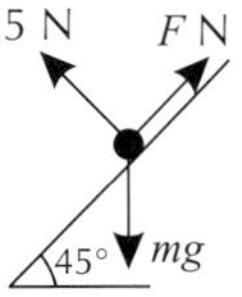

a) Resolving parallel to the plane:
$F = mg \sin 45°$
Resolving perpendicular to the plane:
$25 = mg \cos 45°$
$\cos 45° = \sin 45°$, therefore $F = 25$ N

b) Resolving perpendicular to the plane:
$mg \cos 45° = 25$
$m = \frac{25}{g \cos 45°} = \frac{25}{9.8 \times \cos 45°} = 3.61$ kg (to 3 s.f.)

Q5

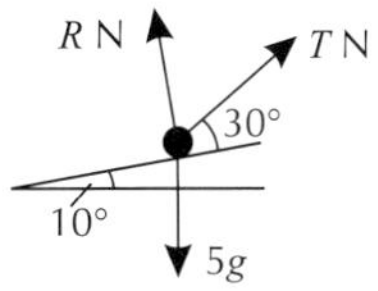

a) Magnitude of the tension in the string = T
normal reaction force = R
Resolving parallel to the plane:
$T \cos 30° = 5g \sin 10°$
$T = \frac{(5 \times 9.8 \times \sin 10°)}{\cos 30°} = 9.83$ N (to 3 s.f.)

b) Resolving perpendicular to the plane:
$R + T \sin 30° = 5g \cos 10°$
$R = 5g \cos 10° - T \sin 30° = 43.3$ N (to 3 s.f.)

Q6

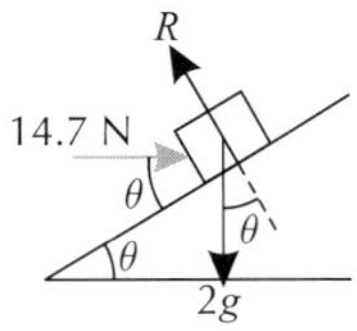

Resolving the forces on the block parallel to the plane:
$14.7 \cos \theta = 2g \sin \theta$
$\frac{\sin\theta}{\cos\theta} = \frac{14.7}{2g} \Rightarrow \tan \theta = \frac{14.7}{(2 \times 9.8)} = 0.75$

Q7

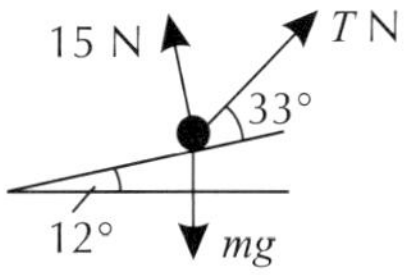

a) The string is inclined at
$45° - 12° = 33°$ to the plane.
Resolving parallel to the plane:
$T \cos 33° = mg \sin 12° \Rightarrow mg = \frac{T \cos 33°}{\sin 12°}$ ①
Resolving perpendicular to the plane:
$15 + T \sin 33° = mg \cos 12°$ ②
Substituting ① into ②:
$15 + T \sin 33° = \frac{T \cos 33°}{\sin 12°} \cos 12°$
$15 = T\left(\left(\frac{\cos 33°}{\sin 12°} \cos 12°\right) - \sin 33°\right)$
$T = \frac{15}{\left(\frac{\cos 33°}{\tan 12°} - \sin 33°\right)} = 4.41$ N (to 3.s.f)

b) Rearrange ① and substitute the values for T and g (= 9.8 ms^{-2}):
$m = \frac{T \cos 33°}{g \sin 12°} = 1.82$ kg (to 3 s.f.)

Q8

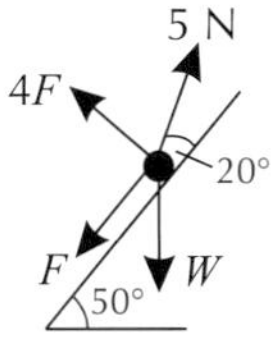

The string makes an angle of
$70° - 50° = 20°$ with the plane.
Resolving perpendicular to the plane:
$R + 5 \sin 20° = W \cos 50°$
$R = 4F$, so:
$4F + 5 \sin 20° = W \cos 50°$ ①
Resolving parallel to the plane:
$F + W \sin 50° = 5 \cos 20°$
$F = 5 \cos 20° - W \sin 50°$ ②
Substituting ② into ①,
$20 \cos 20° - 4W \sin 50° + 5 \sin 20° = W \cos 50°$,
$W = \frac{20 \cos 20° + 5 \sin 20°}{4 \sin 50° + \cos 50°} = 5.53$ N (to 3 s.f.)

2. Friction

Exercise 2.1 — Friction

Q1 The particle is on the point of slipping, so $F = \mu R$.
Resolving vertically: $R = W$, so using $F = \mu W$:

a) $F = 0.25 \times 5 = 1.25$ N

b) $F = 0.3 \times 8 = 2.4$ N

c) $F = 0.75 \times 15 = 11.25$ N

Q2

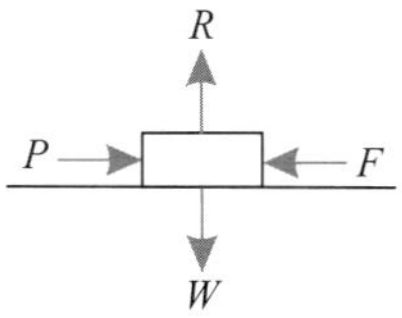

Resolving vertically: $R = W$
Resolving horizontally: $P = F$, so using $F = \mu R$, $\mu = \frac{P}{W}$

a) $\mu = \frac{5}{20} = 0.25$

b) $\mu = \frac{8}{15} = 0.53$ (to 2 d.p.)

Q3 **a)**

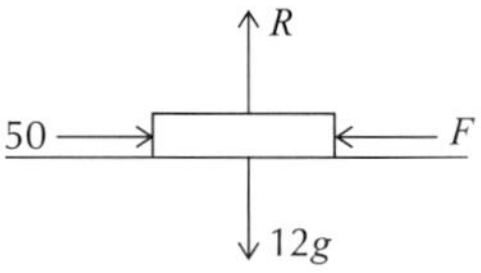

Resolving vertically: $R = 12g$
Find the frictional force, F:
$F \leq \mu R$
$F \leq 0.5 \times 12g$
$F \leq 58.8$ N
50 N isn't big enough to overcome friction — so the particle won't move.

b) Maximum force is when $F = \mu R$, so max force is 58.8 N

Q4

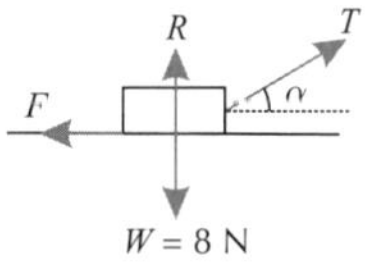

Resolving vertically: $R + T \sin \alpha = 8$

Resolving horizontally: $T \cos \alpha = F$

T will be greatest when the object is in limiting equilibrium: $F = \mu R$

a) $\alpha = 0 \Rightarrow R = W$ and $T = F = \mu R = 8\mu$
$T = 0.61 \times 8 = 4.88$ N

b) $\alpha = 35°$
Resolving vertically: $R + T \sin 35° = 8$
$R = 8 - T \sin 35°$ ①
Resolving horizontally:
$T \cos 35° = F$
$F = \mu R \Rightarrow T \cos 35° = 0.61R$ ②
Substitute the value of R from ① into ②:
$8 - T \sin 35° = \frac{T \cos 35°}{0.61}$
$T \cos 35° = 0.61(8 - T \sin 35°)$
$T(\cos 35° + 0.61 \sin 35°) = 4.88$
$T = \frac{4.88}{\cos 35° + 0.61 \sin 35°} = 4.17$ N (to 3 s.f.)

Q5

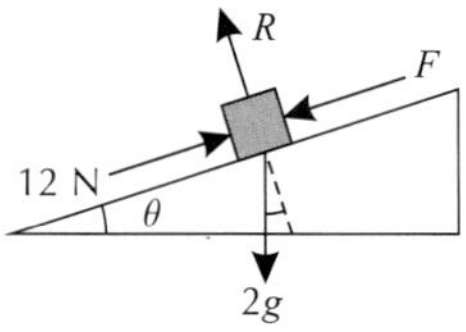

Resolving parallel to the plane:
$12 = F + 2g \sin \theta$
$F = 12 - 2g \sin \theta$
Resolving perpendicular to the plane:
$R = 2g \cos \theta$

The object is on the point of moving, so $F = \mu R$.
$\mu = \frac{F}{R} = \frac{(12 - 2g \sin \theta)}{2g \cos \theta}$

$\cos \theta = \frac{4}{5}$, so, drawing a right-angled triangle, the side adjacent to the angle will be 4, and the hypotenuse will be 5. Then, by Pythagoras' theorem, the opposite side will be 3.

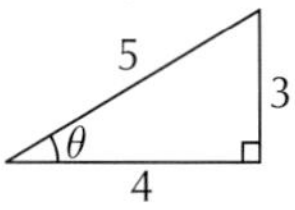

So $\sin \theta = \frac{3}{5}$

$\mu = \frac{\left(12 - \left(2 \times 9.8 \times \frac{3}{5}\right)\right)}{\left(2 \times 9.8 \times \frac{4}{5}\right)} = 0.02$ (2 d.p.)

Q6 **a)** Resolving horizontally:
$F = 4 \cos 25°$
Friction is limiting, so, using $F = \mu R$:
$4 \cos 25° = 0.15R$
$\Rightarrow R = 24.168... = 24.2$ N (3 s.f.)

b) Resolving vertically:
$R = mg + 4 \sin 25°$
$mg = 24.168... - 4 \sin 25° = 22.477...$
So $m = 22.477... \div 9.8 = 2.29$ kg (3 s.f.)

Q7

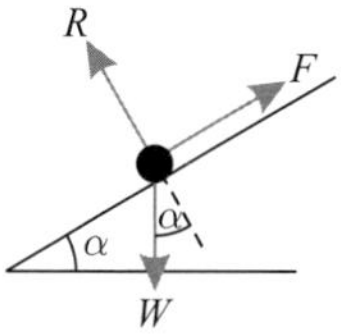

Resolving perpendicular to the plane:
$R = W \cos \alpha$ ①
Resolving parallel to the plane:
$F = W \sin \alpha$ ②
$F \leq \mu R$ when the object is at rest on the plane.
Substituting in values for F and R from ① and ②:
$W \sin \alpha \leq \mu W \cos \alpha$
$\frac{W \sin \alpha}{W \cos \alpha} \leq \mu \Rightarrow \tan \alpha \leq \mu$

Q8

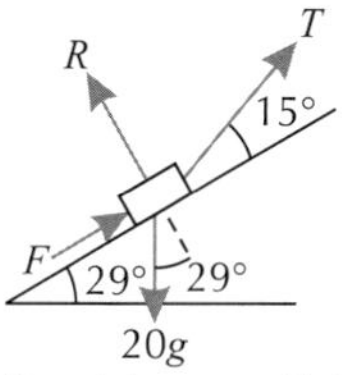

Resolving parallel to the plane:
$T \cos 15° + F = 20g \sin 29°$ ①
Resolving perpendicular to the plane:
$R + T \sin 15° = 20g \cos 29°$
$R = -T \sin 15° + 20g \cos 29°$ ②

T is only just large enough to hold the sledge at rest, so the sledge is in limiting equilibrium and $F = \mu R = 0.1R$. Substituting this and ② into ①:
$T \cos 15° + 0.1(-T \sin 15° + 20g \cos 29°) = 20g \sin 29°$
$T = \frac{20g \sin 29° - 2g \cos 29°}{\cos 15° - 0.1 \sin 15°} = \frac{77.880...}{0.9400...} = 82.8$ N (3 s.f.)

3. Newton's Laws of Motion

Exercise 3.1 — Newton's laws of motion

Q1 **a)** Resolving horizontally:
$F_{net} = ma$
$10 \cos 45° = 3a$
$a = 7.071... \div 3 = 2.357...$
$a = 2.36$ ms^{-2} (3 s.f.)

b) Resolving vertically:
$F_{net} = ma$
$R + 10 \sin 45° - 3g = 0$
$R = 3g - 10 \sin 45° = 22.328...$
$R = 22.3$ N (3 s.f.)

Q2 **a)** Resolving horizontally:
$F_{net} = ma$
$16 \cos 20° - 7 = (49 \div 9.8) \times a$
$a = 8.035... \div 5 = 1.607...$
$a = 1.61$ ms^{-2} (3 s.f.)

b) Use a constant acceleration equation.
$u = 0,\ v = v,\ a = 1.607...,\ t = 7$
$v = u + at$
$v = 0 + (1.607... \times 7) = 11.2\ \text{ms}^{-1}$ (3 s.f.)

c) Resolving vertically:
$R + 16 \sin 20° - 49 = 0$
$R = 49 - 16 \sin 20° = 43.527...$
$R = 43.5$ N (3 s.f.)

Q3 a) Resultant, $\mathbf{R} = (6\mathbf{i} - 10\mathbf{j}) + (4\mathbf{i} + 5\mathbf{j}) = (10\mathbf{i} - 5\mathbf{j})$ N
$|\mathbf{R}| = \sqrt{10^2 + (-5)^2} = \sqrt{125}$ N
$F_{net} = ma$
$\sqrt{125} = 5a$
$a = \sqrt{125} \div 5 = 2.236... = 2.24\ \text{ms}^{-2}$ (3 s.f.)

b) $\mathbf{R}_{new} = (10\mathbf{i} - 5\mathbf{j}) + (-6\mathbf{i} + 2\mathbf{j}) = (4\mathbf{i} - 3\mathbf{j})$ N
$|\mathbf{R}_{new}| = \sqrt{4^2 + (-3)^2} = 5$ N
$F_{net} = ma$
$5 = 5a \Rightarrow a = 1\ \text{ms}^{-2}$
So the acceleration decreases by $1.24\ \text{ms}^{-2}$ (3 s.f.)

Q4 a) $7\mathbf{i} + \mathbf{j} = (6\mathbf{i} - 2\mathbf{j}) + (5\mathbf{i} + 5\mathbf{j}) + \mathbf{x}$
$\mathbf{x} = (7 - 6 - 5)\mathbf{i} + (1 + 2 - 5)\mathbf{j} = (-4\mathbf{i} - 2\mathbf{j})$ N

b) $\boldsymbol{F}_{net} = m\boldsymbol{a}$
$7\mathbf{i} + \mathbf{j} = 10\boldsymbol{a} \Rightarrow \boldsymbol{a} = 0.7\mathbf{i} + 0.1\mathbf{j}\ \text{ms}^{-2}$
List the variables: $\boldsymbol{u} = 0,\ \boldsymbol{v} = \boldsymbol{v},\ \boldsymbol{a} = 0.7\mathbf{i} + 0.1\mathbf{j},\ t = 8$
$\boldsymbol{v} = \boldsymbol{u} + \boldsymbol{a}t$
$\boldsymbol{v} = 0 + 8(0.7\mathbf{i} + 0.1\mathbf{j}) = (5.6\mathbf{i} + 0.8\mathbf{j})\ \text{ms}^{-1}$

c) The new resultant force is
$(7\mathbf{i} + \mathbf{j}) + (-7\mathbf{i} - \mathbf{j}) = 0\mathbf{i} + 0\mathbf{j}$ N
So there is no resultant force, and hence the particle does not accelerate (i.e. it continues to move at whatever velocity it had when the fourth force began to act).

Exercise 3.2 — Friction and inclined planes

Q1 a) Resolving parallel to the plane (↙):
$F_{net} = ma$
$1.5g \sin 25° = 1.5a$
$a = 9.8 \sin 25° = 4.141... = 4.14\ \text{ms}^{-2}$ (3 s.f.)

b) $u = 0,\ v = v,\ a = 4.141...,\ t = 3$
$v = u + at$
$v = 0 + (4.141... \times 3) = 12.4\ \text{ms}^{-1}$ (3 s.f.)

Q2 a)

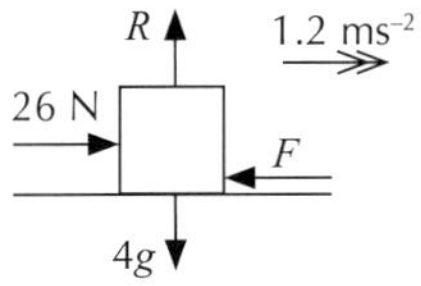

b) Resolving vertically:
$R - 4g = 0$
$R = 4g = 39.2$ N

c) Resolving horizontally:
$F_{net} = ma$
$26 - F = 4 \times 1.2$
$F = 21.2$ N

d) The box is moving, so friction takes its maximum value, i.e. $F = \mu R$:
$21.2 = 39.2\mu \Rightarrow \mu = 0.54$ (2 d.p.)

Q3 a)

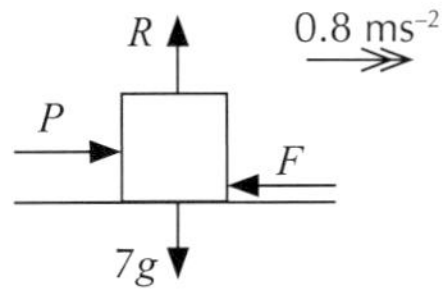

b) Resolving vertically:
$R - 7g = 0 \Rightarrow R = 7g = 68.6$ N

c) Resolving horizontally:
$F_{net} = ma$
$P - F = 7 \times 0.8$
$P - \mu R = 5.6$
$P = 5.6 + (0.2 \times 68.6) = 19.3$ N (3 s.f.)

Q4 a)

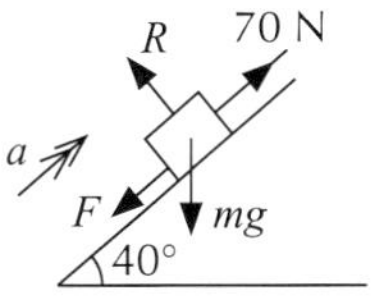

b) Resolving perpendicular to the plane (↖):
$R - mg \cos 40° = 0$
$\Rightarrow R = mg \cos 40°$
Resolving parallel to the plane (↗):
$F_{net} = ma$
$70 - F - mg \sin 40° = 3.2m$
$70 - \mu R - mg \sin 40° = 3.2m$
$70 - 0.35(mg \cos 40°) - mg \sin 40° = 3.2m$
$70 = m(3.2 + 0.35g \cos 40° + g \sin 40°)$
$m = 70 \div 12.126... = 5.77$ kg (3 s.f.)

Q5 a)

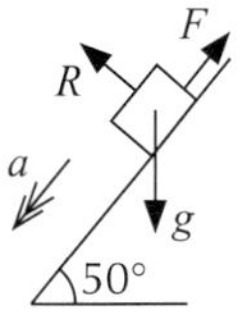

b) Resolving perpendicular to the plane (↖):
$R = g \cos 50°$
Resolving parallel to the plane (↙):
$F_{net} = ma$
$g \sin 50° - F = a$
$g \sin 50° - \mu R = a$
$g \sin 50° - (0.4 \times g \cos 50°) = a$
$\Rightarrow a = 4.9875...$
$s = 2,\ u = 0,\ a = 4.9875...,\ t = t$
$s = ut + \frac{1}{2}at^2$
$2 = \frac{1}{2}(4.9875...)t^2$
$t^2 = 0.8020... \Rightarrow t = 0.896$ s (3 s.f.)

Q6 Resolving perpendicular to the plane (↖):
$R = 5000 \cos 10°$
Resolving parallel to the plane (↗):
$F_{net} = ma$
$T - F - 5000 \sin 10° = (5000 \div 9.8) \times 0.35$
$T - \mu R - 5000 \sin 10° = 178.57...$
$T - (0.1 \times 5000 \cos 10°) - 5000 \sin 10° = 178.57...$
$\Rightarrow T = 1539.21... = 1540$ N (3 s.f.)

Q7

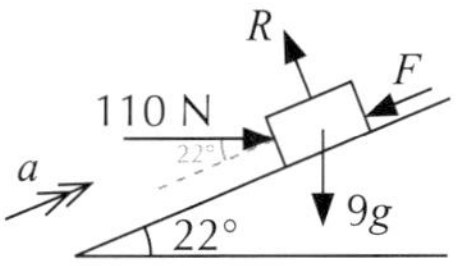

Resolving perpendicular to the plane (↖):
$R = 9g \cos 22° + 110 \sin 22° = 122.98...$
Resolving parallel to the plane (↗):
$F_{net} = ma$
$110 \cos 22° - F - 9g \sin 22° = 9 \times 1.3$
$110 \cos 22° - \mu 122.98... - 9g \sin 22° = 11.7$
$\mu 122.98... = 110 \cos 22° - 9g \sin 22° - 11.7$
$\Rightarrow \mu = 57.24... \div 122.98... = 0.47$ (2 d.p.)

Q8 You need to consider the situations where the block is on the point of sliding up the plane, and when it is on the point of sliding down the plane.
If the block is on the point of sliding up the plane, then friction acts down the plane:

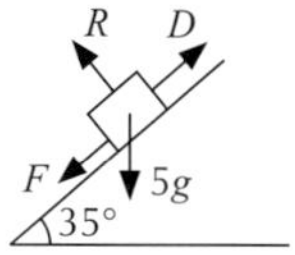

Resolving perpendicular to the plane (↖):
$R = 5g \cos 35° = 40.138...$ N
The block is in limiting equilibrium, so $F = \mu R$:
$F = 0.4 \times 40.138... = 16.055...$ N
Resolving parallel to the plane (↗):
$D - F - 5g \sin 35° = 0$
$D = 16.055... + 28.105... = 44.160...$ N
If the block is on the point of sliding down the plane, then friction acts up the plane:

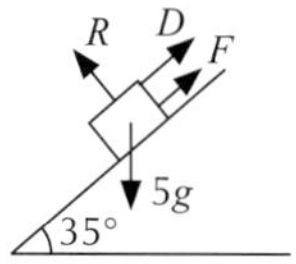

Resolving perpendicular to the plane (↖):
$R = 5g \cos 35° = 40.138...$ N
The block is in limiting equilibrium, so $F = \mu R$:
$F = 0.4 \times 40.138... = 16.055...$ N
Resolving parallel to the plane (↗):
$D + F - 5g \sin 35° = 0$
$D = 28.105... - 16.055... = 12.049...$ N
So the range for which the block is in equilibrium is $12.049...\text{ N} \leq D \leq 44.160...$ N

Exercise 3.3 — Connected particles, pegs and pulleys

Q1 a) Resolving vertically (↓) for B:
$F_{net} = ma$
$8g - T = 8 \times 0.5 \Rightarrow T = (8 \times 9.8) - 4 = 74.4$ N

b) Resolving vertically (↑) for A:
$R - 10g = 0 \Rightarrow R = 98$ N
A is sliding, so F takes its maximum value, i.e. $F = \mu R = 98\mu$
Resolving horizontally (→) for A:
$F_{net} = ma$
$T - F = 10 \times 0.5$
$98\mu = 74.4 - 5 \Rightarrow \mu = 0.71$ (2 d.p.)

Q2 Resolving parallel to the plane (↗) for the lower block:
$F_{net} = ma$
$20 - T - F - mg \sin \theta = ma$ ①
Resolving parallel to the plane (↗) for the upper block:
$F_{net} = ma$
$T - F - mg \sin \theta = ma$ ②
Substituting ① into ②:
$T - F - mg \sin \theta = 20 - T - F - mg \sin \theta$
$2T = 20 \Rightarrow T = 10$ N

Q3 a) $s = 10, u = 0, a = a, t = 30$
$s = ut + \frac{1}{2}at^2$
$10 = \frac{1}{2} \times a \times 30^2 \Rightarrow a = 0.0222...$
Resolving parallel to the plane (↗) for the car:
$F_{net} = ma \Rightarrow T - 800g \sin 25° = 800 \times 0.0222...$
$\Rightarrow T = 3331.10... = 3330$ N (3 s.f.)

b) Resolving vertically (↑) for the truck:
$R_T = 2000g$
You're told you can assume that friction is at its maximum value, so using $F = \mu R$:
$F = 0.8 \times 2000g = 1600g$ N
Resolving horizontally (→) for the truck:
$F_{net} = ma$
$W - F - T = 2000 \times 0.022...$
$W = 44.44... + 1600g + 3331.10...$
$= 19\,055.54... = 19\,100$ N (3 s.f.)

c) Resolving horizontally (→) for the truck:
$F_{net} = ma$
$W - F = 2000a$
$19\,055.54... - 1600g = 2000a$
$\Rightarrow a = 1.69$ ms^{-2} (3 s.f.)

Q4 a) Resolving perpendicular to the plane (↖) for P:
$R = 60g \cos 50°$
Using $F = \mu R$:
$F = 0.1 \times 60g \cos 50° = 6g \cos 50°$
Resolving parallel to the plane (↙) for P:
$F_{net} = ma$
$60g \sin 50° - F - T = 0$
$T = 60g \sin 50° - 6g \cos 50° = 412.63...$ N
Resolving vertically (↑) for Q:
$F_{net} = ma$
$T - 20g - K = 0$
$K = 412.63... - 20g = 216.63...$
$K = 217$ N (3 s.f.)

b) Resolving vertically (↑) for Q:
$F_{net} = ma$
$T - 20g = 20a$ ①
Resolving parallel to the plane (↙) for P:
$60g \sin 50° - F - T = 60a$
$60g \sin 50° - 6g \cos 50° - T = 60a$ ②
Adding ① and ②:
$60g \sin 50° - 6g \cos 50° - 20g = 80a$
$\Rightarrow a = 2.707...$ ms^{-2}
$u = 0, v = v, t = 5$
$v = u + at$
$v = 0 + (2.707... \times 5) = 13.539...$
$v = 13.5$ ms^{-1} (3 s.f.)

Q5 a)

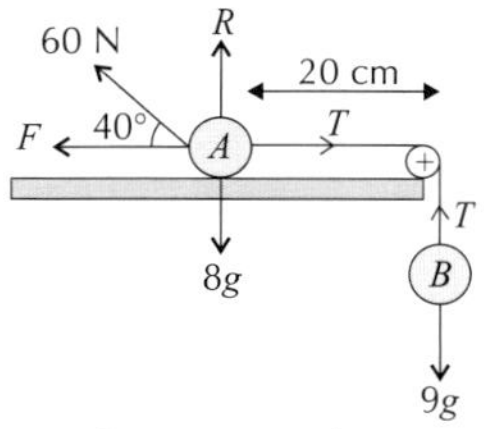

Resolving vertically (↓) for B:
$F_{net} = ma$
$9g - T = 9 \times 0.4$
$\Rightarrow T = 84.6$ N
Resolving vertically (↑) for A:
$R + 60 \sin 40° = 8g$
$\Rightarrow R = 39.832...$ N
Resolving horizontally (→) for A:
$F_{net} = ma$
$T - F - 60 \cos 40° = 8 \times 0.4$
$F = 84.6 - 60 \cos 40° - 3.2 = 35.437...$ N
$F = \mu R \Rightarrow \mu = 35.437... \div 39.832... = 0.89$ (2 d.p.)

b) $s = 0.2, u = 0, v = v, a = 0.4$
$v^2 = u^2 + 2as$
$v^2 = 2 \times 0.4 \times 0.2 = 0.16 \Rightarrow v = 0.4$ ms^{-1}

c) When A hits the pulley, A and B are both travelling at 0.4 ms^{-1}.
$s = 0.6 - 0.2 = 0.4$, $u = 0.4$, $v = v$, $a = 9.8$
$v^2 = u^2 + 2as$
$v^2 = 0.4^2 + (2 \times 9.8 \times 0.4) = 8$
$v = 2.83 \text{ ms}^{-1}$ (3 s.f.)

Q6 a) Resolving perpendicular to the 40° plane (↖) for A:
$R_A = 0.2g \cos 40°$
Using $F = \mu R$:
$F_A = 0.7 \times 0.2g \cos 40° = 0.14g \cos 40°$
Resolving parallel to the 40° plane (↗) for A:
$T - F_A - 0.2g \sin 40° = 0.2a$
$T - 0.14g \cos 40° - 0.2g \sin 40° = 0.2a$
$\Rightarrow 5T - 0.7g \cos 40° - g \sin 40° = a$ ①
Resolving perpendicular to the 20° plane (↗) for B:
$R_B = 0.9g \cos 20°$
Using $F = \mu R$:
$F_B = 0.05 \times 0.9g \cos 20° = 0.045g \cos 20°$
Resolving parallel to the 20° plane (↘) for B:
$F_{net} = ma$
$0.9g \sin 20° - F_B - T = 0.9a$
$0.9g \sin 20° - 0.045g \cos 20° - T = 0.9a$
$\Rightarrow g \sin 20° - 0.05g \cos 20° - (1.11...)T = a$ ②
Solving the simultaneous equations ① and ②:
$5T - 0.7g \cos 40° - g \sin 40° =$
$g \sin 20° - 0.05g \cos 20° - (1.11...)T$
$\Rightarrow (6.11...)T = g \sin 20° - 0.05g \cos 20° + 0.7g \cos 40° + g \sin 40°$
$\Rightarrow T = 2.363... = 2.36$ N (3 s.f.)

b) Using ① from part a):
$a = 5T - 0.7g \cos 40° - g \sin 40°$
$= (5 \times 2.363...) - 0.7g \cos 40° - g \sin 40° = 0.2648...$
$u = 0$, $v = v$, $s = 0.35$
$v^2 = u^2 + 2as$
$v^2 = 2 \times 0.2648... \times 0.35 = 0.185...$
$\Rightarrow v = 0.431 \text{ ms}^{-1}$ (3 s.f.)

Chapter 16: Moments

1. Moments

Exercise 1.1 — Moments

Q1 a) Moment = Force × Perpendicular Distance
$= 3 \times 2.5 = 7.5$ Nm anticlockwise

b) Moment $= 6 \sin 30° \times 4 = 12$ Nm clockwise

c) Moment $= 23 \times 0.8 \cos 18°$
$= 17.5$ Nm (3 s.f.) clockwise

Q2 a) Taking clockwise as positive:
$(2 \times 9) - (7 \times 2) = 4$ Nm clockwise

b) Taking clockwise as positive:
$(19 \times 1.5) - (14 \times 2) = 0.5$ Nm clockwise

c) Taking clockwise as positive:
$(7 \sin 30° \times 3) - (15 \times 1.5) = -12$ Nm
i.e. 12 Nm anticlockwise

d) Taking clockwise as positive:
$(10 \sin 30° \times 3) + (8 \sin 45° \times 4) = (15 + 16\sqrt{2})$ Nm
$= 37.6$ Nm (3 s.f.) clockwise

e) Taking clockwise as positive:
$(18 \sin 25° \times 13) - (24 \sin 32° \times 13)$
$= -66.4$ Nm (3 s.f.), i.e. 66.4 Nm anticlockwise

f) Taking clockwise as positive:
$(6 \cos 15° \times 10) + (2 \sin 50° \times 8)$
$- (3 \sin 28° \times 10) - (1 \times 8)$
$= 48.1$ Nm (3 s.f.) clockwise.

Q3 Find the sum of the moments about O, taking clockwise as positive:
$(13 \times 3) + (X \times 1) - (21 \times 3) = X - 24$
For the rod to rotate clockwise, the sum of the moments must be positive:
$X - 24 > 0 \Rightarrow X > 24$

Q4 Find the sum of the moments about O, taking clockwise as positive:
$4(d + 1) + (38 \times 2) - (25 \times 1) -$
$(24 \cos 60° \times 5) = 4d - 5$
For the rod to rotate anticlockwise, the sum of the moments must be negative:
$4d - 5 < 0 \Rightarrow d < 1.25$
Also, d must be greater than or equal to zero, otherwise the 25 N force will be applied to a point which isn't on the rod, and so the rod will rotate clockwise. So:
$0 \le d < 1.25$

Exercise 1.2 — Moments in equilibrium

Q1 a) Taking moments about A:
Moments clockwise = Moments anticlockwise
$1.5y + (2 \times 8) = 16 \times 5.5$
$1.5y = 72$
$y = 48$ N
Resolving vertically:
$x + 16 = y + 2$
$x = 50 - 16 = 34$ N

b) Resolving vertically:
$10 + 6 = 13 + x \Rightarrow x = 3$ N
Taking moments about C:
$10y + 2x = 6 \times 5$
$10y = 30 - 6 \Rightarrow y = 2.4$ m

Q2 a) Mass acts at centre of rod, i.e. 2.5 m from A, 1 m from C and 0.5 m from D.
Taking moments about the midpoint:
$0.5R_D = 49 \times 1$
$R_D = 98$ N

b) Resolving vertically:
$Mg = 49 + R_D$
$M = (49 + 98) \div 9.8 = 15$ kg

Q3 When the rod is about to tilt about D, $T_C = 0$.
The rod is uniform, so its mass acts at its midpoint, x m from D.
Taking moments about D:
$3g \times 1.5 = 9g \times x$
$9x = 4.5 \Rightarrow x = 0.5$
So the distance from B to the midpoint is
$1.5 + 0.5 = 2$ m, and so the length of the rod is
$2 \times 2 = 4$ m

Q4 a)

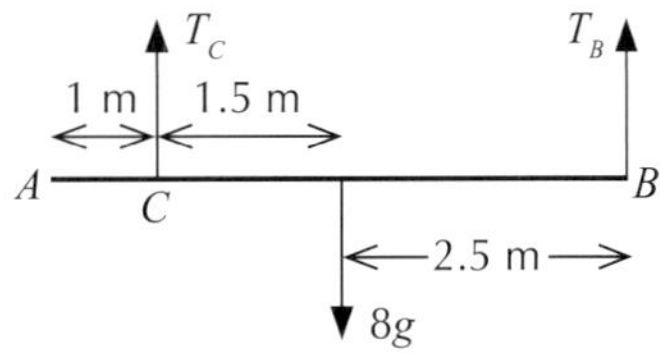

Taking moments about C:
$8g \times 1.5 = 4T_B$
$T_B = 3g = 29.4$ N
Resolving vertically:
$T_C + T_B = 8g$
$T_C = 78.4 - 29.4 = 49$ N

b) When the beam is about to tilt about C, $T_B = 0$.
Taking moments about C:
$8g \times 1.5 = 16g \times x$
$1.5 = 2x \Rightarrow x = 0.75$
So the distance of the particle from A is
$1 - 0.75 = 0.25$ m

Q5 a) The painter stands at the plank's COM, so the weight acting at this point is his weight plus the plank's weight, i.e. $80g + 20g = 100g$.

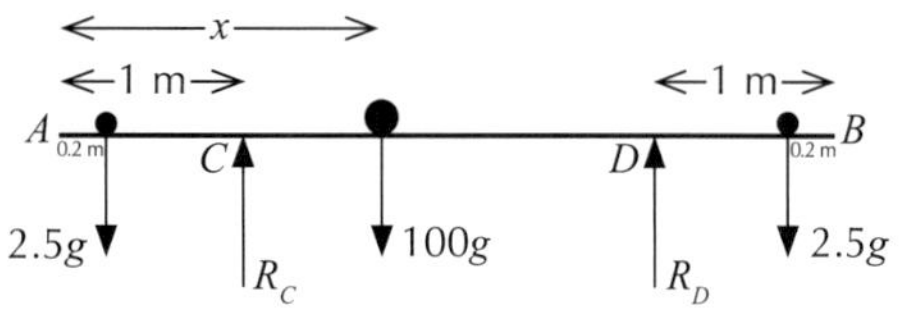

The reaction at C is 4 times the reaction at D, i.e. $R_C = 4R_D$
Resolving vertically:
$2.5g + 100g + 2.5g = R_C + R_D$
$105g = 4R_D + R_D$
$105g = 5R_D$
$\Rightarrow R_D = 21g \text{ N} \Rightarrow R_C = 84g \text{ N}$
Taking moments about A:
$(2.5g \times 0.2) + (100g \times x) + (2.5g \times 3.8) = R_C + 3R_D$
$10g + 100gx = 84g + 63g$
$10g + 100gx = 147g$
$x = (147 - 10) \div (100) = 1.37$ m

b) The distance between the plank's COM and the point D is $4 - 1 - 1.37 = 1.63$ m

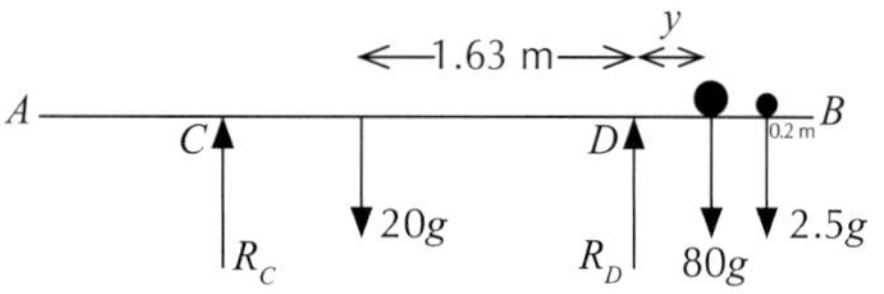

When the plank is about to tilt about D, $R_C = 0$.
Taking moments about D:
$20g \times 1.63 = (80g \times y) + (2.5g \times 0.8)$
$\Rightarrow 32.6 = 80y + 2 \Rightarrow y = 30.6 \div 80 = 0.3825$
So the distance from B is:
$1 - 0.3825 = 0.6175$ m

Q6 Taking clockwise as the positive direction:

a) Sum of moments $= (6 \times 5) - (10 \times 10)$
$= 30 - 100 = -70$ Nm
(i.e. 70 Nm anticlockwise)

b) Sum of moments $= (6 \times 0.5) + (10 \times 5) = 3 + 50$
$= 53$ Nm clockwise

c) Sum of moments $= (6 \times 2) + (10 \times 4) = 12 + 40$
$= 52$ Nm clockwise

d) Sum of moments $= (6 \times 3.5) - (10 \times 6)$
$= 21 - 60 = -39$ Nm
(i.e. 39 Nm anticlockwise)

For parts a)-d), F acts either vertically or horizontally, so you can just use x or y respectively as the perpendicular distance when taking moments. Drawing a diagram will help you to see whether it acts clockwise or anticlockwise.

e) Sum of moments $= (6 \times 2.5) + (10 \cos 30° \times 5) - (10 \sin 30° \times 5)$
$= 15 + 43.301... - 25$
$= 33.3$ Nm clockwise (3 s.f.)

f) Sum of moments $= (6 \times 0.5) - (10 \sin 60° \times 1) - (10 \cos 60° \times 11)$
$= 3 - 8.6602... - 55 = -60.7$ Nm
(i.e. 60.7 Nm anticlockwise) (3 s.f.)

Q7 a)

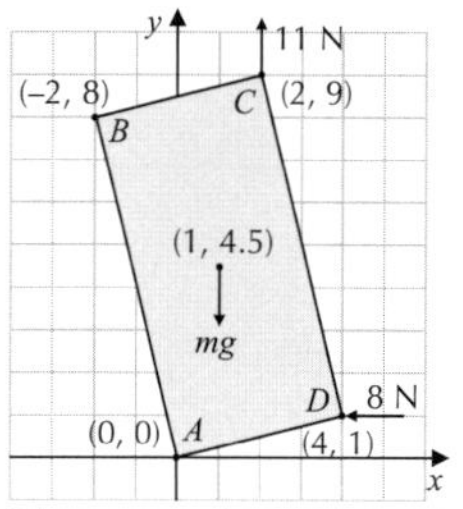

The centre of mass is at the centre of the lamina, so to find its coordinates, find the midpoint of AC (or BD).

b) Taking moments about A:
Clockwise moments = anticlockwise moments
$(mg \times 1) = (8 \times 1) + (11 \times 2) \Rightarrow 9.8m = 8 + 22 = 30$
$\Rightarrow m = 30 \div 9.8 = 3.06$ kg (3 s.f.)
You don't need to split the forces into components — it's much easier to find the perpendicular distance from the line of action of the force to the pivot, since all of the forces act in either the x- or y-direction.

2. Reaction Forces and Friction

Exercise 2.1 — Reaction forces

Q1 a) Taking moments about A:
$3g \times 2 = T \times 1 \Rightarrow T = 58.8$ N

b) Resolving vertically:
$T = 3g + R \Rightarrow R = 58.8 - 29.4 = 29.4$ N

Q2 a) Taking moments about A:
$0.6g \times 0.4 = T \sin 60° \times 0.6$
$\Rightarrow T = 2.352 \div 0.519... = 4.526... = 4.53$ N (3 s.f.)

b)

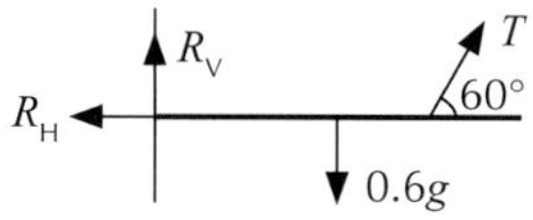

Resolving horizontally:
$R_H = T \cos 60° = (4.526...) \cos 60° = 2.263...$ N
Resolving vertically:
$R_V + T \sin 60° = 0.6g$
$\Rightarrow R_V = 0.6g - (4.526...) \sin 60° = 1.96$ N

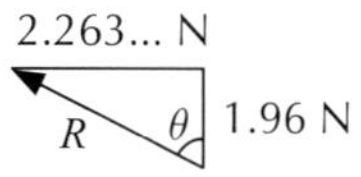

Using Pythagoras' theorem:
$R = \sqrt{(2.263...)^2 + 1.96^2} = 2.993... = 2.99$ N (3 s.f.)
Using trigonometry:
$\theta = \tan^{-1}\left(\frac{2.263...}{1.96}\right) = 49.106...°$
(measured anticlockwise from the upward vertical)
So direction $= 90° + 49.106...° = 139°$ (3 s.f.).

Q3 a) Taking moments about A:
$(1.5g \times 0.6) + (0.5g \times 1.2) = T \sin 20° \times 2.4$
$\Rightarrow T = 14.7 \div 0.820... = 17.908...$
$= 17.9$ N (3 s.f.)

b)

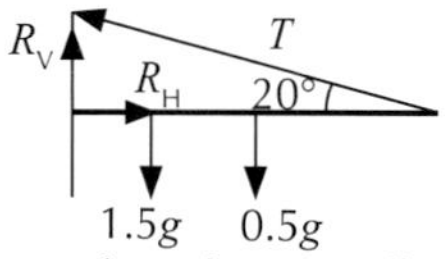

Resolving horizontally:
$R_H = T \cos 20° = (17.908...) \cos 20° = 16.828...$ N
Resolving vertically:
$R_V + T \sin 20° = 1.5g + 0.5g$
$\Rightarrow R_V = 19.6 - 6.125 = 13.475$ N

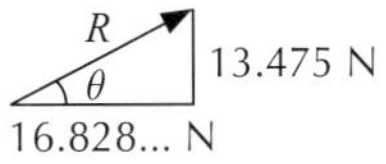

Using Pythagoras' theorem:
$R = \sqrt{(16.828...)^2 + 13.475^2} = 21.558...$
$= 21.6$ N (3 s.f.)
Using trigonometry:
$\theta = \tan^{-1}\left(\frac{13.475}{16.828...}\right) = 38.7°$ (3 s.f.)

Q4 a) Let the distance from A to the COM be x.
Taking moments about A:
$(1.8g \times x) + (3.5 \sin 80° \times 7) = 95 \cos 71° \times 3.5$
$\Rightarrow x = 84.123... \div 17.64 = 4.768... = 4.77$ m (3 s.f.)

b)

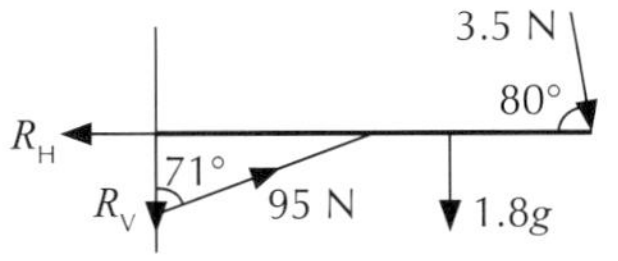

Resolving horizontally:
$R_H = 95 \sin 71° + 3.5 \cos 80° = 90.432...$ N
Resolving vertically:
$R_V + 1.8g + 3.5 \sin 80° = 95 \cos 71°$
$\Rightarrow R_V = 9.842...$ N

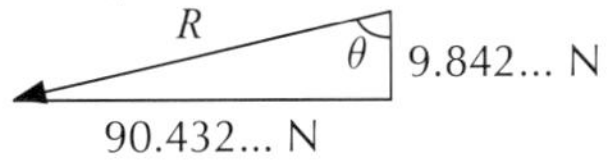

Using Pythagoras' theorem:
$R = \sqrt{(90.432...)^2 + (9.842...)^2} = 90.966...$
$= 91.0$ N (3 s.f.)
Using trigonometry:
$\theta = \tan^{-1}\left(\frac{90.432...}{9.842...}\right) = 83.788...°$
(measured clockwise from the downward vertical)
So direction = 270° – 83.788...° = 186° (3 s.f.)

Q5 a) Taking moments about A:
$(110 \cos 30° \times 2) + (T \sin 30° \times 4) = (1.75g \times 4) + (2.25g \times 8)$
$\Rightarrow T = 54.474... \div 2 = 27.237... = 27.2$ N (3 s.f.)

b)

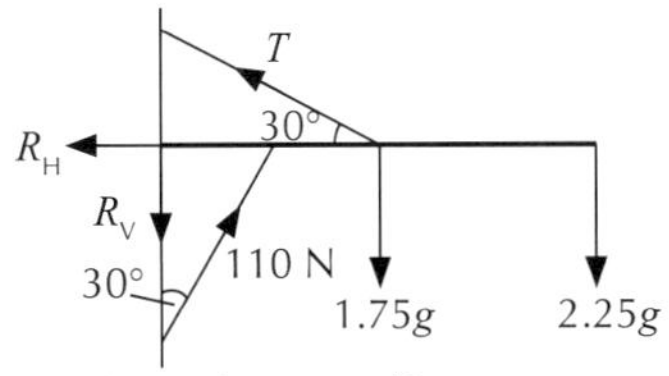

Resolving horizontally:
$R_H + (27.237...) \cos 30° = 110 \sin 30°$
$\Rightarrow R_H = 31.411...$ N
Resolving vertically:
$R_V + 1.75g + 2.25g = (27.23...) \sin 30° + 110 \cos 30°$
$\Rightarrow R_V = 69.681...$ N

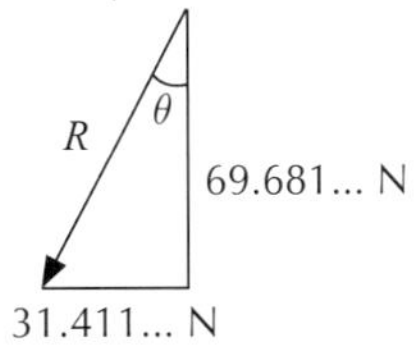

Using Pythagoras' theorem:
$R = \sqrt{(31.411...)^2 + (69.681...)^2} = 76.434...$
$= 76.4$ N (3 s.f.)
Using trigonometry:
$\theta = \tan^{-1}\left(\frac{31.411...}{69.681...}\right) = 24.265...°$
(measured clockwise from the downward vertical)
So direction = 270° – 24.265° = 246° (3 s.f.)

Exercise 2.2 — Friction

Q1 a) Taking moments about A:
$12g \times 8 = T \sin 20° \times 16$
$\Rightarrow T = 940.8 \div 5.472... = 171.919... = 172$ N (3 s.f.)

b)

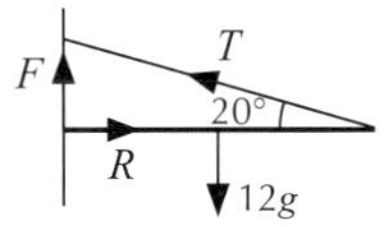

Resolving horizontally:
$R = T \cos 20° = (171.919...) \cos 20° = 161.551...$ N
Resolving vertically:
$F + T \sin 20° = 12g$
$\Rightarrow F = 117.6 - (171.919...) \sin 20° = 58.800...$ N
Using $F = \mu R$:
$58.800... = \mu \times 161.551... \Rightarrow \mu = 0.36$ (2 d.p.)

Q2 a)

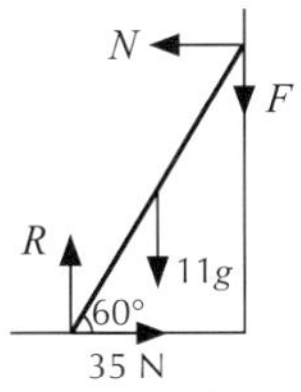

Resolving horizontally: $N = 35$ N

b) Resolving vertically: $F + 11g = R$
Taking moments about top of ladder:
$(11g \cos 60° \times 3.5) + (35 \sin 60° \times 7) =$
$R \cos 60° \times 7$
$\Rightarrow R = 400.826... \div 3.5 = 114.521...$ N
So $F = 114.521... - 11g = 6.721... = 6.72$ N (3 s.f.)

c) $F = \mu N$
$6.721... = \mu \times 35 \Rightarrow \mu = 0.19$ (2 d.p.)

Q3 a)

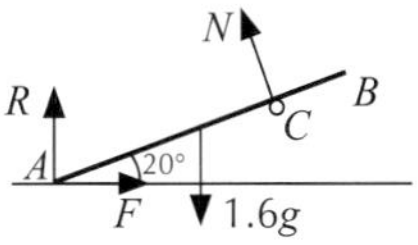

Taking moments about A:
$1.6g \cos 20° \times 0.75 = N \times 1.1$
$\Rightarrow N = 10.046... = 10.0$ N (3 s.f.)

b) Resolving vertically:
$R + (10.046...) \cos 20° = 1.6g$
$\Rightarrow R = 6.239... = 6.24$ N (3 s.f.)

c) Resolving horizontally:
$F = N \sin 20° = (10.046...) \sin 20°$
$= 3.435... = 3.44$ N (3 s.f.)

d) $F = \mu R$
$3.435... = \mu \times 6.239... \Rightarrow \mu = 0.55$ (2 d.p.)

Q4

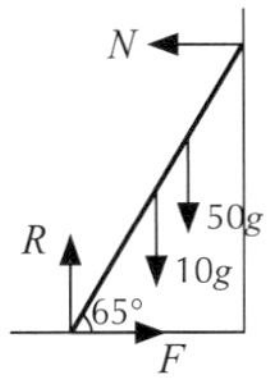

Resolving vertically: $R = 10g + 50g = 60g$
Resolving horizontally: $F = N$
If the ladder is on the point of slipping, then $F = \mu R$.
$\Rightarrow \mu R = N \Rightarrow N = 0.3 \times 60g = 176.4$ N
Let x be the girl's distance up the ladder from the base.
Taking moments about base of ladder:
$176.4 \sin 65° \times 6 = (10g \cos 65° \times 3) + (50g \cos 65° \times x)$
$\Rightarrow x = 834.986... \div 207.082... = 4.03$ m (3 s.f.)

Q5 a)

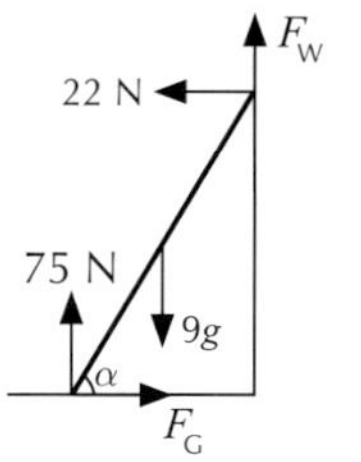

Resolving horizontally: $F_G = 22$ N
Resolving vertically:
$75 + F_W = 9g \Rightarrow F_W = 13.2$ N
Taking moments about base of ladder:
$9g \cos\alpha \times 2.4 = (22 \sin\alpha \times 4.8) + (13.2 \cos\alpha \times 4.8)$
$148.32 \cos\alpha = 105.6 \sin\alpha$
$\tan\alpha = 1.4045... \Rightarrow \alpha = 54.550...° = 54.6°$ (3 s.f.)

b) $F = \mu R$
$F_W = \mu_W \times 22 \Rightarrow \mu_W = 13.2 \div 22 = 0.6$

c) $F_G = \mu_G \times 75 \Rightarrow \mu_G = 22 \div 75 = 0.29$ (2 d.p.)

Q6 If the ladder is on the point of slipping downwards:

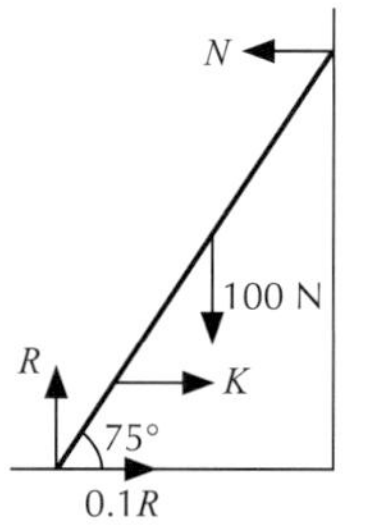

Resolving vertically: $R = 100$ N
Resolving horizontally
$K + 0.1R = N \Rightarrow N = K + 10$
Taking moments about the base of the ladder:
$(K \sin 75° \times 1) + (100 \cos 75° \times 4) = (N \sin 75° \times 8)$
$K \sin 75° + 400 \cos 75° = 8(K + 10) \sin 75°$
$400 \cos 75° - 80 \sin 75° = 8K \sin 75° - K \sin 75°$
$400 \cos 75° - 80 \sin 75° = K(7 \sin 75°)$
$K = 26.253... \div 6.761... = 3.882...$ N

If the ladder is on the point of slipping towards the wall:

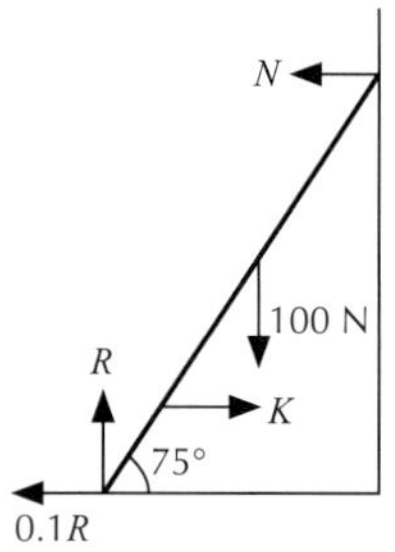

Resolving vertically: $R = 100$ N
Resolving horizontally: $K = N + 0.1R \Rightarrow N = K - 10$
Taking moments about the base of the ladder:
$(K \sin 75° \times 1) + (100 \cos 75° \times 4) = (N \sin 75° \times 8)$
$K \sin 75° + 400 \cos 75° = 8(K - 10) \sin 75°$
$400 \cos 75° + 80 \sin 75° = 8K \sin 75° - K \sin 75°$
$400 \cos 75° + 80 \sin 75° = K(7 \sin 75°)$
$K = 180.80... \div 6.761... = 26.73...$ N

So the possible range of values for K is
$3.882...\text{ N} \le K \le 26.73...\text{ N}$

Q7 a)

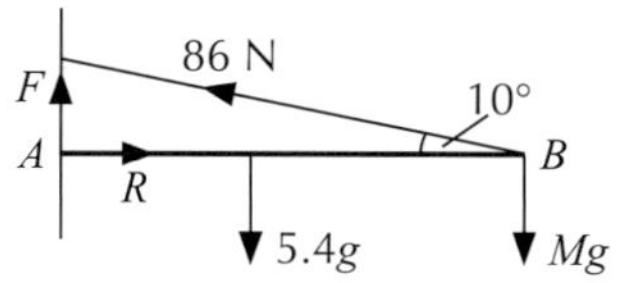

Resolving horizontally:
$R = 86 \cos 10° = 84.693...$N
Using $F = \mu R$:
$F = 0.55 \times 84.693... = 46.581...$ N
Resolving vertically:
$46.581... + 86 \sin 10° = 5.4g + Mg$
$\Rightarrow M = 8.595... \div 9.8 = 0.8770...$
$= 0.877$ kg (3 s.f.)

b) Let x be the distance of the COM from A.
Taking moments about A:
$(5.4g \times x) + (Mg \times 11) = 86 \sin 10° \times 11$
$\Rightarrow x = 69.724 \div 52.92 = 1.32$ m (3 s.f.)

Q8

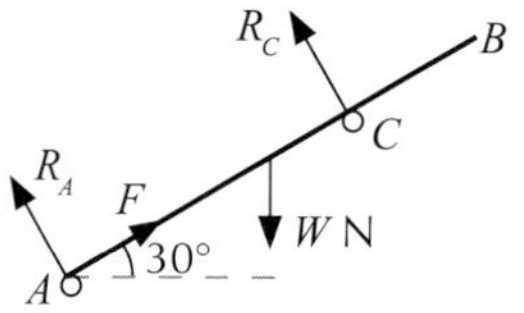

Let l be the length of the beam.
Taking moments about A:
$W \cos 30° \times 0.5l = R_C \times 0.75l$
$\Rightarrow R_C = \frac{\sqrt{3}}{2} W \times \frac{0.5l}{0.75l} = \frac{\sqrt{3}}{3} W$ N
Resolving perpendicular to the beam:
$R_A + R_C = W \cos 30°$
$\Rightarrow R_A = \frac{\sqrt{3}}{2} W - \frac{\sqrt{3}}{3} W = \frac{\sqrt{3}}{6} W$ N
Resolving parallel to the beam:
$F = W \sin 30° = \frac{1}{2} W$ N
Using $F = \mu R$:
$\frac{1}{2} W = \mu \times R_A$
$\Rightarrow \mu = \frac{1}{2} W \div \frac{\sqrt{3}}{6} W = \frac{1}{2} \times \frac{6}{\sqrt{3}} = \frac{3}{\sqrt{3}} = \sqrt{3}$
$= 1.73$ (2 d.p.)

Usually, the coefficient of friction between a body and a surface is between 0 and 1, but it can actually be any number greater than 0 — as in this case.

Q9

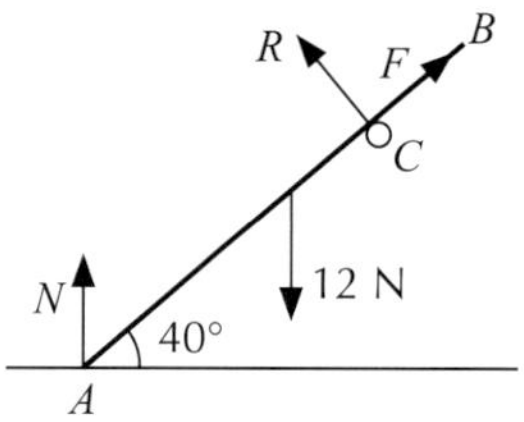

The beam is in equilibrium,
so resolving forces horizontally:
$R \sin 40° = F \cos 40°$
$F = \tan 40°\, R$
$F \le \mu R$, so:
$\tan 40°\, R \le \mu R$
$\tan 40° \le \mu$
$\mu \ge \tan 40°$
So the range of possible values is $\mu \ge 0.84$ (2 d.p.)

Glossary

A

Absolute value
Another name for the **modulus**.

Acceleration
The rate of change of an object's **velocity** with respect to time.

Addition law
A formula linking the probability of the **union** and the probability of the **intersection** of events A and B.

Algebraic division
Dividing one algebraic expression by another.

Algebraic fraction
A fraction made up of algebraic expressions.

Alternative hypothesis
The statement that you will accept instead if you decide to reject the **null hypothesis** in a **hypothesis test.** It gives a range of values for the **parameter** and is usually written H_1.

Arc
The curved edge of a **sector** of a circle.

Arccos
The **inverse** of the cosine function, also written as arccosine or $\cos^{-1}$.

Arcsin
The **inverse** of the sine function, also written as arcsine or $\sin^{-1}$.

Arctan
The **inverse** of the tangent function, also written as arctangent or $\tan^{-1}$.

Arithmetic sequence/series
A **sequence** or **series** where successive terms have a common difference.

Assumption
A simplification of a real-life situation used in a **model**.

B

Beam
A long, **thin**, straight, **rigid** body.

Binomial coefficient
The number of orders in which x objects of one type and $(n - x)$ objects of a different type can be arranged.
Equal to $\binom{n}{x} = \frac{n!}{x!(n-x)!}$.

Binomial distribution B(n, p)
A discrete probability distribution which models the number of successes x in n independent trials when the probability of success in each trial is p.

Binomial expansion
A method of expanding functions of the form $(q + px)^n$. Can be used to give a finite expansion or to find an approximation of a value.

C

Cartesian equation
An equation relating the perpendicular axes x and y in 2D (or x, y and z in 3D).
Cartesian coordinates are given in the form (x, y) in 2D (or (x, y, z) in 3D).

Chain rule
A method for **differentiating** a function of a function.

Coefficient of friction
A number greater than or equal to zero which measures the effect of **friction** between an object and a surface.

Collinear points
Three or more points are collinear if they all lie on the same straight line.

Component
The effect of a **vector** in a given direction.

Composite function
A combination of two or more functions acting on a value or set of values.

Compression
See **thrust**.

Concave curve
A curve with a negative second derivative — i.e. $f''(x) < 0$ for all x.

Conditional probability
A probability is conditional if it depends on whether or not another event happens.

Constant of integration
A constant term coming from an indefinite **integration** representing any number.

Continuity correction
A correction made in order to approximate a discrete distribution with a continuous distribution.

Convergence
A sequence converges if the terms get closer and closer to a single value.

Convergent sequence/series
A **sequence/series** that tends towards a **limit**.

Convex curve
A curve with a positive second derivative — i.e. $f''(x) > 0$ for all x.

Correlation
A linear relationship between two variables showing that they change together to some extent.
(A correlation does not necessarily mean a causal relationship.)

Cosec
The **reciprocal** of the sine function, sometimes written as cosecant.

Cot
The **reciprocal** of the tangent function, sometimes written as cotangent.

Critical region
The set of all values of the **test statistic** that would cause you to reject the **null hypothesis**.

Critical value
The value of the **test statistic** at the edge of the **critical region**.

D

Degree
The highest power of x in a **polynomial**.

Derivative
The result after **differentiating** a function.

Differential equation
An equation connecting variables with their rates of change.

Differentiation
A method for finding the rate of change of a function with respect to a variable — the opposite of **integration**.

Divergence
A sequence diverges if the terms get further and further apart.

Divergent sequence/series
A **sequence/series** that does not have a **limit**.

Divisor
The number or expression that you're dividing by in a division.

Domain
The set of values that can be input into a **mapping** or **function**. Usually given as the set of values that x can take.

E

e
An **irrational number** for which the gradient of $y = e^x$ is equal to e^x.

Equating coefficients
Making the coefficients of equivalent terms on each side of an identity equal in order to calculate the value of unknowns in the identity.

Equilibrium
A state where there is no **resultant force** or **moment** acting on a body, hence the body is at **rest** (or moving with constant **velocity**).

Exponential function
A function of the form $y = a^x$. $y = e^x$ is known as 'the' exponential function.

F

Factorial
n factorial, written $n!$, is the product of all **integers** from 1 to n.
So $n! = 1 \times 2 \times ... \times n$.

Finite sequence
A **sequence** that has a 'last term'.

Force
An influence which can change the motion of a body (i.e. cause an **acceleration**).

Friction
A frictional force is a resistive **force** due to **roughness** between a body and surface. It always acts against motion, or likely motion.

Function
A type of **mapping** which maps every number in the **domain** to only one number in the **range**.

G

g
Acceleration due to gravity.
g is usually assumed to be 9.8 ms^{-2}.

Geometric sequence/series
A **sequence** or **series** in which you multiply by a **common ratio** to get from one term to the next.

H

Hypothesis
A statement or claim that you want to test.

Hypothesis test
A method of testing a **hypothesis** using observed sample data.

I

i unit vector
The standard horizontal **unit vector** (i.e. along the x-axis).

Identity
An equation that is true for all values of a variable, usually denoted by the '≡' sign.

Implicit differentiation
A method of differentiating an **implicit relation**.

Implicit relation
An equation in x and y written in the form $f(x, y) = g(x, y)$, instead of $y = f(x)$.

Indefinite integral
An **integral** which contains a constant of integration that comes from integrating without limits.

Independent events
If the probability of an event B happening doesn't depend on whether or not an event A happens, events A and B are independent.

Inextensible
Describes a body which can't be stretched (usually a **string**).

Infinite sequence
A **sequence** that has no 'last term', so it goes on forever.

Integer
A whole number, including 0 and negative numbers. The set of integers has the notation $\mathbb{Z}$.

Integral
The result you get when you **integrate** something.

Integration
Process for finding the equation of a function, given its **derivative** — the opposite of **differentiation**.

Integration by parts
A method for **integrating** a product of two functions. The reverse process of the **product rule**.

Integration by substitution
A method for **integrating** a function of a function. The reverse process of the **chain rule**.

Intersection (of events A and B)
The set of outcomes corresponding to both event A and event B happening.

Inverse function
An inverse function, e.g. $f^{-1}(x)$, reverses the effect of the function $f(x)$.

Irrational number
A number that can't be expressed as the **quotient** (division) of two **integers**. Examples include surds and π.

Iteration
A numerical method for solving equations that allows you to find the approximate value of a **root** by repeatedly using an iteration formula.

Iteration sequence
The list of results x_1, x_2, ... etc. found with an iteration formula.

J

j unit vector
The standard vertical **unit vector** (i.e. along the y-axis).

K

k unit vector
The standard **unit vector** used in 3D to represent movement along the z-axis.

Kinematics
The study of the motion of objects.

L

Lamina
A flat two-dimensional body whose thickness can be ignored.

Light
Describes a body which is modelled as having no mass.

Limit (sequences and series)
The value that the individual terms in a **sequence**, or the sum of the terms in a **series**, tends towards.

Limiting equilibrium
Describes a body which is at **rest** in **equilibrium**, but is on the point of moving.

Limits (integration)
The numbers between which you integrate to find a **definite integral**.

Logarithm
The logarithm to the base a of a number x (written $\log_a x$) is the power to which a must be raised to give that number.

Lower bound
The lowest value a number could take and still be rounded up to the correct answer.

M

Magnitude
The size of a **vector**.

Many-to-one function
A function where some values in the **range** correspond to more than one value in the **domain**.

Mapping
An operation that takes one number and transforms it into another.

Model
A mathematical approximation of a real-life situation, in which certain **assumptions** are made about the situation.

Modulus
The modulus of a number is its positive numerical value.
The modulus of a function, f(x), makes every value of f(x) positive by removing any minus signs.
The modulus of a **vector** is the same as its **magnitude**.

Moment
The turning effect a **force** has about a pivot point.

Mutually exclusive
Events are mutually exclusive (or just 'exclusive') if they have no outcomes in common, and so can't happen at the same time.

N

Natural logarithm
The **inverse function** of e^x, written as $\ln x$ or $\log_e x$.

Natural number
A positive integer, not including 0. The set of natural numbers has the notation $\mathbb{N}$.

^nC_r
The **binomial coefficient** of x^r in the **binomial expansion** of $(1 + x)^n$.
Also written $\binom{n}{r}$.

Newton-Raphson Method
An **iterative** method for finding a root of an equation, using the formula:
$$x_{n+1} = x_n - \frac{f(x_n)}{f'(x_n)}$$

Non-uniform
Describes a body whose mass is unevenly distributed throughout the body.

Normal distribution
A 'bell-shaped' continuous probability distribution where the further from the mean a value is, the less likely it is to occur.

Normal reaction
The reaction **force** from a surface acting on an object. It acts perpendicular to the surface.

Null hypothesis
A statement which gives a specific value to the **parameter** in a **hypothesis test**. Usually written H_0.

O

One-tailed test
A **hypothesis test** is 'one-tailed' if the **alternative hypothesis** is specific about whether the **parameter** is greater or less than the value specified by the **null hypothesis**.
E.g. it says $p < a$ or $p > a$ for a parameter p and constant a.

One-to-one function
A function where each value in the **range** corresponds to one and only one value in the **domain**.

P

Parameter (hypothesis testing)
A quantity that describes a characteristic of a **population**.

Parameter (parametric equations)
The variable linking a set of **parametric equations** (usually t or θ).

Parametric equations
A set of equations defining x and y in terms of another variable, called the **parameter**.

Partial fractions
A way of writing an **algebraic fraction** with several linear factors in its denominator as a sum of fractions with linear denominators.

Particle
A body whose mass is considered to act at a single point, so its dimensions don't matter.

Particular solution
A solution to a **differential equation** where known values have been used to find the constant term.

Peg
A fixed support which a body can hang from or rest on.

Percentage error
The difference between a value and its approximation, as a percentage of the real value.

Plane
A flat surface.

Point of inflection
A **stationary point** on a graph where the **gradient** doesn't change sign on either side of the point.

Polynomial
An algebraic expression made up of the sum of constant terms and variables raised to positive **integer** powers.

Product law
A formula used to work out the probability of two events both happening.

Product moment correlation coefficient
A measure of the strength of the **correlation** between two variables.

Product rule
A method for **differentiating** a product of two functions.

Progression
Another word for **sequence**.

Projectile
A body, projected into the air, which moves only under the influence of gravity.

Proof
Using mathematical arguments to show that a statement is true or false.

Proof by contradiction
Assuming that a statement is false, then showing that this assumption is impossible, to prove that the statement is true.

Pulley
A wheel, usually modelled as fixed and **smooth**, over which a **string** passes.

***p*-value**
The probability under H_0 of the **test statistic** taking a value at least as extreme as the observed value in a **hypothesis test**.

Q

Quotient
The result when you divide one thing by another, not including the **remainder**.

Quotient rule
A method of **differentiating** one function divided by another.

R

Radian
A unit of measurement for angles. 1 radian is the angle in a **sector** of a circle with radius r that has an **arc** of length r.

Range
The set of values output by a **mapping** or **function**. Usually given as a set of values that y or $f(x)$ can take.

Rational expression
A function that can be written as a fraction where the numerator and denominator are both **polynomials**.

Rational number
A number that can be written as the **quotient** (division) of two **integers**, where the denominator is non-zero. The set of rational numbers has the notation $\mathbb{Q}$.

Real number
Any positive or negative number (or 0) including all **rational** and **irrational numbers**, e.g. fractions, decimals, integers and surds. The set of real numbers has the notation $\mathbb{R}$.

Recurrence relation
A **function** that tells you how to work out a term in a **sequence** from the previous term.

Remainder (algebraic division)
The expression left over following an **algebraic division** that has a **degree** lower than the **divisor**.

Resolving
Splitting a **vector** up into **components**.

Rest
Describes a body which is not moving. Often used to describe the initial state of a body.

Resultant force/vector
The single **force/vector** which has the same effect as two or more forces/vectors added together.

Rigid
Describes a body which does not bend.

Rod
A long, **thin**, straight, **rigid** body.

Root
A value of x at which a **function** $f(x)$ is equal to 0.

Rough
Describes a surface for which a **frictional force** will oppose the motion of a body in contact with the surface.

S

Sec
The **reciprocal** of the cosine function, sometimes written as secant.

Second order derivative
The result of **differentiating** a **function** twice.

Sector
A section of a circle formed by two radii and part of the circumference.

Sense
The direction of a rotation (clockwise or anticlockwise).

Separating variables
A method for solving **differential equations** by first rewriting in the form $\frac{1}{g(y)}\,dy = f(x)\,dx$ in order to **integrate**.

Sequence
An ordered list of numbers (referred to as terms) that follow a set pattern. E.g. 2, 6, 10, ... or –4, 1, –4, 1, ...

Series
An ordered list of numbers, just like a **sequence**, but where the terms are being added together (to find a sum).

Sigma notation
Used for the sum of **series**. E.g.
$\sum_{n=1}^{15}(2n+3)$ is the sum of the first 15 terms of the series with n^{th} term $2n + 3$.

Significance level (α)
Determines how unlikely the observed value of the **test statistic** needs to be (under H_0) before rejecting the **null hypothesis** in a **hypothesis test**.

Significant result
The observed value of a **test statistic** is significant if, under H_0, it has a probability lower than the **significance level**.

Small angle approximations
Functions that approximate sin x, cos x and tan x for small values of x in **radians**.

Smooth
Describes a surface for which there is no **friction** between the surface and a body in contact with it.

Standard normal variable, Z
A random variable that follows a **normal distribution** with mean 0 and variance 1.

Static
Describes a body which is not moving. Often used to describe a body in **equilibrium**.

Stationary point
A point on a curve where the gradient is 0.

Statistic
A quantity that is calculated using only known observations from a **sample**.

String
A **thin** body, usually modelled as being **light** and **inextensible**.

Sum to infinity
The sum to infinity of a **series** is the value that the sum tends towards as more and more terms are added. Also known as the **limit** of a series.

T

Taut
Describes a **string** or **wire** which is experiencing a **tension force** and is tight and straight.

Tension
The **force** in a taut **wire** or **string**.

Test statistic
A **statistic** calculated from **sample** data which is used to decide whether or not to reject the **null hypothesis** in a **hypothesis test**.

Thin
Describes a body which is modelled as having no thickness.

Thrust
The **force** in a compressed **rod**.

Trajectory
The path followed by a **projectile**.

Trapezium rule
A way of estimating the area under a curve by dividing it up into trapezium-shaped strips.

Turning point
A **stationary point** that is a (local) maximum or minimum point of a curve.

Two-tailed test
A **hypothesis test** is 'two-tailed' if the **alternative hypothesis** specifies only that the **parameter** doesn't equal the value specified by the **null hypothesis**. E.g. it says $p \neq a$ for a parameter p and constant a.

U

Uniform
Describes a body whose mass is evenly spread throughout the body.

Union (of events A and B)
The set of outcomes corresponding to either event A or event B (or both) happening.

Unit vector
A **vector** of **magnitude** one unit.

Upper bound
The upper limit of the values that a number could take and still be rounded down to the correct answer.

V

Vertex
Turning point of a graph — the maximum or minimum point for a quadratic graph.

Vector
A quantity which has both a **magnitude** and a direction.

Velocity
The rate of change of an object's **displacement** with respect to time.

W

Weight
The **force** due to a body's mass and the effect of gravity: $W = mg$.

Wire
A **thin** body often modelled as being **light**. It can be bent to form a shape.

Index

Formula Sheet

These are the formulas you'll be given in the exam, but make sure you know exactly **when you need them** and **how to use them.**

Pure Mathematics

Trigonometry

Trigonometric Identities:

$\sin(A \pm B) \equiv \sin A \cos B \pm \cos A \sin B$

$\cos(A \pm B) \equiv \cos A \cos B \mp \sin A \sin B$

$\tan(A \pm B) \equiv \dfrac{\tan A \pm \tan B}{1 \mp \tan A \tan B} \quad (A \pm B \neq (k + \frac{1}{2})\pi)$

Small Angle Approximations:

$\sin\theta \approx \theta$

$\cos\theta \approx 1 - \frac{1}{2}\theta^2$

$\tan\theta \approx \theta$

where θ is measured in radians.

Differentiation

First Principles: $f'(x) = \lim\limits_{h \to 0} \dfrac{f(x+h) - f(x)}{h}$

f(x)	**f′(x)**
$\tan x$	$\sec^2 x$
$\operatorname{cosec} x$	$-\operatorname{cosec} x \cot x$
$\sec x$	$\sec x \tan x$
$\cot x$	$-\operatorname{cosec}^2 x$

For $y = \dfrac{f(x)}{g(x)}$, $\dfrac{dy}{dx} = \dfrac{f'(x)g(x) - f(x)g'(x)}{(g(x))^2}$

Integration

f(x)	$\int$ **f(x) dx**
$\tan x$	$\ln\lvert\sec x\rvert + c$
$\cot x$	$\ln\lvert\sin x\rvert + c$

$\displaystyle\int \frac{f'(x)}{f(x)}\,dx = \ln\lvert f(x)\rvert + c$

$\displaystyle\int u\frac{dv}{dx}\,dx = uv - \int v\frac{du}{dx}\,dx$

Numerical Methods

Trapezium rule: $\displaystyle\int_a^b y\,dx \approx \frac{1}{2}h[y_0 + 2(y_1 + y_2 + \ldots + y_{n-1}) + y_n]$, where $h = \dfrac{b-a}{n}$

The Newton-Raphson formula iteration for solving $f(x) = 0$: $x_{n+1} = x_n - \dfrac{f(x_n)}{f'(x_n)}$

MAT62